Lecture Notes in Electrical Engineering

Volume 905

The book series *Lecture Notes in Electrical Engineering* (LNEE) publishes the latest developments in Electrical Engineering - quickly, informally and in high quality. While original research reported in proceedings and monographs has traditionally formed the core of LNEE, we also encourage authors to submit books devoted to supporting student education and professional training in the various fields and applications areas of electrical engineering. The series cover classical and emerging topics concerning:

- Communication Engineering, Information Theory and Networks
- Electronics Engineering and Microelectronics
- Signal, Image and Speech Processing
- Wireless and Mobile Communication
- Circuits and Systems
- Energy Systems, Power Electronics and Electrical Machines
- Electro-optical Engineering
- Instrumentation Engineering
- Avionics Engineering
- Control Systems
- Internet-of-Things and Cybersecurity
- Biomedical Devices, MEMS and NEMS

For general information about this book series, comments or suggestions, please contact leontina.dicecco@springer.com.

To submit a proposal or request further information, please contact the Publishing Editor in your country:

China

Jasmine Dou, Editor (jasmine.dou@springer.com)

India, Japan, Rest of Asia

Swati Meherishi, Editorial Director (Swati.Meherishi@springer.com)

Southeast Asia, Australia, New Zealand

Ramesh Nath Premnath, Editor (ramesh.premnath@springernature.com)

USA, Canada:

Michael Luby, Senior Editor (michael.luby@springer.com)

All other Countries:

Leontina Di Cecco, Senior Editor (leontina.dicecco@springer.com)

** **This series is indexed by EI Compendex and Scopus databases.** **

More information about this series at https://link.springer.com/bookseries/7818

J. Dinesh Peter · Steven Lawrence Fernandes ·
Amir H. Alavi
Editors

Disruptive Technologies for Big Data and Cloud Applications

Proceedings of ICBDCC 2021

Springer

Editors
J. Dinesh Peter
Department of Computer Science and Engineering
Karunya Institute of Technology and Sciences
Coimbatore, Tamil Nadu, India

Steven Lawrence Fernandes
Department of Computer Science
Creighton University
Omaha, NE, USA

Amir H. Alavi
Civil and Environmental Engineering
University of Pittsburgh
Pittsburgh, PA, USA

ISSN 1876-1100 ISSN 1876-1119 (electronic)
Lecture Notes in Electrical Engineering
ISBN 978-981-19-2176-6 ISBN 978-981-19-2177-3 (eBook)
https://doi.org/10.1007/978-981-19-2177-3

This Springer imprint is published by the registered company Springer Nature Singapore Pte Ltd.
The registered company address is: 152 Beach Road, #21-01/04 Gateway East, Singapore 189721, Singapore

Preface

This work comprises the proceedings of the International Conference on Big Data and Cloud Computing (ICBDCC'21). This conference was organized with the primary theme of promoting ideas that provide technological solutions to the big data and cloud computing applications. ICBDCC provided a unique forum for the practitioners, developers and users to exchange ideas and present their observations, models, results and experiences with the researchers who are involved in real-time projects that provide solutions for research problems of recent advancements in big data and cloud computing technologies. In the last decade, a number of sophisticated and new computing technologies have been developed. With the introduction of new computing paradigms such as cloud computing, big data and other innovations, ICBDCC provided a high-quality dissemination forum for new ideas, technology focus, research results and discussions on the evolution of computing for the benefit of both scientific and industrial developments. ICBDCC is supported by a panel of reputed advisory committee members both from India and from all across the world.

This proceedings includes topics in the fields of big data, data analytics in cloud, cloud security, cloud computing and big data and cloud computing applications. The research papers featured in this proceedings provide novel ideas that contribute to the growth of the society through computing technologies. The contents of this proceedings will prove to be an invaluable asset to the researchers in the areas of big data and cloud computing.

We appreciate the extensive time and effort put in by all the members of the organizing committee for ensuring a high standard for the papers published in this volume. We would like to express our thanks to the panel of experts who helped us to review the papers and assisted us in selecting the candidate for the Best Paper Award.

We would like to thank the eminent keynote speakers who have shared their ideas with the audience and all the researchers and academicians who have contributed their research works, models and ideas to ICBDCC'21.

Coimbatore, India — J. Dinesh Peter
Pittsburgh, PA, USA — Amir H. Alavi
Omaha, NE, USA — Steven Lawrence Fernandes

Contents

About the Editors

J. Dinesh Peter is currently working as an associate professor, Department of Computer Sciences Technology at Karunya University, Coimbatore. Before this, he was a full-time research scholar at the National Institute of Technology, Calicut, India, from where he received his Ph.D. in computer science and engineering. His research focus includes big data, image processing, and computer vision. He has several publications in various reputed international journals and conference papers which are widely referred to. He is a member of IEEE, CSI and IEI and has served as session chairs and delivered plenary speeches for various international conferences and workshops.

Steven Lawrence Fernandes is an assistant professor in the Department of Computer Science at Karunya University, Coimbatore. After earning his Ph.D. in Electronics and Communication Engineering at Karunya Institute of Technology and Science and his Masters in Microelectronics at Manipal Institute of Technology, he began his postdoctoral research at the University of Alabama, Birmingham. There, he worked on NIH-funded projects. He also conducted postdoctoral research at the University of Central Florida. This research included working on DARPA, NSF, and RBC-funded projects. His publications include research articles in highly selective artificial intelligence venues. Dr. Fernandes's current area of research is focused on using artificial intelligence techniques to extract useful patterns from big data. This includes robust computer vision applications using deep learning and computer-aided diagnosis using medical image processing.

Dr. Amir H. Alavi is an assistant professor in the Department of Civil and Environmental Engineering and holds a courtesy appointment in the Department of Bioengineering at the University of Pittsburgh. Dr. Alavi's research interests include structural health monitoring, multifunctional structures, advanced sensors, low-power energy harvesting, and engineering informatics. His research activities involve the implementation of self-sustained and multifunctional sensing and structural systems enhanced by engineering informatics in the fields of civil infrastructure, construction, aerospace, and biomedical engineering. Dr. Alavi has authored seven books and over 200 publications in archival journals, chapters, and conference proceedings. He has received several award certificates for his journal articles. He is among the Google Scholar 200 Most Cited Authors in Civil Engineering, Web of Science ESI's World Top 1% Scientific Minds in 2018, and the Stanford University list of Top 1% Scientists in the World 2020.

A Statistical Performance Analysis of GPU WAH Range Querying

Mitchell Nelson, Joseph M. Myre, and Jason Sawin

1 Introduction

As science, industry, and entertainment continue to generate staggering amounts of data, database managers have been seeking hardware/software hybrid solutions to assist in processing the deluge of information. Bitmap indices are commonly employed software solutions for large read-only scientific databases. Such indices are coarse binary representations of the underlying data. As these representations are generally sparse matrices, they are highly compressible using hybrid run-length compression schemes. The compressed results of such schemes can be queried directly using native hardware-enabled bitwise operators, significantly increasing query processing efficiency.

Recent work has shown that porting some or all of the bitmap query processing to graphical processing units (GPUs) can significantly increase efficiency [1, 2, 10, 11, 14, 15]. GPUs are massively parallel computational accelerators initially designed to assist in the execution of graphic-intensive software packages. However, they are now commonly used to execute a wide array of applications, including scientific simulations, neural networks, etc. GPUs are now commonplace in many computing systems. This ubiquity means that incorporating GPUs into query execution adds benefits beyond their fast processing rate. The use of GPUs frees the CPU and system RAM to manage less specialized processes. It also means that a more complete use of the system resources is realized, thus saving money and reducing the environmental cost of query processing.

This paper presents a statistical performance analysis of the GPU bitmap query engine we introduced in Nelson et al. [11]. The engine was specifically designed to process range queries over bitmaps compressed using Word Aligned Hybrid codes

M. Nelson · J. M. Myre · J. Sawin (✉)
Department of Computer and Information Sciences,
University of St. Thomas, Saint Paul, MN, USA
e-mail: jason.sawin@stthomas.edu

J. D. Peter et al. (eds.), *Disruptive Technologies for Big Data and Cloud Applications*,
Lecture Notes in Electrical Engineering 905,
https://doi.org/10.1007/978-981-19-2177-3_1

Table 1 Left table shows a relation which records the number of different models of cars sold in a year. The right shows a potential bitmap index for that relation

Cars Sold		Model Bins						Volume Bins			
Model	# sold	m_0	m_1	m_2	m_3	m_4	m_5	v_0	v_1	v_2	v_3
Tiago	51K	1	0	0	0	0	0	0	0	0	1
Nexon	48K	0	1	0	0	0	0	0	0	1	0
Alitroz	47K	0	0	1	0	0	0	0	0	1	0
Harrier	15K	0	0	0	1	0	0	0	1	0	0
Tigor	10K	0	0	0	0	1	0	0	1	0	0
Zest	220	0	0	0	0	0	1	1	0	0	0

(WAH) [17]. Here, we present an ANOVA analysis [7] applied to the results of a rigorous empirical study of our engine. The products of this analysis indicate potential features of the query framework that can be tuned to increase efficiency. Further, the results provide guidance for the creation of a more generalized framework.

2 Background

Bitmap Indices: Bitmap indices are created by binning the tuples of the relation being indexed. First, attribute domains are partitioned into sets of discrete values bins or ranges of values bins depending on the domain. Then, each tuple is analyzed, creating a row in the bitmap. The value of each attribute is inspected, and a 1 is placed in the bitmap bin that corresponds to that value. The remaining bins in that set are assigned 0.

A possible example of this binning process is shown in Table 1. The relation on the right records the yearly sales of Tata car models. Since there are two attributes, there are two sets of bins in the bitmap shown on the left. As the values of the **Model** attribute are discrete, each value is assigned its own bin. These are the bins prefixed with an m. A 1 in the m_0 bin indicates the corresponding tuple had the `Tiago` value for the **Model** attribute, m_1 indicates a value of `Nexon`, and so on. The **# sold** attribute holds values that fall in a continuum. Thus, it can be binned using range bins. In this example, the bin v_3 represents values $\geq$50 k, v_2 represents the range [20 k–50 k), v_1 is [10 k, 20 k), and v_0 is <10 k. The first tuple in **Cars Sold** results in a bitmap row of 1 in m_0 representing that it records the number of `Tiagos` sold. All other m bins are assigned 0. Since there were over 50 k `Tiagos`, a 1 is placed in the v_3, and the remaining v bins are assigned 0. This process is applied to all the tuples in **Cars Sold**.

A significant benefit of bitmap indices is the ability to query the bitmap directly using hardware-enabled bitwise operations. Again, consider the example shown in Table 1. An answer to the query "Which models had a sales volume between 15 and 55 k cars?" can be derived by bitwise ORing columns of the bitmap. Specifically,

$v_1 \wedge v_2 \wedge v_3 = R$, where R is the result vector of the operation. Each row in R that contains a 1, corresponds to a tuple in the database that has a **# sold** value that needs to be further inspected to see if it falls into the exact range.

The other main benefit of bitmap indices is their compressibility. The matrices generated by the binning process are generally very sparse. As such, the columns of the bitmaps are amenable to run-length compression. Numerous compression techniques have been developed for bitmap indices. One of the most prominent is Word Aligned Hybrid codes (WAH) [17]. WAH uses a hybrid encoding of literal values and compressed homogeneous runs tuned to the system's word length. It has been shown that WAH can achieve tremendous compression with the added benefit that the compress results can be queried directly without requiring explicit decompression.

Graphics Processing Units (GPU): NVIDIA's CUDA provides a programming platform for NVIDIA GPUs. With CUDA, thousands of threads can be organized into Cartesian structures (1-, 2-, or 3-dimension). These structures intuitively map to many computational problems. In CUDA, threads are organized hierarchically as threads, thread blocks, and thread grids. Threads are executed in warps (groups of 32 threads). Thus, thread blocks are most often a multiple of 32 threads. The memory hierarchy present in NVIDIA GPUs consists of global memory, L2 cache, L1/shared memory, and registers, in descending order of access time. The shared memory can act as a fast access memory for data storage on chip.

GPU Range Queries: NVIDIA GPUs have been fundamental in enhancing the performance of WAH queries [1, 2, 10, 11, 14, 15]. All of these approaches transfer the bins to the GPU in their compressed form. To fully make use of the parallel processing power of the GPUs, the bins are then decompressed. In this paper, we will focus on the highest performing method published thus far, the ideal hybrid approach of Nelson et al. [10]. The hybrid query approach computes a range query of the form $A_1 \vee \ldots \vee A_n$, where A_i is a decompressed bitmap bin using a parallel reduction. Parallel reductions exploit the presence of independent operations to enhance performance. For example, $R_3 = A_5 \vee A_6$ and $R_4 = A_7 \vee A_8$ could be solved simultaneously. This reduces the time of this operation from $\mathcal{O}(n)$ to $\mathcal{O}(\log n)$, where n is the number of bins in the query.

3 Evaluation Methodology

In this section, the testing methodology used to produce our results is described. All tests were executed on a machine running Ubuntu 16.04.5 LTS equipped with dual 8-core Intel Xeon E5-2609 v4 CPUs (each at 1.70 GHz) and 32 GB of RAM. All tests were developed using CUDA v9.0.176 and run on two different GPUs: an NVIDIA GeForce GTX 1080 with 8 GB of memory and an NVIDIA Titan X with 12 GB of memory.

Parameters related to the GPU implementation and the characteristics of the queried dataset can affect the performance of WAH range queries on GPUs. Varying these factors can induce a nonlinear response in performance. This necessitates analysis to determine the parameters that reliably predict the performance of WAH range queries on GPUs.

We examine two sets of performance parameters. The first set is intrinsic to the architectural implementation of the GPU WAH range-query algorithm. These include the use of shared memory, CUDA streams for decompression, CUDA streams query execution, GPU base clock frequency, and global memory bandwidth. The second set of parameters is intrinsic to the dataset/query. These parameters include the number of bins being queried, the number of rows in the dataset, and the dataset compressibility.

We use the ideal hybrid GPU-based range-query method described in Nelson et al. [10] to execute a range queries of 4, 8, 16, 32, and 64 random bit vectors.

Each experiment was run six times, and the execution time of each was recorded. To remove transient program behavior, the first result is discarded and the remaining execution times are used for analysis.

3.1 Dataset Creation

We use the Zipf synthetic datasets for our evaluation. These datasets are created using a Zipf distributions which represent a clustered approach to discretization. This essentially creates a skewed distribution of 1's in our bitmap, simulating the types distribution seen in real data. The Zipf distribution generator assigns each bit a probability of: $p(k, n, \text{skew}) = (1/k^{\text{skew}})/\sum_{i=1}^{n}(1/i^{\text{skew}})$, where n is the number of elements determined by cardinality, k is their rank, and the coefficient skew creates an exponentially skewed distribution. We set $k = 10, n = 10$ and set skew $= 0, 1, 2,$ and 3. These different skew values create datasets of varying bit density. Using these parameters, we generate 16 different datasets containing 100 bins (i.e., ten attributes discretized into ten bins each) and 8, 16, 32, and 64 million rows.

3.2 Statistical Performance Analysis

We use a statistical ANOVA [7] approach to analyze our test results. This approach quantifies the impact of each factor (from the beginning of this section) on performance. An ANOVA analysis determines whether a statistically significant difference exists among the means of each test. This is done by separating the total observed measurement variation into two components: (1) the variation within a system (assumed to be measurement error) and (2) the variation between systems (assumed to be due to both actual differences between systems and to measurement error). Statistically significant differences between systems are determined via an F-test that compares variances across the systems.

An m-factor ANOVA analysis is required as more than two factors are present in the sets of dataset and architectural factors outlined at the beginning of this section. When two or more factors are present in an ANOVA analysis, the interactions between factors can be considered. Interactions are important to consider as the combination of factors can be more impactful than simply summing the impact of each factor independently (the whole can be greater than the sum of the parts). The impact of each unique factor and interaction of factors are called effects. From the results of the ANOVA analysis, it is possible to compute the percent impact of each effect by forming the ratio of total variation in measurement due to each effect to the sum of total variation of all effects and measurement errors.

4 Results

Here, we present the results of the experiments and ANOVA analyses described above. We first present the results of the ANOVA analysis of all factors, then analyzes of factors linked strictly to architectural details and dataset details. All results are reported as a rank ordering of percent impact of each effect.

Rankings of effects for all factors are shown in Fig. 1. The two most significant factors contributing to variations in performance are the use of shared memory on the GPU and the number of rows in the dataset, accounting for 40.26% and 27.32% of the overall variation in performance, respectively, and the interaction thereof accounting for 20.66% of the overall variation in performance. The sum of the stand-alone and interaction effects of these two factors accounts for 88.24% of the overall variation in performance. All remaining effects each account for less than 3.5% of the overall variation in performance.

Rankings of effects for architecturally linked factors are shown in Fig. 2. The two most significant architectural factors are the use of shared memory on the GPU and the base clock rate of the GPU, accounting for 90.79% and 5.04% of the overall variation in performance, respectively, and the interaction thereof accounting for 3.2% of the overall variation in performance. The sum of the stand-alone and interaction effects of these two factors accounts for 99.03% of the overall variation in performance. The sum of all remaining effects accounts for less than 1% of the overall variation in performance.

Rankings of effects for dataset linked factors are shown in Fig. 3. The two most significant factors associated with the dataset are the number of rows in the dataset and the number of columns in the query, accounting for 63.06% and 23.68% of the overall variation in performance, respectively, and the interaction thereof accounting for 12.17% of the overall variation in performance. The sum of the stand-alone and interaction effects of these two factors accounts for 98.81% of the overall variation in performance. The sum of all remaining effects accounts for just over one percent of the overall impact on performance.

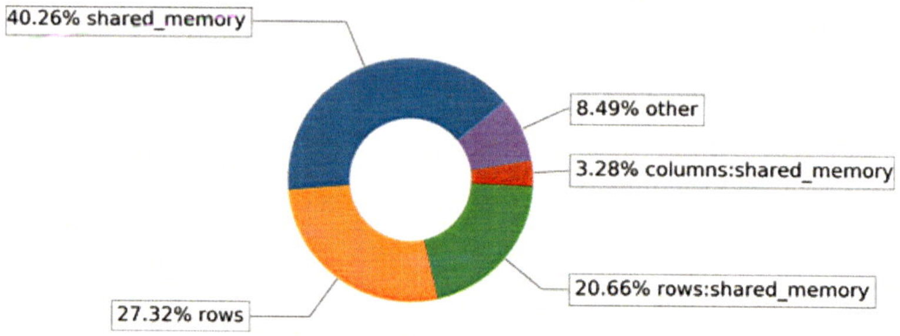

Fig. 1 Five most influential factors to variations in performance ranked by percent of overall variation

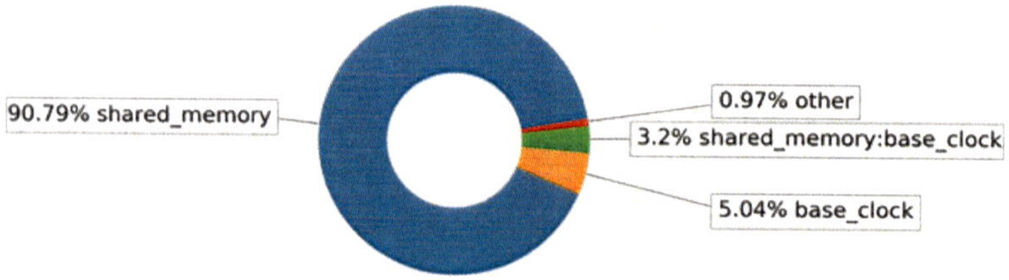

Fig. 2 Four most influential architectural factors to variations in performance ranked by percent of overall variation

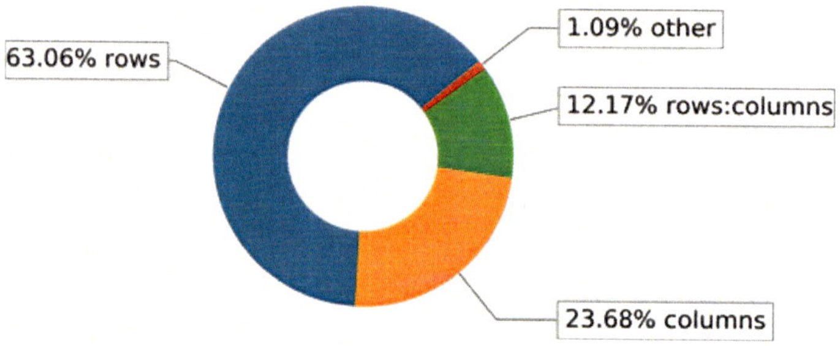

Fig. 3 Four most influential dataset and query factors to variations in performance ranked by percent of overall variation

5 Discussion of Results

As seen in Fig. 1, the two most significant factors are the use of shared memory and the number of rows (88.24% of the total variation in performance). Interestingly, bit density has no significant effect on performance for the GPU query method tested here. Figure 4 presents the approximate profiles of GPU execution time when varying the two most significant factors, which demonstrate these effects in practice. As shown, (A) does not use shared memory and has a large number of rows, (B) does not use shared memory and has a small number of rows, (C) uses shared memory

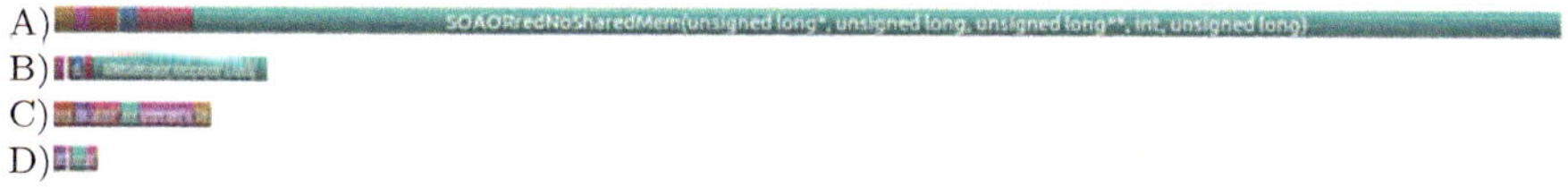

Fig. 4 Shown are profiles of four query executions when the two most significant factors to performance (the use of shared memory and the number of rows in the database) are varied

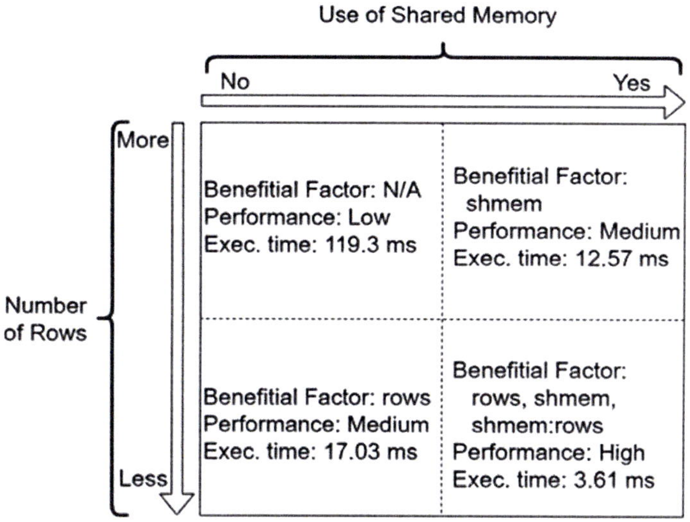

Fig. 5 Effects of the three most significant factors (the use of shared memory on the GPU, the number of rows in the dataset, and the interaction thereof) on execution time

and has a large number of rows, and (D) uses shared memory and has a small number of rows.

Query performance derived from the primary factors (use of shared memory and the number of rows) can also be visualized in Fig. 5. Following the arrows in this figure results in performance enhancement. E.g., using shared memory, decrease the total number of rows, or both. We can see the importance of examining interactions of factors in this figure, as exploiting both factors to gain performance is more beneficial than only using one.

6 Related Work

GPUs and CUDA have enabled the acceleration of many general-purpose computing problems. Many times, these come from a focus on core mathematical routines [6, 12, 13] or parallel programming primitives [5, 8]. With these, researchers have been able to accelerate numerous other applications including computational fluid dynamics models [3], finite element methods [16], and traditional relational databases [4].

Among the many tools that are useful for performance analysis, ANOVA has used to identify critical factors to focus on for performance enhancements [7, 9].

7 Conclusion and Future Work

In this paper, we present a statistical performance analysis of WAH range queries using GPUs and the method of Nelson et al. [10]. We use an ANOVA analysis to quantify the effects of architectural and dataset factors on the variation of overall performance. Focusing only on architectural factors, our analysis finds that the use of shared memory and the rate of the base clock are responsible for 99.03% of the total variation in performance. Focusing only on dataset factors, our analysis finds that the number of rows and columns is responsible for 98.81% of the total variation in performance. When considering all factors, our analysis finds that the use of shared memory and the total number of rows accounts for 88.24% of the total variation in performance.

These results suggest that for GPU WAH range queries, the use of shared memory is critical for performance. Secondary to shared memory is the rate of the base clock. Intuitively, our results also show that datasets with more rows will be slower to query than datasets with fewer rows, all else being equal.

We plan to continue investigating the performance space of the hybrid method. This includes the effect of additional data characteristics, data layout in shared memory, and the distribution of data/queries across multiple GPUs.

References

1. W. Andrzejewski, R. Wrembel, GPU-WAH: applying GPUs to compressing bitmap indexes with word aligned hybrid, in *International Conference on Database and Expert Systems Applications* (Springer, Berlin, Heidelberg, 2010), pp. 315–329
2. W. Andrzejewski, R. Wrembel, GPU-PLWAH: GPU-based implementation of the PLWAH algorithm for compressing bitmaps. Control Cybern. **40**, 627–650 (2011)
3. P. Bailey, J. Myre, S.D. Walsh, D.J. Lilja, M.O. Saar, Accelerating lattice Boltzmann fluid flow simulations using graphics processors, in *2009 International Conference on Parallel Processing* (IEEE, 2009), pp. 550–557
4. P. Bakkum, K. Skadron, Accelerating SQL database operations on a GPU with CUDA, in *Proceedings of the 3rd Workshop on General-Purpose Computation on Graphics Processing Units* (ACM, New York, NY, 2010), GPGPU-3, pp. 94–103. http://doi.acm.org/10.1145/1735688.1735706
5. N. Bell, J. Hoberock, Thrust: a productivity-oriented library for CUDA, in *GPU Computing Gems Jade Edition* (Elsevier, 2012), pp. 359–371
6. J. Dongarra, M. Gates, A. Haidar, J. Kurzak, P. Luszczek, S. Tomov, I. Yamazaki, Accelerating numerical dense linear algebra calculations with GPUs, in *Numerical Computations with GPUs* (2014), pp. 1–26

7. D.J. Lilja, *Measuring Computer Performance: A Practitioner's Guide* (Cambridge University Press, 2005)
8. D. Merrill, CUB: CUDA unbound (2016). http://nvlabs.github.io/cub
9. J. Myre, S.D. Walsh, D. Lilja, M.O. Saar, Performance analysis of single-phase, multiphase, and multicomponent lattice-Boltzmann fluid flow simulations on GPU clusters. Concurr. Comput. Pract. Exp. **23**(4), 332–350 (2011)
10. M. Nelson, Z. Sorenson, J.M. Myre, J. Sawin, D. Chiu, GPU acceleration of range queries over large data sets, in *2019 IEEE/ACM International Conference on Big Data Computing, Applications, and Technologies (BDCAT19)* (IEEE/ACM, 2019). https://doi.org/10.1145/3365109.3368789
11. M. Nelson, Z. Sorenson, J.M. Myre, J. Sawin, D. Chiu, Parallel acceleration of CPU and GPU range queries over large data sets. J. Cloud Comput. **9**(1), 1–21 (2020)
12. S. Tomov, J. Dongarra, M. Baboulin, Towards dense linear algebra for hybrid GPU accelerated manycore systems. Parallel Comput. **36**(5–6), 232–240 (2010a)
13. S. Tomov, R. Nath, H. Ltaief, J. Dongarra, Dense linear algebra solvers for multicore with GPU accelerators, in *2010 IEEE International Symposium on Parallel & Distributed Processing, Workshops and Phd Forum (IPDPSW)* (IEEE, 2010b), pp. 1–8
14. B. Tran, B. Schaffner, J. Sawin, J.M. Myre, D. Chiu, Increasing the efficiency of GPU bitmap index query processing, in *International Conference on Database Systems for Advanced Applications* (Springer, 2020), pp. 339–355
15. B. Tran, B. Schaner, J.M. Myre, J. Sawin, D. Chiu, Exploring means to enhance the efficiency of GPU bitmap index query processing. Data Sci. Eng. **6**(2), 209–228 (2021)
16. S.D. Walsh, M.O. Saar, P. Bailey, D.J. Lilja, Accelerating geoscience and engineering system simulations on graphics hardware. Comput. Geosci. **35**(12), 2353–2364 (2009)
17. K. Wu, E.J. Otoo, A. Shoshani, Compressing bitmap indexes for faster search operations, in *Proceedings 14th International Conference on Scientific and Statistical Database Management* (IEEE, 2002), pp. 99–108

Anonymized Questionnaire Analysis with Differential Privacy for Large-Scale Crowdsourcing

Yuichi Sei and Akihiko Ohsuga

1 Introduction

Several studies for crowdsourcing have been proposed recently [3, 4, 18]. In crowdsourcing, a *requester* posts tasks, such as questionnaires, programming, and proofreading, and a *worker* selects and conducts a task while considering the complexity and fee of the task. Crowdsourcing systems receive money from requesters in advance and pay fees to workers for completed tasks. Therefore, personally identifiable information (PII) of each requester and worker is registered to crowdsourcing systems.

In this article, we focus on questionnaires as tasks. A requester posts a questionnaire, which a worker can answer if he/she wishes to. Because the crowdsourcing system has the PII of workers, it is preferable that workers send the answers of the questionnaire to the requester directly when questionnaires need sensitive information of workers, such as salary and religion. However, it is possible that requesters can identify workers. Most questionnaires contain questions about the answerers' attributes, such as age and sex, because the requester analyzes the results of the questionnaire in terms of the attributes of the answerers in many cases. The requester may identify the worker from these basic items of questionnaires, such as age and sex. Sweeney [17] found that 87% of the US population is uniquely identified by {date of birth, sex, 5-digit ZIP}. Of course, other attributes can also be used for identification. Rocher et al. [14] reported that 99.98% of Americans can be identified using 15 attributes. This problem can cause workers to avoid answering questionnaires.

Y. Sei (✉) · A. Ohsuga
The University of Electro-Communications, Tokyo, Japan
e-mail: seiuny@uec.ac.jp

A. Ohsuga
e-mail: ohsuga@uec.ac.jp

Y. Sei
JST, PRESTO, Kawaguchi, Saitama, Japan

J. D. Peter et al. (eds.), *Disruptive Technologies for Big Data and Cloud Applications*, Lecture Notes in Electrical Engineering 905,
https://doi.org/10.1007/978-981-19-2177-3_2

Moreover, requesters can obtain other information about workers from crowdsourcing systems. The requester can see the IDs of workers in crowdsourcing systems because the requester should check the work result of each worker and tell the crowdsourcing system that the requester accepts to pay an agreed fee for the work if the requester is satisfied with it. Based on the IDs, requesters can check several attribute values registered in the crowdsourcing system of workers, such as state/province of address, career, and skills. This information can also be used to identify workers. In this study, we assume that a *worker agent* exists in each worker's personal computer or smartphone. The worker agent anonymizes the worker's attributes.

Our previous work can collect information from workers while protecting their privacy. However, it assumes that the number of answers to be collected is only one; moreover, it assumes that the privacy-protection level among workers should be the same. In this study, we propose a technique that can collect multiple answers under the assumption that each worker can set a different privacy-protection level for each answer.

The rest of this article is organized as follows. In Sect. 2, we introduce our application and attack model. In Sect. 3, we define the privacy used in this work. In Sect. 4, we describe the related methods. We present the design of the proposed algorithm in Sect. 5, and we present the simulation results in Sect. 6. In Sect. 7, we conclude the paper.

2 Assumptions

We assume that requesters want to analyze the results of questionnaires containing several sensitive information, such as salary and religion. For instance, a requester may want to know the relationship among educational background, age, sex, and salary. The requester can post a questionnaire that contains questions about these attributes. If we can ensure that the requester cannot identify workers from educational background, age, sex, and other public attribute values registered in the crowdsourcing system, workers can answer their salary without worries. However, if we cannot ensure that, the worker is hesitant to answer this questionnaire.

In addition, we assume that all sensitive attribute values are categorized ones. In many cases, analyzers (i.e., requesters in this study) create a histogram or a cross tabulation of the data for their analysis. Therefore, analyzers do not need detailed numerical values. If analyzers want to analyze numerical values, such as salary, they categorize it in advance. For example, category 1 includes from 0 to 100, and category 2 includes from 100 to 200.

We assume that requesters generate cross tabulation for a basic analysis. In practice, the applications of cross tabulations involve only less than four variables at a time [8, 19] even if the requesters collect information about many attributes. For instance, even if a questionnaire has questions about age, sex, education, hobby, salary, and disease, it is difficult to analyze a cross tabulation created from all attributes all at once. As usual, analyzers create several cross tabulations from selected attributes for each purpose.

Further, we assume that the crowdsourcing system has the PII of workers. Therefore, workers do not send the answers to the questionnaire to the crowdsourcing system but send them to the requester after being anonymized. We assume that requesters are semi-honest entities. That is, the requesters follow the proposed protocol but try analyzing the individual information from each disguised data.

3 Privacy Metric

There are many important privacy metrics, such as k-anonymity [2, 10] and l-diversity [11, 13]. In this study, we use differential privacy [6]; in particular, we focus on local differential privacy [5], which has been widely studied recently.

Definition 1 [ε-local differential privacy] Let $\mathcal{X}$ be a set of sensitive values. A randomized mechanism $\mathcal{A}$ satisfies ε-local differential privacy if for any $x, x' \in \mathcal{X}$ and $y \in \mathcal{Y} \subset \text{Range}(A)$,

$$P(\mathcal{A}(x) = y) \leq e^{\varepsilon} P(\mathcal{A}(x') = y'). \tag{1}$$

We consider that each question is a database with only one record having only one column. Each worker can set a privacy level for each question. Moreover, a requester can set different fees for different questions and privacy levels. For instance, the fees of questions about gender with high- and low-privacy levels are 2 and 3 cents, respectively, and the fees of questions about disease names with high- and low-privacy levels are 5 and 10 cents, respectively.

Let q be the number of questions in a questionnaire. When each database (i.e., each answer) of a worker satisfies ε_1, ..., ε_q differential privacies, the set of answers satisfies $(\varepsilon_1 + \cdots + \varepsilon_q)$ differential privacy [9]. Workers can consider not only each privacy level of each answer but also the privacy level of the set of all answers if needed.

4 Related Work

Many privacy-protection data collection methods using differential privacy have been proposed. We proposed single to randomized multiple dummies with Bayes (S2Mb) previously [16]. In S2Mb, we assumed that there was only one value to be collected from participants, and the privacy-protection level was common among all participants. In this study, we use S2Mb as a basic approach for data collection.

There are several other methods on privacy-preserving data collection, such as [1, 7, 12]. Erlingsson et al. [7] proposed a randomized aggregatable privacy-preserving ordinal response that provides local differential privacy, and it was implemented in Google Chrome. Murakami and Kawamoto [12] assumed that there were sensitive and nonsensitive data and proposed a local differential privacy mechanism that could enhance data utility by protecting only sensitive data.

5 Proposed Method

5.1 System Overview

An overview of the proposed system is shown in Fig. 1. Each worker can determine his/her privacy-protection level (i.e., the value of ε). Because it is difficult for a layperson to understand the meaning of ε, we assume that the requester can determine several values of ε in advance. For example, the requester can prepare four levels: nonanonymization ($\varepsilon = \infty$), high-anonymization ($\varepsilon = 0.1$), middle-anonymization ($\varepsilon = 1$), and low-anonymization ($\varepsilon = 10$). How to determine the privacy-protection level is outside the scope of this study.

5.2 The Worker Protocol

The worker protocol is based on S2Mb [16] although S2Mb does not assume multiple data collection and different privacy levels. Let F_i be the number of options of a question i of a questionnaire, S_i be a set of options of question i, and $t_{i,j}$ be the selected option of the true answer of worker w_j for question i.

When worker w_j determines to answer a questionnaire of the crowdsourcing, he/she first specifies privacy levels $\varepsilon_{1,j}, \ldots, \varepsilon_{q,j}$ for each of q questions of the questionnaire. Then, he/she calculates a set of parameters $s_{i,j}$ and $p_{i,j}$ for all $i = 1, \ldots, q$ on the basis of $\varepsilon_{i,j}$ and F_i, using the following equation:

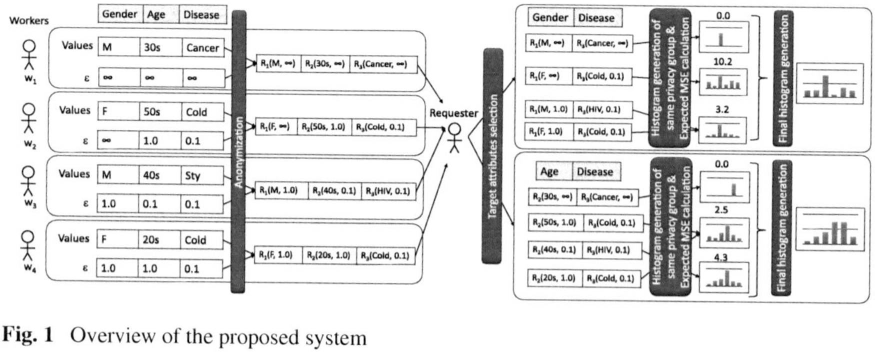

Fig. 1 Overview of the proposed system

$$s_{i,j} = \max\left(\left\lfloor \frac{F_i}{1 + e^{\varepsilon_{i,j}}} \right\rfloor\right), \; p_{i,j} = \frac{e^{\varepsilon_{i,j}}}{F_i - s_{i,j} + e^{\varepsilon_{i,j}} s_{i,j}} \tag{2}$$

Worker w_j creates an empty set $R_{i,j}$ for question i. Next, the worker with probability $p_{i,j}$ adds $t_{i,j}$ and $s_{i,j} - 1$ elements from $S_i \setminus \{t_{i,j}\}$ to set $R_{i,j}$ randomly. With probability $1 - p_{i,j}$, the worker adds $s_{i,j}$ elements from $S_i \setminus \{t_{i,j}\}$ to set $R_{i,j}$ randomly. Finally, worker w_j sends $R_{i,j}$ for all j to the requester.

5.3 *The Requester Protocol*

5.3.1 Histogram Generation for Same Privacy Group

First, the requester divides answers from the workers into several groups according to the privacy-protection levels. Within the same group, the privacy-protection level is the same. Let N represent the number of workers who answered the target questionnaire, and $N_1, \ldots, N_g, \ldots, N_\delta$ represent each number of workers in each group. By calculating the Cartesian product of S_i for all i and calculating the combinations of $s_{i,g}$ and $p_{i,g}$, we can reconstruct a histogram using the Bayes approach proposed in [16].

The estimated mean squared error (MSE) of group g can be calculated by

$$E[\text{MSE}_g] = \frac{(1 - F)(p_g^2 F + s_g - (F + 2p_g)s_g + s_g^2)}{N_g F (s_g - p_g F)^2}, \tag{3}$$

where $F = \prod_i F_i$.

5.3.2 Final Histogram Generation

Integrating method of different results was proposed by [15]; however, they did not use differential privacy but used their privacy definition.

The requester calculates the weighted summation of each reconstructed value based on its MSE. Specifically, let δ denote the number of different privacy-protection levels among all workers.

We use $X_{g,i}$ to represent the number of requesters whose true category is C_i in requester group G_g. Let $\widehat{X_{g,i}}$ denote the estimated value of $X_{g,i}$.

First, we calculate $\widehat{X_{g,i}}$. We focus on each value $\widehat{X_{g,i}}$ of C_i; then, we determine the final reconstructed value of $\widehat{X_i}$.

Let $\sigma_{g,i}^2$ denote the MSE of category C_i in requester group G_g. The weighted summation of each $\widehat{X_{g,i}}$ based on its MSE can be expressed as follows:

$$\widehat{X_i} = N \cdot \frac{\sum_{g=0}^{\delta-1}(1/E[\sigma_{g,i}^2])\widehat{X_{g,i}}/N_g}{\sum_{g=0}^{\delta-1} 1/E[\sigma_{g,i}^2]}, \tag{4}$$

where the expected MSE of category i of each group g is calculated from Eq. 3.

6 Evaluation

We used two baseline methods: the simple summation method and middle epsilon method. The simple summation method uses Eq. 5 for integrating each group's result.

$$\widehat{X_i} = \sum_{g=0}^{\delta-1} \widehat{X_{g,i}}. \tag{5}$$

In the middle epsilon method, the requester sets one value for ε. If a worker thinks the privacy-protection level is enough, the worker participates in the task. Contrariwise, if the worker does not consider the privacy-protection level as being enough, the worker does not participate in the task. In the experiment, we assumed that there were three privacy-protection levels, and each worker can choose one value from $\varepsilon = 0.1$, 0.5, or 1.0, randomly. In addition, the requester sets ε to 0.5 for all workers.

We used the Adult dataset[1] used in many articles on privacy-preserving data mining. Owing to the space reason, we show one experimental result using the age attribute of the Adult dataset. We divided the age values into nine groups. Each worker determined the value of ε from $\{0.1, 0.5, \text{or } 1.0\}$ randomly. The results of MSEs are shown in Fig. 2a; the MSE of the proposed method was the smallest compared with other methods.

Figure 2b shows the example results of the original and reconstructed histograms. The histogram of the proposed method is closest to the original histogram.

7 Conclusions

We proposed a differentially private questionnaire analysis method for large-scale crowdsourcing. The proposed method can handle multiple questions, and each worker can determine his/her privacy-protection level for each question. We showed that the proposed method can reduce MSE by approximately 65% compared with baseline approaches. As a future task, we should conduct experiments with other datasets. We also plan to devise a method that can be used with location data.

---

[1] https://archive.ics.uci.edu/ml/datasets.html.

Fig. 2 Experimental results

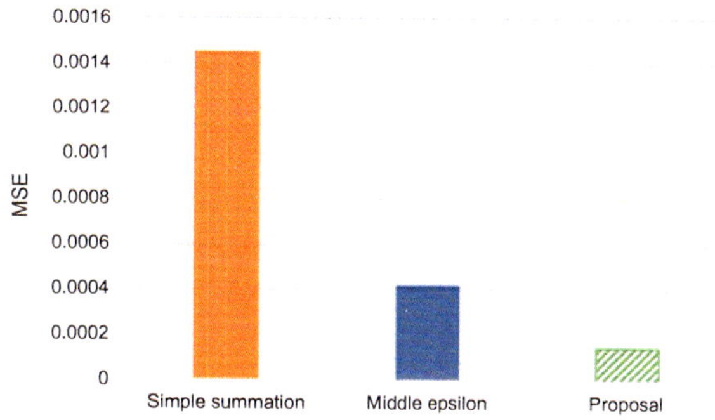

(a) Results of MSEs.

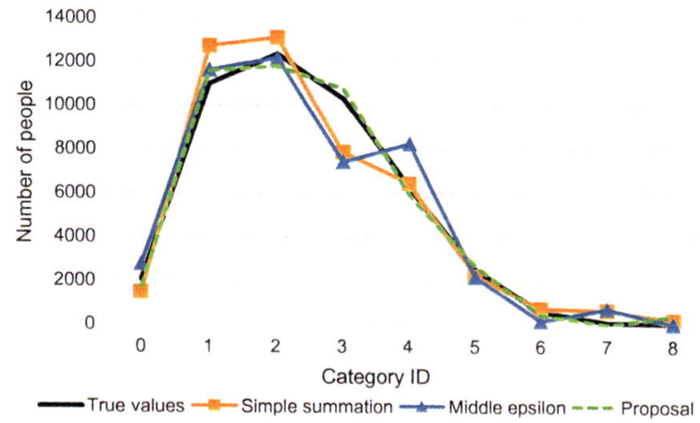

(b) Example of output histograms.

Acknowledgements This work was supported by JSPS KAKENHI Grant Numbers JP17H04705, JP18H03229, JP18H03340, JP18K19835, JP19K12107, JP19H04113, JP21H03496, and JST, PRESTO Grant Number JPMJPR1934.

References

1. J. Acharya, K. Bonawitz, P. Kairouz, D. Ramage, Z. Sun, Context-aware local differential privacy, in *Proceedings of ICML* (2020). http://arxiv.org/abs/1911.00038
2. J. Andrew, J. Karthikeyan, J. Jebastin, Privacy preserving big data publication on cloud using Mondrian anonymization techniques and deep neural networks, in *Proceedings of International Conference on Advanced Computing & Communication Systems (ICACCS)* (IEEE, 2019), pp. 722–727. https://doi.org/10.1109/ICACCS.2019.8728384
3. J. Chodera, A.A. Lee, N. London, F. von Delft, Crowdsourcing drug discovery for pandemics. Nat. Chem. **12**(7) (2020). https://doi.org/10.1038/s41557-020-0496-2
4. A. Desai, J. Warner, N. Kuderer, M. Thompson, C. Painter, G. Lyman, G. Lopes, Crowdsourcing a crisis response for COVID-19 in oncology. Nat. Cancer **1**(5) (2020). https://doi.org/10.1038/s43018-020-0065-z
5. J.C. Duchi, M.I. Jordan, M.J. Wainwright, Local privacy and statistical minimax rates, in *Proceedings of IEEE 54th Annual Symposium on Foundations of Computer Science (FOCS)*

(IEEE, 2013), pp. 429–438. https://doi.org/10.1109/FOCS.2013.53
6. C. Dwork, Differential privacy, in *Proceedings of ICALP* (2006), pp. 1–12
7. Ú. Erlingsson, V. Pihur, A. Korolova, RAPPOR: randomized aggregatable privacy-preserving ordinal response, in *Proceedings of ACM CCS* (2014), pp. 1054–1067
8. R. Grover, M. Vriens, *The Handbook of Marketing Research: Uses, Misuses, and Future Advances* (Sage, 2006)
9. S.P. Kasiviswanathan, H.K. Lee, K. Nissim, S. Raskhodnikova, A. Smith, What can we learn privately? SIAM J. Comput. **40**(3), 793–826 (2013)
10. K. LeFevre, D. DeWitt, R. Ramakrishnan, Mondrian multidimensional K-anonymity, in *Proceedings of IEEE ICDE* (2006), p. 25
11. A. Machanavajjhala, D. Kifer, J. Gehrke, M. Venkitasubramaniam, D. Kifer, M. Venkitasubramaniam, l-diversity: privacy beyond k-anonymity. ACM TKDD **1**(1), 3 (2007)
12. T. Murakami, Y. Kawamoto, Utility-optimized local differential privacy mechanisms for distribution estimation, in *Proceedings of USENIX Security Symposium* (2019), pp. 1877–1894
13. K. Oishi, Y. Sei, Y. Tahara, A. Ohsuga, Semantic diversity: privacy considering distance between values of sensitive attribute. Comput. Secur. **94**(101823), 1–18 (2020). https://doi.org/10.1016/j.cose.2020.101823
14. L. Rocher, J.M. Hendrickx, Y.A. de Montjoye, Estimating the success of re-identifications in incomplete datasets using generative models. Nat. Commun. **10**(1), 1–9 (2019)
15. Y. Sei, A. Ohsuga, Privacy preservation for participatory sensing applications, in *Proceedings of IEEE AINA* (2016), pp. 653–660. https://doi.org/10.1109/AINA.2016.19
16. Y. Sei, A. Ohsuga, Differential private data collection and analysis based on randomized multiple dummies for untrusted mobile crowdsensing. IEEE Trans. Inf. Forensics Secur. **12**(4), 926–939 (2017)
17. L. Sweeney, k-anonymity: a model for protecting privacy. Int. J. Uncertainty Fuzziness Knowl.-Based Syst. **10**(05), 557–570 (2002)
18. Y. Tong, Z. Zhou, Y. Zeng, L. Chen, C. Shahabi, Spatial crowdsourcing: a survey. VLDB J. **29**(1) (2020). https://doi.org/10.1007/s00778-019-00568-7
19. M.J. Wood, J. Ross-Kerr, *Basic Steps in Planning Nursing Research: From Question to Proposal* (Jones & Bartlett Publishers, 2010)

An Optimized K-means Clustering Approach on Top of MapReduce

Omar Abdul Wahab

1 Introduction

Cluster analysis is an unsupervised (and semi-supervised) learning approach that aims to cluster a group of data points based on some similarity measures [8]. That is, the data points that share some common properties are grouped into the same cluster. Technically speaking, for each data point, a set of features has to be extracted and some similarity/dissimilarity measures (e.g., Manhattan distance) have to be applied to determine how close/far these points are to/from each other. Due to its power in exploring hidden relationships among data points in large-scale datasets, cluster analysis had been adopted in many critical disciplines such as medicine, image segmentation, market analysis, document clustering, and network management [1]. For this purpose, plenty of clustering techniques have been proposed to fit different technical needs and data types. The existing algorithms can be classified into three major categories: partition-based, density-based, and hierarchical [7].

The main idea of partition-based clustering is to minimize a certain criterion (e.g., distance) through iteratively assigning and reassigning data points among clusters until a (local) optimal partition is found [2]. Partition-based clustering techniques are often praised for their efficiency in processing spherical clusters in small to medium datasets. However, these approaches have many drawbacks that limit their applicability in many scenarios. One of these drawbacks lies in the difficulty of finding globally optimal partitions, a problem that is known to be NP-hard [2]. Therefore, local optimal partitions are often generated instead. Another drawback of partition-based clustering refers to its reliance on randomly generated cluster centers in its initial phase, which negatively influences the quality of the generated clusters and increases its time to converge to a stable clustering structure. The most popular partition-based clustering algorithms are K-means and K-medoids.

O. A. Wahab (✉)
Université du Québec en Outaouais, Gatineau, QC, Canada
e-mail: omar.abdulwahab@uqo.ca

J. D. Peter et al. (eds.), *Disruptive Technologies for Big Data and Cloud Applications*, Lecture Notes in Electrical Engineering 905,
https://doi.org/10.1007/978-981-19-2177-3_3

In density-based clustering, clusters are determined through detecting those dense regions that are separated by regions of lower data points density [6]. Then, a cluster is formally defined as a maximal set of density-connected data points. Data points that do not belong to any of the identified clusters are then deemed to be noises. The main advantage of density-based clustering approaches lies in their ability to recognize clusters of arbitrary shapes. However, the main limitation of density-based clustering techniques is that they become less efficient when datasets have large differences in their densities. The most popular density-based clustering algorithm is the density-based spatial clustering of applications with noise (DBSCAN).

Hierarchical clustering techniques try to find a hierarchy of clusters that resembles a tree structure, called a dendrogram [5]. Hierarchical clustering techniques can be either *agglomerative* or *divisive*. Agglomerative techniques follow a bottom-up approach in which each data point is considered to be a separate clustering in the beginning [3]. Then, some greedy techniques are applied to join these clusters together, through taking, at each iteration, the two most similar clusters and merging them together. On the other hand, divisive techniques follow a top-down approach where all data points form one single cluster in the beginning [11]. Thereafter, the cluster is recursively divided into two clusters at a time using some common clustering technique (e.g., K-means). The main advantages of hierarchical clustering are that they do not require a priori knowledge for the number of clusters and that they are easy to visualize and interpret. The main weaknesses of this type of clustering are its sensitivity to noise and outliers and its high complexity in large-scale datasets.

1.1 Contributions

In this paper, we focus on K-means, which is a partition-based clustering technique. In K-means, K data points are initially randomly chosen to be the cluster centers. Then, in the cluster assignment phase, the algorithm finds, for every data point, the cluster center that is the closest to it and adds the data point to the underlying cluster. In the cluster recentering phase, the algorithm replaces each cluster center with the average of the data points that belong to the cluster. This process is continuously repeated until no change in the clusters' members identity takes place. We provide in this work two contributions to improve the performance and applicability of K-means. In the first contribution, we incorporate a statistical method into K-means to optimize the process of electing initial cluster centers. In the second contribution, we explain how K-means can be integrated into the MapReduce programming model and provide the details of the *map* and *reduce* functions that should contain the logic of the distributed K-means algorithm. This aims at enabling the distributed and parallel execution of K-means on top of large clusters using big data analytics techniques such as Hadoop [10] and Apache Spark [13].

2 Proposed Solution

In this section, we discuss our two contributions in detail. First, we explain the statistical method that seeks to improve the selection of initial centorids. Second, we give the details of the MapReduce implementation of K-means and explain the map and reduce functions.

2.1 *Initial Centroids Selection: A Statistical Method*

We begin by explaining the statistical method that we integrate into the original K-means to optimize the process of selecting initial cluster centers (a.k.a centroids) whose methodology is inspired by the work proposed in [12]. Given a certain feature (e.g., salary, energy consumption, etc.), the main idea of the statistical method is to calculate the variance between each data point and the rest of data points in terms of the considered feature and then calculate the variance of the whole set of data points. Thereafter, these two variances are compared, and the data points whose variance is lower than the that of the whole set are marked to be eligible for being selected as initial centroids. This step is important to exclude the outliers that have high variance from being chosen as centroids. The next step is to further filter the set of eligible data points through computing the Euclidean distance between each pair of data points in terms of the chosen feature. The data point which minimizes this distance with respect to all other data points is then selected to be the first centroid. To ensure the cohesion of the clustering process and avoid the case of overlapping clusters, the second centroid is chosen to be the one that maximizes the distance from the first centroid. Having selected two initial centroids, the next step is to augment the number of centroids up to κ (the number of clusters specified by the user). To do so, a repeated search process for κ times is initiated inside each of the two established clusters to find the data points that maximize the distance from each of the chosen initial centroids. In the following, we give the details of the above-described statistical method and illustrate the associated formulae:

- **Step 1**: Calculate the Euclidean distance $d(x_i, x_j)$ between each pair (x_i, x_j) of data points according to a set of K features using Eq. (1).

$$d(x_i, x_j) = \sqrt{\sum_{k=1}^{K} (x_i^k - x_j^k)^2} \tag{1}$$

- **Step 2**: Compute the variance between all data points and their mean using Eq. (2).

$$\sigma = \sqrt{\frac{1}{n-1} \sum_{i=1}^{n} d(x_i - \bar{x})^2} \tag{2}$$

$$\bar{x} = \sum_{i=1}^{n} \frac{x_i}{n} \tag{3}$$

- **Step 3**: Compute the variance between each data point x_i and all other data points using Eq. (4).

$$\sigma_i = \sqrt{\frac{1}{n-1} \sum_{i=1}^{n} d(x_i, x_j)^2} \tag{4}$$

- **Step 4**: Add the data points having a variance that is less (or equal) than the variance of the whole set of data points (proportionally to a stretch factor ω) to the set μ of all eligible candidates to be initial centroids (Eq. (5)).

$$\mu = \{x_i | \sigma_i \leq \omega\sigma, i = 1, \ldots, n\} \tag{5}$$

- **Step 5**: Compute the distance d_i of each data point x_i (from all other data points) using Eq. (6):

$$d_i = \sum_{j=1}^{n} d(x_i, x_j) \tag{6}$$

- **Step 6**: Pick out the first centroid μ_1 to be the one which minimizes the distance (computed as per Eq. (6)) from all other data points using Eq. (7):

$$\mu_1 = \arg\min_{x_i \in \mu} \{d_i | i = 1, \ldots, n\} \tag{7}$$

- **Step 7**: Pick out the second centroid μ_2 to be the one which maximizes the distance from the first centroid μ_1 as depicted in Eq. (8):

$$\mu_2 = \arg\max_{x_i \in \mu} \{d(x_i, \mu_1) | i = 1, \ldots, n\} \tag{8}$$

- **Step 8**: Select κ new centroids within each cluster as the ones that are the farthest from their current center (μ_1 or μ_2) using Eq. (9).

$$c' = \arg\max_{x_l \in a_i \cap P} \{d(x_l, \mu_i) | i = 1, \ldots, n\} \tag{9}$$

After executing the above-described statistical method, κ initial centroids would be generated. The next step is to perform the actual K-means clustering steps to further improve the quality of the initial clusters.

2.2 Optimized K-means Clustering

We give in Algorithm 1 a thorough description of the optimized K-means clustering process. Steps (7) to (13) implement the statistical method proposed to find the set of initial centroids. Steps (14) to (18) implement the K-means clustering process which can be categorized into two phases: cluster assignment phase and centroids reassignment phase. In the cluster assignment phase (Steps (14) and (15)), the algorithm finds, for each data point x_i, the index j of the centroid μ_j that is the closest to x_i, i.e., $z_j \leftarrow \arg\min_j \|\mu_j - x_i\|^2$. In the centroids reassignment phase, the value of each centroid μ_j is updated to be the mean of the data points assigned to cluster j, i.e., $\mu_j = \frac{1}{n_j} \sum_{i:z_i=k} x_i$.

Algorithm 1: Optimized K-means clustering

1: **Input**: Set of data points
2: **Input**: Number κ of clusters
6: **procedure** OptimizedKMeans
7: Use Eq. (1) to derive the distance between each pair of data points
8: Use Eq. (2) to compute the variance of the whole set of data points
9: Use Eq. (4) to derive each data point's variance w.r.t all other data points
10: Use Eq. (5) to determine the set μ of candidate initial centroids
11: Use Eqs. (7) and (8) to determine two initial centroids
12: Assign each data point to the closest centroid
13: Use Eq. (9) to augment the number of clusters up to κ
14: Compute the distance between each data point and each centroid μ_j
15: Assign each data point x_i to the cluster of index z_j whose centroid μ_j gives the minimum distance, i.e., $z_j \leftarrow \arg\min_j \|\mu_j - x_i\|^2$
16: Recompute each centroid μ_j such that $\mu_j = \frac{1}{n_j} \sum_{i:z_i=k} x_i$
17: Recompute the distance between each data point and the new centroids
18: If no data point was reassigned then stop, otherwise repeat from step 14
19: **end procedure**

2.3 MapReduce for Optimized K-means

MapReduce is a programming model which enables the processing of big data in parallel and distributed fashions over a cluster of nodes [4, 9]. A MapReduce program consists of a *map* phase and a *reduce* phase. In the map phase, input data are processed to generate a set of intermediate key-value pairs, i.e., $map(k_1, v_1) \rightarrow \text{list}(k_2, v_2)$. In the reduce phase, intermediate values with the same key received from *mappers* are processed to output a (possibly) smaller set of values for each particular key, i.e., $(k_2, \text{list}(v_2)) \rightarrow (k_2, \text{list}(v_2))$.

To run K-means on top of MapReduce, the cluster assignment phase can be executed in the map phase, where mappers can, for each data point given as an input to the map function, **independently** assign this data point to the centroid with minimal distance from it. The *key* of the map function would be the cluster label, and the data point would represent the *value* of the function. On the other hand, the centroids reassignment phase can be executed in the reduce phase since it consists of aggregating values (to compute the average) **within the same key (i.e., cluster label)**.

Technically speaking, the map phase runs on each data point x_i through computing, for each x_i, the squared distance between x_i and each centroid. It then determines the centroid μ_j that minimizes this distance. The cluster index of the chosen centroid along with a data point assigned to it is then emitted as the intermediate key-value pair to be passed to the reduce phase. The pseudocode of the map phase is given in Algorithm 2.

Algorithm 2: Optimized K-means—map phase

Input: $\mu = [\mu_1, \cdots, \mu_j]$
Input: x_i
procedure MAP(μ,x_i)
    $z_j \leftarrow \arg\min_j \|\mu_j - x_i\|^2$
    $emit(z_j, x_i)$
end procedure

The reduce phase is a simple pairwise summation over the values for each key followed by a simple division operation. The reduce function takes as inputs a cluster label j emitted by the map function, along with the list X of all the data points that are associated with that label. Given a two-value pair of data points belonging to a particular key, the reduce function combines these data points through adding up each corresponding element in the pair. It then computes the new mean μ' of the data points for a particular cluster label through dividing the sum of their values by their count. The output of the reduce function is a cluster label and its associated new mean. The pseudocode of the reduce phase is given in Algorithm 3.

Algorithm 3: Optimized K-means—reduce phase

Input: Cluster index j
Input: Set X_j of data points associated with cluster j
procedure REDUCE(j,X_j)
    $sum = 0$
    $count = 0$
    for each $x_j \in X_j$
        $sum = sum + x_j$
        $count = count + 1$
        $\mu' = sum/count$
    end for
    emit(j,μ')
end procedure

3 Conclusion and Future Work

This paper provides two contributions toward boosting the efficiency and applicability of the K-means clustering technique in big data environments. In the first contribution, we integrate a statistical method on top of K-means to optimize the process of selecting initial centroids, thus improving the quality of the generated clusters and reducing the time to converge to a stable clustering topology. In the second contribution, we illustrate how the optimized K-means can be practically executed in parallel and distributed fashions using the MapReduce programming model. This helps researchers and practitioners run K-means on top of state-of-the-art big data analytics platforms (e.g., Hadoop and Apache Spark) to analyze massive amounts of data.

References

1. M.R. Anderberg, *Cluster Analysis for Applications: Probability and Mathematical Statistics: A Series of Monographs and Textbooks*, vol. 19 (Academic Press, 2014)
2. S. Äyrämö, T. Karkkainen, Introduction to partitioning-based clustering methods with a robust example. Rep. Dep. Math. Inf. Technolo. Ser. C Softw. Eng. Comput. Intell. (1/2006) (2006)
3. I. Davidson, S. Ravi, Agglomerative hierarchical clustering with constraints: theoretical and empirical results, in *European Conference on Principles of Data Mining and Knowledge Discovery* (Springer, 2005), pp. 59–70
4. J. Dean, S. Ghemawat, Mapreduce: simplified data processing on large clusters. Commun. ACM **51**(1), 107–113 (2008)
5. S.C. Johnson, Hierarchical clustering schemes. Psychometrika **32**(3), 241–254 (1967)
6. H.P. Kriegel, P. Kroger, J. Sander, A. Zimek, Density-based clustering. Wiley Interdiscip. Rev. Data Min. Knowl. Discov. **1**(3), 231–240 (2011)
7. F.H. Kuwil, F. Shaar, A.E. Topcu, F. Murtagh, A new data clustering based on critical distance methodology. Expert Syst. Appl. (2019)
8. C. Romesburg, Cluster analysis for researchers. Lulu.com (2004)
9. O.A. Wahab et al., Federated machine learning: survey, multi-level classification, desirable criteria and future directions in communication and networking systems. IEEE Commun. Surv. Tutor. **23**(2), 1342–1397 (2021)
10. T. White, *Hadoop: The Definitive Guide* (O'Reilly Media, Inc., 2012)
11. T. Xiong, S. Wang, A. Mayers, E. Monga, DHCC: divisive hierarchical clustering of categorical data. Data Min. Knowl. Discov. **24**(1), 103–135 (2012)
12. D. Yu, G. Liu, M. Guo, X. Liu, An improved k-medoids algorithm based on step increasing and optimizing medoids. Expert Syst. Appl. **92**, 464–473 (2018)
13. M. Zaharia, R.S. Xin, P. Wendell, T. Das, M. Armbrust, A. Dave, X. Meng, J. Rosen, S. Venkataraman, M.J. Franklin et al., Apache spark: a unified engine for big data processing. Commun. ACM **59**(11), 56–65 (2016)

A Framework to Preserve and Examine Pandemic-Healthcare-Data Using IoMT

Seifedine Kadry and Venkatesan Rajinikanth

1 Introduction

The progress in science and engineering domain is rapidly rising due to a considerable research works, and this improvement helped to create a better living ambience to the human community. The development in technology enhanced the facilities of the healthcare sector, which includes automated diagnosis, disease prediction, remote patient monitoring, disease information preservation, and telemedicine [1, 2]. Further, the usage of eHealth facility enhanced the ability in patient monitoring particularly when the patient condition is extremely ill. The subdivision of eHealth known as the mobile-health (mHealth) also helps to diagnose the patient with high-speed wireless network [3, 4].

The recent development in medical domain; Healthcare 4.0 integrates all the recent innovations, such as big data processing, cloud computing, Internet of medical things (IoMT), and artificial intelligence (AI) schemes to support the better patient monitoring and diagnosis in hospital and remote level. Further, this scheme is timely updated by including the recent innovations in the abovesaid subdivisions.

Even though a considerable number of medical facilities are implemented to enhance the healthcare facilities, the disease occurrence rates in humans are rapidly rising, and this will cause a burden to hospitals, when the number of patients are rising. During the spread of the pandemic disease, large human groups are infected

S. Kadry
Faculty of Applied Computing and Technology, Noroff University College, 4608 Kristiansand, Norway

V. Rajinikanth (✉)
Department of Electronics and Instrumentation Engineering, St. Joseph's College of Engineering, OMR, Chennai, Tamilnadu 600119, India
e-mail: rajinikanthv@stjosephs.ac.in

J. D. Peter et al. (eds.), *Disruptive Technologies for Big Data and Cloud Applications*, Lecture Notes in Electrical Engineering 905,
https://doi.org/10.1007/978-981-19-2177-3_4

with the disease, and this situation will cause a heavy burden in disease diagnosis and treatment implementation [5, 6].

From December 2019, respiratory infection due to COVID-19 affected a large human group irrespective of their race and gender and due to this, World Health Organization (WHO) declared it as pandemic [7]. The diagnosis of COVID-19 infection involves in a series of procedures; (i) recognition of the preliminary disease symptoms, (ii) reverse-transcription polymerase-chain-reaction (RT-PCR) test, (iii) lung imaging (CT/X-ray) supported severity detection, and (iv) treatment planning and implementation.

The availability of the COVID-19-related data (specific symptom, image features, treatment protocol followed, drugs considered and its does level, etc.) is gradually rising with increase in the infection rate, and these data must be preserved to know the vital information about the disease and also to develop the disease model. The proposed research aims to develop a framework to preserve the disease features and examine the COVID-19 based on the preserved features. Further, this research also proposes an IoMT framework to support the remote diagnosis of COVID-19. In this work, the lung images recorded with modalities such as CT and X-ray are considered for the assessment, and the diagnosis of the COVID-19 based on the image features along with related information is discussed with appropriated IoMT scheme.

The other sections are arranged as follows; Sect. 2 presents the related works, Sect. 3 depicts the methodology, Sect. 4 depicts the discussion, and Sect. 5 shows the conclusion of this research.

2 Related Research Works

Due to its significance, a considerable number of advanced monitoring facilities are proposed to examine the COVID-19 infection in humans [8]. The earlier works employed procedures, like AI, IoMT, big data, and cloud computing to support the accurate diagnosis of the disease in patient using the clinically collected and preserved disease information.

The summary of these works is presented in Table 1.

Table 1 presented some recently employed approaches to examine the pandemic data using the recently developed methods. The work of Haleem and Javaid [18] confirms the need for the IoMT and AI techniques in developing the Medical 4.0 for the COVID-19 information assessment. When the modern medical facilities are employed, then the PHD can be effectively assessed and shared to support the fasted diagnosis, decision-making, and treatment implementation. The work by Rajinikanth and Kadry [4] presented a detailed discussion regarding the preservation of the disease information for the future disease prediction and evaluation. In the proposed research, similar architecture is proposed to support the COVID-19 diagnosis using the various clinical informations collected from the infected patients. The proposed work also presents a detailed architecture to implement the IoMT framework by considering the AI and big data.

Table 1 Summary of the recent methods in pandemic data assessment

Reference	Methodology employed
Liu et al. [9]	This work proposed a software model to predict the COVID-19 infection in humans based on different individual and clinical information
Sipior [10]	A detailed methodology to be adopted to predict and diagnose the COVID-19 using the recent AI techniques is presented in this research
Ahuja et al. [11]	A summary of the pre-trained deep learning scheme assisted COVID-19 detection based on the disease features is presented with clinical level data
Dey et al. [12]	Assessment of the disease with a chosen machine-learning system is discussed and the proposed methodology is confirmed using the lung CT slices
Sheng et al. [13]	A detailed procedure regarding the implementation of big data to examine the disease information is presented and discussed
Bragazzi et al. [14]	Employment of the AI and big data is discussed with appropriate examples to predict the disease and its security
Javaid and Khan [15]	The need for the Internet of things (IoT) supported healthcare facility is clearly discussed to examine the pandemic data
Aman et al. [16]	Implementation of IoMT-based COVID-19 diagnosis is discussed with appropriate architecture. This work also presented a suggestion related to the network requirement and its security
Al-Turjman [17]	The need for the AI and IoT in COVID-19 detection and the necessary architecture is clearly presented with appropriate experimental results
Haleem and Javaid [18]	A detailed review regarding the role of Medical 4.0 for COVID-19 pandemic is presented

3 Methodology

This section of the research presents the proposed architecture and the related information to implement IoMT-enabled PHD preservation and examination.

Figure 1 depicts the proposed scheme for IoMT-based COVID-19 data preservation and access methodology with the help of cloud computing process. The data existing in the remote server will be continuously accumulated (big data) based on the received PHD. The preserved data in remote server contain the necessary information and which can be considered for the research purpose, disease modeling, and diagnosing the patients by a doctor existing in a remote location with lesser medical facilities. The suggested scheme consists three basic sections; (i) healthcare center level collection, processing and storage of the disease, and personal information, (ii) storing the data in a remote (cloud) server, and (iii) permitting the access to the remote server to doctors/researchers who need the vital information regarding the COVID-19. The main merit of the proposed scheme is that every part of this scheme is interlinked using a wired/wireless Internet facility and permits the controlled access

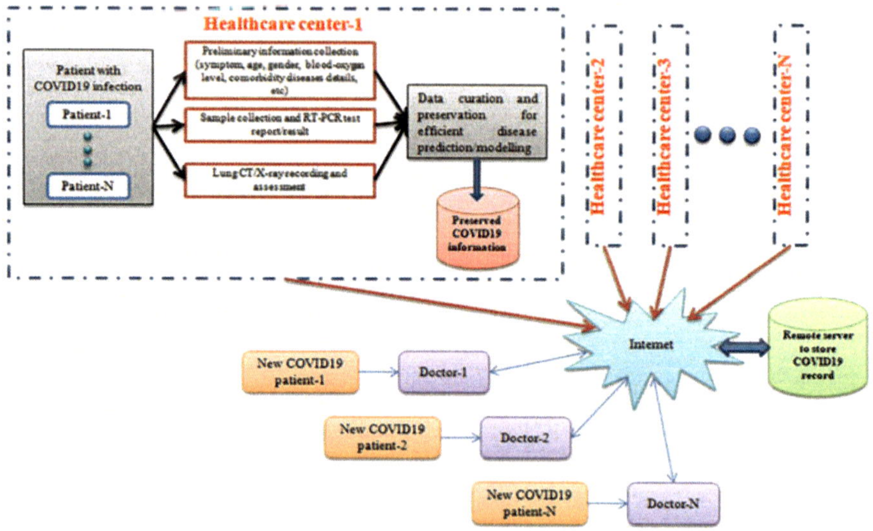

Fig. 1 Proposed IoMT supported scheme for healthcare data preserve and disease detection

to the doctors/researchers to access the preserved data to get the essential information about the disease under study.

3.1 Disease Data Collection

When the patient visits the healthcare center with the disease symptom; essential details, such as the patient's age, gender, fever severity, and comorbidity disease (if any), are initially recorded, and the patient is advised to undergo for an RT-PCR test to confirm the patient is having the COVID-19 infection or not (i.e., positive/negative). When the result turns a negative result, then the patient is treated with the general drugs. When the test result is positive, then the patient is permitted to undergo a lung image examination with a chosen modality (CT/X-ray). Further, infection due to double mutant COVID variant (B.1.617) must be diagnosed only with the imaging procedures due to its rapidity [19, 20]. After collecting the necessary information from the patients, appropriate treatment procedure is employed to treat the patient, and the progress in recovery rate is recorded till the patient is discharged from the hospital. From the current scenario, it can be noted that the number of infected people are more and the available COVID data are large and it must be preserved for current/future use at a remote server with controlled access.

3.2 *Preserving the Disease Information*

Disease data preservation plays a major role in medical domain, particularly during the handling of the PHD. In the current era, the improvement in technology and communication helped to support the remote diagnosis of the disease with the IoMT technique. When the necessary communication facility is initiated, then the preserved data can be shared among the global experts to get the timely suggestion regarding the treatment to be employed for the patient admitted with varied severity level. This process plays a vital role when a large number of severely admitted patients are treated by the doctors. The main motive for having the cloud data storage with the controlled access is to support the remote level assessment, modeling, treatment suggestion, and medicine-related suggestions.

3.3 *Treatment of Disease*

The existing information in the cloud server is continuously improved when a large patient's information is shared, and this information consists the complete information regarding; methods to be followed in disease prediction/diagnosis, severity assessment, recommended drugs and its dosage level to be used, recovering the patient from critical condition, and controlling the disease spread. Further, with the help of the preserved data, any doctor (with prior permission) can able to access the cloud data, to get the essential information to implement the testament for the patients.

When this scheme is employed, then the disease diagnosis will become simple, and every doctor can get the complete protocol to be followed to recover the patient from the disease without further delay when the virus version is continuously changing from conventional, single-mutant, double mutant, and future versions.

4 Results and Discussion

This section of the research presents the procedure to be followed in detecting the infection rate using the lung image recorded using CT/X-ray. In the literature, a considerable number of AI supported image processing for the COVID infection detection is proposed and implemented using CT/X-ray images.

The AI supported assessment of the lung CT/X-ray images is demonstrated with appropriate experimental results, and the disease information to be preserved for future reference is discussed. Initially, the lung CT images available in [21–24] are considered for the assessment, and the sample test images with various infection rates are depicted in Fig. 2. Figure 2a–c depicts COVID-19 infection with categories, mild, moderate, and severe. This work implements the lesion segmentation technique

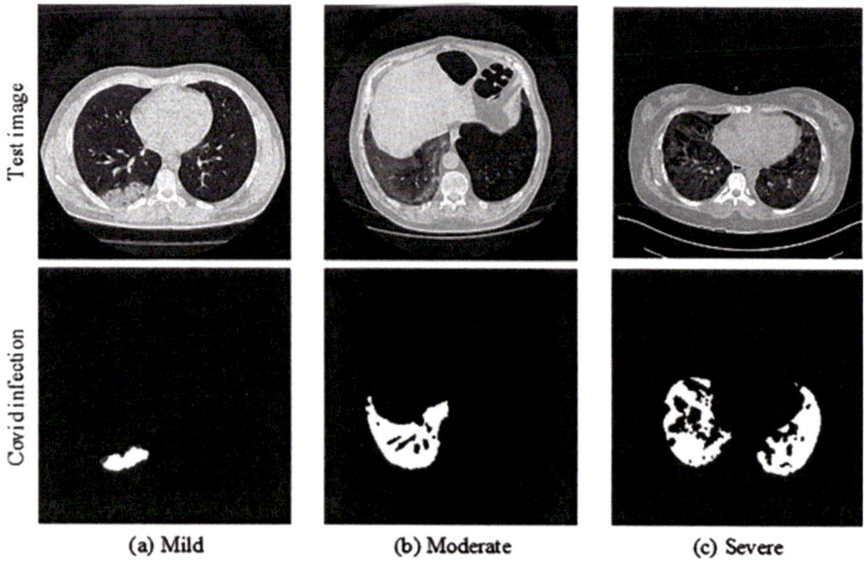

Fig. 2 Sample test images and the extracted COVID lesion

employed in the work of Dey et al. [12], and the essential handcrafted features are extracted and stored as the disease information.

The lung X-ray collected from a benchmark database is presented in Fig. 3 [24], and the visibility of the infection in this image is poor and it needs an appropriate image processing scheme to detect the infected section. After extracting the essential disease information (features), the dominant features are selected and preserved in the database. Figure 3 presents the chosen test image and the result by the deep learning scheme. In this work, the deep learning procedure employed in Rajinikanth and

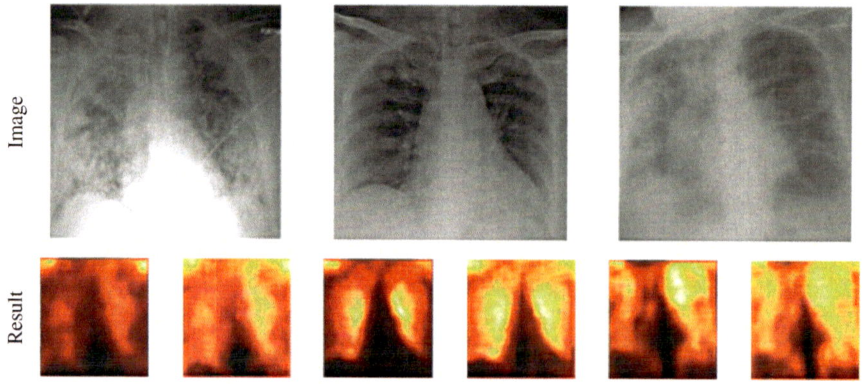

Fig. 3 Sample X-ray and attained result with deep learning

Kadry [4] is adopted, and this technique helped to the deep features. After selecting the essential features from the chosen test image, it is stored in the local/remote server along with the patient information.

When an image processing scheme is employed, it helps to get handcrafted and deep features; which can be processed and preserved for the future reference in a remote/local server. When new patient information is to be examined, then the patient's image is simply compared against the preserved information existing in the server. The comparison of the patient's lung image with the preserved information can be easily performed by a lab staff, and this process considerably reduces the diagnostic burden of the doctor. The doctor is permitted to visually check the lung image and the report provided by the lab staff and based on this assessment, the possible treatment is implemented by the doctor to recover the patient. The report provided by the lab staff includes the personal data of the patient, the nature of lung infection (mild/moderate/severe), and the recommended drug by the computer algorithm. When this scheme is employed, then it is possible to implement the timely treatment for the patients to reduce the impact of the COVID. Further, the critical ill patient will be identified, and the appropriate care will be provided without the delay. After the approval by the ethical committee of the healthcare center, this information can be shared to the local/remote server for further use.

When this scheme is implemented, then every permitted doctor/hospital will have the complete access to the server which is having the essential information about the disease (with varied virus mutants) and its treatment process. When this scheme is employed, every registered healthcare center and the doctor will follow a unique treatment methodology to recover the patient from the disease. Further, the suggestion about the precautionary measures, initial treatment protocol, and the recommended drug and the dose level will be discussed and updated, if necessary to support better treatment.

5 Conclusion

The proposed work of this research aims to develop a scheme to preserve the PHD and support the examination of the disease in new patient using the hospital/remote level data. This scheme helps to support the remote diagnosis, data sharing, and treatment assistance for the patient with varied level of COVID infection. The chief merit of the proposed scheme is the availability of the necessary guidelines about the initial test to be followed, disease diagnosis, severity prediction, drug discovery, treatment implementation, and recovery monitoring. This scheme also supports a networked data handling to exchange the information between hospitals to support the efficient treatment. This scheme also supports the added knowledge about the current variant of the virus and its severity. The preserved data can be considered to prepare a complete COVID infection model, which helps to predict the disease and its progression among humans. The proposed scheme utilizes the efficient communication methodology (Internet); which helps to get the treatment suggestion from the national and global experts.

References

1. J.A. Behar, C. Liu, K. Kotzen, K. Tsutsui, V.D. Corino, J. Singh, … G.D. Clifford, Remote health diagnosis and monitoring in the time of COVID-19. Physiol. Meas. **41**(10), 10TR01 (2020)
2. F. Shi, J. Wang, J. Shi, Z. Wu, Q. Wang, Z. Tang, … D. Shen, Review of artificial intelligence techniques in imaging data acquisition, segmentation and diagnosis for covid-19. IEEE Rev. Biomed. Eng. **14**, 4–15 (2020). https://doi.org/10.1109/RBME.2020.2987975
3. D.N. Vinod, S.R.S. Prabaharan, Data science and the role of artificial intelligence in achieving the fast diagnosis of Covid-19. Chaos Solitons Fractals **140**, 110182 (2020). https://doi.org/10.1016/j.chaos.2020.110182
4. V. Rajinikanth, S. Kadry, Development of a framework for preserving the disease-evidence-information to support efficient disease diagnosis. Int. J. Data Warehous. Min. (IJDWM) **17**(2), 63–84 (2021). https://doi.org/10.4018/IJDWM.2021040104
5. I.F. Miller, A.D. Becker, B.T. Grenfell, C.J.E. Metcalf, Disease and healthcare burden of COVID-19 in the United States. Nat. Med. **26**(8), 1212–1217 (2020)
6. Y. Ji, Z. Ma, M.P. Peppelenbosch, Q. Pan, Potential association between COVID-19 mortality and health-care resource availability. Lancet Glob. Health **8**(4), e480 (2020)
7. R.C. Khanna, M.V. Cicinelli, S.S. Gilbert, S.G. Honavar, G.V. Murthy, COVID-19 pandemic: lessons learned and future directions. Indian J. Ophthalmol. **68**(5), 703 (2020)
8. S. Kadry, F. Al-Turjman, V. Rajinikanth, Automated segmentation of COVID-19 Lesion from Lung CT images using U-Net architecture, in *International Summit Smart City 360°* (Springer, Cham, 2020), pp. 20–30. https://doi.org/10.1007/978-3-030-76063-2_2
9. X.X. Liu, S.J. Fong, N. Dey, R.G. Crespo, E. Herrera-Viedma, A new SEAIRD pandemic prediction model with clinical and epidemiological data analysis on COVID-19 outbreak. Appl. Intell. 1–37 (2021). https://doi.org/10.1007/s10489-020-01938-3
10. J.C. Sipior, Considerations for development and use of AI in response to COVID-19. Int. J. Inf. Manage. **55**, 102170 (2020)
11. S. Ahuja, B.K. Panigrahi, N. Dey, V. Rajinikanth, T.K. Gandhi, Deep transfer learning-based automated detection of COVID-19 from lung CT scan slices. Appl. Intell. **51**(1), 571–585 (2021). https://doi.org/10.1007/s12559-020-09751-3
12. N. Dey, V. Rajinikanth, S.J. Fong, M.S. Kaiser, M. Mahmud, Social group optimization–assisted Kapur's entropy and morphological segmentation for automated detection of COVID-19 infection from computed tomography images. Cogn. Comput. **12**(5), 1011–1023 (2020). https://doi.org/10.1007/s12559-020-09751-3
13. J. Sheng, J. Amankwah-Amoah, Z. Khan, X. Wang, COVID-19 pandemic in the new era of big data analytics: methodological innovations and future research directions. Br. J. Manag. (2020). https://doi.org/10.1111/1467-8551.12441
14. N.L. Bragazzi, H. Dai, G. Damiani, M. Behzadifar, M. Martini, J. Wu, How big data and artificial intelligence can help better manage the COVID-19 pandemic. Int. J. Environ. Res. Public Health **17**(9), 3176 (2020)
15. M. Javaid, I.H. Khan, Internet of Things (IoT) enabled healthcare helps to take the challenges of COVID-19 Pandemic. J. Oral Biol. Craniofac. Res. **11**(2), 209–214 (2021)
16. A.H.M. Aman, W.H. Hassan, S. Sameen, Z.S. Attarbashi, M. Alizadeh, L.A. Latiff, IoMT amid COVID-19 pandemic: application, architecture, technology, and security. J. Netw. Comput. Appl. **174**, 102886 (2020). https://doi.org/10.1016/j.jnca.2020.102886

17. F. Al-Turjman, AI-powered IoT for COVID-19. CRC Press (2020)
18. A. Haleem, M. Javaid, Medical 4.0 and its role in healthcare during COVID-19 pandemic: A review. J. Ind. Integr. Manag. **5**(4), 531–545 (2020). https://doi.org/10.1142/S2424862220300045
19. M. Hoffmann, H. Hofmann-Winkler, N. Krüger, A. Kempf, I. Nehlmeier, L. Graichen, … S. Pöhlmann, SARS-CoV-2 variant B. 1.617 is resistant to Bamlanivimab and evades antibodies induced by infection and vaccination. bioRxiv (2021)
20. P.D. Yadav, S. Mohandas, A.M. Shete, D.A. Nyayanit, N. Gupta, D.Y. Patil, … B. Bhargava, SARS CoV-2 variant B. 1.617. 1 is highly pathogenic in hamsters than B. 1 variant. bioRxiv (2021)
21. https://mosmed.ai/en/
22. M.E. Chowdhury, T. Rahman, A. Khandakar, R. Mazhar, M.A. Kadir, Z.B. Mahbub, … M.T. Islam, Can AI help in screening viral and COVID-19 pneumonia? IEEE Access **8**, 132665–132676 (2020). https://doi.org/10.1109/ACCESS.2020.3010287
23. http://medicalsegmentation.com/covid19/
24. https://ieee-dataport.org/open-access/covid-19-and-normal-chest-x-ray

AR Cloud-Based Indoor Navigation

V. Kiruthika, M. Jagadeeswari, Sneha Prabha, and Sreejaa

1 Introduction

Augmented reality becomes a part of day-to-day life of every humans. With the advance of other technologies such as sensing technologies, graphics, mobile computing, and computer vision systems, augmented reality plays a main role in achieving people's goal effectively and efficiently with utmost satisfaction. This augmented reality intensifies real world in which the user can experience the real-world scenario. AR obtains data from the cloud and provides immediate solution according to the application.

3D indoor navigation can be achieved by using AR technology which provides turn by turn navigation instructions and an arrow pointer for guiding the user to reach the correct destination. The systematic flow of the application will be such that when the user is in need of reaching certain destination; this android application helps the user by providing two types of navigation; one is through Google Map, and another one is through AR, and the database can be stored in AR cloud. The main advantage of AR cloud-based [1] indoor navigation is no additional glasses/wearable devices are needed for the user to achieve their goal.

V. Kiruthika (✉) · M. Jagadeeswari · S. Prabha · Sreejaa
Department of ECE, Sri Ramakrishna Engineering College, Coimbatore, Tamil Nadu, India
e-mail: kiruthika.v@srec.ac.in

M. Jagadeeswari
e-mail: hod-ece@srec.ac.in

S. Prabha
e-mail: snehaprabha.1902224@srec.ac.in

Sreejaa
e-mail: sreejaa.1902229@srec.ac.in

J. D. Peter et al. (eds.), *Disruptive Technologies for Big Data and Cloud Applications*, Lecture Notes in Electrical Engineering 905,
https://doi.org/10.1007/978-981-19-2177-3_5

2 Literature Survey

Prashant Varma et al. [1], focused on creating an indoor navigation system by using augmented reality technology and cloud platform where the administrators can manage the database in the cloud. They compared different indoor navigation systems and concluded that GPS navigation system consumes more power, and the accuracy range is about 6–10 m. The experiment was conducted in a hospital and proved that the augmented reality really works well, and also, the changes can be done simply by the administrator. Dwi Ely et al. [2] created an application which displays 3D objects related to the areas that exist in the map, and it also shows the location in real time by augmented reality. This makes the tourists feel easy to identify the destination. But, the main drawback is that there is no digital advancement. Enis Karaarslan et al. [3], designed a smart tourist guide on mobile application for the tourist people who travel across countries using the technology augmented reality. This helps the people to identify the required destination in real time those who are new to the places. But, the disadvantage is the captured images cannot be processed. Archana Mantri et al. [4] developed an augmented reality navigation in real world without the implementation of 2D map and image processing technology. This navigation system gives accurate results which enables the user to find the location accurately within a short period of time. Wolfgang Narzt et al. [5] created AR system with appropriate information where the user can clearly understand both virtual and real objects alternately by introducing head-mounted displays. Huang et al. [6] who resolved the problems regarding the indoor navigation in a smart intelligent environment. Austin Corotan et al. [7] created AR-based navigation systems using JAQL robot and Google Pixel to android smart phone. While developing this application, they mainly investigated about the current capability of augmented as it is an all-in-one solution for indoor routing, localization, and object detection. Xin Hui Ng et al. [8] built provides an indoor navigation experience using AR technology which uses existing built-in sensors embedded in most of the mobile devices which detects the users' location. They mainly used Indoor Atlas which enables the indoor positioning to detect user's position and obtains the route, and ARCore is used to display AR guidance using the navigation route. Jong-Bae Kim et al. [9] created an AR-based indoor navigation systems where they used personal location method in image sequences which makes the user to navigate to the unknown places in an indoor environment. By identifying the location through image sequences, the indoor navigation uses a color histogram matching method. The results can be view through the augmented reality for the user's view. Georg Gerstweiler [10] basically concentrates on three main topics as why such guiding systems are not still available in real-world situations. Firstly, a tracking solution called HyMo track is presented based visual hybrid tracking approach for mobile phones and tested in a real-world airport scenario. The guiding and tracking parts request a 3D model of the environment for indoor navigation. For this reason, a 3D model generation algorithm is implemented which creates a 3D model of the vectorized 2D floor plan. Finally, the AR guiding system in the humans' aspect is researched, and AR path concept is applied for guiding the people.

3 Motivation

In the older version of this similar application, localization approach is based on trilateration and Wi-Fi access point approach. The distance will be calculated from the starting point of the user through free space path, and it is measured in decibels. The methodology involves two ideas; one is trilateration, and another one is fingerprinting. This will be based on received signal strength indicator (RSSI) measurements. For the localization in indoor positioning and to calculate the distance and location from three starting points the user computes using RSSI. The overall distance between the required points will be calculated by the algorithms whether it is offline or real time. For augmented reality experience, they have used some APIs across android and IOS. It also involves the ARCore which uses three main capabilities to combine the virtual world with the real world which can be visualized through the phone's camera. The key capabilities used in ARCore include:

- Motion tracking which enables the phone to understand and monitor the relevant area with respective to the real world
- Understanding of environment which enables the phone to detect the particular size and location of the types of the surfaces which includes vertical, angled, and horizontal surfaces
- Light estimation which allows the phone to estimates the light with respect to the environment's condition.

ARCore's motion tracking technology generally uses the camera mode of the smart phone which identifies the wave points which is called as features. Also, it tracks on how these points moves over time. By using the measurements detected from the phone's sensors, it can detect the flat surfaces and average lightning in the area. These features enabled ARCore which builds its own understanding with the real world.

4 Proposed Methodology

The AR-based cloud (ARBC) app [6] was developed for 3D indoor navigation. This application can be installed in smart phones for quick and easy navigation. It displays the route for the entire required destination, whatever the user has entered. The ARBC app validates the user, and the current location will get scanned once ARBC is installed on the smart phone. After scanning, the user has to enter the required destination where the user has to reach. The required destination to be reached will be scanned by using bar code.

The bar code gets validated by ARBC app and automatically guides the user by showing 3D arrow marks. Both Google Map and AR mode in ARBC app will be turned on so that the user can switch into any of the mode. All the possible pathways will be shown in the proposed ARBC app where the user can select any of the possible pathways.

The ARBC application is built in such a way that it takes the present location of the user based on the location where the indoor navigation is designed. The outline of the area should be required for the indoor space for reference and based on that the sketch map is designed by using 3D Blender software. After designing the sketch map, this map will be imported into Unity 3D software. It is not necessary to build the sketch maps for each indoor space since bar codes are placed for scanning so that the results will be accurate. As per the destination points, the venue will be plotted with reference of the sketch map. After plotting the venue, the additional floors, 3D arrow marks will be designed where it has to display during the navigation. For identifying the distance between each destination from user's present location unique navigation, floor markers and navigation markers are used, and the points are recognized by AR location path for exact destination. The unique barcodes are created by using barcode generator, and these barcodes are printed using sheet printing, and the barcodes are placed in required pathways. As the user starts navigating through scanning the barcode, the required pathway starts displaying in the real-time environment. This application is completely software oriented and no need of hardware requirement. Here, the latest versions of smart phone can be used for navigation. The smart phones used by the user should have high clarity with high google pixel for accuracy. The time taken to construct the sketch map will be based on the size of the indoor space, and a basic sketch map can be built in two days. The time taken to navigate will be purely based on the user's Internet connectivity. If the user has high bandwidth, it will be navigated in few seconds, and the space required for the entire ARBC app is 33.66 MB. Figure 1 shows the working of proposed ARBC app. Figure 2 shows how the internal design is made while developing ARBC application

Figure 1 shows the working of ARBC app which navigates the user for reaching the required destination. The shortest pathway for reaching the destination is also shown for facilitating the user to reach the required destination within a short period of time.

Figure 2 describes how the application was developed internally, and it shows the internal process of the application.

Fig. 1 ARBC flowchart

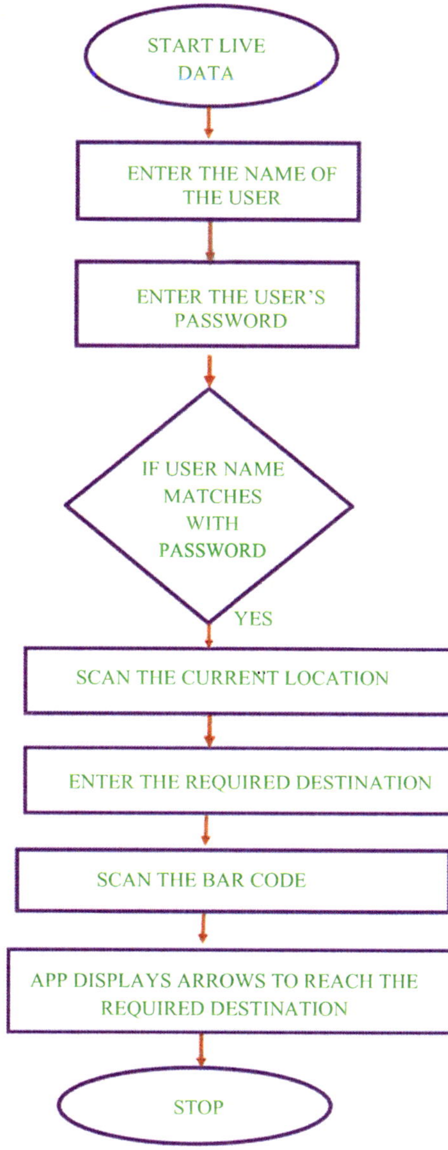

5 Simulation Results

For calculating the distance, this application does not require any algorithms and equations for calculation. There is no need of Wi-Fi access points since original map [11–13] can be implemented. Figure 3 shows the another way of Google Map indication to reach the required destination. Figure 4 shows the bar code indication

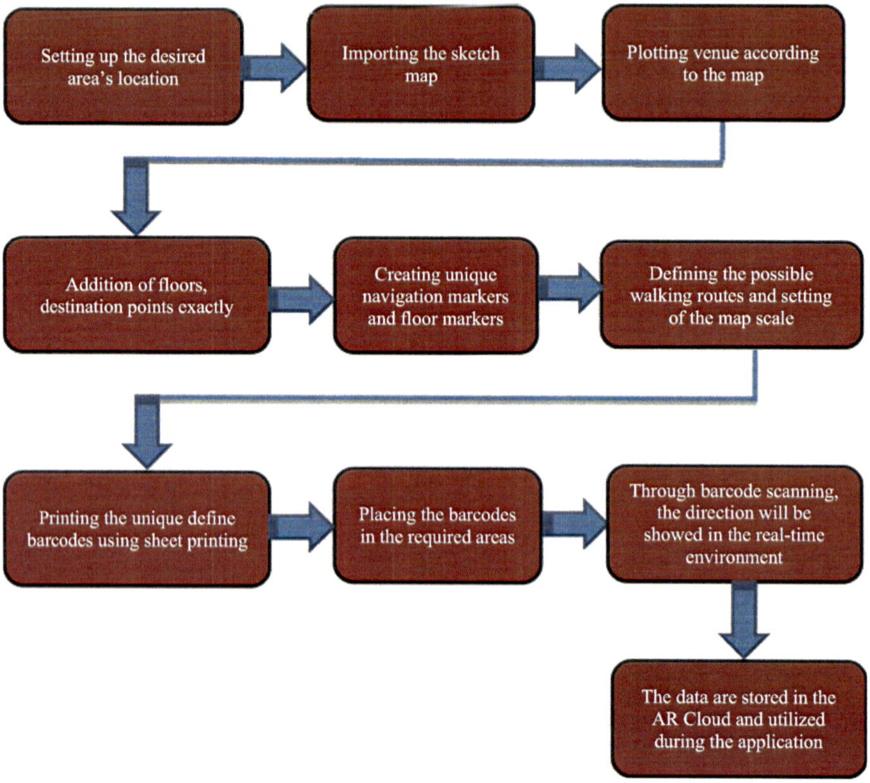

Fig. 2 Internal design on developing ARBC application

for reaching the destination. Figure 5 shows how the arrow marks are displayed in the application to direct the user for reaching the required destination. Figure 6 shows the way how the areas are mapped.

Fig. 3 Google Map indication

Fig. 4 Bar code indication

Fig. 5 Navigation to the destination in AR mode

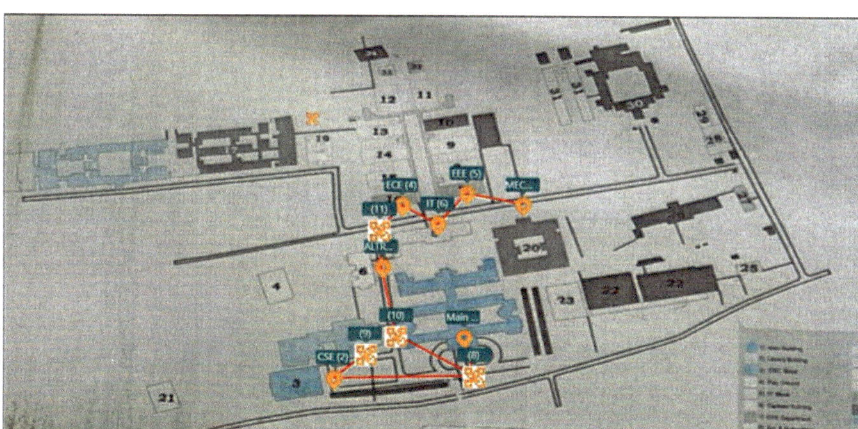

Fig. 6 Possible walking routes

6 Conclusion

This application provides certain information on how to frame a map in real time by using augmented reality (AR) which will help us in path way finding. This application automatically shows the required pathway to reach the destination without any manual guidance. The simulation results are taken from real time, so the exact way can be predicted as in the simulation results. Like Google Maps, this application directs the user to reach the required place using 3D objects. This application makes easier for the users to reach the destination quickly without any confusion. In future, we can develop the same application with higher advancement. Without using GPS and scanning of barcodes, the same application can be built. At some point of time, when the users are using old version smart phones, people find it difficult to scan the barcodes. So, if the directions are inbuilt automatically, then there will be no necessity for barcodes. So, any type of smart phones can be used for this application. These additional features enable the users to handle the application even more easier than the previous application.

References

1. N.S. Sandu, E. Gide, S. Karim, Improving learning through cloud-based mobile technologies and virtual and augmented reality for Australian higher education, in *Proceedings of the 2019 International Conference on Mathematics, Science and Technology Teaching and Learning*. https://doi.org/10.1145/3348400.3348413
2. P. Varma, K. Agrawal, V. Sarasvati, Indoor navigation using augmented reality, in *Proceedings of the 2020 4th International Conference on Virtual and Augmented Reality* (2020), pp. 58–63. https://doi.org/10.1145/3385378.3385387
3. D.E. Kurniawan, A. Dzikri, M. Suriya, Y. Rokhayati, A. Najmurrokhman, Object visualization using maps marker based on augmented reality, in *2018 International Conference on Applied Engineering (ICAE)*. https://doi.org/10.1109/INCAE.2018.8579411
4. E. Karaarslan, O.F. Demir, Augmented reality-application for smart tourism. IEEE Power Energy Soc. https://doi.org/10.13140/RG.2.2.25672.42242
5. A. Mantri, B. Horan, D.P. Kaur, A framework utilizing augmented reality to enhance the teaching–learning experience of linear control systems. IETE J. Res. https://doi.org/10.1080/03772063.2018.1532822
6. M. Chen, C. Ling, W. Zhang, Analysis of augmented reality application based on cloud computing, in*2011 4th International Congress on Image and Signal Processing, IEEE Conference*.https://doi.org/10.1109/CISP.2011.6100311
7. A.F.G. Ferreira, D.M.A. Fernandes, A.P. Catarino, J.L. Monteiro, Localization and positioning systems for emergency responders: A survey. IEEE Commun. Surveys Tuts. **19**(4), 2836–2870 (2017)
8. X. Wang, L. Gao, S. Mao, S. Pandey, CSI-based fingerprinting for indoor localization: a deep learning approach. IEEE Trans. Veh. Technol. **66**(1), 763–776 (2017)
9. H. Moustafa, H. Kenn, K. Sayrafian, W. Scanlon, Y. Zhang, Mobile wearable communications [Guest Editorial]. IEEE Wirel. Commun. **22**(1), 10–11 (2015)
10. U. Rehman, S. Cao, Augmented reality-based indoor navigation using google glass as a wearable head-mounted display, in *Proceedings of IEEE International Conference in Systems, Man and Cybernetics* (2015), pp. 1452–1457

11. J.Z. Liang, E. Turner, A. Zakhor, N. Corso, Image-based positioning of mobile devices in indoor environments, in *Multimodal Location Estimation of Videos and Images* (Springer, New York, NY, USA, 2015), pp. 85–89
12. C. Luo et al., Pallas: Self-bootstrapping fine-grained passive indoor localization using WiFi monitors. IEEE Trans. Mobile Comput. **16**(2), 466–481 (2017)
13. J.L.V. Carrera, Z. Zhao, T. Braun, H. Luo, F. Zhao, *Discriminative learning-based smartphone indoor localization* (2018). arXiv preprint arXiv:1804.03961

Health Record Maintenance Using Cloud Computing and Multi Authority Attribute-Based Encryption

S. Hamsanandhini, Malathi Eswaran, and V. Varanambika

1 Introduction

In these recent years, a health record has been introduced as a centric model of exchanging health information. It allows patient to create and to have a control over their personal data across the web that made storage and retrieval more efficient. In order to store such records in a large repository, there may be a need of building a high cost data centers. Many such services provided by various third party providers for instance, Health vault in Microsoft [1]. Many architectures have been emerged in an efficient storing of patient health record [2]. The patient involved in controlling and sharing their personal health data especially stored in third party service is not truth worthy. Although there may be a chance of implementing many regulations of health care that includes HIPAA which is the most enhanced contribution to various business associates [3]. Encryption serves to be the best way from providing confidentiality to the patient's data that includes their social security and also many sensitive personal information.

An incident stated that department of veteran database that includes health information of about 26.5 million military associates and it was stolen without any proper authorization by an employee [4]. In order to provide a secured access over patient's data there is a need of encrypting the health information files and also allowing only the responsible users in accessing those data. Further, the patients are always given the right to not only granting permission but also in revoking privileges that has been already provided [5]. Here there are two variety of user as personal and professional. Examples of personal users include family and friends and for the professional users it can be doctors, researchers and pharmacist. User's request in accessing data is commonly unpredictable on other hand many of the situations regarding the single

S. Hamsanandhini (✉) · M. Eswaran · V. Varanambika
Department of Computer Technology-PG, Kongu Engineering College, Erode 638060, India
e-mail: hamsanandhini@gmail.com

J. D. Peter et al. (eds.), *Disruptive Technologies for Big Data and Cloud Applications*, Lecture Notes in Electrical Engineering 905,
https://doi.org/10.1007/978-981-19-2177-3_6

Table 1 Detailed view of patient's health record

Record attribute	Content
Target	Personal details, Tracking ID number
Physiological measures	Glucose level, Pulse rate, Blood pressure and so on
Health status	Diagnosis of disease and other health issues stated
Dosage of medication	Amount of injected syringes and medicines per day are taken into consideration
General health measurements	Height, Weight, Cholesterol rate, etc.,
Non physiological status	Pain, insomnia, fatigue, poor appetite, blurry vision and so on
Lifestyle	Diet control, Physical activity, Exercise and so on
Health history of family	Inherited disease encountered like sugar, blood pressure, etc.,

data owner has been considering existing works [6]. Table 1 shows the content of the patient's health record that provides a clear view about the clinical symptoms of a particular patient.

There may be multiple owners who are provided with separating encrypting method that is possibility done by implementing large no of cryptographic keys. This paper secured sharing of health information that has been stored on various semi trusted servers in highly conserved. It mainly focuses on issues generally using various cryptographic keys. Attribute-based encryption is chosen as a main encryption method in providing a secured data access. It ensures trusted access policies based on the data attributes of the users that enables patient to share their personal health information to a secured set of users according to their desires. The main contributions in implementing are

1. The proposed framework works under multi owner sceneries. To main domains namely, public and personal domains are involved in dividing the users of system. This in return helps in addressing the challenges using the key management.
2. By implementing multi authority ABE security is improved in public domain and also avoids the problem during key management (Key escrow problem).
3. The complexity is reduced and the scalability is extended in terms of various metrics in communication, computation, key management and storage.

2 Literature Survey

The proposed survey mainly related with the encrypting standards that accounts for secured data access. The concern of providing access control in fined grained traditional approach of implementing public key encryption [7]. That might incur overhead in key management. In Goyal et al.'s suggested an idea on ABE [8]. The encryption of data is done under a particular set of attributes which in turn decrypted by multiple users whose who are provided with proper matching keys. This action mainly guaranteed efficient encryption and key management [9]. Perversion against user collusion is consider to be the important properties of attribute-based encryption.

Many approaches are done in realizing the fine grained access control has been listed for various outsource data [10]. Narayan et al. suggested an infrastructure based on attribute for electronic health records of patients. These files are encrypted by implementing CP ABE [11]. Akinyele et al. suggested an investigation across generating EMR that could be their stored on cell phones on cloud servers [12]. The reason of storing in cell phones or cloud servers is that though the health provider is in offline EMR could be easily retrieved. Many drawbacks may be listed during when the above work is encountered. First, it may be due to the usage of single authority that is highly trusted in the system. It results not only in bottleneck but also suffers from key escrow issue. Second, user revocation is lacked for ABE due to frequent updates and changes that are essential for secured sharing of Health Records. The idea entirely dividing the system conceptually into two main types is same as in [13] and managing the key for dealing with single authority is concerned for professional domain. Yu et al. [14] suggested an ABE approach in accessing data indulged the cloud server that is dealt with single user in encrypting and data sharing with authorized users. Chase and Chow [15] demonstrated CC multi authority attributed-based encryption (MA–ABE) in governing a subset of different users' request and attributes which in turn helps in generating collective number of secret key of the users.

User revocation and providing on demand service of ABE is the most challenging and well known problem. This is done by implementing periodic updates of key involved to the users those who are unrevoked frequently [16]. This implementation does not help in achieving both the forward and backward security and also considered to be less efficient. The study [17] proposed two different schemes of CP ABE that concerns with immediate revocation of user attributes instead of revoking periodically. Lewko and Waters's [18] provided a solution by providing a separate secret key for each users from the subset of the Trusted Authorities involved in the system. But there exist a gap when a cipher text component is updated to users those who are not revoked. This problem suffers from differentiating the public and personal domains. The above defined problem is solved by implementing a unified framework that is highly securable for sharing of patient's health records. The capturing of personal and public use of health information is done by this proposed framework. By the combination of both the YWRL Attribute-Based Encryption schemes and CC MA Attribute-Based Encryption (ABE), best of granting data access is granted. This scheme allows patients to choose and can create their own data accessing policies without the overhead.

3 Key Management Architecture

The encrypted key is combination of string of bits generated randomly such that key is distinct and cannot be predicted easily. The process of generating such type of encrypted keys are suggested using attribute-based encryption method (ABE). First step is to generate a secret key for the users' data in order to protect it from vulnerable attacks. Second, cipher text is used in the process of generating secret key. Such type of cipher text depends on the attributes based on the country code and the subscription available with the user.

The main objective of the proposed work is governing an efficient management of keys in encryption and secured data access. First, the system is divided into multiple user domains such as public and personal domains, in agreement with data requirements suggested. In both domains, cryptography is ensured by utilizing attribute-based encryption (ABE). Especially, multi authority encryption standard is used in governing multiple set of disjoint attributes. Role attributes represents the role of personal domains and user obligations. To have a controlled access over the data, fine grained is specified for separate user files. This scheme does not require the authorized users list while implementing encryption. Each user is trusted authority of their personal health information since they use secret key and have the right in accessing only their own data. Data attributes refers to inherent properties of user's personal data that includes the category of the data file. Also, the multi-domain approach is best suited for different types of users and in accessing the requirements involved in the system (Figs. 1 and 2).

The algorithm for encrypting the user's health record while uploading is given below.

1. First, the alphabets are arranged as tree from a to n.
2. Start from a, assign the left sub tree to be 0 and right sub tree to 1.
3. Repeat the steps up to n.
4. Arrange them according to the odd and even positions.
5. In odd position, alphabets are arranged at the first and even at the last.
6. Assign the value such as $a = 0$, $b = 00$ and so on.

Fig. 1 Tree representation of alphabets including their positions

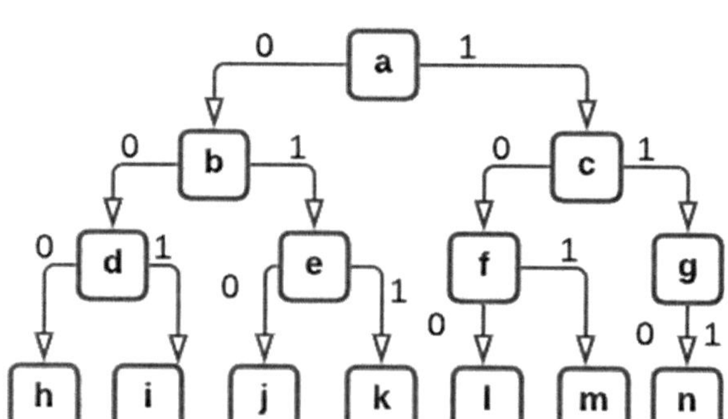

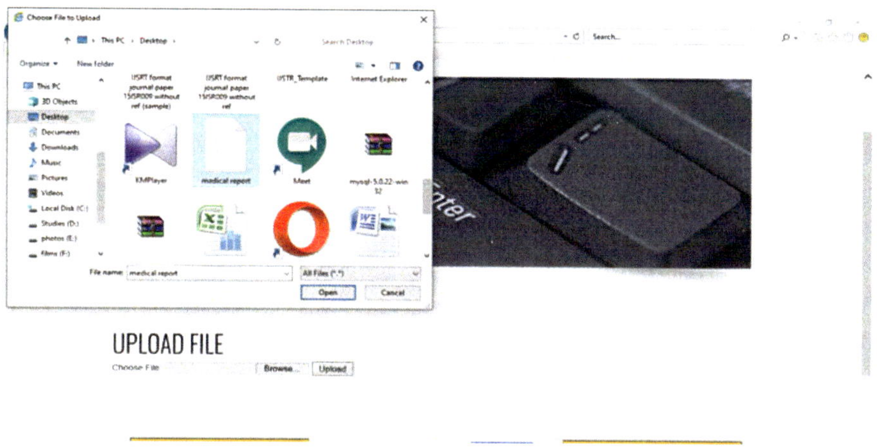

Fig. 2 Uploading patient's health record

7. The redundancy bits are inserted between the values of the user files.
8. The file will be uploaded after the addition of redundant bits (Fig. 3).

The algorithm for decrypting the user file is demonstrated below.

1. After receiving the data record, start from *a*, assign the left sub tree to be 0 and right sub tree to 1.
2. The added redundant bits (111, 111) helps in representing the start and end.
3. The actual data values are taken out after the redundant bits have been detected.
4. The alphabets are assigned to the numerical values.
5. According to odd and position rearrange and take out the original value of received message (Fig. 4).

4 Data Owner Framework

This framework consists of secured manner of patient centric data access and also ensures an efficient management of key at the corresponding time. This module first divides the entire system into multiple user domains. It includes public and personal domains (PUDs and PSDs) according to the user requirements for accessing the data. PUDs include users who could be given accessible based on their roles considered. The roles might include doctors, pharmacists and many medical researchers. The mapping of PUD to an independent department in the society that consists of government, healthcare or it might be insurance sector. But in case of PSD, the users are linked personally to the data owner (it includes family and friends) and given access to their personal health information based on the rights owned. The generation of secret key as shown in Fig. 5 is used in managing data of each users and has the privilege in accessing their own data and also has the right to share their access to

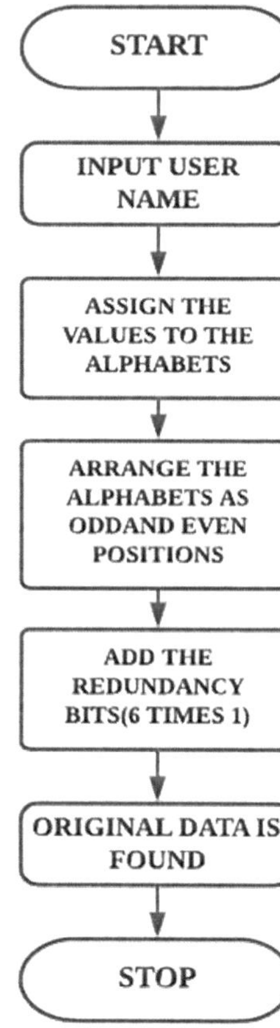

Fig. 3 Flow for encrypting user's health record

other individuals. Each data file is labeled with the representative data attributes and the corresponding key is linear in size. The number of data users in PSD is quite often small when compared to PUD, the burden is completely reduced. Intrinsic data properties are known to the user when the encryption is done.

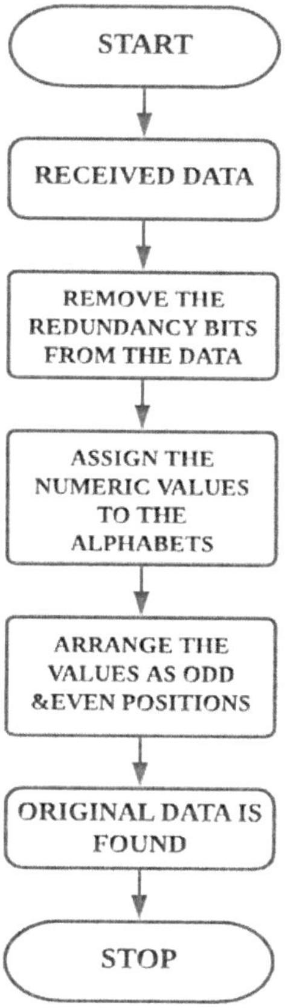

Fig. 4 Flow for decrypting user's health record

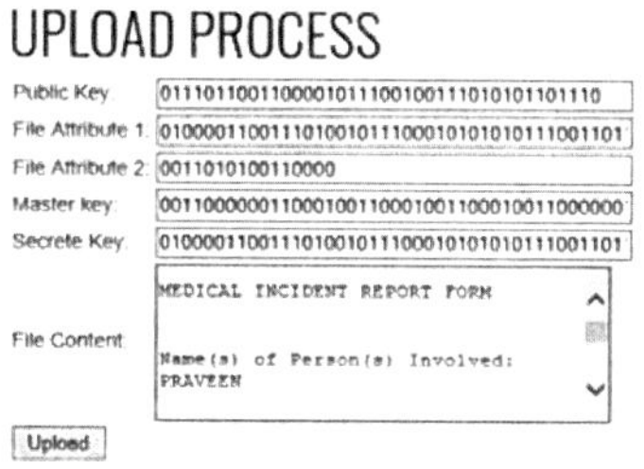

Fig. 5 Generation of Master key and Secret key

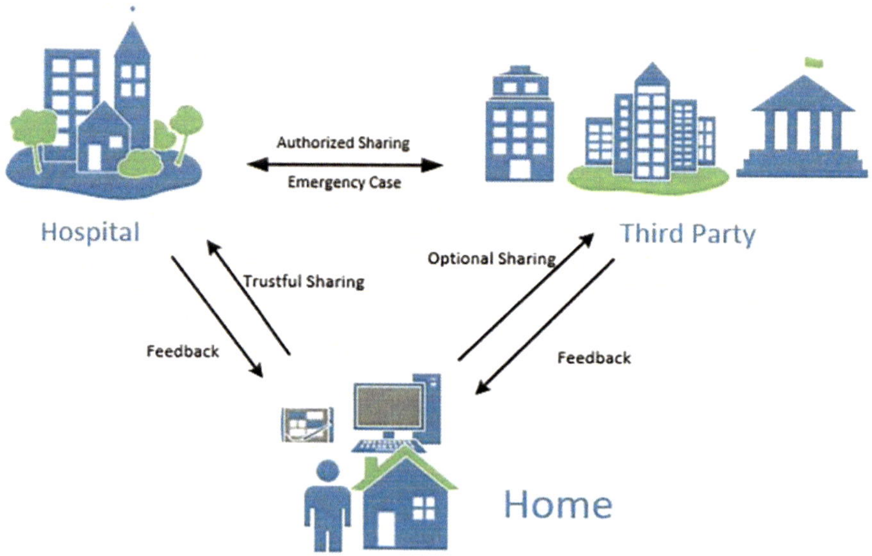

Fig. 6 Conceptual view of sharing health information

5 Cloud Server Architecture

In this module, the server is prompted to determine as much private information in the stored files but works with a protocol. Some of the data user may try in accessing the data files beyond the given level of privileges. It could be well explained with a practical example. The patient's prescriptions help in marketing for the pharmacy department and also in increasing the profit level. In order to gain those information, they may collude with the other data users including the server. Each party in the entire system is pre-loaded using private and public pair of keys. Authentication for each and every entity is done by implementing traditional protocol called challenge—response. Figure 6 [19] demonstrates the scenario of managing patient he himself at home and efficient sharing of health details.

6 Attribute-Based Data Access Policy

In accessing policy, it is implemented with multiple Safety Datasheets (SDs), owners, attributes and data users. The user those who have the access in reading the data is stated as data readers and write access is termed as data contributors. The proxy encryption is adopted in delegating two operations. One, the component of public key for the attributes that are affected. Two, all the ciphers that contains the attributes shall be updated by the server. In order to revoke a user, minimum number of attributed

Fig. 7 Generation of encrypted key

subsets is to be discovered. In delegating the updates of the secret key, an additional inclusion of dummy attribute is required. In lazy revocation, only the ciphers that are affected and secret keys of the users are updated during an unrevoked user who is affected logging into the system another time. Using rekey, the aggregation of all the formal updates are done from the rearmost login to the recent one (Fig. 7).

7 Data Confidentiality

The data owners are then allowed to upload the attribute-based encrypted files (health records of the patients) to the cloud server. The uploaded file undergoes both fine grained and role-based data accessing policies for the data users from the public domains. The set of data attributes that are selected helps in allowing the data access from the private sector. And hence only the authorized users alone had given the right to decrypt the personal health information files, eliminating the server. Figure 8 depicts the enrollment of secret key generated using ABE technique which in turn

Fig. 8 Enrollment of secret key for decryption

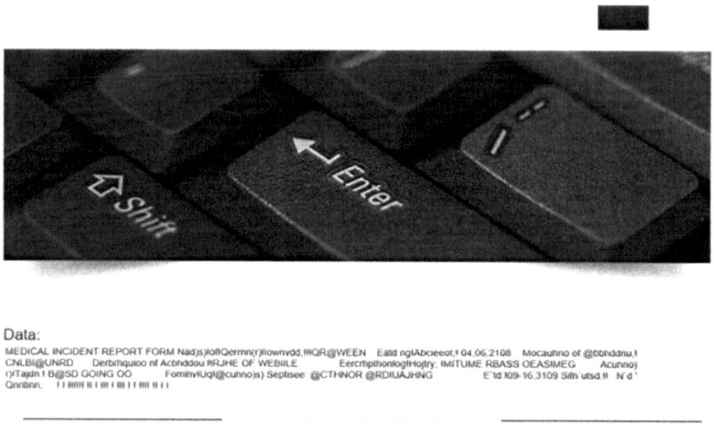

Fig. 9 Original Users' Data after Unlocking using Secret Key

used in viewing the actual data content of the user. The original data content after unlocking using the secret key is show in the Fig. 9.

8 Result and Discussion

The data record thus encrypted proves to be helpful in secured sharing of data. The health record encryption using ABE overcomes the complexity in key management, storage computation and communication. The storing of such enormous amount of health records is done efficiently under secured mechanism. Only the authorized users are allowed to access the health records using the key generated. Since the accessing of data documents and files is done through the centralized server, the user can have an access to other multiple users' data simultaneously. The performance analysis of various ABE schemes has been analyzed and comparison is provided in Table 2. Among those, MA ABE is used in the proposed method which provides more efficiency and scalable than other methods.

Table 2 Performance analysis of various ABE schemes

ABE schemes	Fine grained access control	Benefits	Computational overhead
Key-policy attribute-based encryption (KP-ABE)	Comparatively low but high when decryption considered	Efficiency of the mechanism is average, sometimes high for most of the broad cast type structure	High computational overheads
Cipher text-policy attribute-based encryption (CP-ABE)	Realization of complex access control is relatively average	Average efficiency but mostly not efficient for modern enterprise organization	Average computational overheads
Hidden attribute-based encryption (HABE)	Access control is good	Governs flexibility and scalable	Some of overhead
Hierarchical attribute-set-based encryption (HASBE)	Access control is better	High Efficiency and flexible	Relatively Less overhead
Multi Authority Attribute-Based Encryption (MA ABE)	Provides higher access control	More efficient	Mostly, Lesser overhead than other schemes

9 Conclusion

In this paper, a framework has been proposed in secured sharing and accessing of personal health information of patients in cloud storage. In partially trusted cloud server, a patient centric and secure model has been implemented and also in achieving the privacy of the data users. The privacy has been successfully given to the user by encrypting the user files. It addresses various issues and challenges faced during multiple data owners and users involved. The key management is greatly reduced thus enhances the users' privacy when compared with the existing works. ABE (A schemes helps user to access only their personal data not the other user's particulars and also contributes in providing on demand user revocation by enhancing the existing MA ABE schemes. By this implementation, it is evidently shown that the proposed work is both efficient and scalable. Some of the mathematical models can be also implied in the encryption technique. Rivest-Shamir-Adleman (RSA) is one of the main highlighting technique involving mathematical computation. The public key consists of pair of integers such that (e, n) where n represents the product of distinct pairs that is quite large. For each key generation where $k = (e, n)$ the encryption function is depicted as $E_k = P_{e,n}$ and the corresponding decryption function will be $D_k = P_{d,n}$. The private key computational formula is $ed = 1 \bmod \Phi(n)$. Such mathematical models would also help in generating the secret key in future ahead.

References

1. M. Li, S. Yu, K. Ren, W. Lou, Securing personal health recordsin cloud computing: patient-centric and fine-grained data access control in multi-owner settings, in *Proceedings of Sixth International ICST Conference Security and Privacy in Communication Networks (SecureComm '10)*, pp. 89–106 (2010)
2. H. Lo¨hr, A.-R. Sadeghi, M. Winandy, Securing the E-Health cloud, in *Proceedings of First ACM International Health Informatics Symposium (IHI '10)*, pp. 220–229 (2010)
3. *The Health Insurance Portability and Accountability Act* (2012). http://www.cms.hhs.gov/HIPAAGenInfo/01_Overview.asp
4. *At Risk of Exposure—in the Push for Electronic Medical Records, Concern Is Growing About How Well Privacy Can Be Safeguarded* (2006). http://articles.latimes.com/2006/jun/26/health/he-privacy26
5. K.D. Mandl, P. Szolovits, I.S. Kohane, Public standards and patients' control: how to keep electronic medical records accessible but private. BMJ **322**(7281), 283–287 (2001)
6. J. Benaloh, M. Chase, E.Horvitz, K. Lauter, Patient controlled encryption: ensuring privacy of electronic medical records, in *Proceedings of ACM Workshop Cloud Computing Security (CCSW '09)* (2009), pp. 103–114
7. C. Dong, G. Russello, N. Dulay, Shared and searchable encrypted data for untrusted servers. J. Comput. Secur. **19**, 367–397 (2010)
8. V. Goyal, O. Pandey, A. Sahai, and B. Waters, Attribute-based encryption for fine-grained access control of encrypted data, in *Proceedings of 13th ACM Conference in Computer and Communications Security (CCS '06)* (2006), pp. 89–98
9. M. Li, W. Lou, K. Ren, Data security and privacy in wireless body area networks. IEEE Wirel. Commun. Mag. **17**(1), 51–58 (2010)
10. A. Boldyreva, V. Goyal, V. Kumar, Identity-based encryption with efficient revocation, in *Proceedings of 15th ACM Conference in Computer and Communications Security (CCS)* (2008), pp. 417–426
11. S. Narayan, M. Gagne´, R. Safavi-Naini, Privacy preserving EHR system using attribute-based infrastructure, in *Proceedings of ACM Cloud Computing Security Workshop (CCSW '10)* (2010), pp 47–52
12. J.A. Akinyele, C.U. Lehmann, M.D. Green, M.W. Pagano, Z.N.J. Peterson, A.D. Rubin, *Self-Protecting Electronic Medical Records Using Attribute-Based Encryption*, Cryptology ePrint Archive, Report 2010/565 (2010). http://eprint.iacr.org/
13. L. Ibraimi, M. Asim, M. Petkovic, *Secure Management of Personal Health Records by Applying Attribute-Based Encryption*, technical report, University of Twente, 2009
14. S. Yu, C. Wang, K. Ren, W. Lou, Attribute based data sharing with attribute revocation, in *Proceedings of Fifth ACM Symposium Information, Computer and Communications Security (ASIACCS '10)* (2010)
15. M. Chase, S.S. Chow, Improving privacy and security in multi-authority attribute-based encryption, in *Proceedings of 16th ACM Conference in Computer and Communications Security (CCS '09)* (2009), pp. 121–130
16. X. Liang, R. Lu, X. Lin, X.S. Shen, *Ciphertext Policy Attribute Based Encryption with Efficient Revocation*, technical report, University of Waterloo (2010)
17. J. Hur, D.K. Noh, Attribute-based access control with efficient revocation in data outsourcing systems. IEEE Trans. Parallel Distrib. Syst. **22**(7), 1214–1221 (2011)
18. Encryption, *EUROCRYPT: Proceedings 30th Annual International Conference in Theory and Applications of Cryptographic Techniques: Advances in Cryptology* (2011), pp. 568–588
19. C. Peng, *A Hybrid Cloud Approach for Sharing Health Information in Chronic Disease Self-Management*, Master's thesis—Computer Science, Thesis no: MCS-2013–18, School of computing at Blekinge Institute of Technology, December 2013

20. B.S. Elger, J. Iavindrasana, L. Lo Iacono, H. Müller, N. Roduit, P. Summers, et al., Strategies for health data exchange for secondary, cross-institutional clinical research. Comput. Methods Programs Biomed. **99**(3), 230–251 (2010)

Ensemble DNN for the Brain Tumor Segmentation—A Hybrid Framework Centric on Layer Level and Decision Level Fusion of Multimodal Medical Images

S. Sandhya, M. Senthil Kumar, and B. Chidhambararajan

1 Introduction

Image processing techniques are now widely used for performing tasks like classification and segmentation of images in the field of medical applications. Performing segmentation of MRI images obtained from multimodalities is widely used to develop better medical image acquisition systems for diagnosis of brain tumor [1]. Brain tumor is the condition of unusual development of cells in brain. Gliomas are the casual type of brain tumor which is categorized as high-grade gliomas (HGG) and low-grade gliomas (LGG) based on the assessment of the tumor. Different MRI modalities are available in the field of neuro-imaging and the images from multimodalities are integrated as a whole in order to overcome the limitations imposed by single modal imaging techniques. MRI images from multimodalities like $T1$-weighted, $T2$-weighted, and fluid attenuated inversion recovery (FLAIR) afford complement evidence about the nature of the gliomas. $T1$-weighted images provide a virtuous contrast between the tissues of white matter and gray matter, $T2$-weighted images help in visualizing the abnormalities in the tissues like lesions, and FLAIR images help in enhancing the contrast of white matter [2, 3]. As brain tumor segmentation from multimodal MRI, brain images possess a vital role in grounds of radiology and medicinal diagnosis, manual segmentation leads to time-consuming and overwhelming task. Thus, automation is required for brain tumor segmentation in

S. Sandhya (✉)
Department of IT, SRM Valliammai Engineering College, Chennai, India
e-mail: sanrsn@gmail.com

M. Senthil Kumar
Department of CSE, SRM Valliammai Engineering College, Chennai, India

B. Chidhambararajan
SRM Valliammai Engineering College, Chennai, India

J. D. Peter et al. (eds.), *Disruptive Technologies for Big Data and Cloud Applications*, Lecture Notes in Electrical Engineering 905,
https://doi.org/10.1007/978-981-19-2177-3_7

detecting gliomas from multimodalities in order to progress the exactness of the results [4, 5].

2 Materials and Methods

Over the past few decades, imaging techniques gained advancement with which huge volumes of medical images are growing tremendously. Such images differ not only in the volumes but also vary in the facets of dimension, resolution, modalities, and quality. Medical images are obtained from modalities like magnetic resonance imaging (MRI), X-ray, SPECT, ultrasound, PET, and computed tomography (CT). Machine learning techniques gain its significance in the field of medical image analysis as it handles the structured and labeled data which is helpful in improving the outcomes of diagnosis and survival of the victims. Extracting important and significant features from the images are the critical task in which the conventional machine learning may degrade the performance when the data representation is poor [6]. Deep learning is the division of machine learning techniques which is used to extract the high-level details of the images by creating several levels of disseminated demonstration, and also, it overcomes the difficulties encountered by the existing machine learning techniques. Deep learning is deployed by developing a neural network with several hidden layers which is utilized as decision-making nodes. Convolutional neural network (CNN) is the most popular technique available in deep learning which consists of three major neural layers such as convolutional, pooling, and fully connected. CNN is majorly imparted in the field of image recognition because it performs the task of dimensionality reduction by preserving the significant image details. CNN is capable of handling both the 2D and 3D images, and it exhibits excellent results in brain tumor segmentation by identifying normal versus abnormal [2, 7]. Segmentation from multimodalities involves the fusion of multi-information to improve the results of segmentation. Multimodal medical images are widely used in the field of segmentation for better diagnosis by reducing the uncertainty in information that is being extracted from the images [5]. In Ref. [1], Ahmed et al. suggested a segmentation-based network on U-Net architecture. Modalities like X-ray, MRI, CT, and ultrasound were considered for comparison of the results of segmentation. Wu Fei et al. used encoder to learn the high-level dense features by designing an inception module for batch normalization, and a decoder is used to engross the multilevel features to increase the truthfulness of segmentation densely. Li Sun et al. in Ref. [4] proposed a brain tumor segmentation system that is based on an ensemble model which consists of three various CNN architectures like CA-CNN, DFKZ Net, and 3D U-Net. In Ref. [8], Jiawei Lai et al. deployed a learning-based integrated framework known as GMMD-U by using statistical models to perform the brain tumor segmentation. Zhe Guo et al. discussed the three different schemes for the fusion of multimodal medical images, namely fusion at feature level, classifier level, and decision level [9]. Zhe Guo et al. utilized three different patch-based CNNs with equivalent variations in the network structure for the segmentation of multimodal medical images [10]. Ikram Brahim et al. discussed brain tumor segmentation of multimodal medical images by using two-dimensional U-Net, three-dimensional U-Net, and finally cascaded anisotropic network. The performance of the three variations of CNN was compared with K-Means—an unsupervised technique for MRI segmentation in Ref. [3]. Jose Dolz et al. in Ref. [11] proposed a study to investigate the hyper densely connected 3D fully CNN and Hyper-Dense-Net for the segmentation of brain tumor by involving the concept of multi-path architectures [12].

3 Proposed Method

3.1 Problem Definition

The proposed method aims at developing an automated system for the segmentation of brain MRI images which are obtained from various modalities like $T1$, $T2$, and FLAIR. Equivalent portion of information is obtained from the images that correspond to the modalities but vary with weights associated with each pixel. Therefore, these methods will help in computing the mask for performing segmentation through the available information automatically. The task of segmentation should consider the chore of segmenting the sub-regions of tumor as a whole (WT), core (CT), and enhancing (ET).

3.2 3D U-Net

The 3D U-Net handles the input which is available as 3D volumes and performs the processing like 3D—convolutions, pooling, and up convolutional layers. This network architecture utilizes batch normalization in order to avoid bottlenecks and to perform faster convergence. Figure 1 shows the architecture which consists of paths for the functions of encoding and decoding [3]. The analysis of image is carried out in a contrast trail as a whole and further expansions yield the final results of segmentation. Dice loss function is used in order to train the network.

3.3 Cascaded Anisotropic Network (CAN)

The cascaded anisotropic network (CAN) shown in Fig. 2 is a combination of three CNN architectures which segments the sub-structures of brain tumor in the fashion of hierarchy and sequence [3]. It comprises W-Net to segment the tumor as a whole from the volumes of 3D multimodal images from which a bounding box is obtained

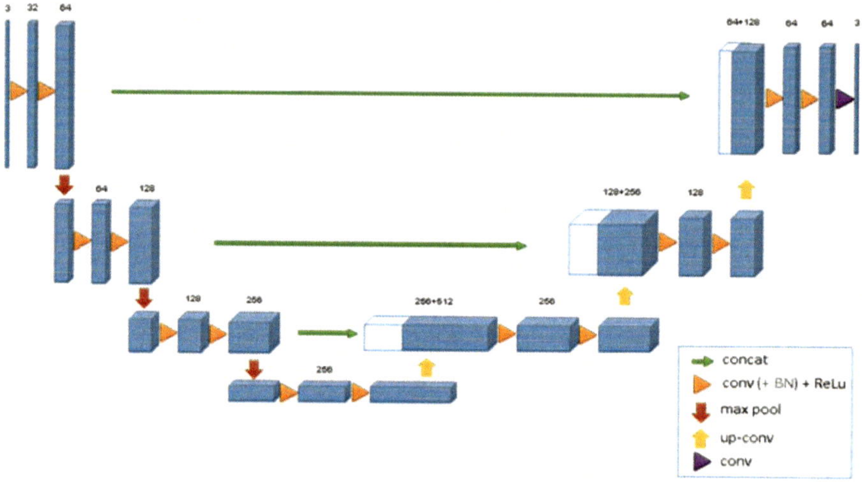

Fig. 1 A 3D U-Net architecture

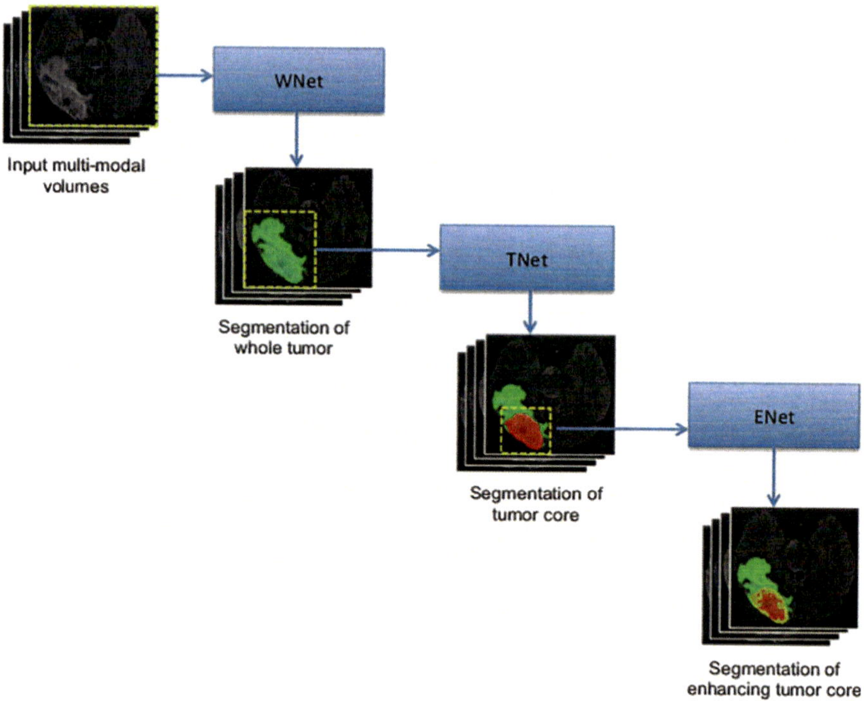

Fig. 2 Cascaded anisotropic network (CAN) architecture

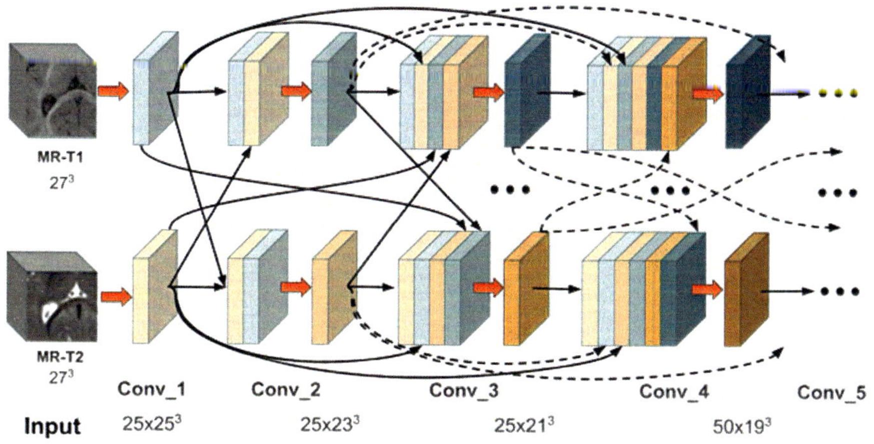

Fig. 3 H-DenseNet architecture

which is then utilized as source for the next level T-Net for the segmentation of core tumor. Input for the third level to enhance the tumor is obtained from the image portions that are lying inside the bounding box. Keeping the ground truth as base, the training phase generates bounding boxes automatically. Bounding boxes for the whole and core tumor are generated during the phase of testing by keeping the results of binary segmentation as a base.

3.4 H-DenseNet

Hyper-DenseNet (H-DenseNet) is an extension of dense connected convolutional networks (DenseNet) illustrated in Fig. 3 [11] that works well on the multimodal. In each modal, it consists of path and dense connectivity exists between layers in the same path as well across layers of different path. H-DenseNet yields more prevailing feature representation from the multimodals, and also, it is useful in learning intricate associations across all levels of abstractions from the modalities.

3.5 Proposed Architecture for Layer Level Fusion of Multimodal Medical Images

In the proposed system, images from multimodals are considered as single input with which the segmentation network is trained, and then with the layers present in the network, the feature representations are fused together which is then fed to decision layer to obtain the final result of segmentation. The proposed system ensemble

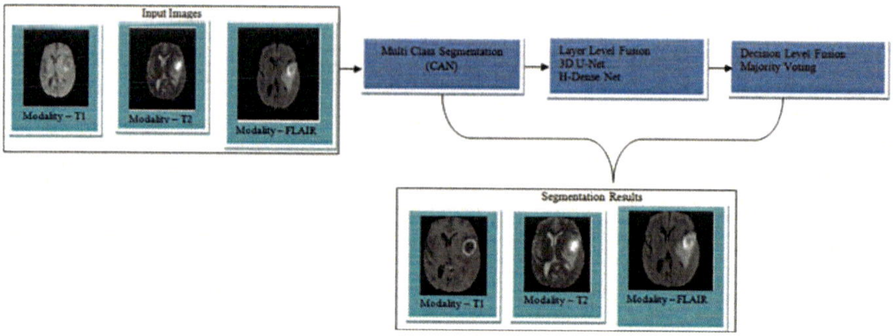

Fig. 4 Proposed ensemble framework for brain tumor segmentation

the networks like cascaded anisotropic network, 3D U-Net, and H-DenseNet which are, respectively, responsible for performing the segmentation of brain tumor. The cascaded anisotropic network is involved to putrefy the multi-class segmentation of multimodal images into the sequences to segment the sub-regions. 3D U-Net and H-DenseNet both represent the layer level architecture for the fusion of images from multimodals like $T1$, $T2$, and FLAIR. Finally, the segmentation results are obtained by decision level by applying majority voting where the label of the networks majorly depends on the voxel of the final label. Figure 4 represents the ensemble model of the proposed segmentation framework.

4 Experimental Results

To perform the segmentation of brain tumor, multimodal images were obtained from BraTS 2017, online dataset was used, and the automatic segmentation system was trained and tested using MATLAB2016a and TensorFlow framework. The experiments were carried out with the networks for the multi-class segmentation using cascaded anisotropic network and layer level using 3D U-Net and H-DenseNet of the images from modals $T1$, $T2$, and FLAIR. Finally, the segmentation results are decided based on majority voting using decision level fusion. All the networks were trained using dice loss function given in Eq. (1).

$$\text{Dice Loss Function} = 1 - \frac{2\sum_i P_i \times T_i}{\sum_i P_i + \sum_i T_i + \in} \quad (1)$$

In Eq. (1), P represents the predicted labels and T represents the set of ground truth, respectively. When the values of P and T are equal to zero, it is handled with the arbitrary value $\in = 1$.

Table 1 Comparative results of segmentation framework

Network	Dice score			Sensitivity			Specificity			Hausdorff distance		
	WT	CT	ET	WT	CT	ET	WT	CT	ET	WT	CT	ET
2D U-Net	0.78	0.65	0.65	0.8	0.74	0.77	0.83	0.81	0.81	5.18	4.23	3.46
CAN	0.75	0.59	0.58	0.75	0.66	0.73	0.78	0.74	0.8	4.92	4.39	3.61
3D U-Net	0.73	0.61	0.46	0.81	0.65	0.72	0.87	0.77	0.86	5.77	4.64	4.13
H-DenseNet	0.81	0.73	0.68	0.81	0.79	0.77	0.84	0.84	0.8	5.18	4.35	3.38
Majority voting	0.81	0.69	0.66	0.82	0.73	0.76	0.86	0.8	0.82	5.06	4.32	3.44

4.1 Estimation Metrics

The experimental results shown in Table 1 present the efficiency of the methods considered in the proposed system through the metrics like Dice Index, Sensitivity, Specificity, and Hausdorff distance. These metrics are given in the equations as follows,

$$\text{Dice} = \frac{2\,\text{card}(P \cap T)}{\text{card}(P) + \text{card}(T)} \tag{2}$$

where card represents the cardinality

$$\text{Sensitivity} = \frac{\text{TP}}{\text{TP} + \text{FN}} \tag{3}$$

$$\text{TP} = \text{True - Positive},\ \text{FN} = \text{False - Negative}$$

$$\text{Specificity} = \frac{\text{TN}}{\text{TN} + \text{FP}} \tag{4}$$

$$\text{TN} = \text{True - Negative},\ \text{FP} = \text{False - Positive}$$

$$\text{Haus}(Y, T) = \max\left\{\sup_{y\in Y}\inf_{t\in T} d(y, t), \sup_{t\in T}\inf_{y\in Y} d(y, t)\right\} \tag{5}$$

In Eq. (5), Haus represents the Hausdorff distance which is used commonly to calculate the extreme and to identify the outliers and $d(y, t)$ represents the Euclidean distance between y and t, the pixel coordinates.

5 Results and Discussion

The outcome of the proposed ensemble segmentation framework is presented in Table 1 which depicts the dice score of the three sub-regions under consideration.

The experimental results were compared with the 2D U-Net, and it is observed that the proposed framework yields better results with the metrics under consideration.

The proposed ensemble framework segments the three sub-regions while performing the task of segmenting the brain tumor by using multi-class segmentation technique and layer level integration of the significant features from the multi-modals, and finally, with the majority voting decision level fusion is achieved to obtain better results of segmentation. In multimodal segmentation of medical images, fusion strategy plays a significant role to achieve accurate results of segmentation. In this regard, deep learning networks are used to learn the complex and hierarchical feature representations from the volume of images. Next, layer level fusion is used to understand the relationships between various medical modalities by applying network layers in hierarchical fashion. In this sense, deep learning techniques gain the area of interest for better fusion than the conventional techniques. The efficiency of the proposed fusion network is evaluated using the parameters dice score, sensitivity, specificity, and Hausdorff distance.

The proposed ensemble network is compared with the existing 2D U-Net and CAN with the help of majority voting rule, and the graphical representations of those results are depicted in Fig. 5 for dice score, Fig. 6 for the metric sensitivity, Fig. 7 represents the specificity, and Fig. 8 shows the Hausdorff distance. From the evaluation results of the proposed system, it is evident that the ensemble fusion network yields better results combined with majority voting rule than the 2D U-Net and CAN.

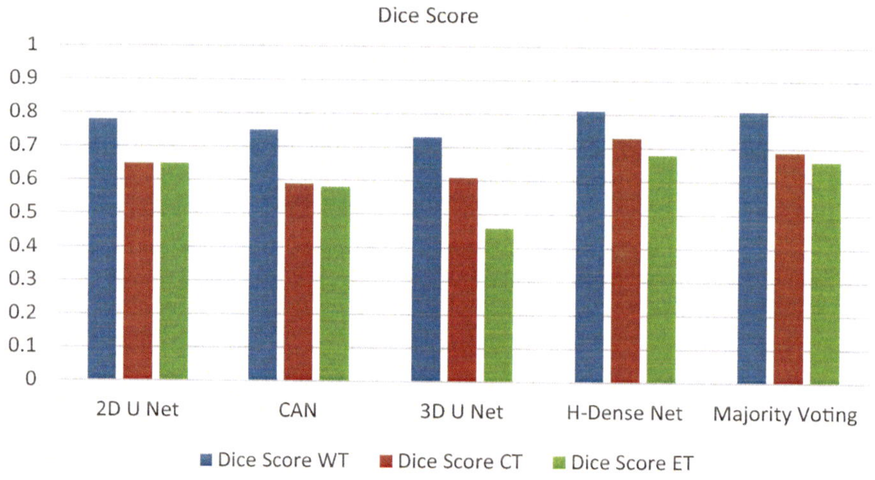

Fig. 5 Evaluation of dice score

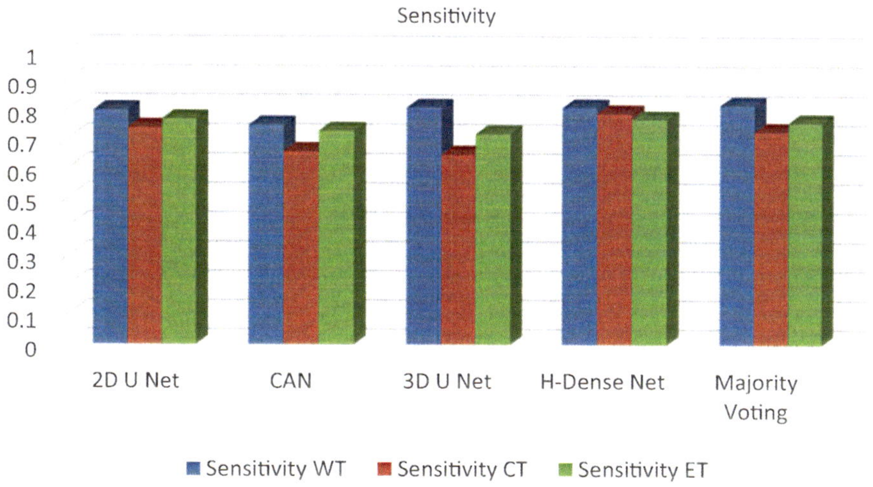

Fig. 6 Evaluation of sensitivity

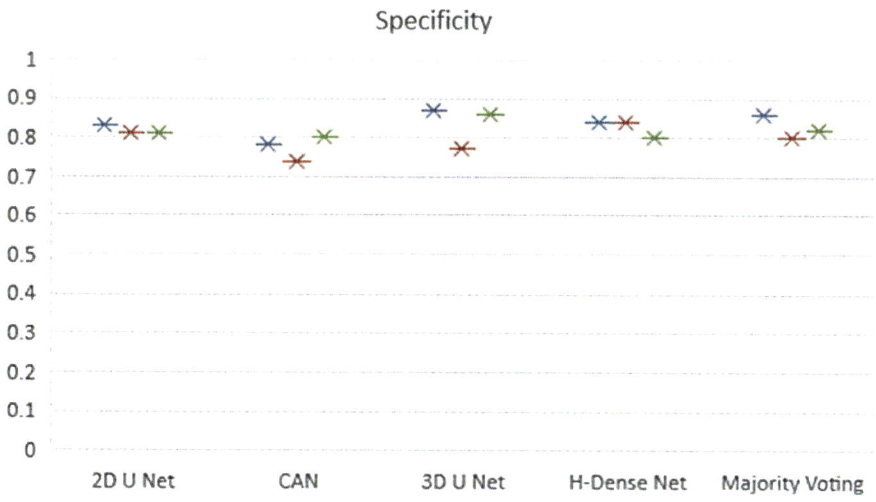

Fig. 7 Evaluation of specificity

6 Conclusion

In the proposed system, a new ensemble fusion framework for segmenting the MRI brain images was presented and the experimental results were compared with other deep neural network-based techniques which are available for the segmentation of brain tumor. The recommended system obtains sub-regions of the tumor such as whole, core, and enhancing accurately. High-level feature representation is obtained

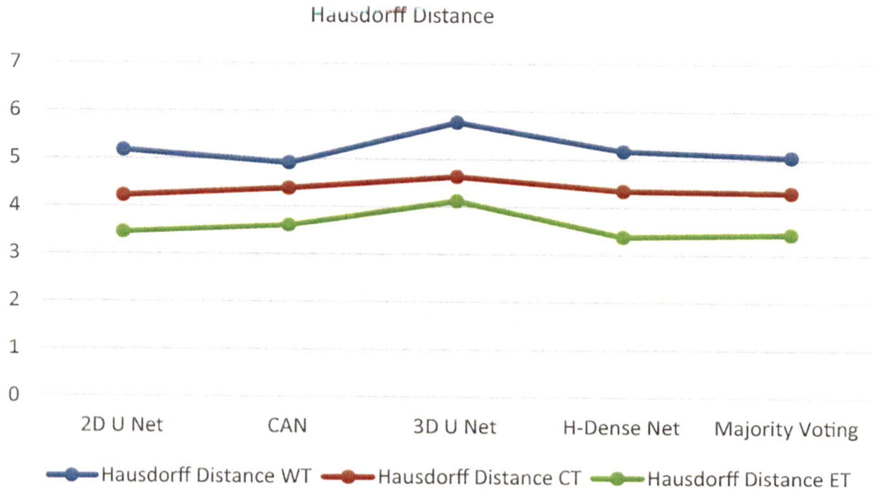

Fig. 8 Evaluation of Hausdorff distance

by fusion of hybrid feature layers present in the H-DenseNet for the tumor segmentation. In the future, other deep learning networks will be integrated with labeled configurations by adopting different architectures and techniques for acquiring better segmentation results.

References

1. A. Harouni, A. Karargyris, M. Negahdar, D. Beymer, T. Syeda Mahmood, Universal multimodal deep network for classification and segmentation of medical images, in *IEEE International Symposium on Biomedical Imaging* (2018), pp. 872–876
2. H. Sun, Z. Liu, G. Wang, W. Lian, J. Ma, Intelligent analysis of medical big data based on deep learning. Deep Learn. Algorithms Internet Med. Things **7**, 142022–142037 (2019)
3. I. Brahim, D. Fourer, V. Vigneron, H. Maaref, Deep learning methods for MRI brain tumor segmentation: a comparative study, in *IEEE* (2019), pp. 1–6
4. L. Sun, S. Zhang, H. Chen, L. Luo, Brain tumor segmentation and survival prediction using multi-modal MRI scans with deep learning. Front. Neurosci. **13**, 1–9 (2019)
5. T. Zhou, S. Ruan, S. Canu, *A Review: Deep Learning for Medical Image Segmentation using Multi-modality Fusion* (Elsevier, 2019), pp. 1–11
6. P. Rastogi, V. Singh, M. Yadav, Deep learning and big data technologies in medical image analysis, in *IEEE International Conference* (2018), pp. 60–63
7. J. Ker, L. Wang, J. Rao, T. Lim, Deep learning applications in medical image analysis, in *IEEE—Soft Computing Techniques for Image Analysis in the Medical Industry Current Trends, Challenges and Solutions*, vol. 6 (2018), pp. 9375–9389
8. J. Lai, H. Zhu, X. Ling, Segmentation of brain MR images by using fully convolutional network and gaussian mixture model with spatial constraints. Hindwai Math. Probl. Eng. **2019**, 1–14 (2019)

9. Z. Guo, X. Li, H. Huang, N. Guo, Q. Li, Deep learning based image segmentation on multimodal medical imaging, in *IEEE Transactions on Radiation and Plasma Medical Sciences* (2019), pp. 1–9
10. Z. Guo, X. Li, H. Huang, N. Guo, Q. Li, Medical image segmentation based on multi-modal convolutional neural network—study on image fusion schemes, in *IEEE International Symposium on Biomedical Imaging* (2018), pp. 903–907
11. J. Dolz, K. Gopinath, J. Yuan, H. Lombaert, C. Desrosiers, I. Ben Ayed, HyperDense-Net: a hyper densely connected CNN for multimodal image segmentation, in *IEEE* (2019), pp. 1–13
12. W. Fei, L. Yifei, Z. Jian, Z. Bin, K. Youyong, S. Huazhong, Brain tissue segmentation integrating multilevel features, in *International Conference on Advanced Cloud and Big Data* (2019), pp. 235–239

Deep Learning-Based BDMSF Resource Sharing—A Systematic Approach for Analysis and Visualization

K. Elaiyaraja, M. Senthil Kumar, and B. Chidhambararajan

1 Introduction

The conversion of medical data into digital data was initiated earlier and it has advanced gradually equated with other technical fields [10]. The medical data becomes one of the focused area with the rapid growth of technology, economy and science. In the legacy medical system, medical data has terrible pressure on raw data as it requires huge amount of memory storage, data structure and data storing and retrieving time [7]. Data loss leads to loss of accuracy. Data integrity is one of the biggest challenge and low data processing which cannot be acquired the end user's satisfaction for visualizing data. Medical Health Information System has accompanied the opportunity of medical development recently [13].

The Internet of Medical Things and the latest technology Cloud Computing have transformed the formation of medicinal data services and increases the efficiency of medical data distribution. Deep learning and machine learning has tremendous improvements in health care systems [11]. Deep learning collects the significant data about patient health from the medical reports. These data can diagnose diseases easily. The *n* number of techniques proposed to embed DL with clinical diagnosing system [15]. Convolutional neural network (CNN) constructs a medical information system framework to increase the diagnosing system and also elaborates the concepts of clinical diagnosis through Sparse DL technique [9]. The data visualization has formed to emphasis complicated and abstract data. The extraction of data supports the CNN

K. Elaiyaraja (✉)
Department of IT, SRM Valliammai Engineering College, Chennai, India
e-mail: k.elaiyaraja@rediffmail.com

M. Senthil Kumar
Department of CSE, SRM Valliammai Engineering College, Chennai, India

B. Chidhambararajan
SRM Valliammai Engineering College, Chennai, India

J. D. Peter et al. (eds.), *Disruptive Technologies for Big Data and Cloud Applications*, Lecture Notes in Electrical Engineering 905,
https://doi.org/10.1007/978-981-19-2177-3_8

for the betterment and performs highly reasonable in multiple visual recognition oriented tasks. CNNs have done on various medical image classification related issues.

The state-of-the-art methods in Medical-based Big Data (MBD) maintains in different formats. The results, test reports and medical history documents are maintained in the image format and audio format and video format. Any kinds of data may adopt by the medical officers and technical members. The modality and devices like biometric data are in the form of shades, special kinds of formats like figures and pictures and statistics. These kinds of data applied to interpret the results obtained through reports and specific persons only can extract significant information.

Data visualization invented at the methodological level and can be categorized based on the usage of technology. Data visualization is an important tool which mainly used for analyzing data and plays a vital part on finding unwanted medical data [3]. Generally, the health information resources have data including medical expenditures, digital record data, laboratory test data, scanned images information, etc. The representation of visual consists of these information which makes data visualization more informative [12]. The increasing scale of the chart lead to the rapid data variations, increasing complexity and storage volume. So, the requirement of memory space is rapidly increasing and algorithms to access data gets complicated which decreased the computational storage ratio.

The techniques used for medical data visualization and combining with medical big data environment are covered here. The progress of DBN, DL methodologies and computational features are summarized. The BDMSF Information system was implemented from the clinical data integration, mining and knowledge supervision. In addition to that the *E* and *H* Charts of data visualization, optimized DL methodologies for Medical Big Data (MBD), resource sharing frameworks and MBD visualization are discussed in the following sections.

2 Materials and Methods

The Deep Belief Network (DBN) is a reproductive graphical representation system consists of several Restricted Boltzmann Machines (RBM). The training method is classified into preprocess and Refinement process. Preprocess is a method to preserve more details and each section in the RBM system requires separate training. The output generated by the RBM is given as input for the next RBM layer [1]. The encoding and decoding processes are done in each section. From the huge amount of training illustrations, the flaw rate between the exact information (E_d) and reformed information (R_d) is reduced [17]. The weight (W_{tm}) is produced in each section after the RBM is pre-processed. In the Refinement process, the Tag is used to compute the performance of DBN once the preprocessing is over [8]. The back-spread technique is used for the error backhaul. It is mandatory to refine the DBN and weight modifications of RBM system to obtain the accuracy. Figure 1 depicts the DL algorithm implement sequence.

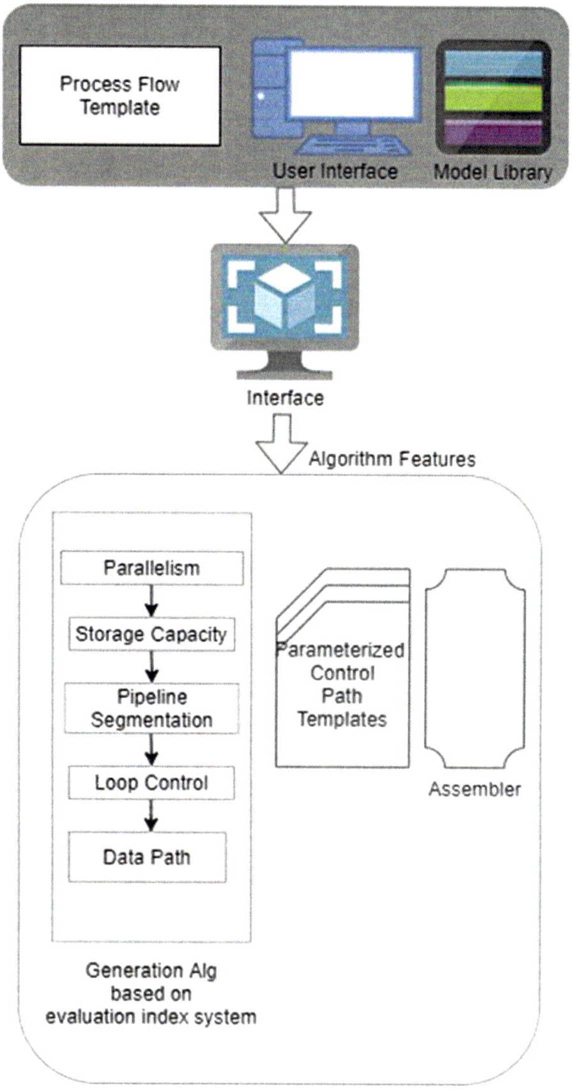

Fig. 1 Deep learning algorithm implement series

Several RBM systems perform unsupervised training in a sequential order while the preprocessing is on. The value of offset is usually set to 0 and the W_{tm} is set along with Gaussian computation. The probability of each section is computed with the value of 1. The unknown layer module is estimated as $(v \rightarrow h)$ using known layer module [2].

$$a = \sum_{m=1}^{Z} n_m W_{tm} + y_1 \quad (1)$$

W_{tm} represents the weight and the y_n represents the offset value. $x(m_1 = 1|N)$ is the activity of m_1 under the progression of each known level module.

$$x(i_1 = 1|N) = \text{Sigmoid}(a) \tag{2}$$

Equated with (1) in the (2), the weight is depending up on the transposition of the matrix and the offset output is set from the value mth value of a which is considered as visible node.

$$a = \sum_{m=1}^{Z} n_m W_{tm} + y_n \tag{3}$$

The network parameter values are updated as $x(i_n = 1|N)$ where i_n is represented as the Gaussian random value.

$$x(i_n = 1|N) = \text{Sigmoid}(a) \tag{4}$$

The guideline for the network constraints W_{tm} and counterweight values α and β are given below, where £ is the learning-rate.

$$W_{tm} = W_{tm} + \pounds(n_m i_n - n_m i_n) \tag{5}$$

Cloud computing is the technology which means of analyzing big data and utilizing during the implementation of big data. In the medical process, the following important scenarios need to be considered for big data processing [4].

- Comparison and analysis of patient's characteristic data and treatment processes.
- Data-based diagnose system which increases the effectiveness of medical treatments.
- Data visualization of BDMSF for optimizing medical processes and reduce the treatment cost.
- Remote observation and recording of patients leads to collect and analyze the data for the betterment of treatment.
- Predictions based on the patient's data for various diseases.

3 DL Distribution Technique for BDMSF

DL comprises a huge number of iteration functions which are dependent and it is tedious process to make independent. It is mandatory to utilize the top most GPU parallel accelerate method [5]. Data parallelism is distributed evenly to exercising at the same duration. The parameters are made to orchestrate as per the certain period

[16]. This one can be extremely error forbearing and the whole training progression will not end as a node off. The Downpour SGD is a technique to manage the asynchronous parallel gradient method in the yesteryears.

The data is partitioned to the consistent training modules and this development is equivalent to the legacy big data technique [14]. This information distributed evenly and warehoused in DFS and each module uses a duplicate method independently. The separate training process and allocation of hardware possessions are done to obtain straight extension. This structure leads to the modifications of existing system including efficient compliance and convenient extension. The server coordinates the data interchange and data update along with the performance blockage of entire system. The effective design required for information update method of this service provider is significant for the act of entire system. One of the efficient algorithm used for data update is Downpour SGD which is based on asynchronous gradient.

4 BDV Methodology

The main idea of Big Data Visualization (BDV) is representing the analysis of data in an instinctive way. The users can easily understand the complex data when representing visually. The open source BDV technologies are available and many large scale organizations supports these open sources in this modern era. Because, the BDV is based on the web-based technology like HTML5, PHP and Java Script. The *H* and *E* Charts, Google Charts and D3.js are some popular open source BDV.

H and *E* Charts are chart library and it is developed by the Java Script and HTML5 Canvas, respectively. In this chart, inserting various charts and supports user interactions. These technology provides data in a graphical representation which are instinctive, customable and interactive. Supports drag and drop, range setting, different data representations and integration. A sample health care statistics shown in Fig. 2.

5 Parallel Optimization of BDMSF Processing

Refinement Process Analysis

Step 1: Initialize parameter in random manner for the RBM.

Step 2: Evaluation of the accelerative transfer routines. For every sample inputs, the output generated by the RBM is calculated and the generated DBN is retrieved. The tag value is deducted from the output generated by the DBN to obtain error value.

Step 3: Error back-spread is calculated from each level and adjust the weight constraint accordingly.

Step 4: Regulate the weight between every layers with respect to the learning-degree.

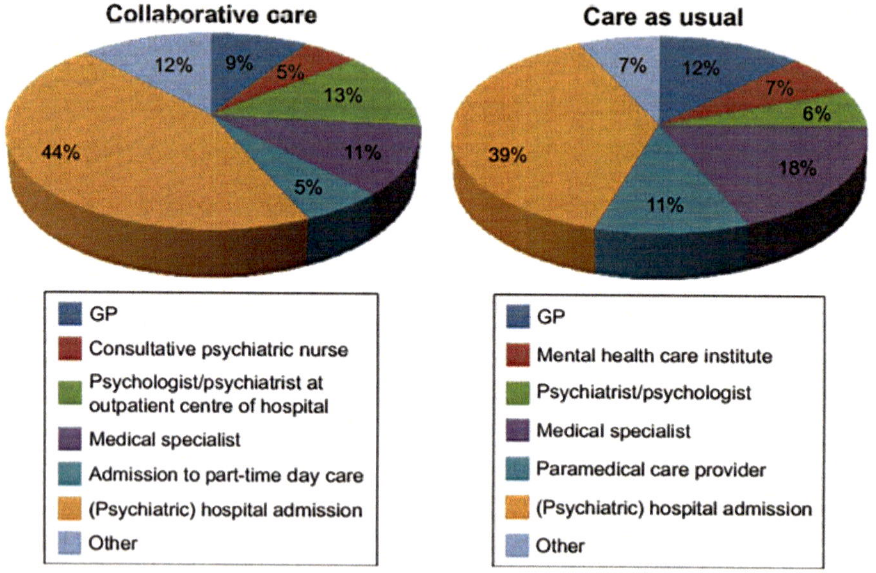

Fig. 2 Statistic analysis of health care system

Step 5: Govern the accuracy. Repeat step 2 if need to compute under the recalibrated weight or abort.

In this work, parallel optimization is used for the refinement process. The refinement process consists of parameter initialization, Assess Catalog Error and Back-spread. The Back-spread technique executes a reverse propagation with respect to the error value. The RBM value is regulated to decrease the error value. The lowest error value is achieved by executing linear search thrice. According to the result calculation, if the iteration <6, then the uppermost RBM is refined. If the iteration >6, then the RBM weight is unchangeable.

Code Authentication

In this work, first encoding the classification process code by using assembly language and simulator is used to verify the process. The FPGA on-chip and DSP resources are compelling into account for code design [6]. Here, dense matrix is used for calculation instead of sparse matrix. For matrix computation, the super-vector coding instruction is used for encryption. The following diagram depicts the super-vector coding. Each coding is classified into Code segmentation, Destination address, Source address 1 and 2 and data-block size.

The universal store/load controller is used to read the externally stored data. The stream buffer writes the on-chip local DRAM under the vector memory control. The multiple calculations based on the time is retrieved from the scalar register through the data which is interchanging over network. Each scalar computations requires one clock iteration to complete pipelined programs. The execution time is depending upon

Name	Code Segment	DA	SA_1	SA_2	Volume
VSUB	21	R_i	R_j	R_k	Size
VADD	20	R_i	R_j	R_k	Size
VISG	30	R_i	R_j	/	Size
VACC	23	R_i	R_j	I_{mm}	Size
VSTAS	31	R_i	R_j	/	Size
VLOAD	80	R_i	R_j	/	Size
VSADD	22	R_i	R_j	/	Size
VMAC	40	R_i	R_j	R_k	Size
VDIV	24	R_i	R_j	I_{mm}	Size
VSMUL	25	R_i	R_j	I_{mm}	Size
VSTORE	81	R_i	R_j	/	Size
VVMC	41	R_i	R_j	R_k	Size

Fig. 3 Super-vector coding

the fundamental block size and the memory control. Here, DDR2 memory regulator is used to extract the matrix block to the external DRAM (Fig. 3).

6 Analysis of Resource Sharing Based on BDMSF

Internet of Medical Things information collection: The medicinal diagnosis establishes the automatic gadgets for observing information and upload through the sensor network. The transmitted data to the server is analyzed then and there. The BP gadget is connected with terminal and accessed the data through the Bluetooth. For pipeline interconnection, the socket is used for transferring the BP data. The gateway uploads the transferred data to the server and it will be stored in the database. The user can remove the improper data in the data base.

Resource Integration and Distribution: The resource integration and data distribution is the fundamental function of BDMSF. It makes the availability of data and plays an intense role for the superior quality of data extraction and knowledge supervision. The BDM service framework entails a proficient with highly stabled system. The medical data is accomplished and managed in a superior way. The medicinal cloud is implemented to achieve the knowledge finding of medical data. The entire BDM service framework is classified into Information-Integration and Sharing (IIS), Extracting Data and Knowledge Supervision (EDKS) and Medical Services (MS). This system adopts Hadoop DFS and it is an open source software by the Apache.

7 BDMSF Visualization

The BDMSF visualization is implemented based on the Java Enterprise Edition and it is an open source framework. The Hibernate is used to obtain the MVC pattern. The model and the view are integrated with the equal level and the supervisor or controller is interacted with the HTTP. The JSON string process the information. The controller requests the facility to function the business logic as well as retrieves the data from the org.json.JSONObject. The Hibernate acts as an intermediate layer between the DAO and DB. For the project supervision, the Maven and its central repository is used. The EclipseMars.2Release integrates with Maven Plugin for creating Maven projects. Json, Weka, log4j, junit, etc. are the tools used for the implementation of framework.

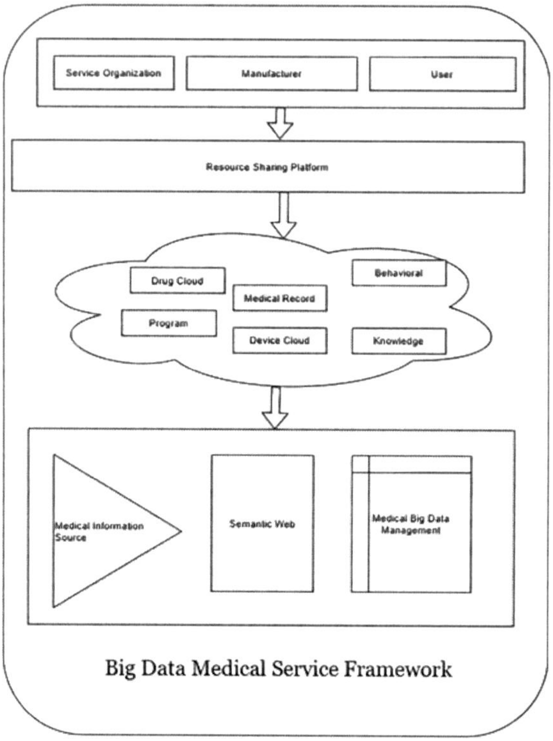

SPARK framework computes the parameter for the patient's age chart who discharged from the clinic. The MySQL DB stores the computed data and flask framework obtains required data from this DB. The flask framework send the requested data for expenditure calculations if a user willing to calculate. Similarly, the user can select the Age Height and Weight (AHW) details from the drop-down bar. The user can also obtain the sickness details by clicking the "Medical Service" feature.

Regional level medical information can be retrieved by selecting "Medical Survey" in the menu bar.

8 Conclusion

The main agenda of this work is to learn the usage of BDMSF visualization to serve the betterment of Big Data Platform which allows the users to share the visual data analysis subjectively. The progress of DBN, DL technique and operational characteristics are analyzed and summarized. The classification computations of DBN was converted into an assembler instruction and the coding was evaluated for the performance. The BDMSF was proposed based on the actual requirements and applications, functional objects and operational methods are technically justified. The implementation of BDMSF for integration, sharing and extracting data and knowledge supervision are analyzed and the operational requirements are increasing accordingly. The practical evaluation and performance investigation can be comprehensive to the scheme of extremely robust and more practical for other kinds of medical imaging tasks. This platform has increased data processing and can use easily. The medical image processing platforms are utilized for processing audiograms which improves the audiograms quality. Overall the cost of medical apparatuses is reduced and increases the efficiency.

References

1. Z. Akkus, A. Galimzianova, A. Hoogi, D.L. Rubin, B.J. Erickson, Deep learning for brain mri segmentation: State of the art and future directions. J. Digit. Imag. **30**(4), 449–459 (2017)
2. X. Fan, Q. Yao, Y. Cai, F. Miao, F. Sun, Y. Li, Multiscaled fusion of deep convolutional neural networks for screening atrial fibrillation from single lead short ECG recordings. IEEE J. Biomed. Health Inform. **22**(6), 1744–1753 (2018)
3. M.H. Hesamian, W. Jia, X. He, P. Kennedy, Deep learning techniques for medical image segmentation: Achievements and challenges. J. Digit. Imag. **32**(4), 582–596 (2019)
4. E. Hosseini-Asl, M. Ghazal, A. Mahmoud, A. Aslantas, A.M. Shalaby, M.F. Casanova, G.N. Barnes, G. Gimel'farb, R. Keynton, A. El-Baz, Alzheimer's disease diagnostics by a 3d deeply supervised adaptable convolutional network. Front. Biosci. (Landmark Ed) **23**, 584–596 (2018)
5. B.A.H.I. Kaggle, *Kaggle Data Science Bowl 2017* (2017). www.kaggle.com/c/data-science-bowl-2017
6. W.J. Kim, Knowledge-based diagnosis and prediction using big data and deep learning in precision medicine. Invest. Clin. Urol. **59**(2), 69–71 (2018)
7. S. Korolev, A. Safiullin, M. Belyaev, Y. Dodonova, *Residual and Plain Convolutional Neural Networks for 3d Brain MRI Classification* (2017). arXiv preprint arXiv:1701.06643
8. M. Liu, J. Shi, K. Cao, J. Zhu, S. Liu, Analyzing the training processes of deep generative models. IEEE Trans. Vis. Comput. Graphics **24**(1), 77–87 (2018)
9. X. Liu, T. Qiu, T. Wang, Load-balanced data dissemination for wireless sensor networks: A nature-inspired approach. IEEE Internet Things J. **6**(6), 9256–9265 (2019)

10. X. Luo, Q. Yang, P. Li, T. Miyazaki, X. Wang, Grouping strategy for rfid-based activity recognition in smart home, in *2019 International Conference on Internet of Things (iThings) and IEEE Green Computing and Communications (GreenCom) and IEEE Cyber, Physical and Social Computing (CPSCom) and IEEE Smart Data (SmartData)* (IEEE, 2019), pp. 96–101
11. Z. Ning, P. Dong, X. Wang, X. Hu, L. Guo, B. Hu, Y. Guo, R.Y.K. Kwok, Mobile edge computing enabled 5g health monitoring for internet of medical things: A decentralized game theoretic approach. IEEE J. Sel. Areas Commun. (2020)
12. Z. Ning, K. Zhang, X. Wang, L. Guo, X. Hu, J. Huang, B. Hu, R.Y.K. Kwok, Online edge computing and caching for internet of vehicles: An imitation learning based solution. IEEE Trans. Intell. Transp. Syst. (2020)
13. Y. Pang, Z. Yang, Y. Yang, X. Wu, Y. Yang, T.-L. Ren, Graphene based wearable sensors for healthcare, in *2019 International Conference on IC Design and Technology (ICICDT)* (IEEE, 2019), pp. 1–4
14. P. Rajpurkar, J. Irvin, K. Zhu, B. Yang, H. Mehta, T. Duan, D. Ding, A. Bagul, C. Langlotz, K. Shpanskaya, et al., *Chexnet: Radiologistlevel Pneumonia Detection on Chest X-rays with Deep Learning* (2017). arXiv preprint arXiv:1711.05225
15. A. Sharif, J. Ouyang, Y. Yan, A. Raza, M.A. Imran, Q.H. Abbasi, Low-cost inkjet-printed rfid tag antenna design for remote healthcare applications. IEEE J. Electromag. RF Microw. Med. Biol. **3**(4), 261–268 (2019)
16. W. Wei, X. Xia, M. Wozniak, X. Fan, R. Damaševiˇcius, Y. Li, Multisink distributed power control algorithm for cyber-physical-systems in coal mine tunnels. Comput. Netw. **161**, 210–219 (2019)
17. P. Zhang, X. Xu, H. Wang, Computer-aided lung cancer diagnosis approaches based on deep learning. J. Comput. Aided Des. Comput. Graph. **30**(1), 90 (2018)

Mitigation and Swift Curative Procedure on Alluring Smart City Using Falcon Technology

G. S. Nivethini, R. Yokesh Srenevas, R. Rahfar Nisha, and G. Ignisha Rajathi

1 Introduction

The great enhancement by the advances in both hardware and software technologies, smart environments already perforate at most spaces of our daily lives. Smart phones, Smart Homes, Smart Buildings like airports, hospitals or universities are already accustomed with a multitude of networks embedded devices as well as connected sensors. People living in these Smart cities have a high standard of living, thereby increasing the amount of waste as much as they use. Updating the features of these requirements, helps to avoid these accidents. Connecting and establishing communication with people is not easy in this modern world. Some activities already envision the "Smart City" which uses the opportunities provided by prevalent computing technologies for the benefits of their public. Hence, smart environments are expected to play a crucial role for coping up with the challenges of urbanization, industrialization and demographic change. Public service, public health, infrastructure development, economic growth, people's life adaptation, safety measures which connect common people securely [1]. We provide a business overview of a large metropolitan supplier of infrastructures and solutions. This includes application scenarios for smart cities and an outline of the involved business and upcoming challenges to be faced by the metropolitan people. A particular focus will be the setup and energy efficient operations of smart infrastructures and data centres. The Internet of Things (IOT) is being used to acquire traffic data quickly and simultaneously send it for processing [2]. The technologies to be brought up for smart cities are real time streaming data (i.e.) Big Data analytics. Traffic density can be analysed best by using predictive analysis. It is connected to a mobile application which acts as an interface to explore and

G. S. Nivethini (✉) · R. Yokesh Srenevas · R. Rahfar Nisha · G. Ignisha Rajathi
Computer Science and Business System, Sri Krishna College of Engineering and Technology, Coimbatore, Tamil Nadu, India
e-mail: nivethinisarav@gmail.com

J. D. Peter et al. (eds.), *Disruptive Technologies for Big Data and Cloud Applications*, Lecture Notes in Electrical Engineering 905,
https://doi.org/10.1007/978-981-19-2177-3_9

control the traffic density in smart cities. Also, it is the inclusion of technologies such as artificial intelligence, machine learning and Robotics. Information and communication technologies (ICTs) and geographical information technologies (GIS) are developing the communication in high tech smart moves in cities. People's interaction and population increase in the city make some sense to develop it smartly. Technology updates always give a change in day-to-day life. Statistically the data refers to the cities to measure the dos and don'ts of the required smart technologies. Smart cities always make people learn innovatively. Experience makes people distinguish between do's and don'ts. The concept of smart cities will always take it to a high level of project [3]. Also, an efficient smart bin system [4] and smart education system [5] will also help in the clean corner of the entire city. In spite of the laws and rules, people are violating those rules and that becomes the reason for road accidents, traffic and also some safety issues for men as well as women. Their views are likely to be more accurate than the views of the driving public, which should be more reliant on stereotypic perceptions of young and older drivers. Traffic density related issues are more prevalent and analytical scriptures aid in addressing these through various predictive analysis. This falcon device can sense the man's body language, nature and lifestyle by specific algorithms. By frequently monitoring human activity, they tend to develop a sense of receptivity by continuous analysis. They differentiate these as recurring natural and non-recurring activities. They also monitor the healthy actions and in case of any deviation from it, they can recognize any possible anomaly. These falcons are not trained; rather they just learn by distinguishing the actions as labelled and unlabelled data centric on machine learning. In fact, the views of police officers about some accident causes may be more accurate than official reports based on road accident records. This is because some factors are difficult to verify or substantiate at the roadside and thus may be underreported in accident records. All of these systems require management effort like monitoring, reporting and interaction to ensure efficient performance of all activities. This work is arranged based on the following sections. The various literature related to the smart city development are discussed in Sect. 2. The proposed research methodology and the working mechanism are discussed in Sect. 3. The results of the planned system and the corresponding inferences are deliberated in Sect. 4. The conclusion part is provided in Sect. 5.

2 Related Works

2.1 Development of Smart Cities

In the 1990's smart city development was addressed. The concept of smart city was to develop the infrastructure and communication network. The information of communication technology for superior infrastructures and upgrades in networks.

The use of information technologies also enables cities to empower the advancement of indispensable services for safety, health, governance and delivery. Global System for Mobile Communication (GSM) develops the latest reports as traffic information systems and intelligent transportation plays an important role in smart city projects. Smart environments host vital technologies to enhance the communication processes in complex cities to ensure better security purposes. They are achieved using intelligent machines paired up with smart technologies to make our spaces more perceptive. The main goal of this smart environment revolves around providing more adequate services available to all people at affordable costs. In case of hazards, it involves 3 vital stages, namely, detection followed by generating alerts and finally notifying respective officials. But even today we have these services available to us but it would be more astonishing if they are made much easier, faster and affordable. To achieve this state of hybrid metropolis, pervasive utilization of sensors, wireless communication, robust monitoring systems are significant towards this giant leap of development. Due to the increase of population in urban areas normal understanding also leads to violence due to lack of patience. Policy makers, researchers and service quality improver in India will have more understanding on information on smart city development [6]. A series of challenges in the economy and demand of cities have emerged in recent years, complicating the promotion of the smart city concept. Furthermore, there is a lack of clarity in the literature on urban planning for smart city programmes [7]. With a clear literature gap in relation to the smart city agenda, theoretical development and evaluation related challenges are included to implement in a country [8].

2.2 *Insecure of Information System*

An attack could affect supervisory control and data acquisition systems for controlling different urban facilities, Internet of Things (IoT) sensors and controllers and communication networks [9, 10]. Cybercrime is the country's leading cause of damage and risk. Every year, according to McAfee Inc. (2014), global cybercrime causes economic losses ranging from USD 375 billion to USD 575 billion.

3 Methods and Materials

Falcon bird is a robotic system to control and monitor the city. Public residing in the city can call this falcon bird for their emergency, so that they can afford their support. Falcon birds can divide into several segments based on people's needs in their surroundings. Falcons have the ability to attach and detach its structure and have its own language to communicate to its sub falcons and make appropriate decisions for the problem. The structure of the falcon is the inclusion of falcon1, falcon2, falcon3 and so on and its structure is based on the intensity of the problem. Public can use these falcons through an application in case of emergency. The falcon, in par with the new modern era, can give a close eye-watch on transportation by scanning the transport details. By doing so, it greatly reduces accidents by pre-determining them through algorithms. It also surveys climatic conditions, usually the smog and pollution impact and its associated risk capturing traffic activities to prevent the accidents and emergency scenarios before-hand. All the above-mentioned processes are made possible by extensive application of data analysis through Big data Analysis. More access to public utilities is made available to the people through websites and related applications along with contact assistance. In case of calamity, the falcon can direct the essential information from source to respective administrative places including hospitals, control centres. This also includes pollution detection, water service, waste management, public confrontations acting as a source of evidence in investigation processes and also addressing environmental issues. For instance, if a woman is caught up alone at midnight and if she is having any emergency, she can call this falcon bird to guide her. Not only in such safety cases but also falcons give their maximum support during emergency cases mentioned in the block diagram. It is not a complicated process to call this falcon, it is very simple and easily accessible by each and every individual who is really in need of guidance and support. As falcons are from the digital family, they will not harm the society and there will be enormous benefits for the people by these human friendly falcons. In case to find or recognize a stolen vehicle or an unofficial vehicle (not registered), falcons are able to create a QR code for that particular vehicle and fix it adjacent to the number panel. Thus, from that QR code it can trace the entire details of that vehicle and can give an alert to the transport office or give an alert message in the application. And if there are any traffic issues, people who are affected can access falcons, so that they can activate signals and clear traffic instantly. AI and Big data are often used synonymously. AI is a method of reducing human interference and resulting in activities that resemble human-like intelligence. This involves processing huge data through which predictions and quality information can be obtained by incorporating Big Data. AI is involved in giving information about public transports, common delays through swift communication methods. It also aids in ensuring public safety by quick mitigation techniques. Additionally, it also contributes to the smart automation systems of the city. Having in hand this huge amount of data not only processing the data is important but also maintaining it is a tedious task. But that is made even simpler by utilizing the Internet of Things (IoT). They aid in making better use of the smart city infrastructure and efficient monitoring.

3.1 Block Diagram of Falcon

Stage wise working behaviour of falcons is given below:

- **Stage 1**: Public should intimate the problem to falcon.
- **Stage 2**: Falcon should make a decision to do an easy process.

- **Stage 3**: Falcon releases its subdivision to the appropriate places.
- **Stage 4**: Finally, it calls its sub falcons by giving a trigger and alerting about relevant problems.

The above four stages are apt to social services, road accidents/traffic, theft vehicle, emergency courier. The technology of both hardware and software with high complication to overcome with people secure. This statement justifies that this method encapsulates hardware analysis with software systems to solve complex issues that pose great threat to the people's security. The falcon here is the hardware device which plays a major role in this smart space which is driven by artificial intelligence (AI), machine learning (ML), data analysis through Big data and equipped with IoT sensors. The software part comprises the mobile application and websites which fetches data to the falcon assisting in its activities. Due to network receiving, in real time birds get affected by these network signals leading to danger for them. To overcome this situation whilst in flight or during standing time, the function of sharing network data will shut down and automatically it starts to leave. By using Big data, traffic control, route planning, intelligent transport systems, congestion management (to predict traffic conditions). Artificial intelligence and the Internet of Things (IoT) are also included in falcon, to make decisions and to control. Data will be accessed by the government. After completion of the process, at the end of the day the appropriate department can retrieve all the data from the falcons. Figure 1 depicts a velar picture of a falcon block diagram. When coming to charging of falcons, its gains charge automatically by solar power and its sub falcons acquire charge when they all combine with the head falcon—this is the process behind charging of falcons. People's privacy is important for this society where there are types of privacy, individual, identity, bodily, territorial, communications and transactions where people should focus on daily life and to protect themselves [11]. The cloud layer has some flexibility and scalability support in smart cities. Today, our cities are being controlled by a network of control centres with much dependence on officials working day and night to maintain and protect our spaces. The advent of

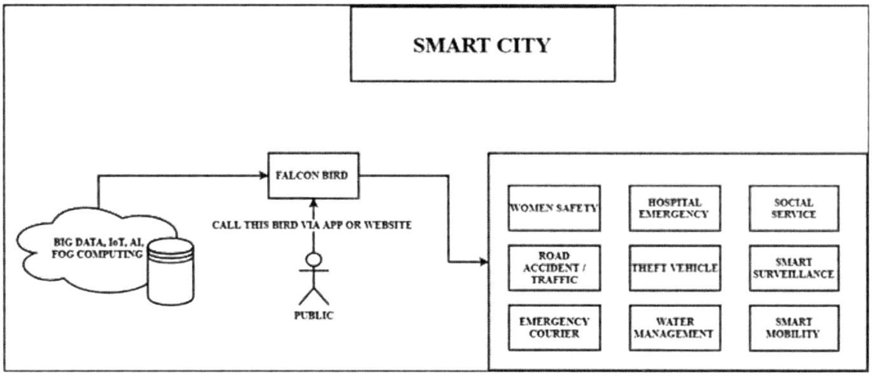

Fig. 1 Falcon general blocks

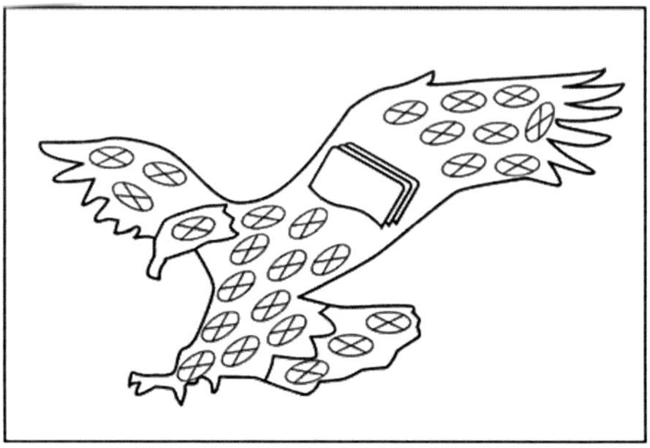

Fig. 2 Falcon device structure

this falcon will make our places much smarter by reducing manpower-dependence. This will also make communication and sharing of vital data much quicker thereby saving lives. It reduces the number of networks connected through control centres by minimizing it in the form of primary and secondary falcon birds. As these falcons can perform all the functions by divide and conquer method, it will be able to solve more than one issue taking place in a smart city in a short span of time. When it comes to dealing with huge data, it surely has a role to play with our privacy and security issues. The virtuous data can be extracted only by respective officials only if there is a need.

3.2 Device Structure

As shown in Fig. 2, the falcon device is easy to detach from its other subdivisions of falcon. Main controller is placed at the centre of the falcon so that it will be a starting point to connect like a chain. Half of the falcon subdivision will contain a camera and without a camera to make captures at critical situations for proves. Falcon can attack the stranger with its two wings if it is about to break, automatically it calls another one. Normally police use 20–30 screens to watch the cities but with the use of falcons it can be reduced. Falcon calculates location with latitude and longitude of the public call with high precision and accuracy.

3.3 Emergency Service

Location of the incident should be known by the falcon immediately. The main process of the user should be reported and determine the problem via application.

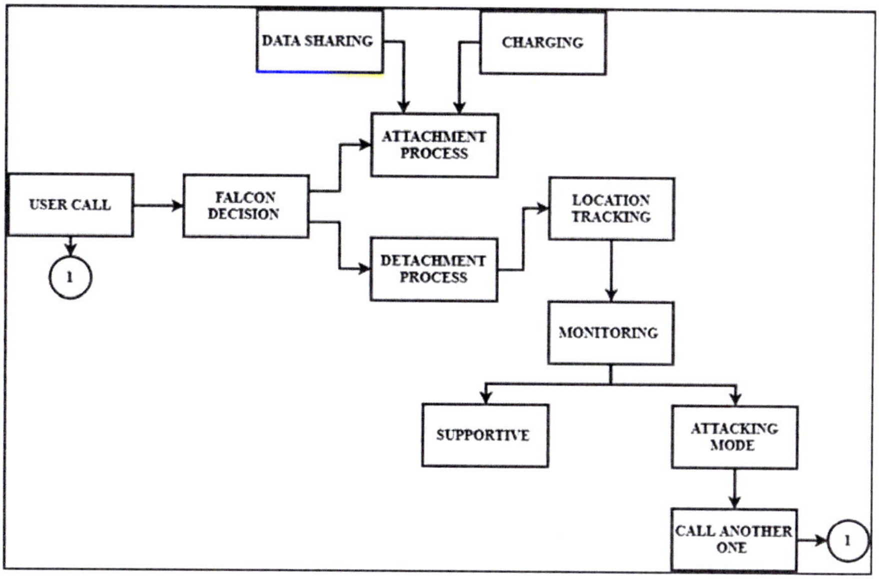

Fig. 3 Process flow of emergency service

And further the falcon receives those data, further the process proceeds as shown in Fig. 3.

The falcon bird device can be accessed through mobile application or through the website. The device then through the data stored in the cloud will just track the location of the incident precisely by processing the data available through fog computing. It is the technological process where numerous devices are connected to a cloud with continual data fetching. After determining and recognizing the required information, the falcon reaches the location and serves to assist the problems including women safety, hospital emergency, social service activities, road accidents or traffic related issues, vehicle theft, smart surveillance systems, emergency delivery services, water management and smart mobility centres. Then from there gathering data it directs it to the cloud and also thereby alerting respective officials about the anomaly and taking appropriate actions accordingly.

4 Results and Discussion

One of the innovative technologies implemented in cities to give a new experience to the public. Smart cities will be the future technology for the upcoming generation. Through this people should enjoy their life without any struggles facing in the streets. Nowadays our populations get more and so wastage and usage gets increased. By introducing this falcon bird, crime and unusual activities can be controlled. The actual sensitivity of the falcon is a natural phenomenon (i.e.) air pollution, fog can slightly affect the functionality of the falcon and there are chances that rays from the falcons can affect the birds. And these falcons are specifically used to enhance cities to smart cities and the precision of devices (i.e.) falcons is based on the difficulty level of that particular problem. Falcons have the capacity to be reused from one city to another. This was the main advantage of using falcons. Falcons' language is highly encrypted. The main authorized falcon gets to know the sensibly used by someone. In metropolitan cities, commercial centres are the places where more security is needed for people. In those places' security plays a vital role and managing a large network and taking suitable steps at the time of hazards seems much more difficult. This can be evaluated based on its sensitivity, specificity, precision and recall, as mentioned below.

- Sensitivity = No. of true positives/[No. of true positives + No. of false negatives]
- Specificity = No. of true negatives/[No. of true negatives + No. of false positives]
- Precision = No. of true positives/[No. of true positives + No. of false positives]
- Recall = No. of true positives/[No. of true positives + No. of false negatives]

Based on these parameters, our performance is analysed. So, if safety and security is ensured with less human intervention in an automated way then it will aid in solving hazards with ease (Fig. 4).

The implementation of this technology has been ignited in the prototype format as software application. The screenshots shown in Fig. 5 holds the acquisition of the login credentials in the leftmost screenshot, followed by the location tracking and identification in the middle and rightmost screenshots.

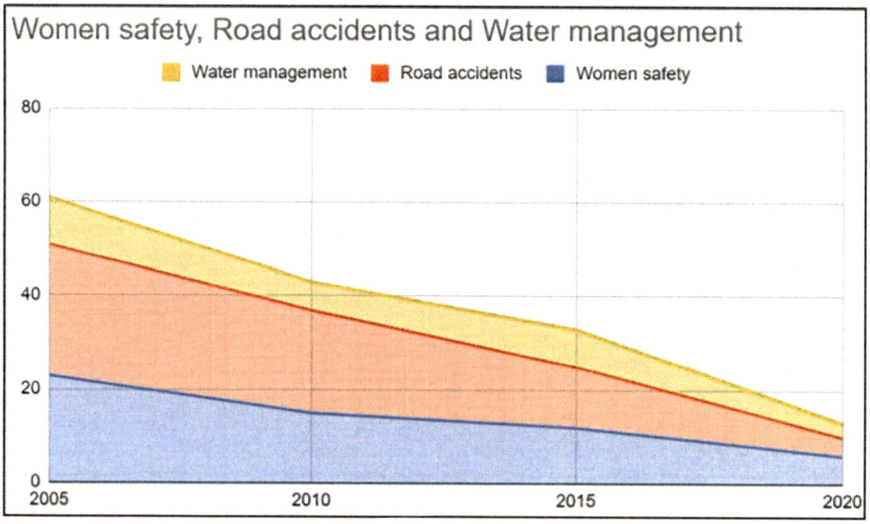

Fig. 4 Aspects supported by falcon in the past years

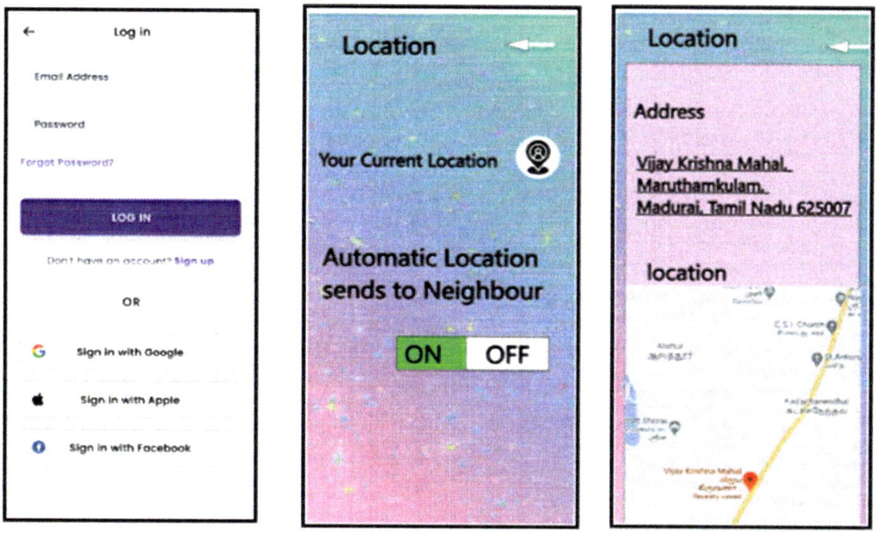

Fig. 5 Prototype screenshots

5 Conclusion

In this fast-paced world, no one has time to even concentrate on one's own work, so it is not possible to guide or help others for a long term. So here we have introduced a term FALCON, which can guide and support humans in all possible ways and its human friendly. The main benefit over falcons is that there will not be any time lag between the incidents, falcons connect themselves as fast as possible to complete their target. And also, they are placed with emotional intelligence so that they can understand the needs of humans prior to the humans themselves. There will not be any direct disaster from the falcons to the society and it is user friendly which greatly matches the Industry 4.0.

References

1. H. Elhoseny, M. Elhoseny, A.M. Riad, A.E. Hassanien, A framework for big data analysis in smart cities, in *Advances in Intelligent Systems and Computing* (2018), pp. 405–414.https://doi.org/10.1007/978-3-319-74690-6_40
2. P. Rizwan, K. Suresh, M.R. Babu, Real-time smart traffic management system for smart cities by using Internet of Things and big data, in *2016 International Conference on Emerging Technological Trends (ICETT)* (2016). https://doi.org/10.1109/icett.2016.7873660
3. V. Fernandez-Anez, Stakeholders approach to smart cities: a survey on smart city definitions, in *Lecture Notes in Computer Science* (2016), pp. 157–167.https://doi.org/10.1007/978-3-319-39595-1_16

4. G. Ignisha Rajathi, R. VedhaPriyaVadhana, L.R. Priya, Robotic Dustbin on Wheels. Int. J. Innov. Technol. Explor. Eng. **9**(1) (2019). ISSN No. 2278-3075
5. G. Ignisha Rajathi et al., Holistic education by breaking the binaries beyond the boundaries. Turk. J. Comput. Math. Educ. **12**(1s), 709–717 (2021)
6. N.P. Rana, S. Luthra, S.K. Mangla, R. Islam, S. Roderick, Y.K. Dwivedi, Barriers to the development of smart cities in Indian context. Inf. Syst. Front. (2018). https://doi.org/10.1007/s10796-018-9873-4
7. M. Angelidou, Smart cities: A conjuncture of four forces. Cities **47**, 95–106 (2015). ISSN 0264-2751
8. E.P. Trindade, M.P.F. Hinnig, E.M. da Costa, et al., Sustainable development of smart cities: a systematic review of the literature. J. Open Innov. **3**, 11 (2017)
9. P.T.I. Lam, R. Ma, Potential pitfalls in the development of smart cities and mitigation measures: an exploratory study. Cities (2018). https://doi.org/10.1016/j.cities.2018.11.014
10. R. Kitchin, The ethics of smart cities and urban science. Phil. Trans. R. Soc. A. Math. Phys. Eng. Sci. **374**, 20160115
11. T.A. Butt, M. Afzaal, Security and privacy in smart cities: issues and current solutions, in *Advances in Science, Technology & Innovation* (2019), pp. 317–323.https://doi.org/10.1007/978-3-030-01659-3_37
12. M. Condoluci, F. Sardis, T. Mahmoodi, Softwarization and virtualization in 5G networks for smart cities, in *Lecture Notes of the Institute for Computer Sciences, Social Informatics and Telecommunications Engineering* (2016), pp. 179–186.https://doi.org/10.1007/978-3-319-47063-4_16

A Review of Security Analysis of Wearable Implantable Medical Devices Using Biometric Encryption

G. Arun Jeba Kumar and V. Evelyn Brindha

1 Introduction

The Wearable and Implantable Medical Devices has been utilized in the healthcare sector in the recent years to monitor the vital signs of a patient as well as in the diagnoses of tachycardia, bradycardia, with the help of an electrocardiogram (ECG or EKG) and similarly in treating arrhythmia with pacemakers. These monitoring devices can be used as handheld devices or implanted in the patient's body such as pacemaker. The signals from the WIMDs has been sent to a Wireless Network Area with the help of an Bluetooth device which is inbuilt in the WIMDs (handheld device) and the signal has been received by the Medical Practitioner's personal computer or by the server of the Hospital, so that the data and the status of the patient can be accessed by the physician at any point of time. Since the data is traveling in a Wireless Area Network and it contains confidential information about the patient and the data should be protected. The cryptographic method using fuzzy vault and fuzzy commitment techniques have been elaborates in this paper. The biometric methods for automated user recognition have been widely deployed in the last ten to fifteen years. A variety of applications, including automatic teller machines, border checkpoints and cell phones, use biometric authentication [1, 2]. Biometric innovations contributed to the efficiency of these processes. The success of these systems was due to advancement in biometrics. The biometric features are unique to each person and unlike traditional security features, they are extremely difficult to lose, forget, or guess [3]. Biometric-based identification schemes provide the customer with greater ease and efficiency than conventional authentication mechanisms. Regardless, the advantages of biometric identification, still there is a serious concerns about biometric authentication and privateness. The image retrieval module and device database can

G. Arun Jeba Kumar · V. Evelyn Brindha (✉)
Department of EEE, Karunya Institute of Technology, Coimbatore, India
e-mail: evelynbrindha@karunya.edu

J. D. Peter et al. (eds.), *Disruptive Technologies for Big Data and Cloud Applications*, Lecture Notes in Electrical Engineering 905,
https://doi.org/10.1007/978-981-19-2177-3_10

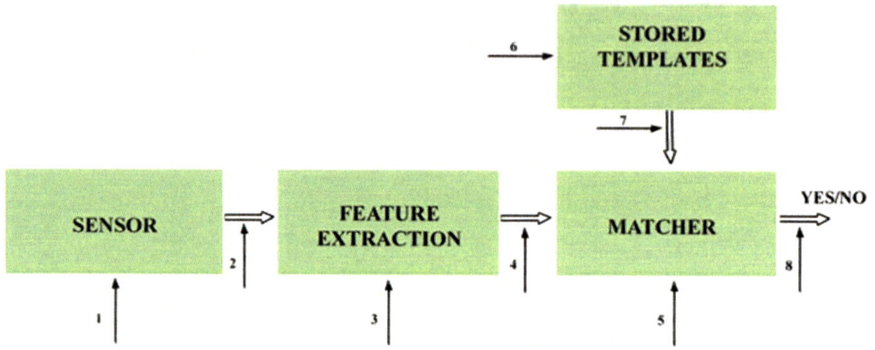

Fig. 1 Attack in biometric system

also be hacked in order to obtain the created genuine template, it can be changed or substitute it with a fake templates (see Fig. 1).

2 Cutting Edge Technology

Biometric mutability issue poses a significant challenge in the development of attribute-based encryption BTP systems for the protection of biometric information standard cryptographic methods should be considered [4, 5]. The biometric template protection classified as follows, Feature Extraction, Biometric Cryptography, Homomorphic Encryption (see Fig. 2), Visual Cryptography, Hybrid Methods and other methods like Stenography and Watermarking. The Visual Cryptography method permits the visual information (images, text, etc.) can be encrypted using a cryptographic approach that allows the decrypted information to display as a visual image. The hybrid method can be achieved by information transfer using a separate

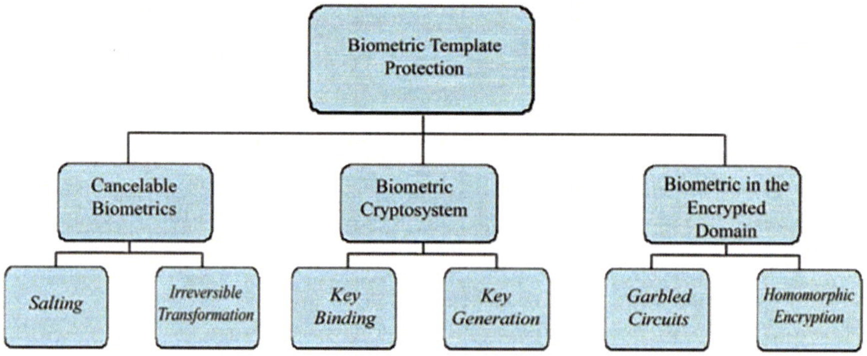

Fig. 2 Cryptography and encrypted domain of biometrics

unique key aside with symmetrical encryption. Meanwhile the stenography technique conceals and entraps the data. The watermarking technique hides the data into a carrier signal. Digital watermarks can be used to confirm the carrier signal's validity or integrity, as well as to identify the signal's owners. Feature extraction is a technique for identifying key characteristics or features in data. Biometric information may be protected using standard secure protocols, where the input data differs significantly. This algorithmic encryption produces significant variations in the output. The intraclass heterogeneity renders these procedures useless biometrics, since encrypted models cannot be contrasted successfully. The models may also be decrypted before being compared. However, any authentication attempt fails since the models are unprotected, this procedure is insecure, and standard cryptographic algorithms are ineffective at protecting biometric templates [4].

3 Proposed Method

The security of a physiological data is totally dependent on the cryptography even though the source data is hidden, once it travels in a Wireless Area Network, it can be easily hacked, so some of the researches used the technique of adding chaff points to the actual data. In this paper, we are proposing to actually increasing the number of chaff points to the real physiological data, so that if the data is hacked, the actual source cannot be accessed by the hacker. Moon [27] mentioned three possible solutions for the protection of biometric data like (i) One-time template method, (ii) Securing Fuzzy fingerprint vault method, (iii) Geometric has table method any one of the methods will be followed in this paper for protection of the physiological data.

3.1 Scheme for Protecting Biometric Templates: Categories and Requirements

In the recent decade, researchers focused on developing BTP methods to address the shortcoming of conventional cryptographic methods. These schemes must comply with all of the standards specified in the ISO/IEC 24,745 2011 standard on biometric data security [6]. Retrieving the initial biometric template from a secure biometric template must be irreversible.

- Irreversibility: The original BT must be algorithmically not possible to extract from a consistent biometric template.
- If a protected blueprint is breached, the scheme must be able to revoke it and produce a new one using the same biometric data.
- Unlink ability: It is trickier to figure out whether two or more templates, each with its own key are linked to the same biometric feature and this property protects users' anonymity by preventing cross-comparisons with other databases.

Simultaneously, precision verification time and storage requirements must all be handled in a way that is equivalent to an unregulated method [7]. It is difficult to design BTP schemes that fully fulfill the above requirements without sacrificing verification performance. In the literature various methods have been suggested and they can be divided into three groups:

Cancelable Biometrics, Biometric Cryptosystems and Encrypted Domain of Biometrics (see Fig. 2).

A transformation function specified by a consumer key is applied to the exposed template in the Cancelable Biometrics category. For verification considerations, both the revised template and the key (also known as auxiliary data (AD)) are saved for verification purposes. The template probe is then modified in the same way, and the comparison takes place in the altered domain. There are two subcategories based on the transformation function's characteristics:

- Salting is a transformation that uses a reversible function. As a result, the security of these schemes is dependent on the key's security.
- Irreversible transformation involves applying a one-way function on an unsecured template, resulting in a modified template that is difficult to reverse even though the code is available.

The second block indicates the Biometric Cryptography, which combines biometrics and cryptography both to create a safe design that does not disclose any data about biometric data or the key. The comparison is done indirectly during authentication by utilizing the AD to evaluate the authenticity of a key retrieved from the biometric probe. Obtaining the key from the AD must be technically challenging if the authentic users' biometric data is unclear. However, if the template probe is far enough to the registered template, decoding the AD and recovering the key should be simple. Intraclass variability is often handled using error correction coding methods. These schemes provide for the secure administration of biometric data while also allowing for the protection of biometric data. There are two types of biometric cryptosystems:

- Key Binding is a form of Asymmetric Decryption in which the key is linked to biometric data. The fuzzy vault [8] and fuzzy commitment [9] were the two popular schemes in this group.
- Key Generation, this group contains a stable sketch and fuzzy extractor [10] principles, where the auxiliary data is obtained solely from the biometric data.

To maintain system protection, the stored AD in Biometric Cryptosystem and Cancelable Biometrics approaches need not disclose substantial sensitive details [4, 11] have a more in-depth look at these two groups.

Alternatives to the above solution that have recently been used as BTP approaches include privacy-preserving computation [12] and homomorphic cryptosystem [13]. In the encoded format, these results are divided into a new field called Biometrics. The most often used methods in this group are Garbled Circuits [14] and Homomorphic (HE) [15].

Garbled circuit is a method of "encrypting a calculation" by revealing just the algorithm's result while concealing the inputs and intermediate values. We use the terminology "circuit" because garbled circuits operate by expressing the calculation as a circuit and then performing cryptographic operations on each operation (AND, OR, NOT) in the circuit.

Most notably, efficient biometric device implementations based on Homomorphic Encryption have been established, ensuring that only protected biometric data is managed. HE schemes perform recognition in the encrypted domain with results that are entirely equal to those obtained by unregulated programs, but they only perform a limited number of operations in this format as a consequence, applying advanced comparative approaches to this mechanism is more complex. Furthermore, this authentication method necessitates a built on the presumption that encryption keys are kept hidden at all times [12], and it has a higher computing cost and bandwidth.

Homomorphic encryption is a type of encryption that preserves data encrypted even when it is being analyzed and modified. It allows user or a third party to perform operations on encrypted information without having to divulge the data's values.

3.2 Similar Work

Several works focused on the security of biometric models using dynamic signatures have been proposed. The authors proposed an irreversible transformation (no well-defined set of values for its state variables) method to protect signature time sequences in [16, 17], in which Hidden Markov Models (HMMs) local features extracted from complex signatures are converted and used in the comparisons. The models were constructed using maximum and minimal signatures, with event points such as ending points, crossing points and high curvature points taken into account. In a dynamic signature verification (DSV) scheme based on UBM, to secure fixed-length dynamic signature models, the fuzzy commitment technique was used, and the solution presented in [18] makes use of helper data information and error correction methods.

3.3 Background of Fuzzy Vault

In general, the security weakness inherent in cryptographic key storage, the fuzzy vault is a technique that enhances standard cryptographic security solutions by combining it with biometric authentication. Although the efficiency of the fuzzy vault will depend on the minutiae information with high number of chaff points added to it and the authenticated key generation. When we talk about fuzzy vault the other terminology comes to mind is the fuzzy logic. The fuzzy logic (FL) is a reasoning approach that is similar to human thinking. FL's method is modeled after how people make decisions, which includes all options in between the digital values YES and NO. The fuzzy logic deals with the method of computation that uses "degrees of truth" rather than the traditional "true or false" approach.

Two popular cryptographic primitive techniques of key generation and key agreement such as FC and FV using biometric data used to secure the physiological values (PVs) such as ECG signals, heart rate, blood pressure, temperature and make the signal less prone to vulnerabilities. A fuzzy vault encryption and decryption shown in Fig. 3.

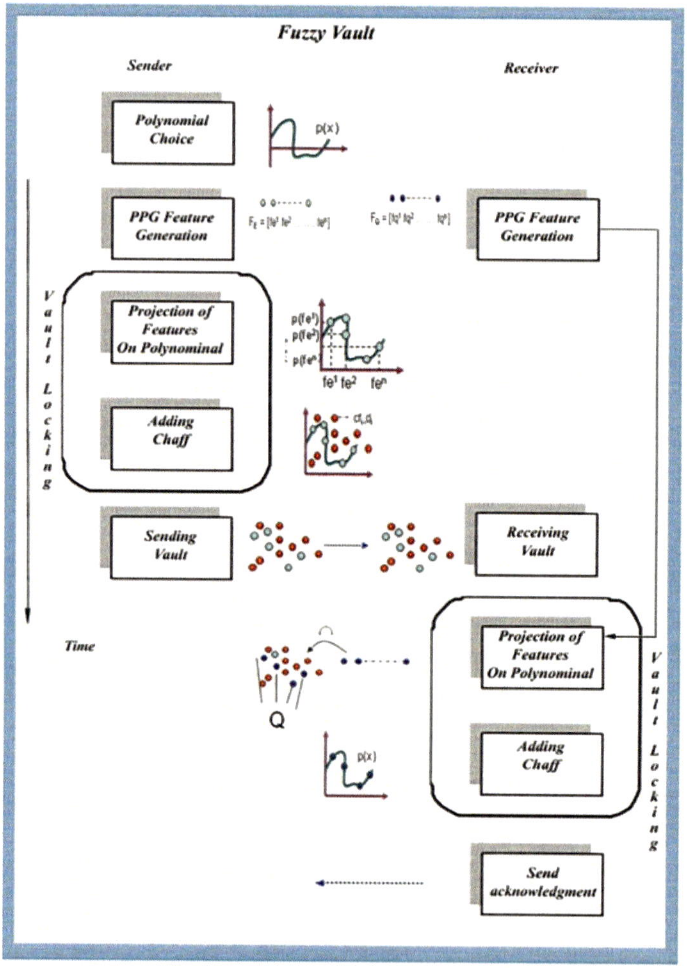

Fig. 3 Fuzzy vault encryption and decryption

3.4 Applications of Fuzzy Vault

Many studies have been published in which the FV method has been used to cover various biometric modes. A fingerprint template-based FV scheme is introduced [19]. The author suggest a tweaked version of the FV mode proposed [8] in this case, owing to the challenges experienced in using error correction to reconstruct the polynomial. There is no error corrector code added during vault encoding. Instead, the user key is used to produce bits or CRC. The encoded key is used to represent the L-degree polynomial used in vault construction, and these bits are appended to the key. The remainder of the vault coding procedure is identical to the FV system. Using the probe sample points of chaff are picked during the authentication process from vault. Authentication fails if fewer than L1 points are detected, otherwise, they find all feasible L1 point combinations and generate a candidate polynomial for each one, which is then checked using CRC. The tested polynomial is discarded in the case of error detection and then does it all over again for the next point combination. If errors not observed, it ensures that the right polynomial was discovered and that the corresponding key is correct. The disadvantage of this scheme is that it would evaluate multiple candidates' polynomial for each authentication attempt, which wastes a lot of time. This FV scheme on the other hand has been one of the most common method. The fuzzy vault method can also be updated to include face [20] and palmprint [21] modalities as per the researchers. The protected template is derived from human fingerprint and iris templates in a multibiometric research-based FV [22]. In this case, the multibiometric vault outperforms the unibiometric vault in terms of efficiency and security. To protect iris biometric info, a version of the original FV scheme [23] is suggested. A comparison of various researches and techniques is seen in the tabular column.

The prototype in this paper's fuzzy vault scheme is based on fixed period models with application to dynamic signature authentication were formed using only fifteen global signature features according to **Ponce-Hernanz** [24]. A proprietary list of signatures along with the public signature databases-MCYT and biosecure are used to verify device performance. As a result, safe references are more resistant to threats, although authentication performance is not significantly worsened when compared to a defenseless device. They believe that in applications where a key is not needed, a simpler version of the fuzzy vault's technique may be used. Otherwise, only the Reed-Solomon encoding would be added to the reference prototype during enrollment, and only the disused symbols received would be stored for each enrolled person. The Reed-Solomon decoder will then decide if the user is right during verification, using the associated redundancy.

In comparison to fuzzy vault, **Nguyen** [25] proposes a non-random chaff point generator that is secure enough to prevent reported attacks. Unfortunately, it relies on a complicated equation, which slows down the implementation process. The Author also states that this article introduces a modified variant of non-random chaff point generator. The suggested solution makes use of the circle packing procedure to restore the Euclidian distance estimate in the chaff point generator. This method speeds up

Galois Field execution while still cutting down on the time it takes, where Galois Field refers to the field with least number of elements feasible, and in the future, the mobile payment system will use this upgraded version. Another authentication stage process should be tested in order to include a fuzzy vault system for mobile payment that is not only reliable but also fast.

Bakhteri [26], indicated that in this implementation of a fingerprint-based biometric encryption scheme on a field programmable gate array development board as a standalone system operating on software is defined in this paper. The memory of the FPGA board was found to be adequate for the RTOS and the fingerprint template database. There is an opportunity for improvement for this system when it comes to increasing accuracy by introducing better image processing algorithms. Through putting compute-intensive operations on hardware accelerators, the system's tempo and efficiency can be increased, decreasing the hand-shaking procedures between the different process blocks and introducing USB communication between FPGA and host personal computer.

The blend replacement attack is a major flaw in the current Cycle Redundancy Check-based fuzzy vault scheme, according to **Moon** [27]. As a result, a new non-random chaff point generator was developed, as well as verification modules. The latest modules will detect any changes to the original vault using constant hashing and linear projection, which eliminates the blend substitution attack on standard Cycle Redundancy Check-based fuzzy vaults. We provide a security analysis of our novel scheme as well as partial results on FVC2002 DB1 datasets.

According to **Ghouzali** [28], the risk of attacks on the template database has recently increased as a result of the increased use of biometric systems. Furthermore, the author believes that the results reveal a connection between key length and process operation, and that using an iris shuffling algorithm to assess revocability merely enhanced the separation of authentic and pretender distributions. The proposed method employs two user-specific keys, one for the 128-bit shuffling algorithm and the other for the Reed-Solomon encoding algorithm, both of which are sufficiently robust to guarantee the system's revocability and security.

To test the feasibility of the proposed algorithm, **Moon** [27] identified three methods of solutions for the FFV and performed few tests on the FVC2002 databases. There are three options: (I) a geometric hashing method for automatic alignment of fingerprint; (II) a secure FFV that can withstand the attack on the correlation; and (III) the fingerprint vault with fuzziness uses the OTT to create an unique prototype of biometrics for each person. With these solutions, the safety of biometric information is enhanced.

- Geometric hashing method: Automatic alignment of fingerprint is presented using the geometric hash table approach. This is the process of determining true minutia by looking at the direction of minutia in a fingerprint. While recording geometrics data on the orientation and position, it provides a point of reference to the actual or chaff points. The geometric hash table is used in verification to align verification details when they are transferred to the reference point one by one.

- Protection of correlation attack: [27] To solve the correlation attack, using minutiae orientation information, the author uses a vector circle with a radius of (r) to calculate the quantity of false minutiae. The angle between the minutiae and the regular minutiae, using calculations based on the tangent circle added to the normal minutiae, additional fake minutiae are generated by selecting a point on the tangent of the angle between the reference minutiae.
- One-time template: The template which is generated only once for the FV runs like an single time password. When a user is properly approved, it creates a new template called one-time template, which is used to ensure that the fingerprint minutiae do not match the old template. The OTFV also expands the size of a fingerprint picture, allowing for the addition of 100 or even 1000 of chaff minutiae. The OTFV can be used to insert number of chaff spots without affecting the quality of the verification. As a result, OTFV can significantly enhance the safety of the fuzzy fingerprint vault when it comes to brute-force or correlation attacks.

According to **Moon** [27], biometric access control in mobile technology is usually based on biometric data with traces, as a result, hacker not able to identify or tamper copies of real biometric information. This paper, according to the source, looked into the safety of FV method under this assumption, analyzing a variety of less-resolution degrees. As a result, the author appears to have tested this attack against FV biometric cryptosystems that use fingerprints and face images, even if the degree of alteration is high, this scheme is vulnerable to attacks particularly in the case of Sface images (Table 1).

4 Conclusion

This paper performed the security analysis of WIMD's of physiological data's like as ECG, EEG, pacemaker signals, etc. The comparative analysis shows that the researches had different kinds of security methods for the appropriate WIMDs which they had been using in their research. Some of the researchers have used the Cancelable Biometrics techniques (salting and Irreversible Transformation), others used the technique of Key Binding and Key Generation in Biometrics Cryptosystem and some of the researchers used the Biometric in the Encrypted Domain techniques of Garbled Circuits or the Homomorphic Encryption. In analyzing all the mentioned researches, it has been concluded that the security of a physiological data is totally depend on the encryption of the data as well as the number of increased chaff added to the data, so that if any attacker tries accesses the data, he/she cannot get the actually hidden data and in the future, most of the WIMDs will transmit their data in the air (wirelessly) in Wireless Area Networks, so the data will be more vulnerable and according to the analysis, it indicated that any of the methods indicated in this paper which has high number of chaff points will protect the data.

Table 1 Comparative analysis of fuzzy vault and fuzzy commitment

Topic	Author	Method	Summary
Implementation and performance of fingerprint-based fuzzy vault	Karthik Nandakumar, Anil K. Jain,	The author implemented fingerprint-based fuzzy vault that secures secrets as a 128 bit advanced encryption standard encryption keys	The approach is quite helpful in securing the keys up to 128 bits
mHealth (Remote health care using mobile devices) application security by focusing on heart rhythm	Chris I. De Zeeuw, Robert M. Seepers, Christos Strydis, Ioannis Sourdis, and Robert M Seepers	It is heartbeat based security (HBBS) method which forms a biometric security key based on time interval between two consecutive heart beats	The researcher introduced multi key generator based on Inter-multi-pulse Interval (ImPI)
Heartbeat timing is used as an independent marker to protect a body sensor network	Shu-Di Bao, Carmen C Y. Poon, Yuan-Ting Zhang, Lian Feng Shen,	The method is generation of binary entity identifier (EIs) based on IPI data and it secures wireless BSNs	Author proposes generation of binary entity identifier (EIs) to secure wireless BSNs
EKG-based key agreement body sensor networks	Ayan Banerjee, Sandeep K. S Gupta, Krishna Kumar Venkatasubramanian	EKG signals generating and distributing keys to enable secure inter-sensor communication	The proposed method is to secure inter-sensor communication by EKG signal distributing keys
A review of ECG-based key distribution for securing of WIMDs	Guanglou Zheng, Rajan Shankara Wencheng Yang Craig Valli, Li Qiao, Mehmet A. Orgun, Subhas Chandra Mukhopadhyay	The fuzzy vault and fuzzy commitment merits and demerits are analyzed by polynomial computation and binary sequences generation	In the analysis the author says that the fuzzy commitment has a better false acceptance rate because of random ECG binary sequence
A study of key distribution schemes in wireless body area networks using fuzzy commitment and fuzzy vault	Rajan Shankaran, Guanglou Zheng, Gengfa Fang, Mehmet A. Orgun	Analysis done with the help of sequence generation and polynomial computation and	Researches says that the analysis between fuzzy vault and fuzzy commitment shows that the FC has the most complicated process in EKG measurement but its key concealing and reveling is easier than FV

(continued)

Table 1 (continued)

Topic	Author	Method	Summary
Based on the fuzzy vault system, a new method for generating an invariant Iris private key	Youn Joo Lee, Jaihie Kim	The intravariations of iris images reduced by two techniques pattern-clustering and local shift matching method	Proposed solution improves results without the need for realignment

References

1. A.K. Jain, A. Kumar, *Second Generation Biometrics: The Ethical, Legal and Social Context*, vol. 11 (2012)
2. A.K. Jain, K. Nandakumar, Biometric authentication: system security and user privacy. Computer (Long. Beach. Calif). **45**(11), 87–92 (2012)
3. S. Prabhakar, S. Pankanti, A.K. Jain, Biometric recognition: security and privacy concerns. IEEE Secur. Priv. **1**(2), 33–42 (2003)
4. K. Nandakumar, A.K. Jain, A. Nagar, Biometric template security. EURASIP J. Adv. Signal Process. **2008** (2008)
5. A.K. Jain, A. Nagar, K. Nandakumar, Fingerprint template protection: from theory to practice. Secur. Priv. Biometrics 187–214 (2013)
6. ISO/IEC JTC1 SC27 Security Techniques, ISO/IEC 24745:2011. Information Technology—Security Techniques—Biometric Information Protection (2011)
7. K. Simoens et al., Criteria towards metrics for benchmarking template protection algorithms, in *Proceedings—2012 5th IAPR International Conference on Biometrics, ICB 2012*, pp. 498–505 (2012)
8. A. Juels, M. Sudan, A fuzzy vault scheme, in *IEEE International Symposium on Information Theory—Proceedings*, vol. 38, p. 408 (2002)
9. A. Juels, M. Wattenberg, Fuzzy commitment scheme, in *Proceedings of the ACM Conference on Computer and Communications Security*, pp. 28–36 (1999)
10. Y. Dodis, R. Ostrovsky, L. Reyzin, A. Smith, Fuzzy extractors: how to generate strong keys from biometrics and other noisy data. SIAM J. Comput. **38**(1), 97–139 (2008)
11. C. Rathgeb, A. Uhl, A survey on biometric cryptosystems and cancelable biometrics. EURASIP J. Inf. Secur. **3**, pp. 1–25 (2011)
12. J. Bringer, H. Chabanne, A. Patey, Privacy—preserving biometric identification using secure multiparty computation: an overview and recent trends. IEEE Signal Process. Mag. **30**(2), 42–52 (2013)
13. C. Aguilar-Melchor, S. Fau, C. Fontaine, G. Gogniat, R. Sirdey, Recent advances in homomorphic encryption: a possible future for signal processing in the encrypted domain. IEEE Signal Process. Mag. **30**(2), 108–117 (2013)
14. A.C.C. Yao, How to generate and exchange secrets, in *Annual Symposium on Foundations of Computer Science (Proceedings)*, pp. 162–167 (1986)
15. C. Fontaine, F. Galand, A survey of homomorphic encryption for nonspecialists. EURASIP J. Inf. Secur. **10**, 1–10 (2007)
16. E. Maiorana, P. Campisi, J. Fierrez, J. Ortega-Garcia, A. Neri, Cancelable templates for sequence-based biometrics with application to on-line signature recognition. IEEE Trans. Syst. Man Cybern. Part A Syst. Humans **40**(3), pp. 525–538 (2010)
17. E. Maiorana, M. Martinez-Diaz, P. Campisi, J. Ortega-Garcia, A. Neri, Template protection for HMM-based on-line signature authentication, in *2008 IEEE Computer Society Conference on Computer Vision and Pattern Recognition Workshops, CVPR Workshops* (2008)

18. M.R. Freire, J. Fierrez, J. Ortega-Garcia, Dynamic signature verification with template protection using helper data, in *ICASSP, IEEE International Conference on Acoustics, Speech and Signal Processing—Proceedings, 2008*, pp. 1713–1716 (2008)
19. U. Uludag, S. Pankanti, A.K. Jain, Fuzzy vault for fingerprints. LNCS **3546**, 310–319 (2005)
20. Y. Wang, K.N. Plataniotis, Fuzzy vault for face based cryptographic key generation, in *2007 Biometrics Symposium, BSYM*, pp. 1–6 (2007)
21. A. Kumar, A. Kumar, Development of a new cryptographic construct using palmprint-based fuzzy vault. EURASIP J. Adv. Signal Process. **2009**(1), 967046 (2009)
22. W. Ponce-Hernandez, R. Blanco-Gonzalo, J. Liu-Jimenez, R. Sanchez-Reillo, Fuzzy vault scheme based on fixed-length templates applied to dynamic signature verification. IEEE Access. https://doi.org/10.1109/ACCESS.2017
23. Y.J. Lee, K. Bae, S.J. Lee, K.R. Park, J. Kim, Biometric key binding: fuzzy vault based on iris images, in *Advances in Biometrics, International Conference, ICB 2007—Proceedings*, vol. 4642 (Seoul, Korea, 2007), pp. 800–808
24. R. Bakhteri, M.K. Hani, Biometric encryption using fingerprint fuzzy vault for FPGA-based embedded systems, in *TENCON 2009—2009 IEEE Region 10 Conference* (2009). https://doi.org/10.1109/TENCON.2009.5396047
25. T.K. Dang, M.T. Nguyen, Q.H. Truong, Chaff point generation mechanism for improving fuzzy vault security. IET Biometrics **5**(2), 147–153. https://doi.org/10.1049/iet-bmt.2015.0023
26. M. Fouad, Towards template security for Iris-based biometric systems. https://ruor.uottawa.ca/handle/10393/22736
27. K.Y. Moon, D. Moon, J.-H. Yoo, H.-S. Cho, Biometrics information protection using fuzzy vault scheme, in *2012 Eighth International Conference on Signal Image Technology and Internet Based Systems* (2012)
28. S. Ghouzali, M. Lafkih, W. Abdul, M. Mikram, M. El Haziti, D. Aboutajdine, Trace attack against biometric mobile applications. Mobile Inf. Syst. **2016**, Article ID 2065948 (2016)

G. Arun Jeba Kumar Research Scholar (RPK19EE009)—Department of Electrical and Electronics Engineering, Karunya Institute of Technology, Coimbatore 641035, Tamil Nadu, India.

Blockchain for CCTV Surveillance

D. Dharani, K. Anitha Kumari, and R. Vasanthan

1 Introduction

In order to provide better services while ensuring effective and optimum use of available resources, a smart city uses information technology to connect and maintain physical, social, and business infrastructures. Smart cities will have creative solutions and more direct connectivity through technology such as the Internet of Things (IoT), cloud computing, and integrated networks. Since IoT devices are heterogeneous in nature, the design of IoT architecture needs to ensure goals such as energy, latency, throughput, scalability, topology, security, and security from a technology perspective. Energy denotes that the nominal power supply would work for the lifetime of an IoT system. Latency of a network is the amount of time taken for data transmission and delivery of messages. Despite a range of possible IoT advantages, there are many information security and privacy issues. Cybercriminals can hack IoT devices, and they can get access to users' personal and financial information [1]. The hazards posed by exploitable devices may be threat to their users and threat to those that they face. If the technology continues to evolve, the amount of data that IoT devices collect and transfer will increase, which will in turn add to the need for increased bandwidth [2]. Surveillance cameras are a part of the smart city application. Since crime and injury rates in densely populated areas are typically high, the protection and security of people are among the key objectives of most smart cities. Video surveillance uses streaming cameras to solve this issue. There is a necessity for a technology which ensures the integrity and confidentiality of an IoT application. CCTV footages are vulnerable against different video altering attacks. The recorded videos are stored in a single place, so they face single-point failure (DVR, NVR, hard disk), and they are stored temporally (per month cycle). To

D. Dharani (✉) · K. Anitha Kumari · R. Vasanthan
Department of Information Technology, PSG College of Technology, Coimbatore, India
e-mail: dharani0609@gmail.com

J. D. Peter et al. (eds.), *Disruptive Technologies for Big Data and Cloud Applications*, Lecture Notes in Electrical Engineering 905,
https://doi.org/10.1007/978-981-19-2177-3_11

solve these issues, the proposed system uses blockchain-based ledger to guarantee the trustworthiness of the stored recordings which would provide increased integrity and permanent storage solution [3]. However, it is still very costly to store large documents, as the 1 MB size limit per block in the blockchain of Bitcoin will limit the file size that can be uploaded. Using decentralized storage systems such as the InterPlanetary File System (IPFS), the need to store massive files was addressed. The IPFS is content-addressable, peer-to-peer, open-source, a globally distributed file system that can be used to store and transfer a large volume of high-throughput files. The system integrates IPFS to blockchain in order to provide immutable and decentralized storage system for preserving the video recordings [4, 5].

2 Related Works

As recent research interest continuously focuses on integrating the Internet of Things with blockchain for providing security for the data, many sensitive fields like health care, surveillance, and smart grid systems are getting benefits out of it. This integration cryptographically secures the data and maintains the distributed ledger in every peer node to help maintain the integrity of the system. Regardless of the potential benefits of this integration, there are some limitations in using the blockchain to all the IoT systems. Not all the IoT systems require the high-end security. Without analyzing the feasibility of the integration and requirements, the entire system will become invalid [2]. The systems like CCTV surveillance generate huge streaming data. It is highly required to be stored and protected for future purposes as it might be an evidence of some criminal activities. If the video data are tampered, entire case will go off in the wrong direction. Hence, this system is completely fit for integrating with blockchain irrespective of the scalability issues present in Blockchain Network. Ethereum like public blockchain frameworks has an issue to maintain the scalability of the system. As the streaming video data are huge in size and it continuously grows drastically, it would be infeasible to handle all the data entirely in blockchain [3]. Blockchain holds sound security mechanisms, and it is going to rule the world without any doubt. It also changes the entire world with the help of digital currencies. Due to the potential benefits of the blockchain, it would sustain for long period of time and serve the world for providing immutability to the sensitive records. However, it also has the limitations such as time taken for mining, power consumption, and financial constraints [4].

3 Materials and Methods

Video: A CCTV video's metadata contains various information like date, time, logs, device name, and location details. Every video uploaded into blockchain using this preservation system will store the metadata also into the block, which is used to retrieve the CCTV information's data very effectively. Each and every video and metadata will be associated with Ethereum smart contract.

IPFS: The CCTV videos are stored in distributed, immutable content-addressable, high-throughput, peer-to-peer system such as the InterPlanetary File System. This file system uses a decentralized storage protocol. The stored files can be of any type, which are permanently stored, and it will calculate a unique cryptographic hash value, which will allow us to identify the files from the system quickly. IPFS does not support duplication. The file system does not allow to repetitive file uploads; hence, it uses minimal bandwidth [6].

Blockchain: A peer-to-peer distributed ledger technology, which records transactions, agreements, contracts, and sales, originally created for support cryptocurrency and can be utilized for any form of transactions without an intermediary. The benefit of blockchain is that an attacker has to compromise 51% of the systems to surpass the hashing power of the target network which is infeasible to compute and attack the network. In this system, the IPFS-calculated hash value is stored into the block of blockchain. Using this method, cost of storage can be reduced and provides efficiency and accessibility [7].

DApp: Decentralized applications are software that connects with the blockchain, which keeps track of all network actors' current states. The user interface of decentralized applications is identical to that of any Web site or mobile app. A decentralized application's core logic is represented by a smart contract. Smart contracts are essential components of blockchain, as they process data from external sensors or events and assist the blockchain in keeping track of the state of all network actors.

Truffle framework: The Truffle framework consists of three primary development frameworks for Ethereum smart contract and decentralized application (DApps) development called Truffle. Truffle platform is developed for modeling DApps using the Ethereum virtual machine (EVM). It also comes up with a development environment, testing framework, and asset pipeline. Truffle is a collection of three development platforms for Ethereum smart contracts and decentralized applications (DApps). Truffle is a development environment, testing platform, and asset pipeline for Ethereum virtual machine (EVM) DApps. The features that make Truffle Ethereum one of the most widely used IDEs for Ethereum blockchain development is it has the built-in support for compiling, deploying, and linking the smart contracts.

Ganache: Ethereum Ganache is a local in-memory blockchain designed for development and testing. It simulates the features of a real Ethereum network, including the availability of a number of accounts funded with test. Ethereum Ganache is available in two versions, as a graphical application with a user interface and as a command line version. It is a readymade private blockchain with some basic default values for the parameters, such as EVM used, gas price and limit of RPC interface to be presented to the local host.

MetaMask: MetaMask is an extension for browser, which used to access Ethereum Dapp. It also serves a wallet for holding tokens and used to allow user to connect the local blockchain network ethers. MetaMask allowing users to access services built on the network via the wallet. MetaMask is a browser plugin that serves as an Ethereum wallet. MetaMask is one of the simpler Ethereum wallets and DApp browsers to use and can be set up in a couple of minutes in most cases.

Smart contract: A smart contract is a series of self-verifying, self-executing, and tamper-resistant systems. Smart contracts that use blockchain technologies are capable of performing tasks in real time at a low cost and with a higher level of security. A smart contract is a computer application that runs on the blockchain and consists of a series of rules. It helps execute the code without the having an association with any third-party applications. The meaning, address, functions, and state of a smart contract are all included. It takes a transaction as an object, runs the code associated with it, and then activates the output events. This reduces the need for any trustworthy third party for smart contract communication while still ensuring the system's security against any malicious attack [2].

ReactJS: React JS is a JavaScript library for creating dynamic elements on the Web sites. A JavaScript library focuses on assisting developers in the development of user interfaces (UI). UIs are the set of on-screen menus, search bars, keys, and everything else that someone communicates with the system to access a page of Web sites and mobile apps. React comes with two main features that contribute to its appeal for JavaScript developers: JSX and virtual DOM, in addition to having reusable React library code. The purpose of the library codes are for saving production time and reducing the risk of coding errors. HTML records are the foundation of every Web site. These documents are read by Web browsers and viewed as Web pages on the computer, tablet, or phone. Browsers construct a document object model (DOM) during this process, which is a representational tree of how the Web page is structured. Developers will then use languages like JavaScript to change the DOM to add dynamic content to their designs. JSX (short for JavaScript extension) is a React extension that allows Web developers to easily change their DOM using simple HTML-style code. The DOM (the mechanism that causes items to "change" on screen without requiring a user to manually reload a page) of the Web site will be updated using HTML. This is good for plain, static Web sites, but it can be problematic for interactive Web sites that need a lot of user interaction. React JS builds a virtual DOM, while a developer uses JSX to manipulate and edit their DOM.

4 Experimental Design

The system is intended to provide scalable environment to access the video files which is in turn an alternate to the existing traditional way of storing and keeping track of the video files. It is achieved by incorporating the following steps.

A. *Implementation of IPFS and development of client-side application.*
The system contains the client-side application called DApp providing GUI which communicates with the Web3 browser, and metaMask using personal ID, the wallet can also be used to interact with decentralized applications or DApps. By connecting to Ethereum-based DApps, users can accept the transaction request and video player with all the videos uploaded.

B. *Integration of Blockchain and IPFS*
Ipfs-http-client is a smaller library that controls an IPFS node that is already running via its HTTP API. js-ipfs actually uses this library internally if it detects that another node is already running on your computer. Interaction can happen with the HTTP API directly when using fetch() function in a browser or a module like request in Node.js. It can be done by integrating IPFS http gateway using infura API.

C. *Testing with Ethereum and smart contract*
The Truffle configuration file that provides the network communication with the Ethereum has to be configured. Using this communication, user can deploy and migrate smart contract through author ID. The smart contract triggers event which in return store the video hash value and upload it to blockchain [8].

This system provides a decentralized application which captures the file uploaded by the user, and then, it compiles the smart contract. The smart contract is been created and connected with Truffle framework. The framework then migrates the smart contract to the local blockchain network called Ganache. Ganache, which in turn run in the local host under your network, then adds smart contract, which may comply by truffle framework to blockchain. MetaMask connects with the local blockchain and the DApp using Web3 technology. The react framework provides the predefined libraries and user interface where the document or the video can be uploaded. It connects with the Ganache framework that is local blockchain. On uploading the video, it is forwarded to IPFS using js-ipfs-http-client. After the video is segmented and uploaded into the IPFS, it returns the hash value of that video file, and it is then forwarded to ganache so that it stores the hash value in the smart contract by the event triggers from the smart contract. This process triggers to send the approvement notification to the DApp. On approving the transaction using MetaMask, this triggers the event to store the hash value of video and the user address using Ganache framework as shown in Fig. 1.

5 Implementation

The implementation flow is done with the sequence of steps as shown in Fig. 2 where the entire process flow is depicted along with the required dependencies and its installations. After installing the required components, deploying the smart contract written in solidity is the essential portion of the proposed work. Blockchain test

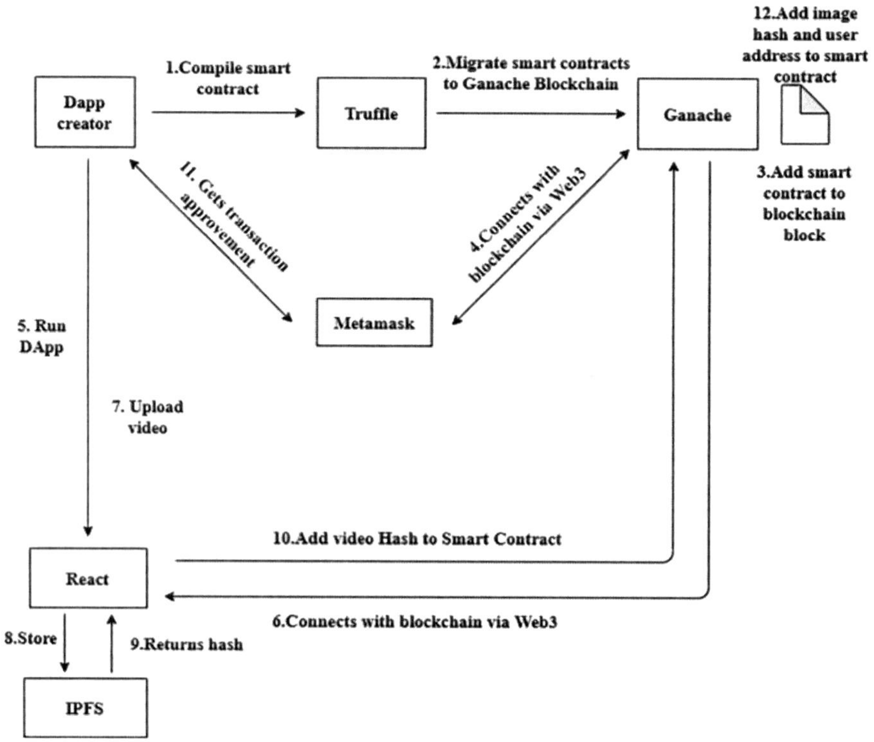

Fig. 1 Architecture of the proposed system

network is the final process which helps to test the entire blockchain network and its functionalities prior to the actual deployment of blockchain network.

5.1 Installation of NPM and Dependencies

A run-time environment executes JavaScript on a standalone machine. Using node.js, a standalone application can be built. It has many sub-modules and server-side components. Node packet manager (NPM) can be used for configuring all the required dependencies. The intermediate installation procedure is shown in Fig. 3.

5.2 Deployment of Smart Contract

The proposed system contains a smart contract file in the contract folder where the heart of the proposed system is coded using solidity programming language. This

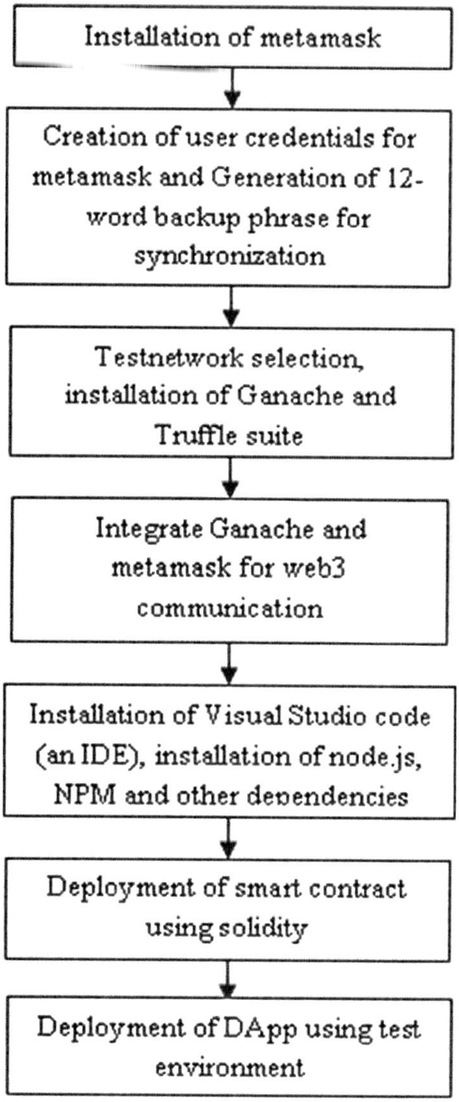

Fig. 2 Ethereum network setup and process flow

language uses the (.sol) as its extension for storing the code. Struct keyword is used to store values in solidity programming. Struct datatype along with the video object to store the parameters such as video ID, hash, title and the author (address) is defined. This information is used as an event reference. Mapping data structure similar to key value pair is used as database, and this would return the video value with the ID which will help us to access the video. The code creates an event function to trigger the event in the smart contract. On occurrence of an event, those values are then passed to the blockchain and will be stored in the block. On adding of the

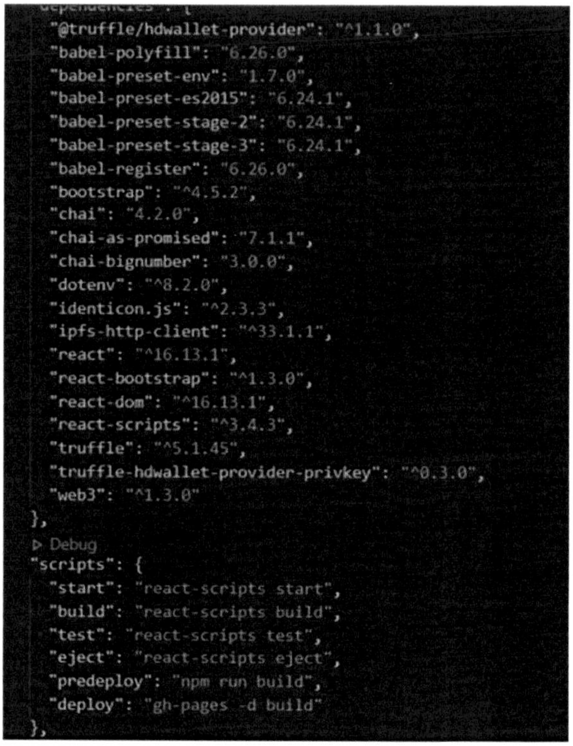
```
"@truffle/hdwallet-provider": "^1.1.0",
"babel-polyfill": "6.26.0",
"babel-preset-env": "1.7.0",
"babel-preset-es2015": "6.24.1",
"babel-preset-stage-2": "6.24.1",
"babel-preset-stage-3": "6.24.1",
"babel-register": "6.26.0",
"bootstrap": "^4.5.2",
"chai": "4.2.0",
"chai-as-promised": "7.1.1",
"chai-bignumber": "3.0.0",
"dotenv": "^8.2.0",
"identicon.js": "^2.3.3",
"ipfs-http-client": "^33.1.1",
"react": "^16.13.1",
"react-bootstrap": "^1.3.0",
"react-dom": "^16.13.1",
"react-scripts": "^3.4.3",
"truffle": "^5.1.45",
"truffle-hdwallet-provider-privkey": "^0.3.0",
"web3": "^1.3.0"
},
▷ Debug
"scripts": {
"start": "react-scripts start",
"build": "react-scripts build",
"test": "react-scripts test",
"eject": "react-scripts eject",
"predeploy": "npm run build",
"deploy": "gh-pages -d build"
},
```

Fig. 3 NPM installation

video there, it increases the count of the video in block. The smart contract makes sure that the video hash is not empty and also verifies that the user address is same as migrated. The smart contract for uploading a video in blockchain is shown in Table 1.

5.3 *Development of Client-Side Application*

An interactive graphical user interface is an essential thing to deal with any kind of systems. The blockchain along with IPFS and GUI which is created using JavaScript with a suitable video player would help interact with the system in a better way. In the proposed system, created DApp for providing user interface consist of three components:

1. App.js (To create a GUI)
2. Main.js (To create video player)
3. Navbar.js (For styling purposes)

The interactive Web page to handle the blockchain along with the video player is shown in Fig. 4.

Table 1 Smart contract for uploading video to blockchain

```
pragma solidity ^0.5.0;
contract CCTV_Video {
CCTV_VideoCount = 0;
name = "CCTV_Video";
mapping(uint => Video) public vid;

public {
require(bytes(_vidHash)> 0);
require(bytes(_heading)> 0);
require(msg.sender!=address(0));

struct Vid {id;hash; heading; author;}
event VidUploaded(id, hash, heading, author);
constructor() public {}
function uploadVid(Hash, heading)

CCTV_VideoCount ++;

vid[CCTV_VideoCount] = Vid(CCTV_VideoCount, _vidHash,
_heading, sender);
emit VidUploaded (CCTV_VideoCount, _vidHash, _heading,
sender);
}
}
```

Fig. 4 GUI with video player

6 Result Analysis

Objective of the proposed system is to obtain the integrity of the information stored in a centralized platform and integrate IPFS file system to avoid the bottle neck of storage in blockchain. IPFS does not support duplication on their file system. For proving this, for a sample, we uploaded three videos as shown in Table 2. They are, namely, original video, tampered video, and duplicate video.

Upon uploading the three videos, new blocks are created and the hash value of them stored in those three new blocks, respectively. The IPFS returned a hash value of the tampered video which is not same as the hash of the original video. So, the hash value of original video and tampered video is different from each other. But,

Table 2 Result analysis

	Duration (s)	ID (Title of the video)	Hash value
Original video	24	Original	QmR57ECGeYvoZhCUT6B9xixjA9S4PHPH7UFjRpsAB9PMpH
Tampered video	17	Demo1	QmZjgRnH2gc6aK7KNoqbrLnqWwt9dWoJk1B48a7WxeUtkn
Duplicate video	24	Duplicate	QmR57ECGeYvoZhCUT6B9xixjA9S4PHPH7UFjRpsAB9PMpH

the hash values of duplicate video and original video are of the same. This proves that the CCTV footages uploaded to this application are immutable. Since the hash value is stored in the blockchain, that would help attain the integrity and also proves that IPFS does not support duplication or modification of CCTV footage files.

7 Future Scope

Blockchain has become one of the hyped technologies these days, but they are not scalable as their counterpart centralized systems. To solve these problems, new emerging technology called Parachain can be proposed. Parachain solves the interoperability and scalability problem of blockchain technology. Our future work is to implement Parachain and develop decentralized application by integrating various blockchain networks and it features. Most of the applications like businesses and medical data storage management systems require a system that allows only the trusted parties who can access the network data. Hence, a private blockchain solution is required. Our future work is also to focus on developing the private permissioned blockchain framework.

8 Conclusion

The main objective of this project is to provide distributed and immutable data preservation system for smart city application using blockchain technology. Thus, created a decentralized application by integrating IPFS and blockchain where the CCTV video is uploaded and viewed on the basis of the date of the video frame. This solves the problem of storing large documents in a centralized platform. The system also stores large volume of data with high throughput. This can be used for various smart city applications. It also ensures the immutability of the data which help safeguard the system from various video altering attacks.

References

1. L. Zarpala, F. Casino, A blockchain-based forensic model for financial crime investigation: the embezzlement scenario. Available at SSRN 3740042 (2020)
2. K. Christidis, M. Devetsikiotis, Blockchains and smart contracts for the internet of things. Ieee Access **4**, 2292–2303 (2016)
3. D. Lee, N. Park, Blockchain based privacy preserving multimedia intelligent video surveillance using secure Merkle tree. Multimedia Tools Appl. 1–18 (2020)
4. K. Biswas, V. Muthukkumarasamy, Securing smart cities using blockchain technology, in *2016 IEEE 18th International Conference on High Performance Computing and Communications; IEEE 14th International Conference on Smart City; IEEE 2nd International Conference on Data Science and Systems (HPCC/SmartCity/DSS)* (IEEE, 2016)
5. P. Gallo, S. Pongnumkul, U.Q. Nguyen, BlockSee: blockchain for IoT video surveillance in smart cities, in *2018 IEEE International Conference on Environment and Electrical Engineering and 2018 IEEE Industrial and Commercial Power Systems Europe (EEEIC/I&CPS Europe)* (IEEE, 2018)
6. J. Sun et al., Blockchain-based secure storage and access scheme for electronic medical records in IPFS. IEEE Access **8**, 59389–59401 (2020)
7. R. Almadhoun et al., A user authentication scheme of IoT devices using blockchain-enabled fog nodes, in *2018 IEEE/ACS 15th International Conference on Computer Systems and Applications (AICCSA)* (IEEE, 2018)
8. E. Reilly et al., An IoT integrity-first communication protocol via an Ethereum blockchain light client, in *2019 IEEE/ACM 1st International Workshop on Software Engineering Research & Practices for the Internet of Things (SERP4IoT)* (IEEE, 2019)

MS D. Dharani is working as an Assistant Professor in Department of IT in PSG College of Technology, India. She received her M.Tech degree in Information Technology in the year 2016 and B.Tech degree in Information Technology in the year 2014 from Anna University, Chennai and is currently pursuing Ph.D. in Information and Communication Technology under Anna University, Chennai. She had published papers in International/National Journals/Conferences. She had also co-authored a book chapter entitled COVID-19: AI-Enabled Social Distancing Detector using CNN and has been published by Springer. She had completed PSG CARE Sponsored One year Certification Course on Data Science. She had successfully completed NPTEL and Coursera certification courses on blockchain and cybersecurity for Internet of Things. Her research interest includes areas of blockchain technology, edge computing, and data analytics. You may contact her at ddi.it@psgtech.ac.in or dharani0609@gmail.com.

Dr K. Anitha Kumari is working as an Associate Professor in Department of IT for past 10 Years in PSG College of Technology, India. She is highly passionate and curious about learning new stuffs in New Generation Computing Technologies. As an independent researcher, she had an opportunity to present her UGC sponsored research paper based on quantum cryptography in USA and visited a few Foreign Universities. To her credit, she had published a PATENT and 60 technical Papers in refereed and Impact Factored International/National Journals/Conferences published by IEEE, Elsevier, Springer, T&F, etc. Out of her research interest, she published a book in "Edge Computing: Fundamentals, Advances and Applications" by CRC Press and contributed several book chapters published by CRC Press, Springer, IGI Global, etc. Currently, she's being the Program Coordinator of ME [Biometrics and Cybersecurity] program and coordinated PSG CARE Sponsored One Year Certification Course on Cyber Security [Jan–Dec 2020]. She is also serving as Editorial Member for Scopus-indexed/Open-access Journals like International Journal of Education, Science, Technology, and Engineering, International Journal of Artificial Intelligence, Decision-making: Applications in Management and Engineering, etc., and acting as SPOC for the MoU signed between PSG and SETS Laboratory, Chennai. Also, she's been an active reviewer for Prestigious Journals published by IEEE (IEEE Communications Surveys and Tutorials (IF: 20.230), IEEE Transactions on Industrial Informatics (IF: 5.43)), IEEE Access, Springer, Wiley, etc., and Technical Program Committee (TPC) for many International/National conferences. Her areas of interest include cloud and IoT Security, design and analysis of security protocols, attacks and defense, security in computing, bioinformatics, quantum cryptography, web service security, network security, cognitive security, and analysis of algorithms. She delivered ample guest lectures in her area of interest. Her security project is sanctioned and granted by AICTE for a sum of Rs. 1,180,000/–. She has been the mentor for Technovator Projects (2019, 2018 and 2014) and "MEDROIDZ," an ICICI—Trinity 2014 funded project that was selected as one among the six projects in India. Academically, she has secured RANK-I and awarded gold medal in ME (SE) and in BE (CSE) from Anna University and from Avinashilingam University. As a supervisor, she is currently guiding Ph.D. scholars in her research area. You may contact her at anitha.psgsoft@gmail.com/ kak.it@psgtech.ac.in.

Mr. R. Vasanthan is a PG Scholar of PSG College of Technology. He is doing his Master of Engineering in Biometrics and Cybersecurity specialization. He has secured "All-rounder award" in ME (BCS) from Department of IT, PSG CT for his outstanding performance. His research interest includes blockchain and cybersecurity.

Intelligent Traffic Management System Using YOLO Machine Learning Model

B. Gomathi and G. Ashwin

1 Introduction

Conventional traffic control techniques for smooth traffic are ineffective with ever growing traffic congestion around the world. As a result, there is a need for a widely acknowledged solution that improves traffic control. In the existing static techniques, the traffic lights change within pre-determined normal time period, but the total vehicles on the road are vary from one signal to another. In this situation, the densely populated side of the road will still be filled when the signal does not turn at the usual interval. Many projects are being developed to transform existing transportation systems into "cleverer" ones by implementing intelligent transportation systems. For detecting vehicles on the road, some methods installed RF-ID sensors and others relied on ultrasonic sensors. Some researchers have attempted to use vehicle detection in the past, but their efforts have failed due to the inefficiency of the underlying machine learning models. Therefore, our aim is to build a system which analyzes the current road situation and also deals with traffic problems.

As a result, You Only Look Once (YOLO) machine learning model is utilized to detect the number of vehicles thereby performing dynamic signal allocation based on the vehicle count. YOLO is a neural network model that helps to detect an object. The proposed model is a combination of region-based convolutional neural networks (RCNN) and single shot detector (SSD), which makes the algorithm much faster, more effective, and powerful. In addition to object identification using OpenCV, YOLO model removes the backgrounds and diminishes the noise in the input picture.

B. Gomathi (✉)
Department of Information Technology, Hindusthan College of Engineering and Technology, Coimbatore, India
e-mail: gomathi.babu@gmail.com

G. Ashwin
Hexaware Technologies, Chennai, India

J. D. Peter et al. (eds.), *Disruptive Technologies for Big Data and Cloud Applications*, Lecture Notes in Electrical Engineering 905,
https://doi.org/10.1007/978-981-19-2177-3_12

The captured images from traffic surveillance CCTV cameras will be used to feed an input picture into the pre-trained model. To capture the shot, camera separates individual road side into separate frames with equal size. The obtained vehicle count [1] is fed into a program within the model. Switching time will be allocated for each side of the road based on the count. The system will first determine the count of vehicles in all lanes, and then the signal will be switched dynamically, with the lane with the most vehicles opening first.

2 Literature Survey

Hole filling technique [2] was proposed to detect and track the vehicles by processing images, respectively. In this method, vehicle detection based on sector/area was the location detection technique that detects vehicles when they enter the centroid of the detection area. Background subtraction algorithm utilized to perform the blob analysis and image segmentation. Hole filling technique was proposed to fill out blank spaces enclosed by white border for the detected object's border, thereby highlighting the overall body of the vehicle with Laplacian operator technique. Then, the adaptive morphological dilation technique was used to recover the missing parts of the border from input. The above-mentioned methods were relied upon to accomplish more accuracy in detecting the object's location. So, whenever a vehicle crossed the predefined centroid area, the vehicles are detected and counted thereby obtaining vehicle count information. Gaussian mixture model and Kalman filter [3] along would have provided better accuracy in these conditions.

A novel method based on digital image processing (DIP) [4] was utilized the multiple CCTV cameras connected over the Internet to track various roads at the junction and automate the traffic. Edge detection in spatial domain was performed to detect the perimeter of closed figures that helped to identify the vehicle. Concurrently, machine learning model was applied to categories the vehicles as motorcycle, light motor vehicle, and heavy motor vehicle. The dynamic time slot was generated by using longest remaining job first and utilized by hybrid round robin traffic scheduling algorithm to update the traffic signal lights periodically to relax the traffic.

Android-based mobile application [5] was utilized to switch the traffic signals by giving priority to congested lanes based on real-time traffic data. The CCTV cameras at each lane were utilized to acquire traffic images that were transmitted to Raspberry Pi 3 microcontroller to estimate traffic density using image processing technique. Based on the Android-based application, it controlled the traffic light operations. From the literature study, we conclude that there is a need for machine learning model providing better accuracy and speed for traffic management systems. Our goal is to build and create a traffic management system with a machine learning model of higher efficiency that can manage traffic signal flow by detecting the number of vehicles on the road.

3 Proposed System

The proposed model analyzes the presence of vehicles on the road and tackles traffic flow problems, resulting in better controlled, more organized, and intelligent traffic management system. This can be achieved by using a qualified object detection machine learning model called YOLO [6] to analyze vehicle count based on input from CCTV cameras installed in highways or traffic signals. IDD dataset is utilized to train our proposed model. The traffic data input is passed into the YOLO machine learning model that counts the vehicles based on the object detection in input video stream. On the basis of density of vehicles in each lane, the dynamic signal switching happens.

Initially, the model enters the vehicle detection phase [7] as shown in Fig. 1. In this phase, the input image is resized into 416 * 416 Resolution, which helps in better performance of the model to detect inputs. This configuration is built-in already in YOLO's configuration file. The resized image is then returned as a tensor variable into the model. Now, the model uses the neural network weights to detect the input with bounding boxes initially and then performs non-max suppression to avoid duplicate bounding box detections before object detection. Our model counts the vehicle and performs dynamic signal switching based on vehicle count as shown in Fig. 2.

3.1 YOLO

YOLO model [8] consists of two main mechanisms such as feature extractor and detector. Initially, feature extractor processes input picture and produces either one of the three different scales. These scaling information are then passed into detector that produces bounding boxes of the input images.

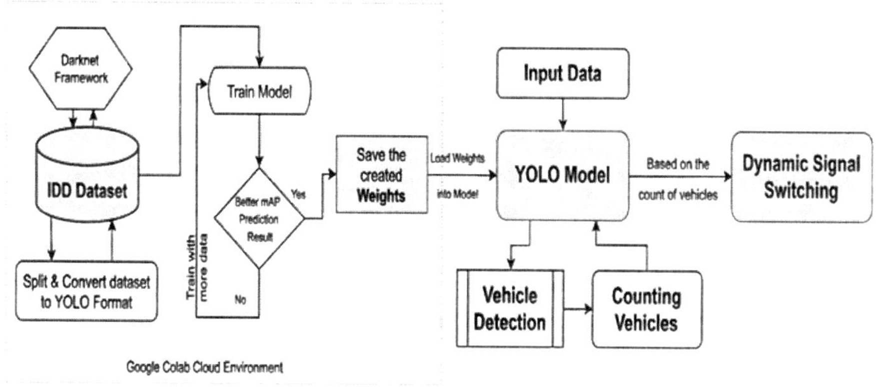

Fig. 1 Proposed system—architecture diagram

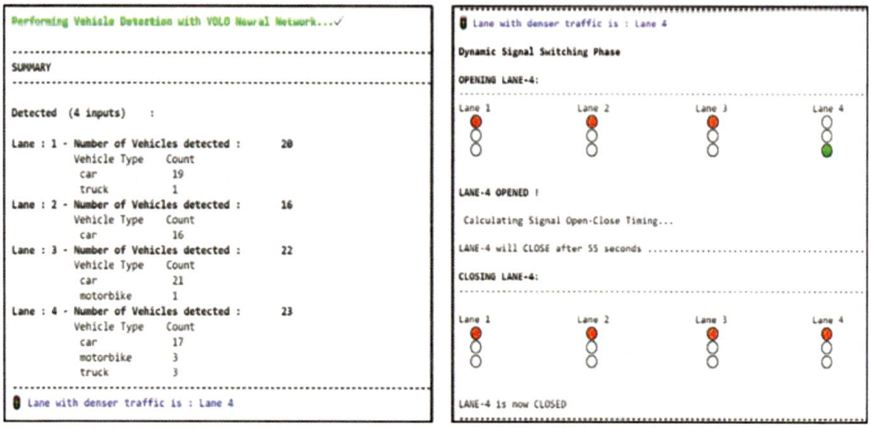

Fig. 2 Vehicle detection with dynamic signal switching with YOLO

3.2 Darknet-53

YOLO uses Darknet-53 [9] as its backbone. Darknet-53 is (an ImageNet trained neural network) built with consecutive 3 × 3 and 1 × 1 convolution layers followed by a skip connection, resulting in a total of 53 convolution layers which are further stacked with 53 more layers for the detection head, making YOLO a total of a 106 layer fully convolutional underlying architecture.

3.3 Dataset

While several datasets are already available to develop machine learning models, **IDD—India Driving Dataset** [10], a novel dataset is chosen to recognize the road sights which are unstructured as well as deeply overcrowded Indian roads. IDD dataset is composed from 182 different driving patterns and comprises of 25,000 pictures that are annotated with 34 different categories.

We will be using 67% of the IDD dataset for training our YOLO machine learning model, whereas 22% of the dataset is used for validation and then the rest of the 11% is used for testing our developed machine learning model. Annotations in dataset are labeled as billboard, auto-rickshaw, animal, etc. The annotation stamps for the dataset are planned as a four-level hierarchy. The distinctive integers are specified for each of these levels. As YOLO divides an image/frame into 13 × 13 grids and predicts at least five bounding boxes on each grid—we must convert any dataset into our YOLO format. At the end of training, we will be able to notice the "weights" necessary for functioning of our machine learning model.

3.4 YOLO Architecture

The proposed YOLO model [11] has twenty-four convolutional layers followed by two fully connected layers. It utilizes 1×1 reduction layers followed by a 3×3 convolutional layer. The 7×7 layer (rightmost) is one of the many bounding boxes that is classified by our YOLO model as shown in Fig. 3. Our model applies its algorithm in each of these many bounding boxes that our model has already classified. The proposed model splits the received input picture into $G * G$ grid. Each individual partition cells predict C bounding boxes and confidence scores for individual boxes. The confidence score specifies the way in which box contains object. The confidence score is calculated using (Eq. 1):

$$\mathrm{CS} = \Pr(\text{object}) * \mathrm{IoU} \tag{1}$$

Each individual cell in grid also predicts CC conditional class probabilities using (Eq. 2):

$$\Pr(\text{Class i}|\text{Object}) \tag{2}$$

It only predicts one set of class probabilities per grid cell, regardless of the number of boxes C. During testing, these conditional class probabilities are multiplied by individual box confidence predictions which give class-specific confidence scores for each box using (Eq. 3). These scores show both the probability of that class and how well the box fits the object.

$$\Pr(\text{Class i}|\text{Object}) * \Pr(\text{Object}) * \mathrm{IoU} = \Pr(\text{Class i}) * \mathrm{IoU} \tag{3}$$

The last predictions of a confidence score are encoded as shown in (Eq. 4),

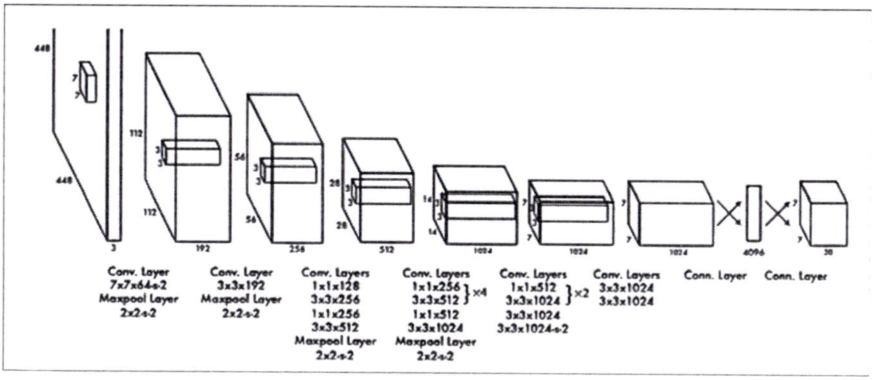

Fig. 3 YOLO model—convolution layers

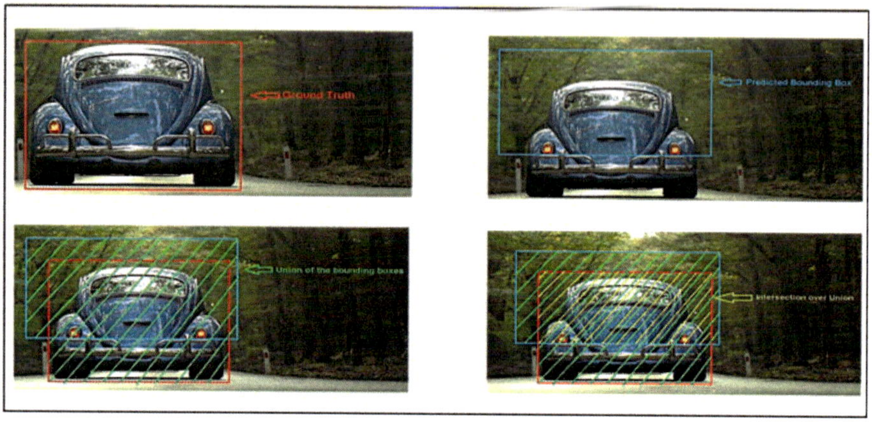

Fig. 4 Demonstration of IoU used for YOLO confidence score

$$G \times G \times \left(C^{*}5 + \mathrm{CS}\right) \tag{4}$$

3.5 *Intersection Over Union (IoU)*

Intersection over union (IoU) [12, 13] is utilized to assess the object detection algorithm that intersects among the ground truth and the predicts the bounding box as shown in Fig. 4. Typically, the threshold value of IoU is reserved as > 0.5. If the IoU of a bounding box < the identified threshold, that bounding box is not engaged into deliberation. The threshold is utilized by proposed model to predict whether the box location is correct or not. If the predicted box with IoU value $\leq$ threshold, then it indicated that high overlap between predicted and ground truth boxes. It implies that object is detected by proposed model successfully. The detected region is called as positive by using (Eq. 5). On the other hand, detected region is called as negative which does not have an object.

$$\begin{aligned} \text{Class(IoU)} = [&\text{Positive when IoU} \geq \text{Threshold} \\ &\text{Negative when IoU} < \text{Threshold}] \end{aligned} \tag{5}$$

3.6 *Non-maximum Suppression*

A technique through which the proposed model identifies the object only once. For example: Consider the Fig. 5a where the probabilities of the boxes are 0.7, 0.9, and

Fig. 5 **a** Multiple bounding boxes of the same object. **b** Single bounding box after non-max suppression

0.6, respectively. To eliminate the replicas, the box with the highest probability is chosen and produces that as a prediction. Later, it excludes any bounding box with IoU > 0.5 (or any threshold value) with the predicted outcome. Hence, the resultant output is shown in Fig. 5b.

Our model can accept multiple inputs (i.e., traffic input image from each lane) all at once and process them one by one. Once feature extractor does image processing, YOLO can now use the trained weights along with the class file of our dataset to detect objects from the input and the vehicles are classified from the detected output. Based on the obtained vehicle count data [1, 14], the dynamic signal switching takes place.

3.7 *Dynamic Signal Switching Approach Using Vehicle Count*

This process can be repeated for every traffic interval cycle by connecting the system with the existing traffic management system's CCTV cameras as inputs. If two or more lanes with equal vehicle count are detected, the lane which is detected first will be opened first as shown in Fig. 7. Since the machine learning models can learn to adapt to different scenarios with continuous use. Thereby, round the clock safety and hassle-free traffic management can be obtained using proposed intelligent traffic management system. In addition, the main objective of this proposed work helps to classify whether the object is vehicle or not. When related with other object detector models, YOLO provided better results than other detectors as shown in Fig. 6.

4 Conclusion

The major goal of intelligent traffic management system is to resolve the traffic problem that almost all cities in urban and sub-urban areas are suffering. Hence, the proposed model focuses on reducing vehicle congestion. The setup requires traffic

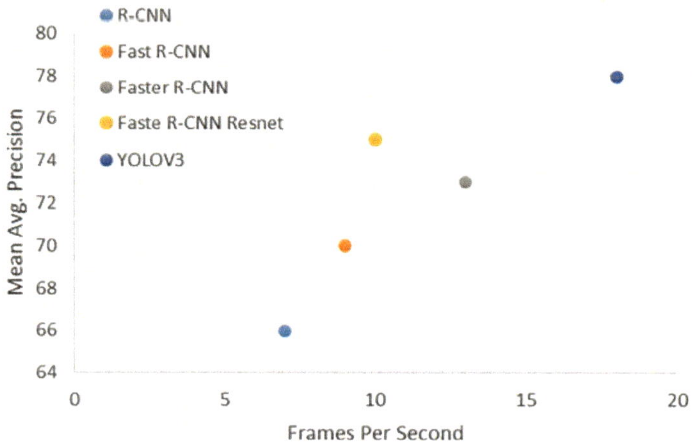

Fig. 6 Comparison between performances of different object detector

Signal Switching Approach	Signal Closed - , Signal Open – , Wait time – 5s, Signal Open Time – Varies with each approach. *Dynamic Switching Time = (Avg Vehicle Count x 2s +15s)			
	Lane A	Lane B	Lane C	Lane D
Static Timer Approach (worst - case scenarios)	- 30+5			
		- 30+5		
			- 30+5	
				- 30+5
Dynamic Approach* - I (Avg. Vehicle Count > 60)	- 135+5			
		- 135 +5		
			- 135 +5	
				- 135+5
Dynamic Approach* - II (Avg. Vehicle Count > 45)	- 105 +5			
		- 105 +5		
			- 105 + 5	
				- 105 +5
Dynamic Approach* - III (Avg. Vehicle Count > 20)	- 55 +5			
		- 55 +5		
			- 55 +5	
				- 55 +5
Dynamic Approach - IV (Avg. Vehicle Count < 20)	- 20 + 5			
		- 20+5		
			- 20+5	
				- 20+5

Fig. 7 Dynamic signal switching

data as input, which is to be used with our machine learning model for intelligent traffic management on the road. Even though the proposed model may yield extra training time but the responce time is less. The proposed model decides smart switching timing for the signal on all sides of the road. Hence, people do not want to wait for extended interval of time on the road and provide smooth flow of traffic on the road. Since the system learns from time to time, the system can be updated in future to become fully autonomous with training and learning. To enable for improved road management in the future, our system may be enhanced with capabilities such as automatic ambulance vehicle recognition, cloud-based traffic data logging, monitoring of vehicles that violate traffic regulations, and detection of civilian presence on the road.

References

1. P. Soundarya Lahari et al., Density based traffic control with emergency override, in *2018 3rd IEEE International Conference on Recent Trends in Electronics, Information and Communication Technology, RTEICT 2018—Proceedings* (Institute of Electrical and Electronics Engineers Inc., 2018), pp. 2094–2099. https://doi.org/10.1109/RTEICT42901.2018.9012488
2. R. Roopa Chandrika et al., Vehicle detection and classification using image processing, in *2019 International Conference on Emerging Trends in Science and Engineering, ICESE 2019* (Institute of Electrical and Electronics Engineers Inc., 2019). https://doi.org/10.1109/ICESE46178.2019.9194678
3. S. Maqbool et al., Vehicle detection, tracking and counting, in *2018 IEEE 3rd International Conference on Signal and Image Processing, ICSIP 2018* (Institute of Electrical and Electronics Engineers Inc., 2019), pp. 126–132. https://doi.org/10.1109/SIPROCESS.2018.8600460
4. R. Krishnamoorthy, S. Manickam, Automated traffic monitoring using image vision, in *Proceedings of the International Conference on Inventive Communication and Computational Technologies, ICICCT 2018* (Institute of Electrical and Electronics Engineers Inc., 2018), pp. 741–745. https://doi.org/10.1109/ICICCT.2018.8473086
5. C.S. Asha, A.V. Narasimhadhan, Vehicle counting for traffic management system using YOLO and correlation filter, in *2018 IEEE International Conference on Electronics, Computing and Communication Technologies, CONECCT 2018* (Institute of Electrical and Electronics Engineers Inc., 2018). https://doi.org/10.1109/CONECCT.2018.8482380
6. Z. Tian et al., FCOS: fully convolutional one-stage object detection, in *Proceedings of the IEEE International Conference on Computer Vision* (Institute of Electrical and Electronics Engineers Inc., 2019), pp. 9626–9635. https://doi.org/10.1109/ICCV.2019.00972
7. K. Dimililer, Y.K. Ever, S.M. Mustafa, Vehicle detection and tracking using machine learning techniques, in *Advances in Intelligent Systems and Computing* (Springer, 2020), pp. 373–381. https://doi.org/10.1007/978-3-030-35249-3_48
8. B. Chethan Kumar, R. Punitha, Mohana, YOLOv3 and YOLOv4: multiple object detection for surveillance applications, in *Proceedings of the 3rd International Conference on Smart Systems and Inventive Technology, ICSSIT 2020* (Institute of Electrical and Electronics Engineers Inc., 2020), pp. 1316–1321. https://doi.org/10.1109/ICSSIT48917.2020.9214094
9. W. Tarimo, M. Sabra, S. Hendre, Real-time deep learning-based object detection framework, in *2020 IEEE Symposium Series on Computational Intelligence, SSCI 2020* (Institute of Electrical and Electronics Engineers Inc., 2020), pp. 1829–1836. https://doi.org/10.1109/SSCI47803.2020.9308493
10. G. Varma et al., IDD: a dataset for exploring problems of autonomous navigation in unconstrained environments, in *Proceedings—2019 IEEE Winter Conference on Applications of*

Computer Vision, WACV 2019 (Institute of Electrical and Electronics Engineers Inc., 2019), pp. 1743–1751. https://doi.org/10.1109/WACV.2019.00190

11. A.M. Ghoreyshi, A. Akhavanpour, A. Bossaghzadeh, Simultaneous vehicle detection and classification model based on deep YOLO networks, in *Iranian Conference on Machine Vision and Image Processing, MVIP* (IEEE Computer Society, 2020). https://doi.org/10.1109/MVIP49855.2020.9116922
12. A. Dubey et al., Internet of Things based adaptive traffic management system as a part of intelligent transportation system (ITS), in *2017 International Conference on Soft Computing and its Engineering Applications: Harnessing Soft Computing Techniques for Smart and Better World, icSoftComp 2017* (Institute of Electrical and Electronics Engineers Inc., 2018), pp. 1–6. https://doi.org/10.1109/ICSOFTCOMP.2017.8280081
13. J.T.G. Nodado et al., Intelligent traffic light system using computer vision with android monitoring and control, in *IEEE Region 10 Annual International Conference, Proceedings/TENCON* (Institute of Electrical and Electronics Engineers Inc., 2019), pp. 2461–2466. https://doi.org/10.1109/TENCON.2018.8650084
14. R.A. Asmara et al., Prediction of traffic density using yolo object detection and implemented in raspberry pi 3b + and intel ncs 2, in *4th International Conference on Vocational Education and Training, ICOVET 2020* (Institute of Electrical and Electronics Engineers Inc., 2020), pp. 391–395. https://doi.org/10.1109/ICOVET50258.2020.9230145

IoT-Based Security Camera Bot Using Raspberry Pi

M. Rufus, J. John Paul, S. Merlin Gilbert Raj, and G. Shine Let

1 Introduction

In today's society, security has become an essential as well as an important aspect. Year after year criminal activities are going on increasing, and this causes a lot of fear and problems among the people. Especially in domestic and sensitive areas, it is very difficult to provide conclusive evidences and to ensure security. These are some of the reasons why lot of crimes are being committed in domestic and sensitive areas.

Security systems such as radar-based system and surveillance cameras are not portable. In surveillance systems, the cameras used were fixed in a given location and can cover only a limited distance. So, it is necessary to develop a prototype which can move as well as to provide the video feed over the Internet. IoT-based security camera bot using Raspberry Pi fulfils this need. The main goal and vision of the prototype are to create a crimeless society by monitoring the criminal activities, taking the necessary actions based on the conclusive evidences and providing security to the people. The prototype can also be used in remote areas and sensitive areas, to eliminate crimes or providing the authorities with the necessary information and evidences. The bot needs to be connected to the Internet continuously in order to monitor and to get the live video feed which can be possible nowadays as we are moving towards a fully connected community.

M. Rufus · J. John Paul (✉) · S. Merlin Gilbert Raj · G. Shine Let
Department of ECE, Karunya Institute of Technology and Sciences, Coimbatore 641114, India
e-mail: johnpaul@karunya.edu

J. D. Peter et al. (eds.), *Disruptive Technologies for Big Data and Cloud Applications*, Lecture Notes in Electrical Engineering 905,
https://doi.org/10.1007/978-981-19-2177-3_13

2 Literature Review

The author [1] discusses the surveillance system developed using an Arduino, Team Viewer, etc. DTMF controller is used to power up the robot. VNC server is used to login into the remote PC. The author has used GSM which turns on the system, Internet to make a connection with the PC and RF to control the bot. But the disadvantage is the huge lag between the systems when implemented. The author has used RF technology which means the transmitter and the receiver must be in line of sight and cannot be used in huge halls. It is impossible to control the robot from another room. The author [2] suggests that the bot is created for trespassing and obtrusion. It possesses the facial recognition system and provides valuable information which can prevent a lot of criminal activities. The author has used VNC Viewer to login into the Raspberry Pi. Webpage is not created, and the user needs to control the bot by logging in the Raspberry Pi. The live feed cannot be viewed, and the bot cannot be controlled from other devices which makes the bot unreliable. The author [3] explains the need of constructing virtual environment which can give valuable information, evidences, video feed, etc., without the loss of human life. The author has used Raspberry Pi, GSM module, motion detection, etc. The author specifies that the bot sends SMS and email alerts to the user. The robot is controlled by logging into the Raspberry Pi through VNC Viewer. The web application developed is used to turn on and turn off the bot. Live video is not available, and the bot cannot be controlled through the webpage or any interface. The author [4] discusses the model developed to perform surveillance in domestic areas. Arduino, Wi-Fi module along with CAYENNE software, etc., is used in the developed module. The author uses a wireless system connected via Internet to control the bot. The user controls the bot using the software design at the user end which is complex. The author did not specify the outputs gathered from the bot. The author [5] proposes a surveillance system using IoT, Raspberry Pi, Webpage, etc. The author has developed a webpage to control the bot and has made an option to view the video feed. The bot is equipped with PHP for backend operations but the webpage is used as a local host and is not implemented in the cloud where only local host can access. The webpage cannot be opened in other devices. The author [6] of this research paper explains the need of creating a spy robot surveillance system using IoT. The author has used raspberry pi to construct the spy bot. The system sends data through IoT. The system proves to be successful but the advantage of using the proposed security camera bot is that the robot can be used anywhere in the building. The author [7] of this research paper talks about the scaling of horizontal and vertical surfaces using a surveillance robot. The author has used raspberry pi for the controlling the robot. Zig-bee is used for data transmission. The research paper describes the mathematical process but the proposed prototype can be used anywhere in the building as the webpage is interfaced to the cloud using python. The movement and controlling of the bot for the proposed prototype is referred from [8]. The webpage used in the proposed prototype is referred from [9]. The basic idea of surveillance and its working is referred from [10].

To overcome the disadvantages of the modules listed in the literature survey, a webpage is created and is interfaced into the cloud environment using PythonAnywhere platform. The webpage can be opened in any device supporting HTML. It is possible that anyone with proper authorization can view the live video feed and control the bot through the webpage. The security camera bot can be used anywhere in the house because of the in-built Wi-Fi module present into the Raspberry Pi. Lag is dependent upon the Internet speed which can be easily rectified.

3 Novelty of the Proposed Model

The proposed model was implemented through a webpage, which provides live video feed from the bot. The webpage also has options to control the bot. As the webpage was hosted on the cloud platform using PythonAnywhere, the webpage can be accessed by anyone with proper authentication. The webpage can be opened in any device that supports HTML application. The user can control the Raspberry Pi remotely using VNC Viewer to initialize the setup for usage. So, it is possible to monitor the domestic as well as sensitive areas and provide the conclusive evidences.

4 Idea of the Proposed Model

In this paper, IoT-based security camera bot using Raspberry Pi is proposed as shown in Fig. 1. The bot is constructed using a Raspberry Pi module, webcam, motor driver

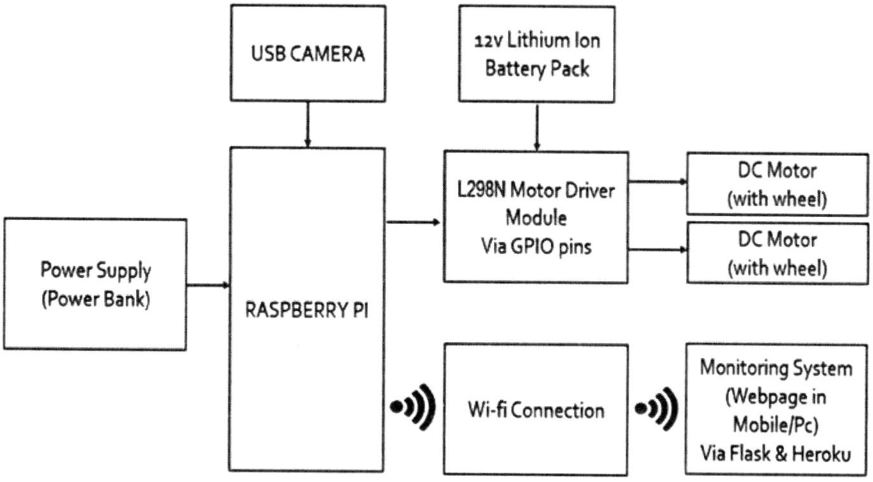

Fig. 1 Block diagram of the proposed model

ICs and DC motors. The controlling core of the system is the Raspberry Pi. The webcam monitors the given environment, and the bot is able to move along via DC motors. Raspberry Pi is connected to the Internet through Wi-Fi module. A user can access the bot through the HTML page created using Flask application. The bot and the pi are connected to the cloud platform using PythonAnywhere.

5 Components Used

The components used both hardware and software are discussed in the upcoming modules.

5.1 *Hardware Components*

(a) **Raspberry Pi 3B +**: The Raspberry Pi 3B + is one of the versions of the Raspberry Pi. The Raspberry Pi 3B + is the size of a credit card. It is the controlling core of the bot developed. It has various USB ports, Ethernet ports, GPIO pins, etc. It also supports both Wi-Fi and Bluetooth.

(b) **Motor Driver Module**: The L298N motor driver module is used to drive DC motors and stepper motors. It requires a steady 12 V power supply.

(c) **Dc Gear Motors**: DC gear motor has both motor and gearbox. It also has a gearhead which increases the speed and reduces the output torque. Two DC gear motors are used for the prototype.

(d) **Webcam**: The Webcam used is a simple USB webcam developed by TVS electronics. It is a high-quality webcam with 1080p pixel quality.

(e) **Rechargeable Battery Cell**: 12 V rechargeable battery cell is made up of lithium ions. Three 18,650 lithium-ion batteries are taken and connected to the 3S 12 V lithium-ion protection board. It can be recharged with the help of 12 V DC power adapter and can be used up to 3–6 h continuously. It is used to power the L298N motor driver module.

(f) **Power Bank**: A Veger 10,000 mAh li-polymer portable power bank is used. The discharge output is 5 V–2.1 A which is used to power the Raspberry Pi.

(g) **Balancing Wheel**: A balancing wheel manufactured by touch companies is used to balance the bot and to provide an extra support to the bot as the bot is heavier.

5.2 *Software Components*

(a) **Motion Detection Library**: Motion detection library should be installed on the Raspberry Pi OS. It is a well-developed library for the camera module.

As the webcam is interfaced to the Raspberry Pi, through the motion library the user can enhance the quality of the live feed and add other audio/quality enhancing features. This library contains lots and lots of features which enhance the quality of the video, audio and the motion detection.

(b) **Python 3 Idle**: Python 3 Idle is the ideal platform for coding the Raspberry Pi and the components used in this bot. Idle is nothing but the integrated development and learning environment which was created for Python 3. Idle is fully equipped Text-Editor which is used to create, manipulate and execute python scripts. It also contains a debugger which debugs the created python scripts.

(c) **HTML**: Hyper-Text Markup Language (HTML) is used to create a webpage. The webpage is displayed on World Wide Web. The webpage created is used to control the bot and display the live video feed from the bot.

(d) **Flask Application**: Flask is nothing but a lightweight WSGI web application framework. It does not require any specific tools or libraries. The Python code is integrated into server using Flask application and is not meant for production purpose.

(e) **PythonAnywhere/Heroku**: PythonAnywhere/Heroku is a cloud-based development server (PaaS—Platform as a Service). It supports several programming languages such as Python, JavaScript since Flask application is not meant for production purpose, Heroku/PythonAnywhere solves this problem as it can be deployed as a webpage provided with a link.

6 Working

The first and the foremost element is the Raspberry Pi 3B+ which acts as the heart core and brain of the system. It is necessary to install the Raspbian OS available on the Raspberry Pi platform and sets up the Raspberry Pi. To power the Raspberry Pi, a power bank with discharge output 5 V–2.1 A is used. The USB camera is interfaced to the Raspberry Pi controller. Following that, installation of motion library packages is done which gives live feed to the IP address: 8081 port. To control the bot, two DC motors are used via L298N motor driver module. The L298N motor driver module requires a steady 12 V supply, so a 12 V lithium-ion battery is used to power the module. The Raspberry Pi is connected to the Internet using Wi-Fi (USB dongle). A Python file is created with code to run the bot through the Flask application and a webpage is developed using HTML with the options to control the bot and view the live video feed. The Python file and HTML code for webpage are interfaced into the cloud environment Through PythonAnywhere platform. Further details of the connection are explained with Figs. 2, 3 and 4.

Figure 2 shows the circuit connection between Raspberry Pi, the L298N motor driver module and two DC motors. The 18th and the 25th GPIO pins of the Raspberry Pi are connected to enable 1 and enable 2 pins of the motor driver module. The 14th, 15th, 24th and the 23rd GPIO pins of the Raspberry Pi are connected to the input

Fig. 2 Connection between Raspberry Pi, L298N motor driver module and two DC motors

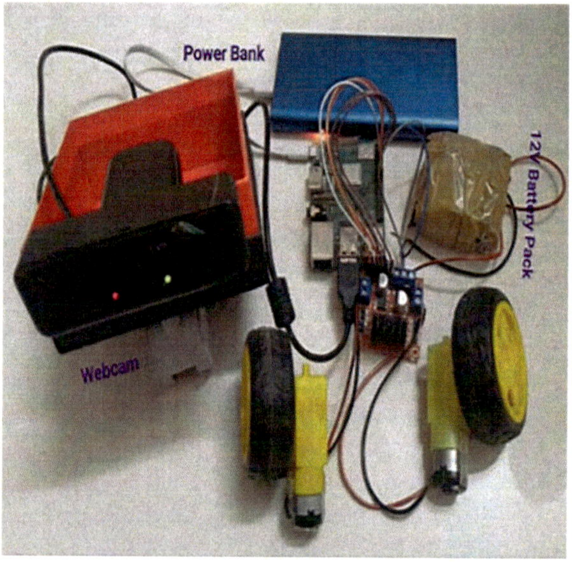

Fig. 3 Circuit connection between Raspberry Pi and Webcam, L298N motor driver module and 12 V battery pack, Raspberry Pi and power bank

pins IN1, IN2, IN3, IN4 of the motor driver module. The output pins OUT1, OUT2, OUT3, OUT4 of the L298N motor driver module are connected to the two DC motors. Figure 3 shows the circuit connection between Raspberry Pi and webcam, L298N motor driver module and 12 V battery pack, Raspberry Pi and power bank. The webcam is interfaced to USB port of the Raspberry Pi and Motion library is enabled. 12 V battery pack made up of lithium-ion batteries is connected to power

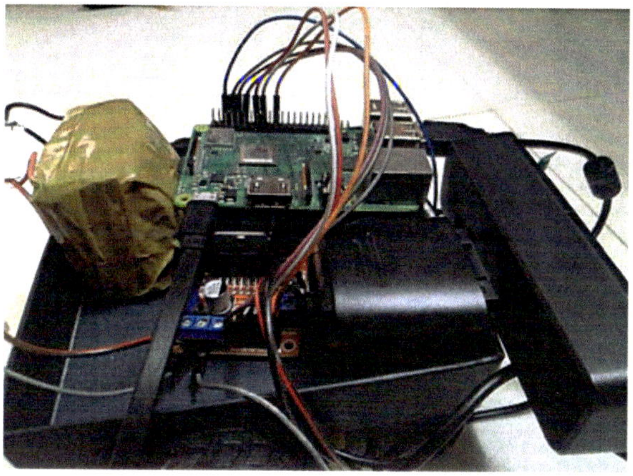

Fig. 4 Overall circuit connection of the bot

the L298N motor driver module. A power bank of discharge output 5 V–2.1 A is used to power the Raspberry Pi. Figure 4 shows the overall circuit connection of the bot. The components such as Raspberry Pi, L298N motor driver module, power bank, 12 V battery pack, and webcam are mounted onto a cardboard box. Since the bot has only two wheels, a balancing wheel is attached at the front portion of the cardboard box to make the bot stable in uneasy pathways such as the roads and uneven surfaces.

7 Results

The results obtained from the developed prototype such as the webpage created, bot created and the pictures taken from the bot are discussed in the upcoming modules.

7.1 Webpage Created Using HTML

Figure 5 shows the webpage created using HTML. On the left half of the webpage, the live feed from the webcam is shown. To the right half, the webpage there are four arrows which controls the movement of the bot. When forward arrow is pressed the bot moves in forward direction, when reverse arrow is pressed the bot moves in reverse direction, when left arrow is pressed the bot turns towards left direction, when right arrow is pressed the bot turns towards right direction. The webpage along with python file is interfaced into cloud environment using PythonAnywhere platform. The interfaced webpage in the PythonAnywhere platform eliminates the disadvantages of Flask application as it is not meant for production process. The motion service is turned on using the command [sudo/etc./init.d/motion restart] to

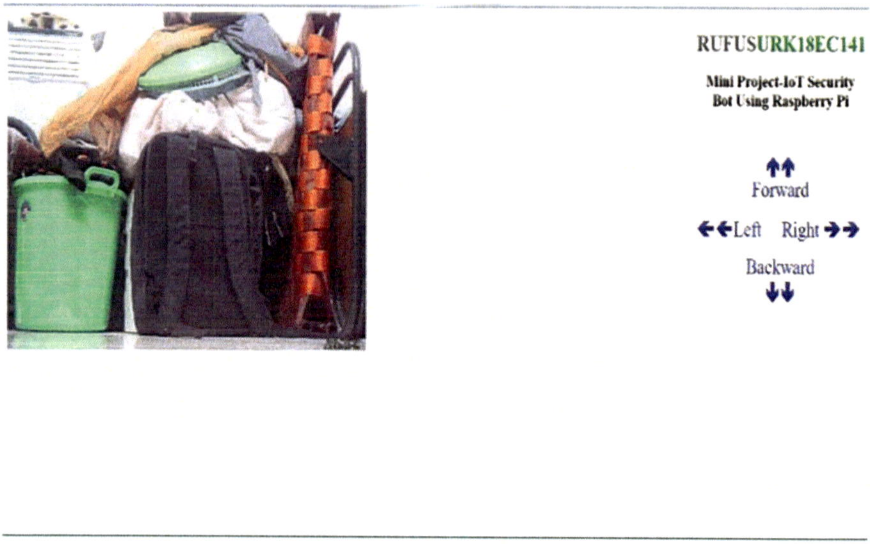

Fig. 5 Webpage created using HTML

visually see the live video feed. The end-user may just enter the link with proper authentication and can monitor the given environment by moving the bot.

7.2 Bot Created and Python IDLE Output

Figure 6 depicts the full structure of bot in the front view, and Fig. 7 depicts the full structure of bot in the back view. Figure 8 shows the response from webpage in the python IDLE output. If any of the arrows are being pressed in webpage, it is shown as an output in python. It also shows the time at which it is operated and IP address of the device using the webpage in order to ensure security The IP address part in the left side is not shown for security and privacy purposes. As the IP address, date and time are visible in the python IDLE window, it is possible to track the user using the webpage. The python IDLE window not only shows the IP address, date and time, it also shows the buttons which is being pressed in the webpage. The HTML file needs to be placed in the same folder as the python code to make the python IDLE output available. The Raspberry Pi is operated by VNC viewer using a laptop. In the VNC viewer, it is necessary to provide proper authentication and security by using a legal username and a strong password.

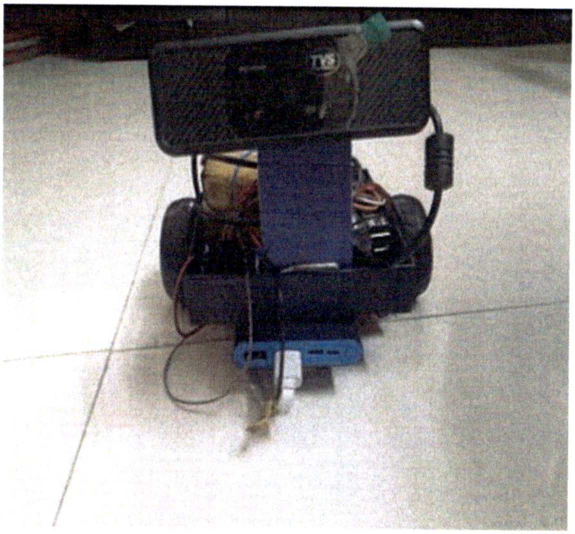

Fig. 6 Bot created (front view)

Fig. 7 Bot created (back view)

7.3 Pictures Taken from the Bot

Figure 9 shows the picture taken when forward arrow is pressed in the webpage. Figure 10 shows the picture taken when right arrow is pressed in the webpage. Figure 11 shows the picture taken when left arrow is pressed in the webpage. Figure 12 shows the picture taken when backward arrow is pressed in the webpage. The bot is

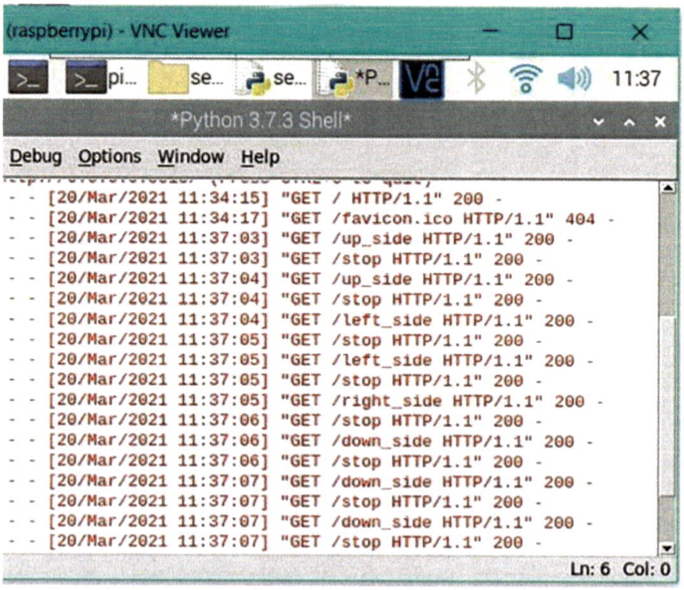

Fig. 8 Python IDLE output

Fig. 9 Picture taken when forward arrow is pressed

present in the terrace of the building and is controlled by the webpage opened in the laptop which is in another room

Fig. 10 Picture taken when right arrow is pressed

Fig. 11 Picture taken when left arrow is pressed

8 Conclusion

The bot was created with a prime motive of eradicating the criminal activities and to provide a safe and secure environment to the society. The developed prototype can also be used for spying purposes. We come across lot of criminal activities such as thefts, murders, fights, quarrels and riots which disturb the society and lead to lot of problems. These types of situations can only be avoided under constant surveillance and monitoring through the cameras and providing conclusive evidences. This project will surely be helpful for the society, and in fact it, can also be used as a prevention tool to avoid massive upcoming losses.

Fig. 12 Picture taken when backward arrow is pressed

8.1 Future Advancements

The proposed model can be improved further more by adding some advancements in future which make it even more reliable and unique such as using USB Wi-Fi dongle to improve the range of connectivity. It is possible to add face recognition system by open CV to recognize the person seen on the feed. With the help of several sensors, it is possible to make an object detection module which detects the object in front of it. By introducing several propellers and their systems, it is possible to develop a drone that come security camera bot which can move as well as fly.

References

1. S. Kumar, S.S. Solanki, Remote home surveillance system, in *2016 International Conference on Advances in Computing, Communication, and Automation (ICACCA)* (2016)
2. R. Harshitha, M.H.S. Hussain, Surveillance robot using Raspberry pi and IoT, in *2018 IEEE International Conference on Design Innovations for 3Cs Compute Communicate Control.* https://doi.org/10.1109/ICDI3C.2018.00018
3. S.M. Ashish, M. Manjunath, L. Ravindra, M. Nadeem, K Neelaja, Automated hybrid surveillance robot. Int. J. Innov. Eng. Res. Technol. [IJIERT] **5**(5) (2018). ISSN: 2394-3696
4. G. Anandravisekar, A. Anto Clinton, T. Mukesh Raj, L. Naveen, IoT based surveillance robot. Int. J. Eng. Res. Technol. (IJERT) **07**(03) (2018). https://doi.org/10.17577/IJERTV7IS030061
5. D. Amriithha, G. Apsara, R. Ramya, R. Chitra, Spy robot surveillance system using IoT **8**(5), 128–133 (2020). https://doi.org/10.17148/IJIREEICE.2020.8627
6. T. Saravanakumar, D. Keerthana, D. Santhiya, J. Sneka, D. Sowmiya, Surveillance robot using Raspberry Pi–IoT. Int. J. Electron. Electr. Comput. Syst. (IJEECS) **7**(3) (2018). ISSN 2348-117X
7. Z. Shaikh, P. Gaikwad, N. Kare, S. Kapade, M.V. Korade, An implementation on—surveillance robot using Raspberry-pi technology. Int. Res. J. Eng. Technol. (IRJET) **04**(04) (2017)

8. https://circuitdigest.com/microcontroller-projects/web-controlled-raspberry-pi-surveillance-robot
9. https://iotdesignpro.com/projects/web-controlled-raspberry-pi-surveillance-robot
10. https://en.wikipedia.org/wiki/Surveillance

Performance Analysis of Different Deep Learning Models for Forest Fire Classification

S. Harshaw Kamal, R. K. Ragul Raj, T. Sabari, and R. Karthika

1 Introduction

Early fire detection is vital for the safety of humans, animals, and vegetation. Several fire detection devices have been developed to avoid damage caused by fire. There are numerous technological solutions one can find. Most of them are based on sensors, which are often limited to indoor environments in general. These strategies, however, have a fatal flaw in which they can only operate to achieve a certain condition. In the worst-case scenario, in the case of real fire, the sensors are damaged or not being properly calibrated will cause heavy casualties. These sensors detect the particles generated by ionization from smoke and fire, requiring proximity to the fire. Consequently, it is not practical to use them to cover broad areas. Video-based fire detection systems are used to circumvent certain constraints. Classic video-based fire systems are however inefficient due to the repetitive nature of it. So, we lean toward deep learning as the models can detect features without human intervention which is possible due to availability of large data. The humongous amount of video data available in today's system supports this cause. The traditional image processing algorithms are more domain-specific rather than, deep learning which provides superior flexibility because deep learning models like CNN can be retrained using a custom dataset for any use case. Due to the limitations of traditional image processing-based techniques, we turn to a deep learning-based system for fire detection. The reason to analyze various deep learning method is to get a clear idea about the accuracy

S. Harshaw Kamal · R. K. Ragul Raj · T. Sabari (✉) · R. Karthika
Department of Electronics and Communication Engineering, Amrita School of Engineering, Amrita Vishwa Vidyapeetham, Coimbatore, India
e-mail: t.sabari007@gmail.com

R. Karthika
e-mail: r_karthika@cb.amrita.edu

J. D. Peter et al. (eds.), *Disruptive Technologies for Big Data and Cloud Applications*, Lecture Notes in Electrical Engineering 905,
https://doi.org/10.1007/978-981-19-2177-3_15

provided by each model and to know which deep learning model will be good for classification.

2 Related Work

In [1], a real-time fire detection method is proposed. To decrease the complexity, the RGB fire picture elements were chosen independently from an image containing fires. [2] demonstrated how to detect fires from photographs by combining the benefits of deep learning and spatial texture analysis. The detected fire regions using spatial characteristics analysis by Linear Dynamical Systems (LDS) were also proposed. [3] proposed a new model architecture called MobileNet, which was built on depthwise separable convolutions. It showed how to use a width multiplier and a resolution multiplier to make a smaller and quicker MobileNet by sacrificing a fair amount of precision to reduce size and latency. Then, it was compared to different MobileNet to popular models outperforming them in terms of scale, speed, and accuracy. In [4], a convolutional neural network was implemented. This network system achieved an accuracy of 95.4% which was greater than all other network systems.

In [5], a deep learning-based Visible Fire Detection was proposed. The proposed architecture produced better results than the state-of-the-art systems by giving an accuracy of 99% on a frame basis. In [6], fire detection using deep learning models with Visual Geometry Group (VGG) model and MobileNet model was proposed. The results produced were better than traditional networks. In [7], an IOT (Internet of Things)-based Forest Fire Prediction and Detection was proposed. The suggested method has shown that K-Nearest Neighbors (KNN) is more efficient than all other algorithms for predicting fire incidence in the forest and cost effective. In [8], a smoke type classification concept for video fire detection was proposed. Different approaches provided a smoke type classification concept due to different temporal and spatial behavior and visual properties in a video sequence. In [9], a real-time video-based smoke detection was proposed. The proposed algorithm was based on a motion accumulation model. The algorithm could detect even minute smoke presence in forest areas. In [10], Flame Detection Using Deep Learning was presented. Shallow learning and deep learning were the two main approaches used to classify fire flames from the perspective of visual images and videos. Both methods performed well in flame detection; however, if the train dataset is large enough, deep learning outperformed shallow learning.

In [11], an indoor fire detection system using a vision sensor derived from computer vision approaches is used. In [12], as vision-based fire detection is cost efficient and has the capacity to detect forest fire in the beginning stages, a lightweight pixel-based model is used. In [13], a Support Vector Machine (SVM)-based approach is used to predict forest fires at the beginning stages by learning process using sensor and image data. In [14], a Wireless Sensor Network (WSN) system that continuously monitors forest fires in real time helps notify forest fires earlier than normal methods and also predicts the directional flow of the fire. In [15], a new architecture with three layers is proposed and the performance is compared with VGG architecture. [16] analyzed the existing algorithms and suggested to use standardized datasets to enhance the performance of training algorithms.

3 Proposed Methods

3.1 VGG-16

This model replaces conventional convolutional networks' large-sized kernels with smaller 3 × 3 kernels. Since deep non-linear layers can learn more complex characteristics, this improves model efficiency. VGG-16 has different layers of convolution, max-pooling layers, and dense layers. 3 dense layers are accompanied by a series of convolutional layers. The final layer is Softmax. The hidden layers have the activation feature of the ReLU.

3.2 ResNet-50-V2

Deeper networks make it more difficult to practice. Model accuracy begins to saturate and degrade rapidly with the depth of the network increasing. Yet over fitting is not the cause of the accuracy deterioration. In image recognition, however, deeper models yield better performance. One of the solutions to the degradation problem is residual networks, or ResNets. There are feedforward shortcut connections between the layers. The feedforward connections perform the transformation of identity that is applied to the stacked layer outputs.

3.3 MobileNet-V2

A residual bottleneck depth-separable convolution is the fundamental block of this architecture. Traditional convolutional layers are more costly in terms of computation than successful convolution in depth. MobileNet-V2 performs 3 × 3 depthwise separable convolution. This reduces the computational cost while losing only a small amount of accuracy. Dense layers with 32 filters make up the first layer.

3.4 Inception-V2

Inception performs convolution on the input using three different filters of sizes 1 × 1, 3 × 3 and 5 × 5. Max pooling is performed, and the outputs are concatenated and sent to the next module. In Inception-V2, the 5 × 5 convolution is replaced by two 3 × 3 convolutions. This is done to decrease computation time, and therefore, the performance was improved.

3.5 Xception

The Xception architecture is a more advanced variant of the Inception architecture. Convolutions are factorized into several blocks in Inception networks. While Xception is like depthwise separable convolution, there are a few key differences. Xception performs 1 × 1 convolution followed by spatial convolution. In Xception, there is a ReLU non-linearity after the operation, but not in the depthwise separable convolution. There are 36 convolutional layers in the Xception architecture. All the other modules, except for the corner modules, have linear residual links.

4 Results and Discussion

4.1 Dataset

The dataset used is from Mendeley Data [15] which was created for this specific problem statement. The images in the dataset are color images with a resolution of 250 × 250. Each image has relevant fire region. It contains 1900 images where 950 images are of the fire class and 950 images are of the no-fire class. The dataset is partitioned for training and testing purposes. 1520 images are for training and validation and 380 images for testing (Figs. 1 and 2).

4.2 Metrics

The quantitative metrics used to evaluate the performance of these algorithms are discussed here: Forest fire classification has been evaluated based on the confusion matrix using: True Positive (TP), True Negative (TN), False Positive (FP), and False Negative (FN). These parameters aid in analysis for classification is discussed here.

We have used four metrics to compare our models: $F1$ score, Precision, Recall, and Accuracy.

Fig. 1 Non-fire images in the dataset [15]

Fig. 2 Fire images in the dataset [15]

Accuracy: It can be described as the ratio of total number of predictions that were correct.

$$\text{Accuracy} = \frac{(\text{TP} + \text{TN})}{(\text{TP} + \text{TN} + \text{FP} + \text{FN})} \tag{1}$$

Precision: It represents the percentage of valid positive predictions out of all positive predictions.

$$\text{Precision} = \frac{\text{TP}}{\text{TP} + \text{FP}} \tag{2}$$

Recall: It represents the percentage of valid positive predictions out of all possible positive predictions.

$$\text{Recall} = \frac{\text{TP}}{\text{TP} + \text{FN}} \tag{3}$$

$F1$ score: As looking at Precision and Recall as isolated metrics does not give us a complete picture, we use the $F1$ score. It encapsulates the properties of both precision and recall.

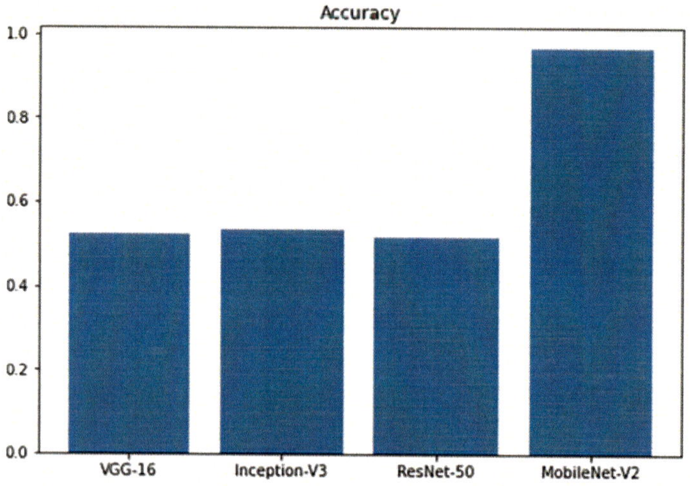

Fig. 3 Accuracy comparison of different models

$$F1\ \text{Score} = 2 * \frac{\text{Precision} * \text{Recall}}{\text{Precision} + \text{Recall}} \quad (4)$$

4.3 Experiment Analysis

To achieve the goal of real-time fire classification, a model must be decided upon as there are several deep learning models available but the best model for the dataset is chosen by comparing the metrics mentioned above of the selected Convolutional Neural Network (CNN) models, namely—VGG-16, Inception-V3, ResNet-50, Xception, and MobileNet-V2. As we need real time, classification accuracy is not the only requirement and the model should be fast too. Accuracy fails on classification problems with a skewed class distribution because of the intuitions developed by practitioners on datasets with an equal class distribution (Figs. 3, 4, 5, and 6).

4.4 Inference

From the above results, we can infer that for limited epochs for all the architectures, MobileNet-V2 model achieved the highest accuracy of 96.84% whereas Inception-V3 achieved 53.15%, VGG-16 achieved 52.10%, ResNet-50 achieved 51.57%, and Xception achieved 56.57%. For precision, MobileNet-V2 model achieved the highest 95.4% whereas Inception-V3, VGG-16, ResNet-50, and Xception models achieved 53.09%, 63.68%, 51.56%, and 56.3%, respectively. For recall,

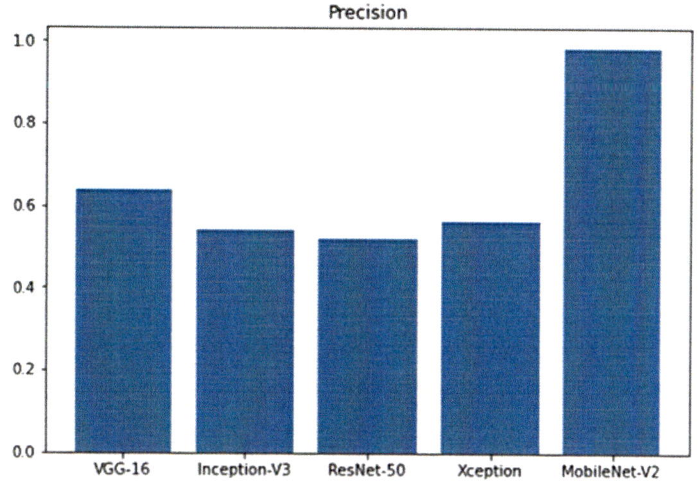

Fig. 4 Precision comparison of different models

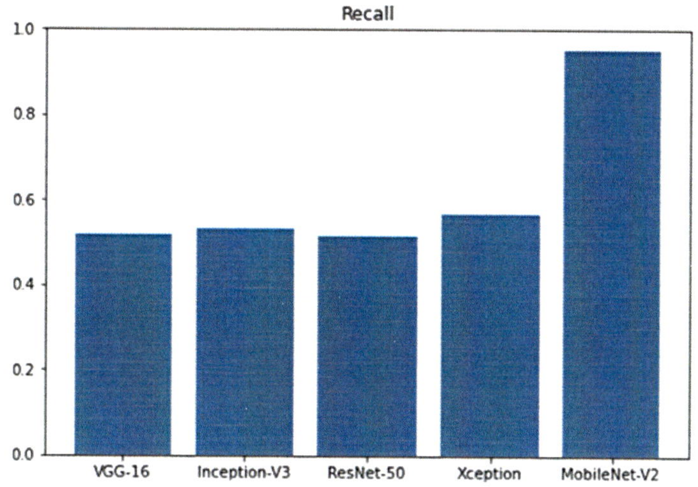

Fig. 5 Recall comparison of different models

MobileNet-V2 model attained the highest 98.4% whereas Inception-V3, VGG-16, ResNet-50, and Xception models attained 54.21%, 51.70%, 52.10%, and 54.18%, respectively. For $F1$ score, MobileNet-V2 model achieved the highest accuracy 96.87% whereas Inception-V3, VGG-16, ResNet-50, and Xception models achieved 53.641%, 57.06%, 51.82%, and 55.21%, respectively. So by observing all this values, it is inferred that MobileNet-V2 architecture provide high accuracy, precision, recall, and $F1$ score than the other models. It is evident that VGG-16, Inception-V3,

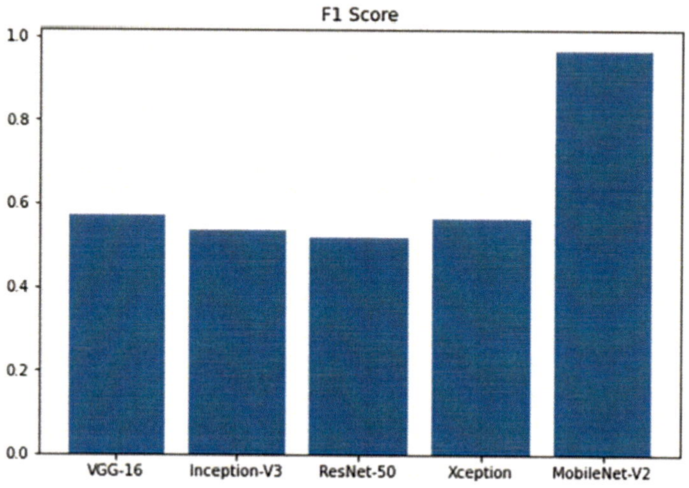

Fig. 6 *F*1 score comparison of different models

ResNet-50, and Xception need to be trained for higher epochs to achieve similar metrics.

Reason for better accuracy of MobileNet-V2 is linear bottlenecks between the layers and shortcut connections between the bottlenecks. The bottlenecks of the MobileNet-V2 encode the inputs and outputs at the intermediate, while the inner layer encloses the model's ability to rework to higher level descriptors like image categories from lower-level concepts like pixels. Faster training and better accuracy can be achieved with traditional residual connections. Therefore, MobileNet-V2 is the best model for the dataset and it is also a small size model which is capable of being deployed in mobile devices. It has high performance which is essential in real-time classification purposes.

5 Conclusion

Forest fires are a grave issue in many countries as once they spread it is impossible to stop it and it leads to heavy losses in vegetation and wildlife. The only solution for this is early detection of fires so that it can be countered as soon as possible before the fire breaks out. Using sensors is not practical as the cost for setup and deployment is not practical over the forest but a simple surveillance camera or drone which can process the video using state-of-the-art deep learning models can easily detect the fire at an early stage and is also cost effective compared to a sensor system and it is easy to setup. From our experiments, we observed that despite the small size of the dataset, MobileNet-V2 achieved the highest accuracy of 96.84% on the test data. MobileNet-V2 is also has reduced network size and hence is best for mobile applications and high performance required for the real-time classification. Therefore, it would be the best model for this purpose. With a dataset with more training images, it can be even more enhanced and it can be a viable for real-world use.

References

1. T.-H. Chen, L.K. Cheng, S.-M. Chang, An intelligent real-time fire-detection method based on video, in *Proceedings of IEEE 37th Annual 2003 International Carnahan Conference on Security Technology*, pp. 104–111 (2003)
2. Q. Zhang, J. Xu, L. Xu, H. Guo, Deep convolutional neural networks for forest fire detection, in *International Forum on Management, Education and Information Technology Application* (Aviation University of Air Force, Changchun, 2016)
3. A.G. Howard, M. Zhu, B. Chen, D. Kalenichenko, W. Wang, T. Weyand, M. Andreetto, H. Adam, MobileNets: efficient convolutional neural networks for mobile vision applications; Cornell University, arXiv.org, Computer Vision and Pattern Recognition (cs.CV); arXiv:1704.04861 (2017)
4. H. Yang, H. Jang, T. Kim, B. Lee, Non-temporal lightweight fire detection network for intelligent surveillance systems. IEEE Access **7**, 169257–169266 (2019)
5. V. Hüttner, C.R. Steffens, S. Silva da Costa Botelho, First response fire combat: deep leaning based visible fire detection, in *2017 Latin American Robotics Symposium (LARS) and 2017 Brazilian Symposium on Robotics (SBR)* (Curitiba, Brazil, 2017), pp. 1–6
6. M. Dua, M. Kumar, G. Singh Charan, P. Sagar Ravi, An improved approach for fire detection using deep learning models, in *2020 International Conference on Industry 4.0 Technology (I4Tech)* (Pune, India, 2020), pp. 171–175
7. M. Trinath Basu, R. Karthik, J. Mahitha, V. Lokesh Reddy, IoT based forest fire detection system. Int. J. Eng. Technol. **7**(2.7), 124 (2018)
8. A. Wellhausen, A. Stadler, A smoke type classification concept for video smoke detection, in *Proceedings of the 16th International Conference on Automatic Fire Detection AUBE'17* (2017)
9. A. Stadler, T. Ike, Real time video based smoke detection using double optical flow estimation. Int. J. Comput. Electr. Autom. Control Inf. Eng. **10**(9) (2016)
10. D. Shen, X. Chen, M. Nguyen, W.Q. Yan, Flame detection using deep learning, in *IEEE 4th ICCAR*, pp. 416–420 (2018)
11. Sridhar, T. Senthil Kumar, L. Parameswaran, A new approach for fire pixel detection in building environment using vision sensor, in *Innovations in Computational Intelligence and Computer Vision*, vol. 1189 (Springer Singapore, Singapore, 2021), pp. 392–400
12. M. Manickam, M. Yogesh, P. Sridhar, T. Senthil Kumar, L. Parameswaran, Video-based fire detection by transforming to optimal color space, in *Proceedings of 3rd International Conference On Computational Vision and Bio Inspired Computing*, vol. 1108 (Springer International Publishing, Cham, 2019)
13. K. Sukumar, V. Joevivek, T. Hemalatha, K.P. Soman, Prediction of forest fire using SVM—a novel approach, in *10th ESRI India User Conference* (2009)
14. Arjun, A. Hanumanthaiah, Wireless sensor network framework for early detection and warning of forest fire, in *2020 International Conference on Inventive Computation Technologies (ICICT)*, pp. 186–191 (2020)
15. B. Saiharsha, B. Diwakar, R. Karthika, M. Ganesan, Evaluating performance of deep learning architectures for image classification, in *2020 5th International Conference on Communication and Electronics Systems (ICCES)* (IEEE, 2020), pp. 917–922
16. S. Durga, R. Nag, E. Daniel, Survey on machine learning and deep learning algorithms used in internet of things (IoT) healthcare, in *2019 3rd International Conference on Computing Methodologies and Communication (ICCMC)* (IEEE, 2019), pp. 1018–1022

Secure GEDAR Routing Protocol for Underwater Data Collection Using WSN

R. Joshua Samuel Raj, Bhupender Singh, D. R. Ganesh, and N. Muthukumaran

1 Introduction

Submerged correspondence procedures, for example, optical, radio recurrence have been proposed for short reach joins where their exceptionally high-transmission capacity can be abused. These signs weaken high force or huge receiving wires procedures. It is suggest longer ranges, however, are constrained by three factors, for example, distance subordinate data transmission, time changing multi-way, and afterward low speed of sound [1–3]. Extensively higher piece rates have been illustrated; however, these outcomes are as yet in the area of exploratory examination. With the advances in acoustic modem innovation, research has moved into the space of organizations [4–6]. The submerged world drives the advancement of USN, and the applications can have unique necessities, for example, fixed, short, and afterward best exertion prerequisites at various plans [4, 7]. The mobility of two boundaries is to be distinctive for sorts of arrangements of submerged sensor organizations. In this module, we examined about remote submerged organizations and furthermore critical work in cabled submerged [8]. Logical applications notice the climate from topographical cycles on the sea depths and furthermore to water qualities to checking

R. Joshua Samuel Raj (✉) · D. R. Ganesh
Department of ISE, CMR Institute of Technology, VTU, Bengaluru, India
e-mail: joshuasamuelraj@gmail.com

D. R. Ganesh
e-mail: ganesh.dr@cmrit.ac.in

B. Singh
Faculty of Water Supply and Environmental Engineering, Arbaminch Water Technology Institute, Arbaminch University, Arba Minch, Ethiopia

N. Muthukumaran
Electronics and Communication Engineering, Francis Xavier Engineering College, Tirunelveli, Tamil Nadu, India

J. D. Peter et al. (eds.), *Disruptive Technologies for Big Data and Cloud Applications*, Lecture Notes in Electrical Engineering 905,
https://doi.org/10.1007/978-981-19-2177-3_16

creature life. Modern applications screen and control business exercises, for example, submerged gear identified with oil extraction, submerged pipelines or business [9]. Military and country security applications include getting or checking port offices or boats in unfamiliar harbors, de-mining, and correspondence with submarines and jumpers [10, 11].

The remainder of this paper is coordinated as follows. In Sect. 2, the identification of issues definition identified with the current techniques. Area 3 presents point-by-point portrayal of proposed methods for tackling plant-related issue. Analysis results and conversations are portrayed in Sect. 4. At last, the end and further improved are given in Sect. 5.

2 Identification of Problems

The recognizable proof of submerged assortment methods geographic and opportunistic routing having the detriment of there is no security given in this convention and vulnerable to different sorts of assaults, for example, egotistical assaults and dark opening assaults [7, 8, 12]. The profundity controlled directing convention having its disadvantage can give higher piece mistake and limitation blunder. In the VOID, mindful pressing factor directing organizations having its detriment can give restriction issue in submerged organizations and a portion of the hubs around a void region will in the end course parcels to neighborhood maxima [4, 5]. The multi-externalization-based restriction of submerged sensor having the detriment of the blemished roundabout shape and the thickness of the curl causes blunders in the limitation results [10, 11]. The versatile pressing factor in steering plan for submergence has the downside of high-organization traffic. The adaptable restriction with portability forecast having the burden of defenseless against assailants [2, 9].

Submerged sensor networks (UWSNs) comprise of submerged sensor hubs ousted in various profundities and furnished with a low-transfer speed acoustic modem and with different sensors that act cooperatively to help applications like submerged strategic reconnaissance, undersea investigation, and submerged catastrophe anticipation [1, 8]. UWSNs are portrayed by having hubs controlled by battery and high-engendering delay because of the sign in water, available transfer speed seriously restricted, and high-powerful parcel misfortune, which prompts the lower ratability [2, 3, 9].

3 Problem Solutions

The fundamental approach of GEDAR uses the voracious sending technique to drive the group at each hop, in the direction of the surface sonobuoys. The assessment challenge of OR next-skip forwarder set decision is the best approach to choose a summary of neighbors with the ultimate objective that the mysterious terminal issue is diminished. The accompanying ricochet forwarder set decision segment of GEDAR considers the circumstance of the neighbors and known sonobuoys to pick the top up-and-comer neighbors. Right, when a center is in a correspondence void region, GEDAR moves it to another significance to proceed with the voracious sending framework. To the furthest extent that we might actually know, GEDAR is the principal coordinating show proposed for compact lowered sensor associations to consider the significance change limit of the sensor center points to oversee correspondence void district issue. The motivations for the usage of this new perspective are triple.

In any case, the center significance change advancement is as of now available. Second, the correspondence task in the lowered sensor network is uncommonly exorbitant. Third, the cost expected to move the centers to new profundities is debilitated along the association action when differentiated and the circumstance where the center point should course data packages along more hops.

3.1 System Model

Submerged sensor hubs can change its profundity through inflatable floats or winch-based contraption. In a lightness-based profundity change framework, a float can be expanded by a siphon, bladders, or other gadget to change the lightness of the buoy comparative with the water. This framework does not utilize drive components, diminishing the energy cost to the profundity change. In winch-based mechanical assembly, sensor hubs are connected to surface floats or anchors through links. A link is then acclimated to move and keep a hub in a decided profundity. A few propositions, which consider profundity change ability of the hubs for inclusion upgrades and restriction frameworks, for example, did not think about the expense comparative with this undertaking. In this work, as we consider that sensor hubs can uninhibitedly float with sea flow, Drogue is a best contender to be utilized as a sensor hub.

3.2 Packet Delivery Estimation

In this part, we gage the submerged parcel conveyance likelihood p(m, d) of m pieces for any pair of hubs with removed, which is utilized in the following bounce subset sending choice strategy of the proposed geo deft steering convention.

4 Results and Discussion

Figure 1 shows the center sending review. The centers are passed on in about 1500 × 500. The center course of action portrayal has a colossal number of adaptable lowered sensor center points at the ocean base and sonobuoys, moreover named sinks centers, at the ocean surface. Each center point is outfitted with various sensor devices and with a low-bandwidth acoustic modem which is used to at times report the distinguished data to the complaints. Figure 2 shows that the presentation of centers. Here, the lowered sensor networks are passed on which contains lowered sensor center points, sonobuoys, and noticing center. In the review, blue tone tends to lowered sensor centers, and green concealing center points address sonobuoys.

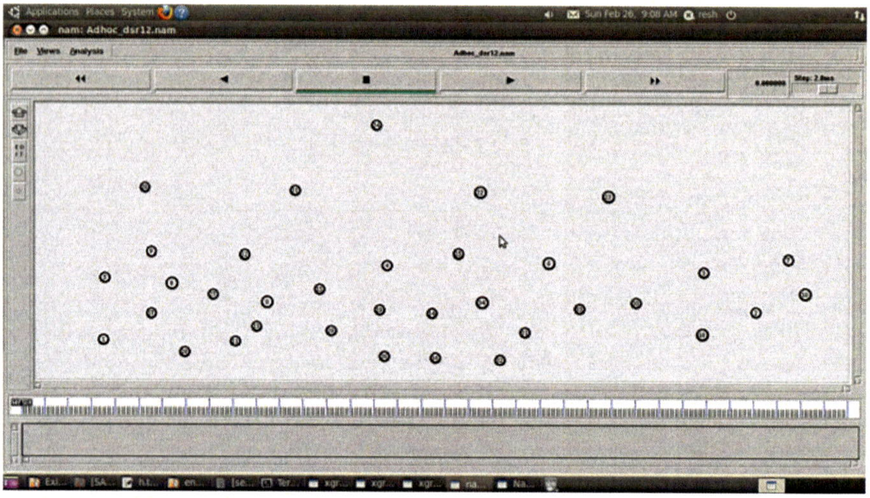

Fig. 1 Node deployment

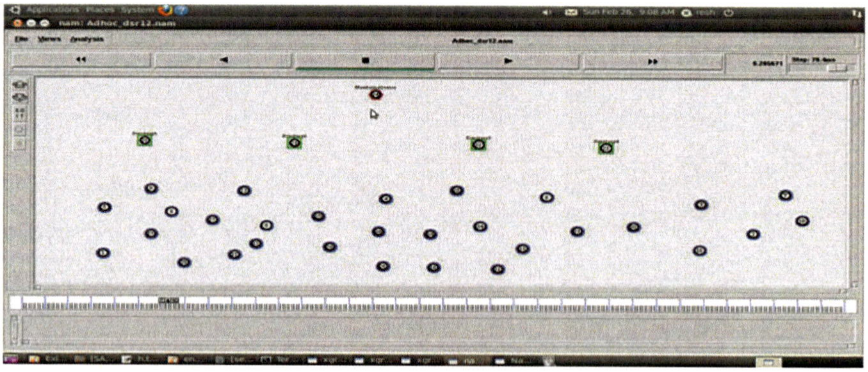

Fig. 2 Initialization of nodes

Here, there are 4 sonobuoys are passed on. Hearty hued tone tends to noticing center. Figure 3 shows that discontinuous beaconing. Discontinuous beaconing accepts a critical part in GEDAR. It is through occasional turning into that every hub gets the area data of its neighbors and reachable sonobuoys. Figure 4 shows that the depiction of control message transmission from checking focuses to sonobuoys. The checking focus sends the control message about detecting the submerged region to submerged sensor networks through sonobuoys. Figure 5 shows that the depiction of control message transmission from sonobuoys to sensor hubs. The observing focus sends the control message about detecting the submerged region to submerged sensor networks through sonobuoys. Figure 6 shows that depiction of submerged sensor hub detecting the climate. Subsequent to gaining power messages got from sonobuoys, the submerged sensor begins detecting the climate. The detected worth is communicated to checking focus through close by sonobuoys. Figure 7 shows source hub introduction. Here, the source hub is 16. This hub is instated in yellow

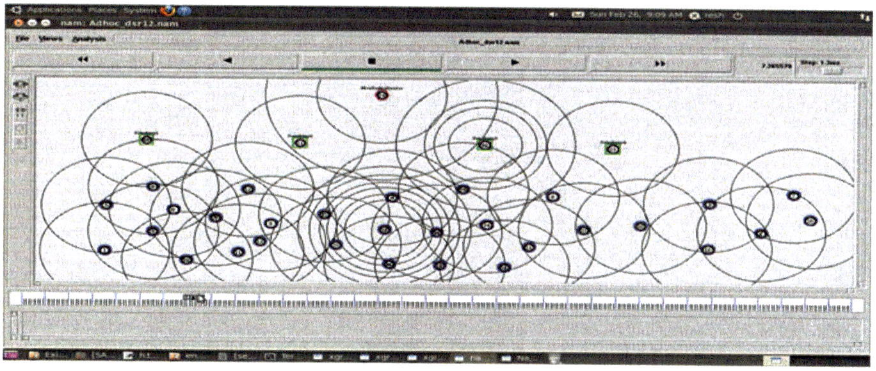

Fig. 3 Periodic beaconing

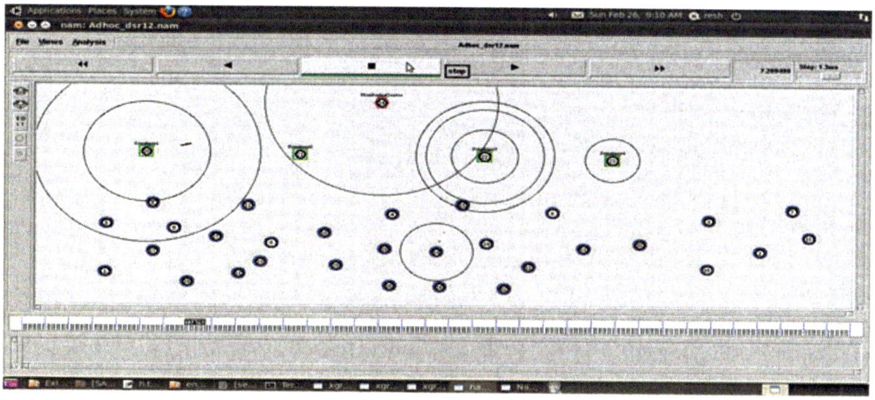

Fig. 4 Control message from monitoring center

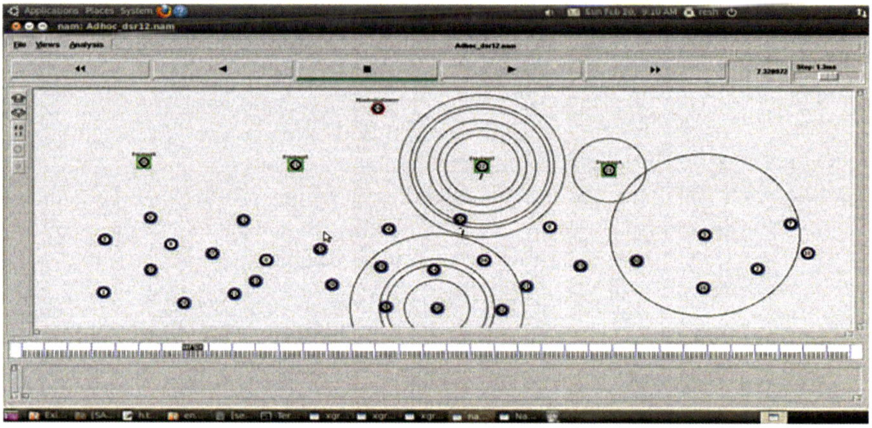

Fig. 5 Control message from sonobuoys

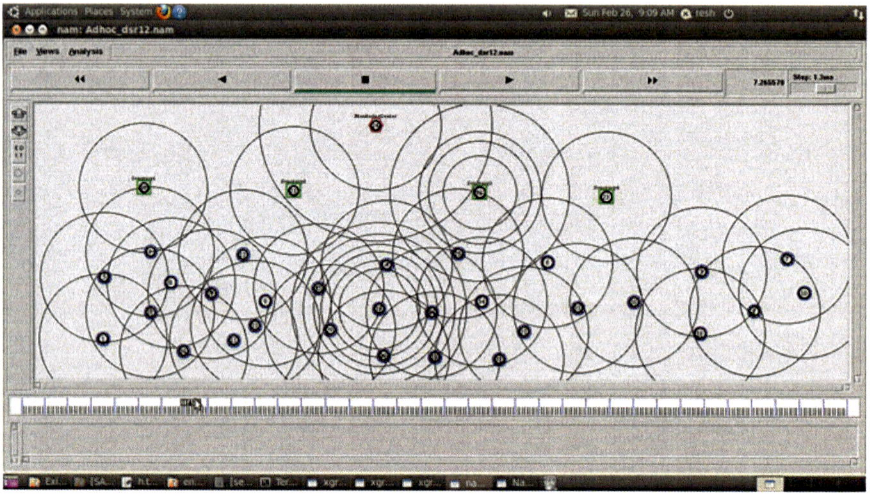

Fig. 6 Underwater sensor start sensing

tone since it needs to send the information to observing focus by means of close by sonobuoys. For that it needs neighbor hub for transmission of messages by means of neighbor hubs. Figure 8 shows that neighbor hub computation.

Figure 9 shows that preview of source hub 16 transmission of message to sonobuoys 22. The source hub 16 ascertains the neighbor hub and sending set to communicate the detected worth to sonobuoys 22 through hub 14 and 2. Figure 10 shows source hub in statement. Here, the source hub is 28. This hub is instated in yellow tone since it needs to send the information to observing focus through close by sonobuoys. For that it needs neighbor hub for transmission of messages

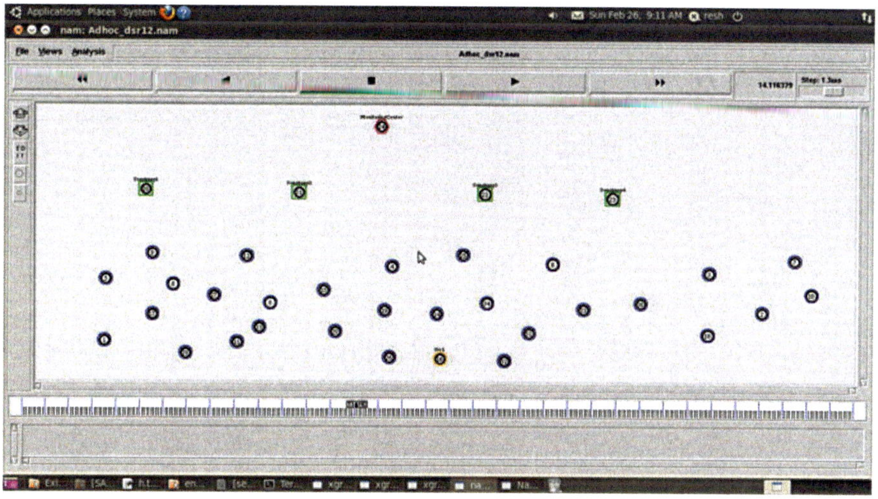

Fig. 7 Source node initialization

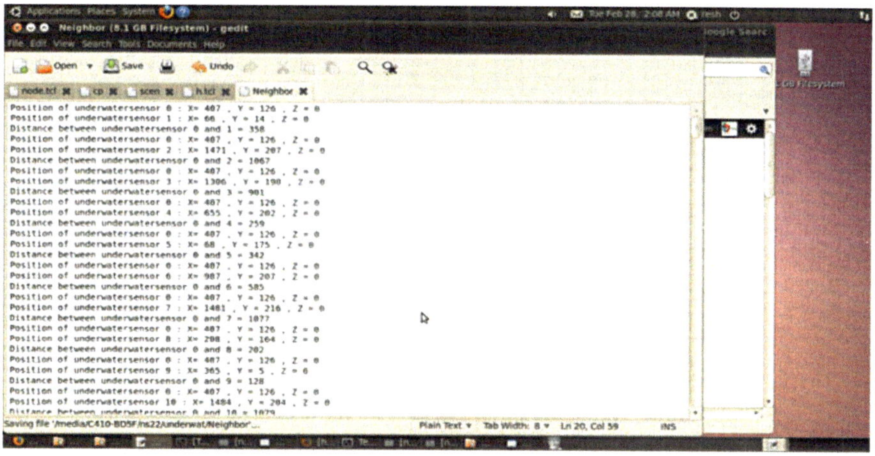

Fig. 8 Neighbor node initialization

through neighbor hubs. Figure 11 shows that preview of source hub 28 transmission of message to sonobuoys 20.The source hub 28 ascertains the neighbor hub and sending set to communicate the detected worth to sonobuoys 20. Figure 12 shows that the sonobuoys 20 sends the message to observing focus. The observing focus gets the data from sonobuoys and sends control message to sensor nodes. Figure 13 shows source hub introduction. Here, the source hub is 18. This hub is instated in yellow tone since it needs to send the information to observing focus through close by sonobuoys. For that it needs neighbor hub for transmission of messages

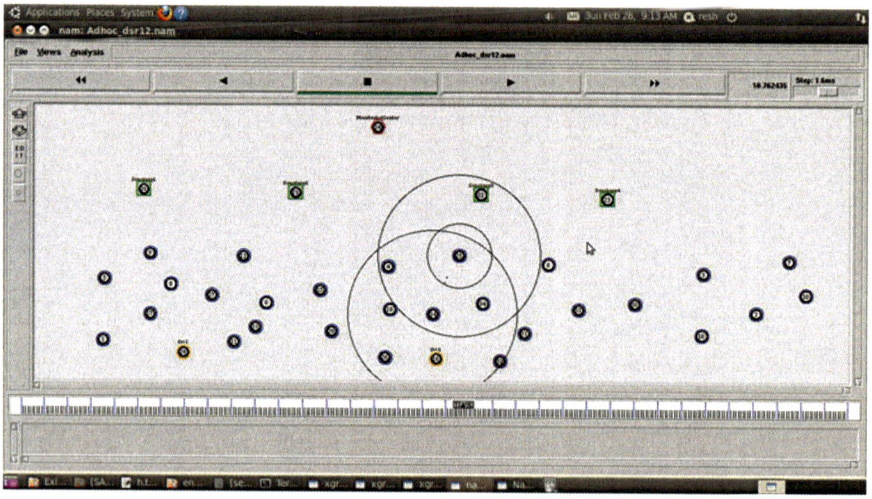

Fig. 9 Transmission of message to sonobuoys 22 from source node 16

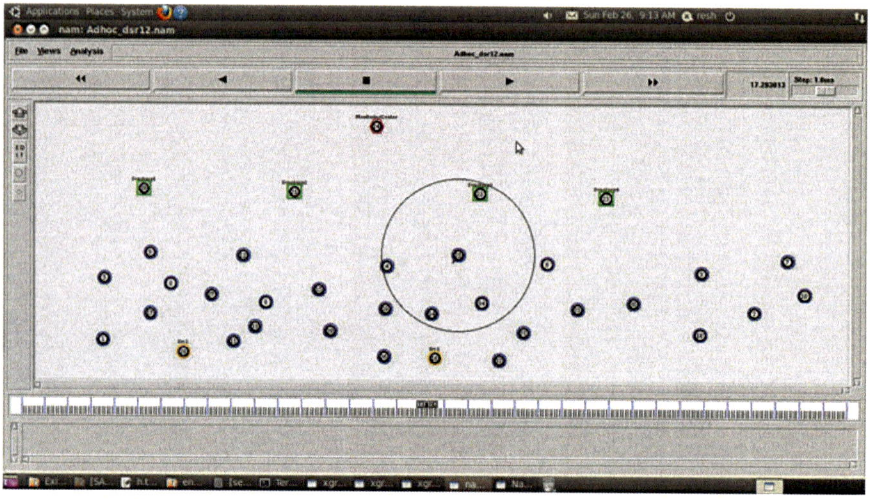

Fig. 10 Source 2 initialization

through neighbor hubs. Figure 14 shows that preview of source hub 18 transmission of message to sonobuoys 23.The source hub 18 ascertains the neighbor hub and sending set to communicate the detected worth to sonobuoys.

Figure 15 shows the hub 14 entering void area. At the point when the hub neglects to advance the information parcels by utilizing the covetous sending system. Rather

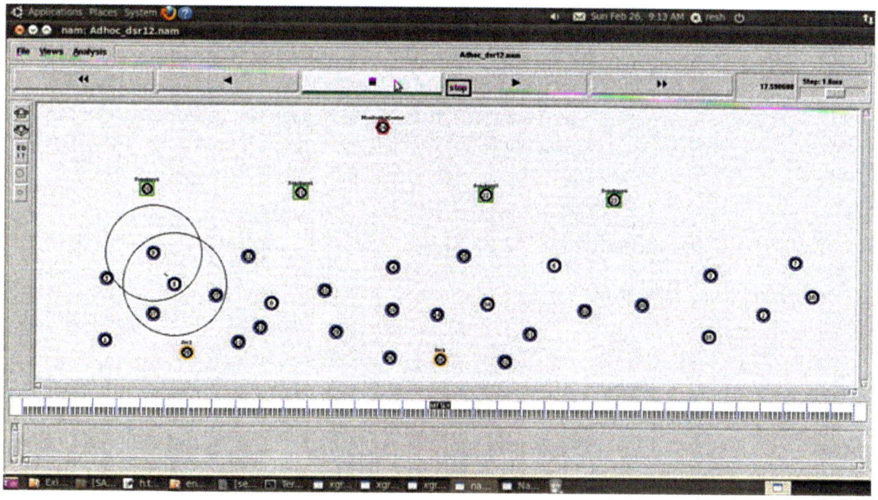

Fig. 11 Source 2 transmitting information to sonobuoys 20

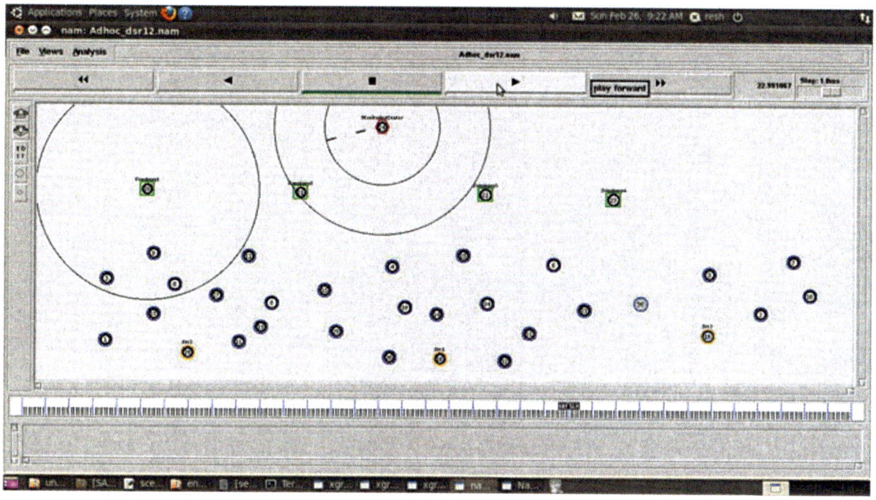

Fig. 12 Sonobuoys 20 sends message to monitoring center

than message-based void hub recuperation methodology, GEDAR exploits the generally accessible hub profundity change innovation to move void hubs for new profundities attempting to continue the ravenous sending. We advocate that profundity change-based geography control for void hub recuperation is more successful as far as information conveyance and energy utilization than message-based void hub recuperation strategies in UWSNs.

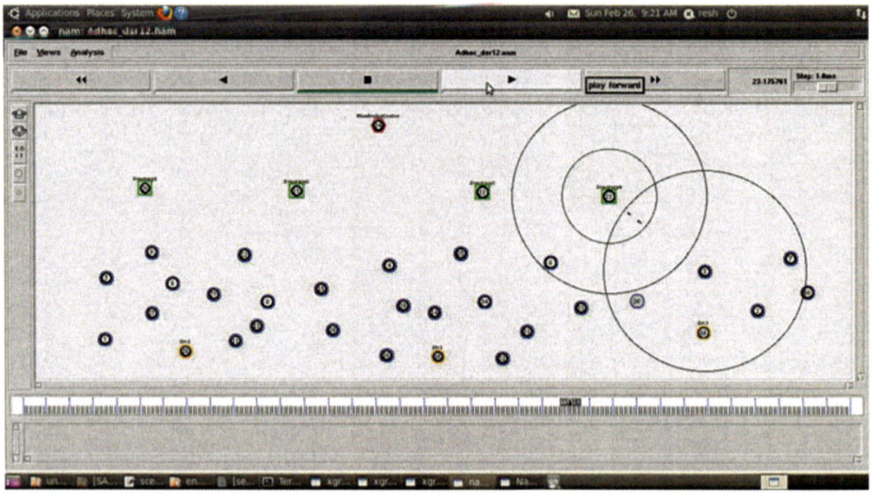

Fig. 13 Source 3 initialization

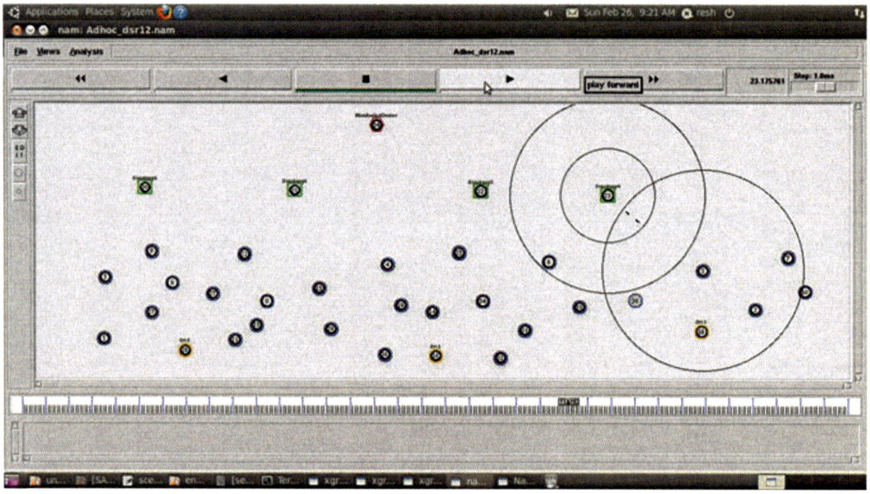

Fig. 14 Source node 18 sending information to sonobuoys 23

In Fig. 16, the recuperation mode technique, the void hub changes its status, stops the beaconing, sends a void hub declaration message to report its void hub condition to the area, and timetables the strategy to compute its new profundity. Figure 17 shows the throughput chart. The diagram is plotted between times versus throughput. Toward the beginning of reproduction, the throughput is diminished due to the presence of pernicious hub. At that point, subsequent to tracking down the substitute secure steering way, the bundle is safely communicated. Along these

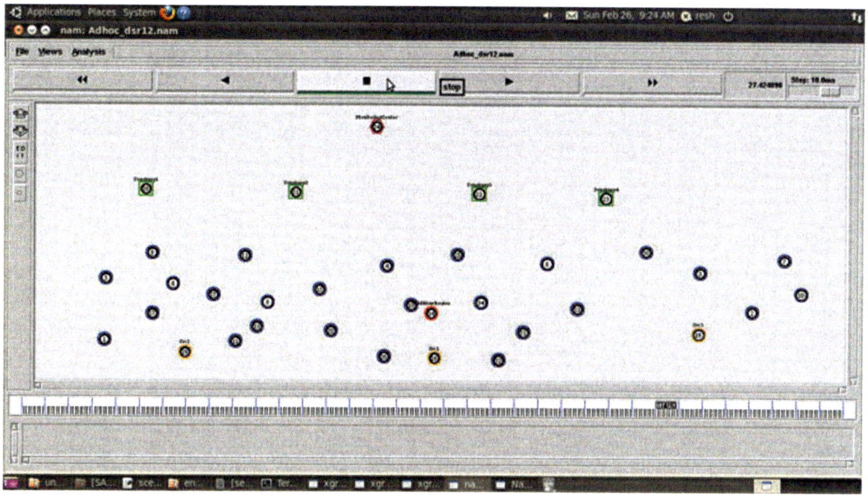

Fig. 15 Void region at node 14

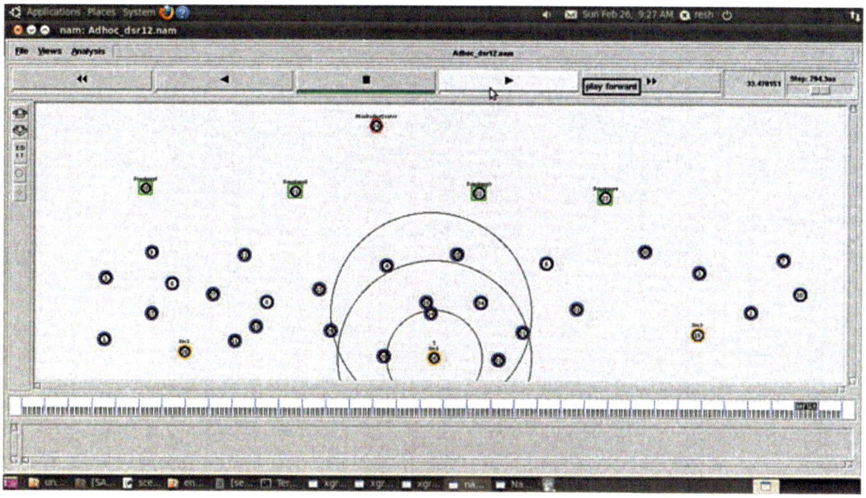

Fig. 16 Void node recovery

lines, the throughput is expanded after the pernicious hub discovery. Figure 18 shows the defer diagram. The diagram is plotted between times versus delay. The diagram shows the correlation of existing framework delay and proposed framework. The green shading line demonstrates the deferral of existing framework. The red shading line shows the proposed framework delay. In this, the deferral can be diminished as the time increments when contrasted with existing framework. Figure 19 shows the parcel misfortune chart. The diagram is plotted between times versus bundle misfortune. The

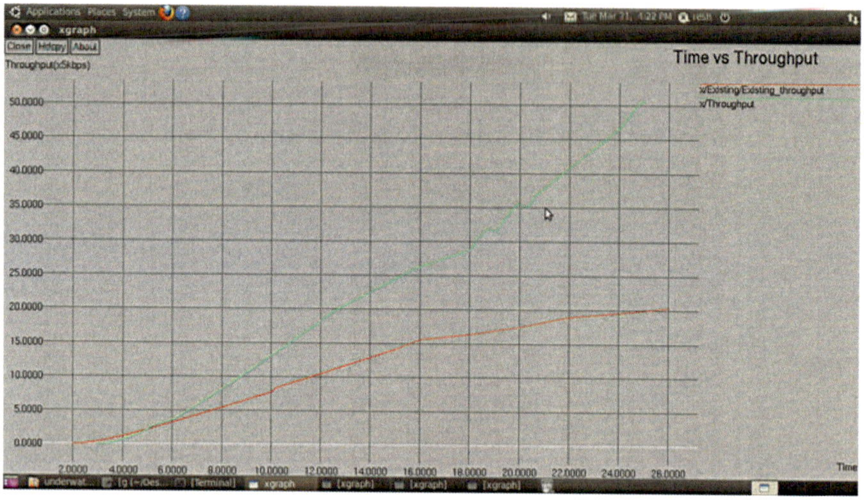

Fig. 17 Throughput graph

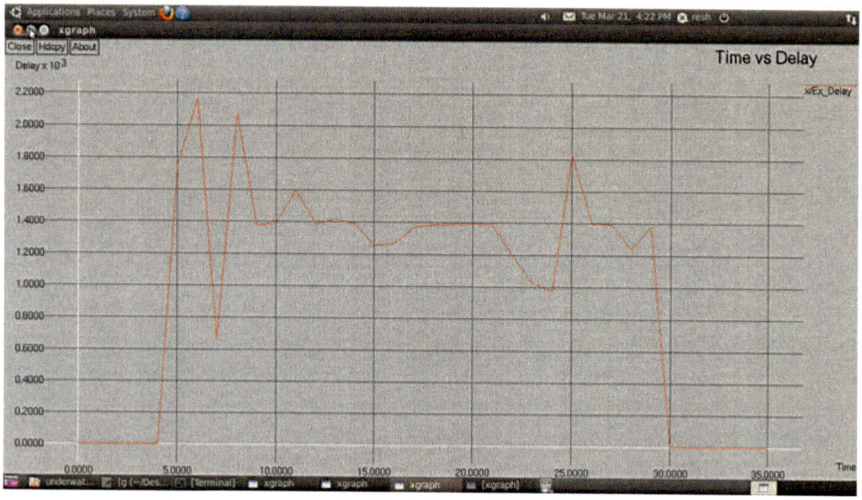

Fig. 18 Delay graph

diagram shows the correlation of existing framework bundle misfortune and proposed framework. The green shading line shows the bundle loss of existing framework. The red shading line shows the proposed framework parcel misfortune. In this, the bundle misfortune can be diminished as the time increments when contrasted with existing framework.

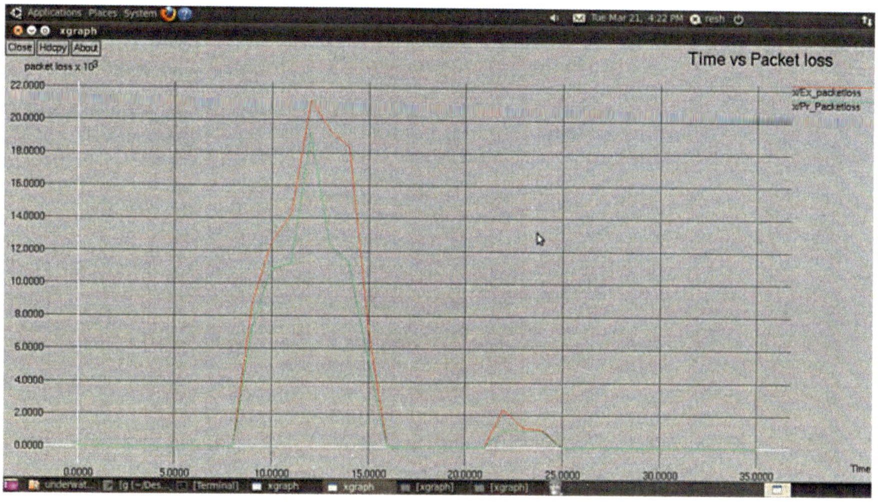

Fig. 19 Packet loss graph

5 Conclusion and Future Enhancement

This investigation work proposes and surveyed the GEDAR guiding show with lightweight obscure affirmed show to improve the data coordinating and insurance from feeble attacks in lowered sensor associations. GEDAR with lightweight show is a fundamental and versatile guiding that uses the position information of the transmission correspondence medium to the sea surface sonobuoys. GEDAR with lightweight show gives a novel significance change-based topography framework used to move centers to new profundities to overcome the correspondence areas. Our reenactment results showed that geographic coordinating shows subject to the circumstance of the center points are more beneficial than pressure directing shows. Here, when differentiated and existing work, the proposed work is a ton of improved in the strategy for throughput and restricting the deferment and bundle disaster in Table 1. In future, this topography control of significance change norms is applied to the arrangement of clever coordinating shows with a considered powerless attacks for UWSNs. By then, unmistakable QoS necessities for data movement and cost for shows up at a neighbor center, and the lifetime of the association is in like manner considered to improve execution.

Table 1 Comparison of existing and proposed methods

S. No	Parameters	Existing system	Proposed system
1	Throughput	100 kbps	250 kbps
2	Delay	2.15 ms	1.05 ms
3	Packet loss	21 kbps	19 kbps

References

1. M. Erol-Kantarci, H. Mouftah, S. Oktug, A survey of architectures and localization techniques for underwater acoustic sensor networks. IEEE Commun. Surv. Tuts. **13**(3), 487–502 (2011)
2. Z. Yu, C. Xiao, G. Zhou, Multi-objectification-based localization of underwater sensors using magnetometers. IEEE Sens. J. **14**(4), 1099–1106 (2014)
3. Z. Zhou, Z. Peng, J.-H. Cui, Z. Shi, A.C. Bagtzoglou, Scalable localization with mobility prediction for underwater sensor networks. IEEE Trans. Mobile Comput., vol. 10, no. 3, pp. 335–348 (2011); M. Zuba, M. Fagan, Z. Shi, J.H. Cui, A resilient pressure routing scheme for underwater acoustic networks, in *Proceedings of 57th IEEE global telecommunication conferences* (2014), pp. 637–642
4. N. Muthukumaran, R. Ravi, The performance analysis of fast efficient lossless satellite image compression and decompression for wavelet based algorithm. Wirel. Pers. Commun. **81**(2), 839–859 (2015)
5. N. Muthukumaran, R. Ravi, Hardware implementation of architecture techniques for fast efficient loss less image compression system. Wirel. Pers. Commun. **90**(3), 1291–1315 (2016)
6. Y. Noh, U. Lee, P. Wang, B.S.C. Choi, M. Gerla, Void-aware pressure routing for underwater sensor networks. IEEE Trans. Mobile Comput. **12**(5), 895–908 (2013)
7. R.W. Coutinho, A. Boukerche, L.F. Vieira, A.A. Loureiro, Geographic and opportunistic routing for underwater sensor networks. IEEE Trans. Comput. **65**(2) (2016)
8. L. Freitag, M. Grund, S. Singh, J. Partan, P. Koski, K. Ball, The WHOI micro-modem: an acoustic communications and navigation system for multiple platforms, in *Proceedings of MTS/IEEE Oceans* (2005), pp. 1086–1092
9. L.F.M. Vieira, Performance and trade-offs of opportunistic routing in underwater networks, in *Proceeding IEEE wireless communications and networking conference* (2012), pp. 2911–2915
10. R.W.L. Coutinho, A. Boukerche, L.F.M. Vieira, A.A. Loureiro, GEDAR: Geographic and opportunistic routing protocol with depth adjustment for mobile underwater sensor networks, in *Proceedings under IEEE International Conference on Communications* (2014), pp. 251–256
11. R. Sudhashree, N. Muthukumaran, Analysis of low complexity memory footprint reduction for delay and area efficient realization of 2D FIR filter. Int. J. Appl. Eng. Res. **10**(20), 16101–16105 (2015)
12. J. Heidemann, M. Stojanovic, M. Zorzi, Underwater sensor networks: applications, advances and challenges. Philos. Trans. Roy. Soc. A Math. Phys. Eng. Sci. **370**(1958), 158–175 (2012)

A Comprehensive Review on Automatic Image Captioning Using Deep Learning

P. V. Kavitha and V. Karpagam

1 Introduction

In our day-to-day life, we come across a lot of images that are being collected from different sources like Internet resources, articles, television, advertisements, etc. These sources do not have a textual description or caption of an image. Humans can understand them easily without a description, whereas it is very tough for a machine to understand. Automatic image captioning comes into focus as machines need a description to recognize the image. Generation of high accuracy and quality captions from an image is the main idea behind the research work. The motivation of this work is to generate quality captions from an image automatically using deep learning technique.

2 Applications and Challenges

Image captioning is used in various applications such as social media platforms, biomedical, e-commerce, military forces, educational sector, web surfing, etc. Image captioning is a hot research area of artificial intelligence that covers both the technologies of computer vision and natural language processing. It focuses on an understanding of images and generating captions for that image. Understanding an image depends on obtaining image features. Generating captions depend on producing syntactically and semantically correct sentences. Automatic interpretation of images

P. V. Kavitha (✉) · V. Karpagam
Information Technology, Sri Ramakrishna Engineering College, Coimbatore, TN, India
e-mail: kavitha.krishna@srec.ac.in

V. Karpagam
e-mail: karpagam.vilvanathan@srec.ac.in

J. D. Peter et al. (eds.), *Disruptive Technologies for Big Data and Cloud Applications*, Lecture Notes in Electrical Engineering 905,
https://doi.org/10.1007/978-981-19-2177-3_17

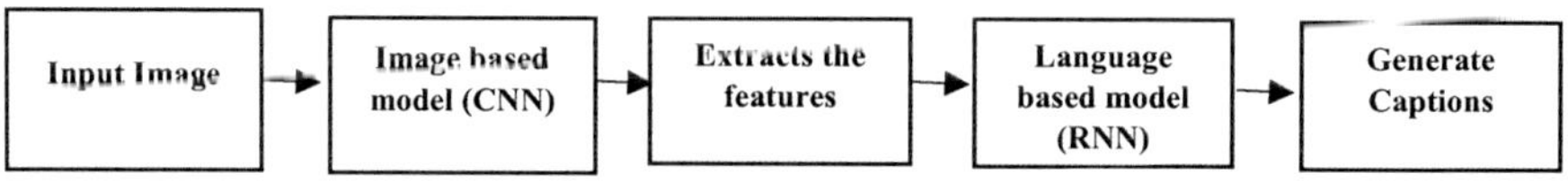

Fig. 1 Image captioning

with high accuracy, relevance, and support for long-term captions is a major challenging task. Though extra research works have been carried out in supervised, unsupervised, and reinforcement learning techniques, the quality of sentence description for an image is still a challenging task.

3 Role of Machine Learning Versus Deep Learning in Image Captioning

In machine learning, the hand-crafted features being task-specific in nature are extracted from input data. The extracted hand-crafted features are given to a classifier to classify an image object. Hence, handling a large set of images and videos becomes complex in machine learning-based techniques. Deep learning technique extracts the features of an input image automatically and handles a large amount of data. It uses convolutional neural networks (CNN) to classify an object and recurrent neural networks (RNN) to generate the textual description of an object.

4 Image Captioning Workflow

The workflow of image captioning is illustrated in the below Fig. 1

Image captioning is a two-step process. Image-based model extracts the features of the input image and language-based model converts the features and objects produced by image-based model to a natural sentence.

5 Image Captioning Techniques

The techniques used for generating automatic image captioning are:

- Template-based image captioning
- Parsing-based image captioning
- Retrieval-based image captioning
- Deep learning based
- Image captioning.

5.1 Template-Based Image Captioning

Farhadi et al. [1] have introduced a fixed template method for image captioning. This method uses a template with a blank number of slots for producing captions of an image. Template slots are filled with a triplet of scene elements such as < object, action, scene > and it generates grammatically correct captions. This method limits the generation of variable length captions since the templates are predefined. The captions generated are not always easy to read, and they are not natural.

5.2 Parsing-Based Image Captioning

Kuznetsova et al. [2] have proposed a parsing-based language model for image captioning. This model produces only limited captions from an image.

5.3 Retrieval-Based Image Captioning

Hodosh et al. [3] have used retrieval approaches for image captioning. It first select a set of similar images from a large database, and based on the query image, it passes the captions of retrieved images. It is impossible to generate new captions. The generated caption is like a human-generated language.

5.4 Deep Learning-Based Image Captioning

The commonly used techniques of deep learning for image captioning are:

- Supervised learning: It uses labeled data.
- Unsupervised learning: It deals with unlabeled data.
- Reinforcement learning: It is a reward-based learning technique that finds suitable action.

6 Methods Used in Deep Learning-Based Image Captioning Techniques

Deep learning image captioning techniques are categorized into encoder–decoder architecture, attention-based, semantic concept-based, and novel object-based image captioning. The following methods are used in supervised, unsupervised, and reinforcement learning.

Table 1 CNN architectures

Architecture	#Params	Size	Top-1 Accuracy	Top-5 Accuracy	Year
AlexNet	61 M	-	57.1	80.2	2012
VGG	138 M	528 MB	70.5	91.2	2013
Inception-V1	7 M	92 MB	69.8	89.3	2013
ResNet – 50	25.5 M	98 MB	75.2	93	2015
Xception	23 M	88 MB	79	94.5	2016
InceptionResNetV2	56 M	215 MB	80.3	95.3	2016
ResNetXt50	25 M	96 MB	77.7	93.8	2017

6.1 Encoder–Decoder Architecture

It is defined as a structure of CNN and RNN. Convolutional neural network (CNN) represents the encoder that reads the given input image and extracts a high-level feature representation. Recurrent neural network (RNN) represents the decoder that generates word captions from the encoded image.

Due to its simplicity and computational power, VGGNet is the first popular choice for extracting the features of images. A clear comparison of CNN architectures associated with params, size, top-1 accuracy, top-5 accuracy, and the year was made [4], and the results are presented in Table 1.

Vinyals et al. [5] put forth a neural image caption generator model that uses CNN for image interpretation and LSTM for caption generation. But this model does not support the generation of lengthy sentences.

Jia et al. [6] have proposed an extension of LSTM and it is called guided LSTM (gLSTM). This model generates long-length sentences. It adds global information to each gate and cell state of LSTM to manage the generation of long-term sentences.

Wang et al. [7] have introduced a deep, hierarchical LSTM-based method for image captioning. The proposed method consists of a CNN and two LSTM networks. It can utilize both past and future information to learn lengthy captions.

6.2 Attention Mechanism

In this approach, CNN is used to extract image features and RNN is used to generate captions. It consists of two types. They are soft and hard attention mechanisms.

In hard attention mechanism, the random sampling method is used for selecting the information. The functional relationship between the final loss function and the attention distribution is not achievable, and backpropagation algorithm cannot be used. Hard attention only focuses on one location and randomly selects that location. It samples the hidden state of the input by probability and does not sample on the hidden state of the entire encoder. Soft attention mechanism focuses on calculating the weighted sum of all regions. It emphasizes all the encoder inputs when calculating each decoder state and the amount of calculation is large.

Xu et al. [8] have introduced an attention-based model for image captioning. It uses both hard and soft attention models. CNN considers the top layer of ConvNet for extracting the features of the salient objects from the image. To preserve the loss of information, this approach uses features from the bottom convolutional layer rather than a fully connected layer.

Jin et al. [9] have proposed an attention-based mechanism to generate captions of an image. It considers the scene as a whole and does not consider the spatial aspects of the image. Language-based model dynamically generates the captions as output. It considers the salient regions of the image to generate sentence descriptions.

6.3 Semantic Concept

In this method, CNN-based encoder encodes the features and semantic concepts of an image. The encoded part is passed into the language model to generate captions. Gan et al. [10] have introduced a semantic compositional network (SCN) for image captioning. Semantic concept vector is constructed from the image and has more potential than visual feature vector and scene vector. It can generate captions emphasizing the entire meaning of the image.

6.4 Novel Object Based

In novel object-based method, a lexical classifier and a language model are used for training unpaired image and text data. A deep caption model is used for training paired image data. Eventually, both these models train jointly to generate captions for novel objects. Generating captions for unseen images in the test images is a challenging task.

Yao et al. [11] have introduced a copying mechanism to generate image captions for novel objects. It uses classifiers for novel objects. A decoder RNN with a copying mechanism to produce captions. It adds a new network to identify the unseen objects from unpaired images and it uses LSTM to generate captions. Venugopalan et al. [12] proposed a novel object captioner (NOC) which uses external resources for generating captions for unseen objects in the image.

7 Unsupervised and Reinforcement Learning

Generative adversarial networks (GANs) are unsupervised learning techniques that learn deep features from unlabeled data. It is suitable for real-time valued data.

It has limitations in backpropagating the discrete data. To overcome this discrete data problem, the Gumbel sampler is used [13]. Shetty et al. [14] generate multiple captions for a single image. But a lot of research works are going in this unsupervised learning by researchers.

Ren et al. [15] have introduced a novel reinforcement learning-based image captioning method. It comprises two networks. The "policy network" acts as local guidance and helps to predict the next word based on the present state. The "value network" acts as global guidance and calculates the reward value considering all the possible extensions of the current state. This mechanism helps to adjust the networks in predicting the right words.

Shi et al. [16] proposed architecture for image captioning with deep reinforcement learning to optimize image captioning tasks. This model consists of two networks such as "policy network" and the "value network." Policy network consists of CNN-RNN framework, whereas value network consists of CNN-RNN-linear mapping framework for generating image captions. CNN-RNN framework faces two problems such as information inadequateness problem in the training phase and misrecognition problem in the caption generation phase. RNN suffers from vanishing gradient and exploding gradient problems.

Even though a lot of research works have been introduced in supervised, unsupervised, and reinforcement learning, still the quality of sentence description for an image is a challenging task. Most of the image captioning methods have used LSTM which faces misrecognition of semantic objects in the caption generation phase.

To improve the quality of captions, image captioning integrates deep reinforcement learning with the CNN-RLSTM framework. In this CNN-RLSTM framework, different weights are assigned to the words to distinguish the importance of different words in the training phase. RLSTM reduces the misrecognition of semantic objects in the generation phase by maximizing the consensus score between the captions and the reference information from the neighboring images of the target image [17]. The aforementioned ideas illustrate how RLSTM outperforms the other frameworks.

8 Comparison of Image Captioning Techniques

Table 2 depicts the comparison of various deep learning-based techniques for the evaluation metrics such as BLUE-1, BLUE-2, and METEOR on different datasets such as Flickr8k and MSCOCO. Flickr8k dataset consists of 8000 images and 5 captions for each image. MSCOCO dataset consists of more than 300,000 images, more than 2 million instances, and 5 captions per image. BLUE-1 g and BLUE-2 g are metrics used to measure the quality of a machine-generated text. METEOR is another metric used to measure the machine-translated language. From Table 2, RLSTM outperforms the best language model for image captioning.

Table 2 Comparison of image captioning techniques

Dataset	Researchers	Techniques and method	Image-based model	Language-based model	BLUE-1	BLUE-2	METEOR
Flickr8k	Jia et al. [6]	Supervised, encoder–decoder	VGGNet	LSTM	0.647	0.459	0.201
	Xu et al. [8]	Supervised, attention	AlexNet	LSTM	0.670	0.457	0.203
MSCOCO	Xu et al. [8]	Supervised, hard attention	AlexNet	LSTM	0.718	0.504	–
	Xu et al. [8]	Supervised, soft attention	AlexNet	LSTM	0.707	0.492	–
	Jin et al. [9]	Supervised, attention	VGGNet	LSTM	0.697	0.519	0.235
	Venugopalan et al. [11]	Supervised, novel object based	VGGNet	LSTM -C	0.638	0.486	–
	Gan et al. [12]	Supervised, semantic concept	ResNet	LSTM	0.741	0.578	0.236
	Shetty et al. [13]	Unsupervised + GAN	GoogLeNet	LSTM	–	–	0.239
	Ren et al. [15]	Reinforcement learning, encoder–decoder	VGGNet	LSTM	0.713	0.539	0.251
	Ding et al. [17]	Supervised, encoder–decoder	ResNet-101	RLSTM	0.765	0.603	0.264

9 Conclusion

In this paper, machine learning techniques such as supervised, unsupervised, and reinforcement learning commonly used for image captioning were discussed. This paper highlights the methods used in deep learning techniques such as encoder–decoder architecture, attention-based, semantic concept-based, and novel object-based image captioning. A comparison of deep learning-based techniques with image and language models was made for the evaluation metrics like BLUE-1, BLUE-2, and METEOR on Flickr8k and MSCOCO datasets. Recently, deep learning-based techniques made remarkable progress, but the quality of generating captions from the image remains a challenging task. Future work may be extended by a novel image captioning method integrating deep reinforcement learning with CNN—RLSTM framework which is yet to be achieved for improving the quality of captions of an image.

References

1. A. Farhadi, M. Hejrati, M.A. Sadeghi, P. Young, C. Rashtchian, J. Hockenmaier, D. Forsyth, Every picture tells a story: Generating sentences from images, in *European conference on computer vision* (Springer, 2010), pp.15–29
2. P. Kuznetsova, V. Ordonez, Tamara L Berg, and Yejin Choi", TREETALK: composition and compression of trees for image descriptions. TACL **2**(10), 351–362 (2014)
3. M. Hodsh, P. Young, J. Hockenmaier, Framing image description as a ranking task: data, models and evaluation metrics. J. Artif. Intell. Res. **47**, 853–899 (2013)
4. Koustubh, ResNet, AlexNet, VGGNet, Inception: Understanding Various Architectures of Convolutional Networks, Available online: https://cv-tricks.com/cnn/understand-resnet-alexnet-vgg-inception/. Accessed on 24 May 2019
5. O. Vinyals, A. Toshev, S. Bengio, D. Erhan, Show and tell: A neural image caption generator, in *Proceedings of the IEEE Conference on Computer Vision and Pattern Recognition* (2015), pp. 3156–3164
6. J. Xu, E. Gavves, B. Fernando, T. Tuytelaars, Guiding the long-short term memory model for image caption generation, in *Proceedings of the IEEE International Conference on Computer Vision* (2015), pp. 2407–2415
7. C. Wang, H. Yang, C. Bartz, C. Meinel, Image captioning with deep bidirectional LSTMs, in *Proceedings of the 2016 ACM on Multimedia Conference* (ACM, 2016), pp.988–997
8. K. Xu, J. Ba, R. Kiros, K. Cho, A. Courville, R. Salakhudinov, R. Zemel, Y. Bengio, Show, attend and tell: Neural image caption generation with visual attention, in *International Conference on Machine Learning* (2015), pp. 2048–2057
9. J. Jin, K. Fu, R. Cui, F. Sha, C. Zhang, Aligning where to see and what to tell: image caption with region-based attention and scene factorization (2015). arXiv preprint arXiv:1506.06272
10. Z. Gan, C. Gan, X. He, Y. Pu, K. Tran, J. Gao, L. Carin, D. Li, Semantic compositional networks for visual captioning, in *Proceedings of the IEEE conference on computer vision and pattern recognition (CVPR)* (2017) pp. 1141–1150
11. T. Yao, Y. Pan, Y. Li, T. Mei, Incorporating copying mechanism in image captioning for learning novel objects, in 2017 *IEEE Conference on Computer Vision and Pattern Recognition (CVPR)* (IEEE, 2017), pp. 5263–5271

12. S. Venugopalan, L.A. Hendricks, M. Rohrbach, R. Mooney, T. Darrell, K. Saenko, Captioning images with diverse objects, in *Proceedings of the IEEE Conference on Computer Vision and Pattern Recognition* (2017) pp. 1170–1178
13. [58] E. Jang, S. Gu, B. Poole, Categorical reparameterization with gumbel-softmax, in *International Conference on Learning Representations (ICLR)* (2017)
14. R. Shetty, M. Rohrbach, L.A. Hendricks, M. Fritz, B. Schiele, Speaking the same language: matching machine to human captions by adversarial training, in *IEEE International Conference on Computer Vision (ICCV)* (2017), pp. 4155–4164
15. Z. Ren, X. Wang, N. Zhang, X. Lv, L.-J. Li, Deep reinforcement learning-based image captioning with embedding reward, in *Proceedings of the IEEE Conference on Computer Vision and Pattern Recognition (CVPR)* (2017), pp. 1151–1159
16. H. Shi, P. Li, B. Wang, Z. Wang, *Image Captioning based on Deep Reinforcement Learning* (ACM Computing, 2018)
17. G. Ding, M. Chen, S. Zhao, H. Chen, J. Han, Q. Liu, *Neural Image Caption Generation with Weighted Training and Reference* (Cognitive Computation, 2018)

A GAN-Based Triplet FaceNet Detection Algorithm Using Deep Face Recognition for Autism Child

R. Joshua Samuel Raj, S. Anantha Babu, A. Jegatheesan, and V. M. Arul Xavier

1 Introduction

The most essential steps to take to limit the impact of Autism Spectrum Disorder and enhance the quality of life of ASD sufferers are early diagnosis and treatment. There is, still, no medical screening for the diagnosis of autism. Observation is commonly used to identify ASD symptoms. ASD signs are typically detected by parents and educators in older and adolescent kids attending kindergarten. The signs of ASD are then assessed by the school's special education staff. This becomes necessary to monitor emotional differences in an infant by analysis when they can be seen as early as 6 months of age, as in contrast to autism-specific neural scans, which can be seen after two years of age [1].

A machine learning algorithms are innovative analytical methods that accurately detect the relevant features in vast quantities of data using mathematical learning, statistical inference, and knowledge theories. This approach has the advantages of precise and consistent estimation with vast numbers of variables and appropriate sample in non-experimental datasets. These approaches have been successfully used in diagnostics of ASD [2], classification by manipulated issue with potentials in Attention Deficit Hyperactivity Disorder, and classification by free expression review

R. Joshua Samuel Raj
Department of ISE, CMR Institute of Technology, VTU, Bengaluru, India

S. Anantha Babu (✉)
Department of CSE, St. Mother Theresa Engineering College, Tuticorin, Tamil Nadu, India
e-mail: ananthmtech@gmail.com

A. Jegatheesan
Department of CSE, Swarnandhra College of Engineering and Technology, Narsapur, Andhra Pradesh, India

V. M. Arul Xavier
Department of CSE, Karunya Institute of Technology and Sciences, Coimbatore, TN, India

J. D. Peter et al. (eds.), *Disruptive Technologies for Big Data and Cloud Applications*, Lecture Notes in Electrical Engineering 905,
https://doi.org/10.1007/978-981-19-2177-3_18

in Schizophrenia, and according recent studies in psychology. Bishop et al. [3] examined adult ASD patients' potential health conditions using machine learning techniques and found that these methods accurately forecast health problems in the digestive, urinary, and respiratory systems.

The major accepted research on the efficacy of a parent-mediated social contact intervention in children between the ages of 9 and 14 months with a high familial risk of ASD found that the therapy effect on symptom duration persisted for another 24 months after the intervention concluded [4, 5]. Moreover, implementing cognitive care to all the children with the risk of ASD depends solely on characteristics that would be economically unfeasible, and the risk/benefit ratio for infants would have grown normally. As a result, early detection of early symptoms of ASD and individual prediction of later progression may help to accurately target early intervention strategies.

The significant variation of the syndrome and the comparatively late development of the main traits of ASD make it difficult to diagnose ASD until the age of 24 months [6, 7]. As a result, examining various forms of data is essential to capture various facets of the condition. Machine learning has the ability to provide a reliable algorithm for predicting future health outcomes by efficiently contributing significant data from various sources and identifying the most predictive combination of steps.

Research work's major contributions can be summarized in the following points:

- We suggest the triplet distillation approach for transferring facial recognition information from an instructor model to a student model.
- In relation to the fixed margin of the original triplet loss, we boost the triplet loss with dynamic margins by using similarity structures among different identities.

The following report is structured as: The second section looks at how connected research has been used to help connected models to investigate the emergence of irregular behavior using face recognition schemes in a variety of classification methods. Theoretical Foundations of improved Triplet Loss FaceNet, GAN generator, and proposed algorithm are covered in Section three. Section four introduces the findings and discusses GTBFaceNet Analyses, and Section five ends with possible term guidance.

2 Background Study

In the field of artificial intelligence, cognitive psychology, cognitive science, neuroscience, and philosophy of mind, conceptual paper is a collection of approaches that models psychological or behavioral phenomena as emergent systems of interconnected networks of simple structures. The most common types of cognitive neuroscience use neural network models, which are cognitive processes essentially based on neural processing of information. Many machine learning researchers recognize process of analytic representation as choosing after the information is retrieved, considering that the valuable features would be found and representative data has been collected. As a result, the researchers decided to concentrate on algorithm creation and machine learning algorithms suggesting the probabilistic neural network (PNN) as trained artificial neural approach to the back-propagation neural network [8].

Fake faces are typically produced and displayed on a computer monitor, mirror, or some other kind of display system using images or pages from books or printed media. The properties of these false faces that appear in various media vary significantly from those of a live face, since the appearance of a live face is found to be distinct from that of a fake face. A fake face, on the contrary, is a two-dimensional portrait. Let $I(x, y)$ be the image quality at a point $P(x, y)$ on the face's surface in the plane (x, y), and $K(x, y)$ be the face's albedo at $P(x, y)$. When the light source directions are unknown, a set of M images can be created by marking the light source directions as $I = 1, \ldots, M$, and the set is specified by,

$$I(x, y) = A(x, y)N(x)S(f_j), \quad (1)$$

where $N(x)$ seems to be the face's surface normal (3D shape), and $S(n)$ is any light source path [9].

Deep learning employs neural network architecture with many hidden layers that represent the brain's neural interactions. Deep learning has been used to research predictors of challenging behavior and assess neuro imaging of people with ASD in the literature studied [10–12].

Unsupervised feed-forward neural networks or autoencoders were used to reduce the number of dimensions. The algorithm then applied supervised learning to a feed-forward neural network by using two identical AdaBoost as weights. With 70% of accuracy, the deep neural network (DNN) outperformed SVM and RF classifiers. However, processing time took more than 32 hours (due to GPU-accelerated computing), although the SVM and RF versions took less than 21 minutes. This research suggests that deep learning analysis of rs-fMRIs may be used to identify ASD biomarkers [11, 12].

The problem identified and applied triplet-based GAN-generated dataset with reduced noise to segregating the long-time (i.e., 8 min) data series into many overlapped short-time (i.e., 7 s) sub-sequences to evaluate the feasibility of using Triplet-dependent FaceNet for separate sets of autistic and non-autistic children, and then sent them as input to the GAN generator to reduce the loss function for training and classification. The findings showed that even with the Triplet-based GAN generator, we were able to obtain a high level of precision with the Autism Spectrum Disorder classification dataset, with 98.30% accuracy [13].

3 Theoretical Foundations

3.1 Improved Triplet Loss FaceNet

The suggested technique persuades the FaceNet network to generate an encoding $F(x)$ in a d-dimensional space from an input image x. The embedding's second rule is set to 1 to ensure that it remains on either the d-dimensional hypersphere.

$$M_i^x\ \text{: anchor image, } M_i^y\ \text{: Positive image and}$$
$$M_i^N\ \text{: Negative image (various image dataset)}$$

$$\left\| f\left(M_i^x\right) - f\left(M_i^y\right)\right\|_2^2 \propto < \left\| f\left(M_i^x\right) - f\left(M_i^y\right)\right\|_2^2 \tag{2}$$

The principle of hard positive and hard negative mining is used to do this. It can be expressed mathematically as follows:

$$\arg\max_{M_i^p} \left\| f\left(M_i^k\right) - f\left(M_i^l\right)\right\|_2^2 \tag{3}$$

$$\arg\min_{M_i^q} \left\| f\left(M_i^k\right) - f\left(M_i^l\right)\right\|_2^2 \tag{4}$$

The problem seems to be that predicting the argmin and argmax of the entire dataset is challenging due to the fact that the majority of the dataset is usually occupied by low-quality images and often mislabeled instances, resulting in a collapsed model $F(x) = 0$. As a result, the triplet is chosen to meet the following correlation.

$$\left\| f\left(M_i^r\right) - f\left(M_i^s\right)\right\|_2^2 < \left\| f\left(M_i^r\right) - f\left(M_i^s\right)\right\|_2^2 \tag{5}$$

Semi-hard negative outcomes find the ideal combination between simple and difficult, facilitating adequate learning and rapid convergence to global optimum (minimum) values. The loss is described as follows:

$$L_1 = F_2 - F_1 - \text{constant-value} \tag{6}$$

where F_2 and F_1 reflect the difference between the anchor and negative image clustering algorithms, alongside the variance between the anchor and positive image classifiers, respectively.

3.2 Improved GAN Generator

Generative Adversarial Networks are part of the generative framework. It implies that they have the ability to create/generate new content. G is a generator that maps the latent space vector (z) to information. The Vector D is a binary classification network that takes an image as input and returns a scalar probability that the image is autistic. For D and G setup, we can decide how they learn through the loss functions and optimization techniques. D takes a $3 \times 64 \times 64$ input image, filters it through a sequence of Conv2d, BatchNorm2d, and Leaky ReLU layers, and outputs the maximum possibility through a Sigmoid activation function.

$$\begin{aligned} K(m, n) &= L = \{i_1 \ldots i_N\}, l_p \\ &= -\left|n'_q . \log m_{nq} + \left(1 - m_q\right). \log(1 - m_q)\right| \end{aligned} \tag{7}$$

We assign a value of 1 to our true label and a value of 0 to our false label. While measuring the losses of DD and GG, these labels will be used and this is also the principle used in the performance of the Triplet-based FaceNet Model that uses GAN [14–17].

3.3 Proposed Procedure

The completed proposed GTLBFaceNet for training and testing procedure is presented in Fig. 1, shows the lifecycle of proposed model.

3.3.1 Proposed Algorithm

GTLBFaceNet Training Algorithm

Step 1. Get the autistic and non-autistic dataset 3d facial image.

Step 2. Encoding all collected image divided with 128 size vectors of entire images set

Step 3. After pre-processing steps the model involving to apply CNN with Triplet-based FaceNet with GAN generator

Step 4. The triplet method finding the picture value with hard positive and hard negative mining of entire sets of images

Step 5. Then the FaceNet model correlate and batch process of autistic and non-autistic face images

Step 6. Next, the proposed model combines GAN-generated output of triplet method. To rectify any loss in triplet method, the GAN decimator predicts exact autistic children face images

Step 7. Train the multiclass images with different classes

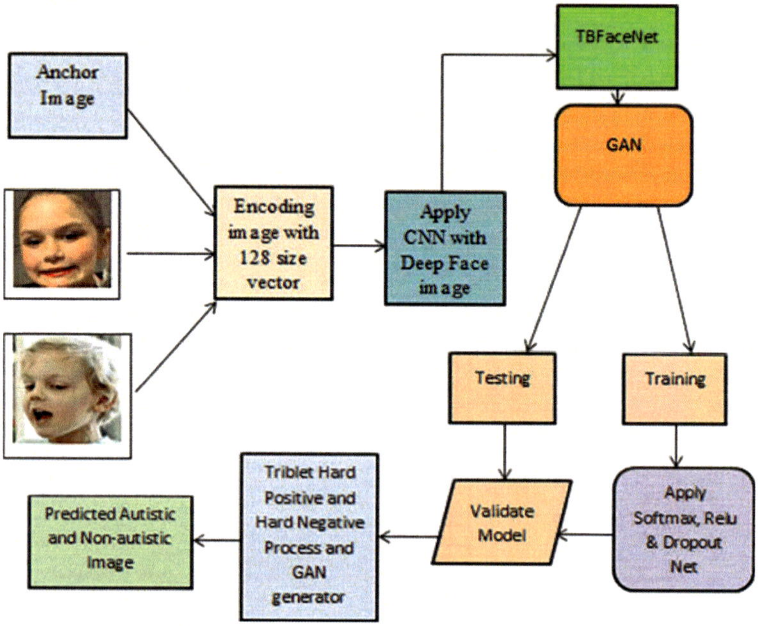

Fig. 1 Proposed triplet-based GAN generator

Step 8. Process the TBFaceNet with GAN model in 40 EPOCS
Step 9. Generating Best Model (FaceNet)
Step 10. Plot the entire samples with accuracy and loss of image
Step 11. Predicting label with autistic and non-autistic images

3.3.2 Proposed Diagram

See Fig. 1.

4 Results and Discussion

The GTLBFaceNet modified the model parameters and split the data by a sliding window of 100 and a stage of 50 to prevent fewer available training samples from fitting. Finally, 7-s data are labeled as a sub-sequence. Each sub-sequence was associated with a distinct name.

The filter size, namely kernel size, describes each node parameter filtered. For example, slides at frequent intervals over the existing layer's neural network matrix and performs dot product. In other terms, from each point, the element of the filter is weighted by the corresponding element of the input sample, and the total is added.

The input sample is [a, b, c, d], one of the filters is [2, 3], and the interval is one. By calculating the dot product, we can get [$a_2 + b_3$, $b_2 + c_3$, $c_2 + d_3$]. The RELU active function, the binary cross-entropy loss function, and the Adam optimizer with default parameter values were also used to train the Triplet-based FaceNet model. The training dropout value was fixed at 0.5%. The rate of learning was 0.01. The default settings were used to start the models. There were 128 filters in all, each of which was an 88-dimensional vector. Then, there is the training, validation, and test sets. They have a 3:1:1 ratio. The GTLBFaceNet was trained in the training set, the model's generalization potential were tested in the verification set, and the best solution with the smallest loss function was saved. Finally, the model was put to the test on a collection of data. For an autism dataset, Fig. 2 and 3 show the success assessment estimates for training error and validation precision.

In the analysis plot, the accuracy and loss measured with the 40 EPOCS GTLB-FaceNet model predict a high accuracy rate as well as a low loss rate. Positive class is referred to in ASD, while negative class is referred to in TD. The percentage of

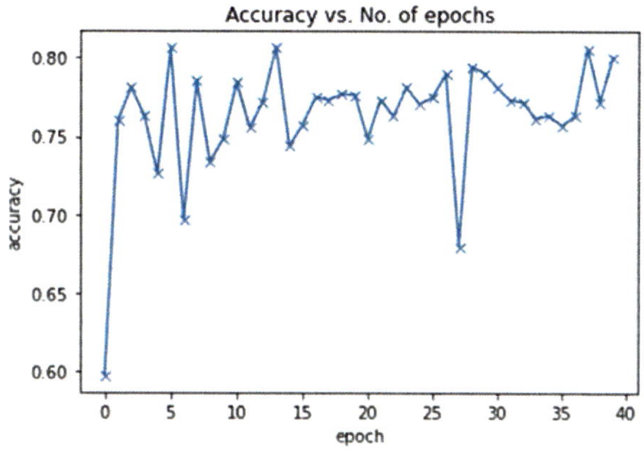

Fig. 2 Accuracy versus No. of EPOCS

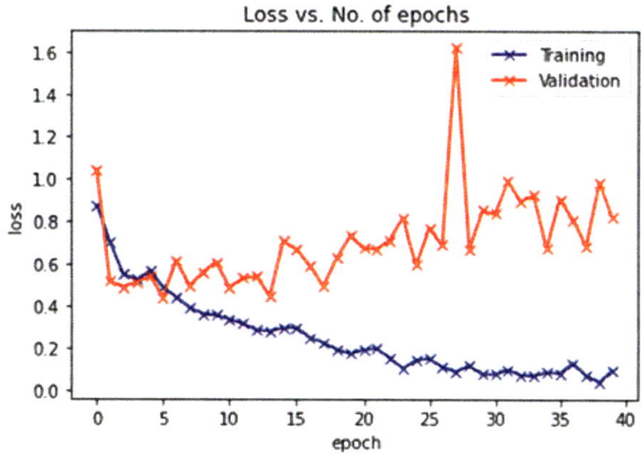

Fig. 3 Loss versus No. of EPOCS

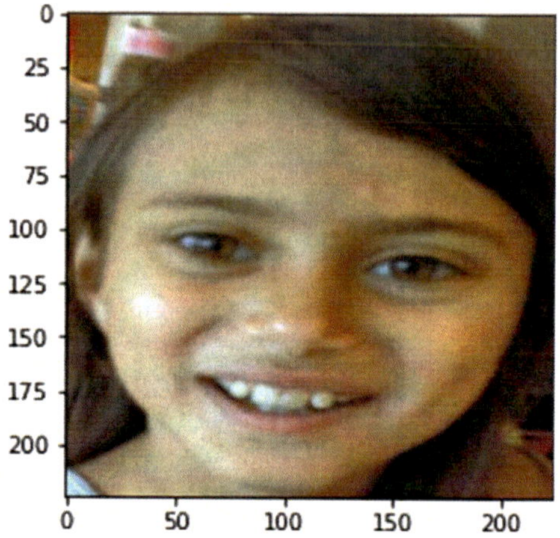

Fig. 4 Label: Autistic, predicted: Autistic (90%)

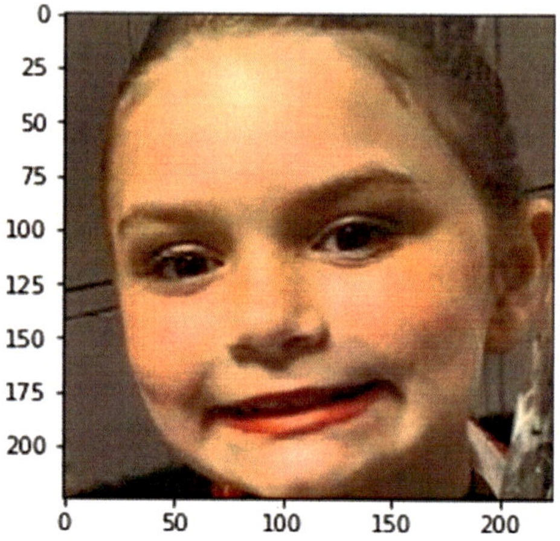

Fig. 5 Label: Non-autistic, predicted: Non-autistic (90%)

individual positives that are properly identified in all positives is referred to as sensitivity. The percentage of accurate diagnosis of all diagnoses is known as accuracy. For example, at the age of 2, an 18-month-old child that loses a key behavior is predictive of autism which can begin to randomly display the behavior. Figures 4 and 5 depict the autism and non-autistic learning analysis outcome sets [18, 19].

4.1 *Quality Measurement GTLBFaceNet with Other Existing Methods*

Quality management is important for determining how well a classification model does in achieving a goal. On the research sample, performance assessment metrics

Table 1 Comparison with other existing methods

Method	Sensitivity	Accuracy	Loss	Time
Naïve Bayes	62	72	40	5 min 02 s
SVM	63	77	30	5 min 33 s
KNN	64	78	28	8 min 10 s
Random forest	66	83	28	8 min 30 s
GTLBFaceNet	79	98	20	10 min 56 s

are used to evaluate the classification model's efficacy and performance. It is critical to choose the right metrics for evaluating the model's results. GTLBFaceNet is compared to other approaches in Tables 1 and 2 and plotted in Figure 6.

Table 2 Comparison with different models

Methods	Model	Sensitivity	Accuracy	Loss	Time
Naïve Bayes	MobileNet	54	61	49	5 m 2 s
	VGGNet	55	62	48	5 m 2 s
	ResNet	58	68	45	5 m 2 s
	FaceNet	62	72	32	5 m 2 s
SVM	MobileNet	56	72	40	5 m 33 s
	VGGNet	56	74	38	5 m 33 s
	ResNet	58	74	38	5 m 33 s
	FaceNet	63	77	30	5 m 33 s
KNN	MobileNet	60	62	42	8 m 10 s
	VGGNet	62	62	35	8 m 10 s
	ResNet	62	65	30	8 m 10 s
	FaceNet	64	78	28	8 m 10 s
Random forest	MobileNet	56	70	44	8 m 30 s
	VGGNet	58	75	40	8 m 30 s
	ResNet	64	78	35	8 m 30 s
	FaceNet	66	83	28	8 m 30 s
GTLB FaceNet	MobileNet	68	56	42	10 m 53 s
	VGGNet	72	88	42	10 m 56 s
	ResNet	74	90	30	10 m 54 s
	FaceNet	79	98	20	10 m 56 s

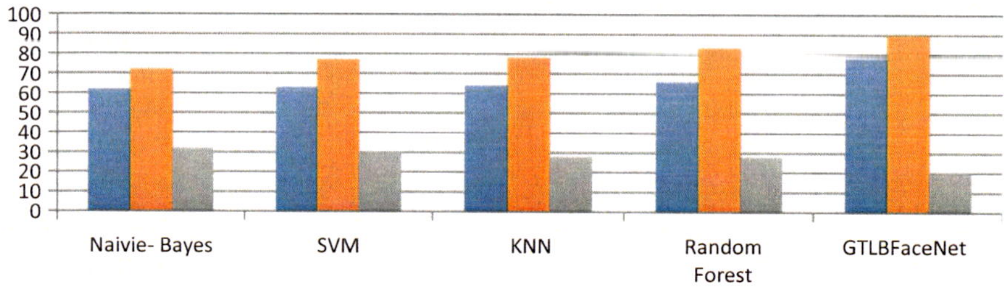

Fig. 6 Comparison plot with other existing methods

5 Conclusion and Future Scope

In this study, multiple machine learning and deep learning methods were used to diagnose Autism Spectrum Disorder. On a non-clinical dataset from three age groups, Child, Adolescents, and Adults, various performance assessment metrics were used to assess the performance of the models applied for ASD identification. Both the SVM and CNN-related models display the same accuracy of estimation of around 98.3% for the ASD Child dataset after treating missed values in this study. However, for the remaining two datasets, the CNN-based approach outperformed all other deemed model-building strategies in terms of precision. These findings clearly indicate that a Triplet-based GAN generator algorithm, rather than the other traditional machine learning classifiers suggested in previous studies, may be used to predict Autism Spectrum Disorder.

References

1. S. Raj, S. Masood, Analysis and detection of autism spectrum disorder using machine learning techniques. Procedia Comput. Sci. **167**, 994–1004 (2020)
2. Q. Junfei, et al., A survey of machine learning for big data processing. EURASIP J. Adv. Sign. Process. **1**(2016), 1–16 (2016)
3. R.J. Landa, Efficacy of early interventions for infants and young children with, and at risk for, autism spectrum disorders. Int. Rev. Psychiatry **30**(1), 25–39 (2018)
4. D. Bzdok, A. Meyer-Lindenberg, Machine learning for precision psychiatry (2017). arXiv preprint arXiv:1705.10553
5. J. Lee, et al., Quantifying performance of machine learning methods for neuroimaging data. NeuroImage **199**, 351–365 (2019)
6. S. Efstratios, et al., Predicting the diagnosis of autism spectrum disorder using gene pathway analysis. Mol. Psychiatry **19**(4), 504–510 (2014)
7. J.N. Constantino, T. Charman, Diagnosis of autism spectrum disorder: reconciling the syndrome, its diverse origins, and variation in expression. Lancet Neurol. **15**(3), 279–291 (2016)
8. S.H. Lee, M.J. Maenner, C.M. Heilig, A comparison of machine learning algorithms for the surveillance of autism spectrum disorder. PLoS ONE **14**(9), e0222907 (2019)
9. A. Rachid, et al., Euclidean and geodesic distance between a facial feature points in two-dimensional face recognition system. Hum. Comput. Interact. **1**, 5 (2017)

10. H. Anibal Sólon, et al., Identification of autism spectrum disorder using deep learning and the ABIDE dataset. NeuroImage Clin. **17**, 16–23 (2018)
11. T. Rajat Mani, et al., Classifying autism spectrum disorder using the temporal statistics of resting-state functional MRI data with 3D convolutional neural networks. Front. Psychiatry **11**, 440 (2020)
12. A. Nicola, et al., Deep learning and multiplex networks for accurate modeling of brain age. Front. Aging Neurosci. **11**, 115 (2019)
13. M. Schrimpf, Brain-inspired recurrent neural algorithms for advanced object recognition. Diss. Master's thesis, Technical University Munich, LMU Munich, University of Augsburg, 2017
14. A. Hermans, L. Beyer, B. Leibe, In defense of the triplet loss for person re-identification (2017). arXiv preprint arXiv:1703.07737
15. X. Liu, et al., Hard negative generation for identity-disentangled facial expression recognition. Pattern Recognit. **88**, 1–12 (2019)
16. A. Zhang, et al., EEG data augmentation for emotion recognition with a multiple generator conditional Wasserstein GAN. Complex Intell. Syst. **021**, 1–13
17. S.A. Babu, R.J.S. Raj, A.X. VM, N. Muthukumaran, DCT based enhanced tchebichef moment using huffman encoding algorithm (ETMH), in *2021 3rd International Conference on Intelligent Communication Technologies and Virtual Mobile Networks (ICICV)* (2021), pp. 522–527. https://doi.org/10.1109/ICICV50876.2021.9388504
18. S. Babu Anantha, P. Eswaran, C. Senthil Kumar, Lossless compression algorithm using improved RLC for grayscale image. Arab. J. Sci. Eng. **41**(8), 3061–3070 (2016)
19. S.A. Babu, E. Perumal, Efficient approach of run length coding technique using lossless grayscale image compression (E-RLC), in *2018 3rd International Conference on Inventive Computation Technologies (ICICT)* (2018), pp. 680–686. https://doi.org/10.1109/ICICT43934.2018.9034377

Rapid Efficient Loss Less Color Image Compression Using RCT Technique and Hierarchical Prediction

R. Joshua Samuel Raj, T. Sudarson Rama Perumal, N. Muthukumaran, and D. R. Ganesh

1 Introduction

In the color image compression, the color transform is used to de-correlate the color components. Each of the transformed components is compressed separately [1]. For example, RGB to YCbCr transform is used for lossy compression of color image and video [2, 16]. In the loss less color image compression, the most of the color transform cannot be used due to their un-invariability with integer arithmetic [3–5]. In this paper, we design hierarchical prediction method for the use of lower row, upper and left row pixels on the way to be encoded. For the loss less compression of color images, the RGB is first transformed based on the RCT techniques and then the Y channel is encoded by a conventional gray scale image compression algorithm [6–8].

For more precise expectation of these signs and furthermore for exact displaying of forecast blunders, we utilize the progressive plan. When the even line picture is encoded then every one of the pixels in even line pixel is added to odd line picture. Followed by the odd and even picture we can pack upper and lower column of pixel picture [9]. In this expansion, since the arithmetical properties of two sub pictures are

R. Joshua Samuel Raj (✉) · D. R. Ganesh
Department of ISE, CMR Institute of Technology, VTU, Bengaluru, India
e-mail: joshuasamuelraj@gmail.com

D. R. Ganesh
e-mail: ganesh.dr@cmrit.ac.in

T. Sudarson Rama Perumal
Department of ECE, Rohini College of Engineering and Technology, Kanyakumari, Tamil Nadu, India

N. Muthukumaran
Electronics and Communication Engineering, Francis Xavier Engineering College, Tirunelveli, Tamil Nadu, India

J. D. Peter et al. (eds.), *Disruptive Technologies for Big Data and Cloud Applications*, Lecture Notes in Electrical Engineering 905,
https://doi.org/10.1007/978-981-19-2177-3_19

very little unique pixel esteems so the Probability Density Function of expectation mistakes of a picture can be precisely demonstrated. Investigations on different sorts of pictures are performed, and it is shown that the proposed strategy gives higher pressure proportion and PSNR esteem than JPEG2000 [10, 11].

The tone change which is utilized for misfortune less shading picture pressure ought to be reversible thus the reversible tone change method is utilized. The encoding of the chrominance picture is finished by progressive plan that utilization the upper, lower, left, odd and even estimations of pixels for the pixel expectation [5, 12]. The inspiration of the modules is to expand the pressure proportion and the top to flag proportion. The remainder of this paper is coordinated as follows. In Sect. 2, the identification of issues detailing identified with the current strategies. Segment 3 presents the proposed calculations for settling the organizations issue. Test results and conversations are depicted in Sect. 4. At last, the ends and further upgraded are given in Sect. 5.

2 Problem Identification

The major plan of the leaving calculation is principally to address the pressure models of the framework and these techniques are essentially associated with following space. In the misfortune, less pressure of shading channel exhibits pictures by setting demonstrating large expectation mistakes which are inescapable close to the edges and inside finished regions [17]. The adaptive determination of shading changes for the reversible picture pressure conspires including the issue that the tone change can spread clamor from one shading segment into another. Hence, it is upsetting the resulting step of spatial de-connection [15]. The progressive of shading channel exhibit pictures utilizing full scale pixel spatial change includes the different limit, for example, the pressure first methodology does not give a net decrease of number of pieces required for the packed picture [18]. The pressure first methodology can diminish the amount of information that should be handled as contribution to the pressure encoder and replicated from the resulting pressure decoder. In the productive pressure for scrambled grayscale pictures have the limitation of customary encoder side stylistic layout connection techniques such change are not relevant here on the grounds that the encryption work has veiled the source reliance [13].

A misfortune less pressure plot for Bayer shading channel cluster pictures have the constraint of the records just one of the three shading parts at every pixel area [14]. Forecast of the missing green examples in the non-green sub picture is excessively troublesome. The high quality straight insertion for Bayer designed shading pictures have the restriction for each CCD pixel we should relate a full RGB esteem [14]. Bilinear introduction creates huge high recurrence content, since it does not consider the connection among the RGB esteems. To defeat this, proposed strategy is considered.

3 Problem Solution

This segment momentarily depicts about the new method of quick productive misfortune less shading picture pressure calculation utilizing RCT strategy and afterward the various leveled expectation coding. The presentation of these modules is contrasted and the current frameworks like JPEG 2000. The boundaries that are thought about for the current framework and proposed framework are pressure proportion, top sign to commotion proportion and mean square mistake.

3.1 Flow Chart Description

The information RGB picture is de-corresponded by utilizing RCT procedures and it is changed over into dark scale picture. The dim scale picture technique separates the light and dull pixels in the picture. The dim scale picture is changed over into RCT method. RCT strategy is utilized to change the RGB picture for the pressure reason. The chrominance picture is deteriorated into set of odd numbered lines and set of even numbered columns. On the off chance that the even column sub picture is encoded, the pixels of even line sub picture is utilized to anticipate the pixels in odd numbered line. For encoding the chrominance picture the progressive plan was created for the pixel expectation.

The proposed framework builds up a progressive forecast conspire for misfortune less picture pressure depend on the raster filter expectation technique. Be that as it may, the raster examine forecast technique is once in a while in productive in the high recurrence district. The various leveled forecast for the pressure interaction just a single pixel insertion is utilized. The chrominance picture is packed utilizing various leveled forecast plot. After pressure the picture is quantized and encoded to deliver the effective misfortune less packed picture.

3.2 RCT Technique

The misfortune less pressure works in a luminance chrominance shading space rather than the standard RGB tone. It is very much perceived in picture pressure that this ordinarily empowers better coordinated pressure because of the de-connection of the RGB shading channels. For misfortune less pressure, the shading space changes should be actually reversible interaction. RCT procedure is utilized to change the RGB picture for the pressure reason. The RCT for 16 digit shading picture coding, the work is propelled by the expanding needs of media applications on low end gadgets. To improve coding great association on the present circumstance, RCT methods tweaked for high shading frameworks (Fig. 1).

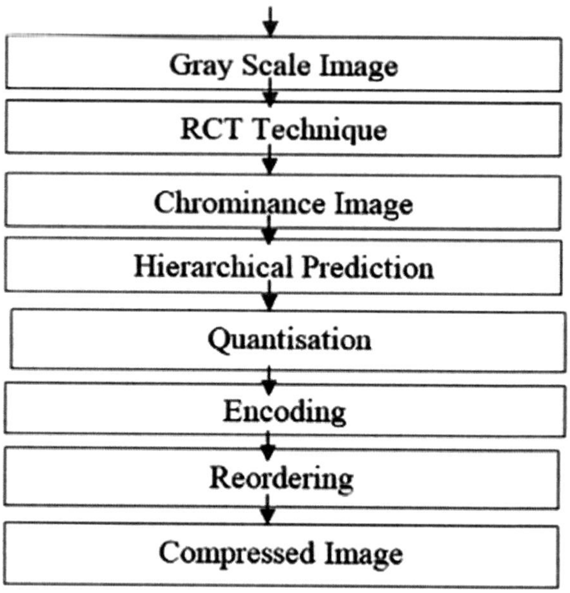

Fig. 1 Flow chart

3.3 *Hierarchical Decomposition*

For the efficient lossless pressure, it is imperative to precisely gage the pdf of expectation mistake for better setting demonstrating, alongside the exact forecast. For this, we propose a various leveled decay conspire as portrayed in Fig. 2, which shows that pixels in an information picture X is isolated into two sub pictures: an even sub picture Xe and an odd sub picture Xo. At that point, Xe is encoded first and is utilized to anticipate the pixels in Xo. Likewise, Xe is additionally used to appraise the measurements of expectation mistakes of Xo.

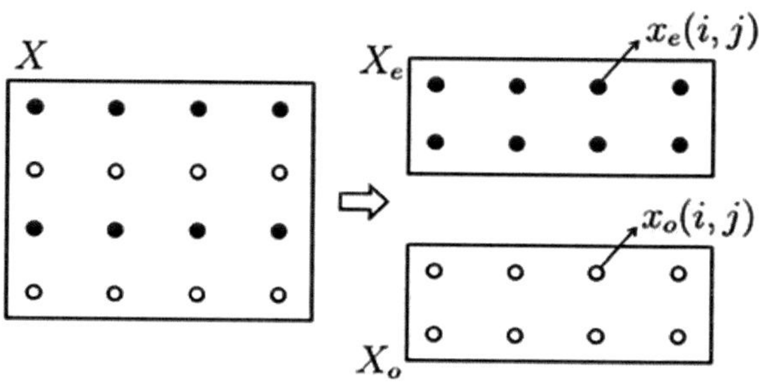

Fig. 2 Input images and its decomposition

3.4 Proposed Algorithm

```
Step 1
Calculation of dir(i, j)
    if |xo(i,j) – x^h (i,j)| + T1 < |x^v (i,j)| then
dir (i,j) ← H
    else
    dir (i,j) ← V
    end if

Step 2
if  (i – 1,j) = H or dir (i, j – 1) = H then
        Calculate dir (i,j) by Algorithm 1
            Encode dir(i,j)
    ifdir (i,j) = H then
            x^o (i,j) ← x^h (i,j)
    else
            x^o (i,j) ← x^v (i,j)
    end if
    else
    x^o (i,j) ← x^v (i,j)
        Calculate dir (i,j) by Algoritm 1
    end if
```

3.5 Quantization

Quantization at the encoder side methods facilitate parceling of the info information range into a more modest arrangement of qualities. There are two primary kinds of quantizer: scalar quantizer and vector quantizer. A scalar quantizer parcels the area of information esteems into fewer spans. On the off chance that the yield spans are similarly divided which is the most straightforward approach to do it the interaction is called uniform scalar quantization; in any case, for reasons generally identified with minimization of all out contortion, it is called non-uniform scalar quantization. Vector quantization procedures broaden the essential standards of scalar quantizations to numerous measurements (Figs. 3, 4, 5, 6, 7, 8 and 9).

Fig. 3 Input image

Fig. 4 Grayscale image

Fig. 5 RCT technique Y channel

3.6 *Encoding*

Encoding is a procedure for addressing the quantized information as minimally as could really be expected. After quantization the majority of the great recurrence

Fig. 6 Chrominance image (Cu Cv channel)

Fig. 7 Even channel

Fig. 8 Odd channel

Fig. 9 Quantization

coefficients are zero. To misuse the quantity of zero a crisscross output of the lattice is utilized respecting long series of zeros. When a pixel building block has been changed to a range and quantized. Encoder makes a fixed length code to address the quantize yield and guides the yield. An encoder packs the compacted values get by the quantized to give more productive pressure.

4 Results and Discussion

4.1 Results and Discussion for RCT Method

In this segment, in view of proposed framework investigation of different boundaries is finished. In proposed framework, the RGB picture is compacted by utilizing misfortune less picture pressure technique and it is contrasted and the current framework JPEG 2000. The boundaries pressure proportion and the pinnacle sign to commotion proportion of the current framework and the proposed framework are contrasted and one another. Figure 10 shows the encoded image. An entropy encoder is used to compress the image values in an efficient compression. Figure 11 shows the reordering of pixels in the image to produce the compressed output image. Figure 12 shows the subplot of images of compressed image. The original image occupies the first position. The noisy image is shown in the second position. The final compressed image is shown in the third position. The final image shows the squared error image. Figure 13 shows the PSNR calculation for the final compressed image. Figure 14 shows the compression ratio of our proposed method.

Fig. 10 Encoding

Fig. 11 Reordering

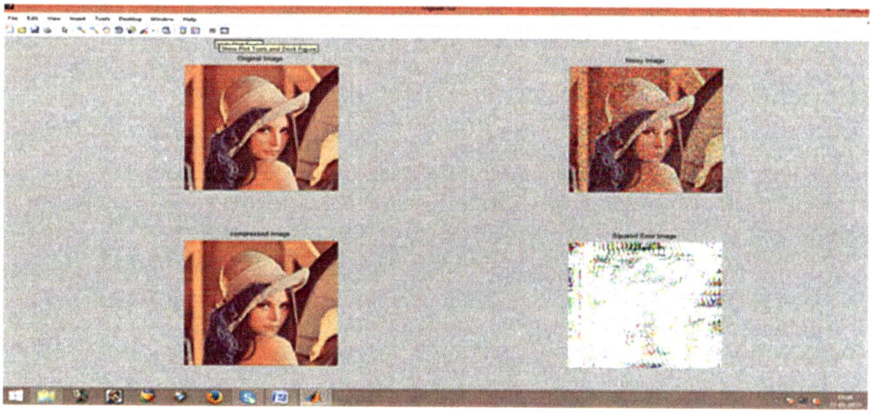

Fig. 12 Compressed image

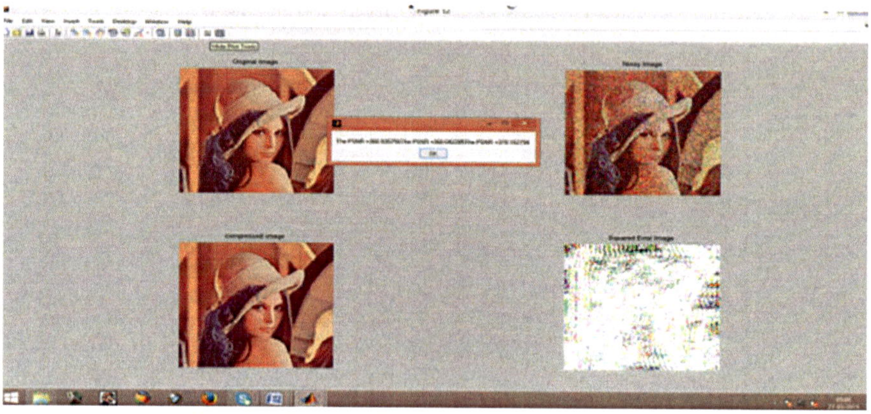

Fig. 13 PSNR calculation

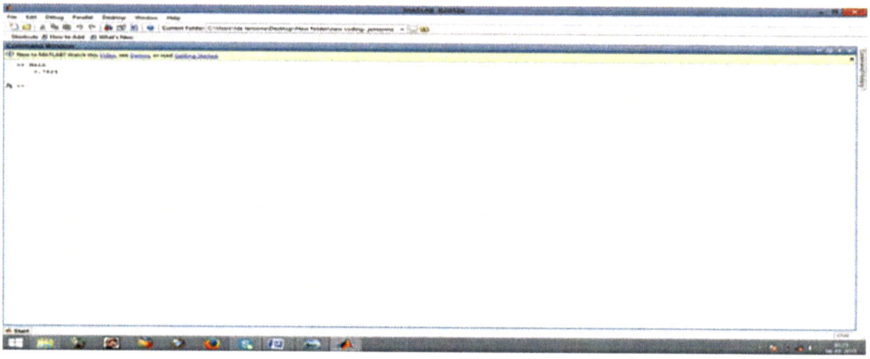

Fig. 14 Compression ratio

4.2 Results and Discussion for JPEG 2000

In this section the proposed loss less color image compression method using Hierarchical prediction is compared with the existing compression method called JPEG 2000.

Figure 15a, b show the single level decomposition. In JPEG 2000 the image has been decomposed on wavelet decomposition technique with different levels of decomposition. Figure 16a, b shows the double level decomposition using 2-D wavelet transform. Figure 17 shows the transform encoding. Figure 18 shows the uniform quantization. Quantization is a methods for optimally reducing a large number scale into a smaller one. The quantized coefficients are then sequenced and loss lessly packed into the output bitstream. Figure 19 shows the zigzag reordering and runlength encoding. Zigzag reordering shuffles transformed samples in a zigzag pattern after transforms and quantization. Figure 20 shows the subplot of images.

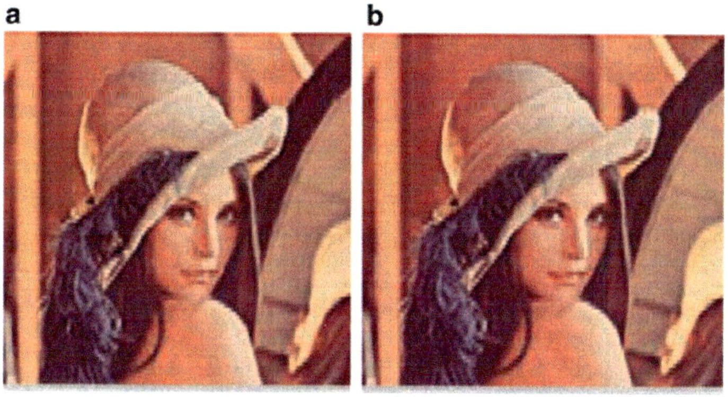

Fig. 15 **a** Single level decomposition input image **b** decompressed image

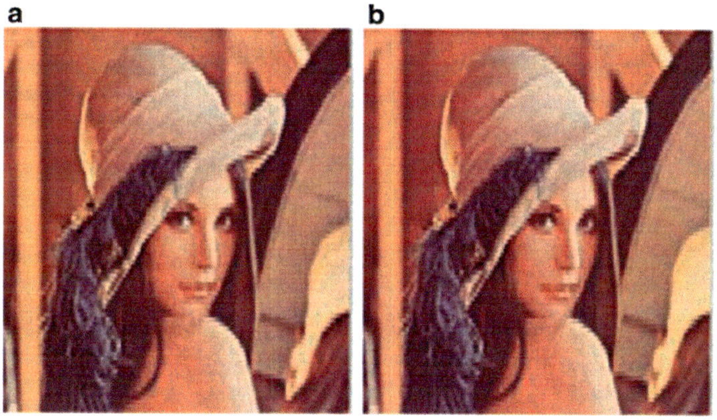

Fig. 16 **a** Double level input image **b** double level compressed image

Fig. 17 Transform encoding

Fig. 18 Uniform encoding

Fig. 19 Zigzag reordering and runlength encoding

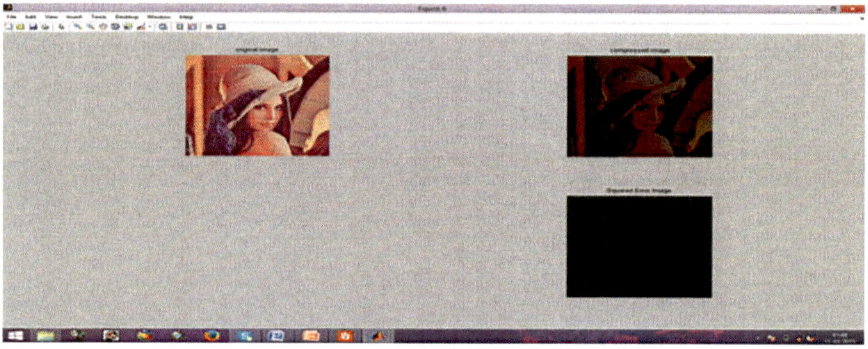

Fig. 20 Compressed images

Table 1 Comparison between proposed method and existing method

Parameters	Proposed method	Existing method
Peak signal to noise ratio (PSNR)	370.16	315.73
Compression ratio (C.R)	78	75

The original image occupies the first position. The final compressed image is shown in the second position. The final image shows the squared error image. Compression ratio is the ratio of the size of the original image to the size of the compressed image (Table 1).

5 Conclusion and Future Enhancement

The proposed system of the project is based on hierarchical prediction scheme and RCT technique. The reversible color transform is applied on the RGB image for the color transformation of the given image. The compression of the RGB image is done by de-correlating the original image by using RCT. After decorrelation of image, each Y component is encoded by convention less loss less grayscale image compression. The encoding of the chrominance image is done by a hierarchical scheme. The horizontal scheme uses the upper, lower pixels in the image. The compressed image obtained is loss less. The PSNR and compression ratio is measured for the compressed image. The coding is developed for JPEG 2000 also. The PSNR and compression ratio is calculated for the existing method JPEG 2000. Then both the measured values are compared with each other. From the comparison the proposed method shows better performance than the existing JPEG 2000 method. The future scope is to process with the MPEG standard in the same loss less compression field and provide better performance measurement.

References

1. H.S. Malvar, G.J. Sullivan, Progressive to loss less compression of color filter array images using macro pixel spectral spatial transformation, in *Proceedings of DCC* (2012), pp. 3–12
2. P. Roos, M.A. Viergever, M.C.A. van Dijke, J.H. Peters, Reversible intra frame compression of medical images. IEEE Trans. Med. Imag. **7**(4), 328–336 (1998)
3. K.-H. Chung, Y.-H. Chan, A loss less compression scheme for Bayer colour filter array images. IEEE Trans. Image Process. **17**(2) (2008)
4. W. Liu, W. Zeng, L. Dong, Q. Yao, Efficient compression of encrypted gray scale images. IEEE Trans. Image Process. **19**(4) (2010)
5. N. Zhang, X. Wu, Loss less compression of color mosaic images. IEEE Trans. Image Process. **15**(6), 1379–1388 (2006)
6. S. Pei, J. Ding, Improved reversible integer to integer color transforms, in *Proceedings of 16th IEEE ICIP* (2009), pp. 473–476

7. S.-T. Hsiang, J.W. Woods, Embedded image coding using zero blocks of sub band/wavelet coefficients and context modelling, Troy, NY 12180–3590.
8. N. Muthukumaran, R. Ravi, The performance analysis of fast efficient loss less satellite image compression and decompression for wavelet based algorithm. Wirel. Pers. Commun. **81**(2), 839–859 (2015)
9. Z. Lei, X. Wua, P. Baob, Real time loss less compression of mosaic video sequences, in *Real-time Imaging*, vol. 11, ed. by L. Zhang et al. (2005), pp. 370–377
10. T. Strutz, Adaptive selection of color transformations for reversible image compression, in *Proceedings of the 20th European IEEE Signal Processing Conference* (2012), pp. 1204–1208
11. X. Wu, N. Memon, Context-based, adaptive, loss less image coding. IEEE Trans. Commun. **45**(4), 437–444 (1997)
12. M. Weinberger, G. Seroussi, G. Sapiro, The LOCO-1 loss less image compression algorithm: principles and Standardization into JPEG-LS. IEEE Trans. Image Process. **9**(8), 1309–1324 (2000)
13. H.S. Malvar, G.J. Sullivan, S. Srinivasan, Lifting-based reversible color transformations for image compression, in Proceedings of SPIE, vol. 707307 (2008), pp. 1–10
14. Z. Mai, P. Nasiopoulous, R. Ward, A wavelet based intra prediction loss less image compression scheme, in *Proceedings of International Conference on Consumer Electronics* (2009), pp. 1–2.

Design and Development of Web-based Photoplethysmogram Signal Monitoring and Human Vital Parameters Measurement

W. S. Nimi, P. Subha Hency Jose, and R. Jegan

1 Introduction

In healthcare network, the usage of noninvasive device has been increased drastically due to monitoring patient health condition, measuring pulse rate, and oxygen saturation, etc. Therefore, the development and analysis of pulse sensor becomes very important which leads to accurate biomeasurement. In many cases, photoplethysmogram (PPG) signal has been widely considered for monitoring patient health condition [1]. The various researchers from different domains of science have become increasingly interested in PPG signal processing for vital parameters measurement because of its advantages as noninvasive, inexpensive, and convenient diagnostic tool for physiological parameter measurement [2]. The measurement of human physiological parameters such as pulse rate and systolic and diastolic blood pressure has been considered most important to analyze the heart-related issues. Recently, the interest on developing the sensor which can be used in patient's home provides an effective solution to monitor patient chronic diseases at home than hospital-based analysis method [3]. Monitoring patient vital parameters such as pulse rate and blood pressure form PPG signal leads more essential to predict and identify the diseases such as cardiovascular problem, sleep apnea, and respiration problems.

The research reported that people suffering from cardiovascular diseases are increasing, nearly 17.9 million people died due to this problem in the year 2016 [4]. Therefore, there is a huge demand to develop a home-based biosensor as initial screening at home for monitoring health conditions. In recent years, the large numbers of medical devices have been developed with low cost for monitoring many vital parameters. Also, there is an evidence for such wearable devices used at home reduced

W. S. Nimi (✉) · P. Subha Hency Jose · R. Jegan
Department of Biomedical Engineering, Karunya Institute of Technology and Sciences, Coimbatore, India
e-mail: nimiwsrec@gmail.com

J. D. Peter et al. (eds.), *Disruptive Technologies for Big Data and Cloud Applications*, Lecture Notes in Electrical Engineering 905,
https://doi.org/10.1007/978-981-19-2177-3_21

the mortality rate because of timely monitoring of patient health [5]. It is very important to develop a device with multi-parameter monitoring capabilities and extend to monitor daily life activities. This can be done with support of developing an efficient algorithm to estimate the vital parameters.

The signal processing algorithm plays a major role and provides a wide flexibility to biomedical signal for extracting the desired information at a particular interval for proper estimations. The previous research reported that the ability of various algorithms such as machine learning and advanced signal processing techniques for successful detection of many diseases in advance and also for vital parameters measurement [6–16]. This will help the care taker to take necessary timely treatment procedure effectively. Due to the aging problems, the wearable devices with support of Internet of medical things and advance signal processing methods provide an alternative solution at home in terms of pre-hospital health monitoring [17, 18]. But, the care must be taken to develop light weight wearable devices which are essential to feel patient more comfort. It is also very important to design a device which consumes very low power. This will provide an efficient continuous long-term health parameter monitoring with very low power. The output response of these devices produced a real-time signals for possible detection of diseases. The designer should ensure the quality of the signal because of its importance to extract correct features. In many cases, the real-time signals have been affected by many kinds of noises which lead to wrong diagnosis. The wide presence of noises makes very difficult to process the signal. Therefore, proper techniques must be adopted in the design phase to maintain the quality of the signals. Researchers have been investigated the possible reasons for these noise signals which are coupled with original real-time signals [18–23]. Therefore, it is necessary to design a system with support of advanced signal processing methods to measure and estimate the patient vital parameters.

The proposed paper targets the development of reliable smart system which consists of PPG sensor for easy acquisition of real-time signal to measure physiological parameters. Also, this paper provides estimation of blood pressure from the features of PPG signal. Here, sensor-based physiological system has been developed for acquisition of a real-time PPG signals. The signal processing methods which include band-pass filtering and wavelet techniques have been studied and implemented for removing various noises, thereby improving the quality of the signal. The feature such as PPG peak and peak time is extracted, and the pulse rate and blood pressure are estimated. This paper comprises of three sections: Sect. 2 of this paper explains the system architecture for vital parameters monitoring, Sect. 3 provides experimental results and their discussions. Section 4 concludes the paper.

2 Materials and Methods

The graphical representation of the proposed system architecture is shown in Fig. 1. The main modules of this system include PPG sensor, an amplifier, and programming environment to process and extract the desired information for vital parameters measurement.

2.1 System Architecture

The sensing module of the proposed system is made up of PPG sensor for sensing the pulse rate of physiological signal. The pulse rate of the patient is measured using PPG sensor, which works on the principle of light-based technology. It consists of LED and photodetector for sensing pulse signal of the patient. The low-power amplifier is used to amplify the signal which can be processed by LabVIEW environment through myDAQ analog device.

2.2 Signal Acquisition and Its Representation

The recording of PPG signal with PPG sensor has been an integral part of medical engineering for measuring patient's vital parameters. The comprehensive literature review in this research incorporated the use of PPG sensor for acquiring the pulse signal. A PPG is a standard method used for the acquisition of photoplethysmogram

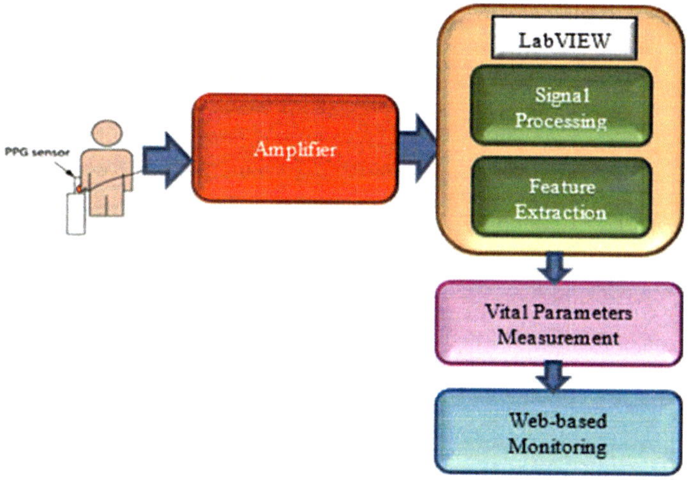

Fig. 1 Proposed system architecture

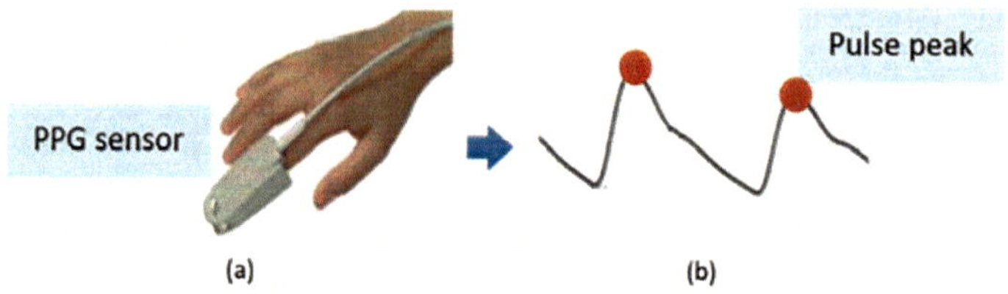

Fig. 2 Schematic representation of pulse sensor: **a** PPG sensor **b** PPG signal

signal. PPG sensor is used to illuminate the skin, and the discrepancy of the light absorbed is estimated for obtaining the PPG signal in real time. The waveform obtained using PPG signal shows the variation in blood volume for every cardiac cycles. Photodetector is used to estimate this variation in blood volume. In this research, PPG signal is acquired from the index finger. Figure 2 shows the typical PPG sensor and their output signal representation.

2.3 *Signal Processing and Vital Parameters Measurement*

Advanced signal processing techniques are incorporated with the acquired PPG signal. While acquiring the PPG signal in real time, it may also include baseline drift due to the occurrence of noise signal during signal acquisition. Low-pass filtering is applied to the signal to eliminate the baseline drift. Also, wavelet-based analysis method is applied on the signal for feature extraction. The advantage of using wavelet-based signal analysis approach is that it provides time and frequency information of the real-time signal. The features such as PPG peak and peak time are extracted from the signal for measuring the pulse rate and estimating the blood pressure. In pulse rate measurement, peak detection approach is designed to count the number of PPG peaks for 60 s. In blood pressure measurement, regression approach is used. PPG signals from 30 subjects have been acquired, and the system is trained with peak voltage extracted from the signals. The subject's PPG signal peak voltage and their standard systolic and diastolic blood pressure values are used in this paper to form the regression equations to estimate blood pressure.

2.4 *Web-based Vital Parameters Monitoring*

In healthcare network, monitoring of patient signal becomes very much essential because of early detection of various diseases and patient health status. In case of critical patients, the 24 h monitoring is very crucial. The majority of the biomedical equipment utilized in the hospitals are very difficult to monitor from remote areas by the physicians due to insufficiency of wireless monitoring enabled devices. This paper proposes an online web server which can be used to receive the vital physiological parameters remotely from a medical device connected with the patient. Through this web server, the client requests are being served on a real-time basis.

3 Simulation Results and Performance Analysis

The measurement of pulse signal interfered with the noise signal generates low peak voltages. Noises due to the power line interference can also be included in the data acquisition system. The peak generated with the power line interference is tremendous. To eliminate such noises, appropriate preprocessing is to be carried out. Analog front-end system plays the major role in the measurement system that comprises the data acquisition unit and the processing unit. In this paper, the analog device and the software threshold-based methods are proposed for human physiological signal recording, preprocessing the signal, and for continuous monitoring. The signal should be preprocessed, and the features are extracted accurately for the estimation of pulse rate and blood pressure.

The proposed method is implemented using LabVIEW environment for signal analysis. For the performance evaluation of the proposed method, real-time PPG signals from the real subject are used. The PPG sensor is properly placed at correct location of human body (finger), and the signals are acquired from different subjects.

The PPG signal is acquired in real time from 30 subjects using PPG sensors as shown in Fig. 3. In order to ensure the quality of the signal, the band-pass filtering and wavelet detrend are applied on the signal to remove the noises and baseline drift of the signal as shown in Fig. 4, and smoothening is performed by moving average filter as shown in Fig. 5.

After filtering, the signal is reconstructed using the wavelet decomposition. Wavelet decomposition is done using Daubechies 4 wavelet, and the decomposition process is carried out in four levels. The reconstructed signal is as shown in Fig. 6. The threshold-based peak detection method is implemented to detect the peak of the signal.

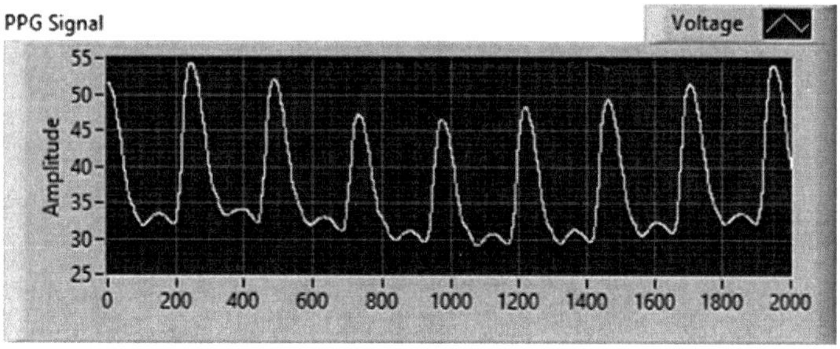

Fig. 3 Real time acquired PPG signal

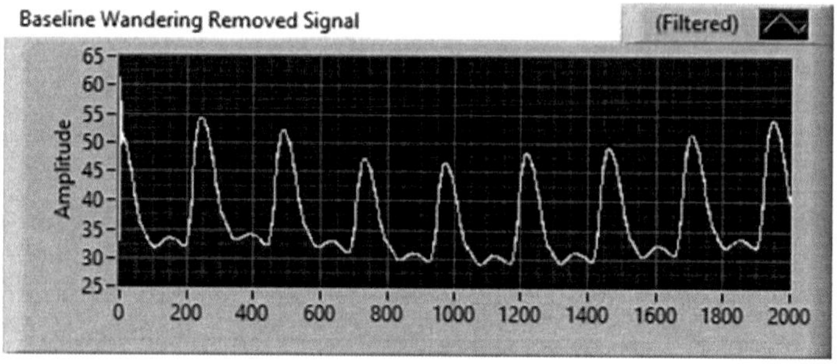

Fig. 4 Baseline wanderinge removed PPG signal

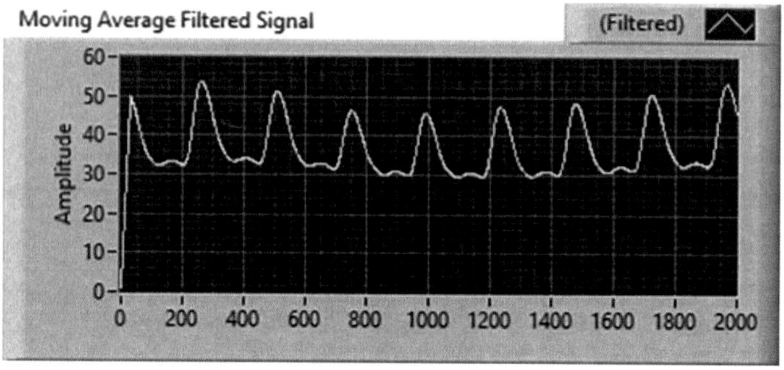

Fig. 5 Moving average filtered signal

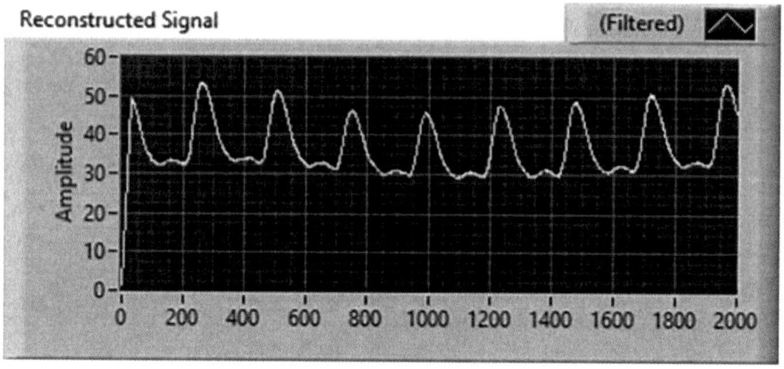

Fig. 6 Reconstructed signal

Comparison is made between the filtered signal, and the reconstructed signal is shown in Fig. 7. PPG signal peak detected in real time is shown in Fig. 8. From the results, it was shown that wavelet analysis method is an optimal method for processing biomedical signals.

The web-based PPG signal is displayed to the user and the physician as shown in Fig. 9 for real-time monitoring. The number of peaks and peak amplitude are

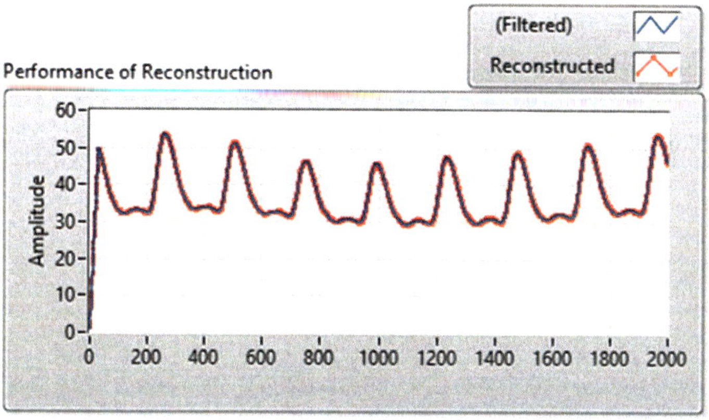

Fig. 7 Comparison of filtered and reconstructed signal

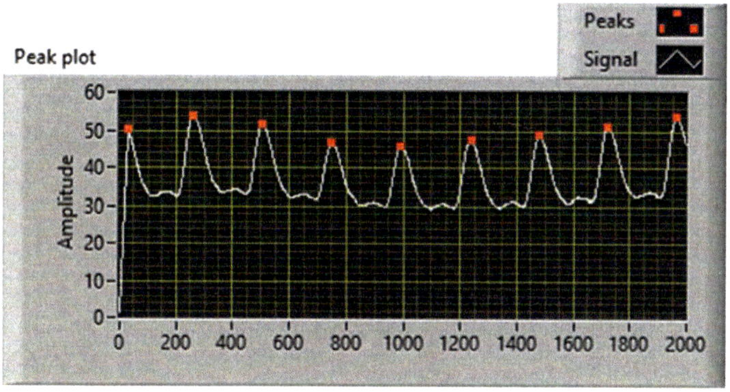

Fig. 8 Real-time PPG signal peak detection

further used to estimate the blood pressure through regression principle by forming differential equations. The measurement of blood pressure is done based on offline analysis. The signals from 30 subjects have been collected, and their features such as number of peaks and peak amplitude are recorded. These features are correlated with their original systolic and diastolic blood pressure which have been done with commercially available digital blood pressure device.

Table 1 shows the performance assessment of pulse rate and the systolic and diastolic blood pressure measurement. Blood pressure is estimated in terms of the systolic blood pressure (SBP) and the diastolic blood pressure (DBP). The comparison of the performance of the proposed method is analyzed by digital hospitalized blood pressure device. The number of peaks of PPG signal for 1 min is calculated. The proposed system is designed and considered as abnormal pulse rate when the number of PPG peaks goes below 70 bpm and exceeds beyond 100 bpm. The major application of the proposed method is real-time monitoring of persons pulse rate and blood pressure from home. By using this proposed method, the pulse and blood pressure can be easily monitored.

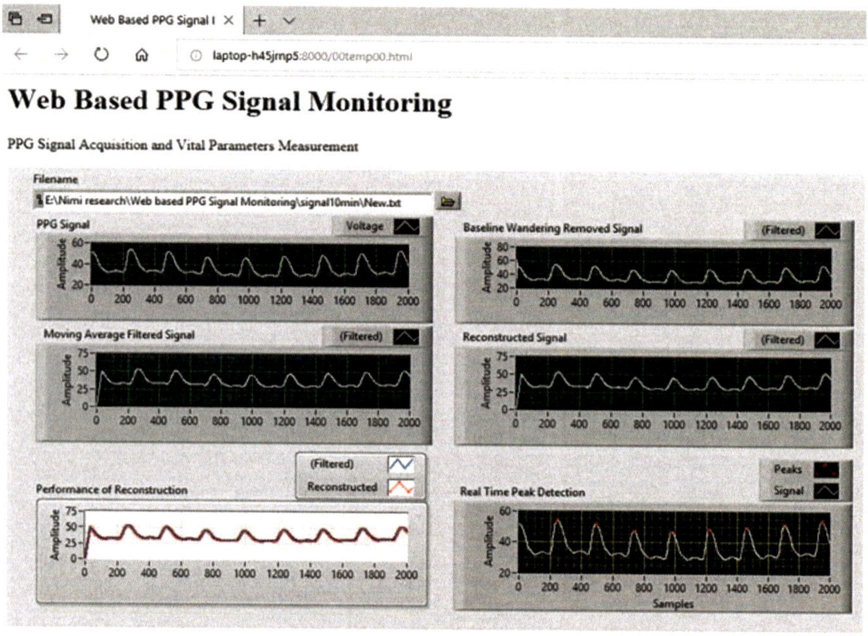

Fig. 9 Web-based real-time monitoring of PPG signal

Table 1 Performance assessment of proposed method

S. No	Pulse Rate (bpm)		Error %	Measured Blood Pressure (mmHg)		Estimated Blood Pressure (mmHg)		Error %	
	Standard Value	Estimated Value		SBP	DBP	SBP	DBP	SBP	DBP
1	77	76	-1.31	128	83	125.4	86.6	-2.07	4.15
2	74	75	1.33	129	84	127.5	86.7	-1.17	3.11
3	72	75	4.00	123	84	128.6	87.2	4.35	3.66
4	76	78	2.56	125	85	127.4	86.7	1.88	1.96
5	73	76	3.94	125	84	128.4	87.1	2.64	3.55
6	76	79	3.79	123	84	127.3	86.6	3.37	3.00
7	77	82	6.09	126	84	127.5	86.6	1.17	3.00
8	75	79	5.06	122	84	127.4	86.7	4.23	3.11

4 Conclusions

The PPG sensor-based measurement system is proposed, and advanced signal processing method is applied on the PPG signal for pulse rate and blood pressure measurement. The performance of proposed method is evaluated with standard hospitalized device. Based on the experimental results, it is concluded that the proposed method exhibits a maximum of 6.09% of error rate on pulse rate measurement and a maximum error percentage of 4.35% and 4.15% for systolic and diastolic blood pressure, respectively. The proposed method is based on noninvasive approach and can be used at home for monitoring patient's health conditions remotely. The integration of web-based technologies with the medical monitoring system effectively helps the physician for efficient remote monitoring and timely disease diagnosis.

References

1. M. Blanco-Velasco, F. Cruz-Roldan, J.I. Godino-Llorente, K.E. Barner, Wavelet packets feasibility study for the design of an ECG compressor. IEEE Trans. Biomed. Eng. **54**(4), 766–9 (2007)
2. B. Sun, Z. Zhang, Photoplethysmography-based heart rate monitoring using asymmetric least squares spectrum subtraction and bayesian decision theory. IEEE Sens. J. **15**(12), 7161–7168 (2015)
3. S. Song, M. Konijnenburg et al., A 769 μW battery-powered single-chip soc with ble for multi-modal vital sign monitoring health patches. IEEE Trans. Biomed. Circ. Syst. **13**(6), 1506–1517 (2019)
4. World Health Organization, Cardiovascular diseases (CVDs). (Online). Available: https://www.who.int/news-room/fact-sheets/detail/cardiovascular-diseases-(cvds) Accessed: 15 Nov. 2020
5. L. Bui, G.C. Fonarow, Home monitoring for heart failure management. J. Amer. College Cardiol. **59**(2), 97–104 (2012)
6. M.R. Ram, K.V. Madhav, E.H. Krishna, N.R. Komalla, K. Sivani, K.A. Reddy, ICA-based improved DTCWT technique for MA reduction in PPG signals with restored respiratory information. IEEE Trans. Instrum. Meas. **62**(10), 2639–2651 (2013)
7. R. Jegan, W.S. Nimi, Sensor based smart real time monitoring of patients conditions using wireless protocol. Int. J. e-Health Med. Commun. (IJEHMC) **9**(3), 79–99 (2018)
8. R. Jegan, K.V. Anusuya, Real-time ECG peak detection for heart rate measurement using wavelet packet transform. Int. J. Biomed. Eng. Technol. **19**(3), 244–254 (2015)
9. P.D. Chazal, M. O'Dwyer, R.B. Reilly, Automatic classification of heartbeats using ECG morphology and heartbeat interval features. IEEE Trans. Biomed. Eng. **51**(7), 1196–1206 (2004)
10. R.J. Nimi, Sensor based biomedical frame work for monitoring patient vital parameters, in *IEEE International Conference on Circuits and Systems* (2020), pp. 96–100
11. A. Hina, W. Saadeh, A noninvasive glucose monitoring SoC based on single wavelength photoplethysmography. IEEE Trans. Biomed. Circ. Syst. **14**(3), 504–515 (2020)
12. R. Jegan, V.H. Ashwin, P. Rajalakshmy, P.S.H. Jose, EEG based internal stress observation: an overview of methods involved from EEG acquisition to system design. J. Critical Rev. **7**(9), 947–952 (2020)
13. S. Acharya, D. Swaminathan, S. Das, K. Kansara, S. Chakraborty, D. Kumar, T. Francis, K.R. Aatre, Non-invasive estimation of hemoglobin using a multi-model stacking regressor. IEEE J. Biomed. Health Inf. 1717–1726 (2020)

14. R. Jegan, W.S. Nimi, Low cost and improved performance measures on filtering techniques for ECG signal processing and TCP/IP based monitoring using LabVIEW, in *2017 4th International Conference on Advanced Computing and Communication Systems (ICACCS)* (Coimbatore, 2017), pp. 1–7
15. M.A. Motin, C.K. Karmakar, M. Palaniswami, PPG derived heart rate estimation during intensive physical exercise. IEEE Access **7**, 56062–56069 (2019)
16. R.J. Melvin, Performance measures on blood pressure and heart rate measurement from PPG signal for biomedical applications, in *Proceedings of IEEE International Conference on Innovations in Electrical, Electronics, Instrumentation and Media Technology* (2017), pp 31–35
17. A.R. Jegan, Embedded based patient's vital sign monitoring system using wireless methods. Int. J. Eng. Res. Technol. **3**(4), 835–837 (2014)
18. S. Vishnu, S.R.J. Ramson, R. Jegan, Internet of medical things (IoMT)-an overview, in *2020 5th International Conference on Devices, Circuits and Systems (ICDCS)* (2020), pp. 101–104
19. T.H. Huynh, R. Jafari, W.Y. Chung, Noninvasive cuffless blood pressure estimation using pulse transit time and impedance plethysmography. IEEE Trans. Biomed. Eng. 967–976 (2019)
20. R. Jegan, K.V. Anusuya, Heart signal monitoring for remote patient assistant based on optical remote sensing. Optoelectr. Adv. Mater. Rapid Commun. **9**(5), 570–574 (2015)
21. R. Jegan, K.V. Anusuya, High-performance ECG signal acquisition for heart rate measurement. Int. J. Biomed. Eng. Technol. **12**(4), 371–381 (2013)
22. R. Jegan, K.V. Anusuya, Bio sensor based feature extraction and physiological parameters measurement for bio medical applications. Int. J. Biomed. Eng. Technol. **28**(1), 67–80 (2018)
23. R. Jegan, P.S.H. Jose, P. Rajalakshmy, P.A.C. Raj, H.J. Kanmani, W.S. Nimi, Methodological role of mathematics to estimate human blood pressure through biosensors. Int. J. Integr. Eng. **13**(1), 240–248 (2021)

Smart Fetal Health Monitor

J. Anu Shilvya and P. Subha Hency Jose

1 Introduction

Every year, over 2.8 million pregnant women and newborns die or one every 11 s. According to UNICEF, the World Health Organization (WHO), the United Nations Population Division, UNFPA, and the World Bank Group, recent mortality numbers were provided by UNICEF, the World Health Organization (WHO), the United Nations Population Division, UNFPA, and the World Bank Group. The vast majority of them can be avoided [1]. Both the mother and fetus appear vulnerable during the period leading up to and following birth. Complications that can lead to maternal mortality in expectant mothers can happen any time throughout pregnancy or childbirth, without any kind of indication.

Almost most of these deaths occurred due to the lack of resources. Women die because of many complications that happen during pregnancy and delivery. These complications are curable if happened to diagnose them during the pregnancy itself. Some problems are there in the women's body even before pregnancy itself. They get worsen during the pregnancy due to lack of proper care and facilities. According to researchers, severe bleeding (typically after childbirth), infections (usually after childbirth), high blood pressure during pregnancy (preeclampsia and eclampsia), delivery problems, and botched abortion result for about 75% of all maternal deaths [2].

Poverty, remoteness, a lack of information, insufficient resources, and cultural customs are just a few of the other factors that contribute to pregnant women receiving

J. Anu Shilvya (✉) · P. Subha Hency Jose
Department of Biomedical Engineering, Karunya Institute of Sciences and Technology, Coimbatore, India
e-mail: anushilvya@karunya.edu.in

P. Subha Hency Jose
e-mail: hency20002000@karunya.edu

J. D. Peter et al. (eds.), *Disruptive Technologies for Big Data and Cloud Applications*, Lecture Notes in Electrical Engineering 905,
https://doi.org/10.1007/978-981-19-2177-3_22

reduced care. So, the requisite measures should begin with delivering effective and high-quality health care to women who are pregnant, which will result in the delivery of healthy babies.

Generally, pregnant women are advised to undergo ultrasound examinations at least twice during the pregnancy to learn about fetal growth. Furthermore, effective and timely health assessments will ensure safe childbirth for the women. Rural women are unaware of the importance of appropriate medicines. India has made significant strides in enhancing the health of its citizens, but it is still far from fulfillment. Availability to a healthcare facility with advanced maternity services is critical for rural women's pregnancy outcomes. Women in rural communities have a high death rate due to a lack of awareness. Also, they sometimes suffer from anemia, fatigue, nausea, migraines, and dizziness, among other health problems.

Generally, doctors are using ultrasound scanning to monitor the development and health of the fetus in the womb of the maternal. Chances of miscarriage, pregnancy confirmation, development of anomalies, multiple pregnancies, and many other complications can be diagnosed by ultrasound scanning. But this type of scanning method is not much cost-effective, and also some studies show that long-term exposure to ultrasound waves can cause serious side effects. This could be the reason why they are not suggested for an extended duration of scanning. Therefore, we have come up with a model that uses some of the latest sensors that may not cause any harm to both the fetus and the maternal.

2 Related Works

Alexandre et al. [3] developed a system that monitors mother and fetal signals incorporating Internet of Things (IoT) integrated sensors and convolutional neural network (CNN) classifier. Various parameters like blood pressure, uterine tonus activity, heartbeat rate, temperature, and saturation of oxygen were measured and produced an accuracy of 92.59% in pregnancy emergencies. Zhao et al. [4] developed an IoT-based wearable system for fetal health monitoring with accelerometers and machine learning. The unit was build using Arduino and medical sensors along with an alarm to alert if the fetal movements are weakened. It also used a cloud computing platform to collect and store medical data for research purposes. Kumaresh et al. [5] developed an android application in which the maternal and fetal heart rate can be transmitted using heartrate sensor and Bluetooth technology. The parameters are also used to alert the hospitals in case of emergency through the hospital URL.

Ramla et al. [6] proposed a method to predict the high-risk pregnancy using a decision tree algorithm called CART. For classification, a CTG dataset with 2126 observations from the UCI machine learning repository was used. The conceptual CART methodology was computed using precision, recall, and F-score after undergoing fivefold classification algorithm. Saniya Ansari et al. [7] briefly summarized the various technologies available in the market for fetal health monitoring and came up with a simplified model using low-cost sensors and Arduino. The parameters like heart rate, fetus heart rate, blood pressure, and temperature of the maternal were measured and displayed on the LCD display. The data was also sent to the doctors so that prescriptions can be provided based on the health conditions. Zhihua Wang et al. proposed a wireless intelligent sensor-based system for fetal heartrate that is obtained through body sound monitoring. It could measure the fetal and maternal heart rate through the sound monitoring of the maternal [8].

Roham et al. [9] developed a wireless wearable system that could measure fetal heart rate and uterine contractions. Doppler ultrasound transducer and tocodynamometer were used to implement the model. The system produced a concordance correlation coefficients (CCC) of 88%. Glenda Puco et al. [10] proposed a system that could detect and control preeclampsia in pregnant women. They integrated the model using Arduino and Wi-Fi technology to monitor blood pressure and proteinuria of the maternal. It produced good results with low error rate and was developed at a comparatively low cost.

Motika et al. [11] developed a wireless fetal heart rate monitoring system with ZigBee technology and IEEE 802.15.4 standard. They used Doppler probe to measure the heart rate of the fetus. The signals were sent over the ZigBee network and displayed on user screen for manipulation of the data. Troyee et al. [12] came up with a model that could measure several physiological parameters such as anemia and glucose rate detection, sudden fall detection, heart rate, and body temperature of the maternal. The system was developed using Arduino and Wi-Fi, and GSM module was incorporated to alert nearby hospitals in case of emergency. Ertuğrul et al. [13] proposed a fetal heart rate monitoring system with a mobile phone integrated Doppler device. The research used ten sets of fetal heart rate instance values taken from the last ten gestation weeks of labor. Electrofetal monitoring algorithm was implemented to produce effective results (Table 1).

2.1 Novelty of Our Proposed System

The novelty of this paper is the use of IoT technology with embedded system to monitor the physical parameters and transfer the measured values to the smart phone devices. As far as we are concerned, we have discovered that measurement of various parameters of maternal is not present in any of the literature. Most papers emphasis on a single parameter such as measuring heart rate or any particular physical parameter. We are certain that our suggested study is an utterly novel attempt since we presented a system that would monitor and detect various abnormalities in the pregnant women.

Table 1 Comparative study with existing scheme

Author	Methodology applied
Alexandre et al. [3]	Convolutional neural network (CNN) was used to develop a system that monitors parameters of pregnant women
Zhao et al. [4]	A wearable system was developed for pregnant women by implementing machine learning into the system
Kumaresh et al. [5]	Developed an android application that could monitor the heart rate of the mother and fetus
Ramla et al. [6]	Decision tree algorithm and UCI machine learning is used in this paper to predict the high-risk pregnancy
Wang et al. [8]	Proposed a system that measures the fetal and maternal heart rate through the monitoring the sound of the maternal body
Roham et al. [9]	Doppler ultrasound transducer and tocodynamometer is used to detect fetal heart rate and uterine contractions in the developed model
Puco et al. [10]	Developed a model using Arduino that could measure preeclampsia in pregnant women
Motika et al. [11]	ZigBee and IEEE 802.15.4 standard are used to develop a wireless heart rate monitoring system

3 Materials and Method

3.1 Hardware Setup

A variety of sensors are used to assess the important parameters of the human body in the hardware setup. Body temperature, blood pressure, heart rate, and the movement of the baby in the mother are among the parameters. Sensors are connected to the device and used to view the parameters that have been recorded. IoT has been incorporated in the system to make remote health care possible. The parameters can be transmitted to the health specialists for real-time monitoring of the patients.

3.2 Heart Rate Sensor

The frequency of the cardiac cycle is defined by the heartbeat rate of a person. One of the four vital signs is the pulse. It is measured by the number of heart contractions per minute and is expressed as "bpm" (beats per minute). During pregnancy, the cardiac output which is the amount of blood pumped by the heart has chances to increase by 30–50%. When cardiac output increases, the heart rate would be at about 70 bpm to 80 or 90 bpm. In this system, uses a microcontroller-based heartbeat sensor is used. It uses infrared (IR) radiation and works on the concept of light modulation by blood flowing through the finger for every pulse (Fig. 1).

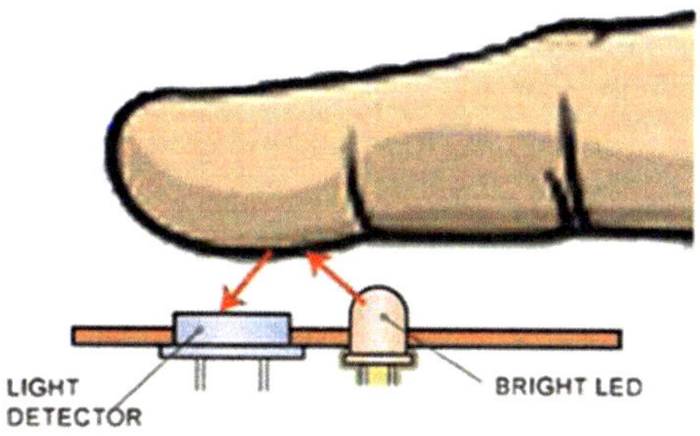

Fig. 1 IR sensor working principle [14]

Module is made of a light detector and LED of bright red color. Device is placed at the fingertip, and when the reflected light falls on the light detector, a signal is produced. This signal varies when blood flows through the vessels because the finger becomes opaquer. This variation in the signals determines the heartbeat rate.

3.3 Blood Pressure Sensor

The next major concern for all human beings is blood pressure. Blood pressure should be maintained normal for the maternal. The reason is that for all pregnant women, the blood pressure will keep altering moderately during pregnancy period. This also makes them feel tired throughout the 10 months. Observation of blood pressure is crucial because delivery of oxygen and essential nutrients to the infant is affected by variations in blood pressure. In most cases, high blood pressure during pregnancy does not cause significant complications. But, hypertension, on the other hand, if left unchecked, may be fatal to both the maternal and the child.

There is a risk of severe complications if elevated blood pressure remains same beyond 20 weeks of pregnancy. Preeclampsia is a condition that may occur during pregnancy. The brain and kidneys, among other organs, can suffer severe damage as a result of this disease. Toxemia of pregnancy-induced hypertension is another name for preeclampsia. As a result, it is crucial to monitor the maternal blood pressure.

3.4 Accelerometer Sensor

The measuring instrument sensor is among the most popular mechanical phenomena detectors. Accelerometer sensors can detect movement along one, two, or three orthogonal axes. The fetus' movements are essentially determined by the uterus's vascular condition of placental insufficiency. The term used for these actions is kicking [15]. The baby will begin kicking about the fourth month, but the mother

Table 2 Measurement criteria for fetal counts

Duration hours	Fetal count (normal)	Fetal count (abnormal)
1	≥3	≤2
2	≥6	<6
12	>10	<10

will be oblivious to it. But clinicians can anticipate the prenatal condition and development by measuring fetal movement. Ultrasound scans are employed to monitor fetal movement, but they are costly. As a result, an accelerometer sensor is utilized.

The ADXL335 is a small, thin, and low-power three-axis sensor with signal-conditioned voltage outputs [16]. The sensor has a comprehensive variety of 3 g and can monitor acceleration. X_{OUT}, Y_{OUT}, and Z_{OUT} are the three voltage output pins of accelerometer sensors. Bandwidths are chosen based on the system's intended use.

The frequency range of the X- and Y-axes is 0.5–1600 Hz, while the frequency range of the Z-axis is 0.5–550 Hz [17]. ADXL335 comes with a 16-lead, low-profile plastic lead frame chip box. The fetal movement is measured using a wireless accelerometer sensor. The output of the accelerometer is communicated to controller via RF modules transmitter and receiver.

Table 2 represents the fetus's kick count which is classified as normal or abnormality [2]. Fetal counts begin during fourth month; yet, the mother would be unaware of it. Only in the fifth month will the mother be able to see the fetus move. It is considered an essential indication of prenatal hypoxia if the fetal movement is reduced to less than 10 times in 12 h. As a result, early action should be done to avoid these problems.

3.5 *Temperature Sensor*

The maternal body temperature is determined by using temperature sensor. During pregnancy, a woman's body temperature is likely to fluctuate. Because of an increase in metabolism, heightened hormone levels, greater stress on the woman's body, and the processing of fetal nutrients, and waste products, the woman's body generates additional heat during pregnancy. Concurrently, dissipation of heat from the body of the woman is caused due to increased peripheral circulation.

Thermistor is a resistor that is measures changes in temperature by changing resistance as temperature changes. The term thermistor is formed by combining the phrases thermal and resistor. The LM35 sensor has a voltage range from 4 to 30 V and has a temperature range of −55 °C to +150 °C. If the resistance-to-temperature relationship is considered to be linear, then

$$\Delta R = k \Delta T \tag{1}$$

where

ΔR difference in resistance
ΔT difference in temperature
k first-order temperature coefficient of resistance

4 Internet of Things

Basically, IoT means Internet of Things. It describes the network of physical objects with the internet. It also refers to a network of interconnected, connected gadgets that can gather and transmit data across a wireless network without the need for human involvement [18]. Devices can create smart decisions using analytical rules, and this comes as a great advantage for IoT devices. Information may be sent, received, stored, and controlled by the devices, which can send it to another device or broadcast it to all devices.

IoT in health care plays a crucial role in providing better health facilities to both patients and doctors by eliminating the communication barriers within them. This has been widely used in hospitals to continuously monitor the patients in ICU and share the parameters with doctors. This aids in the detection of abnormalities in the patients and providing immediate support. Medical equipment has become smarter as IoT technologies have advanced; for example, sensors, actuators, microcontrollers, and boards have made it easier for doctors to access patient data and write prescriptions for their health without having to visit the hospital [19]. On the patient's side, it has considerably aided in stress reduction because patients no longer have to wait in lines in hospitals, and they can also send and receive prescriptions from their doctors via IoT-enabled health monitoring systems. This can be greatly supportive for pregnant women living in remote areas who finds it difficult to get regular checkups by visiting the hospitals located far from their place.

4.1 IoT Board

WEMOS D1 is an ESP8266 12E-based Wi-Fi development board. Its functions are similar to NODEMCU, but the hardware is designed to look like an Arduino UNO. The Boards Manager is the program that is used to set up the D_1 board for use in an Arduino environment. As a result, there is no need for a separate Arduino board because D_1 can be programmed directly using the Arduino IDE which is an open-source programming software. This is very much useful when working with IoT projects.

Its features include.

- 32-bit RISC CPU
- 64 kb instruction RAM
- 96 kb data RAM
- 80 MHz clock speed
- 4–8 MB flash memory.

5 System Description

In this methodology, four basic parameters are measured. They are body temperature, blood pressure, heartbeat rate, and movement of the fetus. The signals are obtained from sensors connected to the maternal body and transferred to the Arduino module. The data obtained from the sensors is transferred to the controller. The information is then analyzed, and results are simulated.

The Arduino UNO is a type of ATMEGA328-based microcontroller. There are 14 digital input/output pins and six analog input pins on this board (Fig. 2).

It also has a Universal Serial Bus (USB), a crystal oscillator with a frequency of 16 MHz, power jack, ICSP header, and reset button. The value that is sensed by the sensor which is analog, so it is linked straight to the analog input pin. Before any additional processing can be done, the Arduino microcontroller has an ADC that will convert the analog signal to digitalized signal. The Arduino produced a digital output signal after further processing. To develop code and upload it to the board, an open-source Arduino software (IDE) is utilized.

IoT is deployed in this system for real-time monitoring of maternal. The parameters that were measured are then transmitted to smart phones that are connected by

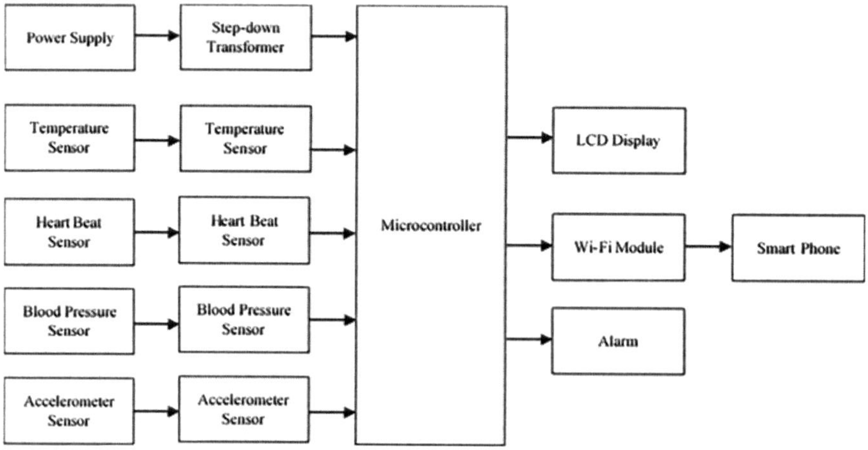

Fig. 2 Block diagram [19]

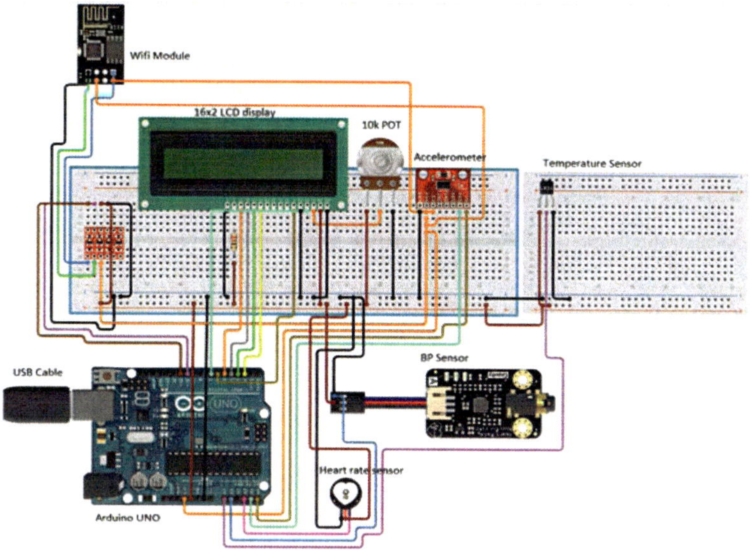

Fig. 3 Circuit design implementation

a Wi-Fi module. The results can be viewed on the mobile phone as well as on the LCD display of the measuring device.

5.1 Electrical Design Implementation

Figure 3 shows the electrical design implementation of smart fetal health monitor. The Arduino microcontroller is connected to the sensors such as heart rate sensor, accelerometer sensor, pressure sensor, and temperature sensor.

6 Result

The device is built using various sensors and a controller to measure body temperature, pulse rate, blood pressure, and movement of the fetus. The movement of the fetus is measured using a compact three axes accelerometer sensor. When the accelerometer sensor is placed on the abdomen of the maternal, the kick counts of the fetus can be detected, and counts are measured to determine the fetal health status. The results obtained are sent to the smart phone with the help of IoT. In addition to that, the results are displayed on the LCD screen connected to the module as shown in Fig. 4. The "HB" represents the real-time monitoring of the heartbeat of the maternal. This is measured through the IR sensor. In Fig. 3, the reading 84 represents the heartbeat

Fig. 4 Output on LCD display

Fig. 5 Output displayed in mobile phone

count and 20 or 34 denotes the time in seconds. "P" represents the blood pressure of the maternal.

All the abnormal conditions will be shown in mobile through IoT. If the temperature value increases or decreases from the normal value, there will be an indication of "ABNORMAL_TEMP". If the pressure value decreases or increases from normal value, there will be an indication of "ABNORMAL_PRESSURE". If the heartbeat value increases or decreases from normal value, there will be an indication of "ABNORMAL_HEARTBEAT". If there is abnormal condition in the position of the fetus it will indicate "ABNORMAL_POS".

Furthermore, the abnormality in the measured values produces an alarming sound through the buzzer enabled in the system (Fig. 5).

7 Discussion

Nine months are the average duration of pregnancy. Trimesters are known as every three-month period during pregnancy. The fetus develops and matures with each trimester. During each trimester, specialized prenatal tests are used to check both the mother's and the fetal's health [20]. Health experts may use contemporary technology to detect birth abnormalities, diagnose problems that could impact childbirth, and treat some fetal disorders before the baby is born.

Ultrasound is used to perform biophysical testing. Because ultrasounds are performed every trimester, it is sufficient if the patient attends the hospital once every three months. Aside from that, the patient must have a monthly health exam, which involves evaluating the mother's and fetus' breathing rates, heart rates, and ECG.

Though ultrasound screening is typically regarded to be safe when done by truly trained and skilled healthcare practitioners, it has some side effects, and the waves has the ability to affect the body biologically. It can sometimes heat the tissues which may result in some long-term impacts to the body. Moreover, long-term complications of these impacts are still unclear. The major drawback of the systems already in place they are expensive and cannot be used at home. It requires skilled professionals to carry out the procedure and also requires a comparatively higher time. Moreover, the scans are not available in rural areas which make pregnant women to travel long distances for the checkup.

An accelerometer sensor is made wireless in the suggested work and is primarily utilized to detect the condition of fetus, as well as certain critical indicators for women like body temperature, pulse rate, and blood pressure. Measured parameters are shown on the LCD display and sent through IoT to the mobile phone [21], where they may be examined. The suggested project entails creating a small assist device for rural pregnant women to monitor maternal and fetal vital signs at a cheap cost utilizing modern sensors and the Internet of Things for individualized treatment.

8 Conclusion and Future Enhancement

Studies show that most of the maternal death are caused in hospitals and are mainly due to poor facilities in the hospitals. But in countries like India and in other developing countries, maternal deaths are happening at home itself. This is due to lack of proper care during the pregnancy period of the women. To address this issue, a regular monitoring of the maternal and fetal is necessary. So, an assistive device is designed in a compact size that is portal and can be used by the maternal without the need of a caretaker. The device can monitor important physiological indicators such as the mother's body temperature, blood pressure, heart rate, and the fetus' movements. The parameters that are measured are displayed on the LCD display. It is also transferred to the mobile device using IoT and Wi-Fi module. This makes the results to be viewed by the person on the smart phone. In case of any abnormalities, an alert would be triggered by a buzzer that is fitted to the device.

In future enhancement, this project can be implemented in more places because it requires low power, low cost, and reduces the personal effort. Furthermore, the system can be enhanced to include other parametric measurements such as oxygen saturation, blood glucose detection, and many more.

References

1. Surviving birth: Every 11 seconds, a pregnant woman or newborn dies somewhere around the world. https://www.unicef.org/press-releases/surviving-birth-every-11-seconds-pregnant-woman-or-newborn-dies-somewhere-around. Accessed 15 May 2021
2. Maternal mortality. https://www.who.int/news-room/fact-sheets/detail/maternal-mortality. Accessed on 10 May 2021
3. J.A.L. Marques, T. Han, W. Wu, J.P. do Vale Madeiro, A.V.L. Neto, R. Gravina, G. Fortino, V.H.C. de Albuquerque, IoT-based smart health system for ambulatory maternal and fetal monitoring. IEEE Internet of Things J. https://doi.org/10.1109/JIOT.2020.3037759. November 2020
4. X. Zhao, X. Zeng, L. Koehl, G. Tartare, J. de Jonckheere, K. Song, An IoT-based wearable system using accelerometers and machine learning for fetal movement monitoring, in *2019 IEEE International Conference on Industrial Cyber Physical Systems (ICPS)*. ISBN: 978–1–5386–8501–3. August 2019
5. S. Kumaresh, M. Sabareesh, R. Srihari, Non-invasive fetus heart rate and growth measurement with abnormality detection using IoT, in *2016 International Conference on Electrical, Electronics, and Optimization Techniques (ICEEOT)*. ISBN: 978-1-4673-9940-1. November 2016
6. M. Ramla, S. Sangeetha and S. Nickolas, "Fetal Health State Monitoring Using Decision Tree Classifier from Cardiotocography Measurements", 2018 Second International Conference on Intelligent Computing and Control Systems (ICICCS), June 2018
7. S. Ansari, M.B. Ansari, Smart health monitoring system for pregnant women. Int. J. Eng. Adv. Technol. (IJEAT). **9**(4) (2020). ISSN: 2249–8958
8. Z. Wang, H. Jiang, Wireless intelligent sensor system for fetal heart rate tracing through body sound monitoring on a pregnant woman, in *2013 IEEE MTT-S International Microwave Workshop Series on RF and Wireless Technologies for Biomedical and Healthcare Applications (IMWS-BIO)*, March 2014
9. M. Roham, E. Saldivar, S. Raghavan, M. Zurcher, J. Mack, M. Mehregany, A mobile wearable wireless fetal heart monitoring system. https://doi.org/10.1109/ismict.2011.5759813. May 2011
10. G. Puco, J. Pallo, C. Granizo, C. Nunez, P. Encalada, C. Gordon, Electronic system for detection and control of preeclampsia in pregnant women, in *2019 6th International Conference on eDemocracy & eGovernment (ICEDEG)*. https://doi.org/10.1109/icedeg.2019.8734368. June 2019
11. G. Motika, A. Prusty, Wireless fetal heartbeat monitoring system using ZigBee and IEEE 802.15.4 standard, in *2011 Second International Conference on Emerging Applications of Information Technology*. March 2011
12. T.G. Troyee, M.K. Raihan, M.S. Arefin, Health monitoring of expecting mothers using multiple sensor approach: preg care, in *2020 2nd International Conference on Advanced Information and Communication Technology (ICAICT)*. ISBN: 978-0-7381-2323-3. February 2021
13. D.Ç. Ertuğrul, H. Kanmaz, M.U. Yüksel, A. Elçi, M. Ertuğrul, Fetal heart rate monitoring system (FHRMS), in *2016 IEEE 40th Annual Computer Software and Applications Conference*. https://doi.org/10.1109/COMPSAC.2016.116. August 2016
14. P.A. Pawar, Heart rate monitoring system using IR base sensor & Arduino Uno, in *2014 Conference on IT in Business, Industry and Government (CSIBIG)*, pp. 1–3. https://doi.org/10.1109/CSIBIG.2014.7057005.
15. N.D. Zakaria, P.E. Numan, M.B. Malarvili, Fetal movements recording system using accelerometer sensor. ARPN J. Eng. Appl. Sci. **13**(3) (2018)
16. S. Shiny Amala, S. Mythili, IoT based health care monitoring system for rural pregnant women. Int. J. Pure Appl. Math. **119**(15), 837–843 (2018)
17. Probots, ADXL335 - Triple Axis Linear Accelerometer Module. https://www.probots.co.in/adxl335-triple-axis-linear-accelerometer-module.html

18. P.V. Dudhe, N.V. Kadam, R. M. Hushangabade and M. S. Deshmukh, Internet of Things (IOT): an overview and its applications, in *2017 International Conference on Energy, Communication, Data Analytics and Soft Computing (ICECDS)*, June 2018
19. S. Banka, I. Madan, S.S. Saranya, Smart healthcare monitoring using IoT, Int. J. Appl. Eng. Res. **13**(15), 11984–11989 (2018). ISSN 0973-4562
20. B.N. Lakshmi, T.S. Indumathi, N. Ravi, A novel health monitoring approach for pregnant women, in *2015 International Conference on Emerging Research in Electronics, Computer Science and Technology (ICERECT)*, ISBN: 978-1-4673-9563-2. June 2016
21. G. Wicahyono, A. Setyanto, S. Raharjo, A. Munandar, Pregnancy monitoring mobile application user experience assessment, in *2019 International Conference on Information and Communications Technology (ICOIACT)*. https://doi.org/10.1109/icoiact46704.2019.8938446. December 2019

Event Location Detection from Online Clustering Algorithms Using Geo-Tagged User Data in Social Streams

Bhuvaneswari Anbalagan

1 Introduction

Social media is the new source of updated user data where social media is considered as the new "Wikipedia" of Internet; then, we are correct as people are more active on the microblogging platforms as compared to any other forum or discussion community. Social media is broad term here as many social sites are all about data in form of images, videos, or music. For broadcasting events, microblogging networks are effective to reach masses and spread awareness. But, our major goal is detecting location of events, so we only need text-based data and geo-tags that can be analyzed to determine the disaster event location [1]. For the significant research findings, we have Twitter, a microblogging site that let people share their views in form of text. These short texts are often referred as tweets [2]. Twitter has around 390 million active users monthly, which clearly articulates that it is a very famous microblogging site among users, and they actively share information over it. Event detection using dynamic post influence is achieved by detecting the real-time event bursts at different time intervals on social media on survey on location recommendation in [3]. These tweets are related to multiple topics, events that happen around the globe and such information gets posted from multiple locations where real-time events are happening along with geo-tags. Conversations are topical and are categorized under some set standards like volume of content of a subject, real-time event occurrence and the natural classification by Twitter where it classifies user content as breaking news. During emergency crisis, events such as breaking news events covered are natural disasters, disease outbreak, protests, rallies, travel recommendation [4], trend analysis [5], and information dissemination [6].

B. Anbalagan (✉)
School of Computing Science Engineering, Vellore Institute of Technology, Chennai, Tamil Nadu, India
e-mail: bhuvana.cse14@gmail.com

J. D. Peter et al. (eds.), *Disruptive Technologies for Big Data and Cloud Applications*, Lecture Notes in Electrical Engineering 905,
https://doi.org/10.1007/978-981-19-2177-3_23

With all such events being covered and detected using social media data streams, it is easy to trace location of events. Also, when we are talking about such events, we also need to validate such events and tweets. These tweets are so powerful and reach out to the world within minutes. People who are affected or are a part of that event post their views and updates as soon as event occurs. Nowadays, these textual data streams (Tweets) reach out so soon that they surpass news and media agencies even. Event detection involves three major components hot events, user-generated post related to the event, and the influential spreader that can help track location and intensity of event. The expense of assortment of portability follow information progressively is diminishing with the headway of innovation, and these days, the accessibility of such information is expanding. Additionally, it is normal that in not so distant future, we would be effectively ready to gather ongoing portability follow information from various sources like cabs, transports, singular advanced mobile phones, and so on.

The remaining section of the paper is organized as follows: Sect. 2 presents the related work on user-behavior analysis using online social media. Section 3 discusses the online k-center clustering algorithm. Section 4 discusses the experiments and results demonstrate location detection techniques. Section 5 concludes the paper with significant future directions.

2 Related Work

The percentage of geo-tagged data, however, is limited compared to non-geo-tagged data, which often contains noises generated by automated robots, location spoofing, and human-made mistakes. Given these challenges, many researchers target to understand the difference in Twitter diffusion characteristics between geo-tagged and non-geo-tagged tweets. A research method [7] developed a geo-social event detection system, which intend to detect the indication of local events via Twitter by witnessing crowd behaviors. The abnormal geo-social events were detected by geographic uniformities inferred from the typical behavior patterns of crowds with a limitation of with geo-tags sparsity.

However, Xu et al. (2015) proposed a method to detect the distribution of user's interest from the travel history on weather condition and season. The probabilistic-based topic model is used to detect the user–user similarity of topic under discussion. Petrović, 2010 on streaming data on Twitter [8] and Flickr dataset was used which contains geo-tagged photos taken in 11 cities of China to determine crowd behavior [9]. Social Internet of Things enabled location framework using geo-tagged tweets and multimedia data [10]. Xu proposed a system head out suggestion motor to eliminate commotions, break down the inner construction of the dispersion interaction, and research the spatial circulation of geo-labeled tweets and their relationship with land-use types [11]. They gathered the tweets utilizing two watchwords "influenza" and film Ted were identified from four designated to portray assorted themes and geological regions in the United States. Various research outcomes deals with spatiotemporal examples of occasions utilizing online media: a contextual investigation of flu patterns [12], semantics-based delicate theme dispersion [13]. The textual and photo features are computed and trained using the microblog—density-based spatial clustering of applications with noise (DBSCAN) clustering algorithm, Naïve Bayes classifier, decision tree (DT), support vector machines (SVMs) classifier, and convolution neural networks (CNNs) classifier. The measure of data (e.g., literary messages, photos, and recordings) that is being produced today surpasses by numerous significant degrees beforehand accessible datasets, making conventional methodologies out of date for current occasion recognition.

Zhang uses a cross-modular master measure to recognize various rotate include in the inquiry window from geo-labeled tweet streams for ongoing neighborhood occasion location with less exactness [14]. Kogan et al. investigated how retweeting movement by the geologically weak varies from the overall Twitter populace. The Twitter-based connections were inspected with the recently shaped group was a direct result of the fiasco [15]. It was tracked down that everyone was retweeting neighborhood tweets and those with locally huge information in for the catastrophe support. Bakerman et al. exhibited a half and half technique that utilized Twitter substance and organization data as model highlights are discussed [16]. They utilized Gaussian blend models to plan the crude spatial circulation of the model highlights to an anticipated area in [16, 17].

Numerous articles center around online media use during different catastrophes occasion in the USA (e.g., 2001 9/11, 2005 Hurricane Katrina, 2012 Hurricane Sandy, 2013 Colorado Flood). Different examinations have given a more global scenery. Geo-labeling administrations support clients to name the media substance like content, photographs, and recordings created in the actual world. Each tweet is related with a period and area, which is a bunch of scope and longitude. Twitter based research work explained disaster, crisis, catastrophe, and emergency management are sometimes used synonymously and sometimes with slight differences, by scholars and practitioners [18]. Many published research papers have mainly focused on the use of Twitter, Facebook, and Weibo networks [19]. Based on the literature survey, an event location detection method for online streams using clustering algorithms is proposed in the research article.

3 Proposed Event Location Detection Method

An event occurs in online social network causes a large number of activities by the OSN users U. It is based on the observation of real-world happening in specific time and at exact location. Generally, the significance of an event E is evaluated by various parameters $E = \langle P_i \rangle$. The parameters P_i uniquely describe the event in multi-dimensions such as what (type) event is happening, who (people) are involved, when (duration) it is happening, where (location) it is happening, and how much (adverse effect) to nearby surroundings. The event detection is either a local or global characteristics depending upon its intensity. The textual information represents an event Rep(e) during a specific time stamp $\{T_{(A(e),\text{start})}, T_{(A(e),\text{end})}\}$ in a location $L(e)$. According to the proposed work, the event detection is defined as follows.

The problem of event location detection is defined as: "Given a stream of various activities $A(n)$ in an online social network, identify all possible attributes $E = \{e_1, e_2, \ldots e_k\}$, where there are K numbers of events." It is represented as $E_i = \{U(e_i), \text{Rep}(e_i), A(e_i), T(e_i), L(e_i)\}$. Given a set of activities $A(e) = \{a_1, a_2, \ldots a_N\}$, the process of identifying a real-world event is done using the following attributes, where $U(e)$—users involved in OSN for the event, Rep(e)—textual burst keywords which represents an event, $A(e)$—set of activities relevant to the event, $T_{(A(e))}$—time stamp for the set of activities between $T_{(A(e),\text{start})}$ and $T_{(A(e),\text{end})}$, $L(e)$—geographical location (latitude, longitude), where the event occurs. The geographic location is extricated from the arrangement of sources S, the time stamp T of the tweets and the legitimate location-based information available from a location-based assistance. The way toward recognizing and detailing of an event is seen is valuable, particularly on account of crisis circumstances like fiascos, for individuals getting early alerts to make suitable events. For instance, Geonames API is used in our research experiments to extract the location from the tweet text. The area is resolved decisively utilizing the greatest probability (aggregate likelihood) of a client who tweets about an area gave in the sources—client profiles, geo-labeled tweet substance, and person-to-person communication registration. The area is additionally granulated at the level, from geographic locale, mainland, country, state, city, time region, road, and milestone. The location passage can have a better degree of granularity like road or neighborhood-level directions.

Proposed online k-centric clustering algorithm

Input: Tweets, Location Threshold T, Geo-tags locations $\{G1, G2, \ldots Gn\}$
Output: Location Clusters $\{L1, L2, \ldots Ln\}$
Begin.

Step 1: Arbitrary select the Geographic Location G
Step 2: Assume G as cluster center
Step 3: Allocate all geo-tags within D distance of G to the cluster.
Step 4: Repeat Step 1 to 3.
Continue until none of the non-clustered geo-tags are left
Step 5: Select the cluster center as Location Cluster Li when
Number of geo-tags > T
Step 6: Return Location Cluster L = {Li **U** Li + 1}

End

We propose following methodologies for detecting location of crisis events: We plan the accompanying standard models to analyze against the proposed models for location identification as follows:

- **Arbitrary Method (AM)**. This technique arbitrarily chooses an event location from the rundown of areas alluded by the client or related with an occasion to be the anticipated location.
- **Frequency-Based Method (FBM)**. The most incessant area from related tweets is chosen as the location of event.
- **Online K-Center Clustering (OK-CC) Method**. For every tweeting user, we select the focal point of the circle with greatest area references inside the circle. The sweep of the circle is the conflict tolerating value and is same as *N* in Accuracy@*N* and 25 miles with value of *N* as zero. The location of a user from a given set of geolocations is intended to predict using the following steps involved in online k-centric clustering algorithm.

The proposed location extraction model in online streams concentrates location references of a client; the area distinguishing proof module recognizes the home location of the client. Location extraction model concentrates geolocations for all area references from DBPedia. The geolocation plot arranges on a Euclidean plane. It determines the area of a client from these directions. The region is considered incorporating these directions to be the dynamic space of the client. The assignment of area distinguishing proof is to anticipate the location area of the client from this dynamic territory. The computational mathematical is utilized and grouping ways to deal with anticipate the area of the client.

The location ID framework is comprised of three principle parts: location extraction from tweets, vagueness goal in area names, and area distinguishing proof of the client from extricated areas. To distinguish area-based sections from tweets, we use gazetteers to dole out a one of a kind identifier to every area-based passage. DBPedia, Geonames, and US evaluation are a portion of the huge scope gazetteers with a huge number of sections for states, urban areas, roads, and milestones. We use DBPedia as the principle asset for distinguishing area-based passages since it settle equivocal area passages that will be clarified in the following segment. We can match each location entry from text to multiple URIs at multiple granularity levels. A granularity level for a location can be street–neighborhood–city–state–country-level information. All locations are transformed to city-level location to retain the accuracy of location identification. A location entry can have a finer level of granularity like street or neighborhood-level coordinates.

The proposed location recognizable framework is comprised of three essential segments: Location extraction from tweets, vagueness goal in area names and location ID of the client from separated areas. To distinguish location-based sections from tweets, gazetteers are used to appoint an extraordinary identifier to every location-based passage. DBPedia, Geonames, and Indian geographical information system (GIS) evaluation are a portion of the enormous scope gazetteers with a large number of sections for states, urban communities, roads, and tourist spots. DBPedia is used in the implementation as the fundamental asset for recognizing area-based passages since it settle uncertain area passages that will be clarified in the following segment.

4 Experiments and Results

The Twitter tweets are collected as a scheduled process using Chorus Analytics by tweet catcher tool which uses Twitter API and stores it as a flat file. Tweet catcher device which uses Twitter API and stores it is anything but a level record. The main contribution of this research implementation is to develop an event detection framework that expedient the challenge of detecting events location in real time. Tweet catcher allows the online clients to move Twitter for explicit hunt information by burst catchphrases happening in Twitter conversations utilizing semantic information-driven approach and find Twitter network dependent on client-driven information. We use the following measures to evaluate the efficiency of location detection models. In the experiment, we used arbitrary method (AM), frequency-based method (FBM), k-center clustering (OK-CC). For each user, we select the center of the circle with maximum location references within the circle. The radius of the circle is the tolerance value and is same as N in Accuracy@N and 25 miles for tolerance value N as zero. Average error distance (AED) is the normal distance between the real areas to the anticipated area of a user with precision (ACC). The level of appropriately anticipated location detection and tracking location is at city level. The resilience esteem is zero (Tolerance value). Accuracy within N miles (ACCURACY@N) is the level of anticipated areas that are inside N miles of the real area. For instance, ACC@75 estimates the level of anticipated areas that are inside 75 miles of area from the user location profile.

The algorithm will coordinate with every location tweet from text to different URIs at various granularity levels. A granularity level for an area can be road–neighborhood–city–state–country-level data. Every location detail is changed to city-level area to hold the exactness of area recognizable proof. We use ACCURACY@25, ACCURACY@50, and ACCURACY@75 to compute precision inside 25, 50, and 75 miles, separately. N is the resilience esteem is processed and noted in Table 1. It is seen with deference the unambiguous area references, the accuracy is acquired sensibly between 50 and 75 miles. Contrasting and all area references, the precision of 83% is gotten most extreme at 75 miles. The impact of time on the spot discovery and level of area detail is appeared in Table 2. It is seen that the area is recognized with most noteworthy precision for KCC calculation. Moreover, the precision of AM and FBM strategies shows less exactness in distinguishing emergency occasion area.

Table 1 Comparison of accuracy of unambiguous location identification methods

Method	Unambiguous location references			
	ACCURACY	ACCURACY@25	ACCURACY@50	ACCURACY@75
Sample—1500 tweets				
AM	0.443	0.573	0.594	0.592
FBM	0.588	0.599	0.577	0.601
OK-CC	0.566	0.676	0.688	0.771

Table 2 Comparison of accuracy of all location identification methods

Method	All location references			
	ACCURACY	ACCURACY@25	ACCURACY@50	ACCURACY@75
AM	0.551	0.561	0.570	0.581
FBM	0.592	0.621	0.629	0.678
OK-CC	0.790	0.810	0.822	0.837

Table 3 Effect of time on location detection and average distance error

Time after event detection (in s)	15 s	30 s	45 s	60 s	75 s	Average distance error (miles)
Sample tweets—2000						
ACCURACY—AM	0.447	0.565	0.616	0.756	0.560	122
ACCURACY—FBM	0.567	0.675	0.766	0.856	0.671	45
ACCURACY—OK-CC@25 miles	0.554	0.745	0.781	0.831	0.854	13
% of tweets with geo-tagged location detail	33	54	66	79	81	–

Table 3 shows the effect of time on location detection and average distance error. The processing time for detecting the event location is shown in Fig. 1 which is deployed in Apache Spark Streaming platform.

The impact of number of tweets, it is seen as that the impact of the measure of tweets on the accuracy of the proposed models. The process of location detection and suspicion increases with the measure of tweets. The precision of the models seems to fix as the amount of tweets increases from 1000 to 2000 tweets. The execution time for time of bearings for the two techniques is equivalent to past models. The accuracy

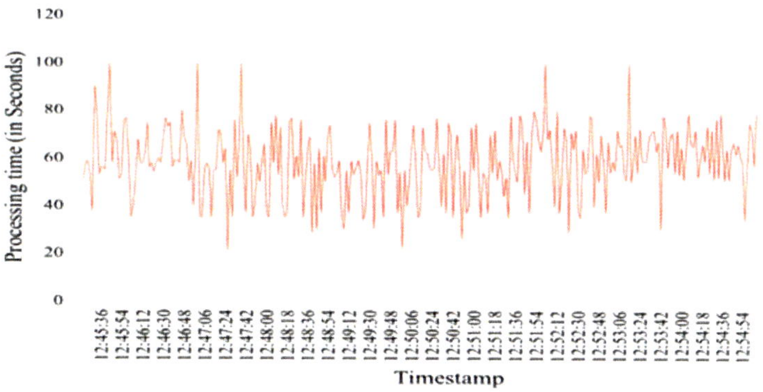

Fig. 1 Processing time (event location prediction)

for ACCURACY-OK-KC@25 miles increases from 72.6 to 85.1% for online k-center clustering algorithm.

5 Conclusion and Future Work

The percentage of geo-tagged data is limited compared to non-geo-tagged data, which often contains noises generated by automated robots. In this paper, an computational mathematical ways is proposed to deal with recognize the location of a client dependent on city level in the tweets. The proposed technique performs better compared to before work around here. The upgrades are accomplished by holding every single uncertain area and anticipating the home area of the client, rather than recognizing the most ideal counterpart for a location from a gazetteer. The proposed model can anticipate the area of the user location with 71% exactness inside 30 miles of the event's current location of the user utilizing 2000 tweets and builds an optimal accuracy to 85% in detecting location inside 25 miles of home area utilizing 2000 tweets. In addition, it is considered that the impact of expansion with minimal average error distance as the value of N increments for k-center grouping. The normal distance blunder increments from 79 miles for N value as 25–120 miles for $N = 75$ value. Online k-center clustering algorithm proves to be an efficient no-regret online algorithm which detects the event location cluster with 85% accuracy.

References

1. S. Xu, X. Fu, J. Cao, B. Liu, Z. Wang, Survey on user location prediction based on geo-social networking data. World Wide Web **23**(3), 1621–1664 (2020)
2. A. Saroj, S. Pal, Use of social media in crisis management: a survey. Int. J. Disaster Risk Reduct. 101584 (2020)
3. C. Bothorel, N. Lathia, R. Picot-Clemente, A. Noulas, Location recommendation with social media data, in *Social Information Access* (Springer, Cham, 2018), pp. 624–653
4. I. Memon, L. Chen, A. Majid, M. Lv, I. Hussain, G. Chen, Travel recommendation using geo-tagged photos in social media for tourist. Wireless Pers. Commun. **80**(4), 1347–1362 (2015)
5. H. Lee, M. Abdar, N.Y. Yen, Event-based trend factor analysis based on hashtag correlation and temporal information mining. Appl. Soft Comput. (2018)
6. A. Bhuvaneswari, C. Valliyammai, Information entropy based disaster event detection framework in online social networks. J. Intell. Fuzzy Syst. (2018). IOS Press
7. L. Palen, A.L. Hughes, Social media in disaster communication, in *Handbook of Disaster Research* (Springer, Cham, 2018), pp. 497–518
8. S. Petrović, M. Osborne, V. Lavrenko, Streaming first story detection with application to twitter, in *Proceedings of Annual Conference of the North American Chapter of the Association for Computational Linguistics Human Language Technologies*, Los Angeles, CA (2010), pp. 181–189
9. R. Lee, K. Sumiya, Measuring geographical regularities of crowd behaviors for twitter-based geo-social event detection, in *Proceedings of the 2nd ACM SIGSPATIAL International Workshop on Location Based Social Networks* (2010), pp. 1–10

10. A. Bhuvaneswari, C. Valliyammai, Social IoT enabled emergency event detection framework using geo tagged microblogs and crowdsourced photos, in *Emerging Technologies in Data Mining and Information Security. Advances in Intelligent Systems and Computing*, ed. by A. Abraham et al., vol. 813, chap. 13 (Springer Nature, Singapore, 2018), pp. 151–162
11. Z. Xu, L. Chen, G. Chen, Topic based context-aware travel recommendation method exploiting geotagged photos. Neurocomputing **155**, 99–107 (2015)
12. Y. Gao, S. Wang, A. Padmanabhan, J. Yin, G. Cao, Mapping spatiotemporal patterns of events using social media: a case study of influenza trends. Int. J. Geogr. Inf. Sci. **32**(3), 425–449 (2018)
13. C. Valliyammai, A. Bhuvaneswari, Semantics-based sensitive topic diffusion detection framework towards privacy aware online social networks. Clust. Comput. 1–16 (2018)
14. J.D. Zhang, C.Y. Chow, CRATS: an LDA-based model for jointly mining latent communities, regions, activities, topics, and sentiments from geosocial network data. IEEE Trans. Knowl. Data Eng. **28**(11), 2895–2909 (2016)
15. M. Kogan, L. Palen, K.M. Anderson, Think local, retweet global: retweeting by the geographically-vulnerable during hurricane Sandy, in *Proceedings of ACM Conference CSCW'15* (2015), pp. 981–993
16. J. Bakerman, K. Pazdernik, A. Wilson, G. Fairchild, R. Bahran, Twitter geolocation: a hybrid approach. ACM Trans. Knowl. Discov. Data (TKDD) **12**(3), 34 (2018)
17. A. Bhuvaneswari, C. Valliyammai, # ChennaiFloods: leveraging human and machine learning for crisis mapping during disasters using social media, in *Proceedings of IEEE 23rd International Conference on High Performance Computing (HiPC)*, Hotel Novatel, Hyderabad, Dec 2016, pp. 50–59
18. A. Bhuvaneswari, J. Timothy Jones Thomas, P. Kesavan, Embedded bi-directional GRU and LSTM learning models to predict disasters on twitter data. Procedia Comput. Sci. **165**, 511–516 (2019)
19. A. Bhuvaneswari, M. Karthikeyan, T. Lakshminarayanan, Improving diversity in video recommender systems and the discovery of long tail. J. Theor. Appl. Inf. Technol. **37**(2), 224–233 (2012)

Smart Cyberbullying Detection with Machine Learning

Shakambhari, Joshua Samuel Raj, and S. Anantha Babu

1 Introduction

Cyberbullying is one example of an online misbehaviour that has had a negative impact on the society. It can simply be described as a deliberate act of harming someone through the use of digital technology. It is been a problem for a long time, but the effect on young people has only recently become more apparent. Cyberbullying occurs on digital platforms such as mobile phones and computers. Cyberbullying happens when people share negative, harmful, inaccurate and mean content through SMS, email, post, tweet, audio, video and other means. Bullies thrive on social networking sites, and teenagers and young adults who use them are particularly vulnerable to attacks [1]. It usually happens on social media sites like Facebook, Twitter and Instagram. We can detect language patterns used by bullies and their victims, using machine learning and thus create rules to automatically detect cyberbullying.

1.1 Effect of Cyberbullying

Since cyberbullying can reach victims in their homes, where they normally feel happiest, it can be particularly painful for them. Victims of cyberbullying sometimes

Shakambhari
IBM, Bangalore, India

J. S. Raj (✉)
Department of ISE, CMR Institute of Technology, VTU, Bengaluru, India
e-mail: joshuasamuelraj@gmail.com

S. Anantha Babu
Department of CSE, St. Mother Theresa Engineering College, Tuticorin, Tamil Nadu, India

J. D. Peter et al. (eds.), *Disruptive Technologies for Big Data and Cloud Applications*, Lecture Notes in Electrical Engineering 905,
https://doi.org/10.1007/978-981-19-2177-3_24

feel as if there is no way out because they have 24/7 access to social media and the Internet. Because of the relentless and systemic nature of cyberbullying, victims often suffer from depression and low self-esteem, which can contribute to a decrease in grades of their overall progress. Given the effects of cyberbullying on victims, it is critical to find effective ways to identify and thereby avoid it. Machine learning is one of the most promising methods for learning from data and generating a model that automatically classifies appropriate behaviour [2].

1.2 Recent Studies

Today's youth are bringing their bullies home with them thanks to technological advancements and the rise of social media. Home, which used to be a safe heaven, has morphed into a whole new battleground where bullies are assured of anonymity and constant access to their victims. Almost 60% of children aged 10–17 have been bullied or abused online or via social media platforms, but 90% of these victims will never disclose the events. According to a recent study conducted by a research centre, 59% of adolescents have personally observed at least one of the six forms of violent online behaviours. A majority of teens have been target of cyberbullying with name-calling and rumour spreading being the most common forms of harassment as shown in Fig. 1.

Cyberbullying can be found on almost every social media site. The most popular social media network used by bullies is Facebook, which has 1.74 billion active users. In addition to this, all the social media platforms such as Twitter, Instagram, Snapchat, Reddit have been used extensively for cyberbullying. Cyberbullying Research Centre in 2016 reported around 34% of 12–17 years of students are being cyberbullied

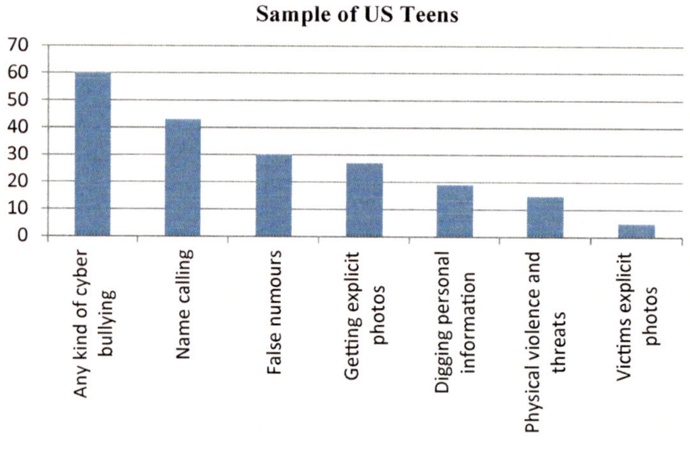

Fig. 1 Sample of US teens who say they have experienced bullying online or on their cell phone

in their life. While approximately twelve percentage of students in this age group revealed that they have been involved in cyberbullying. Moreover, they show that the percentages of people who have experienced cyberbullying at least once during their life increased from 18 to 34% from 2007 till 2020.

In the USA, 95% of adolescents are online, and the vast majority access the Internet via a mobile device, making it the most popular platform for cyberbullying. It has been recorded that 23% of students have said or done something mean or cruel to another person online. It has been recorded that 27% of people have been subjected to online humiliation by others. Girls are more likely than boys to be victims or perpetrators of cyberbullying. Compared to 6% of boys, 15% of girls have been the victim of at least four different types of violent online activity. Furthermore, previous research has shown that 94% of young people with mobile Internet access use social media on a regular basis. Seventy-one per cent of teenagers claim they have multiple social media profiles. According to a study in 2012, nearly all of the participants posted personal information on their profiles, such as their birth date, name, occupation, school and other details, making them more vulnerable to cyberbullying. A survey conducted in 2014 that covered the signs as well that may point towards the child or friend a victim of cyberbullying, as shown in Fig. 2. It stated that 95% of teens who witnessed bullying on any platform have ignored the behaviour and 33% of the young people who reported.

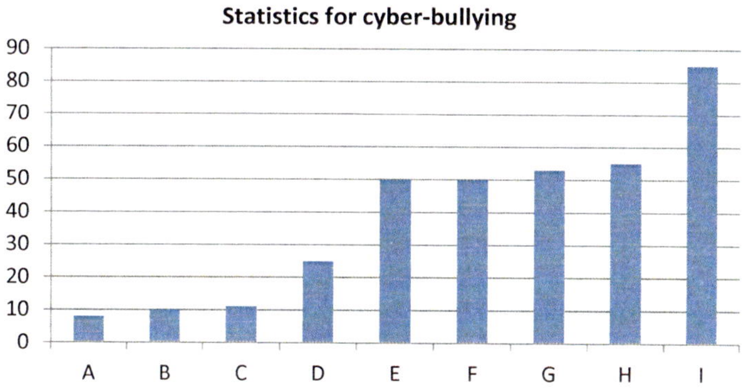

Fig. 2 Statistics of cyberbullying in 2014. Where A—Of parents of adolescents and teens; B—Of all middle school and high school students; C—Adolescents and teens reported; D—Of teenagers report that, they have; E—Of young people surveyed say that; F—Young people have experienced some source of cyberbullying; G—Of young people report being cyberbullied; H—Of all teens who use social media; I—Of teens regularly using cell phones

2 Related Work

Many methods suggest systems that can detect cyberbullying automatically and with high precision. While information retrieval and natural language processing techniques have been successfully applied to a variety of text classification problems, their use in cyberbullying detection is relatively new [3]. Another improvised method employed the Naïve Bayes machine learning approach to suggest a model with 91% accuracy. On still further improvisation, the Naïve Bayes classifier for extracting words and analysing loaded pattern clustering, resulted in 95.79% accuracy. However, this solution has the drawback of the cluster method not running in parallel. Later, using datasets from kaggle, a different method was used, in which two classifiers, Naïve Bayes and SVM, were applied [4]. The average accuracy of the Naïve Bayes classifier was 92.81%, while the SVM classifier was 97.11%. Deep learning and neural networks are another kind of approach [5]. Data from Twitter [6] has been used in this process. Owing to its imbalance, the accuracy of the Twitter dataset could not be determined. Although they achieved 56% precision, 78% recall and 96% accuracy, their dataset was not balanced, resulting in false results, which is reflected in their precision score of 56%. Hybrid approaches were also tested to combine the advantages of the expert system with the machine learning models [7]. The other systems consisting of three major modules have also been made that consists of link filtering, age validation and comments validation. This consists of two phases, training phase and testing phase [8]. Models were made on sentimental informed cyberbullying where it detects cyberbullying from psychological and sociological perspective and later computational methods have been used to detect cyberbullying [9]. Models were made on multi-model context where the approach to detect bullying was via network representation learning [10]. This is another approach as tried by the authors in the detection of road accidents [11] (Table 1).

3 Data Collection

In this section, we describe the collection and labelling of the data we used in our experiments.

3.1 Dataset Origin

As seen in Fig. 3, the social media site that we have selected is Reddit, which is a social news platform that allows users to discuss and vote on content submitted by other users. The beauty of reddit is that it has small communities called sub-reddits that bring people together based on their shared interests. We are using reddit because it is a textual platform where most of its contents are coming from the comments only, and it is mainly used by youngsters which are mostly affected by cyberbullying.

Table 1 History of different methods for identifying cyberbullying

Authors	Year	Methodology	Conclusions/findings	Open issues	Datasets used
Yin et al.	2009	A supervised machine learning approach was used, in which document features (local, contextual and textual) were used to train an SVM classifier	The addition of sentiment and contextual features improves the efficiency of a simple model that only uses local features	Only supervised learning methods are included in the data collection, which consists of stand-alone posts. Pragmatics of communication are not taken into account. Both the predators and the victims have yet to be named	Website data from Kongregate, Slashdot and MySpace
Dinakar et al.	2011	A supervised machine learning approach was used to identify bullying sensitive topics using binary and multiclass classifiers	When it comes to detecting content-based cybercrime, label-specific (binary) classifiers outperform multiclass classifiers	Only supervised learning methods are included in the data collection, which consists of stand-alone posts. Pragmatics of communication are not taken into account. Both the predators and the victims have yet to be named	After clustering comments from various videos into physical appearance, sexuality, race and culture
Dinakar et al.	2012	In addition to machine learning classifiers, a common-sense knowledge base with related reasoning strategies was used	Binary classifiers outperform multiclass classifiers in the identification of textual cyberbullying	The authors did not understand other aspects of the issue, such as pragmatics of communication and discussion	YouTube and Formspring data that has been manually labelled
Nahar et al.	2012	A sentimental research technique for detecting cyberbullying material was proposed	The HITS algorithm is used to find the most influential people (predator or victim)	Indirect cyberbullying is not a priority	Website datasets from Kongregate, Slashdot and MySpace

(continued)

Table 1 (continued)

Authors	Year	Methodology	Conclusions/findings	Open issues	Datasets used
Zhao et al.	2016	A tool for detecting cyberbullying was proposed that combined bullying, latent semantic and BoW features	Semantic capture words conceal information. Text containing bullying can be detected using linear SVM	Only for supervised learning methods, stand-alone articles	http://research.cs.wisc.edu/bullyin
Raisi et al.	2017	Using a machine learning algorithm, a weakly controlled participant vocabulary consistency (PVC) model was developed for simultaneously inferring new vocabulary measures of bullying and user roles in molestation-based bullying	On the basis of qualitative and quantitative assessment, the proposed model was tested on datasets from various social media platforms	To boost accuracy, network features or the sequence of conversations can be considered	Datasets of Ask.fm, Instagram
Van Hee et al.	2018	By modelling texts written by bystanders, preys, and bullies of online bullying, an automatic detection model for cyberbullying in social media was proposed	Experiments show that the proposed method is a promising approach for automatically detecting signals of cyberbullying in social media data, outperforming a word n-gram and keyword-based baseline	Hate, racist terms, curses, and threats can all be detected as fine-grained categories linked to cyberbullying	Datasets of ASK.fm posts in English and Dutch

(continued)

Table 1 (continued)

Authors	Year	Methodology	Conclusions/findings	Open issues	Datasets used
Cheng et al.	2019	Proposed X Bully, a system for detecting cyberbullying that first re-articulates multi-modal data from social networks before training node embedding illustrations on it	Broad experimental results on real-world datasets confirm the proposed framework's feasibility, and multi-modal data may suggest useful visions for representing and identifying cyberbullying behaviours	The identification of cyberbullying would be aided by a better understanding of the different modalities used to describe cyberbullying behaviours	Datasets of Instagram and Vine

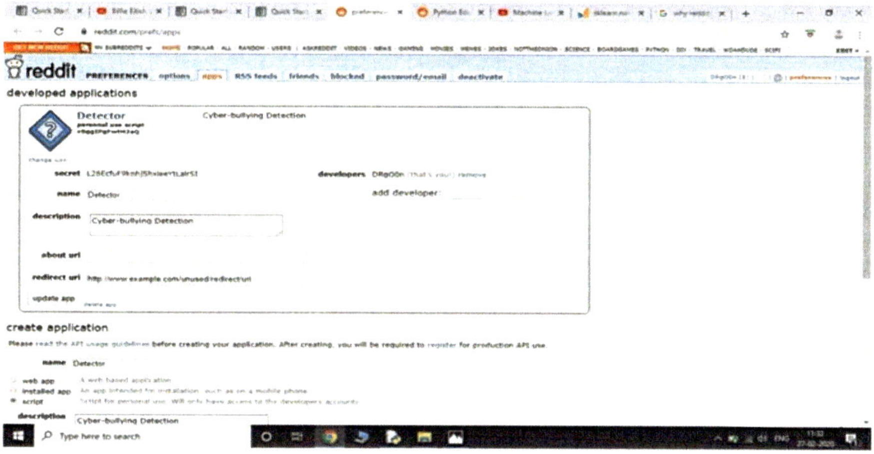

Fig. 3 Reddit platform

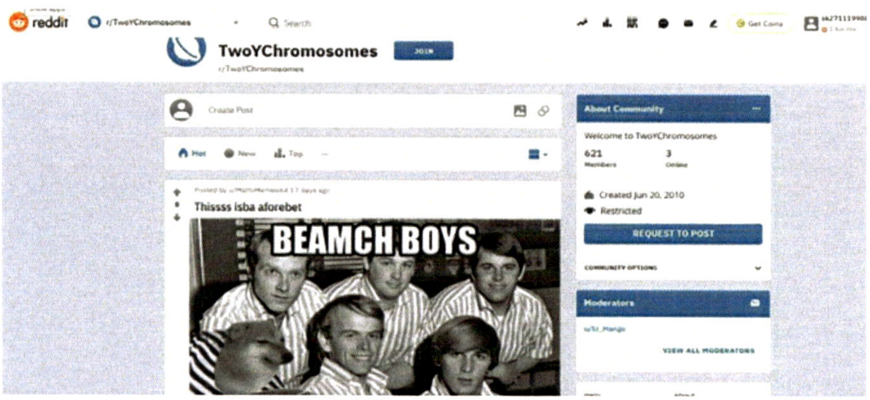

Fig. 4 Sub-reddits

3.2 Labelling of Data

We have analysed the posts and their respective comments on different sub-reddits as shown in Fig. 4. By using PRAW, which is specifically used for fetching the comments of the posts on different sub-reddits and later we classify them using Naïve Bayes classifier. PRAW stands for Python Reddit API Wrapper, a Python package that provides easy access to the Reddit API. It can be used for web application, installed application and script application. And the instances of the reddit class are the gateways to interact with reddit's API through PRAW.

4 Proposed Approach

Preprocessing, feature extraction and classification are the three key steps in the proposed method. We clean the data in the preprocessing step by removing noise and unnecessary text. The following are the steps in the preprocessing process:

- Tokenization: In this section, we take the text as sentences or entire paragraphs and output it as a list of separated words.
- Lowering text: This takes the list of words that were not tokenized and lowers all of the letters, so "THISISAWESOME" becomes "thisisawesome".
- Stop words and encoding cleaning: This is an important step in the preprocessing process because it rids the text of stop words and encoding characters like n and t that give no useful information to the classifiers.
- Stemming: This transforms the list of terms back to their original form, without the use of suffixes or prefixes.

Feature extraction is the second phase in our model. This phase converts textual data into a format that machine learning algorithms can understand. First, we use

TFIDF to extract the input data's features, which we then store in a features list. TFIDF's main concept is that it analyses text and determines the weight of words in relation to the document or sentence. We use a sentiment analysis technique in addition to TFIDF to extract the polarity of the sentence and add it as a function to the feature list containing TFIDF features. The polarity of a sentence indicates whether it is positive or negative in nature. In addition to TFIDF-based function extraction, the proposed method employs N-gram to consider various word combinations during model evaluation. The classification step is the final step in the proposed method, and it involves feeding the extracted features into a classification algorithm to train and evaluate the classifier before using it in the prediction phases. SVM was the classifier, and we used support vector machine. In general, classifiers are evaluated using a variety of evaluation metrics that are dependent on the confusion matrix. Accuracy, recall and F-score are some of the parameters. The following equations are used to calculate them:

$$\text{Accuracy} = \text{TP} + \text{TN}/\text{TP} + \text{TN} + \text{FP} + \text{FN} \tag{1}$$

$$\text{Precision} = \text{TP}/\text{TP} + \text{FP} \tag{2}$$

$$\text{Recall} = \text{TP}/\text{TP} + \text{FN} \tag{3}$$

$$F\text{-score} = 2 * \text{precision} * \text{recall}/\text{precision} + \text{recall} \tag{4}$$

where TP denotes the number of true positives, TN the number of true negatives, FP the number of false positives and FN the number of false negatives groups.

Bag of words is a representation of text that describes the occurrence of words within a document. It involves a vocabulary of known words and a measure of the presence of known words and while extracting the document vectors we use Boolean value 0 for absent and 1 for present. A problem with this algorithm is that when it comes to scoring word frequency, highly frequent words tend to dominate the text, but they do not contain as much information for the model. One another problem with this approach is that each data gets turned into a very long list of word counts which can significantly slow down some machine learning algorithms like SVM. So, to overcome the problem of BoW we have used another algorithm with BoW, that is, term frequency-inverse document frequency (TFIDF). TF is a scoring of the frequency of the word in the current document. IDF is a scoring of how rare the word is across elements in the vocabulary of known words. Support vector machine is a machine algorithm which is used to hyperplane in an N-dimensional space where N is the number of features that distinctly classify the data points.

We have used SVM for our proposed model because:

- It is effective in high-dimensional space.
- It is effective in cases where numbers of dimensions are greater than the number of samples.

- It is memory efficient.
- It is versatile as different kernel functions can be applied for the decision functions. It is possible to specify custom kernels.
- An SVM model is a representation of the examples as a point in space, mapped such that the examples of the different categories are separated by a large distance. With the labelled training data, the algorithm outputs an optimal hyperplane. We have used SVM for the custom feature selection in our model because the word count for each word is its own feature and each additional feature causes a significant increase in computational complexity. So TFIDF has a very high number of features, that means it has high dimensionality, but SVM greatly suffers from the curse of dimensionality. And, as the custom feature extraction is of very low dimensionality, SVM is a deal.

The above-mentioned algorithms have been used for creating stand-alone models from the training dataset. Later, ensemble method of machine learning technique has been used that combines these base models in order to produce one optimal predictive model. Since it can be learned and then used to make predictions, ensemble is a supervised learning algorithm. As a result, the qualified ensemble represents a single hypothesis. This theory, on the other hand, may or may not be found within the hypothesis space of the models from which it is constructed. As a result, ensembles can be seen to have more feature versatility, allowing them to over-fit the training data more than a single model can, resulting in better performance. Ensemble methods are a type of meta-algorithm that incorporates many machine learning techniques into a single predictive model to reduce uncertainty, bias and improve predictions. We used one of their methods called "Voting" to do ensemble, which is one of the easiest ways of integrating predictions from multiple machine learning algorithms. It works by building two or more stand-alone models from the training data. When asked to make predictions for new data, a voting classifier can be used to wrap the models and average the predictions of the sub-models. The sub-models' predictions can be weighted, but manually or heuristically determining the weights for the classifier is difficult. The voting classifier class is used to construct a voting ensemble model for classification. Hence, it produces better outcomes as shown in Fig. 5.

5 Conclusion

In this paper, we have proposed an approach to detect cyberbullying using machine learning techniques on different platforms. We evaluated our model on the two classifiers SVM and Naïve Bayes classifiers and used TFIDF for the feature extraction. In addition, the classifications were evaluated on different n-gram language models. We have used SVM for the custom feature extraction which means that any user who feels offended with any word can specify it and eliminate it. We have used the ensemble approach to combine the different algorithms for better performance. Our efforts would undoubtedly enhance cyberbullying detection, allowing people to use

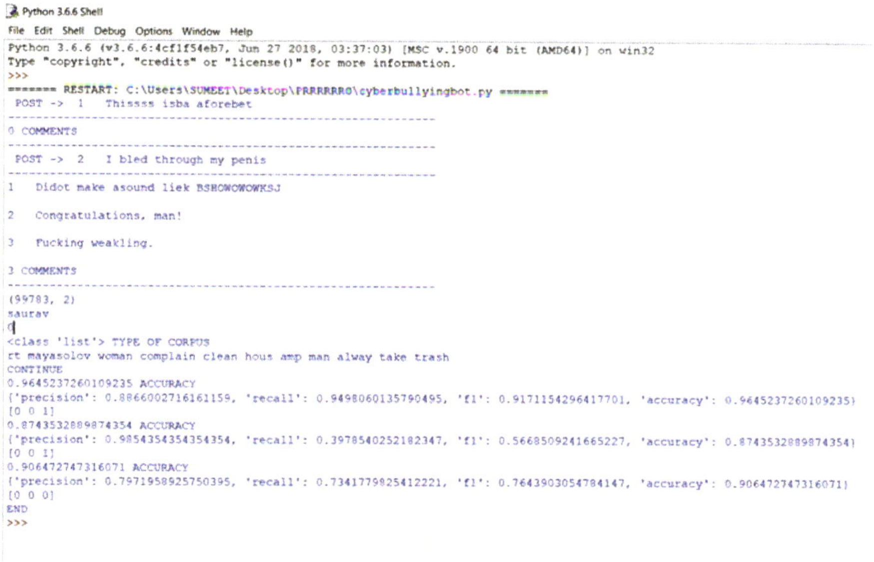

```
Python 3.6.6 Shell
File Edit Shell Debug Options Window Help
Python 3.6.6 (v3.6.6:4cf1f54eb7, Jun 27 2018, 03:37:03) [MSC v.1900 64 bit (AMD64)] on win32
Type "copyright", "credits" or "license()" for more information.
>>>
======= RESTART: C:\Users\SUMEET\Desktop\PRRRRRRO\cyberbullyingbot.py =======
 POST ->  1   Thissss isba aforebet
------------------------------------------------------------
0 COMMENTS
------------------------------------------------------------
 POST ->  2   I bled through my penis
------------------------------------------------------------
1   Didot make asound liek BSHOWOWOWKSJ

2   Congratulations, man!

3   Fucking weakling.

3 COMMENTS
------------------------------------------------------------
(99783, 2)
saurav
d
<class 'list'> TYPE OF CORPUS
rt mayasolov woman complain clean hous amp man alway take trash
CONTINUE
0.9645237260109235 ACCURACY
{'precision': 0.8866002716161159, 'recall': 0.9498060135790495, 'f1': 0.9171154296417701, 'accuracy': 0.9645237260109235}
[0 0 1]
0.8743532889874354 ACCURACY
{'precision': 0.9854354354354354, 'recall': 0.3978540252182347, 'f1': 0.5668509241665227, 'accuracy': 0.8743532889874354}
[0 0 1]
0.906472747316071 ACCURACY
{'precision': 0.7971958925750395, 'recall': 0.7341779825412221, 'f1': 0.7643903054784147, 'accuracy': 0.906472747316071}
[0 0 0]
END
>>>
```

Fig. 5 Results

social media safely. The size of the training data, however, limits the detection of cyberbullying patterns. As a result, more cyberbullying data is needed to enhance results. Thus, deep learning techniques would be appropriate for larger data because they have been shown to outperform machine learning methods on larger datasets.

References

1. K. Reynolds, A. Kontostathis, L. Edwards, Using machine learning to detect cyberbullying, in *2011 10th International Conference on Machine Learning and Applications and Workshops*, vol. 2 (IEEE, 2011)
2. J. Hani et al., Social media cyberbullying detection using machine learning. Int. J. Adv. Comput. Sci. Appl. **10**, 703–707 (2019)
3. H. Rosa et al., Automatic cyberbullying detection: a systematic review. Comput. Hum. Behav. **93**, 333–345 (2019)
4. M. Di Capua, E. Di Nardo, A. Petrosino, Unsupervised cyber bullying detection in social networks, in *2016 23rd International Conference on Pattern Recognition (ICPR)* (IEEE, 2016)
5. M. Dadvar, K. Eckert, Cyberbullying detection in social networks using deep learning based models, in *International Conference on Big Data Analytics and Knowledge Discovery* (Springer, Cham, 2020)
6. V. Nandakumar, B.C. Kovoor, M.U. Sreeja, Cyberbullying revelation in twitter data using Naïve Bayes classifier algorithm. Int. J. Adv. Res. Comput. Sci. **9**(1) (2018)
7. M. Dadvar, D. Trieschnigg, F. de Jong, Experts and machines against bullies: a hybrid approach to detect cyberbullies, in *Canadian Conference on Artificial Intelligence* (Springer, Cham, 2014)

8. V.H. Divyashree, N.S. Deepashree, An effective approach for cyberbullying detection and avoidance. Int. J. Innov. Res. Comput. Commun. Eng. **14** (2016)
9. H. Dani, J. Li, H. Liu, Sentiment informed cyberbullying detection in social media, in *Joint European Conference on Machine Learning and Knowledge Discovery in Databases* (Springer, Cham, 2017)
10. L. Cheng et al., Xbully: cyberbullying detection within a multi-modal context, in *Proceedings of the Twelfth ACM International Conference on Web Search and Data Mining* (2019)
11. R.J.S. Raj, T.S.R. Perumal, N. Muthukumaran, Road accident data analytics using map-reduce concept. Int. J. Innov. Technol. Explor. Eng. **8**(11), 1032–1037 (2019)

Tourist Sentiment Analysis Using Natural Language Processing

T. N. Prabhu, S. Aarthi, and P. Nanthini

1 Introduction

Sentiment analysis can be called as the computational treatment of sentiments. It is the way towards identifying the emotions from particular piece of information to gain an understanding of the opinions expressed by the users. The clients frequently express their sentiments towards the services and products through social media websites. Every day, lot of users provide ratings for the tourist places on various travel industry sites like TripAdvisor, etc.

There were lot of analysis for the hotel reviews in social media which has become a source for doing sentiment analysis. The data utilized for investigation has been gotten from the Web. The data can be separated from the suppositions raised by the clients. In the event that those assets are overseen well, it will likewise be useful for the partners for the development of their business. This examination was utilized to separate between positive, negative and unbiased surveys from the clients at each lodging through their sentiments. Likewise, text mining can have the option to anticipate the key credits, which implies the catchphrase which regularly seems dependent on often happened word from both negative and positive reviews.

This paper analyses the online user-generated customer reviews taken from TripAdvisor website. The reviews were extracted and sentiment analysis was performed which will be helpful for the user to get a clear picture about the hotels that they want. Various machine learning algorithms have been used to implement sentiment analysis. The reviews are then processed using sentiment analysis. Based on the result we get by means of analysis, the user can easily decide the place that they want to visit.

T. N. Prabhu (✉) · S. Aarthi · P. Nanthini
Information Technology, Sri Ramakrishna Engineering College, Coimbatore, India
e-mail: prabhu.natesan@srec.ac.in

J. D. Peter et al. (eds.), *Disruptive Technologies for Big Data and Cloud Applications*,
Lecture Notes in Electrical Engineering 905,
https://doi.org/10.1007/978-981-19-2177-3_25

Classification algorithms like Naïve Bayes Algorithm, Random Forest Algorithm and Decision Tree Algorithm has been implemented, and the results were studied. Along with this, we have implemented feature extraction algorithms like CountVectorization and TFIDFVectorization for better accuracy. The result which was derived my means of the algorithms has been compared to give more accurate output.

The primary point of this paper was to dissect the online surveys of the clients and their view about remaining in the hotel and furthermore to remove the regular words show up on the public reviews. Here, we have performed similar investigations of feature extraction calculations like CountVectorization, TFIDFVectorization, alongside arrangement calculations Naive Bayes (NB), Decision Tree and Random Forest (RF). Execution of calculations has been looked at utilizing different boundaries like precision, review, accuracy and $f1$-score.

2 Literature Survey

In the current time, machines gave exact understandings of what individuals express via online media. Information has been investigated, and the techniques for assessing assumptions has been classified and thought about for better future research [1]. The sources of the reviews are principally survey locales. The interpersonal organization source is viewed as a best wellspring of data on the grounds that numerous individuals impart and examine their insights about a specific theme freely [2].

The TripAdvisor is the informational index, and it is the base for removing all subtleties like hotel name, number of reviews for every lodging, star ratings, hotel address and hotel interface. The Hotel surveys are arranged to discover the Positive words and Negative words, and the level of the Positive and Negative reviews are calculated [3].

Various strategies for performing sentiment analysis were there. In that SentiWordNet for assessment mining from hotel reviews and sentence significance, score-based strategy for assessment rundown of hotel surveys shows 87% of precision of hotel survey arrangement as positive or negative review by AI method [4]. Existing writing in dissecting on the web lodging surveys chiefly centres around a solitary information asset, dictionary and investigation technique and seldom gives and dynamic data to work on business' administration and nature of products [5].

The global meeting about the Tourism surveys gives the information on utilizing feature extraction calculations like CountVectorization and TFIDFVectorization calculation, and three order calculations like Naive Bayes (NB), Decision Tree and Random Forest (RF) have been utilized for opinion arrangement. The adequacy of our methodology is demonstrated against different methodologies accomplishing better results when utilizing Naïve Bayes, Random Forest and Decision Tree order algorithms [6]. A similar examination has been performed for mix of feature extraction and classification algorithms with the assistance of boundaries like execution time, exactness, accuracy, review and $f1$-score [7].

In a most recent investigation, numerous methodologies were presented that consequently perform sentiment detection utilizing different calculations. Hotel surveys have been dissected utilizing different strategies like Naïve Bayes, K-Nearest Neighbour, Support Vector Machine, Logistic Regression and Random Forest. What is more, a learning model was additionally suggested that joins every one of the classifiers, and results were compared [8].

A large portion of the investigations uncovered that it is anything but a worldview to decide the extremity and tendency of gigantic populace towards explicit snippet of data. For instance, individuals rapidly post their reviews online when they go to any spots and afterwards begin presenting remarks on examining about the spot they went. It assists us with assessing, to rate about the allure of any spots and different administrations given by them and to think about if it is anything but a triumph. The data on these locales can be used for advertising, social examinations and exploration papers [7].

The estimation depicts how precise the calculations are utilized to anticipate the notion examination. Before that, the supposition examination of each review in TripAdvisor was determined by annotators. The online reviews and assessments were improved so it separates the important surveys which could be gotten from the not-valuable ones. The point is to separate the essential data in information mining from a lot of the dataset to stay away from information sparsity and performing highlight extraction and arrangements dependent on the same [9].

3 Proposed System

3.1 Problem Statement

Integrity of Reviews cannot be ensured as there are no restrictions about the content to be given. Number of Reviews highly varies which sometimes results in inaccurate ratings of hotel because people tend to give vague reviews despite its features. All these discrepancies were taken into account, and result has been predicted based on the same.

Datasets that were gathered could be analysed and presented in such a way that it becomes easy to identify whether the reviews were positive, negative or neutral. Star rating system might be quick and simple but insightful as the average rating of a five-star system will generate 4.3 usually regardless of the object. To overcome such trouble, the result can be given by terms of the percentage of various parameters of the reviews and also a word cloud has been visualized to get a clear picture of the frequently used positive and negative words. In addition to this, which aspect of the hotel is best is also predicted.

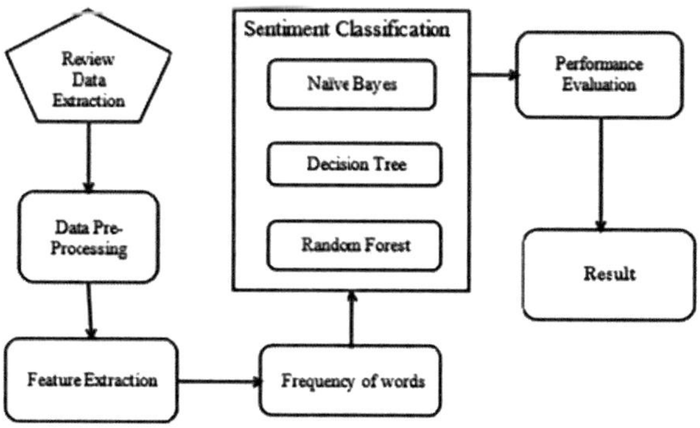

Fig. 1 Process of sentiment analysis

3.2 Block Diagram

See Fig. 1.

3.3 Module Description

Web Scraping

Web scratching or web information extraction is the interaction of information scratching which is utilized for removing information from sites. Here, the information has been removed from a site called TripAdvisor. The gathered information has been changed over into a CSV document to stay away from long runtime during the execution of the paper. Scratching a site page includes bringing and extricating the information from it. Bringing is accomplished by a solicitation from a web worker. The brought information is then separated for additional preparing. The substance of a page might be parsed, looked and reformatted to have its information replicated into a bookkeeping page, etc.

Data Pre-processing

Information pre-handling includes different advances like tokenization, standardization, stop word evacuation, stemming and lemmatization.

Tokenization: It will part the sentence into words utilizing whitespaces and accentuations. Each word is a token, so the interaction is called as tokenization.

Standardization: It is the way towards changing over the whole reviews into lowercase.

Stop Word Removal: In this interaction, words which happen habitually like a, the, this, you, in, is, was and so forth are stop words and furthermore accentuations and unique characters which is of no utilization are taken out. It helps in decreasing the vector size just as incrementing the exhibition.

Lemmatization: Here, tokens in past or future tenses get changed over into current state and tokens in third individual structure gets changed over into first individual structure.

Stemming: It helps in discovering the root expression of the tokens with no "ing" word or different characters.

Text Classification

The Text Classification deals with all pre-handling errands (separating text, tokenization, stop word evacuation, lemmatization) needed for programmed characterization, for instance, with the accompanying hotel surveys:

Text A: "Friendly administration. Unrivalled room! Adored the high roof. Housekeeping administration was top quality. Great breakfast and wellness room".

Text B: "If conceivable I would give the lodging zero stars. The smell is horrendous and gave me a migraine following one evening. Suggest keeping away from".

An assessment investigation classifier will order Text A as certain as there are components that show that the commentator was by and large fulfilled.

Text B would almost certainly be named negative since the analyst is motioning to many negative parts of the lodging and experience.

Feature Extraction

In natural language processing, information is in literary organization. AI classifier expects information to be available in vector design. So, change of tokens into highlights is required. There are different calculations for literary component extraction. CountVectorization and TFIDFVectorization algorithms are utilized for extraction.

CountVectorization: This calculation tallies the frequency of word, i.e. token in a specific archive and change entire report into record term framework. It is practically equivalent to bag-of-word approach. The formula is:

$$\text{Count(word)} = \text{No of occurrence of a word in particular document.}$$

TFIDFVectorization: It figures weight of term specifically record by considering event of term in report just as in entire corpus then it changes entire archive into document term matrix. The equations for TF-IDF calculation are:

$$\text{TF} = (\text{No of times a required word occurs})/(\text{Total no of words in a document})$$

$$\text{IDF} = \log(\text{Total no of document/No of documents contains the required word})$$

$$\text{TF-IDF} = \text{TF} \times \text{IDF}$$

Classification Using Algorithm

Naive Bayes: Naive Bayes is a probabilistic AI model which is broadly utilized for opinion arrangement. MultinomialNB is utilized to execute the Naive Bayes calculation for information having numerous results. Multinomial Naive Bayes considers the probability given class as a multinomial appropriation. It depends on probability likelihood.

Random Forest: Random Forest is a mix of decision tree indicators to such an extent that each tree relies upon the upsides of a random vector examined autonomously and with similar appropriation for all trees in the timberland. It is called Ensemble procedure as it utilizes assortment of results to make last forecast.

Decision Tree: The primary point in utilizing a Decision Tree calculation is to create a training model with the goal that it will be utilized to foresee which class the testing information will have a place with. The Decision Tree has its root to begin foreseeing the class name and afterwards it leaps to the following part of the root dependent on which esteem the information has followed.

Analyse the Sentiment and Provide Ratings

After performing all the above modules, the ratings will be provided to the users as a percentage of positive and negative reviews. In this, each and every word is provided with the ratings and the overall percentage is given. So that user can have a clear idea or opinion about the particular required resort.

Visualization

Anticipated opinion has been envisioned through word cloud for frequently occurred positive and negative words of each tourist place.

Performance Evaluation

Performance evaluation has been performed utilizing boundaries like precision score, accuracy, review, $f1$-score and execution time estimation which will give a clear picture about the result. The four measurements were calculated based on confusion matrix.

4 Results and Discussions

With the expanded utilization of Internet, the need for sentiment analysis is likewise expanding. This is on the grounds that individuals nowadays rely upon the reviews communicated by others on some sort of items, services, subject, issues and so on. Three various algorithms are used. Naive Bayes classifier provides better *precision* whereas Decision Tree provides better *accuracy*. Random forest classifier is almost

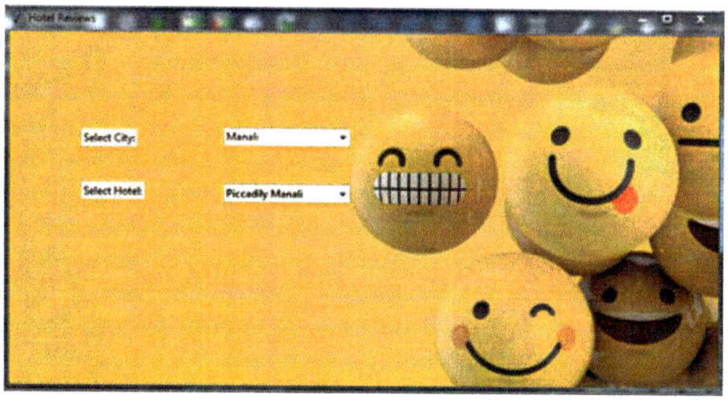

Fig. 2 Selecting hotels

same as the Decision Tree algorithm. Text mining strategies permit us to feature the most every now and again utilized keywords in a section of writings. A word cloud, additionally alluded as text cloud or tag cloud, has made, and a graphical portrayal of different angles are portrayed which is a visual portrayal of information (Figs. 2, 3, 4, 5, 6, 7, 8, 9, 10 and 11).

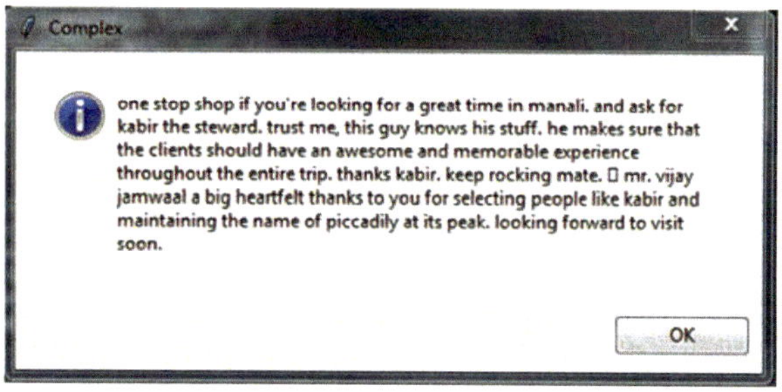

Fig. 3 Collecting reviews

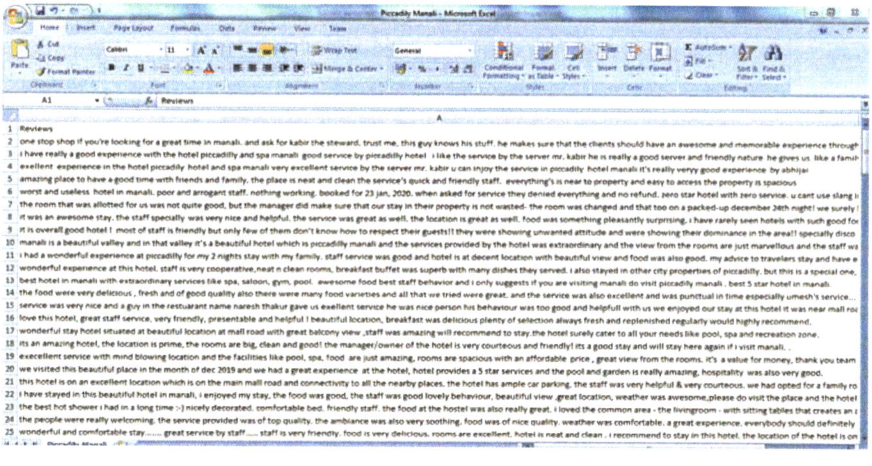

Fig. 4 Dataset

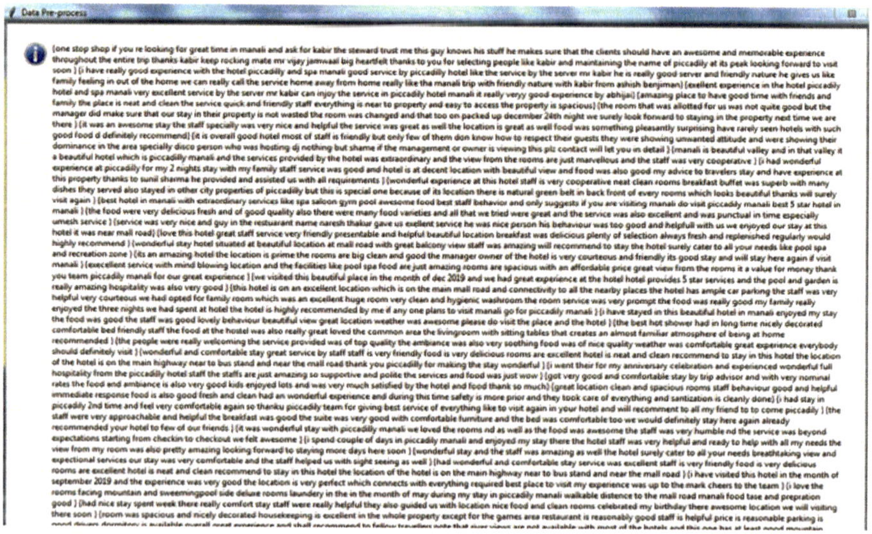

Fig. 5 Data pre-processing

Fig. 6 Wordcloud for positive reviews

5 Conclusion and Future Scope

Here, sentiment analysis is used for extracting and classifying hotel reviews posted by customers on the travel industry sites. The framework visits website pages beginning

Fig. 7 Wordcloud for negative reviews

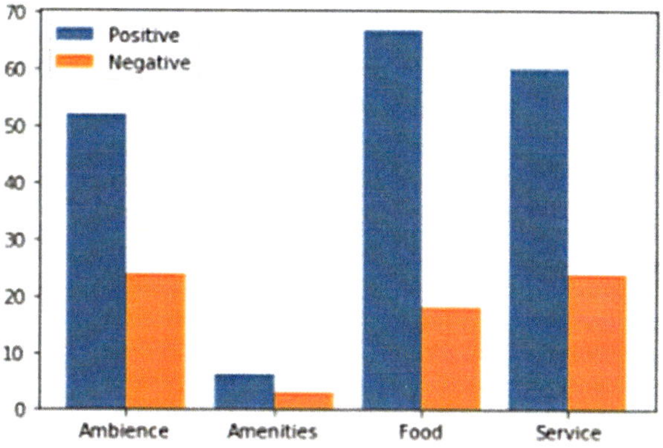

Fig. 8 Comparison of different aspects

Fig. 9 Naive Bayes

```
Naive Bayes
Matrix [[ 0  8]
 [ 0 34]]
Report              precision    recall  f1-score   support

           0       0.00      0.00      0.00         8
           1       0.81      1.00      0.89        34

    accuracy                           0.81        42
   macro avg       0.40      0.50      0.45        42
weighted avg       0.66      0.81      0.72        42

Score 0.8095238095238095
```

Fig. 10 Random Forest

```
RF:
Matrix [[ 4  1]
 [ 1 27]]
Report              precision    recall  f1-score   support

           0        0.80      0.80      0.80         5
           1        0.96      0.96      0.96        28

    accuracy                            0.94        33
   macro avg        0.88      0.88      0.88        33
weighted avg        0.94      0.94      0.94        33

Score 0.9393939393939394
```

Fig. 11 Decision Tree

```
Decision Tree:
Scores: [96.21621621621622, 97.2972972972973, 98.91891891891892,
97.83783783783784, 95.13513513513514]
Mean Accuracy: 97.081%
```

from a given URL, extracts the reviews from page content and then uses feature extraction and classification algorithm to provide positive and negative percentage of reviews. The primary commitment of this paper is the manner by which the process for data collection and data pre-processing, extracting and classifying data is executed with perception of TripAdvisor information. In future work, we will attempt to utilize deep learning-based strategies for feature extraction and include large datasets for better performance.

References

1. K. Chakraborty, S. Bhattacharyya, R. Bag, A survey of sentiment analysis from social media data. IEEE Access **7**(2) (2020)
2. R.A. Laksono, K.R. Sungkono, R. Sarno, Sentiment analysis algorithms and applications: a survey. Ain Shams Eng. J. **5**(4) (2014)
3. N. Divyashree, K.L. Santhosh Kumar, J. Majumdar, Opinion mining and sentimental analysis of TripAdvisor.in for hotel reviews. Int. Res. J. Eng. Technol. **04**(11) (2012)
4. V.B. Raut, D.D. Londhe, Opinion mining and summarization of hotel reviews, in *Sixth International Conference on Computational Intelligence Communication Networks*, Pune (2014)
5. Y.C. Chang, C.H. Ku, C.H. Chen, Social media analytics: extracting and visualizing Hilton hotel ratings and reviews from TripAdvisor. Int. J. Inf. Manage. **48** (2019)
6. A. Kaur, D. Khaneja, K. Vyas, R.S. Saini, Sentiment analysis on Twitter using Apache Spark. ResearchGate (2010). https://doi.org/10.13140/RG.2.2.10737.79200
7. A.A. Wadhe, S.S. Suratkar, Tourist place reviews sentiment classification using machine learning techniques, in *IEEE International Conference on Industry 4.0 Technology (14 Tech)* (2020)
8. S. Anis, S. Saad, M. Aref, Sentiment analysis of hotel reviews using machine learning techniques, in *International Conference on Advanced Intelligent Systems and Informatics*, vol. 1261 (2020)
9. N.Z. Dina, Tourist sentiment analysis on TripAdvisor using text mining: a case study using hotels in Ubud, Bali. ResearchGate **9** (2020). ISSN: 2223-814X

Sentiment Analysis of Twitter Data Using Machine Learning

K. Deepa, H. Sangita, and H. Shruthi

1 Introduction

Twitter which is mostly used social platform for writing statements or messages on different matter in the form of short messages service. Sentimental analysis determines the sensitivity behind a sequence of words, which is used to gain an understanding of the opinions, attitude and emotions expressed [1]. The major challenge of the area of sentiment analysis lies in indicating the emotions expressed in these different texts and google co lab is used to perform this. The prediction is done by people's expression in their vision about the person which assists in analyzing the negative, positive and neutral comments. Sentiment analysis points out the use of natural language processing, computational linguistics and text analysis where this helps to find and then extract subjective details in source materials. Here, judgments of the people are taken on a well-known person. People share their views about the person which will helps out to analyze the negative, neutral and positive comments. The negative, neutral or positive tweets are analyzed with the help of key words. The classification is analyzed to discover the results of sentiment analysis and to check how accurate the results would be.

Sentiment analysis is merely Opinion Mining which recognizes the nature of a particular person by analyzing its Timeline on Twitter [2]. It is an automated process by which the subjective information underlying a text is identified and extracted. The major demand of sentiment analysis occupies the exact space with the development

K. Deepa · H. Sangita (✉) · H. Shruthi
Sri Ramakrishna Engineering College, Coimbatore, India
e-mail: sangita.1805123@srec.ac.in

K. Deepa
e-mail: deepa.senthil@srec.ac.in

H. Shruthi
e-mail: shruthi.1805131@srec.ac.in

J. D. Peter et al. (eds.), *Disruptive Technologies for Big Data and Cloud Applications*,
Lecture Notes in Electrical Engineering 905,
https://doi.org/10.1007/978-981-19-2177-3_26

Fig. 1 Sentimental analysis: the tweets are segregated into three parts say, positive negative and natural. This separation is done using keywords used by the user in the tweet

of social media like Facebook, Quora, Twitter, microblogs, Instagram, Blogs or other social networks. In this survey, the famous microblogging site "twitter" has been used for sentiment analysis. Twitter is the best social media platform where people can express their opinion on different campaigns, events, etc. Twitter is a rich source to learn about people's opinion and sentiment analysis. For each tweet it is important to determine the sentiment of the tweet whether it is neutral, negative or positive [3]. The datasets are collected from Kaggle. The data is analyzed using logistic regression, which gives the output in graphical form. The training accuracy has reached up to 95% and testing has reached up to 94%. Positive, negative neutral tweets have each graphical representation.

Eventually making the process and also extracting only the useful information is a hectic job. The enhancement in the technology has simplified for people around the globe to share, express their points openly, and each potential web user depends on these opinions and reviews to take decisions [4]. Twitter sentiment analysis is not all about number of positive and negative and natural tweets, rather it is about knowing the general feeling of the users. The contrast between these types of tweets might seem easy, though it is easy to understand, it is way more complex [5]. A positive tweet throws a positive sentiment after having all the words analyzed. Sentiment analytics in Twitter is an excellent science because of the accuracy given by the system that analyzes Twitter information and also in a network, where irony plays an important role. Depending on the context each word of the tweet has its own score (Fig. 1).

2 Description

Sentiment analysis is a technique for analyzing and judging somebody else's or feelings or sentiments with esteem to particular topic written in a part of text. It is used to admit and collect the sentiments communicated. Big numbers of clients are drawn for imparting their thoughts in the form of messages or comments in the web-based social networking sites like twitter. The tweets could be then classified into natural, negative or positive. Logistic regression classification is used in this proposed word as a classifier for a feature vector. Using logistic regression the analysis of sentiments in twitter data is a web-based application which gives sentiment value as output by taking tweets as an input.

2.1 Introduction

Sentiment is negative, positive or neutral write up about any topic. Anybody can differentiate tweets by reading whether it is natural, negative or positive. But if there are big numbers of data to be read then it would be a hectic and consumes time work. So, if these all process could be done with the use of an automated program, then it would be simpler and the above manual process could be eliminated. A huge number of data is taken from Kaggle. These datasets are trained and the analysis is done. The final output is displayed.

Logistic Regression

Logistic regression is a classification algorithm. It helps to predict the binary outcome on a set of independent variables.

A binary outcome is one where only two possible situations are possible which are

i. the event happens.
ii. the event does not happen.

Independent variables are those variables which influences the result.

So logistic regression is correct when worked with binary data. It always fits into two categories (such as "yes" or "no"; "pass" or "fail", and so on).

The independent variables fall under any of the following categories:

1. Continuous—like temperature in degree Celsius, Fahrenheit or weight in grams, kilograms.
2. Discrete, ordinal—Ordinal data is the data that will be possibly got from a customer satisfaction inspection.
3. Discrete, nominal—Data which is inserted into named groups, it does not represent any order or scale.

Logistic regression is used to calculate the probability of a binary event that is occurring. For instance, predicting if an incoming email is a spam email or not a spam email, or predicting if a master card transaction is a fraudulent transaction or not a fraudulent transaction, to predict if a tumor is benign or malignant. A web education company may use logistic regression for finding whether a student will complete a course on the given time or not (Fig. 2).

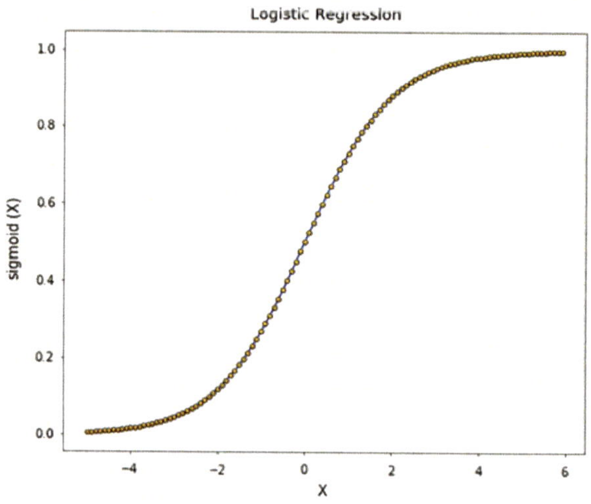

Fig. 2 Sentimental analysis: a sigmoid function fitted to a data

The logistic regression function $p(\mathbf{x})$ is a sigmoid function of $f(\mathbf{x})$: $p(\mathbf{x}) = 1/(1 + \exp(-f(\mathbf{x})))$. As it's often 0 or 1. The function $p(\mathbf{x})$ is interpreted because the predicted probability is 1. Hence, $1 - p(x)$ is that the probability is 0.

3 Methods

There are four modules involved in this project are

1. Data Collection and Preprocessing
2. Word Cloud Generation
3. Graph Generation for Hashtags
4. Model Training (Fig. 3).

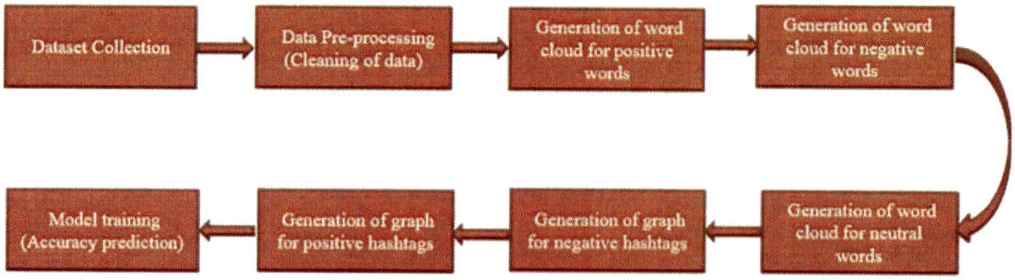

Fig. 3 This flow diagram shows the process of analyzing the twitter data. Here, the dataset is collected and then pre-processed; later it segregates positive, negative and neutral words and then the generation of positive and negative hashtags separation takes place

```
[ ] df['clean_tweet'] = np.vectorize(remove_pattern)(df['tweet'], "@[\w]*")

[ ] df.head()
```

	id	label	tweet	clean_tweet
0	1	0	@user when a father is dysfunctional and is s...	when a father is dysfunctional and is so sel...
1	2	0	@user @user thanks for #lyft credit i can't us...	thanks for #lyft credit i can't use cause th...
2	3	0	bihday your majesty	bihday your majesty
3	4	0	#model i love u take with u all the time in ...	#model i love u take with u all the time in ...
4	5	0	factsguide: society now #motivation	factsguide: society now #motivation

```
[ ] df['clean_tweet'] = df['clean_tweet'].str.replace("[^a-zA-Z#]", " ")
    df.head()
```

	id	label	tweet	clean_tweet
0	1	0	@user when a father is dysfunctional and is s...	when a father is dysfunctional and is so sel...
1	2	0	@user @user thanks for #lyft credit i can't us...	thanks for #lyft credit i can t use cause th...
2	3	0	bihday your majesty	bihday your majesty
3	4	0	#model i love u take with u all the time in ...	#model i love u take with u all the time in ...

Fig. 4 The process where the data is cleaned and pre-processed

3.1 Data Collection and Preprocessing

The data was collected from Kaggle website; this data was gathered from twitter the source was provided as a CSV file. The data contained 31,962 rows and 2 columns of information about the user's name and the tweets which are tweeted by the user. Datasets are divided into two as training data and testing data. Here, in this project 70% of data set are taken for training and 30% data set are used for testing.

Before building the analyzer, the tweets will be pre-processed by using the following steps:

1. Lower Case—Tweets are converted to lowercase letters.
2. URLs—Eliminates all of the URLs via regular expression.
3. @username—Remove "@username" using regex matching or replace it with a common word.
4. #Hashtag—replace the exact word without hashtags, e.g., #boycottUnitedAirlines replaced with "boycottUnitedAirlines".
5. Punctuations and additional spaces—remove the punctuations at the start and end of tweets and hashtags if available (Fig. 4).

3.2 Word Cloud Generation

Now, the cleaned text is ready for identifying sentiment, negative, positive, neutral and hashtags. For all the parameters obtained, new data frames are created. Now word cloud is created using the tweets, so as to find that which words are most used in the tweets. To create a word cloud, define a function, so word cloud can be created

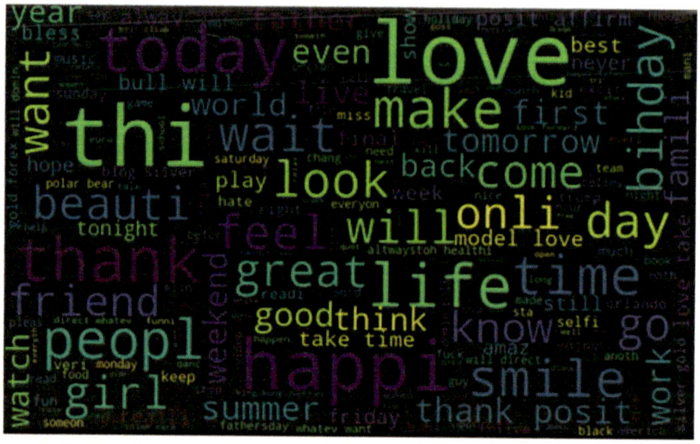

Fig. 5 Word cloud for positive words used in the tweets

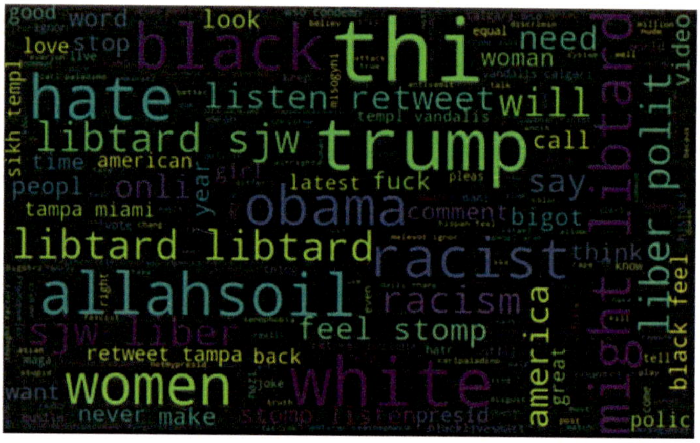

Fig. 6 Word cloud for negative words used in the tweets

for neutral tweets, positive tweets, negative tweets, etc. After defining the function word cloud for all positive, negative and neutral words is obtained (Figs. 5, 6 and 7).

3.3 Graph Generation for Hashtags

A hashtag is a tag that is preceded by # symbol, which is called hash symbol. Hashtags are widely used on social media websites like Twitter, Facebook and Instagram. The hashtags that have been used in the tweets are analyzed and a graph is generated

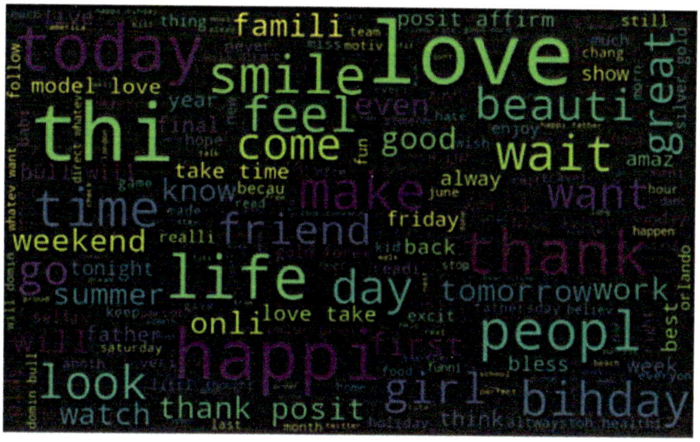

Fig. 7 Word cloud for neutral words used in the tweets

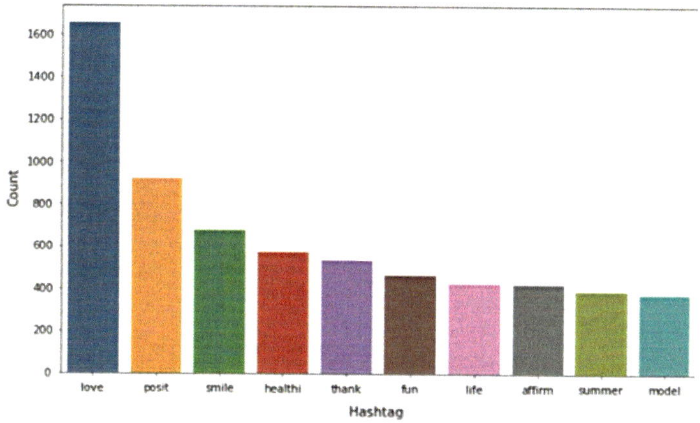

Fig. 8 Graph for positive hashtags used in the tweets

based on the positive and negative hashtags. The graph is generated on the basis of the highly used hashtag to least used hashtag (Figs. 8 and 9).

3.4 Model Training

Training a model means learning good values for all the weights. The model is trained by examining many examples and attempting to seek a model that reduces loss. The dataset is trained using logistic regression which provides an accuracy of 95% and that for testing dataset provided 94% of accuracy (Figs. 10 and 11).

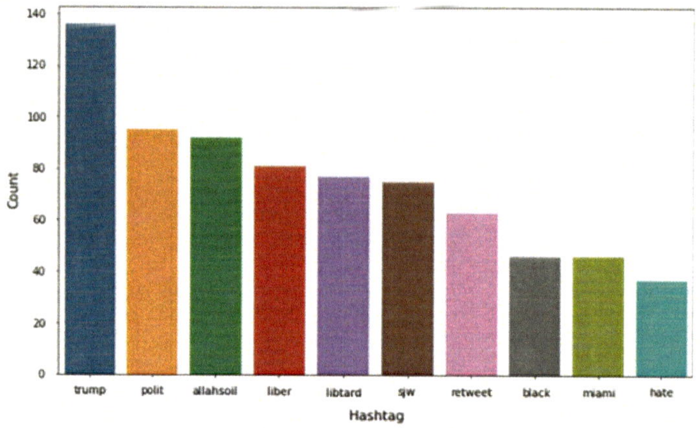

Fig. 9 Graph for negative hashtags used in the tweets

```
[38] from sklearn.linear_model import LogisticRegression
     from sklearn.metrics import f1_score, accuracy_score

[39] # training
     model = LogisticRegression()
     model.fit(x_train, y_train)

     LogisticRegression(C=1.0, class_weight=None, dual=False, fit_intercept=True,
                        intercept_scaling=1, l1_ratio=None, max_iter=100,
                        multi_class='auto', n_jobs=None, penalty='l2',
                        random_state=None, solver='lbfgs', tol=0.0001, verbose=0,
                        warm_start=False)

[40] # training
     pred = model.predict(x_test)
     f1_score(y_test, pred)

     0.49763033175355453

[ ]  accuracy_score(y_test,pred)

     0.9469403078463271
```

Fig. 10 Accuracy produced by training the model

4 Conclusion

Twitter sentiment analysis falls under the category of text and opinion mining. It focuses on analyzing the emotions of the tweets, and using a machine learning model to train it and then get its accuracy, so that one could use this model in the future according to the results. It consists of steps like data collection, text preprocessing,

```
# testing
pred_prob = model.predict_proba(x_test)
pred = pred_prob[:, 1] >= 0.3
pred = pred.astype(np.int)

f1_score(y_test, pred)

0.5545722713864307

[46] accuracy_score(y_test,pred)

0.9433112251282693
```

Fig. 11 Accuracy produced by testing the model

sentiment classification, training and testing the model. Logistic regression helps to get good accuracy for a larger number of datasets.

References

1. O. Adwan, M. Al-Tawil, A.M. Huneiti, Twitter sentiment analysis approaches: a survey. iJET **15**(15), 79 (2020)
2. R. Kamal et al., Emotion classification and crowd source sensing; a lexicon based approach. IEEE Access 27124–27134 (2019)
3. S. Taj, B.B. Shaikh, A.F. Meghji, Sentiment analysis of news articles: a lexicon based approach, in *2nd International Conference on Computing, Mathematics and Engineering Technologies (iCoMET)* (IEEE, 2019), pp. 1–5
4. S. Joshi, Sentiment analysis on WhatsApp group chat using R, in *Data, Engineering and Applications* (Springer, Singapore, 2019), p. 47
5. R. Devika, S. Revathy, *Survey on Clustering Techniques in Twitter Data* (IEEE, 2018), pp. 1073–1077

Comparative Study on Recognition of Food Item from Images for Analyzing the Nutritional Contents

E. S. Sreetha, G. Naveen Sundar, and D. Narmadha

1 Introduction

Nutrition is a vital part of health and development. Individuals with great nourishment are more dynamic and can assemble freedoms to break the patterns of yearning and neediness progressively. Malnutrition poses major threats to human health in any form. Today, the world is facing a double burden of malnutrition, both under nutrition and overweight, especially in countries with low and medium incomes.

According to a WHO report, millions of people suffer from various types of malnutrition. The consumption of high-calorie food can be unhealthy and leads to obesity, which is a preventable medical problem that occurs the body to accumulate abnormal fat. Numerous diseases such as obesity, diabetes, cholesterol, heart attacks, blood pressure, cancers of the breast, colon and prostate, and other diet-related illnesses may result [1]. So, nowadays, people are more interested to follow proper diet mechanisms to avoid such diseases. There are an ever-increasing number of individuals who need to monitor what they eat and how much dietary substance they get every day and check whether they have a sound eating routine. Hence, an exact assessment of dietary caloric admission will be vital.

E. S. Sreetha (✉) · G. Naveen Sundar · D. Narmadha
Department of CSE, Karunya Institute of Technology and Sciences, Coimbatore, Tamil Nadu, India
e-mail: sreethasreedharan@gmail.com

G. Naveen Sundar
e-mail: naveensundar@karunya.edu

D. Narmadha
e-mail: narmadha@karunya.edu

E. S. Sreetha
Department of CSE, Sahrdaya College of Engineering and Technology, Thrissur, Kerala, India

J. D. Peter et al. (eds.), *Disruptive Technologies for Big Data and Cloud Applications*, Lecture Notes in Electrical Engineering 905,
https://doi.org/10.1007/978-981-19-2177-3_27

It is more important than any other time in recent era to create tools for distinguishing and identifying food from pictures to advance the examination of the nutritional ingredients from receipts and track the individual propensity for eating and drinking. The diseases and allergies listed above can be minimized if people become more mindful of their food intake and nutritional value [2].

2 Literature Survey

The survey on nutritional study of food mainly concentrating on the areas such as different methods to recognize the food items, different techniques for analysis of nutrition dietary assessment methods, availability of different datasets. Among these, the most important step is to identify the food. Since the varieties of food images are available nowadays, the item can be identified from image. The identification of the objects in an image can be done in different ways. That is represented in Fig. 1.

2.1 Geometric Method

Lowe suggested the scale-invariant feature transform method (SIFT) [3]. It converts image data into proportional scale-invariant coordinates based on local features. It is used to derive specific invariant features from images that can be used to align different views of the same object reliably. Belongie et al. proposed an object identification method based on shape matching. After matching the shapes, the nearest neighbor classifiers are used with the similarity obtained [4]. The algorithm used is finding point-by-point correspondences between the shape of the image and a shape in the dataset.

Felzenszwalb discusses how to reflect and detect deformable shapes in pictures [5]. In a tree structure, the triangles that decompose a polygon without holes are related.

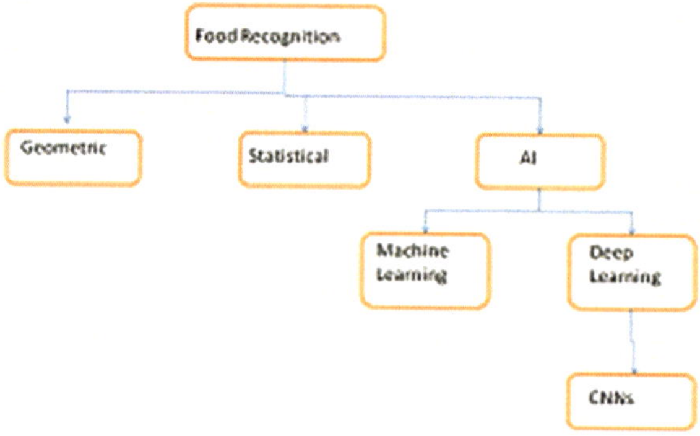

Fig. 1 Categorization of survey on object recognition from images

Here, decompose objects into meaningful pieces by choosing a specific triangulation for the polygons. It represents complex shapes with triangulated polygons by using simple building blocks. Jiang et al. suggest a method for learning and enhancing shape matching by learning a shape prior model for an object class [6]. It can learn the object class's mean shape as well as non-affine and affine transformation variants separately. Shapes are defined by vector fields rather than features in our learning approach, which makes it more general. Shape prior information will be injected during shape matching, ensuring that the matching outcome is compatible with the training examples.

2.2 *Statistical Feature Method*

A method for recognizing food based on the spatial relationships between different ingredients, that is, vegetable and non-vegetable ingredients in an image of single item is suggested by Yang et al. [7]. Pairwise local features are computed. Every pixel in the food picture is allotted a vector addressing likelihood. This describes that each pixel belongs to which food category. Baxter proposed analysis of a food item by marking each pixel with a particular ingredient [8]. Analysis is done using statistics and spatial relationships between the pixel ingredient labels. This method helps to increase the classification accuracy.

Hoashi et al. recommended utilizing multiple kernel learning to join various kinds of picture highlights like color and gradient histogram, Gabor features, and bag of features. Fivefold cross-validation gives a classification accuracy of 62.52% for 85 categories of food items. Bosch et al. proposed a method for defining features that characterize an object so that we can achieve full inter-class discrimination and intra-class robustness during the classification phase [9]. For food identification, a combination of global and local features is used [10].

2.3 *Machine Learning Method*

Machine learning is an artificial intelligence application in which no need to specify the procedure explicitly, but systems will learn and develop the method by their own. As an alternative patch encoding approach, Sanchez et al. suggested using the Fisher kernel system [11]. Fisher kernel is a statistical measure that can be used as a general picture descriptor. By maximizing the probability that two images belong to one of many predefined groups, the Fisher kernel will compute a measure of similarity between them. Fisher vectors are used to maximize a number of parameters that are arranged in a vector shape. An integrated framework for computing dense local features was proposed by Anthimopoulos et al. It uses the scale-invariant function transform on the HSV color space and hierarchical K-means clustering to construct a visual dictionary of 10,000 visual terms. The food images are then classified using a linear support vector machine classifier 78% [12].

A Bayesian framework was proposed by Aizawa et al. for including incremental learning of personal dietary tendencies [13]. A probabilistic model is used in Bayesian estimation to determine the maximum a posteriori likelihood. The naive Bayes classifier will predict the likelihood of a given object falling into a specific category. Random forest classification is a learning algorithm, and it uses a collection of tree predictors. In order to predict the behavior of the trees, values of a random vector are sampled independently. A new discriminative part mining method based on random forests was proposed by Bossard et al. [14]. A patch sampling technique focused on superpixels that avoid the use of multiple detectors on sliding windows. Experiments show that the system outperforms the competition. Fisher vectors classifier has been improved. The methodology and the limitation of various systems are given in Table 1.

2.4 Deep Learning Method

Deep learning is a machine learning technique that constructs artificial neural networks (ANNs) to mimic the structure and function of the human brain. The deep ANN is made by different layers out of neurons. Because of the solid feature learning capacity of profound learning technique, numerous unpredictable issues can be tackled in a quick and powerful manner. Recently, deep learning methods have been used in image recognition. Deep learning strategy begins to be put in the applications of food science, basically alluding to food class identification, products of the food quality recognition, food calorie assessment, etc. [18]. The convolutional neural network satisfies the requirements of the deep learning approach.

Convolutional Neural Networks (CNNs): It is a multi-layer neural network in which the input to the neurons will be taken from the previous layer. Convolutional layers, pooling layers, and completely linked layers make up a CNN structure. Each layer has a number of filters that produce various outputs. The extraction of various features for the image recognition will be done by these filters. The filters are often referred to as (convolution) kernels. Pooling layers sample the measured output from the convolutional layer. The outputs are produced by activating rectangular regions in the pooling layer. A set of convolution's layers and pooling layers is present in A CNN for representing high-level features. The fully connected network (FNN) block, which is made up of fully connected neural units, is typically used for classification [19]. The architecture of a convolution neural network is shown in Fig. 2.

Convolutional neural networks are cutting-edge models for image classification, segmentation, object detection, and a variety of other tasks. In this competitive and ravenous world, knowing some of the popular CNN architectures will keep us on track.

Table 1 Comparison of various object recognition systems

S. No.	Type	Authors	Method	Limitations
1.	Geometric features	Lowe	Image data into scale-invariant coordinates [3]	1. Recognition of the highlights such as edges, counters, and key points is troublesome
2.		Belongie et al.	Finding pointwise correspondences [4]	
3.		Felzenszwalb	Revolve around a particular shape representation using triangulated polygons [5]	
4.		Jiang et al.	It can learn a mean shape of the object class as well as the variations [6]	
1.	Statistical features	Yang et al.	Exploitation of spatial relationships between different ingredients [7]	1. Could not handle cluttered images
2.		Baxter	Calculates pairwise statistics [8]	2. As the size and number of categories grows, need to rigid systems
3.		Hoashi et al.	Fusing various kinds of image features [9]	3. Less accuracy for more categories of food items
4.		Bosch et al.	For recognition of food, the global and local features are combined [10]	
1.	Machine learning	Singh et al.	Discover a set of discriminative patches [15]	1. Less accuracy for more categories of food items
2.		Moosmann et al.	Extremely randomized clustering forests [16]	
3.		Lazebnik et al.	Bag of features method [17]	
4.		Sanchez et al.	Describes patches by their deviation from an universal generative Gaussian mixture model [11]	
5.		Aizawa et al.	The Bayesian framework facilitates incremental learning [13]	

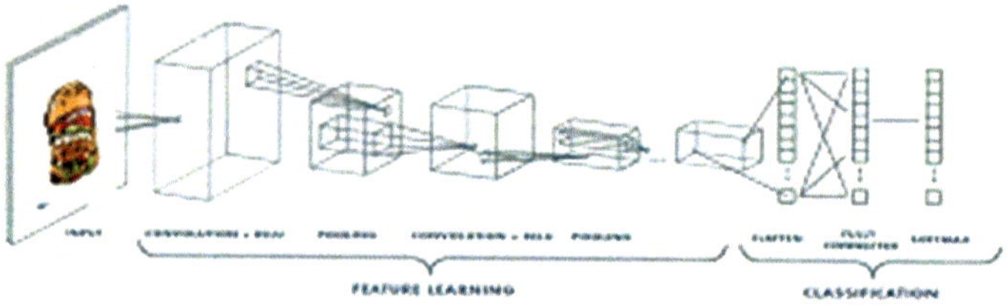

Fig. 2 CNN architecture

Table 2 Comparison of various CNN architectures

S. No.	CNN architecture	Novel method
1	LeNet-5 (1998)	Convolutions and pooling layers stacked on top of one another followed by fully connected layers at the end
2	AlexNet (2012)	Used ReLU as activation functions
3	VGG-16 (2014)	For designing of deeper networks
4	Inception-v1 (2014)	Dense modules are used for network building
5	Inception-v3 (2015)	Uses batch normalization
6	ResNet-50 (2015)	Popularized skip connections, designing deeper CNNs
7	Xception (2016)	Uses depthwise separable convolution layers
8	Inception-v4 (2016)	Change in stem module, uniform choices of Inception-v3 modules
9	Inception-ResNet-V2 (2016)	Uses residual inception blocks
10	ResNeXt-50 (2017)	Reduced the number of hyper parameters

CNN Architectures: The general principles used in almost all the CNN architectures are same. It uses a sequence of convolution and pooling layers followed by fully connected layers. Different CNN architectures differ in the number of convolution and pooling layers, size of each layers, etc. CNN architectures such as LeNet, AlexNet, VGG-16, GoogleNet, ResNet-50, ResNet-152, and others are available for image recognition [20]. Each architecture shows difference in the Top-1 accuracy and Top-5 accuracy. The novel features introduced in different CNN architectures are given in Table 2.

CNN architectures seem to be more accurate for recognition of food items from the images. Different architectures are developed on different dataset for the food identification. Selection of dataset for the training and testing purpose is most impor-

Table 3 Comparison of accuracy of food recognition

S. No.	Approach	Accuracy
1	Statistical method—pairwise local features	78% for 7 categories
2	Ingredient-level features	81.2% for 7 categories
3	Fusion of image features	62.52% for 85 categories
4	Random forest	50.76% on Food-101 dataset
5	Shape reconstruction	6134% on UDA food
6	SIFT detector	78% on PFI dataset
7	CNN	81.55% on ILSVRC dataset
8	Kfoodnet	91.3% for 23 categories
9	AlexNet	70.41% on Food-101 dataset
10	DeepFood	77.40% on Food-101 dataset
11	ResNet	78.5% on Food-101 dataset
12	NutriNet	86.72% on Food-101 dataset

tant in convolutional neural network. There are lots of food image datascts available which includes Food11, Food-5K, Food-101, UEC-Food256, etc. Among this the most widely used dataset in food recognition is Food-101. It contains 101 varieties of food, and each item is having around 1000 images. The food recognition accuracy of various methods with different datasets is given in Table 3.

3 Conclusion

We addressed various strategies proposed by various authors to detect and identify multiple food items from an image in this paper. We looked at how each approach works and how accurate they are, and we came to the conclusion that convolutional neural networks outperform all other base methods for recognition of food. Also, the most suitable dataset for food identification is Food-101. In the future, we expect to develop a system that can identify food and find the nutritional content present in the food item for recommending the diet plan.

References

1. C. Liu, Y. Cao, Y. Luo, G. Chen, V. Vokkarane, Y. Ma, DeepFood: deep learning-based food image recognition for computer-aided dietary assessment. IEEE Access **8**, 47477–47489 (2020)
2. R. Yunus, O. Arif, H. Afzal, M.F. Amjad, H. Abbas, H.N. Bokhari, S.T. Haider, N. Zafar, R. Nawazy, A framework to estimate the nutritional value of food in real time using deep learning techniques. IEEE Access (2018)

3. D.G. Lowe, Distinctive image features from scale-invariant keypoints. Int. J. Comput. Vis. **60**(2), 91–110 (2004)
4. S. Belongie, J. Malik, J. Puzicha, Shape context: a new descriptor for shape matching and object recognition, in *Proceedings of the Advances in Neural Information Processing Systems* (2001), pp. 831–837
5. P.F. Felzenszwalb, P.F. Felzenszwalb, Representation and detection of deformable shapes. IEEE Trans. Pattern Anal. Mach. Intell. **27**(2), 208–220 (2005)
6. T. Jiang, F. Jurie, C. Schmid, Learning shape prior models for object matching, in *Proceedings of the IEEE Conference on Computer Vision and Pattern Recognition*, June 2009, pp. 848–855
7. S. Yang, M. Chen, D. Pomerleau, R. Sukthankar, Food recognition using statistics of pairwise local features, in *Proceedings of the IEEE Computer Society Conference on Computer Vision and Pattern Recognition*, June 2010, pp. 2249–2256
8. J. Baxter, Food recognition using ingredient-level features
9. H. Hoashi, T. Joutou, K. Yanai, *Image Recognition of 85 Food Categories by Feature Fusion* (Department of Computer Science, The University of Electro-Communications, Tokyo, 2010)
10. M. Bosch, F. Zhu, N. Khanna, C.J. Boushey, E.J. Delp, Combining global and local features for food identification in dietary assessment. IEEE Trans. Image Process. **2011**, 1789–1792 (2011)
11. J. Sanchez, F. Perronnin, T. Mensink, J. Verbeek, Image classification with the Fisher vector: theory and practice. Int. J. Comput. Vis. **105**(3), 222–245 (2013)
12. M.M. Anthimopoulos, L. Gianola, L. Scarnato, P. Diem, S.G. Mougiakakou, A food recognition system for diabetic patients based on an optimized bag-of-features model. IEEE J. Biomed. Health Inform. **18**(4), 1261 (2014)
13. K. Aizawa, Y. Maruyama, H. Li, C. Morikawa, Food balance estimation by using personal dietary tendencies in a multimedia food log. IEEE Trans. Multimed. **15**(8), 2176–2185 (2013)
14. L. Bossard, M. Guillaumin, L. Van Gool, Food-101—mining discriminative components with random forests, in *Proceedings of the European Conference on Computer Vision* (Springer, Cham, Switzerland, 2014), pp. 446–461
15. S. Singh, A. Gupta, A.A. Efros, Unsupervised discovery of mid-level discriminative patches, in *Computer Vision ECCV* (Springer, Berlin, Germany, 2012), pp. 73–86
16. F. Moosmann, E. Nowak, F. Jurie, Randomized clustering forests for image classification. IEEE Trans. Pattern Anal. Mach. Intell. **30**(9), 1632–1646 (2008)
17. S. Lazebnik, C. Schmid, F.J. Ponce, *Beyond Bags of Features: Spatial Pyramid Matching for Recognizing Natural Scene Categories* (Beckman Institute, University of Illinois, INRIA Rhone-Alpes Montbonnot, Ecole Normale Superieure, Paris, France)
18. https://onlinelibrary.wiley.com/doi/full/10.1111/1541-4337.12492
19. https://www.analyticsvidhya.com/blog/2020/10/what-is-the-convolutional-neural-network-architecture/
20. https://www.jeremyjordan.me/convnet-architectures/

Shot Boundary Detection and Video Captioning Using Neural Networks

Avantika Balaji, S. Ganesh, T. Abishek Balaji, and K. R. Sarath Chandran

1 Introduction

There has been a surge in the number of videos in cyberspace due to the availability of highly developed communication technologies and multimedia devices. This makes video segmentation and analysis the need of the hour. Shot boundary detection is a substantial process of video browsing and retrieval as it aims to detect transitions and their boundaries between consecutive shots. The factors affecting the performance of a shot boundary detection algorithm are computation complexity and the ability to accurately identify transitions in a video, and this is still very much a domain of research as even state-of-the-art transition models tend to get confused by fast abrupt changes of the visual contents caused by camera/environment/entity. This makes it important to search for more advanced approaches. The shot information can further be used for video captioning, content-based video indexing and retrieval. In this era where an exploding amount of visual data is generated every day, video captioning can have many real-life applications. For example, automatic generation of captions for videos would greatly help users to filter what's relevant to them among the huge number of videos on platforms like YouTube. Additionally, video captioning techniques will make videos accessible to the disabled. But, one of the main challenges faced by video captioning is the scarcity of video datasets that have accompanying

A. Balaji (✉) · S. Ganesh · T. Abishek Balaji · K. R. Sarath Chandran
Department of CSE,SSN College of Engineering,Chennai, India
e-mail: avantika17021@cse.ssn.edu.in

S. Ganesh
e-mail: ganesh174307@cse.ssn.edu.in

T. Abishek Balaji
e-mail: abishek17005@cse.ssn.edu.in

K. R. Sarath Chandran
e-mail: sarathchandran@ssn.edu.in

J. D. Peter et al. (eds.), *Disruptive Technologies for Big Data and Cloud Applications*,
Lecture Notes in Electrical Engineering 905,
https://doi.org/10.1007/978-981-19-2177-3_28

text descriptions, because collecting and annotating videos are significantly more arduous and expensive than images. Due to this, most video captioning models are not very accurate and are very heavy weight and complex. To optimize the system, we extract frames at a regular interval from each shot of the video as images, generate captions for the images, and combine the captions of the images belonging to the same shot using semantic textual similarity. This results in a caption/a set of captions describing each shot of the video.

2 Related Works

Shot boundary detection or techniques used to identify a transition in a sequence of frames are a highly researched topic where various methods have been proposed for the past few decades.

Pixel-based approaches are the simplest form of shot boundary detection algorithms. In the work proposed by [2], the pixel values of two consecutive frames are taken and compared pixel by pixel to evaluate the global change in the pixel values. When the resulting value is greater than a threshold, it detects that a transition has occurred. This involves major drawbacks during the detection of transitions in fast camera movements or pan/zoom shots.

Histogram difference approach, as used in [11], is similar to pixel based, but instead of comparing pixel values, it compares the histograms of two consecutive frames. Since histogram is not as sensitive to minor changes, it results in less false hits. But, one of the major problems here is any two frames of a video sequence can have same histogram values but differ in the content of the frames.

The latest approaches involve using convolutional neural networks due to their ability to extract high-level features from images and video frames. The latest work proposed in [3] employs gradient and color information to make the technique of SBD resilient to illumination effects and abrupt motions. Another approach, proposed by [12], uses a modified CNN called a dilated convolutional neural network. This approach uses multiple 3D convolutions resulting in a wider receptive field as well as fewer trainable parameters.

Describing the content of an image or a video has been a highly researched topic in artificial intelligence. This problem requires two phases to be solved, that is, a convolutional neural network for feature extraction and natural language processing to form a description of the extracted features [7]. Earlier methods consist of visual recognizers coupled with structured formal language like And-Or graphs [13]. These types of methods are extremely hand-built and are subject to errors. Another drawback is that such models are designed exclusively for a specific use case and do not work effectively for other cases beyond the scope.

Recent works on image captioning are dominated by deep neural network techniques that use CNN for feature extraction from a set of images and subsequently use a language model, typically a recurrent neural network like long short-term memory (LSTM) to generate a description for the objects (features) extracted by the CNN layer(s) [14, 15]. The utility of this technique is that the CNN layers can be trained on various different types of image data that can extract very deep features of a given frame thus resulting in a much accurate description generated by combining it with the LSTM layer.

3 Methodology

This section details the overall flow of the system, the architectural design, and the methodology of the proposed system. Figure 1 represents the architecture of the system.

The system can be broken down into three parts—shot boundary detection, video captioning, and semantic textual similarity. The detailed design of each module, its functionality, and the algorithms used are discussed in the following sections.

3.1 Shot Boundary Detection

The proposed architecture follows the standard convolutional architectures, but with the convolutions replaced by dilated convolutions. The input to the network is a series of N successive frames of the video to which a sequence of 3D convolutions is applied. The final layer returns a prediction for each frame in the given input, where the prediction represents the likelihood of a given frame being a shot boundary. Based on a fixed threshold θ, the shot boundaries can be identified. Once the frame numbers of the beginning and end of a shot are obtained, the respective timestamps are computed by dividing the frame number by the frame rate (fps) (Fig. 2).

The principal constituent of the model is the dilated CNN. The dilated CNN network comprises four 3-dimensional 3 * 3 * 3 convolutions, with each convolution

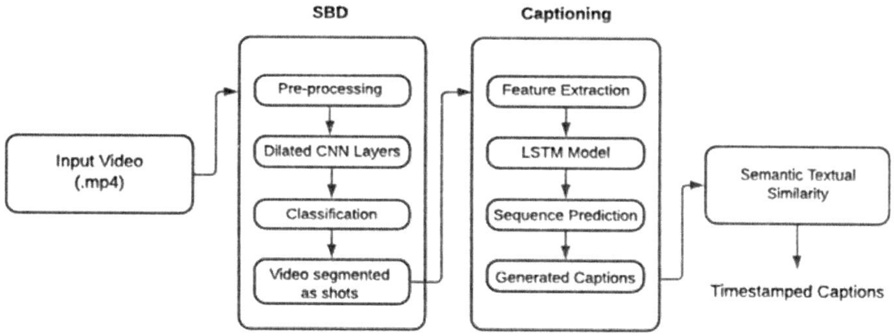

Fig. 1 Proposed system

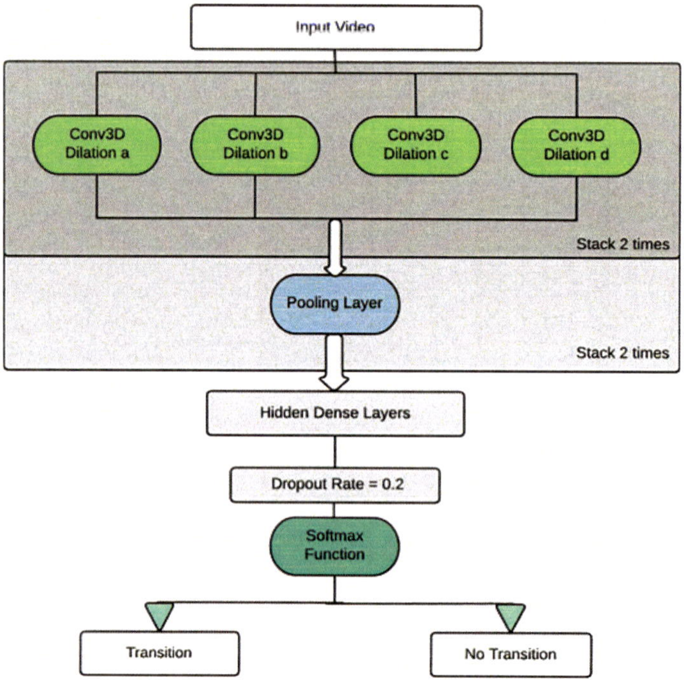

Fig. 2 Architectural design of shot boundary detection

applying a different dilation rate and the output is concatenated in the channel axis. Here, the channel is RGB. Subsequently, this output is passed to the pooling layer which employs the average pooling operation.

The dense layers refine the extracted features, and a dropout of 0.2 is used. The dataset used for training was the TRECVID IACC dataset since it is accompanied by a group of predetermined temporal segments. Training examples are generated by combining any two shots from the set with an arbitrary transition. The network was trained over 30 epochs, each with 300 batches and a batch size of 20. For testing, the BBC dataset [1] was considered.

3.2 Video Captioning

The performance and accuracy of traditional video captioning models are heavily limited by the scarcity of video datasets accompanied by text descriptions. To add to this, video captioning models are also very heavy weight and computationally intensive. To overcome this, we have used an image captioning model, which is lighter, to generate captions for the entire video. The input video is split into an array

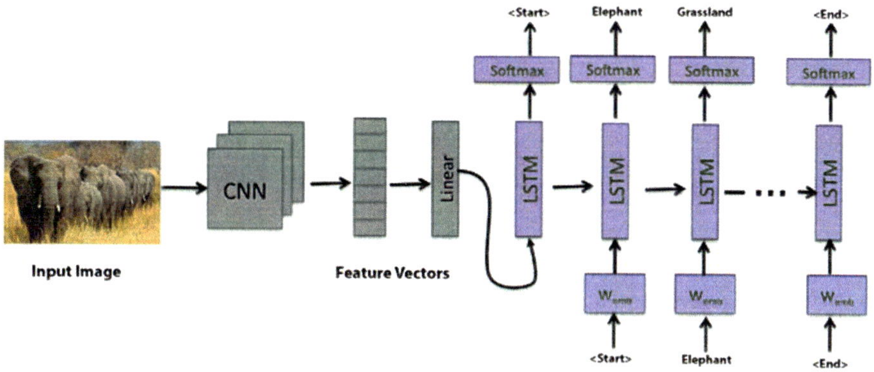

Fig. 3 Architecture of the caption generator

of frames, and at regular intervals (per second of the video), a frame is extracted as an image and given as input to the caption generator.

The architecture comprises two major components: convolutional neural network (encoder) for feature extraction and a recurrent neural network (decoder) for predicting the sequence of words/caption from the feature vectors obtained from the CNN. Here, the RNN used for predicting the sequence is long short-term memory (LSTM) (Fig. 3).

The encoder is a convolutional neural network which extracts the key features of the input image in the form of a vector. A CNN is typically a combination of feature extraction and classification. The visual feature extractor used here is the ResNet-50 [6] CNN architecture prior to feeding the images to the decoder network (LSTM). The ResNet-50 architecture is an image classification model that is 50 layers deep and can classify images into 1000 different categories. However, image classification is not relevant in this scenario, and only the feature vectors are required. Therefore, by applying transfer learning, the final classification layer of the ResNet-50 model is removed and the feature vectors are obtained.

Subsequently, the language model—LSTM—takes as input the vectors obtained from the CNN and generates the caption. During training, the inputs given to the network are the features extracted from the images and sentence embeddings generated from the ground truth, i.e., the captions. For predicting, the only input given is the extracted features. The MS-COCO [8] dataset was used for training and validation. The model was trained for 40 epochs, and a batch size of 120 was used. During the decoding phase of training, the model minimizes the negative log likelihood of the output predicted.

3.3 Semantic Textual Similarity

The output of the video captioning module is a set of captions describing each shot in the video. These captions are generated for the whole video by taking frames from each shot of the video at regular intervals and giving each of these frames to the image caption generator to generate a caption. The frames extracted from the same shot of the video may have repetitive/similar content, thereby producing the same/similar captions.

To eliminate any redundant captions, the semantic textual similarity between every pair of captions describing a shot is computed. First, the captions are encoded into sentence embeddings using a sentence transformers[1] model called "DistilRoBERTa-base". These sentence embeddings are then used for computing the similarity scores, using cosine similarity. A threshold between 0–1 is chosen, and from pairs of captions that have a similarity score higher than the threshold, a caption may be removed. From experimentation, it was found that the threshold value of 0.7 worked the best for retaining all unique captions and reducing redundancy. The outcome is a single caption/set of captions per shot of the video stored in a text file.

4 Results

In this section, we discuss the qualitative results obtained and quantitative evaluation of the results in terms of precision, recall, and $F1$-scores for shot boundary detection, and BLEU, CIDEr, and METEOR for the caption generator. Figure 4 depicts

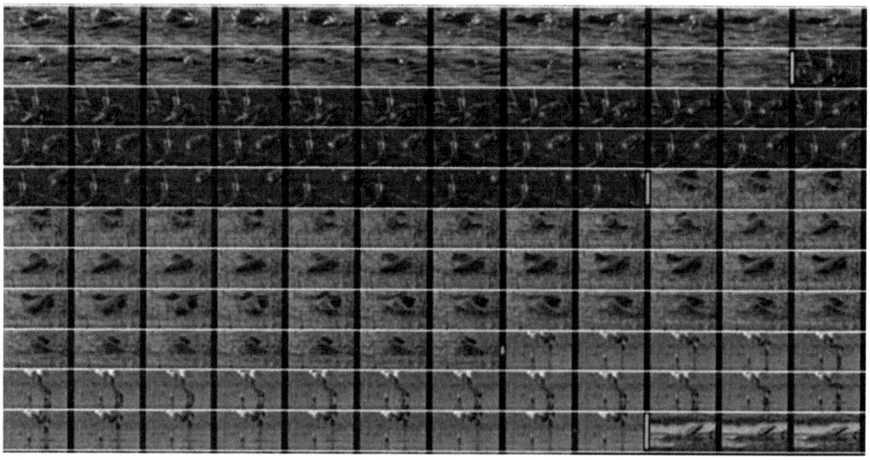

Fig. 4 Visualized results of shot boundary detection

[1] https://pypi.org/project/sentence-transformers/0.3.0/.

the visualized results of the SBD module, where the green lines indicate the shot boundaries.

Figure 5 shows the predictions generated by the captioning module. Figure 6 depicts the overall output of the system, which is in the form of a text file containing a pair of start and end frames for each shot and its description.

In Table 1, T is the total number of transitions, FN indicates false negatives, TP indicates true positives, and FP indicates false positives.

Table 2 discusses the results of the caption generation model in terms of different metrics. These metrics compare the predicted sentences with the ground truth sentences and measure the similarity between them.

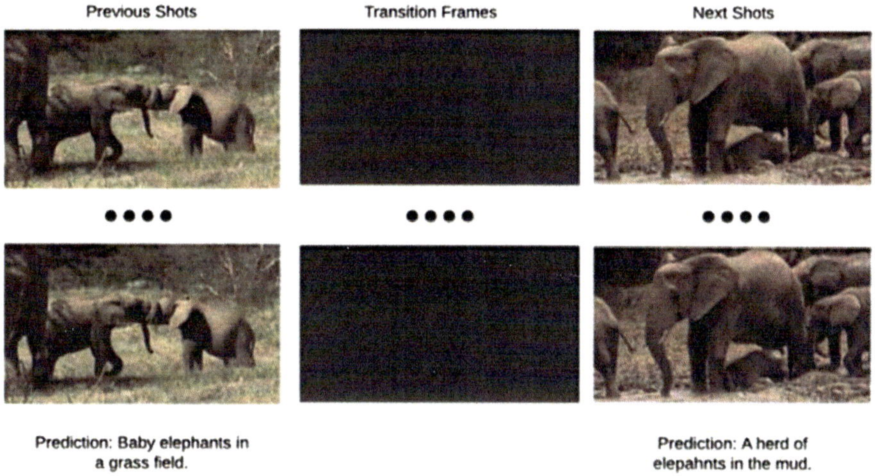

Fig. 5 Qualitative results of the caption generator

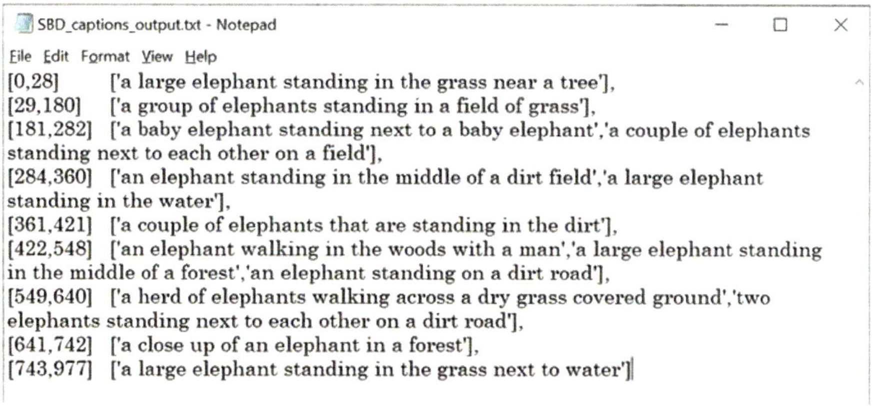
SBD_captions_output.txt - Notepad
File Edit Format View Help
[0,28] ['a large elephant standing in the grass near a tree'],
[29,180] ['a group of elephants standing in a field of grass'],
[181,282] ['a baby elephant standing next to a baby elephant','a couple of elephants standing next to each other on a field'],
[284,360] ['an elephant standing in the middle of a dirt field','a large elephant standing in the water'],
[361,421] ['a couple of elephants that are standing in the dirt'],
[422,548] ['an elephant walking in the woods with a man','a large elephant standing in the middle of a forest','an elephant standing on a dirt road'],
[549,640] ['a herd of elephants walking across a dry grass covered ground','two elephants standing next to each other on a dirt road'],
[641,742] ['a close up of an elephant in a forest'],
[743,977] ['a large elephant standing in the grass next to water']

Fig. 6 Output of the system

Table 1 Results obtained on BBC dataset for SBD

Test video	T	TP	FP	FN	Precision	Recall	$F1$-score
BBC_01	445	389	13	56	0.967	0.874	0.918
BBC_02	383	334	17	49	0.951	0.872	0.909
BBC_03	421	385	11	36	0.972	0.914	0.942
BBC_04	472	422	15	50	0.965	0.894	0.928
Overall	1721	1530	56	191	0.964	0.889	**0.924**

Bold values indicate the final score obtained by our model

Table 2 Results of the caption generation model

CIDEr	METEOR	BLEU 1	BLEU 2	BLEU 3	BLEU 4
0.692	0.213	0.657	0.471	0.327	0.228

5 Conclusion and Future Work

We have created a system that incorporates shot boundary detection with a video captioning model with the help of sentence transformers. Shot boundary detection segments the input video, and the description of these segmented videos is generated using our caption generation model. Finally, these captions are compiled into concise information that corresponds to that segment of the video using sentence transformers. This allows us to index videos or categorize different segments of a video automatically based on the content of the video.

Despite the competent performance of the caption generator, it is still far from ideal. Slight errors in the predicted captions may occur due to specific features in the images not being picked up on. In the future, we would like to experiment with attention mechanisms to overcome this and further improve the performance of our model. Currently, our system is also limited by the availability of large-scale datasets to be trained for a generalized sample space. Thus, generating datasets that focus on a specific topic or a field with captions can increase the accuracy. Our work and source code for this system can be found at our GitHub repository.[2]

References

1. L. Baraldi, C. Grana, R. Cucchiara, A deep Siamese network for scene detection in broadcast videos, in *Proceedings of the 23rd ACM International Conference on Multimedia*, Oct 2015, pp. 1199–1202
2. Y. Bendraou, *Video Shot Boundary Detection and Key-Frame Extraction Using Mathematical Models* (Université du Littoral Côte d'Opale, 2017)

[2] https://github.com/Avantika-Balaji/SBD-with-video-captioning.

3. S. Chakraborty, D.M. Thounaojam, SBD-Duo: a dual stage shot boundary detection technique robust to motion and illumination effect. Multimed. Tools Appl. **80**(2), 3071–3087 (2021)
4. Y. Chu et al., Automatic image captioning based on ResNet50 and LSTM with soft attention. Wireless Commun. Mob. Comput. **2020** (2020)
5. D.S. Guru, M. Suhil, P. Lolika, A novel approach for shot boundary detection in videos, in *Multimedia Processing, Communication and Computing Applications* (Springer, New Delhi, 2013)
6. K. He et al., Deep residual learning for image recognition, in *Proceedings of the IEEE Conference on Computer Vision and Pattern Recognition* (2016)
7. N. Krishnamoorthy, G. Malkarnenkar, R. Mooney, K. Saenko, S. Guadarrama, Generating natural-language video descriptions using text-mined knowledge, in *Proceedings of the AAAI Conference on Artificial Intelligence*, June 2013, Vol. 27, No. 1
8. T.-Y. Lin et al., Microsoft coco: common objects in context, in *European Conference on Computer Vision* (Springer, Cham, 2014)
9. E. Nishani, B. Cico, Computer vision approaches based on deep learning and neural networks: deep neural networks for video analysis of human pose estimation, in *2017 6th Mediterranean Conference on Embedded Computing (MECO)* (2017), pp. 1–4
10. N. Reimers, I. Gurevych, Sentence-BERT: sentence embeddings using Siamese BERT-networks, in *Proceedings of the 2019 Conference on Empirical Methods in Natural Language Processing* (Association for Computational Linguistics, 2019)
11. H. Shao, Y. Qu, W. Cui, Shot boundary detection algorithm based on HSV histogram and HOG feature, in *2015 International Conference on Advanced Engineering Materials and Technology*, Aug 2015 (Atlantis Press, 2015), pp. 951–957
12. T. Souček, J. Moravec, J. Lokoč, TransNet: a deep network for fast detection of common shot transitions (2019). arXiv preprint arXiv:1906.03363
13. C.F. Tsai, Image mining by spectral features: a case study of scenery image classification. Expert Syst. Appl. **32**(1), 135–142 (2007)
14. S. Tsutsui, D. Crandall, Using artificial tokens to control languages for multilingual image caption generation (2017). arXiv preprint arXiv:1706.06275
15. S. Venugopalan, H. Xu, J. Donahue, M. Rohrbach, R. Mooney, K. Saenko, Translating videos to natural language using deep recurrent neural networks (2014). arXiv preprint arXiv:1412.4729
16. O. Vinyals, A. Toshev, S. Bengio, D. Erhan, Show and tell: a neural image caption generator, in *Proceedings of the IEEE Conference on Computer Vision and Pattern Recognition* (2015), pp. 3156–3164
17. H. Wang, C. Gao, Y. Han, Sequence in sequence for video captioning. Pattern Recogn. Lett. **130**, 327–334 (2020)

Ensuring the Presence of a Person During Virtual Classes Using Histogram of Oriented Gradients (HOG) Algorithm

S. Nithya, M. Revathi, A. Sathiya Sree, T. Sivapriya, and P. Vaishnavi

1 Introduction

Nationwide, school closures during the COVID-19 pandemic have created an unprecedented level of activity in the online learning space. At this situation, educators and policymakers have expressed concerns about adequate student engagement through remote learning. E-learning system is introduced to facilitate and enhance the learning process; a problem associated with their usage is that the lack of face-to-face interaction which makes it impossible to get student's attention and response.

Computer can detect a person's face by accessing the webcamera. It can be compared with the image stored in database. The automated management system gives a facility to the faculties to reduce the burden in taking attendance and engaging the students' physical presence. This system takes the attendance automatically using face recognition. There are different advances that are done during the process of face recognition framework; yet, the essential steps of these are face detection using Haar cascades algorithm and face recognition using histogram of oriented gradients algorithm. This promising practices briefly explore the research on monitoring and promoting the presence and engagement in online learning.

2 Related Works

The related work enables attendance marking using face recognition [1, 2] system which uses two cameras installed in and out which will cost more. The methodology

S. Nithya (✉) · M. Revathi · A. Sathiya Sree · T. Sivapriya · P. Vaishnavi
Department of Computer Science and Engineering, Coimbatore Institute of Technology, Coimbatore, India
e-mail: nithya.s@cit.edu.in

J. D. Peter et al. (eds.), *Disruptive Technologies for Big Data and Cloud Applications*, Lecture Notes in Electrical Engineering 905,
https://doi.org/10.1007/978-981-19-2177-3_29

used here involves enrollment, face detection, face recognition, confirmation by class cameras and attendance marking [3, 4]. The camera inside the class will provide the attendance during the lecture and stored in the database. The most important issue was it cannot perform recognition on multiple faces at one instant. The other work is automated criminal identification [5]. In this, they use Haar classifier algorithm for feature extraction. Criminals are automatically identified using face recognition by open computer vision classifier; if they found match in any of the criminal database, timestamp will be noted. In the other, paper class attendance management system using face recognition [6] uses Raspberry Pi to detect and recognize students. Haar Feature-based Cascade algorithm [7] is used to detect and extract the given image from the set of images or from real videos. The result will be directly to MYSQL database whether the student is present or not. This section provides information gathered under different publications to use and rectify the shortcomings over in it.

3 Proposed System

The proposed system exhibits monitoring the students present during virtual class that is made via software application that has been developed for the faculty/staff member. The application is designed with several modules in user-friendly and easily understandable way to access it. Haar cascade and HOG algorithm are implemented for detection and recognition of the student during virtual class. By using webcam, student's presence will be monitored. To monitor the student activeness in between the class time, general questions will be popped. At the end of the class, the generated report will be sent to the corresponding staff mail ID.

4 System Architecture

The system architecture describes the conceptual model of structure, behavior and views of the system. The entire flow of the system is divided into six modules, namely enrollment module, checking module, face detection module, face authentication module, questions popping module and report generation module. These modules are interconnected to each other to generate the report which is used to give the user's presence and activeness during virtual classes with the supreme support of Haar cascade algorithm and histogram of oriented gradients (HOG) algorithm (Fig. 1).

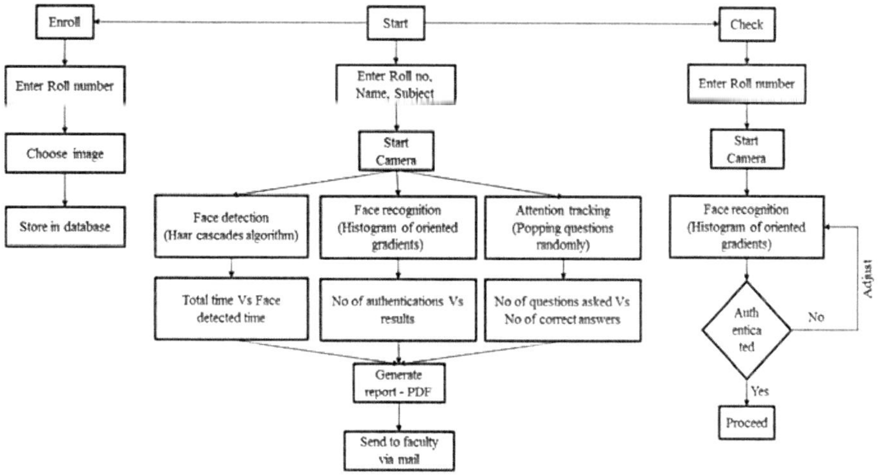

Fig. 1 Architecture diagram for the proposed system

5 Modules

5.1 Enrollment

This module is designed for the user (i.e., students, staffs) whomever attending virtual classes, webinars, meetings, tests to register with their image, if they are new to the system. Initially, the user is requested to enter their roll number and select the image and that will be stored separately for further use.

5.2 Checking

The checking module is used by the enrolled user before the class gets started to check their position using webcam, to know whether the environment does not have any disturbances (e.g., lighting) and are able to authenticated by the system correctly. In this module, the user is asked to enter the registered roll number. The webcam gets accessed and the face is detected and recognized using histogram of oriented gradients (HOG) algorithm. If the face does not get authenticated, he has to change the position and retry. This process continues till the detected face gets authenticated. Then, he can proceed for further functions.

5.3 Face Detection

At the time of logging into the Gmeet for virtual classes, he has to proceed with this module. The user is requested to enter their registered roll number, name and the subject name or code. The camera gets accessed by the software, and the video option in Gmeet platform still remains only in the mute (camera off) condition. Adding to this, our system would not store this live feed from webcam anywhere. So, the privacy of the user would not be affected. The face gets detected with the support of **Haar cascade algorithm**. Total time the face detected out of the total class duration will be calculated and provided for report.

5.4 Face Authentication

Once the face gets detected, authentication process needs to be started. The detected face is authenticated randomly by comparing their image stored in the database with the detected face using the **histogram of oriented gradients algorithm**. Count the total number of times the face gets authenticated out of the total number of times the face gets detected. The result will be used to generate the report.

5.5 Popping Questions

This module is designed to track attention of the user during virtual classes or meetings. The questions are popped at random interval of time during classes, only when the Gmeet platform is active or the student is taking notes in Word. Based on how many questions answered out of the total questions has to be popped, the user's attention will be tracked and provided to report generation.

5.6 Report Generation

This module is used to generate report from the results of the face detection, face authentication and popping questions modules such as total duration versus face detection duration, number of authentications done and number of questions answered out of popped, respectively, and a value for attentiveness will also be calculated from these values. This complete information will be stored in the Google Sheets.

6 Algorithms

6.1 Histogram of Oriented Gradients (HOG)—Face Recognition

The histogram of oriented gradients (HOG) is a feature descriptor which is used in computer vision and image processing for the purpose of detecting objects present. This technique will count the occurrences of gradient orientation in localized portions of an image. The main idea behind the histogram of oriented gradients descriptor is that, from the image, local object appearance and shape can be explained by the distribution of intensity gradients or edge directions. The image will be divided into cells of small connected regions, and for the pixels within each cell, a histogram of gradient directions will be compiled. The concatenation of these histograms is the histogram. In order to improve the accuracy, the local histograms can be contrast-normalized by the calculation of measure of intensity across block, the larger region of image and by using this value to normalize all cells which are present within the block.

The main advantages of HOG descriptor over other descriptors are its ability to be invariant to geometric and photometric transformations which is possible due to its operation on local cells, except for object orientation. These changes will appear only in larger spatial regions. For these reasons, this histogram of oriented gradients can be suited for detection of human in particular from images.

6.1.1 Loading the Image and Converting It to Gray Scale

Figure 2 depicts the original image and corresponding grey scale image.

6.1.2 Computing and Visualizing the HOG Descriptor

Figure 3 shows a role of HOG descriptor in detecting an object with background and without background.

6.1.3 Plotting the Bar Graph for Each Cell and for Its Orientation

Figure 4 shows the detailed graph for each cell and its orientaion.

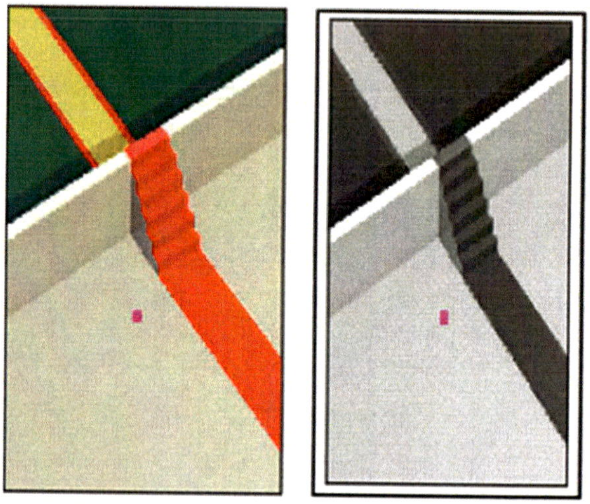

Fig. 2 Original image and grayscale image—interactive

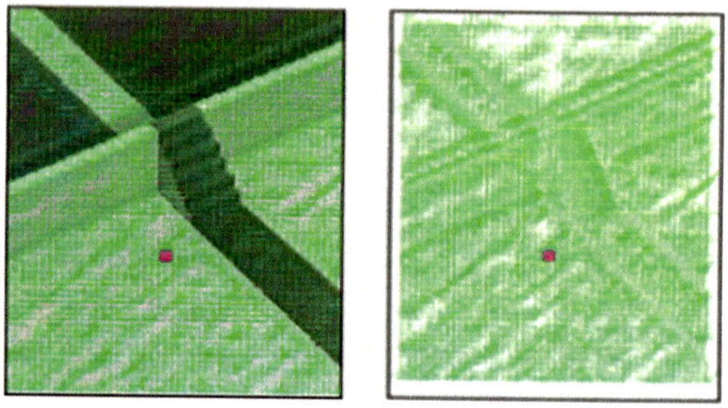

Fig. 3 HOG descriptor with and without background

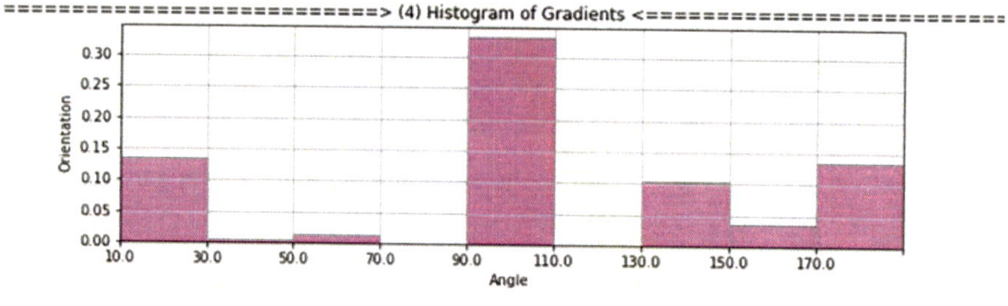

Fig. 4 Graph for each cell and its orientation

6.2 Haar Cascade—Face Detection

Haar cascade classifiers is an effective way for detecting objects. Here, the image to be used will be divided into multiple **sub-windows** and many **Haar-like features** for the computation of that image at various scales and positions for each sub-window. After that the sub-windows will be checked one by one for the **presence or absence of face** using a **cascade of classifiers**.

$$\text{Features} = \text{sum(pixels in the black area)} - \text{sum(pixels in the white area)}$$

Figure 5 shows that when we choose the second image for calculating the features, we can get even more features because the bridge there **is lighter when compared to the nearby area**. But if we use the same feature, we can only get fewer features. The third image, will also give us more features as it can detect the eye region. The reason for this is eye region will be darker when compared to the region below it. The other thing which has to be taken into consideration is that one single feature will not be capable of detecting faces with high accuracy. But using many such features will give a very accurate result in detecting the presence or absence of a face.

The main features which have real importance while detecting the face detection are

- The eye regions which will be darker than cheek regions.
- The nose region which will have bright pixels than the eye region.

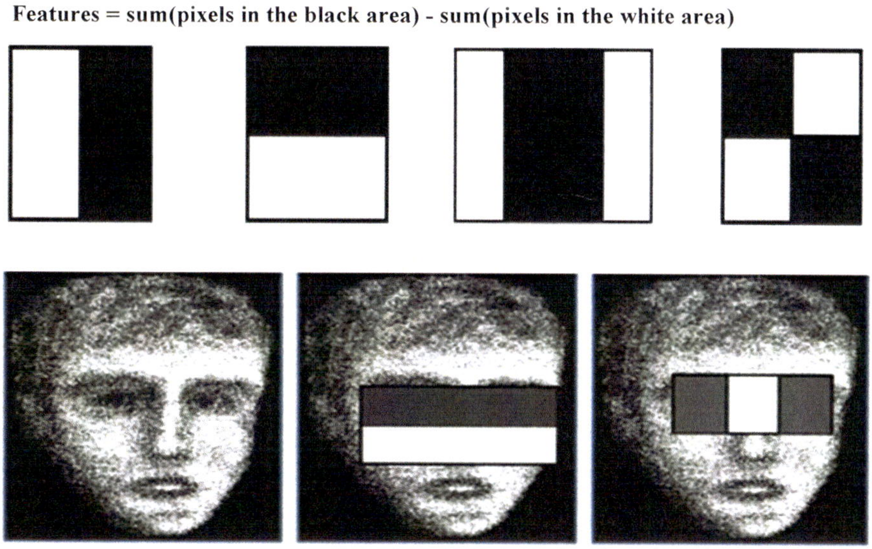

Fig. 5 Example Haar features

The failure in any one of this test will lead to the automatic discard of the window. This Haar cascade approach is good as they are very fast like Haar features. The actual solution we want is to detect the faces in the live feed of webcam which should be independent of the image size and scale, and by using this Haar cascade, we can get all these. So, it is able to detect objects by using this Haar cascade algorithm (Viola–Jones algorithm).

7 Result

Figures 6, 7, 8, 9, 10 and 11 gives the detailed flow of detecting the presence of students by extracting a facial features and by posting oral questions.

Finally the attentiveness of the each student will be generated in terms of .csv file. Sample file will be shown in Fig. 12.

Figure 13 depicts the Accuracy Comparison of HOG algorithm with other face recognition algorithms.

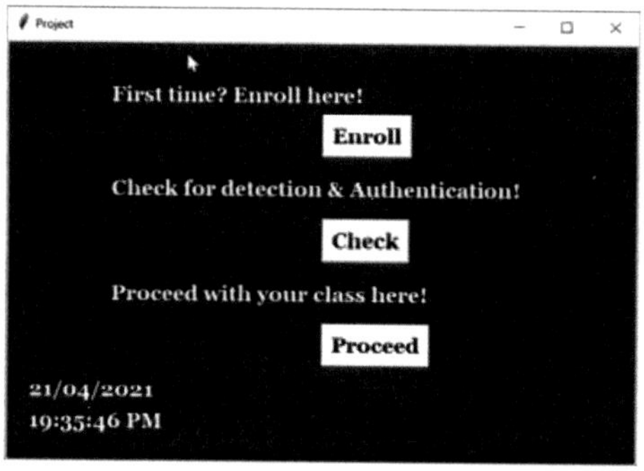

Fig. 6 Home page of the system

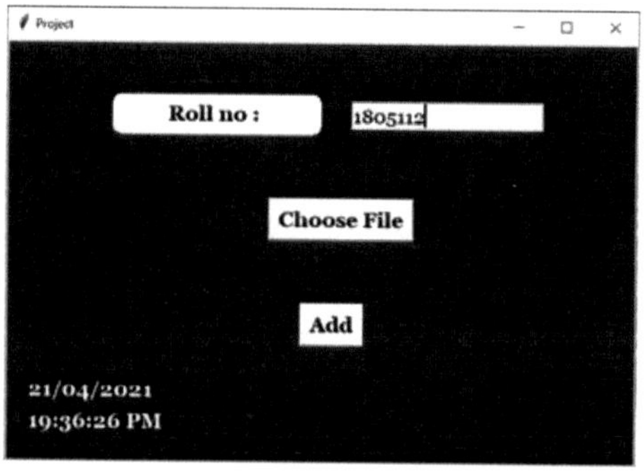

Fig. 7 Enrollment page of the system

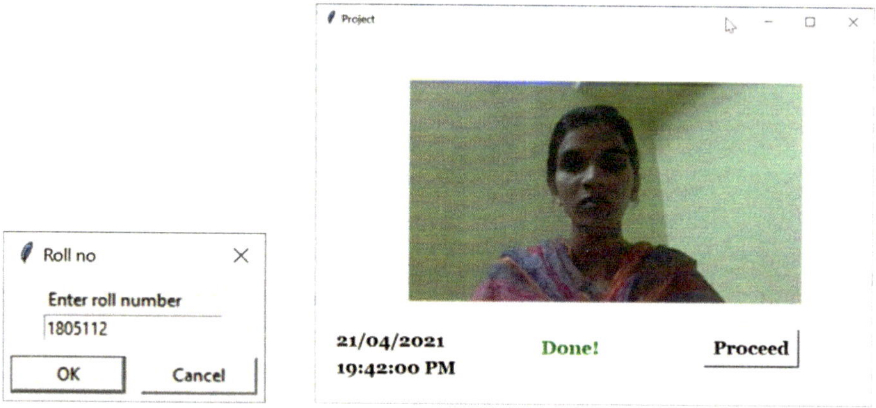

Fig. 8 Checking page of the system

Fig. 9 Start of detection module of the system

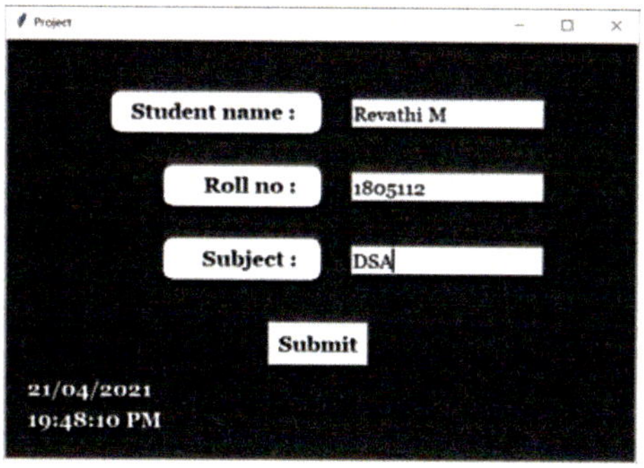

Fig. 10 Authentication by the system

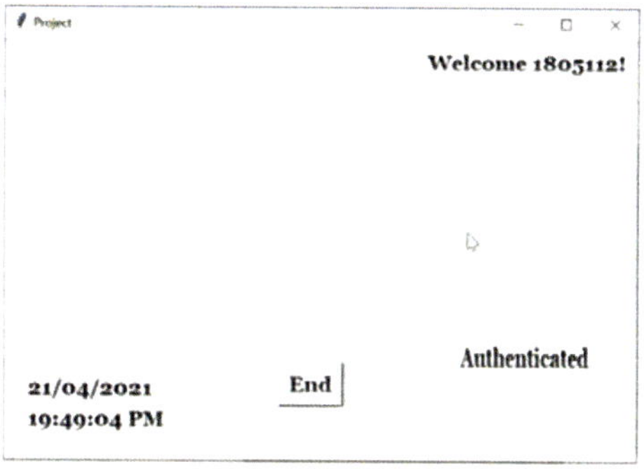

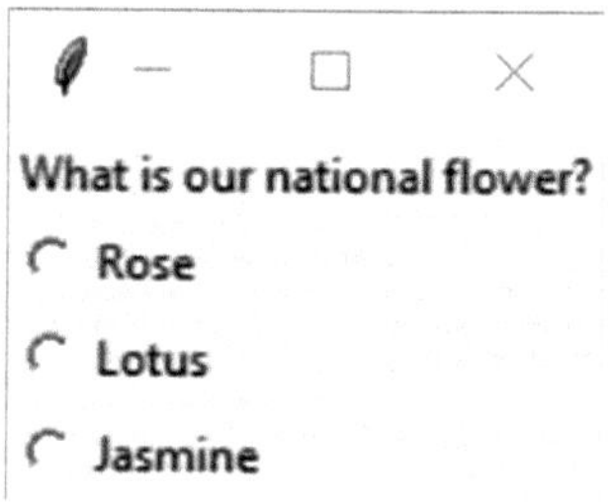

Fig. 11 Question popped by the system

Date	Name	Roll no	Subject	Total time	Face detected time	Authentications done	Results	No. of qns asked	Answered	Attentiveness(in %)
21/04/2021	Revathi M	1805112	DSA	0:00:54	0:00:54	6	6	6	5	94.44444444
21/04/2021	Sathiya Sree A	1805116	DSA	0:00:53	0:00:43	6	5	6	4	77.04402515
21/04/2021	Sivapriya T	1805119	DSA	0:00:54	0:00:51	6	5	6	5	87.03703703
21/04/2021	Vaishnavi P	1805126	DSA	0:00:54	0:00:52	6	6	6	6	98.76543209
21/04/2022	Suwetha S	1805123	DSA	0:00:53	0:00:50	6	5	6	5	87.00209643
21/04/2023	Madhuvarshini R	1805097	DSA	0:00:53	0:00:53	6	6	6	6	100
21/04/2024	Swathi R	1805124	DSA	0:00:54	0:00:52	6	6	6	5	93.20987654

Fig. 12 Result updated in the Google Sheet

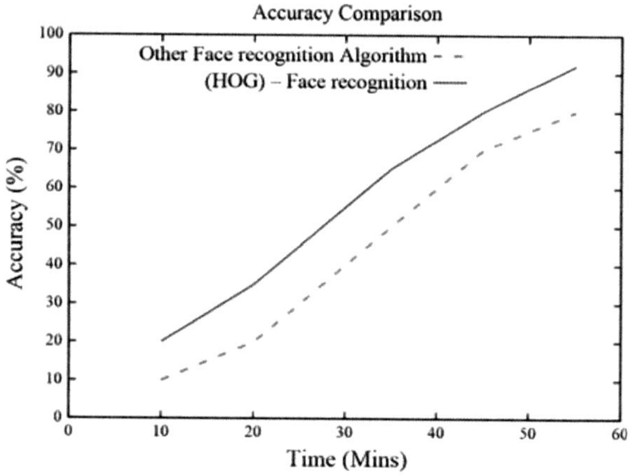

Fig. 13 Accuracy comparison of HOG algorithm with other face recognition algorithm

8 Conclusion and Future Work

In this project, an app using Python code is developed with tkinter for user interface and will be given permission to send email. This app can monitor students' presence during virtual classes by accessing webcamera, which can produce more accurate result. In terms of privacy concern, no information is taken out, because the application runs on user's laptop or PC. So, it is more secure when providing access to webcamera and mail.

It can be further implemented in near future with NLP as mining text from the class and the questions related to this can be popped instead of popping general questions. In addition to this, a database which have all the details of that particular student can be created to get these details from the student while enrolling and stored which will reduce manual typing of these details every time.

References

1. S. Dev, T. Patnaik, Student attendance system using face recognition, in *Proceedings of the International Conference on Smart Electronics and Communication (ICOSEC 2020)—IEEE Xplore Part Number: CFP20V90-ART* (2020). ISBN: 978-1-7281-5461-9
2. S. Sawhney, K. Kacker, S. Jain, S.N. Singh, R. Garg, Real time smart attendance system using face recognition techniques, in *9th International Conference on Cloud Computing, Data Science & Engineering (Confluence)* (2019)
3. O.A.R. Salim, R.F. Olanrewaju, W.A. Balogun, Class attendance management system using face recognition, in *7th International Conference on Computer and Communication Engineering (ICCCE)* (2018)
4. J.H. Lim, E.Y. Teh, M.H. Geh, C.H. Lim, Automated classroom monitoring with connected visioning system, in *APSIPA Annual Summit and Conference* (2017)
5. P. Apoorva, H.C. Impana, S.L. Siri, M.R. Varshitha, B. Ramesh, Automated criminal identification, in *Third International Conference on Computing Methodologies and Communication* (2019)
6. K. Goyal, K. Agarwal, R. Kumar, Face detection and tracking (using OpenCV), in *International Conference on Electronics, Communication and Aerospace Technology—ICECA 2017* (2017)
7. M. Awais, M.J. Iqbal, I. Ahmad, M.O. Alassafi, R.A. AlGhamdi, M. Basheri, M. Waqas, Real-time surveillance through face recognition using HOG and feedforward neural networks. IEEE Access (2019). https://doi.org/10.1109/ACCESS.2019.2937810

Identification of Alzheimer's Disease Using Principal Component Analysis-Based Data Mining Techniques

T. Jemima Jebaseeli, D. Jasmine David, and R. Emilin Renitta

1 Introduction

More than a century ago, Dr. Alois Alzheimer, a German scientist, discovered abnormalities in a dementia patient's brain parts. AD is a fatal brain disorder that affects 1 person in 10 around 65 years of age [1]. As the world's population grows, the number of older individuals with dementia will rise. About 4.2 million people in the US have dementia, at a cost of care of approximately US $200 billion annually, according to the latest figures. The worldwide dementia burden is estimated to be 44.3 million people, with future annual costs of $604 billion. By 2050, a worldwide prevalence of 135.5 million is expected to triple [2]. In India, there are 7.6 million people may get affected according to Dementia in India report published in 2020 by Alzheimer's and Related Disorders Society of India (ARDSI). However, some recent global population studies have found that the age-specific risk of dementia is decreasing, which will help to reduce the projected increase in dementia cases [1]. Rare protein particles such as plaques and tangles accumulate and damage brain cells in everyone [3]. They begin in the hippocampus, which first forms memories in the brain. For many, plaques and encounters kill the hippocampus progressively and fresh memories become more difficult.

Perhaps, there are no simple remembrances from a few hours or days ago. Along with this, more plaques and tangles grow across various areas of the brain that kills cells and compromises their work everywhere [4]. This propagation affects the

T. Jemima Jebaseeli (✉)
CSE, Karunya Institute of Technology and Sciences, Coimbatore, India
e-mail: jemima_jeba@karunya.edu

D. Jasmine David
ECE, Karunya Institute of Technology and Sciences, Coimbatore, India

R. Emilin Renitta
Food Processing Technology, Karunya Institute of Technology and Sciences, Coimbatore, India

J. D. Peter et al. (eds.), *Disruptive Technologies for Big Data and Cloud Applications*,
Lecture Notes in Electrical Engineering 905,
https://doi.org/10.1007/978-981-19-2177-3_30

various phases of AD. The disorder is transmitted from the hippocampus into the brain area where it is articulated. It becomes harder to find the right word when this happens. The disease is sliding into the brain, where rational thought occurs at a breakneck pace, rendering a person incapable of solving issues, comprehending ideas, or forming plans. The plaques and enclosures penetrate the brain portion that regulates emotions. The patient loses control of moods and emotions progressively as this happens. The disorder then travels where the brain understands what it sees and smells. At this point, a person's senses are stuffed by Alzheimer's and it triggers hallucinations [5].

Eventually, the tangles and plaques erase the earliest and most valuable memories of a human located in the back of the brain. At the very final step, the disorder undermines a person's stability and coordination, destroying the portion of the brain that controls breathing and the heart. The development is gradual and steady from mild to death and takes an average of 8–10 years [6]. The continuous and incurable support of relatives and neighbors in families to embrace AD helps lessen societal stigma.

2 Related Works

The insoluble compounds of a peptide such as amyloid beta or Abeta are plaques [7]. They're made when two enzymes, beta and gamma secretase that break a protein called Amyloid Precursor in a certain order. Other molecules may be produced as a result of the breakdown; however, Abeta is the primary cause of the illness. Abeta continues to be misfolded and sticky to form soluble oligomers at last. Certain of these are grouped in massive insoluble fibrils, which are placarded in the brain [8]. In many types of organisms, the oligomers arrive. Although predicting precisely which organisms are toxic is complex, evidence suggests that communications and plasticity are weakened. This prevents the brain from recalling or developing memories. Neurons are not the only cells damaged in AD; astrocytes and microglia also play a role. Microglia are immune cells that remove waste and prune synapses during development. Abeta is taken up by microglia, but it is activated by inflammatory cytokines, which can also damage neurons [2]. The microglia are already beginning to delete phagocytosis synapses. When synapses start to fail and neurons die, irregular patterns of behavior arise and the brain cannot efficiently process and store data.

Another major feature of AD is neurodegeneration [9]. Abeta induces cell death and damage, although certain Abeta effects appear to be mediated by Tau, a tangling factor identified in the patients' brain. A stable neuron has metabolites that Tau remains stable throughout a network of pathways in the axon. However, in AD, Tau has been altered causes the detachment of microtube. This induces an abnormal shape and the cell body moves away from the axon. Tau comes in several ways like Abeta and does not know which ones help the disease. Like Abeta, it is either soluble or sticks and places. Those processes would eventually destroy the neuron.

The misplaced Tau proteins will spread through healthy neurons via the symptoms of animal models. They also appear to misfold healthy Tau proteins, spreading disease across the brain [10]. The sequence of propagation from early to late stages of AD disease through the individual brain regions corresponds to the evolving symptoms. This pattern also shows how certain neurons are more susceptible to death than others. No treatment occurs despite these progressions in AD illness. Although medications are being formulated to target Amyloid Beta or Tau, it is not known whether or not the condition will be treated successfully.

A neural network has been utilized as a multi-class marker for identifying different phases of AD [10]. The three-dimensional cross-sectional corpus-callous and hippocampus visualization was used to assess AD, co-axial to brain characteristics. This method was used for the precision of results up to 90.65% [11]. Voxel-based morphometry (VBM) is commonly used for therapeutic procedures that need less time for completing [12]. VBM is also used in the field of medicines. Most importantly, VBM has no structural skew and can be used on any part of the brain [13]. In Japan, stand-alone VBM software and a voxel-based, dynamic spatial evaluation system for AD are used for clinical diagnosis.

The decrease in the amount of grass in the minor population of neurological disorders may be diagnosed when opposed to a stable elderly individual with AD [14, 15]. Using fractal-based analyses based on MRI-related results, a sophisticated local brain binary pattern focusing on the independent variable analysis of AD study has been widely used [16, 17]. Earlier diagnosis of AD can also be made using MRI findings [18]. Structural changes attributable to neuroanatomical configurations and function discrepancies in the human brain can be overviewed by visual inspections [19]. Furthermore, visual examinations have a great difficulty to detect minute changes in the human brain. The use of MRI 3D brain segmentation provides clinicians with a fresh perspective on AD [20].

Data mining tools for diagnostics and a better understanding of the classification of diseases are used in biological engineering and science [4]. The usage of classifier systems in medical diagnosis are increased nowadays. This new advancement in technology has made it possible to record large amounts of data [1]. To facilitate the analysis of such data for clinical decision making and diagnosis, data mining approaches were introduced. The majority of these techniques were favorable, but they were assessed on a variety of pathologically untested datasets, making a fair assessment [21, 22].

3 Materials

Open Access Series of Imaging Studies (OASIS) consist of MRI images of the human brain that are offered to the scientific society for research purpose [23]. The dataset is the combination of people aged from 18 to 96 years. It consists of mild to moderate AD images that are clinically diagnosed. Also, it consists of 400 individuals MRI images without dementia. CDR is the instrument to measure the stage of AD. The acquisition of MRI image details is as follows. TR—9.7 ms, TE—4.0 ms, flip angle is 10, TI—20 ms, TD—200 ms, orientation is sagittal, thickness gap is 1.25, 0 mm, slice number is 128, and resolution is 256 × 256 (1 × 1 mm) pixels. There are 436 MRI images which are used for the proposed method's experimental setup [24].

4 Methods

The proposed technique suggests a software approach to identify AD brain anomalies. Currently, available equipment relies on a doctor's manual interpretation of an infected person's MRI. To validate the condition, review the patient's previous health records. An MRI image is made up of nearly 120 brain slices. Each slice takes a lot of patience and time to diagnose. This procedure is often vulnerable to human error. The proposed system describes a procedure for automating the role of a doctor examining an MRI image. The morphometric features of the brain image have been used to analyze AD. As shown in Fig. 1, the input MRI image is taken for preprocessing. The MRI image is preprocessed to segment the gray and white matters, and cerebrospinal fluid.

The features of the MRI image are extracted using principal component analysis (PCA) method. It is used to classify the image and to make the decision as normal or abnormal. Then the density and original volume of the brain tissue are calculated. In the end, the abnormal MRI image is graded under the category of mild, moderate, and severe. The following expression is used to find the region of interest (*R*) of the input image required for the feature extraction process.

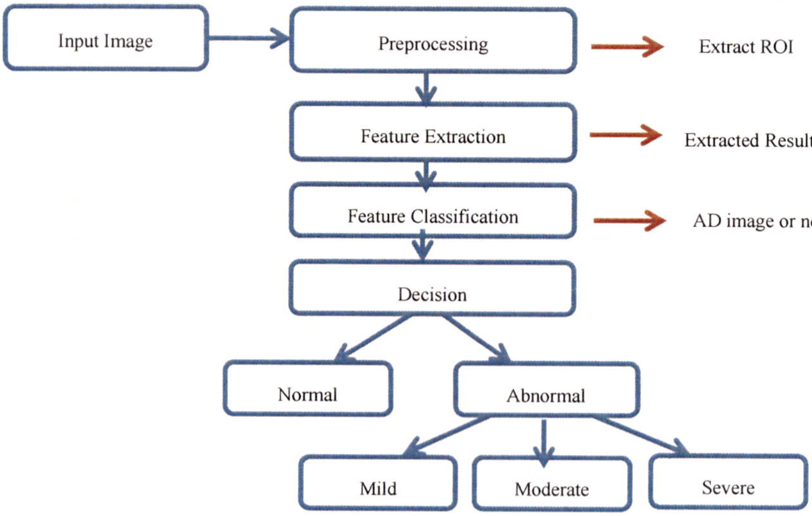

Fig. 1 Schematic diagram of the proposed method to predict Alzheimer's disease

$$R = S^T x \tag{1}$$

where $R = m' \times 1$ and it transforms the data points with m' features.

The transformation matrix is given in Eq. (2).

$$S = m \times m \tag{2}$$

The original data point in the m feature set is described as follows.

$$x = m \times 1 \tag{3}$$

PCA transforms the features of the original dataset and divides them into a small linear subset. The high weight assignments on the features would generate a high variability in classification. Hence, the linear discriminant analysis step is included in the transformation matrices. The mathematical expression is given in Eq. (4).

$$F(S) = \frac{S^{\mathrm{T}} Q_{\mathrm{B}} S}{S^{\mathrm{T}} Q_{\mathrm{S}} S} \tag{4}$$

where Q_{B} represent between class covariance matrix of the expected feature vectors. Q_{S} is within the class covariance matrix of the expected feature vectors. Then the regression model is required to build the meta-dataset attributes, the feature extraction method which extracts the features, and the classical model produces the output. The above said processes enable the system to predict the values of the attribute continuously. It is represented through the following mathematical model.

$$x = \left[V\left(x_{\mathrm{DC}_1}, \ldots x_{\mathrm{DC}_i}\right), V\left(x_{\mathrm{MO}_1}, \ldots x_{\mathrm{MO}_i}\right)\right] \tag{5}$$

where,

$V\left(x_{\mathrm{DC}_1}\right)$ is the value of the attributes.

$V\left(x_{\mathrm{DC}_i}\right)$ represents dataset characteristics.

$V\left(x_{\mathrm{MO}_1}\right)$ is the value of the attributes.

$V\left(x_{\mathrm{MO}_i}\right)$ is the model of output characteristics.

The set of possible categorical output attribute of the class (R) is described as follows,

$$C(x) = \mathrm{range}(R) \tag{6}$$

The abstract model is used to build the final decision tree as shown in Fig. 2. Here, the meta-training dataset is used to analyze the decision tree. It is constructed based on if–then rules for the knowledge base. These rules are establishing the relationship based on between meta-dataset characteristics, feature extraction, and classification model. The proposed technique produces the results and reduces the complexity between the dataset by choosing the characteristics that are needed for the data mining process.

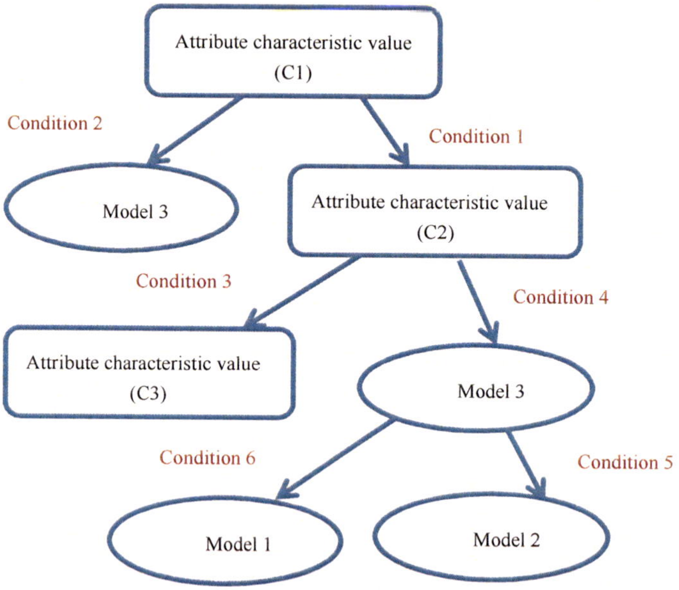

Fig. 2 Abstract model of the decision tree

5 Results and Discussions

The various dimensions such as white and gray matter volume, cortical region, cavity area, and brain density have been used to assess the MRI images from both normal and Alzheimer's patients. According to the results, the proportion of gray material with white matter is higher for an individual with brain abnormalities. There are 416 data samples in the OASIS dataset consider for the proposed method's experimentation. In a 4:1 ratio, the dataset is split into a testing dataset and a training dataset. A validation dataset was created using 10% of the training dataset's results.

As shown in Fig. 3, in a moderate situation, the white matter is compromised without an inflated lateral ventricle. In the moderate case, the white matter is somewhat reduced, with the lateral ventricle somewhat enlarged. The white matter diffuse at the extreme lateral ventricle enlarges in the severe case. The arrows indicate that the ventricular dilation is irregularly dilated posteriorly.

An average citizen is affected by AD which is between the ages of 20 and 45 will have a CDR value with a ratio of 1.1, 50–80 will have a ratio of 1.3, and anyone over 60 will have a ratio of 1.5. The patient is diagnosed with a brain abnormality if the measured ratio differs from the normal ratio. As per the global CDR scale, if it is 0 then there is no dementia, very mild case it is 0.5, mild case it is 1, moderate case it is 2, and 3 represent severe dementia, respectively. The proposed system results are compared with the competitive methods and given in Table 1. Here, the methods such as C4.5, SVM, Naive Bayes, and KNN give an average of 91.70% accuracy

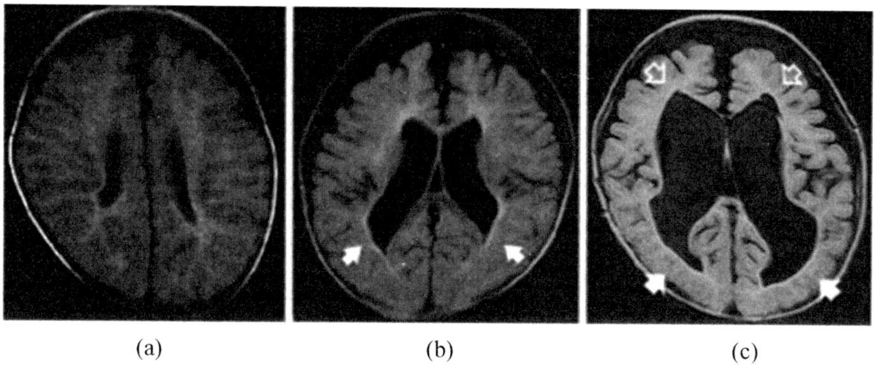

Fig. 3 Grading of Alzheimer's disease: **a** mild stage, **b** moderate stage, and **c** severe stage

Table 1 Comparison with other medical image classifiers

Methods	Accuracy %
C4.5	84.71
SVM	90.45
Naïve Bayes	76.03
KNN	86.61
Random forest	86.4
Proposed (PCA)	**99.76**

Bold values indicates, the proposed method achieved more accuracy (99.76%) than the competitive methods

[24]. The random forest method produces an average accuracy of 86.4% [1, 25]. The proposed system achieved an average accuracy of 99.76%. The proposed technique outperforms the other competitive methods in terms of performance measures. The experts evaluated the results and reported that the quality is not compromised on all MRI images of the OASIS dataset used for the experimentation.

6 Conclusions

Using the brain MRI data analysis, the proposed system is developed as an accurate approach to AD diagnosis. While a majority of current research focuses on binary classification, the model significantly improves multi-class classification. The proposed PCA technique is a very useful tool for detecting early-stage AD. Even though the suggested model has only been validated on the AD dataset, the proposed PCA method can be used for similar problems in the clinical domain. The proposed framework will be updated in the future to evaluate the PCA technique for various AD databases and other brain disease diagnoses.

References

1. J. Escudero, E. Ifeachor, J.P. Zajicek, C. Green, J. Sheare, S. Pearson, Machine learning-based method for personalized and cost-effective detection of Alzheimer's disease. IEEE Trans. Biomed. Eng. **60**(1), 164–168 (2013)
2. K.M. Langa, Is the risk of Alzheimer's disease and dementia declining? Alzheimer's Res. Ther. **7**(1), 34 (2015)
3. A. Khan, L.-S. Liu, M. Usman, S. Fong, Early diagnosis of Alzheimer disease using instance-based learning techniques. J. Med. Imaging Health Inform. **6**(4), 1111–1118 (2016)
4. K. Tejeswinee, G.J. Shomona, R. Athilakshmi, Feature selection techniques for prediction of neuro-degenerative disorders: a case-study with Alzheimer's and Parkinson's disease. Procedia Comput. Sci. **115**, 188–194 (2017)
5. M. Marttinen, J. Paananen et al., A multiomic approach to characterize the temporal sequence in Alzheimer's disease-related pathology. Neurobiol. Dis. **124**, 454–468 (2019)
6. H. Matsuda, MRI morphometry in Alzheimer's disease. Ageing Res. Rev. **30**, 17–24 (2016)
7. R.O. Weller, R.O. Carare, D. Boche, Amyloid: vascular and parenchyma, reference module, Neurosci. Biobehav. Psychol. (2017)
8. D. Sarwinda, A. Bustamam, Detection of Alzheimer's disease using advanced local binary pattern from hippocampus and whole brain of MR images, in *International Joint Conference on Neural Networks (IJCNN)* (2016), pp. 5051–5056
9. J. Zhang, Y. Gao, Y. Gao, B.C. Munsell, D. Shen, Detecting anatomical landmarks for fast Alzheimer's disease diagnosis. IEEE Trans. Med. Imaging **35**(12), 2524–2533 (2016)
10. Y.F. Fathabad, M. Balafar, Application of content-based image retrieval in diagnosis brain disease. Int. J. Tech. Phys. Probl. Eng. **4**(4), 122–128 (2012)
11. R.A. Ben, B. Faouzi, A. Hamid, Diagnosis of Alzheimer diseases in early step using SVM, in *IEEE 13th International Conference on Computer Graphics, Imaging and Visualization (CGiV)* (2016), pp. 364–367
12. M.M. Dessouky, M.A. Elrashidy, T.E. Taha, H.M. Abdelkader, Computer-aided diagnosis system for Alzheimer's disease using different discrete transform techniques. Am. J. Alzheimers Dis. Other Demen. **31**(3), 282–293 (2016)
13. K.S. Nayaki, A. Varghese, Alzheimer's detection at early stage using local measures on MRI: a comparative study on local measures, in *International Conference on Data Science & Engineering (ICDSE)* (2014), pp. 224–227
14. Y. Kaewaramsri, S.A. Alfarozi, K. Woraratpanya, Y. Kuroki, Fractal dimension for classifying 3D brain MRI using improved triangle box counting method, in *International Conference on Information Technology and Electrical Engineering (ICITEE)* (2016), pp. 1–6
15. S. Lahmiri, M. Boukadoum, Automatic detection of Alzheimer disease in brain magnetic resonance images using fractal features, in *6th International IEEE/EMBS Conference on Neural Engineering (NER)* (2013), pp. 1505–1508
16. W. Yang, H. Xia, B. Xia, L.M. Lui, X. Huang, ICA-based feature extraction and automatic classification of AD-related MRI data, in *International Conference on Natural Computation (ICNC)* (2010), pp. 1261–1265
17. K.P. Gunawardena, R.N. Rajapakse, N.D. Kodikara, I.U. Mudalige, Moving from detection to pre-detection of Alzheimer's disease from MRI data, in *Sixteenth International Conference on Advances in ICT for Emerging Regions (ICTer)* (2016), p. 324
18. M.S. Holia, V.K. Thakar, Mutual information based image registration for MRI and CT SCAN brain images, in *International Conference on Audio, Language and Image Processing (ICALIP)* (2012), pp. 78–83
19. A. Huang, R. Abugharbieh, R. Tam, A hybrid geometric–statistical deformable model for automated 3-D segmentation in brain MRI. IEEE Trans. Biomed. Eng. **56**(7), 1838–1848 (2009)
20. Alzheimer's disease image dataset, https://www.oasis-brains.org/

21. J. Feng, S.-W. Zhang, L. Chen, Identification of Alzheimer's disease based on wavelet transformation energy feature of the structural MRI image and NN classifier. Artif. Intell. Med. **108** (2020)
22. X. Zhao, C.K.E. Ang, U.R. Acharya, K.H. Cheong, Application of artificial intelligence techniques for the detection of Alzheimer's disease using structural MRI images. Biocybern. Biomed. Eng. **41**(2), 456–473 (2021)
23. MRI for Alzheimer's, https://www.kaggle.com/jboysen/mri-and-alzheimers
24. B. Li, W. Li, D. Zhao, Multi-scale feature based medical image classification, in *IEEE 3rd International Conference on Computer Science and Network Technology* (2013), pp. 1182–1186
25. T. Jemima Jebaseeli, R. Venkatesan, K. Ramalakshmi, Fraud detection for credit card transactions using random forest algorithm, in *Advances in Intelligent Systems and Computing* (2021), pp. 189–197

Dew Computing-Inspired Mental Health Monitoring System Framework Powered by a Lightweight CNN

Tanusree Podder, Diptendu Bhattacharya, and Abhishek Majumdar

1 Introduction

During human communication, facial expressions indicate the mindset of a person's emotions like positive (happy, surprise, neutral) or negative (angry, sad, fear, disgust). Several automatic facial expression recognition (FER) systems have already been there in literature due to advancements in big data processing and machine learning methods. The FER system uses images [1], speech [2], video [3], or ECG [4] signals as input to capture the emotional states. Recently, there has been an increasing demand to utilize the FER system in the healthcare industry to capture patients' mental health conditions.

Researchers are currently trying to incorporate a FER system to treat a patient having difficulty in identifying the correct emotional states, e.g., autism [5], schizophrenia [6], Parkinson's [7], and bipolar disorder [8], etc. In this context, Yolcu et al. [9] designed deep learning-based FER system to diagnose and monitor different neurological disorders. Firstly, they formed an iconized image after collecting essential parts from the face image. Then, the system trains the raw and iconized image together with a cascaded deep learning model for emotional state identification. For autism treatment, Sivasangari et al. [10] used a Q sensor to capture the EEG data for

T. Podder
National Institute of Technology Agartala, Agartala, Tripura, India

D. Bhattacharya
Department of Computer Science and Engineering, National Institute of Technology Agartala, Agartala, Tripura, India

A. Majumdar (✉)
Department of Computer Science and Engineering, Karunya Institute of Technology and Sciences, Coimbatore, Tamilnadu, India
e-mail: abhishekmajumdar91@gmail.com

J. D. Peter et al. (eds.), *Disruptive Technologies for Big Data and Cloud Applications*, Lecture Notes in Electrical Engineering 905,
https://doi.org/10.1007/978-981-19-2177-3_31

stress level identification. Further, it utilizes a FER module to alert the caregivers by capturing a 3D shape of the face.

The traditional process of FER is used to follow different handcrafted feature extraction methods to recognize emotion. But recently, researchers started detecting emotion with the deep learning model due to its self-learning ability. These emotion detection methods are applied recently in the industry for smarter healthcare services. For example, for improving the healthcare system in smart cities, Muhammad et al. [11] proposed a facial expression recognition system that applies bandlet transform to obtain the sub-bands from the face image. Afterward, the CS-LBP histogram is extracted from each sub-band and concatenated to classify using the Gaussian mixture model and SVM. Finally, the score of both the classifier is fused to get the final confidence score to determine the final emotional state. Likewise, an emotion-aware healthcare system for smart home scenarios is proposed in [12]. The system receives speech and image as input through different IoT devices and performs the prediction using edge cloud technology. After prediction, the system alters the caregiver in case pain is detected in the remote cloud. Fei et al. [13] proposed a deep CNN-based method for FER for emotion interpretation toward mental health. It modifies AlexNet with fully connected layers to use as a CNN model for emotion detection. Further, experiments are carried out to classify the extracted deep features using *linear discriminant analysis* (LDA), SVM, and *k-nearest neighbors* (KNN). The outcome shows that the LDA outperforms the other two classifiers in terms of recognition accuracy. For emotional health problems, Uddin et al. [14] proposed a system that utilizes depth video data to extract local directional position patterns (LDPP). Then, they filtered the features by employing the PCA and generalized discriminant analysis (GDA) and applied a deep belief network (DBN) for classification. Kim et al. [15] developed a system to monitor facial expression during radiotherapy. The system gives warning signs to therapists regarding the advanced movement of the patient to enhance the stability of the radiation procedure. It uses a deep convolution network to train the system for expression detection using the CK+ dataset. For detecting emotion from audio-visual big data, Hossain and Muhammad [16] proposed a CNN-based method. The audio signal is first converted into Mel spectrogram to process using 2D-CNN, and the corresponding video signal is processed with 3D-CNN. The extracted features are fused to feed into two EML and finally classified with support vector machine (SVM). Furthermore, the advancement of deep learning models like AlexNet [17], VGG [18], Inception [19], DenseNet [20], etc., encourages researchers to design application based on it. But, when applied to the real-time application, these models consume a lot of processing time due to their heavyweight nature.

In this regard, a lightweight CNN has been designed in this study to monitor emotion in the healthcare environment with minimum time. In addition, a dew computing-inspired framework has also been proposed to configure it in the real-time healthcare environment.

2 Methodology

The proposed framework for real-time mental health status monitoring has been depicted in Fig. 1, inspired by the literature [21–24]. It comprises three layers: the dew computing layer, hospital server layer, and cloud layer. All layers have their specific tasks and perform independently, thereby providing efficient service to the next adjacent layer and form an effective system. The dew computing layer is responsible for acquiring the continuous facial expressions of the patient through a real-time camera deployed in front of the patient's bed. Afterward, it will be forwarded to the embedded microcontroller configured with the proposed lightweight CNN model for recognizing the patient's mental health status efficiently and faster. Eventually, these recognized decisions for a patient in a specific period will be sent to the hospital server for analyzing it at the edge level. The hospital server is responsible for analyzing a patient's collective mental health status and notifying the concerned health consultants accordingly. Besides, the hospital server will also send all the data gathered from the dew computing layer to the cloud layer for permanent storage and further in-depth analysis.

2.1 System Model

Figure 2 represents the system model for facial emotion detection. The frames will be processed first to detect the face region. This further goes through preprocessing

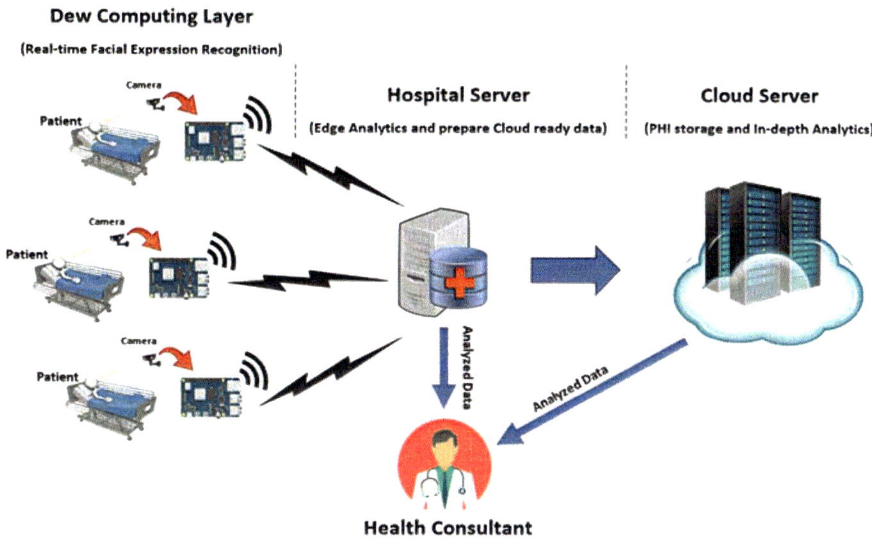

Fig. 1 Proposed framework for real-time mental health status monitoring

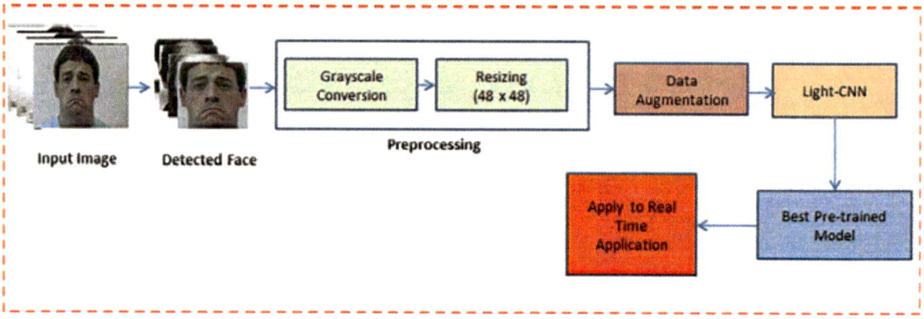

Fig. 2 Proposed system model for real-time application

module for grayscale conversion and resizing. Afterward, on the fly, data augmentation is carried out to give versatility to the dataset. Finally, the augmented images are trained with the light CNN for emotion detection. After training, the best-trained weight is stored to apply in real-time detection.

2.1.1 Preprocessing

Minimal preprocessing steps have been employed here as the application needs faster real-time processing of emotion. The main step toward FER is face detection, and the Haar cascade classifier proposed by Viola and Jones [25] is utilized for faster face detection. In the next step, on the fly augmentation through ImageDataGenerator of Keras is used to improve the accuracy of the CNN model. The methods adopted as a part of data augmentation are rotation, image shifts, brightness adaptation, zooming, horizontal flip, etc. Figure 3 shows some sample augmented images. Further, the images are resized to 48×48 and normalized to [0, 1].

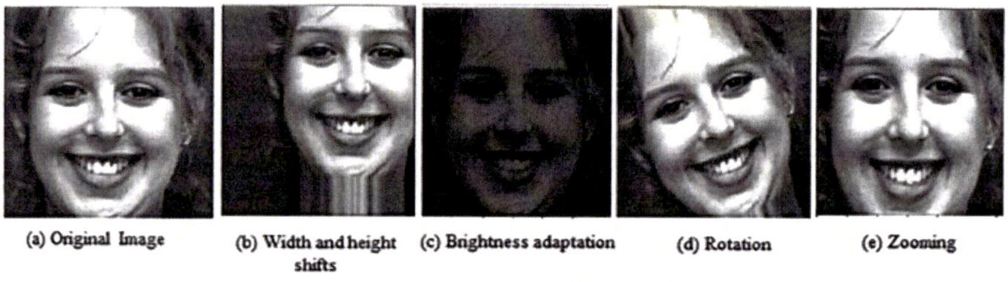

Fig. 3 Augmented images from CK+ dataset

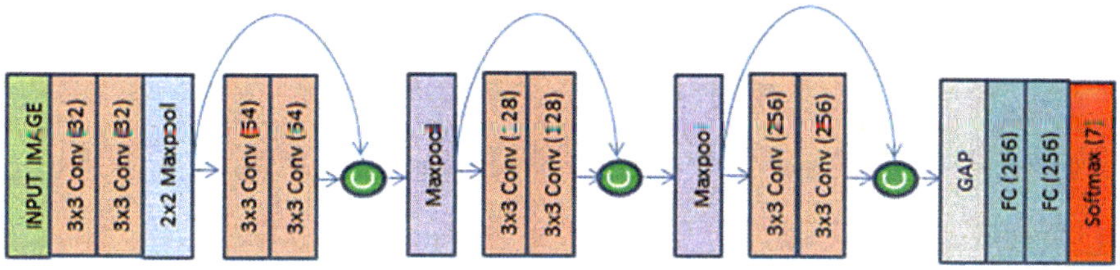

Fig. 4 Proposed CNN architecture

2.1.2 Proposed CNN Architecture

The proposed CNN model is a lightweight variant of VGG and DenseNet, as shown in Fig. 4. The model starts with two convolutional (Conv) layers to extract 32 feature maps with a 3×3 filter and a maxpooling layer with filter size 2×2 for dimensionality reduction. The maxpooling layer down-sample the feature maps to reduce the dimension by preserving the relevant details. The feature further passes to two consecutive Conv layers to extract 64 feature maps with a 3×3 filter size. The output of which gets concatenated with the output of the previous maxpooling layer for better gradient propagation. The concatenated features pass through a maxpooling layer, and the same process is repeated until we reach the global average pooling (GAP) layer. The GAP layer returns a 1D features map of size 480 followed by two fully connected (FC) layers of size 256 before final classification through a softmax layer. Additionally, batch normalization is applied after every layer to accelerate the training with standardizing input. Further, a dropout layer is added after each fully connected layer with a probability of 0.5 to regularize the network.

2.2 Datasets

The detailed overview of the dataset utilized to showcase the effectiveness of the proposed method is given here with processed sample images shown in Fig. 5.

FER-2013 [26]: The research community widely uses the FER-2013 dataset as it is freely available and resembles images from a wild setting. The dataset contains 35,887 grayscale images of size 48×48 belonging to 7 basic facial expressions. Further, it is well distributed into training, validation, and testing set with 28,709, 3589, and 3589 images (Table 1).

CK+ [27]: It is released as an extended version of the Cohn-Kanade (CK) database. The dataset contains a video sequence of expression images starting from neutral to peak expression. Along with seven basic expressions, it contains contempt emotion. The size of the images is either 640×490 or 640×480 pixels. It recorded the expression of 210 adults in a laboratory-controlled environment from varying age group and community (Table 2).

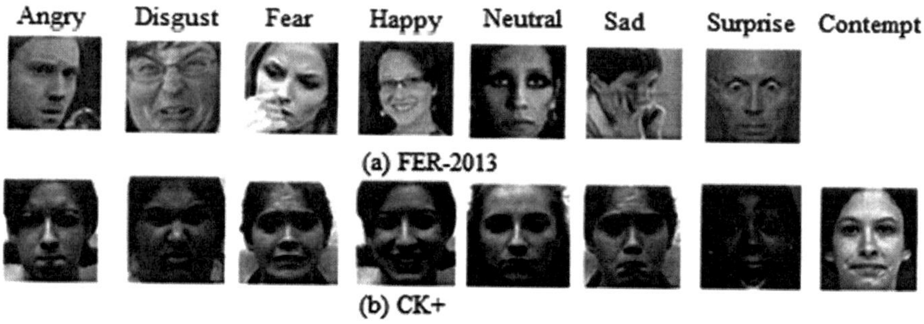

Fig. 5 Processed sample images from FER-2013 and CK+ dataset

Table 1 Performance comparison with literature on the FER-2013 database

Methodology	Accuracy (%)
AlexNet + FC6 + LDA [13]	56.40
MobileNet with softmax + center loss [5]	68.31
AlexNet [17]	58.32
VGG16 [18]	65.80
InceptionV3 [19]	68.86
Proposed method	**68.93**

Table 2 Performance comparison with literature on the CK+ database

Methods	Accuracy (%)
CNN [15]	85.60
AlexNet + FC6 + LDA [13]	94.70
MobileNet with softmax + center loss [5]	95.38
AlexNet [17]	86.40
VGG16 [18]	89.45
InceptionV3 [19]	90.32
Proposed method	**96.12**

3 Results and Discussion

The experiments carried out on the FER-2013 and CK+ datasets are illustrated here to evaluate the performance of the proposed system. The proposed algorithm outperforms all other methods because of the bypass connection introduced in the light CNN for gradient propagation. Further, for smaller databases like CK+, it utilized transfer learning to obtain better accuracy. For the runtime environment, the GPU support of Google Colab is utilized during the execution of the CNN model. Furthermore, for compilation, an Adam optimizer with a fixed learning rate of 0.001 is applied for 150 epochs during the training process.

3.1 Experiments on FER2013 Dataset

It is challenging to obtain a good accuracy while maintaining a minimum time frame as the images are captured in the wild environment. But, the proposed method manages to achieve 68.93% accuracy with minimum time. Figure 6 shows the confusion matric and its normalized version along with the classification report and the training accuracy graph. The normalized confusion matrix shows that the lowest accuracy is obtained in classifying fear. It was caused because out of 528 images, 140 images are classified as sad due to low-interclass variation in both emotions. The classification report shown in Fig. 3a reports that the method obtains 70.34% precision, 68.93% $f1$-score, and 69.04% recall. The graph on training and validation accuracy shows how validation data converges after 50 epochs of running. Further, the best pretrained weight on FER-2013 is saved to apply in healthcare applications.

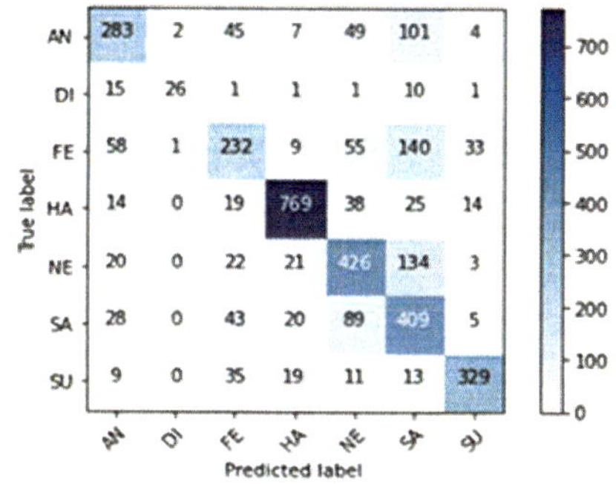

(a) Confusion Matrix

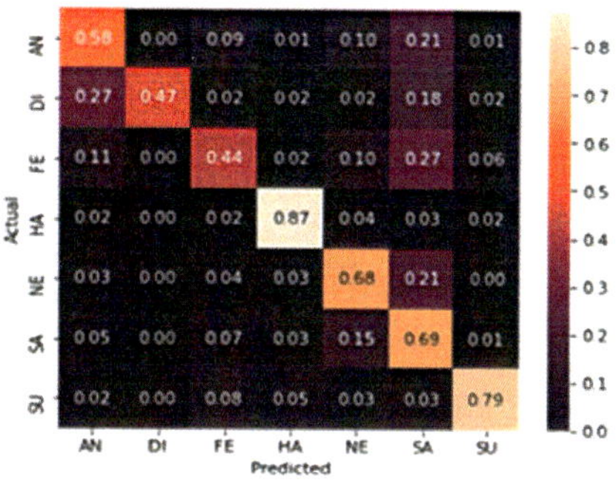

(b) Normalized Confusion Matrix

	precision	recall	f1-score	support
AN	0.6628	0.5764	0.6166	491
DI	0.8966	0.4727	0.6190	55
FE	0.5844	0.4394	0.5016	528
HA	0.9090	0.8749	0.8916	879
NE	0.6368	0.6805	0.6579	626
SA	0.4916	0.6886	0.5736	594
SU	0.8458	0.7909	0.8174	416
accuracy			0.6893	3589
macro avg	0.7181	0.6462	0.6683	3589
weighted avg	0.7035	0.6893	0.6904	3589

(c) Classification Report

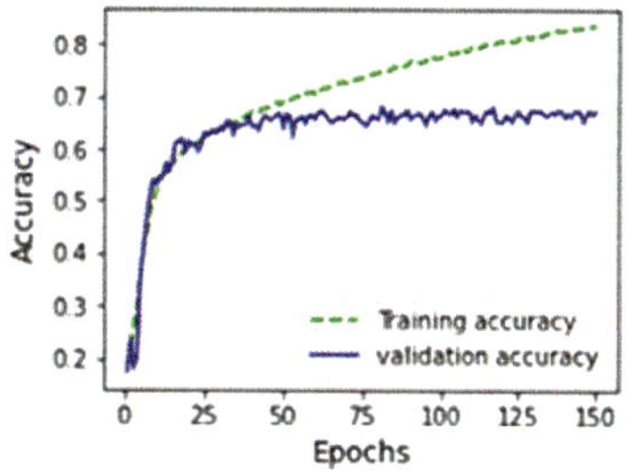

(d) Training vs. validation accuracy

Fig. 6 Confusion matrix, normalized confusion matrix, classification report, and accuracy graph on FER-2013 dataset

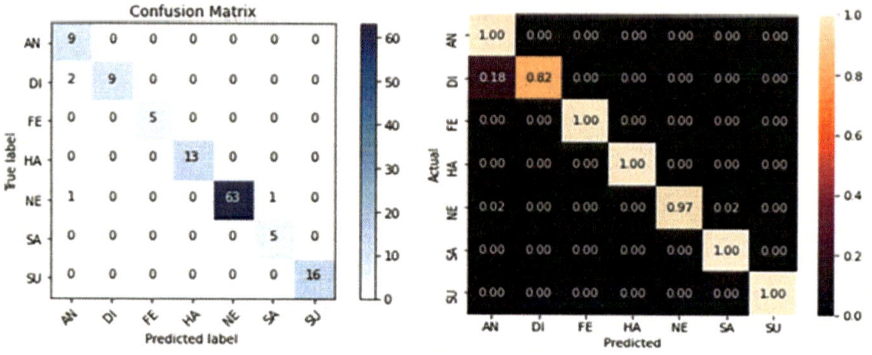

(a) Confusion Matrix

(b) Normalized Confusion Matrix

	precision	recall	f1-score	support
AN	0.7500	1.0000	0.8571	9
DI	1.0000	0.8182	0.9000	11
FE	1.0000	1.0000	1.0000	5
HA	1.0000	1.0000	1.0000	13
NE	1.0000	0.9692	0.9844	65
SA	0.8333	1.0000	0.9091	5
SU	1.0000	1.0000	1.0000	16
accuracy			0.9677	124
macro avg	0.9405	0.9696	0.9501	124
weighted avg	0.9751	0.9677	0.9689	124

(c) Classification Report

Fig. 7 Confusion matrix, normalized confusion matrix, and classification report on CK+ dataset

3.2 Experiments on CK+ Dataset

To evaluate the performance of the CK+ dataset, a fivefold cross-validation technique is adopted as the dataset is not split into training, validation, and testing set. Further, the pretrained weight on FER-2013 is used to fine-tune the CNN model during training of the CK+ dataset as it contains relatively fewer images. The proposed method achieves a mean accuracy of 96.12%. Figure 7 shows the confusion matrix and its normalized version with classification report on one of the folds. It reveals that 18% disgust images are misclassified as angry that means out of nine images, two images are misclassified. The classification report reveals that the particular fold gets 97.51% precision, 96.77% $f1$-score, and 96.89% recall.

3.3 Real-Time Experiment

For the real-time experiment, a webcam is utilized to get the frames which further preprocessed to resized grayscale images after face detection. The images are fed to the pretrained model for classification. Figure 8 shows sample images classified

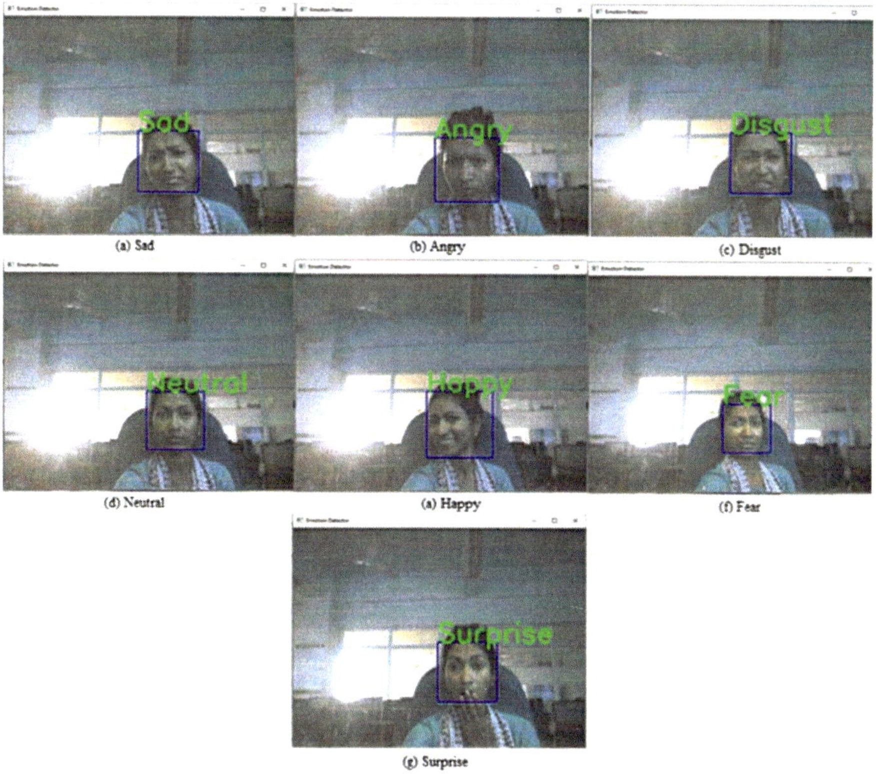

Fig. 8 Real-time image classified using the proposed method

with the proposed system, which reveals that the current system can correctly detect emotion for healthcare monitoring with lesser time. The system can detect faster because the CNN architecture is designed with only 1.6 M parameters compared to literature, as shown in Table 3. The system can alter the healthcare staff when negative emotion is detected after processing the real-time data.

Table 3 Comparison of state-of-the-art CNN architectural parameters

CNN architectures	No of parameters (in million)
AlexNet + FC6 + LDA [13]	41.5
MobileNet with softmax + center loss [5]	3.2
InceptionV3 [19]	24
VGG16 [18]	138
AlexNet [17]	61
Proposed method	**1.6**

4 Conclusion

In this work, a dew computing-based framework to monitor patients' mental health has been proposed using a lightweight CNN. The lightweight CNN is designed with a minimal parameter, so to compensate for it, the features from previous layers are combined for better gradient propagation. Furthermore, the proposed expression detection module outperforms state-of-the-art methods in terms of accuracy despite being lightweight. Further, for real-time detection, the pretrained model of FER-2013 can be deployed in the cloud infrastructure for a faster response. Future work includes the improvement in CNN structure after collecting feedback from the real-time detection module.

References

1. J.C. Hung, K.-C. Lin, N.-X. Lai, Recognizing learning emotion based on convolutional neural networks and transfer learning. Appl. Soft Comput. **84**, 105724 (2019)
2. Z. Peng, J. Dang, M. Unoki, M. Akagi, Multi-resolution modulation-filtered cochleagram feature for LSTM-based dimensional emotion recognition from speech. Neural Netw. **140**, 261–273 (2021)
3. H. Kaya, F. Gürpınar, A.A. Salah, Video-based emotion recognition in the wild using deep transfer learning and score fusion. Image Vis. Comput. **65**, 66–75 (2017)
4. T. Zhang, W. Zheng, Z. Cui, Y. Zong, Y. Li, Spatial–temporal recurrent neural network for emotion recognition. IEEE Trans. Cybern. **49**(3), 839–847 (2018)
5. Y. Miao, H. Dong, J.M. Al Jaam, A. El Saddik, A deep learning system for recognizing facial expression in real-time. ACM Trans. Multimed. Comput. Commun. Appl. (TOMM) **15**(2), 1–20 (2019)
6. N.I. Muros, A.S. García, C. Forner, P. López-Arcas, G. Lahera, R. Rodriguez-Jimenez, K.N. Nieto, J.M. Latorre, A. Fernández-Caballero, P. Fernández-Sotos, Facial affect recognition by patients with schizophrenia using human avatars. J. Clin. Med. **10**(9), 1904 (2021)
7. B. Sonawane, P. Sharma, Review of automated emotion-based quantification of facial expression in Parkinson's patients. Vis. Comput. 1–17 (2020)
8. J. Codina-Filbà, S. Escalera, J. Escudero, C. Antens, P. Buch-Cardona, M. Farrús, Mobile eHealth platform for home monitoring of bipolar disorder, in *International Conference on Multimedia Modeling* (Springer, Cham, 2021), pp. 330–341
9. G. Yolcu, I. Oztel, S. Kazan, C. Oz, K. Palaniappan, T.E. Lever, F. Bunyak, Deep learning-based facial expression recognition for monitoring neurological disorders, in *2017 IEEE International Conference on Bioinformatics and Biomedicine (BIBM)* (IEEE, 2017), pp. 1652–1657
10. A. Sivasangari, P. Ajitha, I. Rajkumar, S. Poonguzhali, Emotion recognition system for autism disordered people. J. Ambient Intell. Humaniz. Comput. 1–7 (2019)
11. G. Muhammad, M. Alsulaiman, S.U. Amin, A. Ghoneim, M.F. Alhamid, A facial-expression monitoring system for improved healthcare in smart cities. IEEE Access **5**, 10871–10881 (2017)
12. L.Y. Mano, B.S. Faiçal, L.H.V. Nakamura, P.H. Gomes, G.L. Libralon, R.I. Meneguete, P.R.G. Filho et al., Exploiting IoT technologies for enhancing health smart homes through patient identification and emotion recognition. Comput. Commun. **89**, 178–190 (2016)
13. Z. Fei, E. Yang, D.D.-U. Li, S. Butler, W. Ijomah, X. Li, H. Zhou, Deep convolution network based emotion analysis towards mental health care. Neurocomputing (2020)
14. M.Z. Uddin, M.M. Hassan, A. Almogren, A. Alamri, M. Alrubaian, G. Fortino, Facial expression recognition utilizing local direction-based robust features and deep belief network. IEEE Access **5**, 4525–4536 (2017)

15. K.H. Kim, K. Park, H. Kim, B. Jo, S.H. Ahn, C. Kim, M. Kim et al., Facial expression monitoring system for predicting patient's sudden movement during radiotherapy using deep learning. J. Appl. Clin. Med. Phys. **21**(8), 191–199 (2020)
16. M.S. Hossain, G. Muhammad, Emotion recognition using deep learning approach from audio–visual emotional big data. Inf. Fusion **49**, 69–78 (2019)
17. A. Krizhevsky, I. Sutskever, G.E. Hinton, Imagenet classification with deep convolutional neural networks, in *Advances in Neural Information Processing Systems* (2012)
18. K. Simonyan, A. Zisserman, Very deep convolutional networks for large-scale image recognition. arXiv preprint arXiv:1409.1556 (2014)
19. C. Szegedy, V. Vanhoucke, S. Ioffe, J. Shlens, Z. Wojna, Rethinking the inception architecture for computer vision, in *In Proceedings of the IEEE Conference on Computer Vision and Pattern Recognition* (2016), pp. 2818–2826
20. G. Huang, Z. Liu, L. Van Der Maaten, K.Q. Weinberger, Densely connected convolutional networks, in *Proceedings of the IEEE Conference on Computer Vision and Pattern Recognition* (2017), pp. 4700–4708
21. A. Majumdar, T. Debnath, S.K. Sood, K.L. Baishnab, Kyasanur forest disease classification framework using novel extremal optimization tuned neural network in fog computing environment. J. Med. Syst. **42**(10), 1–16 (2018)
22. A. Majumdar, N.M. Laskar, A. Biswas, S.K. Sood, K.L. Baishnab, Energy efficient e-healthcare framework using HWPSO-based clustering approach. J. Intell. Fuzzy Syst. **36**(5), 3957–3969 (2019)
23. A. Majumdar, T. Debnath, A. Biswas, S.K. Sood, K.L. Baishnab, An energy efficient e-healthcare framework supported by novel EO-μGA (extremal optimization tuned micro-genetic algorithm). Inf. Syst. Front. 1–18 (2020)
24. A. Biswas, A. Majumdar, S. Nath, A. Dutta, K.L. Baishnab, LRBC: a lightweight block cipher design for resource constrained IoT devices. J. Ambient Intell. Humaniz. Comput. 1–15 (2020)
25. P. Viola, M.J. Jones, Robust real-time face detection. Int. J. Comput. Vis. **57**(2), 137–154 (2004)
26. I.J. Goodfellow, D. Erhan, P.L. Carrier, A. Courville, M. Mirza, B. Hamner, W. Cukierski et al., Challenges in representation learning: a report on three machine learning contests, in *International Conference on Neural Information Processing* (Springer, Berlin, Heidelberg, 2013), pp. 117–124
27. P. Lucey, J.F. Cohn, T. Kanade, J. Saragih, Z. Ambadar, I. Matthews, The extended Cohn-Kanade dataset (CK+): a complete dataset for action unit and emotion-specified expression, in *2010 IEEE Computer Society Conference on Computer Vision and Pattern Recognition-Workshops*, June 2010 (IEEE, 2010), pp. 94–101

Feature Dimensionality Reduction Method on Social Network Dataset

R. Rajkumar

1 Introduction

The curse of dimensionality may be a common problem in social network dataset. Ineffective and unreasonable feature dimensionality reduction will compromise machine learning efficiency, pattern recognition accuracy, and data mining efficiency while increasing the workload of measured data experiments to some extent. A group of high-dimension features possesses the subsequent problems: too many features but few samples, too many features with few or no relations to the mining task, and excessive redundancy among features. Feature dimensionality reduction methods are often classified into two types: feature selection and have projection. Feature projection is additionally called feature reparameterization and transforms the first feature space into one with a smaller dimensionality and more independent dimensions [1]. Feature projection generally uses principal component analysis (PCA), kernel principal component analysis, or other improved principal component analysis methods for linear or nonlinear transformations [2].

Although these methods can realize feature dimensionality reduction to some extent, they require extensive calculations and high-computational complexity. These dimensionality reduction methods also cannot reduce necessary experimental data and are inconvenient when identifying features heavily correlated to mining tasks. Against this, feature selection obtains low-dimension features more effectively compared to machine learning or data processing and carries low-computational complexity. However, existing feature selection methods are still time-intensive with reference to calculations, and a few data dimensionality reduction problems in social network datasets remain difficult to unravel using these methods [3].

R. Rajkumar (✉)
Department of Computer Science and Applications, The Gandhigram Rural Institute, Gandhigram, Dindigul, India
e-mail: rajkumarbdu@gmail.com

J. D. Peter et al. (eds.), *Disruptive Technologies for Big Data and Cloud Applications*, Lecture Notes in Electrical Engineering 905,
https://doi.org/10.1007/978-981-19-2177-3_32

The social network dataset is taken as an example during this paper, capturing feature dimensionality reduction in multivariate statistic as an example, within the social network dataset 51 samples for analysis; this increases the challenges of feature dimensionality reduction.

To solve the matter of excessive dimensions, this paper proposes a replacement feature selection method supported a social network dataset survey method and an exhaustive search algorithm [3]. The overall process is as follows: first, we build a feature-scoring system and a reduced-dimension length-scoring system supported the sampling survey method; second, we sort features and reduced-dimension lengths consistent with their scores and choose the features and reduced-dimension lengths that are ranked within the front; third, to use selection optimization to features and reduced-dimension lengths simultaneously, we redefine the meaning of the knowledge carried by individuals within the exhaustive search algorithm population and improve the algorithm accordingly. The accuracy of FOCUS-RRK algorithm estimation was generally above 8% and effectively reduced the redundancy data within the social network dataset.

2 Dimensionality Reduction

Dimensionality reduction is a crucial technique in various fields like data processing, machine learning, pattern recognition, image retrieval, and text mining, then forth. Various real-world applications in data processing usually have high-dimensional data. So as to handle the info adequately, its dimensionality must be reduced. The most objective of dimensionality reduction is to rework the high-dimensional data samples into the low-dimensional space, such the intrinsic information contained within the data is preserved. As soon because the dimensionality gets reduced, it helps to enhance the robustness of the classifier and reduces the computational complexity. Figure 1 shows the dimensionality reduction process. Generally, dimensionality reduction is performed using various techniques inclusive of principal component analysis, principal feature analysis, fisher criterion, correlational analysis, and classical scaling.

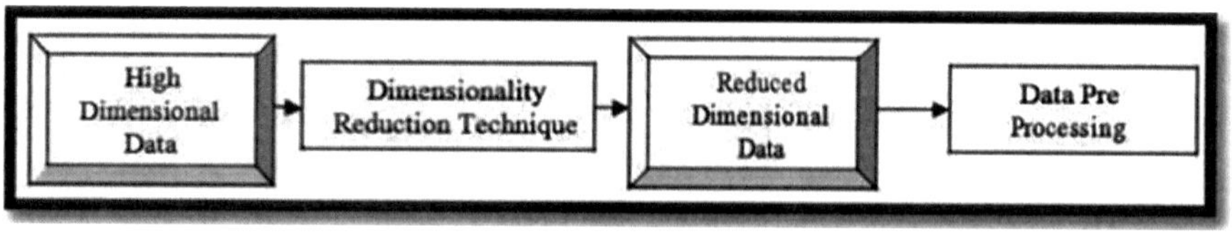

Fig. 1 Dimensionality reduction process

2.1 New Feature Dimensionality Reduction Method Based on Social Network Dataset and Improved FOCUS-RRK Algorithm

Scoring system supported sampling survey method. The sampling survey method takes a part of the population as samples for analysis. During a typical dimensionality problem, given a huge number of features, the ultimate number remains unknown after dimensionality reduction; therefore, using an enumeration to research, all samples require an intensive workload and should fail. By employing the sampling survey method and a particular sample scale, this paper analyzes the features and number of them after dimensionality reduction. All samples must have an equivalent chance of being selected. To obtain useful information for feature dimensionality reduction, I propose a feature-scoring system and a rating system for reduced-dimension length.

2.2 FOCUS-RRK Algorithm

FOCUS is one among the first algorithms within machine learning. FOCUS starts with an empty set and carries out a breadth-first search until it finds a minimal subset that predicts pure classes. If the complete set has three features, the basis is (0 0 0), its children are (0 0 1), (0 1 0), and (1 0 0) where a '0' means the absence of the respective feature and '1' means its presence within the feature subset. It is an exhaustive search in nature and original works on binary and noise-free data with some simple amendment of FOCUS, FOCUS-RRK which will work on non-binary data with noise.

To eliminate the redundant features from the dataset, the redundant features are removed by constructing the minimal spanning tree and separate the tree having the sting distance quite its neighbor to make the separate clusters, from the clusters features which could be strongly related to the target features are selected to make the subset of features. Many feature subset selection algorithms, a couple of can efficaciously do away with irrelevant features, however, fail to deal with redundant features yet a number of the others can eliminate the irrelevant while taking care of the redundant feature [4]. To map out this problem, a replacement method within the research has been proposed, namely the FOCUS-RRK algorithm. FOCUS-RRK algorithm is that the first contribution of this research. This is often the filter-based feature selection algorithm [3].

2.3 *Dataset*

The dataset is that the main input to look at the new algorithm. Within the research, social network dataset is employed. The real-time dataset is employed within the research. Here, the real-time dataset is taken from social network data sites.

2.4 *Social Network Dataset*

Social network dataset is a real-time dataset, which consists of 14 features and 51 instances (Table 1).

3 Novelty of FOCUS-RRK Algorithm

The proposed new algorithm is named the FOCUS-RRK algorithm. The techniques are vital. An honest search strategy can reduce the computational cost while improving the accuracy at an equivalent time. Consistent with different searching nature, the search strategy includes exhaustive search (Fig. 2).

4 Exhaustive Search (ES)

The exhaustive search uses the algorithm to get every possible combination of feature subset and compute the consistency rate for every of them. A threshold is about up at the start consistent with the consistency rate calculated for the primary selected feature subset which is additionally set to be the present best subset. Because the algorithm proceeds, the present best subset could also be replaced by the one with a better consistency rate and a smaller size of the feature set during which case the edge is additionally replaced. The exhaustive search can start with either a group with one feature and perform by adding features into a group or with a full feature set and subtract features from it. It is evident that exhaustive search is time-consuming and computationally expensive because it calculates every combination, and many of which can be redundant and will be avoided if a learning process is involved [4]. The efficiency is abating fast with the dimensions of the search space. For instance, if a testing data have n features, the amount of combinations is $\sum_{i=1}^{n} C_n^i$ which suggests that there will be? $\sum_{i=1}^{n} C_n^i$ Times of calculation for consistency rate. Last, the exhaustive search is inefficient and dear for top feature size. An example of an exhaustive search is FOCUS-RRK. Below is that the exhaustive search algorithm where ConCal() may be a function to calculate the consistency rate for social network dataset D. S is that the full feature set for the D.

Table 1 Social network dataset

Relation: Social_Network_DataAnalysis

No.	1: User id	2: Name	3: Gender	4: Date_of_birth	5: Address	6: Mobile	7: Email	8: Id	9: Join_date	10: Class
	Numeric	String	Nominal	Date	String	Numeric	String	String	Date	Nominal
1	1.0	Sharmi	Female	1984-01-02	Aranthangi	8.344526894E9	sharmi@gmail.com	BPKfUZ	1980-01-28	c0
2	101.0	Rams	Male	1992-11-02	Madurai	9.655446994E9	rams@gmail.com	GBJL7m	2016-03-27	c1
3	601.0	Keerthi	Female	1992-06-02	Pudukkottai	9.87654321E9	keerthi@gmail.com	nFKRtN	2016-03-27	c0
4	701.0	Nivetha	Female	1993-08-25	Thanjavur	8.973946796E9	nivetha@gmail.com	KWj85c	2016-03-28	c0
5	201.0	Anjali	Female	1982-03-04	London	7.654390823E9	Anjali@gmail.com	fiZqN9	2016-03-31	c0
6	901.0	Manju	Female	1993-05-01	Mannargudi	9.988776645E9	manju@gmail.com	gQMjVx	2016-04-13	c0
7	1001.0	Sharmi	Female	1992-01-15	Marathahalli	7.750724707E9	sharmi1@gmail.com	K19dZK	2016-04-14	c0
8	1101.0	Sharmi	Female	1991-01-16	Salem	8.887722334E9	sharmi2@gmail.com	lJAPXn	2016-04-14	c0
9	1201.0	Keerthi	Female	1992-06-03	Ooty	9.987655432E9	keerthi1@gmail.com	tovrGI	2016-04-14	c0
10	1301.0	Reshma	Female	1993-11-01	Bommidi	8.833447721E9	rams1@gmail.com	q7jR2O	2016-04-14	c0
11	1401.0	Roshan	Male	1993-09-12	Namakkal	9.090808077E9	Roshan1@gmail.com	3ZIML6	2016-04-14	c1
12	1401.0	Roshan	Male	1993-09-12	Namakkal	9.090808077E9	Roshan1@gmail.com	3ZIML6	2016-04-14	c0
13	1501.0	Kavya	Female	1994-06-06	Kadathur	9.099021309E9	kavya@gmail.com	xfgZdE	2016-04-14	c0
14	1601.0	Sundar	Male	1992-04-18	Namakkal	9.67788569E9	sundar@gmail.com	pVHmcP	2016-04-14	c1
15	1701.0	Sutha	Female	1992-06-20	Puduviduthi	9.715167295E9	sutha@gmail.com	qKEbXF	2016-04-14	c0
16	301.0	Simbu	Male	1992-01-04	Puthupatti	9.977886655E9	simbu@gmail.com	3nVuAS	2016-04-15	c1
17	401.0	Simbu	Male	1992-01-04	Puthupatti	9.977886655E9	simbu1@gmail.com	qUf7ty	2016-04-15	c1
18	501.0	Simbu	Male	1992-01-04	Puthupatti	9.977886655E9	simbu2@gmail.com	nFMdyG	2016-04-15	c1
19	801.0	Mounika	Female	1992-01-07	Marathahalli	9.750060023E9	mounika1@gmail.com	Az4kJH	2016-04-16	c0

(continued)

Table 1 (continued)

Relation: Social_Network_DataAnalysis

No.	1: User id	2: Name	3: Gender	4: Date_of_birth	5: Address	6: Mobile	7: Email	8: Id	9: Join_date	10: Class
	Numeric	String	Nominal	Date	String	Numeric	String	String	Date	Nominal
20	1801.0	Mounika	Female	1991-01-06	Vengatharahalli	9.750060023E9	mounika@gmail.com	I6s9Tz	2016-04-17	c0
21	1901.0	Mounika	Female	1993-01-08	Salem	9.750060023E9	mounika2@gmail.com	ZOIt7c	2016-04-17	c0
22	2001.0	Mahalakshi	Female	1993-02-14	Pappireddipatti	9.988776645E9	mahalakshi@gmail.com	GtMbxn	2016-04-17	c0
23	2101.0	Mahalakshi	Female	1993-02-14	Madurai	9.988776645E9	mahalakshi1@gmail.com	6mEpsP	2016-04-17	c0
24	2201.0	Mahalakshi	Female	1993-02-14	Bommidi	9.977886655E9	mahalakshi2@gmail.com	EwXYNZ	2016-04-17	c0
25	2301.0	Vinoth	Male	1984-11-24	Kadathur	9.750060023E9	vinoth@gmail.com	B4xGIU	2016-04-17	c1
26	2401.0	Vinoth	Male	1984-04-14	Salem	9.750060023E9	vinoth1@gmail.com	BM4HRd	2016-04-17	c1
27	2501.0	Vinoth	Male	1985-02-04	Ooty	9.750060023E9	vinoth2@gmail.com	1C4j7D	2016-04-17	c1
28	2601.0	Hari	Male	1992-07-10	Namakkal	8.384203557E9	hari@gmail.com	8CLP6q	2016-04-17	c1
29	2701.0	Hari	Male	1992-06-10	Namakkal	9.9445566E9	hari1@gmail.com	YReS8G	2016-04-17	c1
30	2801.0	Hari	Male	1991-03-07	Aranthangi	8.833447721E9	hari2@gmail.com	IyHSWR	2016-04-17	c1
31	2901.0	Vasanth	Male	1993-12-07	Nellai	7.708075245E9	vasanth1@gmail.com	suEkGt	2016-04-17	c1
32	3001.0	Selvam	Male	1992-08-27	Kamalam	9.988776645E9	selvam1@gmail.com	UtZ8Ed	2016-04-17	c1
33	3101.0	Deepa	Female	2008-07-13	Chennai	8.765453672E9	deepa2@gmail.com	DEZc76	2016-04-17	c0
34	3201.0	Selvam	Male	1993-02-15	Kadathur	9.988776645E9	selvam2@gmail.com	6zlP9o	2016-04-17	c1
35	3301.0	Ganesh	Male	1993-03-05	Madurai	9.911223344E9	ganesh2@gmail.com	UA4HQt	2016-04-17	c1
36	3401.0	Siva	Male	1994-04-05	Chennai	9.988332211E9	siva@gmail.com	OuqT2V	2016-04-17	c1

(continued)

Table 1 (continued)

Relation: Social_Network_DataAnalysis

No.	1: User id	2: Name	3: Gender	4: Date_of_birth	5: Address	6: Mobile	7: Email	8: Id	9: Join_date	10: Class
	Numeric	String	Nominal	Date	String	Numeric	String	String	Date	Nominal
37	3501.0	Siva	Male	1991-01-01	Chennai	9.988776655E9	siva2@gmail.com	K9WqvJ	2016-04-17	c1
38	3601.0	Arun	Male	1992-02-04	London	9.977886655E9	arun@gmail.com	CtQDVq	2016-04-17	c1
39	3701.0	Arun	Male	1991-01-03	London	9.988776645E9	arun1@gmail.com	zJBpUC	2016-04-17	c1
40	3901.0	Ganesh	Male	1992-03-12	Salem	9.977665544E9	ganesh@gmail.com	I51lxs	2016-04-17	c1
41	4001.0	Ganesh	Male	1992-06-08	Kadathur	9.988776645E9	ganesh1@gmail.com	vE7kcj	2016-04-17	c1
42	4101.0	Siva	Male	1993-01-07	Chennai	9.988777448E9	siva1@gmail.com	51rVaH	2016-04-17	c1
43	4201.0	Arun	Male	1993-01-07	London	9.988776645E9	arun2@gmail.com	aSQZHq	2016-04-17	c1
44	4301.0	Vignesh	Male	1993-12-27	Karur	9.677453667E9	vignesh@gmail.com	QEIB4b	2016-04-18	c1
45	4401.0	Vignesh	Male	1993-12-27	Karur	9.677453667E9	vignesh1@gmail.com	GgyzbR	2016-04-18	c1
46	4501.0	Vignesh	Male	1993-12-25	Malaikovil	9.677453667E9	vignesh2@gmail.com	VOXr5D	2016-04-18	c1
47	4601.0	Divya	Female	1992-12-29	Thanjavur	9.988776655E9	divya@gmail.com	7mBbKi	2016-04-18	c0
48	3801.0	Karthi	Male	1993-12-11	Trichy	8.344060645E9	karthi2@gmail.com	Ji2DsH	2016-04-25	c1
49	4701.0	Karthi	Male	1993-12-11	Trichy	8.344060645E9	karthi@gmail.com	tHuh0C	2016-04-25	c1
50	4801.0	Karthi	Male	1993-12-11	Trichy	8.344060645E9	karthi1@gmail.com	HjsEKk	2016-04-25	c1
51	4901.0	Arul	Male	1986-06-04	Trichy	8.973946796E9	Arul@gmail.com	fQS4gD	2016-04-28	c1

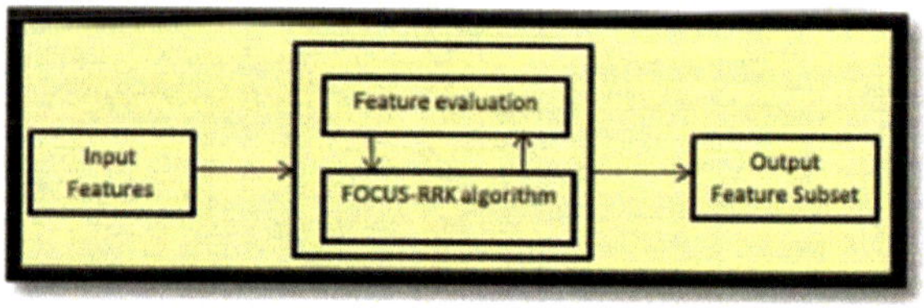

Fig. 2 FOCUS-RRK searching algorithms

5 FOCUS-RRK Algorithm for Removing Redundant Features and Calculate Consistency

Input: *Data D, Feature set fs*

Output: *Remove Redundant Feature subset. BestSet=* [

Feature Space = $-\infty$

For i = 1 *to* |*fs*|

Begin

⋆*fs'* *where* |*fs'*| = i *and* *fs'* ∈ *fs*

CCONR = *ConCal* (*fs'*, *D*)

If CCONR ≥ *Removed Redundant Features*

If |*BestSet*| < |*fs*|

BestSet = *fs'*

Removed Redundant Features = *CCONR*

End

Return BestSet

In the current situation, the feature sets are a mix of noisy, irrelevant, and redundant features in instances [4]. Finding suitable and appropriate features may be a tedious task within the machine learning and data processing process. There are many algorithms that have currently evolved within the context.

Among the varied feature selection algorithms, the main target algorithm is taken into account to be efficient in terms of reducing redundant features [4]. This algorithm deals with dependent and independent datasets.

It uses exhaustive search techniques for locating suitable attributes. Here, the utilization of consistency measure for feature selection and calculate confusion matrix is employed to define the precision, recall, and *F*-measure is enhancement the accuracy in social network data using feature selection techniques [3].

The social network dataset may be a real-time dataset collected from social network Websites. This social network dataset has been taken for feature selection experiment with FOCUS-RRK and FOCUS methods, and therefore, the results are presented in Table 2.

Table 2 Comparison of FOCUS-RRK and FOCUS feature selection algorithms on social network dataset

Feature selection		Features									
Algorithm name		F1	F2	F3	F4	F5	F6	F7	F8	F9	F10
FOCUS-RRK	Selected times	30	28	29	12	9	26	7	3	25	5
	Selected probability (%)	90	82	78	11	23	76	45	9	74	30
	Relevant features	F1 F2 F3 F6 F9									
FOCUS	Selected times	4	10	2	12	22	5	14	9	18	3
	Selected probability (%)	74	47	76	63	9	82	17	85	3	78
	Relevant features	F8 F6 F10 F3 F1									

Table 3 Accuracy results between FOCUS-RRK and FOCUS in Naive Bayes classifier on social network dataset

Algorithm	Precision	Recall	F-measure	Accuracy (%)
FOCUS-RRK	0.823	0.8611	0.8415	84
FOCUS	0.761	0.784	0.771	76

The results depict that the main target-RRK algorithm has extracted the highly relevant features compared to the FOCUS algorithm. The feature-selected probabilistic percentage of the main target-RRK algorithm is higher within the feature selection compared to the FOCUS algorithm. This means that the FOCUS-RRK performs better within the redundant nature of the feature space of the social network dataset. The amount of times the actual feature is being selected is additionally higher within the FOCUS-RRK algorithm than the main target algorithm. The results depict that the main target-RRK algorithm has extracted the highly relevant features compared to the FOCUS algorithm. The feature-selected probabilistic percentage of the main target-RRK algorithm is higher within the feature selection compared to the FOCUS algorithm. This means that the FOCUS-RRK performs better within the redundant nature of the feature space of the social network dataset. The amount of times the actual feature is being selected is additionally higher within the FOCUS-RRK algorithm than the main target algorithm.

From Table 3, it is inferred that the FOCUS-RRK method leads to better accuracy. It also weights a lesser rating for irrelevant features. This helps to eliminate the redundant features from the feature space.

6 Conclusion

Feature selection may be a vital a part of the classification task of knowledge mining. Among the varied feature selection algorithms, there are many exhaustive search strategies in exercise which include is that the most time-efficient, on the brink of optimal.

FOCUS-RRK algorithm is employed while there could also be no known manner to locate an optimal solution during which case the goal is to expand a simple process with demonstrable better time period and an improved solution. Also, the FOCUS-RRK algorithm removing the redundant features, and targeted data results are often achieved.

FOCUS-RRK algorithm is straightforward and efficient. It is an extended version of the main target algorithm. This algorithm is partly stimulated by instance-based learning techniques. FOCUS-RRK employs the exhaustive approach for appropriate feature estimation within the larger feature space. The efficiency of the algorithm is measured in terms of the accuracy of classification. The accuracy of classification is measured in terms of precision, recall, and F-measures. By employing the exhaustive search approach, the redundancy and incomplete portion of features within the feature space are managed efficiently in social network dataset. From the experiment, it is been inferred that the main target-RRK algorithm outperforms the FOCUS algorithm.

In the huge volume of datasets, the redundancy of features or datasets is extremely common within the data processing field. The choice of relevant features from a huge volume of feature spaces may be a complex process. To beat the redundancy issue, a replacement exponential search strategy approach has been incorporated within the FOCUS-RRK algorithm. The FOCUS-RRK algorithm more efficiently selects the features within the redundant nature of the features. The proposed FOCUS-RRK algorithm has been compared with the main target algorithm. The results of the proposed FOCUS-RRK algorithm showed enhanced classification accuracy than the main target algorithm. The proposed FOCUS-RRK algorithm enhances the classification accuracy by 8% (84–76%), because the redundant features are deported.

References

1. J. Liu, Feature dimensionality reduction for myoelectric pattern recognition: a comparison study of feature selection and feature projection methods. Med. Eng. Phys. **36**(12), 1716–1720 (2014)
2. H.D. Gangurde, Feature selection using clustering approach for big data, in *ITCCE* (2014)
3. R. Rajkumar, Behavior analysis and feature selection in online social network. IJSEAS **1**(8). ISSN: 2395-3470 (2015)
4. K. Sutha, J. Jebamalar Tamilselvi, A review of feature selection algorithms for data mining techniques. Int. J. Comput. Sci. Eng. **7** (2015)
5. A. Jalilvand, N. Salim, Feature unionization: a novel approach for dimension reduction. Appl. Soft Comput. **52**, 1253–1261 (2017)
6. J. Liu, Y. Lin, M. Lin, S. Wu, J. Zhang, Feature selection based on quality of information. Neurocomputing **225**, 11–22 (2017)
7. W. Zhao, S. Du, Spectral–spatial feature extraction for hyperspectral image classification: a dimension reduction and deep learning approach. IEEE Trans. Geosci. Remote Sens. **54**(8), 4544–4554 (2016)
8. G.F. Khamis, T. Ramadas Naik, Identifying and removing irrelevant and redundant features in high dimension data using feature subset. Int. J. Mag. Eng. Technol. Manag. Res. **2** (2015)

9. A. Tommasel, Integrating social network structure into online feature selection, in *IJCAI* (2016)
10. R. Rajkumar, Anbuselvi, Set cover-RRK feature selection technique for enhancing the accuracy in social network data. WJERT **3**(1), 97–102 (2017). ISSN 2454-695X

Internet of Things (IoT) for Coronavirus (COVID-19) Pandemic: A Survey on Trailblazing Techniques

Salomi Selvadass, J. John Paul, I. Thusnavis Bella Mary, and A. Diana Andrushia

1 Introduction

The coronavirus or the COVID-19 with varied mutants known as the SARS-CoV and SARS-CoV-2 has shuffled the entire globe not only in the medical space but also in discrete facets of human life. The health, education, IT, politics, transportation sector, etc. have had a perspicacious effect on the global economy. Being an orderly respiratory sickness that can be effectually treated with proper medical ailments, what poses a precarious menace is the human-to-human transmission which drastically happens. At the time at which this paper was recorded, the World Health Organization (WHO) showed that a total of 220 countries and territories have fallen to this virus with around 160 million cases. The USA with the highest number of cases stating 32 million is backed by India with 24 million confirmed cases and nearly 242,362 deaths (May 10, 2021) [1].

The second wave was expected to have a higher peak than the first wave governing epidemiological and social conditions which are locked in a mutual feedback loop [2]. With India experiencing a deadly second wave at the time of this writing, the researchers are facing deliberate challenges to predict its infection with high exactitude. Frequent tests are undergone by a COVID-infected person to clinch that he has recovered thoroughly. But with the mutations and asymptomatic infections spiralling across various continents, the basic reverse transcription-polymerase chain reaction (RT-PCR) test also fails to indicate if the virus is present inside one's body [3]. Figure 1 depicts the demography of the COVID situation in India with the first graph showing the new cases encountered every month since the beginning of this pandemic and the second graph showing the death case data.

S. Selvadass · J. John Paul (✉) · I. Thusnavis Bella Mary · A. Diana Andrushia
Department of ECE, Karunya Institute of Technology and Sciences, Coimbatore 641114, India
e-mail: johnpaul@karunya.edu

J. D. Peter et al. (eds.), *Disruptive Technologies for Big Data and Cloud Applications*, Lecture Notes in Electrical Engineering 905,
https://doi.org/10.1007/978-981-19-2177-3_33

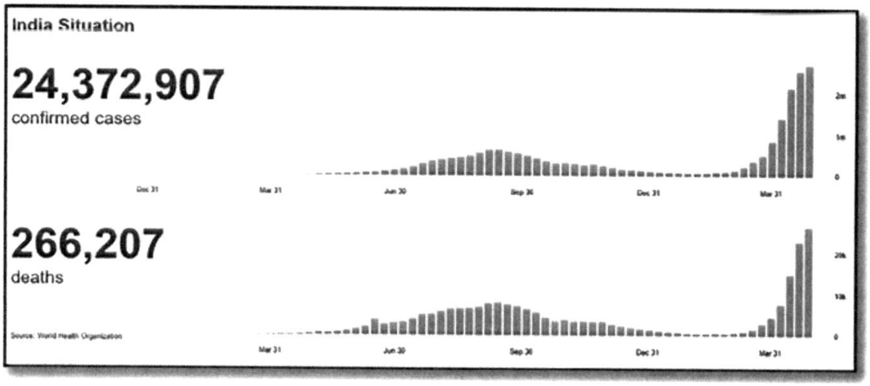

Fig. 1 Statistics in regards to the COVID-19 in India. *Data Source* WHO Situation Report—17 May, 2021 [4]

Far and wide, this paper assays the assorted methodologies used for the detection of COVID-19 using AI, IoT, cloud, block chain, etc. It also addresses the various ways implemented currently to prevent transmission and raise awareness among the people [5]. These technologies assist to extenuate the hostile effects of this pandemic and also quicken the recovery process. A comprehensive review is also provided of a proposed idea which can prove to be a low-cost system for the continuous remote monitoring of an infected person.

2 Related Work

The health system of manifold nations has arrived at a collapse point with the increase in casualties due to COVID. A weak immune system is prone to be infected which is intimidating for people above the age of 60 and young infants. The availability of rapid diagnostic tests (RDT) is very limited with the ongoing surge, and with a low sensitivity rate, it proves to be costly for healthcare sectors to rely upon. The chest radiology scan acts as an extensive tool for early diagnosis as this virus targets the respiratory system. This system uses the CNN architecture for classification [6].

A new intelligent security system was proposed to recognize elderly people in public with the help of deep learning and IoT [7]. The system works by identifying the age of the person through the face image realized using deep learning and machine learning algorithms connected to the Internet for real-time updates. According to WHO, the best proceeding to block the air droplets from spreading person to person via the nose or mouth is by cultivating appropriate social distancing practices and wearing a mask. Motion detection, computer vision and human contour detection are some of the extensively used detection processes for human presence. Ahamad et al. proposed [8] a new method to detect people using the MobileNet Single Shot

Table 1 A definite correlation of the different related works undertaken

Author name	Methodology	Principle	Notability	Constraints
Alghamdi et al. [6]	Deep learning approach	Detection of COVID from chest X-ray images	Multi-class classification	Disease progression assessment and prognosis analysis
Othman et al. [7]	DL and IoT-based surveillance system	Recognition of elderly people in public	Face alignment algorithm with 89.45% accuracy	Use of two deep learning models
Ahamad et al. [8]	Segmented ROI system	Person detection for social distancing and safety violation	Object detection framework	Difficult detection in outdoors with distant scenes
Somaldo et al. [9]	YOLO implemented ROS	Social distancing surveillance drone	Localization and navigation scheme with crowd detection performance	Implementation of methodology in real drone

Multibox Detector (SSD) object tracking model along with OpenCV library for image processing. A drone surveillance system was recommended by [9] to detect people and find the ones violating the norms. These contemplated schemes identify a local area network and update information through the cloud where the data is stored and analysed for a longer run, thereby learning and making new strategies [10]. Table 1 summarizes these researches in the form of a concise representation.

3 Proposed Idea

The prime advantage of the Internet of Things (IoT) is the real-time continuous updates received from the sensors connected to a server for the monitoring of a person or environment. The advancement of IoT along with artificial intelligence (AI) for healthcare applications is the new era of science. Inspired by the smart surveillance systems, this paper suggests a novel smart COVID mask for telehealth (Fig. 2).

The components used are ESP32 microcontroller, temperature and ultrasonic sensor which have been preferred to make it highly affordable for every person. The temperature sensor used is LM35 to determine the body temperature along with other important health factors such as respiratory rate and heart rate. The ultrasonic sensor (HC-SR04) determines the distance between two persons, and an alert message is directed if the minimum distance is not maintained. The ThingSpeak cloud platform is preferred for data storage, analysis and visualization of data. The proposed system

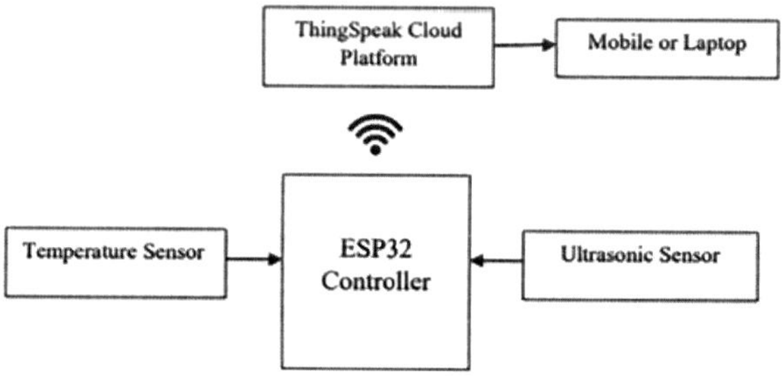

Fig. 2 Block diagram of the proposed system

defers from the other systems discussed in literatures in terms of how the data is directed to the person and analysis is done. The aggregated data is regularly sent to the person wearing the mask or the doctor who examines the person's regular health in comparison with the conventional approach. The proposed prototype is integrated with a mask worn by the person as it has become an integral part of human life coevally. The outflow of how this system is implemented has been drafted, which is the base in obtaining a fully demonstrating telehealth mask to fight the COVID-19. Figure 3 displays the flow chart of the proposed system. Initially, the proposed system is connected with a Wi-Fi network for the transfer of data collected from the sensors to the cloud to reach the person for analysis. The system calculates the respiratory rate every time a person exhales. Since air droplets inhaled through the nostrils are the main reason for the COVID spread, the ultrasonic sensor mounted in front of the nostrils would calculate the distance between any two or more people and ensure that an appropriate distance is being maintained as ordered by the social distancing norms. The LED is the immediate alert system for social distancing. The data is then sent to the cloud, which is viewed by the user. Since ThingSpeak's cloud platform has better visualization, the data can be viewed in the form of graphs or discrete values on laptop's Web browser or a mobile platform.

4 Conclusion and Future Scope

Reviewing the trailblazing technique literature, we treasure that the Internet of Things along with AI and the cloud plays a compelling part in withstanding the COVID-19 pandemic. A brief introduction to the coronavirus with the current medical scenario and needs to be catered has been provided with this paper. The various approaches taken by researchers throughout nations to detect and monitor this disease have been reviewed. The proposed system will assist people to monitor their health parameters such as temperature, respiratory rate and heart rate, remotely. The orientation and rundown given for the new mask prove to be a low-cost and highly efficient system. The proposed novel smart COVID mask for telehealth protects an individual from being exposed to infection as well as monitor their health remotely. This research

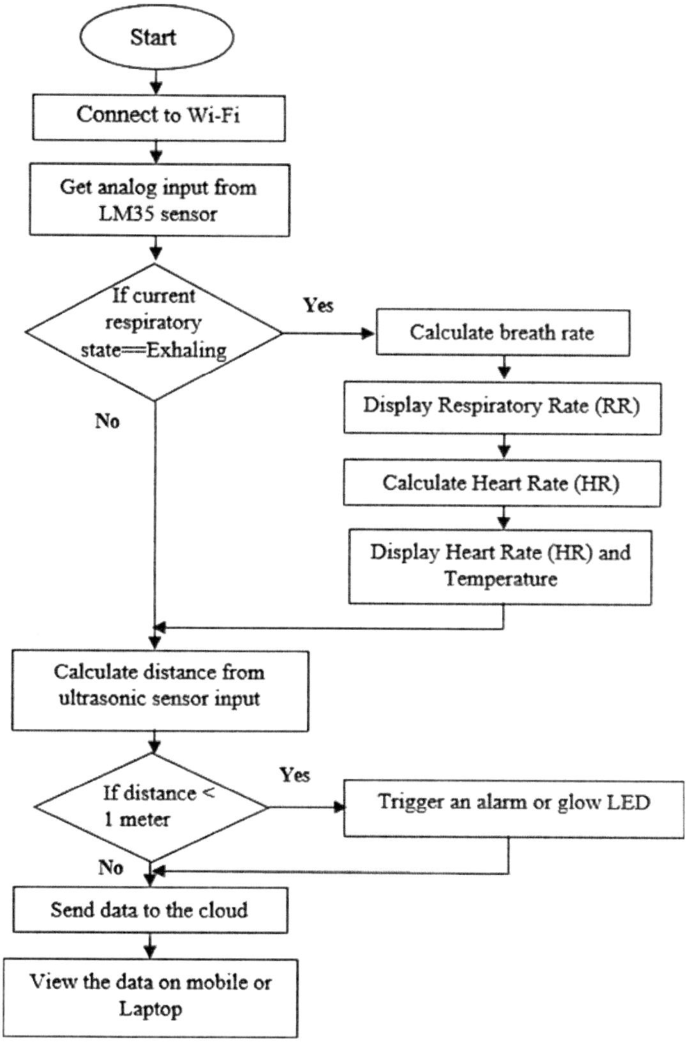

Fig. 3 Flow chart of the proposed system

can further be enhanced with the use of relevant machine learning and deep learning techniques.

References

1. World Health Organization novel coronavirus 2019 (COVID-19)
2. S.A. Pedro et al., Conditions for a second wave of COVID-19 due to interactions between

disease dynamics and social processes. Front. Phys. **8**, 428 (2020). https://doi.org/10.3389/fphy.2020.574514. ISSN: 2296-424X
3. C. Jiang et al., Molecular detection of SARS-CoV-2 being challenged by virus variation and asymptomatic infection. J. Pharm. Anal. (2021). ISSN 2095-1779. https://doi.org/10.1016/j.jpha.2021.03.006.
4. May 2021. [Online]. Available: https://www.who.int/countries/ind/
5. V. Chamola, V. Hassija, V. Gupta, M. Guizani, A comprehensive review of the COVID-19 pandemic and the role of IoT, drones, AI, blockchain, and 5G in managing its impact. IEEE Access **8**, 90225–90265 (2020). https://doi.org/10.1109/ACCESS.2020.2992341
6. H.S. Alghamdi, G. Amoudi, S. Elhag, K. Saeedi, J. Nasser, Deep learning approaches for detecting COVID-19 from chest X-ray images: a survey. IEEE Access **9**, 20235–20254 (2021). https://doi.org/10.1109/ACCESS.2021.3054484
7. N.A. Othman, M.Z.N. Al-Dabagh, I. Aydin, A new embedded surveillance system for reducing COVID-19 outbreak in elderly based on deep learning & IoT, in *2020 International Conference on Data Analytics for Business and Industry: Way Towards a Sustainable Economy (ICDABI)* (2020), pp. 1–6. https://doi.org/10.1109/ICDABI51230.2020.9325651
8. A.H. Ahamad, N. Zaini, M.F.A. Latip, Person detection for social distancing and safety violation alert based on segmented ROI, in *2020 10th IEEE International Conference on Control System, Computing and Engineering (ICCSCE)* (2020), pp. 113–118. https://doi.org/10.1109/ICCSCE50387.2020.9204934
9. P. Somaldo, F.A. Ferdiansyah, G. Jati, W. Jatmiko, Developing smart COVID-19 social distancing surveillance drone using YOLO implemented in robot operating system simulation environment, in *2020 IEEE 8th RIO Humanitarian Technology Conference (RIO-HTC)* (2020), pp. 1–6. https://doi.org/10.1109/RIO-HTC49770.2020.9357040
10. S.S. Vedaei et al., COVID-SAFE: an IoT-based system for automated health monitoring and surveillance in post-pandemic life. IEEE Access **8**, 188538–188551 (2020). https://doi.org/10.1109/ACCESS.2020.3030194

Comparison of Stock Market Prediction Using Deep Learning Algorithms

S. Revathi, Regina Begam, Radhika, and R. Akila

1 Introduction

Big data analytics developed with machine learning and deep learning algorithms are gaining importance in a variety of fields of application such as stock market investment, agricultural field, credit card fraud detection, weather forecasting, and so on. It is due to the advancement of machine learning and deep learning algorithms. The main reason is that an algorithm does not require any assumptions about the information, but it frequently produces higher accuracy than economics and applied mathematical models. The stock market refers to a large collection of shares from various companies that are combined, and the regular and individual activity of shopping for and selling those shares is listed. The economical market suggestion defines that this stock values reflect all reachable information within the market. For the investment ways, stock value prediction is required.

Many methodologies are used to predict the daily returns of the stock, including auto-regression integrated moving average—ARIMA, support vector regression—SVR, and various statistic formulas [1]. Formalized paraphrase stock values are determined by musician Kimura, Vinicius Amorim Sobreiro and shut Bruno Miranda Henrique using SVR on daily and up-to-the-minute costs. The SVM—support vector machine—is a machine learning algorithm that can perform both regression and classification. This technique employs training observations to construct a linear model based on nonlinear classification thresholds, which maps variables across a much wider range of dimensions. SVR produces a large number of correct results when using the kernel function. The most common kernel functions are linear, polynomial, and sigmoid. For prediction, the sigmoid function is used.

S. Revathi (✉) · R. Begam · Radhika · R. Akila
B.S. Abdur Rahman Crescent Institute of Science and Technology, Chennai, India
e-mail: srevathi@crescent.education

J. D. Peter et al. (eds.), *Disruptive Technologies for Big Data and Cloud Applications*, Lecture Notes in Electrical Engineering 905,
https://doi.org/10.1007/978-981-19-2177-3_34

Reference [2] the stock value is predicted using the ARIMA method in another methodology. Adebiyi et al. proposed the ARIMA model prediction method. ARIMA models have been found to be more robust and efficient in statistical analysis, particularly for short-term prediction. The ARIMA model defines three parameters: p, q, and d. The best ARIMA model is determined by the relatively low standard error rate and the absence of any significant pattern within the autocorrelation and thus the partial autocorrelation function.

The goal of the project is to forecast a stock market daily value that will help businesses increase their profits. A reasonable prediction system could help stock investors by providing future market values or information. And by knowing the market behavior, prediction system facilitates the investor to take the trading decision. A stock market dataset is collected from the Yahoo finance that contain the numerous share market index data value. A proposed system takes the two-stock data such as SPDR S&P 500 ETF (Standard and Poor Depository Rate 500 Exchange Traded Fund) and BSE-100 (Bombay Stock Exchange Rate 100) [3].

SPDR S&P 500 ETF and BSE-100 daily return direction is to be predicted by SRNN and LSTM. This method starts with feature selection and then preprocesses the data to manage missing values and, as a result, mismatched samples. The entire dataset is preprocessed or transferred then employed for prediction using SRNN and LSTM algorithm. Finally, the results of prediction by LSTM and SRNN algorithm are compared with the results of prediction by standard SVR model, ARIMA, and neural network. Pang et al. [4] proposed a method called an innovative neural network for stock market prediction. The method demonstrates the concept of "stock vector" where input is multi-stock high dimensional historical data. The model uses LSTM with an embedded layer and automatic encoder to analyze stock market values. Ghosh et al. [5] proposed a method of predicting the Indian share market prices using the LSTM algorithm. Wang and Wang [6] have proposed work of predicting indexes of stock market movement using PCA - Principal Component Analysis and STNN - Stochastic Time effective Neural Networks. Rajput and Bobde [7] have proposed a new approach based on the hybrid approach of opinion mining and clustering techniques. The objective was to predict the stock movement of NSE - National Stock Exchange exploitation an opinion mining and the clustering methodology. Qiu and Song [8] have proposed the work based on predicting Stock Market Index values and trends by Optimized ANN - Artificial Neural Network Model. The author defined the prediction of the next day's value by an optimized ANN model. Optimize the prediction accuracy by the GA - Genetic Algorithms. Hiransha et al. [9] developed a new approach based on Deep-Learning Models for prediction. The frequency of day-wise closing price of stock markets two different companies and various sectors of each NSE - National Stock Exchanges of Asian nation and NYSE - New York Stock Exchange was taken for prediction. Zhang et al. [10] have proposed the work based on GAN - Generative Adversarial Network. Using the GAN - Generative Adversarial Network, the categories of the neural network which is of the unsupervised learning methodology. Chiang et al. [11] has proposed a work based on a stock index trading decision support system. An effective approach for recognizing trading or commerce signals and suggesting to the capitalist a trading call or trading decision by utilizing a system that adapts an input and also the prediction model supported the required output. In the adaptive approach, multiple input and modeling techniques are utilized. Selvamuthu et al. [12] have proposed a new approach to the prediction of the tick data. The three different learning algorithms like Levenberg-Marquardt, Scaled Conjugate Gradient, and Bayesian Regularization are supported by the neural network. For a stock market prediction, all three methods supported tick data. Niaki and Hoseinzade [13] designed a method, where S&P 500 index was forecasted using ANN.

2 Problem Definition

Companies have the capacity of gaining the large profit by a stock market. It is quite difficult to predict a stock values since the data of stock are complex nature and nonlinear. Use of a large historical data may also increase the dimension. To help the stock investor by providing the future market direction information, an efficient prediction method is needed and help the investors, to take the trading decision by market behavior and to yield more profits.

The model includes collection of 10 years historical data of SPDR S&P 500. The data are preprocessed using the min–max normalization method in order to eliminate missing values. Then, apply the prediction algorithm such as SVR, ARIMA and multi-layer feed forward network, SRNN—simple recurrent neural network, and LSTM—long short-term memory algorithm. Finally, the error rate of each algorithm is calculated.

Figure 1 depicts the proposed system's system architecture. Stocks' historical data are collected, and the best features are chosen and preprocessed for prediction. Following preprocessing, the data are divided into train and test datasets. The dataset is used as input for the prediction rule. Finally, from the validation results, the optimize algorithm for stock prediction is chosen.

The selection of relevant data for prediction is part of the data collection process. The data used in the proposed method were obtained from a reliable source. A proposed system's data are historical data from the S&P 500 ETF, and historical data from the BSE-100 are obtained from yahoo.finance.com.

It is easier to analyze data when it is represented graphically. A histogram is a graphical representation of the distribution of a process dataset. Figure 2 depicts a data histogram of the stock market close price in bar chart format, which aids in initial data analysis by representing the number of data points that fall within a range of values. The system collects various stock index data in order to find an optimized algorithm to predict the stock and to determine whether an optimized algorithm is capable of predicting any stock index accuracies. The data of a proposed system are the day-wise historical data of the S&P 500 ETF and BSE-100, which were collected from yahoo.finance.com. The dataset contains attribute of open, close, high, low, adj.close and volume values. Dataset dated from 1/4/1995 to 29/4/2020 that contain about 6312 rows.

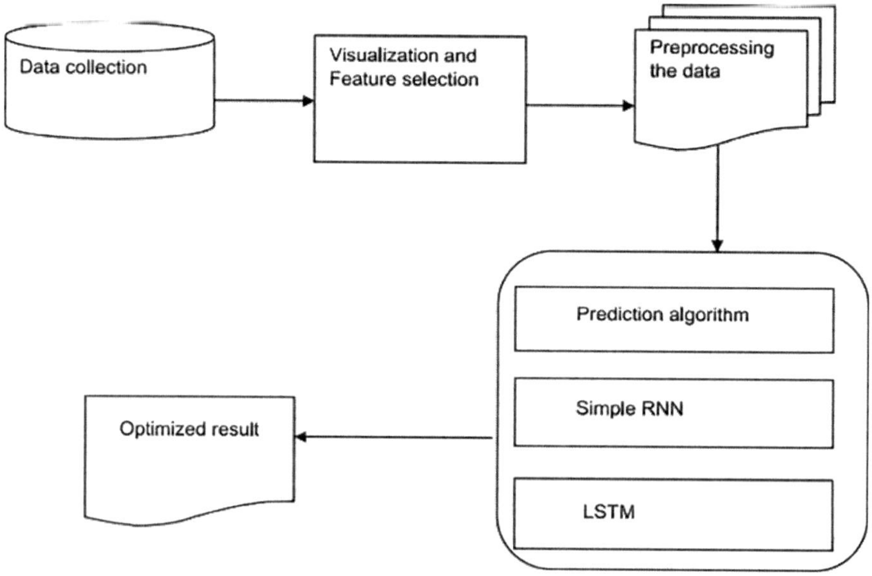

Fig. 1 System architecture

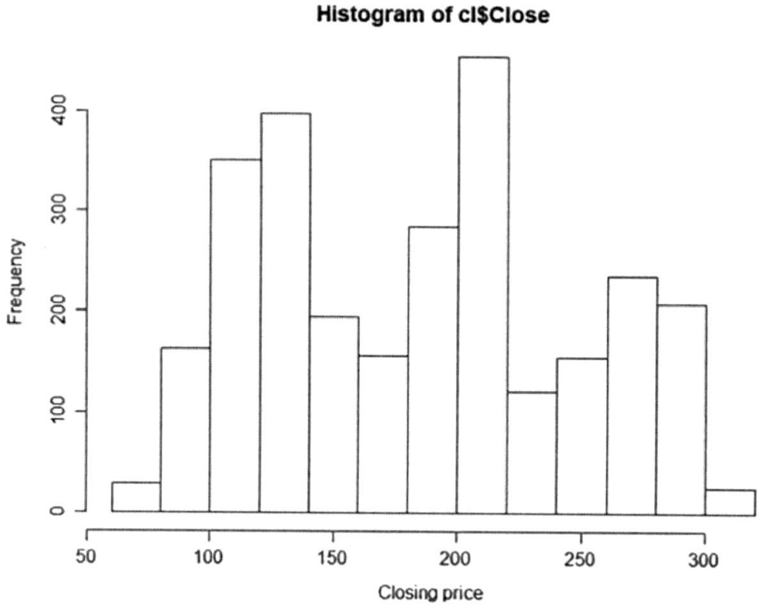

Fig. 2 Histogram of closing price of stock

Feature Selection

Feature selection is the process of selecting the appropriate variable or predictors for model generation. The preprocessed system uses the recursive feature elimination (RFE), which remove the weakest features one by one to reach the exact feature for prediction. The RFE simplifies the model and makes it more computationally efficient. And also, this allows us to create a model with the best dimensions possible. The closing price of the future stock is to be prediction, by using RFE the appropriate attribute required to predict the closing price are found to be high, open, adj.close and low.

Figure 3 represents the output of the feature selection method. By recursive feature elimination method, the top four variables to predict closing price of future stock are high, open, adj.close, and low.

Preprocessing the Data

The most important issue for prediction is data quality. Because the data may contain missing values and mismatched samples, the prediction accuracy may deviate if they are not preprocessed; thus, it is critical to preprocess the data. Missing data can be dealt with in one of three ways:

1. Remove missing data: If the dataset has a minimal number of missing data, this can be done.
2. Substitute other values for missing data, such as the mean or the most common value.
3. Leave any missing data alone.

The data used in the proposed methodology contain information for 5032 trading days. The other significance of the normalization method is to make the data values to have the same scales, so each value is taken as equal important.

```
Recursive feature selection

Outer resampling method: Cross-Validated (10 fold)

Resampling performance over subset size:

 Variables   RMSE Rsquared    MAE  RMSESD RsquaredSD   MAESD Selected
         1 2.4836   0.9973 1.6263 2.07518  4.283e-03 1.45098
         2 1.1002   0.9997 0.6938 0.25188  1.754e-04 0.14615
         3 0.9241   0.9998 0.5884 0.12358  7.753e-05 0.03838
         4 0.8041   0.9998 0.5123 0.07542  4.283e-05 0.02340        *
         5 0.8305   0.9998 0.5427 0.07025  4.255e-05 0.01865

The top 4 variables (out of 4):
   High, Open, Adj.Close, Low

> predictors(results)
[1] "High"      "Open"      "Adj.Close" "Low"
```

Fig. 3 Output of feature selection

Figure 4 depicts the histogram of the SPDR S&P 500's denormalized and normalized closing prices. The denormalized data are ranges widely for the range of 50–300 and difficult to choose the feature range for prediction. After normalization, the data values are within the range, where every values is taken equal importance. Normalized value provides the faster and accurate results.

Figure 5 represents the preprocessed dataset where the data are normalized by min–max method and scaled to the particular range. All the values are between the range of 0 and 1.

Learning

The preprocessed dataset is split into two, i.e., 70% of data as training set and 30% as test set. The training dataset contains an observed output variable and an input variable. Training set makes an algorithm to learn.

Simple RNN Algorithm

A RNN is a feed forward network generalization with internal memory. RNN uses the memory for processing an input layer. Each neuron connected directly with other neuron in the next successive layer. Each layer of RNN generate the output that depends on two input such as sequence of input and the previous output.

Steps for building RNN algorithm:

Step 1: Load the necessary library. Library are Keras and TensorFlow.

Step 2: Load the data, normalize, and set the model parameters such as maximum length of the training samples, batch size, which is the number of unique samples to

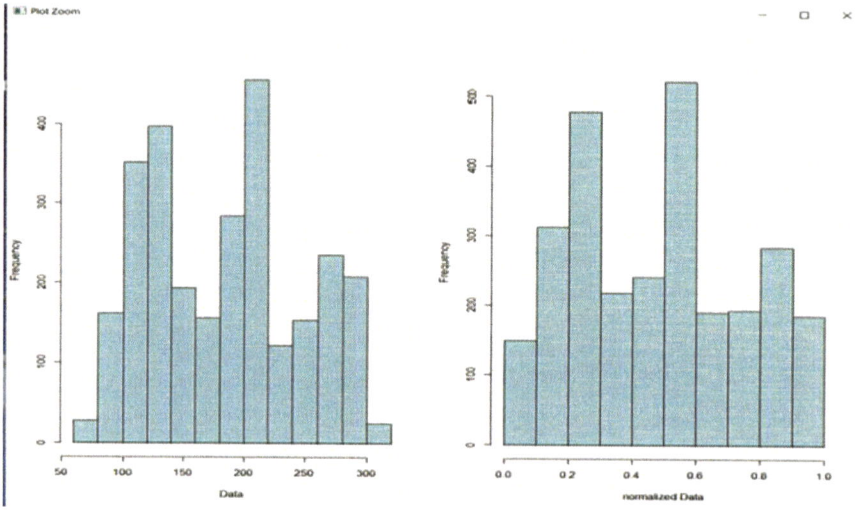

Fig. 4 Histogram of current closed and normalized closed price

Close_norm	Open_norm	High_norm	Low_norm	Adj.Close_norr
0.094764030	0.110173722	0.106154820	0.095443962	0.069871948
0.107260386	0.098495570	0.100146049	0.097273035	0.079315322
0.102136863	0.113290669	0.106530355	0.099018953	0.075443558
0.090223671	0.094796792	0.092384726	0.089582655	0.066440831
0.073770149	0.084282292	0.079073639	0.074908566	0.054007014
0.096055308	0.075554814	0.090673914	0.062312927	0.070847691
0.077102523	0.089186294	0.092050902	0.080728294	0.056525259
0.072312245	0.076593796	0.077446269	0.075074848	0.052905368
0.079018626	0.071482023	0.071854794	0.065721673	0.057973345
0.055775397	0.074640538	0.070394339	0.057449284	0.040408538
0.030574400	0.050619232	0.052201136	0.033047908	0.021364435
0.047527789	0.039522906	0.045483006	0.030096433	0.034175930
0.070479438	0.058058351	0.070895049	0.055121395	0.051520295
0.073103693	0.080417277	0.073064896	0.069504496	0.053503445
0.086891326	0.067949486	0.080075120	0.071250414	0.063922589
0.091556612	0.085944639	0.083997486	0.088875975	0.067448064
0.058316326	0.081290018	0.073231811	0.061356845	0.042328675

Fig. 5 Preprocessed dataset

look at one times during the training and the number of epoch, which is the number of times the model will be exposed on training set.

Step 3: Initializing the model, with an input, hidden, and output layers. The activation function is also defined in the hidden state. Generally, the simple RNN model uses the tanh activation function.

Step 4: Train the model, evaluate, and tune to minimize the loss by increasing the number of neurons and epoch.

Step 5: Predict the output by passing the test dataset.

LSTM Algorithm

LSTM is a type of RNN algorithm capable of learning long-term dependencies across sequences [5]. The data in LSTM algorithm flow through a mechanism called cell states that can store its own information. An algorithm consists of different memory block called cells. Every next cell, an input is being transferred, namely cell state and hidden state. This memory is manipulated by three major mechanisms known as gates. The three gates are referred to as the forget gate, the input gate, and the output gate.

Forget gate (ft)

It is responsible for removing information from the cell gate that is the information that are not required is removed. The gate takes two input such as hidden or output of the previous state and an input of a current cell state. The weight for the forget_gate neurons with the two parameters such as the output of the previous LSTM block and input at the current timestamp is added with bias. It is applied to the sigmoid function.

Input gate (it)

It is in charge of adding new information to the cell's state. Normally, an input gate is three-step process such as

1. Regulate what are the values to add in the cell state by involving sigmoid function called regulatory filter names as it.
2. Create candidate vector ($\hat{c}$) by using the tanh activation function.
3. Multiply the filter value and vector values, then add multiplied ones to the cell state. And the update is made to a current cell state Ct.

Output gate (ot)

The third one is for the output gate, which is used to activate the LSTM block's final output at timestamp 't'. Create the vector after applying the tanh function to a current cell state Ct. Like input gate, it also generates the regulatory filter. Finally, multiply both and generate output.

Evaluation Setup

A test dataset is a set of observations used to estimate the performance of an algorithm using a performance metric. It is significant that the test set contains no members of the trained set. A program that generalizes well will be able to perform a task with new data effectively. For training, 70% dataset has taken, and the remaining 30% of the dataset is fed into an algorithm that predicts the stock's closing price.

3 Implementation of Simple RNN and LSTM ALGORITHM

The closing of stock was predicted for two dataset such as SPDR S&P 500 and for BSE-100 with the same RNN model. The library files required to load and perform the functions of simple RNN algorithm are Keras and TensorFlow. The batch size of a proposed model is 32, and the number of epochs is 50.

The dataset has been enhanced with large attributes such as future economic factors and gold rates. In addition, to determining the impact of stock prices with the above parameters, hybrid algorithms may be used.

Figure 6 defines the output graph of the RNN and the real final price of the stock for the SPDR S&P 500 dataset. The model can be tuned to reduce the error by

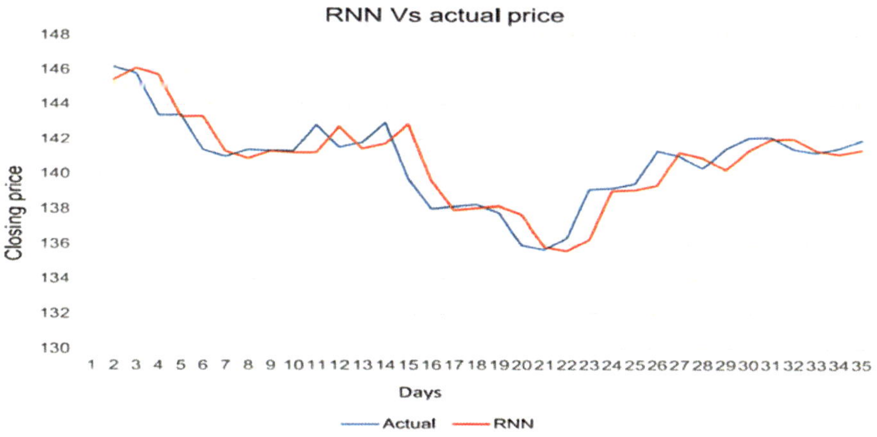

Fig. 6 Output graph of RNN and actual closing price for SPDR S&P 500

increasing the number of hidden layers. Graph represents the comparison of actual price with the predicted price.

Figure 7 defines the output graph of the RNN and the real final price of the stock for BSE-100 dataset. Graph shows the predicted closing price for the particular days and shows the chances of the predicted closing price by RNN with the actual closing price. The model can be tuned to reduce the error by increasing the hidden layers, the neurons, and the number of epochs.

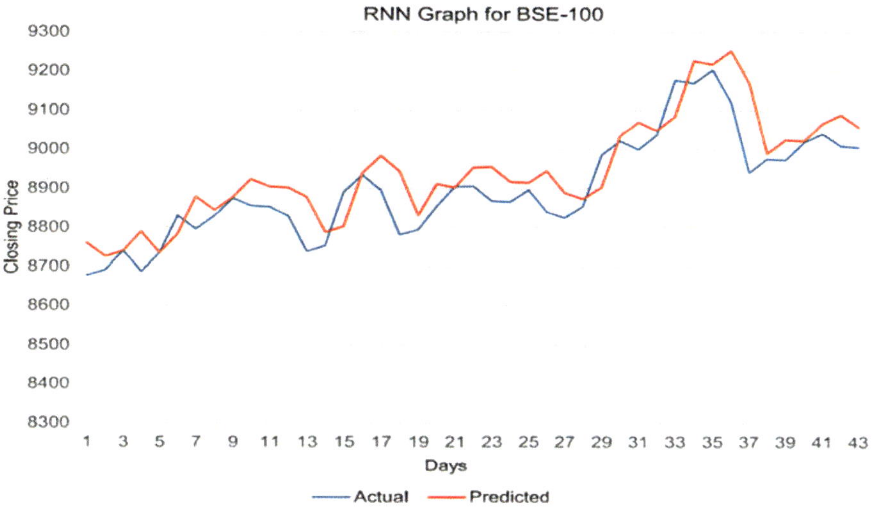

Fig. 7 Output graph of RNN and actual closing price for BSE-100

Implementation of LSTM Algorithm

The libraries required to load, for performing LSTM algorithm, are library (tidyverse), library (forecast), and library (Keras). Keras in the important library to implement the LSTM algorithm In LSTM model, some terminology needed to define. Some terminologies are

Batch size—no. of training sample that must be processed before updating the weights.

Epoch—a complete forward and backward pass of all the training samples. A proposed model uses epoch of 50.

Dropout—a technique where the randomly selected neuron is ignored during training.

Figure 8 shows the output graph of an actual and therefore the expected price for SPDR S&P 500 dataset. And it shows a comparison of actual and predicted closing price.

Figure 9 shows the output graph of the actual and therefore the expected price for BSE-100 dataset. Graph represents the comparison of changes of actual closing price day by day with the change of predicted closing price by LSTM model for BSE-100 dataset.

Comparison of Simple RNN with MLFFNN and ARIMA Model

Figure 10 represents the comparison graph of predicted closing price of SPDR S&P 500 dataset by a proposed simple RNN and existing ARIMA and MLFFNN. The RMSE value of ARIMA gives 1.7094; MLFFNN gives 1.3868, and simple RNN gives 0.6490. From the RMSE value of each model of ARIMA, MLFFNN, and

Fig. 8 Output graph of LSTM and actual stock price for SPDR S&P 500

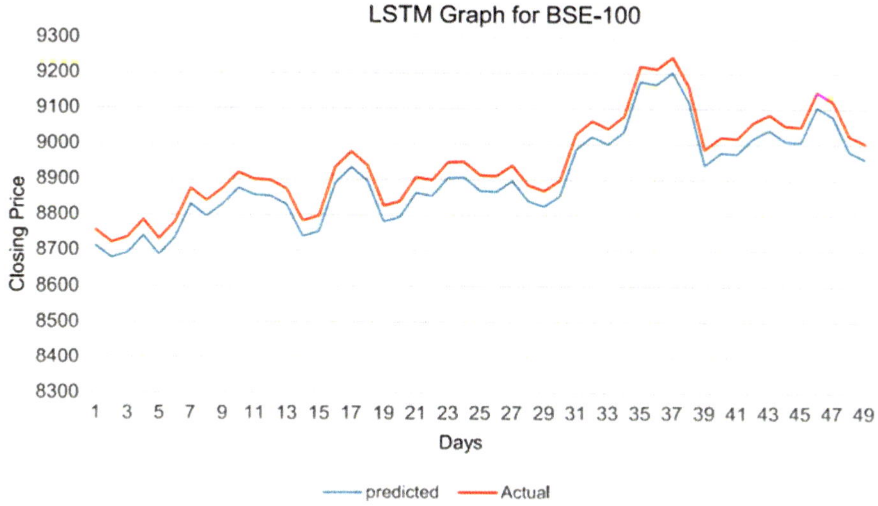

Fig. 9 Output graph of LSTM and actual stock price for BSE-100

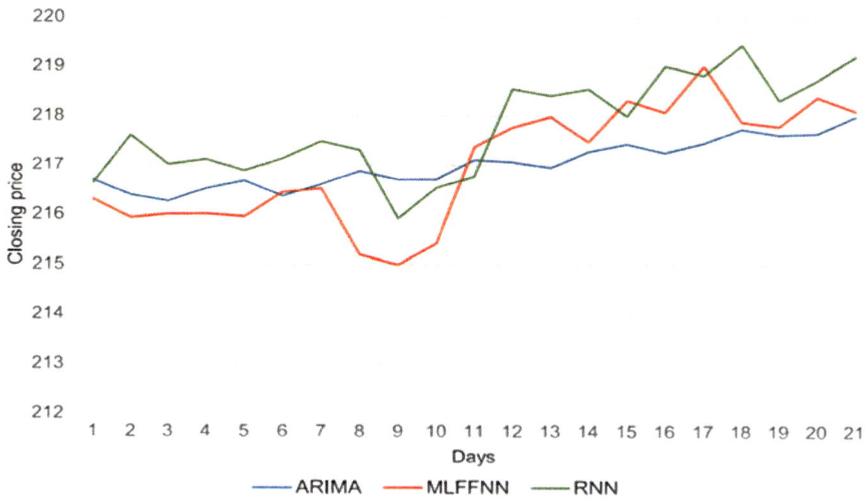

Fig. 10 Comparative graph of RNN with existing MLFFNN and ARIMA model

RNN, simple RNN has less error rate; thus, simple RNN is more optimal model for prediction compared to the existing model.

Comparison of LSTM with MLFFNN and ARIMA Model

Figure 11 represents the comparison graph of predicting the closing price of SPDR S&P 500 dataset by LSTM with ARIMA and MLFFNN.

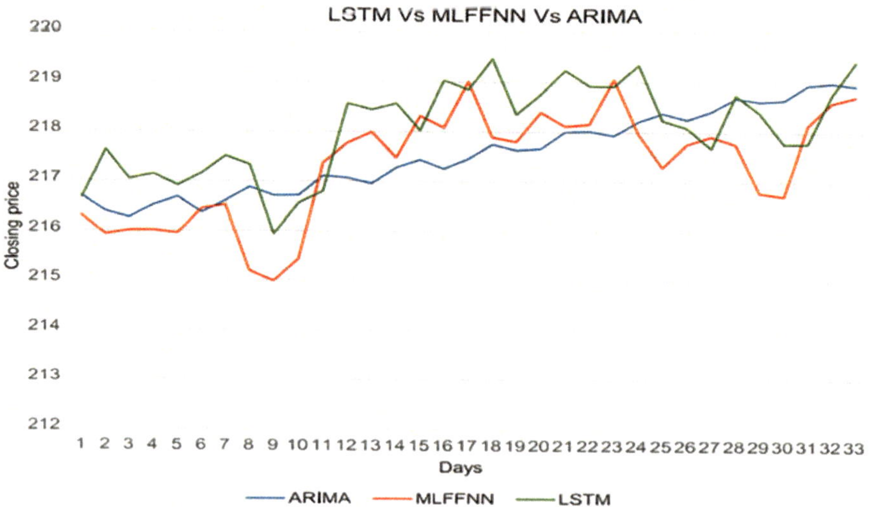

Fig. 11 Comparative graph of LSTM with existing MLFFNN and ARIMA model

The RMSE value of ARIMA gives 1.7094; MLFFNN gives 1.3868, and LSTM gives 0.4032. From the RMSE value of each model of ARIMA, MLFFNN and LSTM, LSTM has less error rate; thus, LSTM is more optimal model for prediction.

Comparison of Simple RNN with LSTM Model

The parameter such as number of neurons, dropout rate, number of epoch and dataset, and training split is taken same for both model of the simple RNN and LSTM. The training split was 70%; dropout rate was set to 0.2, epoch of 50 for each model. A proposed RNN and LSTM are applied to two datasets separately such as SPDR S&P 500 and S&P BSE-100.

Figure 12 shows the comparison graph of prediction of closing price of stock of SPDR S&P 500 by simple RNN and LSTM model. The performance of a proposed system of stock prediction by simple RNN and LSTM for SPDR S&P 500 dataset is calculated. RMSE of simple RNN is 0.64907, and those of LSTM is 0.4032.

Figure 13 shows the comparison graph of prediction of closing price of stock for S&P BSE-100 by simple RNN and LSTM model. The performance of a proposed system of stock prediction by simple RNN and LSTM for BSE-100 dataset is calculated. RMSE of simple RNN is 0.9076, and those of LSTM is 0.5637.

The performance of the model is analyzed on the SPDR S&P 500 dataset by the parameter such as RMSE value, MAE, and MSE. The model with less RMSE, MSE, and MAE is better and optimal algorithm for prediction. The lesser the value of metric, lesser the error rate. Initially, for SPDR S&P 500 dataset, closing price is predicted by both existing and proposed model. Then, the parameter such as RMSE, MSE, and MAE value is calculated for each model.

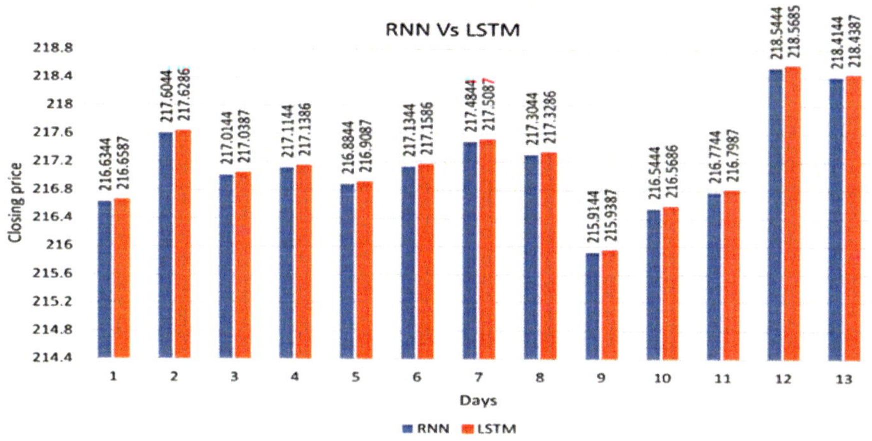

Fig. 12 Comparison of simple RNN and LSTM algorithm for BSE-100

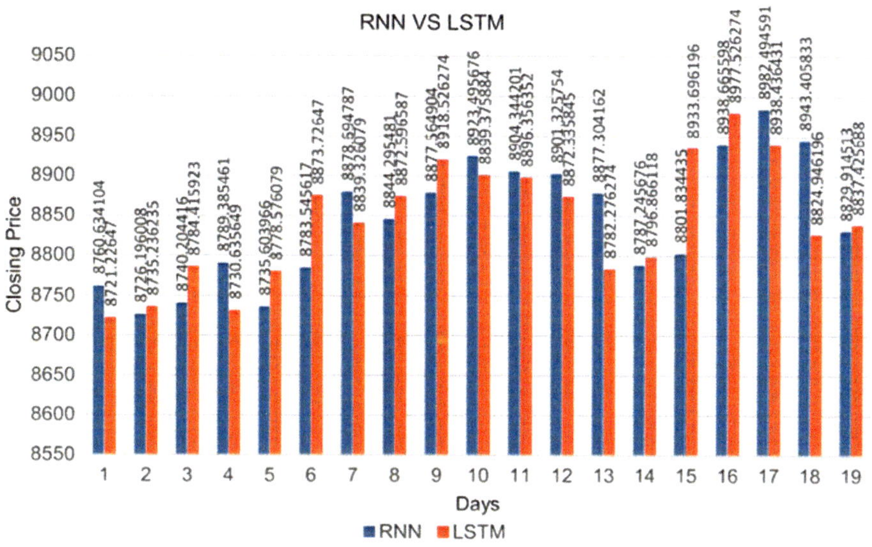

Fig. 13 Comparison of simple RNN and LSTM algorithm for SPDR S&P 500

Table 1 RMSE, MSE, and MAE values of an algorithm

Algorithm	RMSE	MSE	MAE
Simple RNN	0.6490	0.4212	0.045702
LSTM algorithm	0.4032	0.1625	0.02602
SVR	2.4041	5.779	0.08593
ARIMA	1.7094	2.9220	0.07329
MLFFNN	1.3863	1.9232	0.04532

Table 1 describes the comparison of RMSE, MSE, and MAE value of proposed and the existing algorithm. From Table 1, the RMSE, MSE, and MAE value of proposed algorithm LSTM has less value compared to the existing model such as SVR, ARIMA, and MLFFNN. By the comparison of RMSE, MSE, and MAE value of proposed and the existing system, LSTM has a better prediction accuracy. Thus, the LSTM algorithm is an optimized algorithm for predicting a stock market daily returns with higher accuracy.

4 Conclusion and Future Work

A proposed system uses simple RNN and LSTM algorithm for predicting the daily returns of the future stock price. An analysis of the results also indicates that both models of simple RNN and LSTM give better accuracy. From the analysis, an optimized model of prediction is found as LSTM algorithm. The major focus of the system is to present the process of building the RNN and LSTM algorithm for predicting the daily returns of a stock exchange. This method assists companies in increasing profits and guides traders in investing in the market. The dataset has been enhanced with large attributes such as future economic factors and gold rates. In addition, to determining the impact of stock price with the above parameters, hybrid algorithms may be used.

References

1. B.M. Henrique, V.A. Sobreiro, H. Kimura, Stock price prediction using support vector regression on daily and up to the minute prices. J. Finance Data Sci. **4**(3) (2018). https://doi.org/10.1016/j.jfds.2018.04.003
2. A.A. Adebiyi, A.O. Adewumi, C.K. Ayo, Stock price prediction using the ARIMA model, in *16th International Conference on Computer Modelling and Simulation* (Cambridge University, UK, 2014). https://doi.org/10.1109/UKSim.2014.67
3. X. Zhong, D. Enke, Predicting the daily return direction of a stock market using the hybrid machine learning algorithm. Financ. Innov. **5**(1) (2019). https://doi.org/10.1186/s40854-019-0138-0
4. X. Pang, Y. Zhou, P. Wang, W. Lin, V. Chang, An innovative neural network approach for stock market prediction. J. Supercomput. (2018). https://doi.org/10.1007/s11227-017-2228-y
5. A. Ghosh, S. Bose, G. Maji, N.C. Debnath, S. Sen, Stock price prediction using LSTM on Indian share market, in *Proceedings of 32nd International Conference on Computer Applications in Industry and Engineering, EPiC Series in Computing*, vol. 63 (2019). https://doi.org/10.29007/qgcz
6. J. Wang, J. Wang, Forecasting stock market indexes using principle component analysis and stochastic time effective neural networks. Neurocomputing **15** (2015). https://doi.org/10.1016/j.neucom.2014.12.084
7. V.S. Rajput, S.S. Bobde, Stock market prediction using hybrid approach, in *International Conference on Computing, Communication and Automation (ICCCA)* (2016). ISBN (Online): 978-1-5090-1666-2

8. M. Qiu, Y. Song, Predicting the direction of stock market index movement using an optimized artificial neural network model. PLoS ONE (2016). https://doi.org/10.1371/journal.pone.0155133.s001
9. M. Hiransha, E.A. Gopalakrishnan, V.K. Menon, K.P. Soman, NSE stock market prediction using deep-learning models, in *International Conference on Computational Intelligence and Data Science* (2018). https://doi.org/10.1016/j.procs.2018.05.050
10. K. Zhang, G. Zhong, J. Dong, S. Wang, Y. Wang, Stock market prediction based on generative adversarial network, in *International Conference on Identification, Information and Knowledge in the Internet of Things* (2018). https://doi.org/10.1016/j.procs.2019.01.256
11. W.-C. Chiang, D. Enke, T. Wu, R. Wang, An adaptive stock index trading decision support system. Expert Syst. Appl. **59** (2016). https://doi.org/10.1016/j.eswa.2016.04.025
12. D. Selvamuthu, V. Kumar, A. Mishra, Indian stock market prediction using artificial neural networks on tick data. Financ. Innov. (2019). https://doi.org/10.1186/s40854-019-0131-7
13. S.T.A. Niaki, S. Hoseinzade, Forecasting S&P 500 index using artificial neural networks and design of experiments. J. Ind. Eng. Int. (2013). https://doi.org/10.1186/2251-712X-9-1

Automatic Irrigation and Crop Protection System Based on IoT

M. Raja, N. M. Nithish, B. Saravana Shankar, and D. Sadhurwanth

1 Introduction

Automatic irrigation developing model is a real-time monitoring system. Most farmers use many hectares of farmland. It is becoming very difficult to monitor and check every area of large land. Sometimes, there is a possibility of uneven water irrigation and crop damage caused by animal attack. This results in low quality crops, which also leads to a loss of production. At present, automation is among the most essential roles in human life. This not only brings comfort, it also reduces energy consumption or saving time [1]. The industries are now using a control and automation machine that is expensive and not approximate for use in farming. This also constructs a low cost, intelligent irrigation technology that can be used by Indian farmers. This project suggests an automatic irrigation system based on the moisture content of the soil. In the existing farming system, Arduino boards and GSM technology are used to monitor crop yields [2], in which Arduino Uno boards act as microcontrollers but not as servers. Therefore, in order to overcome all of these features, microcontrollers like Node MCU are the latest version that acts as both a microcontroller and a server [3, 4]. This methodology's main attraction is its low-installation cost, which has numerous advantages [5]. Information from the various sensors can be regularly updated to the Blynk android application through Wi-Fi where the farmers can analyze the on/off status of motor pump and detection/not detection of animals. Here, you can access and visualize and control the farming system on your laptop, cell phone, etc. [6, 7].

M. Raja (✉) · N. M. Nithish · B. Saravana Shankar · D. Sadhurwanth
Department of Electronics and Instrumentation Engineering, Kongu Engineering College, Perundurai, Erode, Tamil Nadu, India
e-mail: mraja.eie@kongu.edu

J. D. Peter et al. (eds.), *Disruptive Technologies for Big Data and Cloud Applications*, Lecture Notes in Electrical Engineering 905,
https://doi.org/10.1007/978-981-19-2177-3_35

2 Literature Survey

Before briefing automatic irrigation and crop protection system [8], we shall revise about the existing agricultural system [9]. In few of the existing irrigation system, soil moisture is monitored using a voltage meter and water irrigation controller systems. In few other irrigation systems, fuzzy logic controllers are used for effective irrigation with various crops. Fuzzy logic system rises the standard rate of the values obtained and also helps to make decisions. Wireless sensor network and GPRS module are used for automatic irrigation system in the current system of irrigation [10]. Measuring the temperature and humidity is done using sensors and controlled by a microcontroller.

The WIU also has a GPRS module which transmits data to the server with a public mobile Internet. The information data achieved can be monitored remotely through a graphical representation through server. This irrigation system makes it easy to monitor and cultivate even in places with less water and making the system feasible. But Zigbee protocol makes this system costly [11]. The modern agriculture is based on greenhouse which must be accurately controlled both in terms of temperature and humidity. The atmospheric conditions around the fields vary from place to place, making it nearly impossible to irrigate uniformly in all locations of the field manually. To that end, GSM technology is used to gain the report of irrigation process via farmer's mobile phone or other social communication devices [12]. For crop protection system in the existing method PIR sensor, ultrasonic sensor, it may have some drawback to overcome this, and we proposed this system for motion detection of animals.

3 Proposed System

The main objective of our solutions is to supply water when there is less moisture content in the soil without anyone's presence and also to avoid excess water supply during irrigation and also monitoring soil parameters such as moisture content of the soil and atmospheric temperature. The various parameters of the field can also be remotely controlled, at any time, through mobile phone and Web applications. When the animal entering into the field area, the Infrared sensor which senses the presence of animal and send, the input signal to the controller, the APR board will be ON, and sound will be played to divert the animal. It will also be possible to control various parameters of the field remotely from anywhere, anytime by mobile as well as Web application. It will also be possible to remotely control various field parameters using a mobile or Web application from anywhere, at any time. This tells the phone whether to send water to the field or not (when the field is dry). This gives signals to the mobile phone whether to send water (that is when field is dry) to the field or not. It includes Arduino Uno, Node MCU, sensors like soil moisture, IR sensor and Dht11, relays, LCD display, submersible pump, etc.

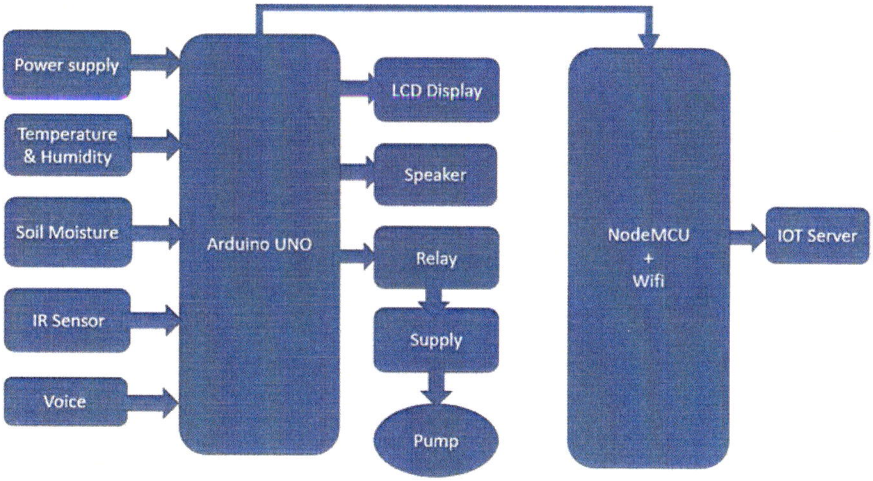

Fig. 1 Block diagram

4 Methodology

The current methodologies include Arduino Uno board and GSM technology. It operates as only a microcontroller, but never as a Wi-Fi server. Thus, to overcome the disadvantages, microcontroller like Node MCU which is the latest version that acts as both a microcontroller and a server. Figure 1 shows the block diagram that gives the basic outcome of the project. It also includes components used in the project such as soil moisture sensor, humidity sensor, and temperature sensor in a single sensor known as DHT 11 sensor, LCD display, relay, submersible pump, IR sensor, voice controller, speaker are used in this experiment. This explanation is as follows.

5 Hardware Description

5.1 Node MCU

Node MCU works on a multi-paradigm, lightweight, high-level programming language designed primarily for embedded applications known as Lua and specifically used for IoT-based applications. It works both as a microcontroller and as a modem. It also works with the help of ESP8266 [Node MCU] Wi-Fi and hardware equipment depending on the ESP-12 module.

5.2 Relay

Relay is the most frequently used switching device in electronics. Relay has an input terminal set and an operating terminal set. The traditional shape of the relay uses an electromagnet to close and open the contacts. Coil 1 is used to trigger the relay. Normally, one end is in contact with 5 v and the other end with ground. Coil 2 is used to trigger the relay, usually one end is connected to 5 v and the other end to the ground. The relay is used to control the submersible pump that is connected to the controller and the sensor.

5.3 DHT 11 Sensor

The DHT 11 sensor is used to measure both the temperature and humidity of the atmosphere. It consists of two parts, the capacitive humidity sensor and also thermistor, which helps to measure both atmospheric humidity as well as atmospheric temperature. The digital temperature and humidity sensor are low cost. It is very easy to use our project. And it has one main advantage that it measures the values of these parameters precisely every 3 to 5 s. Atmospheric humidity alters the capacitance of the DHT11 sensor and gives us an accurate humidity output. There is also a small chip in the sensor that helps convert the analog signal to the digital output. The DHT11 sensor is connected to the Node MCU and the display for the reference of the produced reading. It has three pins for connecting to the circuit, i.e., VCC, data pin, and ground pin.

5.4 Analog Capacitive Soil Moisture Sensor

Capacitive soil moisture sensor which used to measure the soil moisture with respect to the change in capacitance. That is, the moisture of the soil changes the capacitance of the rod and produces certain amount of voltage in respect to the change in capacitance. It basically produces about 1.2–3.0 V. Capacitive soil moisture sensor operates based on the principle of change in capacitance. It measures the change in capacitance rather than change resistance.

5.5 Submersible Pump

The submersible pump used to supply the water to the soil under the control of the microcontroller. The relay helps control the pump when the pump is switched on and off. The submersible pump works with an analog capacitive soil moisture sensor. Once the soil moisture level below a certain level, the motor is automatically switched on, and water is pumped out of the reservoir and supplied to the soil. When the soil moisture level reaches a certain limit, the pump switches off automatically with the support of the microcontroller and the soil moisture sensor connected to a pump. The submersible pump operates on a supply of 3.3–6 V. This can pump approximately to 120 L per hour even when the supply is low (i.e., 220 mA). The pump is submerged in the water, and the tube is connected to the pump outlet and is powered to make the pump work in any condition.

5.6 *APR 9600*

The voice signal is picked up by the microphone. After that, the signal is routed to the APR 9600 multi-section sound record and replay IC. The APR 9600 is a low cost, high performance integrated circuit with flash analog storage. Even after the module's power supply is removed, the recorded sound is preserved. The replayed sound is of high quality and has a low-noise level. By changing the value of a single resistor, the total recording time can be adjusted from 32 to 60 s. The integrated circuit can function in two modes: serial and parallel.

5.7 *IR Sensor*

An infrared sensor that aids in the detection of radiation-interfering objects. The emitter is a simple infrared light emitting diode (LED), and the detector is a simple infrared photodiode. The photodiode is sensitive to infrared light of the same wavelength as the IR LED. When infrared light strikes a photodiode, the resistances and output voltages change proportionally to the magnitude of the IR light. An infrared source, a transmission medium, an optical component, infrared detectors or receivers, and signal processing are the five basic components of an infrared detection system. As infrared sources, specific wavelength infrared lasers and infrared LEDs are used.

5.8 *IoT Cloud*

The IoT cloud used in this project is a Blynk app, and it is an Internet of things (IoT)-based cloud platform that allows to receive and analyze sensor data from the cloud and develop IoT applications. The Internet of things is an emerging trend technology where a more number of built-in devices (things) connected to the Internet. These digital technologies share information with other users and other places but often facilitate sensor information to the cloud storage as well as cloud computing resources where information is processed and evaluated for important insights. Affordable cloud computing has expanded Internet connectivity to make this trend possible. IoT solutions are designed for a wide range of various applications, including health surveillance, vehicle fleet tracking, industrial monitoring and control, environmental monitoring and management, and remote monitoring.

5.9 Setting Up of Blynk Application

The following procedures should be followed to set the Blynk server: create a account on Blynk and login and fill the details in the account. By clicking the "channel" button, we can create new channel by filling these following details. Click the API Key to see the "write API key" button. Insert the API key. We can click the "private view" button and modify the window display as you desire this is all part of Blynk IoT cloud. Now coming to the program section of the IoT-based automatic irrigation and crop protection system source code with ATMEL board. You can upload the code directly to the development board. Before all this, make remember to add OLED Display Library Adafruit GFX Library, SSD1306 Library. API Key, Wi-Fi SSID and password are changed at the Blynk to execute a program.

6 Interfacing and Working

Block diagram of the IoT-based automatic irrigation and crop protection system is shown in Fig. 1. This is where the Fritzing tool is used to design a circuit, and the soil moisture device is connected to A0 pin of Node MCU and DHT11 pin to D4 pin. The relay is controlled by a D5 pin of Node MCU. A LCD display is connected to a Node MCU 12c pi. The relay and motor are controlled by 5v over the Vin pin of Node MCU. The DHT11 sensor, the capacitive soil moisture sensor V1.2, and the LCD display are controlled by a 3.3v supply. Capacitive soil moisture sensor, DHT11 sensor which is interfaced with Node MCU where the air temperature, air humidity, soil moisture, and motor status can be monitored through Blynk app.

The LCD display will start displaying real-time data as soon as the device is powered. When the soil moisture content present in the soil is reduced, the water pump will turn on and irrigates the soil till the required moisture level reached. All of this can be remotely monitored from anywhere in the world through the use of a Blynk server [6]. In this image, it shows that the water level of the soil is being nurtured, and the limit of the soil moisture has been set. So the water pump automatically turns off, and the supply to the motor is shut. The observations along with the soil moisture level as well as the temperature, humidity, and their respective graphs and motor status are also shown on the IoT platform. In Fig. 2, the IoT platform shows us the exact condition of the parameters and the status of the sensor without delay in the readings. Farmers can visualize and analyze online data from any other part of the world [13]. IoT platform not only helps to verify the status of the pump, soil

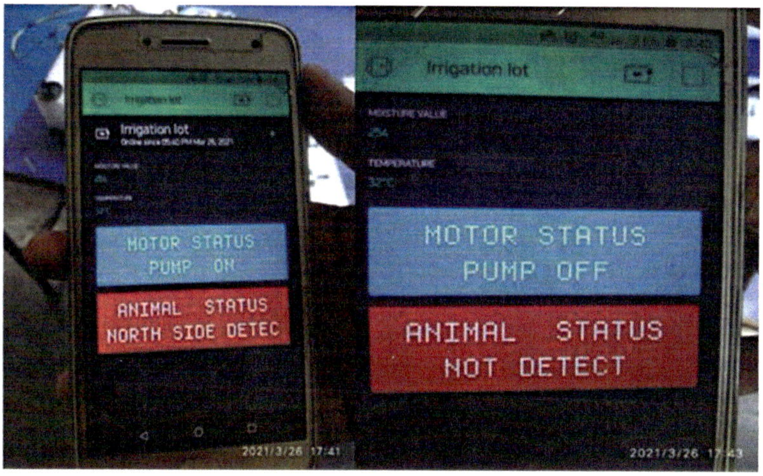

Fig. 2 Server output

moisture, and temperature, and it helps to intimate the intrusion of the animals also it shows in which direction of animals into the field. The platform helps in many ways, and with the help of artificial intelligence, the voice output can be set up based on the animal or bird entering the field and can be driven away [14].

7 Results and Conclusion

Automatic irrigation of water is carried out successfully in all parts of the field whenever a low amount of soil moisture is sensed, the engine switches automatically on and soil becomes wet, the engine turns off, the prototype developed in this project prevents human work and water wastage. By implementing this smart setup, crops will increase productivity. This is more compatible and economical when it comes to setting up on a large scale. A real-time data are shown. Whenever the soil moisture levels are reduced, a water pump switches on and irrigates the area till the required moisture has been reached.

In the Fig. 3, it shows the working prototype on condition in which pumps the water automatically and provides the water to the soil with which water depletion for the plant while there is no manual supply of water. Also the Fig. 3 shows the working prototype off condition in which automatic switching off pump with the help of the microcontroller and relay and maintaining the soil moisture.

The prototype developed in this project prevents the human work, water wastage, and also crop protection system. By implementing this smart setup, crops will increase productivity. This is more compatible and economical when setting up on a large scale. Table 1 displays real-time data. When the moisture content of the soil

Fig. 3 Working prototype on condition

Table 1 Temperature and moisture comparison for field irrigation

S. No.	Crop name	Ideal temperature range	Actual temperature	Setpoint temperature	Actual moisture	Setpoint moisture	Pump status
1	Sugarcane	27–38	32	34	180	80	Off
2	Cotton	20–32	34	26	158	140	On
3	Groundnut	30–35	30	33	120	150	Off
4	Paddy	25–35	31	30	30	10	On
5	Turmeric	25–35	29	28	100	100	Off
6	Tapioca	25–29	28	27	200	170	On

decreases, the water pump activates and irrigates the field until the required moisture is achieved. This is done in comparison with the temperature of the atmosphere. Farmers can use Blynk which allows data to be monitored online from anywhere in the world you can access the Blynk private view. We can monitor the soil moisture, humidity, and temperature, as well as the relay status, from this location. It also shows the wild intrusion into the field and IR detections. Thus with the sound alarm system, crops can be protected from the external interference of animals.

References

1. P.A.V. Deshpande, Design and implementation of an intelligent security system for farm protection from wild animals. Int. J. Sci. Res. (IJSR) 1–4 (2016)
2. D.K. Sheela Sobana Rani, M.R. Lavanya, M. Gayathri, An automatic repelling system to reduce human elephants conflicts using sensors. Int. J. Res. Appl. Sci. Eng. Technol. (IJRASET) 1–5 (2017)
3. L. Yuanguai, An intelligent monitoring system for agriculture based on ZigBee Wireless Sensors Network. J. Adv. Mater. Res. Manuf. Sci. Technol. **383–399**, 435 (2008)
4. M.T. Batte, Changing computer use in agriculture: evidence from Ohio. Comput. Electron. Agric. **47**, 1–13 (2005)

5. P. Divya, P. Usha Kiran, P. Praveen, Iot-based wild animal intrusion detection system. Int. J. Recent Innov. Trends Comput. Commun. 1–3 (2018)
6. S.P. Das, S. Chita, N. Peterson, B.A. Shirazi, M. Bhadkamkar, Home automation and security for mobile devices, in *IEEE PERCOM Workshop* (2011), pp. 141–146
7. I. Gautam, S.R.N. Reddy, Innovative GSM based remote controlled embedded system for irrigation. Int. J. Comput. Appl. **47**(13), 1–7 (2012)
8. M. Yogheshwaran, D. Praveenkumar, S. Pravin, P.M. Manikandan, S. Saravanan, IoT based intelligent traffic control system. Int. J. Eng. Technol. Res. Manage. **4**(4), 59–63 (2020)
9. K. Jambroi, B. Ivanevi, A. Stimac, *The Influence of Ultrasound on Animal Irritation* (2019)
10. K. Subashchandrabose, G. Moulieshwaran, M. Raghul, V. Dhinesh, S. Saravanan, Design of portable sanitary napkin vending machine. Int. J. Eng. Technol. Res. Manage. **4**(4), 52–58 (2020)
11. Y. Kim, R.G. Evans, W.M. Iversen, Remote sensing and control of an irrigation system using a distributed wireless sensor network. IEEE Trans. Instrum. Measur. **57**(7), 1379–1387 (2008)
12. M. Pavithra, S. Pavithra, R. Rama Priya, M. Vaishnavee, M. Ranjitha, S. Saravanan, Fingerprint based medical information system using IoT. Int. J. Eng. Technol. Res. Manage. **4**(4), 45–51 (2020)
13. C. Amardeo, I.G. Sarma, Identity in the future internet of things. Wirel. Pers. Commun. **49**, 353–363 (2009)
14. G. Pooja, M. Umair Bagali, A smart farmland using raspberry pi crop vandalization prevention intrusion detection system. IJARIIE 1–7 (2016)

Visual Question Answering System Using Co-attention Model

D. Karthika Renuka, L. Ashok Kumar, R. Geetha Rajakumari, R. Vinitha, R. Meena, and B. Swathi

1 Introduction

Image Question Answering is a computer vision and a natural language problem that expects models to comprehend the image and the question as well as their relationship. It requires better reasoning abilities and rational knowledge. VQA can help image retrieval system, blind people in navigation, smart transport systems, while handling sensor data, etc. The advent of feature extraction, feature fusion, and attention mechanisms aids in the steady progress of VQA models in various fields.

VQA needs image and question processing to obtain information to answer the question. The image representation is required to understand the image and question features for understanding the question. Then, the proposed system combines these features by fusion methods like concatenation, etc. At last, the model is trained with the fused features to produce an appropriate answer. The requirement of reasoning ability, rational knowledge, and ability to understand the relation between the visual and language features makes VQA a more difficult task. Hence, it becomes difficult to answer the question more accurately. But with the advanced deep learning techniques, better results can be obtained for image question answering. The combination of sensors and VQA can also help machines to obtain human-like abilities and assist people that lack those abilities at low cost.

D. Karthika Renuka (✉) · R. Geetha Rajakumari · R. Vinitha · R. Meena · B. Swathi
Department of IT, PSG College of Technology, Coimbatore, Tamil Nadu, India
e-mail: dkr.it@psgtech.ac.in

L. Ashok Kumar
Department of EEE, PSG College of Technology, Coimbatore, Tamil Nadu, India

J. D. Peter et al. (eds.), *Disruptive Technologies for Big Data and Cloud Applications*, Lecture Notes in Electrical Engineering 905,
https://doi.org/10.1007/978-981-19-2177-3_36

2 Related Works

With the rapid growth of natural language processing and deep learning techniques, visual question answering has become an active area of research. Simonyan et al. demonstrated how VGGNet can be widely utilized for image feature extraction [1]. Antol et al. briefed about the datasets that can be used, the types of questions and answers available in the datasets, and the metrics for analysis the performance of a visual question answering system [2]. Malinowski et al. proposed the use of CNN and LSTM using DAQUAR dataset for image question answering [3]. Ren et al. introduces the method of generating questions and answers from the image for image-based question answering problem [4]. Kiros et al. proposed that the embeddings from questions can be calculated using skip-thought vectors [5]. Kafle et al. suggested the use of Bayesian algorithm to predict the probability of the answer. The probability is then converted as the actual answer [6]. A focused attention dynamic attention model that incorporates question and image understanding filtered out the irrelevant information to improve performance [7]. The most recent technique in the deep learning field is the concept of attention. There are several works that provides an attention-based solutions like question-based image attention, question-guided spatial attention, stacked attention networks, and co-attention [8–12]. Das et al. introduces Human Attention dataset and compares the attention maps generated by visual answering system with human attention [13]. Teney et al. suggested another technique which uses graph-structured representation for the images and questions [14]. Deep learning models for various real world case studies are discussed in Lovelyn Rose et al. [15].

3 Dataset and Data Preparation

3.1 Datasets for VQA

The most commonly used datasets for VQA systems are as follows:

DAQUAR (DAtaset for QUestion Answering on Real-world images): It has images from NYU-DepthV2 dataset with 6795 training and 5673 testing questions and answers. Considering its small size and less variation, DAQUAR cannot be used for complex models.

COCO-QA: COCO-QA consists of COCO images. COCO images are provided with captions about them. The questions for COCO-QA are generated from the captions using NLP techniques. Similarly, answers are also generated for each question. It has 78,736 training and 38,948 testing question and answers.

The VQA Dataset: It is of two kinds: COCO-VQA and SYNTH-VQA. COCO-VQA with images from COCO dataset images is a large dataset with 614,163 questions in total along with 248,349 training and 244,302 testing questions. SYNTH-VQA contains abstract scenes of around 50,000 scenarios with three questions each and ten answers for every question. Unlike DAQUAR, VQA dataset has questions that are open-ended, yes or no, and number based. The diversity and large size of VQA dataset have made VQA as one of the ideal datasets for evaluating complex algorithms with promising accuracy.

3.2 Data Preparation

The proposed system uses VQA dataset which contains COCO images and set of question and answers about the images for both training and validation. The questions and answers are available as JSON files. The first step in data preparation is to create a new JSON file which has the answer along with the question and image path. The questions in the dataset are pre-processed, tokenized, and converted to sequence of integers. On visualizing the answer distribution, it is found that the answer to most of the questions are under few classes of answers. Hence, the proposed model filters out the top 1000 answers from the dataset and encode those answers. These encoded answers can be used for predicting the answers for questions being asked.

4 Proposed Methodology

In general, a VQA system has two main subtasks, namely image feature generation and obtaining text representations. The final task is to develop a model that makes use of these two features to provide a response to the question. A simple approach is utilizing the pre-trained CNN models like VGGNet, GoogLeNet, ResNet, etc., for visual feature extraction. As for question feature extraction, techniques like bag-of-words (BOW) approach, gated recurrent units (GRU), long short-term memory (LSTM) encoders, and skip-thought vectors are available. Considering the question answering task, many algorithms solve it by considering a classification approach. By classifying answers into different answer types, the proposed system can assist the model in classifying the appropriate answer type for the question. Attention mechanism also made its way in making the VQA model to concentrate more on relevant features for question answering. This is achieved by giving different weights to different regions of the image based on their participation in the question being asked. As for textual attention, the key words in the questions are given more importance. It can be said that attention models aim at making the model more focused on relevant details and key features in order to improve the answer predictions. The attention model is represented in Fig. 1.

The main difference between other algorithms and attention mechanism is that features are not considered on the complete image level. In case of images, image representations are considered on a local range with respect to the weightage it holds.

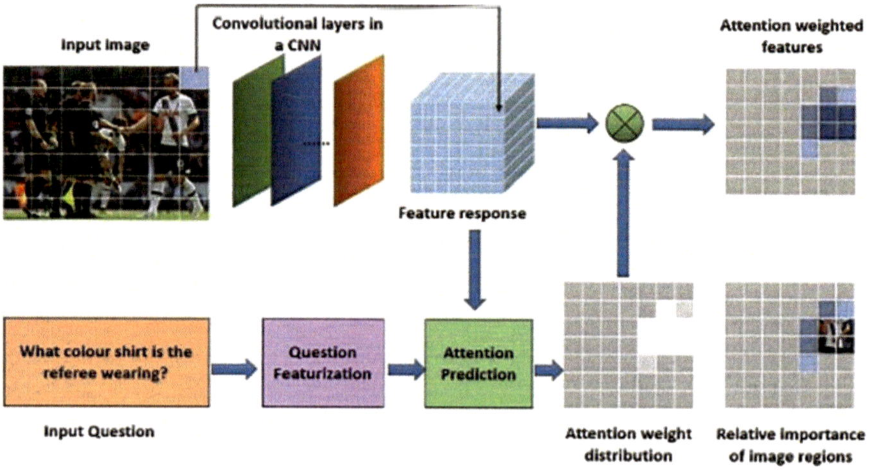

Fig. 1 Incorporating attention into a VQA system

As for the questions, the model gives more weightage on the words that has more weightage in assisting the model to predict the answer. VQA models are made to focus on both important image and question key words. The proposed framework includes a novel multi-modal attention model which focuses on image and question attention together. The model is made to give attention to image and question representations parallel. Hence, it is referred to as parallel co-attention. The image features are extracted using VGGNet19. In the case of questions, the proposed system extracts textual features at different hierarchical levels, namely word, phrase, and sentence level. The model takes these features and apply parallel co-attention at all these levels.

4.1 Question Hierarchy

After data preparation, the question words are embedded to a vector space to get word level features. The word embeddings from the word level is given to 1-D convolution to obtain phrase level features. The system computes the unigram, bigram, and trigrams from the word. After obtaining the convolution result, the model does max-pooling to get the phrase level question features as:

$$q_t^p = \max(\hat{q}_{1,t}^p, \hat{q}_{2,t}^p, \hat{q}_{3,t}^p), t \in \{1, 2, \ldots, T\}. \quad (1)$$

The proposed system uses LSTM on the phrase level features q_t^p to obtain the sentence level features q_t^s.

4.2 Parallel Co-attention

The first step is calculating similarity between image and question features at every region in order to connect the image and question features. V represents the image features, and Q represents the question features. The proposed model calculates an affinity matrix C, as follows:

$$C = \tanh\left(Q^T W_b V\right) \tag{2}$$

where W_b are the weights. Using the affinity matrix, the system try to compute image and question attention maps as follows:

$$H^v = \tanh\left[W_v V + \left(W_q Q\right)C\right] \tag{3}$$

$$H^q = \tanh[W_q Q + (W_v V)C^T)] \tag{4}$$

$$a^v = \text{softmax}\left(w_{hv}^T H_v\right) \tag{5}$$

$$a^q = \text{softmax}\left(w_{hq}^T H_q\right) \tag{6}$$

where W_v, W_q, w_{hv}, and w_{hq} are the corresponding weight parameters. a^v and a^q are the obtained attention probabilities for image and question attention maps, respectively. Using the calculated attention probabilities along with the image and question features, the model calculates the image and question attention vectors using the formula:

$$\hat{v} = \sum_{n=1}^{N} a_n^v v_n \tag{7}$$

$$\hat{q} = \sum_{t=1}^{T} a_t^q q_t \tag{8}$$

where $\hat{v}$ and $\hat{q}$ are the image and question attention vectors. For every level in the question hierarchy, all these steps are followed to achieve co-attention. The parallel co-attention model is depicted in Fig. 2.

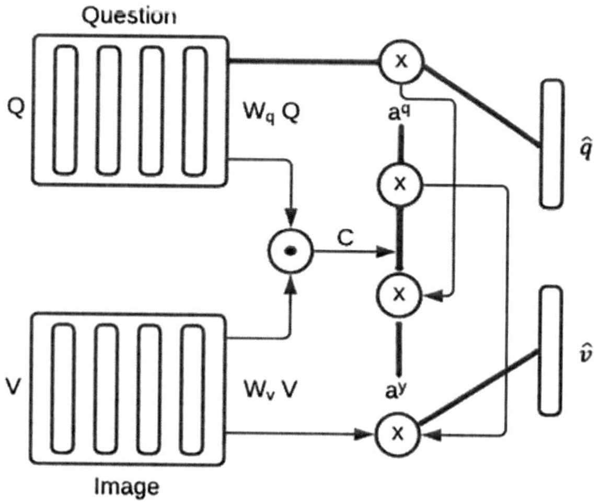

Fig. 2 Parallel co-attention

4.3 *Predicting the Answers*

The top 1000 answers are filtered and encoded. Hence, the system can treat this as a 1000 class classification task. The co-attended vectors obtained from a lower level in the hierarchy is then concatenated with the higher level. After the sentence level, an MLP softmax activation provides us with the probability of the answer as follows:

$$h^w = \tanh\left(W_w\left(\hat{q}^w + \hat{v}^w\right)\right) \tag{9}$$

$$h^p = \tanh\left(W_p\left(\left(\hat{q}^p + \hat{v}^p\right),\ h^w\right)\right) \tag{10}$$

$$h^s = \tanh\left(W_s\left(\left(\hat{q}^s + \hat{v}^s\right), h^p\right)\right) \tag{11}$$

$$\text{answer probability} = \text{softmax}\left(W_h h^s\right) \tag{12}$$

where W_w, W_p, W_s, and W_h are the corresponding weights for each level. The probability from the softmax output is the probability of the most appropriate answer. Figure 3 depicts the answer prediction process.

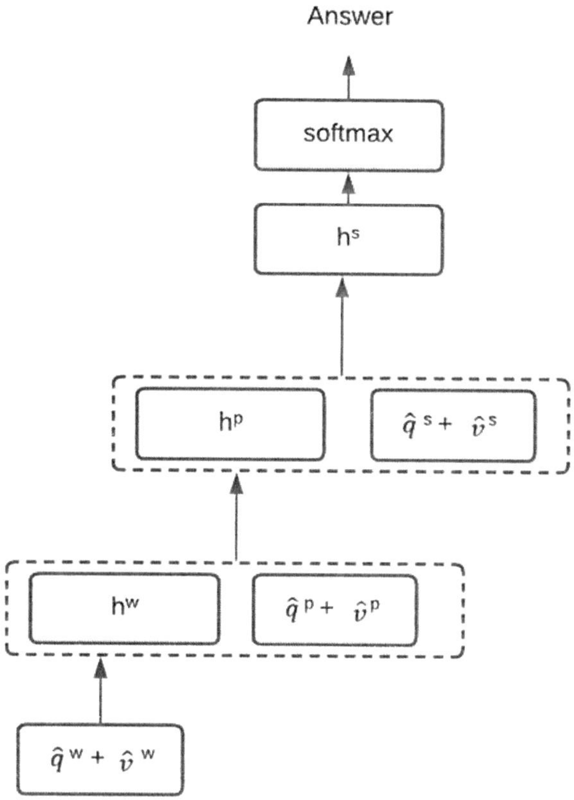

Fig. 3 Answer prediction

5 Experimental Results

5.1 Setup

The proposed system uses python for implementing the model. The model is evaluated using the VQA dataset. The images are resized to a fixed dimension 224 × 224, and VGGNet19 is used for extracting the image features. The maximum length (number of words) of the question is fixed as 22. At a learning rate of 1e−4, dropout rate 0.5, and regularization value of 0.01, the model is trained for 60 epochs. The model uses Adam optimizer, and the batch size is fixed as 300.

Table 1 Performance analysis

Method	Accuracy for each answer type			Overall accuracy
	Yes/No	Numbers	Others	
Per Q-type prior	71.03	35.77	09.38	37.54
Zhang et al.	79.14	–	–	35.25
Nearest neighbor	71.89	24.36	21.94	42.70
SAN	79.3	36.6	46.1	58.7
VGG + LSTM	–	–	–	23.14
VGGNet + Parallel Co-attention (proposed model)	79.5	38.7	48.3	49.2

5.2 *Result Analysis*

The proposed model is trained for open-ended questions from the VQA dataset. Hierarchical Co-attention-based model [12] is the model's baseline. The parallel co-attention model provides an overall training accuracy of 54.78% and test accuracy of 49.28%. The comparison with other existing algorithms showing the accuracy for each class of answer and the overall accuracy is depicted in Table 1. The accuracy is calculated based on the formula,

$$\text{accuracy} = \text{minimum}\ ((\text{no of ground truth answers that match with predictions}/3),\ 1)$$

which implies that the predicted answer is 100% accurate if the answer given by the system matches with three correct answers for the question being asked.

6 Conclusion

VQA is still a newly developing research field which requires deep knowledge and understanding of visual and text information. With the current research advancements, VQA systems are expected to grow with improving accuracy and performance. The performance of the model has to be evaluated in such a way that it is tested with different question and answer types. Other than open-ended questions, multiple choice questions should also be taken into consideration. The scope of VQA in real world is steadily increasing with the improving technology and intelligence systems. Intelligent devices like robots can rely on VQA for assisting people. There has been increasing research and new approaches that can improve the performance and accuracy for image question answering [16–19].

References

1. K. Simonyan, A. Zisserman, *Very Deep Convolutional Networks for Large-Scale Image Recognition* (2014). arXiv preprint arXiv:1409.1556.
2. S. Antol, A. Agrawal, J. Lu, M. Mitchell, D. Batra, C.L. Zitnick, D. Parikh, Vqa: Visual question answering, in *Proceedings of the IEEE International Conference on Computer Vision* (2015), pp. 2425–2433
3. M. Malinowski, M. Rohrbach, M. Fritz, Ask your neurons: a neural-based approach to answering questions about images, in *Proceedings of the IEEE International Conference on Computer Vision* (2015), pp. 1–9
4. M. Ren, R. Kiros, R. Zemel, Exploring models and data for image question answering. Adv. Neural. Inf. Process. Syst. **28**, 2953–2961 (2015)
5. R. Kiros, Y. Zhu, R.R. Salakhutdinov, R. Zemel, R. Urtasun, A. Torralba, S. Fidler, Skip-thought vectors, in *Advances in Neural Information Processing Systems* (2015), pp. 3294–3302)
6. K. Kafle, C. Kanan, Answer-type prediction for visual question answering, in *Proceedings of the IEEE Conference on Computer Vision and Pattern Recognition* (2016), pp. 4976–4984
7. I. Ilievski, S. Yan, J. Feng, *A Focused Dynamic Attention Model for Visual Question Answering* (2016). arXiv preprint arXiv:1604.01485
8. K.J. Shih, S. Singh, D. Hoiem, Where to look: focus regions for visual question answering, in *Proceedings of the IEEE Conference on Computer Vision and Pattern Recognition* (2016), pp. 4613–4621
9. W. Yin, H. Schütze, B. Xiang, B. Zhou, Abcnn: attention-based convolutional neural network for modeling sentence pairs. Trans. Assoc. Comput. Linguist. **4**, 259–272 (2016)
10. H. Xu, K. Saenko, Ask, attend and answer: exploring question-guided spatial attention for visual question answering, in *European Conference on Computer Vision* (Springer, Cham, 2016), pp. 451–466
11. Z. Yang, X. He, J. Gao, L. Deng, A. Smola, Stacked attention networks for image question answering, in *Proceedings of the IEEE Conference on Computer Vision and Pattern Recognition* (2016), pp. 21–29
12. J. Lu, J. Yang, D. Batra, D. Parikh, Hierarchical question-image co-attention for visual question answering. Adv. Neural. Inf. Process. Syst. **29**, 289–297 (2016)
13. A. Das, H. Agrawal, L. Zitnick, D. Parikh, D. Batra, Human attention in visual question answering: do humans and deep networks look at the same regions? Comput. Vis. Image Understand. **163**, 90–100 (2017); L. Ma, Z. Lu, H. Li, Learning to answer questions from image using convolutional neural network, in *AAAI* (2016)
14. D. Teney, Q. Wu, A. van den Hengel, Visual question answering: a tutorial. IEEE Signal Process. Mag. **34**(6), 63–75 (2017)
15. D.S.L. Rose, L. Ashok Kumar, D. Karthika Renuka, *Deep Learning Using Python* (2019)
16. A. Al-Sadi, H. Al-Theiabat, M. Al-Ayyoub, The inception team at VQA-Med 2020: pretrained VGG with data augmentation for medical VQA and VQG, in *CLEF (Working Notes)* (2020)
17. W. Tian, B. He, N. Wang, Z. Zhao, Multi-channel co-attention network for visual question answering, in *2020 International Joint Conference on Neural Networks (IJCNN)* (2020), pp. 1–8. IEEE
18. J. Hong, S. Park, H. Byun, Selective residual learning for visual question answering. Neurocomputing **402**, 366–374 (2020)
19. D. Han, S. Zhou, K.C. Li, R.F. de Mello, Cross-modality co-attention networks for visual question answering. Soft. Comput. **25**(7), 5411–5421 (2021)

Internet of Things and Cloud Computing for Smart Vermicomposting by Using Eisenia Fetida and Its Optimization by ANN

Amar Kumar Das, Saroja Kumar Rout, Srikanta Kumar Dash, and Abhijit Mangaraj

1 Introduction

To feed the world's population, modern agriculture and civilization both require higher food output. In the agricultural area, new technologies and solutions [1–3] are being used to give an optimal alternative for gathering and processing data [4] while increasing net productivity. To execute this aim, automation and intelligent decision-making are becoming increasingly necessary. Internet of Things (IoT) [5–8], ubiquitous computing, wireless ad hoc and sensor networks [9, 10], radio frequency identifier cloud computing [11, 12], remote sensing [13], and other technologies are gaining in popularity in this area.

Motivation

Key assessment causes, as listed below, have impacted the script of this article.

Municipal solid waste (MSW) can be regarded as the waste produced from domestic and industrial usages. India is the second-largest nation accounting for a population of 1.21 billion, with rapid urbanization and prosperity [14]. In developing countries like India, management of solid waste is one of the most daunting problems facing a severe pollution due to the production of enormous amounts of solid waste. This paper provides an evaluation of the current state of municipal solid

A. K. Das (✉)
Department of Mechanical Engineering, GIFT, Bhubaneswar, India
e-mail: amar.das120@gmail.com

S. K. Rout
Department of IT, Vardhaman College of Engineering, Hyderabad, India

S. K. Dash
Department of EEE, REC, Bhubaneswar, India

A. Mangaraj
Department of Civil Engineering, GIFT, Bhubaneswar, India

J. D. Peter et al. (eds.), *Disruptive Technologies for Big Data and Cloud Applications*, Lecture Notes in Electrical Engineering 905,
https://doi.org/10.1007/978-981-19-2177-3_37

waste management (MSWM) in Indian cities. Disposal in landfills requires a large scale area and becomes inevitable due to the increase in population [15]. Landfill also contributes to lowering the soil fertility level, leading to poor crop growth. The incinerating of wet food items may lead to more energy consumption. Potential sites for municipal landfills are likely to generate harmful gases such as methane and carbon dioxide which is contemplated as greenhouse gases subsidizing climate changes globally [16]. Hence, to overcome this crisis, an eco-friendly and hygienic method can be opted which is regarded as composting.

The source of composting and vermicomposting-related materials and biological activity makes it extremely difficult to determine the rate of application of vermicomposting and its suitability as a modification of the soil. MSW is produced generously which is a burden from a waste disposal point of view. The biological conversion of organic waste to bio-fertilizer is termed Vermicomposting. The agent which widely contributes to the conversion of organic wastes into organic compost is earthworm [17, 18]. The earthworm intakes these wastes, and the end product produced is the worm humus which is otherwise termed as vermicast. It has been observed that these castings are rich in nutrients such as nitrogen, phosphorus, potassium, calcium, sodium, magnesium, and iron which can be used directly as a fertilizer to boost up the plant growth [19]. Vermicompost when compared with locally available chemical fertilizer is considerably richer in nutrients, cheap in cost, and an environment-friendly way to sustainably recycle the organic wastes to bio-fertilizer. Vermicomposting is the process of turning organic waste into organic fertilizer rich in NPK, micronutrients, and beneficial soil microorganisms by utilizing earthworms [20]. It is a good growth booster and protector for crop plants [21], and it is a sustainable alternative to artificial fertilizers. It also aids in the conversion of many types of organic wastes into high-nutrient plant and soil fertilizers [22]. Vermicompost adds organic matter to the soil while improving its physical, chemical, and biological qualities. This paper aims to find out the parameters responsible for effective composting and substantially optimizing such parameters for better enrichment in vermicompost as the end product.

2 Materials and Methods

A brief comparison between chemical fertilizer and vermicompost is given below in Table 1. The table shows the increment in potential nutrients values of compost required for the growth of a plant in Vermi based on that of chemical fertilizer.

Figure 1 shows the schematic diagram of worm bin prepared for vermiculture and subsequently producing the compost. It is made of standard size by (6 ft × 4.5 ft × 5 ft) and occupied with maximum volume of 70–80% of total capacity. The preparation of vermicompost is done by providing suitable bedding materials for the earthworms. The vermicompost bed is prepared by the materials such as dry cow dung, watering at proper intervals, cooked rice wastes, shredded cardboard materials or papers, dry neem leaf, and egg shell powders as per the required intake of raw

Table 1 Potential nutrient values

Variables	Chemical fertilizer	Vermicompost
pH	7.8	6.8
Nitrate/nitrogen	156.50 ppm	902.20 ppm
Potassium	0.48%	0.70%
Phosphorus	0.35%	0.47%
Total nitrogen	0.80%	1.94%

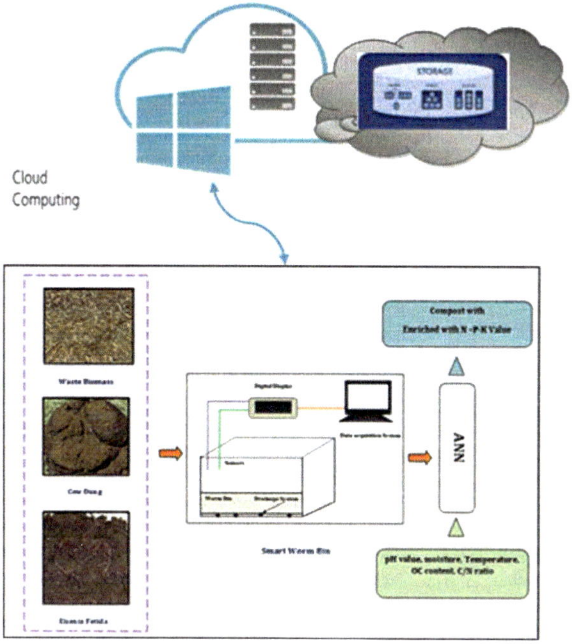

Fig. 1 Schematic experimental worm bin

materials. Earthworms are used for digestion of the bedding materials. Dry cow dung which is free from lumps is also spread evenly. The purpose of using dry cow dung is because it is free from any unusual odor compared to wet cow dung. Water is sprinkled on the dry cow dung for it to retain moisture which will be favorable in digesting the waste. Moisture content is an essential factor, as the earthworms breathe through their skin, they need enough moisture to prosper otherwise their skin dries out and consequentially leading to the death of earthworms. The dung is mixed thoroughly at regular intervals for proper aeration. Rice waste is added to the bin, and it is allowed to decompose for about 2–3 days with proper moisture content. The rice wastes with cow dung proportionately 70:30 are used in our worm bin. The decomposition allows partial digestion of complex materials of the sample, making it suitable for the earthworms to consume.

The worm bin consists of a sample tray having a porous surface at the bottom to drain out excess water. The temperature, pH, and humidity of the bin are measured

Fig. 2 Quality compost prepared

and recorded by sensors assisted data acquisition system. Using temperature and humidity sensors, vermicomposting areas are kept under control. Through the IoT devices and cloud storage platform, optimal levels for both temperature and humidity are preserved which are presented in Fig. 1. In this process, Eisenia fetida was used for preparing compost due to its potential stability with the environment, high reproductive rate, and ease of availability. Earthworm (Eisenia fetida) is introduced to the bin. The moisture content is maintained by spreading shredded cardboard material or papers on top of the bin and sprinkling water at a regular interval of 1–2 days. The shredded cardboard also acts as food material for the earthworm. The earthworm eats the food waste and after digestion excretes a lightweight material which is also called worm humus or vermicompost. This process takes up to 24–28 days, the vermicompost produced is black/brown in color, rich in nutrients required for the healthy growth of plants. The whole process is carried out at different temperature conditions to get optimized values of nutrients.

The vermicompost then undergoes some nutrient tests to measure necessary nutrients (N–P–K) available in the compost as shown in Fig. 2. The nutrient tests for nitrogen (N), phosphorus (P), and potassium (K) in vermicompost are carried out in vermicompost samples. The outcomes observed after testing have been enlisted in Table 2.

The best measures to test the vermicomposting process are to increase its consistency and transformation of food waste into quality manure. In addition to the biochemical properties of waste, microbial biomass and degradation processes in anaerobic digestion are important for determining worm biomass and cocoon generation.

3 Results and Discussion

IoT cloud solutions enable remote cloud servers to perform functions such as real-time data collecting, pay-as-you-go, visualization, data analytics, decision, device

Table 2 NPK values under different conditions

Temperature (°C)	pH Value	C/N ratio	Organic carbon (°C)%	Moisture (%)	N %	P%	K %
30	6.1	5.1:1	6.28	29	1.23	1.53	0.63
31	6	5:1	6.1	30	1.10	1.62	0.72
32	6.2	4.8:1	6.3	31	1.22	1.52	0.68
33	6.2	4.9:1	6.2	35	1.15	1.54	0.64
34	6.17	5.8:1	6.18	31.8	1.19	1.54	0.65
35	6.12	5.4:1	6.23	30	1.21	1.54	0.62
36	6.14	5.3:1	6.32	33	1.24	1.54	0.61
37	6.15	5.5:1	6.35	35	1.28	1.58	0.52

administration, and so on. In the agriculture cum farming business, various cloud service providers are gradually gaining traction. In recent years, new ground-breaking applications have been envisioned thanks to the rise of sensor cloud, big data analytics, and ubiquitous computing. We will quickly go over the fundamentals.

3.1 Sensor Cloud-Enabled Computing

Computing is a relatively new notion on-field IoT applications which is referred to as this is enabled by cloud computing [21] and that finds relevant information from massive volumes of data collected at rapid speeds. Furthermore, the sensor cloud enhances data administration, data access, and device administration chores while allowing for smart vermicomposting in nature.

Below are a few examples of possible applications.

- A cloud-based storage system for tracking the geographic fluctuations of soil and other environmental characteristics over time.
- A mobile sensor cloud is served to monitor a cloud-based infrastructure and estimate the future yield in vermicompost productivity.
- It is vital to create a plan for the remote-controlled vermicomposting system so that the vermicompost field can be plated by an autonomous device.
- The use of the cloud in vermicomposting management systems might be built to produce compost at any time of year. Cloud services could be used to power a real-time livestock monitoring system that automatically improves the farmer's financial situation. Vermicomposting farming is a critical sector where cloud-based services can track and regulate the entire process, from breeding to market sale.

3.2 Process Optimization and Validation Using ANN Technique

The neural network plays an important in machine learning with a backpropagation multilayer feed-forward system. This neural system is most appropriate for different machine learning applications. Neural networks can decide the condition of the brain in the MR image [22]. The neural network is trained and constructed by MATLAB programming, as shown in [23], with the purpose of having the least error in data preparation and the most extreme relationship in a system toolbar coefficient.

The major network was the neural system feeding forward backpropagation with train BR preparation capacity, LEARNGD SSE performance effort, and energy cognitive abilities. This procedure utilizes the ANN apparatus in MATLAB programming to prepare, confirm, and test the system module. Backpropagation is the most generally utilized neural network [24].

Figures 3 and 4 demonstrate the structure of the neural system and describe the starch wastes used for producing organic compost under various parametric conditions. Figures 3 and 4 demonstrate the framework of a neural system and describe the starch wastes used for producing organic compost under various parametric conditions. The figure also explains how different temperature, humidity, C/N ratio, pH value, and retention period are all input into the system and organic carbon content in the waste biomass. The hidden layers of the neural network play a crucial role in establishing a link between input and output and the quality of compost with high nutrient values. Figure 5 depicts the error-correcting technique used to prepare the network and defines the following are the network's features. With both the graph and regression analysis, we defined the parameters and observed the execution. During the preparation of the system's outcome, the figure depicts the prediction error value. Backpropagation approach is used in this execution; MSE, MSCREG, and SSE are

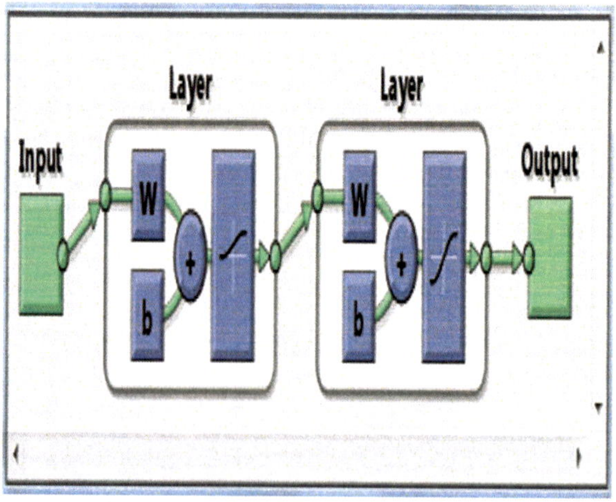

Fig. 3 ANN layout of the system using MATLAB

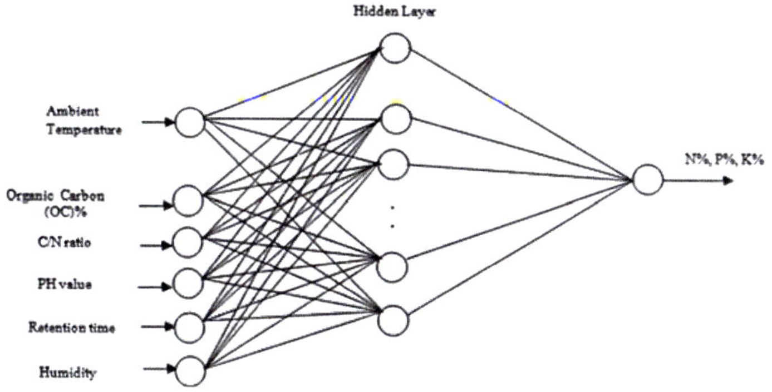

Fig. 4 ANN layout of the vermicomposting

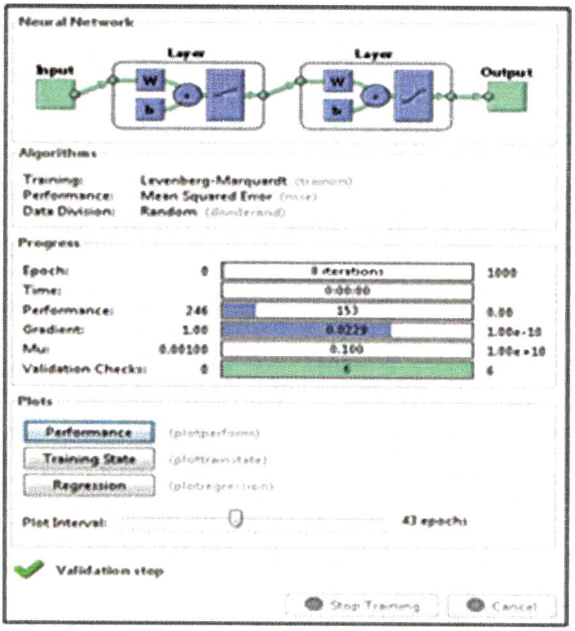

Fig. 5 Neural network training system

examples of different evaluation functions. In addition, various transfer functions, for example Transit, Logsig, and Purelin, are used in the result analysis. We use a large number of iterations to activate and smarten the network's validation tests (Figs. 6 and 7).

The outcome of the training R-value during different scenarios is stated in (Figs. 8, 9, and 10), and iterate obtains the outcome utilizing activation functions. 7.4576 is the MSE precision in implementation work using exchange function of Purelin effort that yields fine evaluation outcomes. Coarse validation findings are expressed as value 0.99571 in the MSEREG execution job with the Tansig exchange function. The SSE

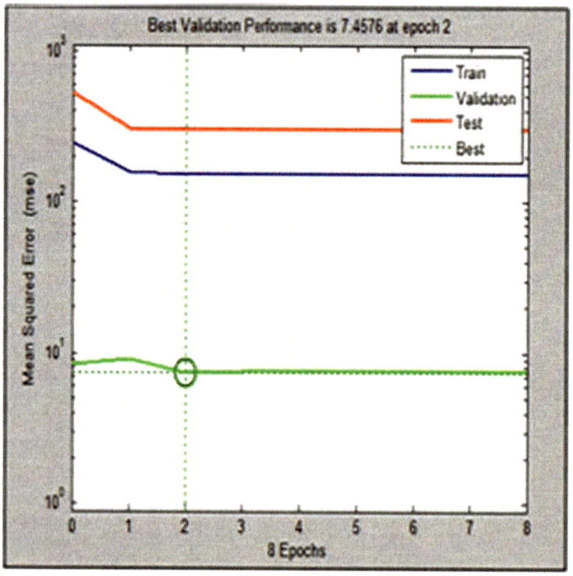

Fig. 6 Validation performance of MSE

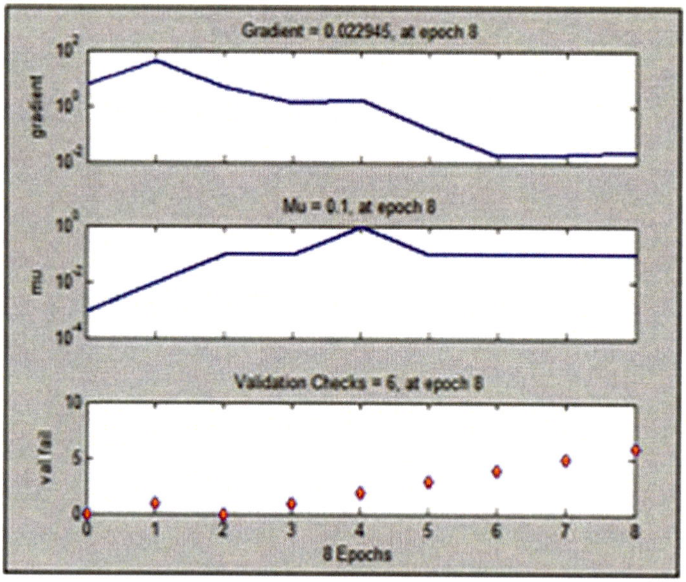

Fig. 7 Training state analysis of MSE

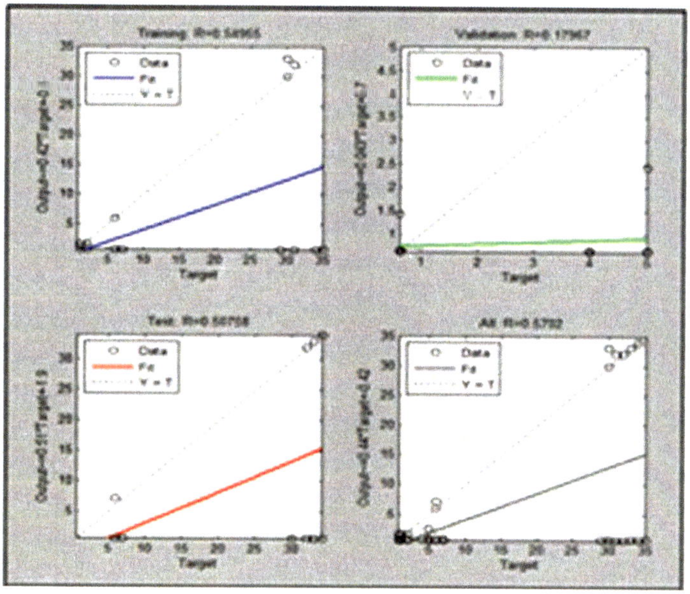

Fig. 8 Regression analysis of system in MSE

Fig. 9 Validation performance of MSEREG

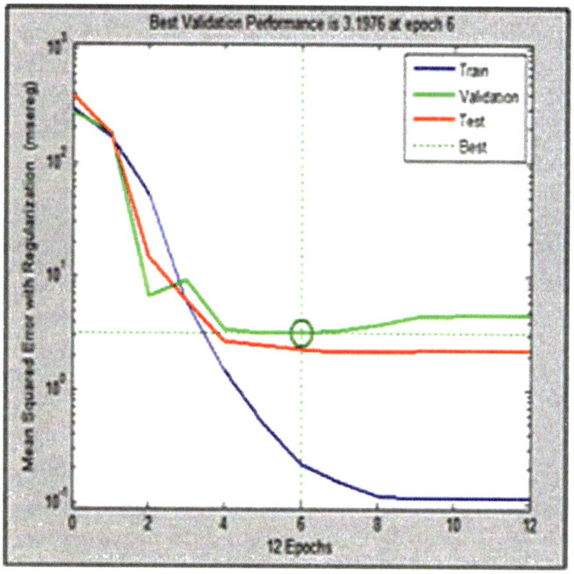

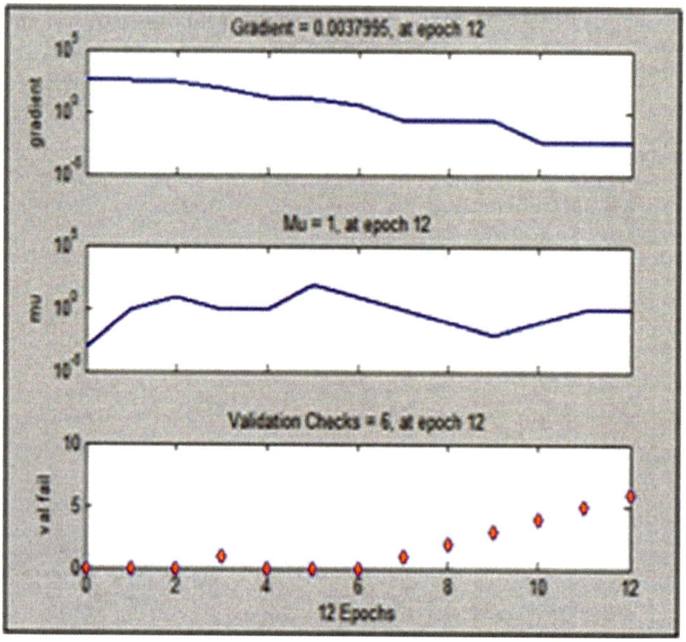

Fig. 10 Training state analysis of MSEREG

performance work combined with Tansig exchange work yields a strong validation result of 98.5565. Similarly, MSE implementation collaborate with Purelin on an exchange basis specifies proper Training outcomes as 0.00007, but it delivers the best validation value when the number of the epoch is increased. When the number of epochs is less than 100, the R-value is low, but when the number of epochs is greater than 100, the R-value improves. The outcome of vermicompost validation performance is described in an efficient manner in (Figs. 11, 12 and 13).

4 Conclusions

By adding new dimensions to the vermicomposting and farming industries, the addition of IoT and the smart cloud is expected to be beneficial. In the agricultural area, a few IoT cloud service companies have recently become prominent. Decentralized composting allows organic waste to be reused where it is made, reducing the amount of waste that must be transported as well as the cost of transportation. It would have a positive effect on the overall cost of municipal waste disposal. Other organic garbage was mixed up with the food trash, such as cattle dung, to improve nutrient content. The findings revealed that Eisenia fetida vermicompost has a higher nutrient content than other vermicomposts and can be used as manure. This research has shown that

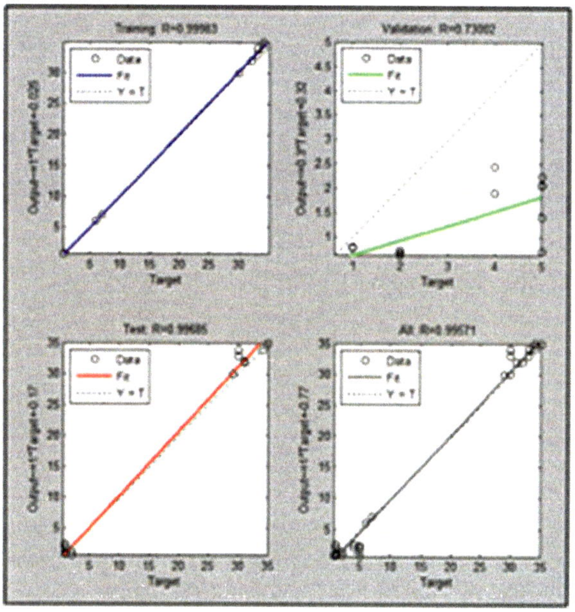

Fig. 11 Regression analysis in MSEREG

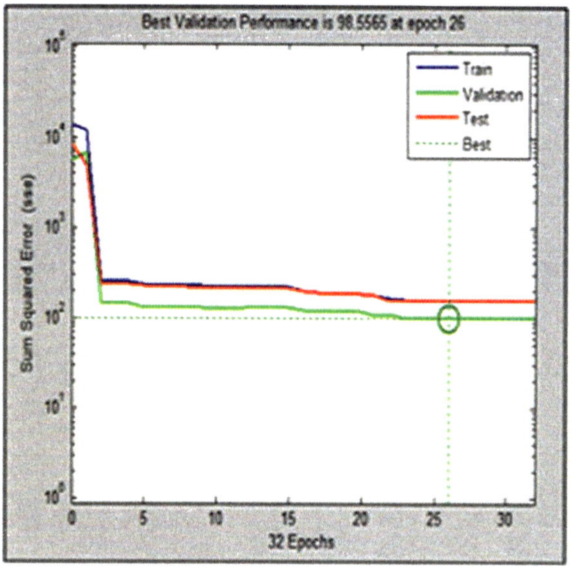

Fig. 12 Validation performance of SSE

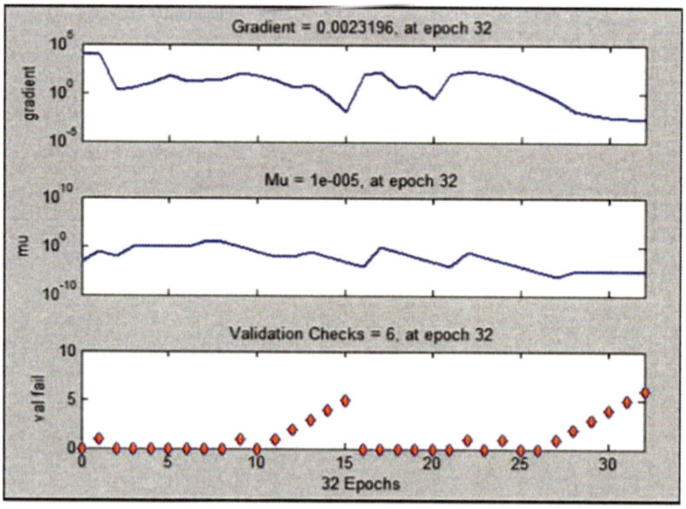

Fig. 13 Training state analysis of SSE

vermicomposting is being used as an efficient technology to turn starch food waste into a valuable commodity. In some challenging areas, such as increased food output, waste recycling, and solid waste management, such as food waste and kitchen waste, vermin technology can be a promising technique that has demonstrated its potential, which could increase the soil's fertility and productivity and generate nutritious and healthy food. The ambient temperature at 30–35 °C, humidity at 29–33%, pH value 6.5–7.0, C/N ratio of 5.3:1 to 5.5:1, the mixture has been maintained to yield good results. The required optimum conditions comprise of the optimization of parameters affecting the yield of high rich nutrient compost has been done and validated using artificial neural network (ANN) techniques. The study reveals that conversion of waste starchy foods into compost exhibits better waste utilization and energy recovery for plants.

References

1. N. Chen, X. Zhang, C. Wang, Integrated open geospatial web service enabled cyber-physical information infrastructure for precision agriculture monitoring. Comput. Electron. Agric. **111**, 78–91 (2015). https://doi.org/10.1016/j.compag.2014.12.009
2. Z. Xu, J. Yin, X. Li, A reliable wireless monitor and control system with low power for greenhouse microclimate, in *New Developments of IT, IoT and ICT Applied to Agriculture* (Springer, Singapore, 2021), pp. 209–216
3. D. Bandyopadhyay, J. Sen, Internet of things: applications and challenges in technology and standardization. Wirel. Personal Commun. **58**(1), 49–69 (2011). https://doi.org/10.1007/s11277-011-0288-5
4. P. Sanjeevi, S. Prasanna, B. Siva Kumar, G. Gunasekaran, I. Alagiri, R. Vijay Anand, Precision

agriculture and farming using Internet of Things based on wireless sensor network. Trans. Emerg. Telecommun. Technol. **31**(12), e3978 (2020)

5. P.P. Ray, Internet of things cloud enabled MISSENARD index measurement for indoor occupants. Measurement **92**, 157–165 (2016). https://doi.org/10.1016/j.measurement.2016.06.014
6. P.P. Ray, Towards an Internet of things based architectural framework for defence, in *IEEE ICCICCT* (Kumaracoil, 2016), pp. 411–416
7. O. Diallo, J.J.P.C. Rodrigues, M. Sene, J.L. Mauri, Distributed database management techniques for wireless sensor networks. IEEE Trans. Parallel Distrib. Syst. **26**(2), 604–620 (2015). https://doi.org/10.1109/TPDS.2013.207
8. J. Lloret, I. Bosch, S. Sendra, A. Serrano, A wireless sensor network for vineyard monitoring that uses image processing. Sensors **11**(6), 6165–6196 (2011). https://doi.org/10.3390/s110606165
9. Y. Cho, K. Cho, C. Shin, J. Park, E.-S. Lee, An agricultural expert cloud for a smart farm, in *Proceedings of Future Information Technology, Application, and Service*, Lecture Notes in Electrical Engineering, vol. 164 (Springer, Berlin, 2012), pp. 657–662. https://doi.org/10.1007/978-94-007-4516-2_69
10. T. Ojha, S. Bera, S. Misra, N.S. Raghuwanshi, Dynamic duty scheduling for green sensor-cloud applications, in *Proceedings of IEEE CloudCom* (Singapore, Berlin, 2014)
11. R. Morais, M.A. Fernandes, S.G. Matos, C. Serôdio, P.J.S.G. Ferreira, M.J.C.S. Reis, A ZigBee multi-powered wireless acquisition device for remote sensing applications in precision viticulture. Comput. Electron. Agric. **62**(2), 94–106 (2008). https://doi.org/10.1016/j.compag.2007.12.004
12. J.M. Irwan, N. Othman, Vermicomposting of Food Waste. Int. J. Integr. Eng. **4**(2), 39–48 (2012)
13. C. Hurst, P. Longhurst, S. Pollard, R. Smith, B. Jefferson, J. Gronow, Assessment of municipal waste compost as a daily cover material for odour control at landfill sites. Environ. Pollut. **135**(1), 171–177 (2005)
14. D.E. Tedesco, C. Conti, D. Lovarelli, E. Biazzi, J. Bacenetti, Bioconversion of fruit and vegetable waste into earthworms as a new protein source: the environmental impact of earthworm meal production. Sci. Total Environ. **683**, 690–698 (2019)
15. W.T. Tsai, Analysis of municipal solid waste incineration plants for promoting power generation efficiency in Taiwan. J. Mater. Cycles Waste Manage. **18**(2), 393–398 (2016)
16. S. Suthar, *Impact of Vermicompost and Composted Farmyard Manure on Growth and Yield Of Garlic (Allium stivum L.) Field Crop* (2009), pp. 27–38
17. A. Das, R. Mohanty, Effective conversion of organic wastes into natural fertilizers using vermicomposting. Int. J. Manage. Technol. Eng. **IX**(I), 463–469 (2019)
18. P.B. Londhe, S.M. Bhosle, Recycling of solid wastes in to organic fertilizers using low cost treatment: vermi-composting. Int. J. Innov. Eng. Res. Technol. 2394–3696 (2015)
19. V.K. Garg, R. Gupta. Effect of temperature variations on vermicomposting of household solid waste and fecundity of Eisenia fetida. Bioremediat. J. **15**, 165–172 (2011)
20. S.R. Rupanagudi, B.S. Ranjani, P. Nagraj, V.G. Bhat, G. Thippeswamy, A novel cloud computing based smart farm ing system for early detection of borer insects in tomatoes, in *Proceedings of International Conference on Communication*, Information & Computing Technology (ICCICT), Mumbai, India (2015)
21. I.G. Smith, The Internet of things 2012 new horizon, in *IERC-Internet of Things European Research Cluster* (2012)
22. S. Suthar, S. Singh, Vermicomposting of domestic waste by using two epigeic earthworms (Perionyx excavatus and Perionyx sansibaricus). Int. J. Environ. Sci. Technol. **5**(1), 99–106 (2008)
23. S.K. Rout, A.K. Rath, C. Bhagabati, Energy efficient dynamic node localization technique in wireless sensor networks. Indian J. Sci. Technol. **10**(15), 1–8
24. S.V. Razavi, M.Z. Jumaat, H. Ahmed, E.I. Shafie, Using feed-forward back propagation (FFBP) neural networks for compressive strength prediction of lightweight concrete made with different percentage of scoria instead of sand. Int. J. Phys. Sci. **6**(6), 1325–1331 (2011)

Time Series Analysis to Forecast Wind Speed

M. Sai Anand and R. Ramalakshmi

1 Introduction

Setting up a wind farm is a very decisive investment. These investments depend upon uncertain factors like wind and energy resources based on natural terrain and geographic locations. Measurement to set up these farms should be effective- particularly optimized to reduce uncertainties and then meticulously observed. These kinds of measurement of value can be calculated by applying a statistical model, and with proper measurement management, a pinpointed optimized value can be obtained, which could help the wind farmers and corporate from a financial perspective.

The objective of this project was to predict the wind speed, which is measured from the anemometer fixed in the windmills located in the coastal regions of the European Union [2–6]. The data fetched from the sources were explored and further analyzed. In the preliminary exploration, it was seen that the data were very irregular. Even though with the advancement of technology day-by-day, given that many algorithms and statistical models were available to predict the wind speed; our focus was to find out the most appropriate and highly accurate model. Since the data are a time series, its stationary was checked by doing a visualization of its correlation.

Once found, first-order differencing was performed over the dataset to tune the stationary, and further on, we applied the ARIMA model over it and which gave fairly good results in return. The Prophet algorithm was also tried out, where the results were not bad, given that the data being volatile. Later on, the long short-term memory network, an artificial neural network architecture used in the field of deep learning was also applied, and the result was yet again satisfactory.

M. Sai Anand · R. Ramalakshmi (✉)
Department of Computer Science and Engineering, Kalasalingam Academy of Research and Education, Krishnankoil, Srivilliputtur, TN, India
e-mail: rama@klu.ac.in

J. D. Peter et al. (eds.), *Disruptive Technologies for Big Data and Cloud Applications*, Lecture Notes in Electrical Engineering 905,
https://doi.org/10.1007/978-981-19-2177-3_38

2 Methodology

2.1 Data Collection

The analysis was performed over the ERA4 dataset is the fifth-gen ECMWF reanalysis for the worldwide environment and climate dating as long since 1950 [7]. The current dataset contained wind speed data ranging from 1st of January 1990 to 30th of April 2020. In addition to the wind speed, there was an additional column of data containing the time and wind direction. The data acquired were fit as a fiddle; thus, no data cleaning was required.

2.2 Data Exploration

The data collected were for a timestamp of each and every minute; it could be seen in Fig. 1 that a whopping 265,871 observations were recorded in the dataset. To narrow down the data in order to get better accuracy, we summarized the data by taking the mean of the wind speed for every month, which could be seen in Fig. 2. Further, we applied the strategy to improve that data for analysis by converting it into a time series [8–11]. Later on, visualization of the data was performed between time and

	Month	Speed		Month	Speed
0	01-01-1990 00:00	3.45	**265867**	30-04-2020 19:00	6.06
1	01-01-1990 01:00	4.22	**265868**	30-04-2020 20:00	6.31
2	01-01-1990 02:00	5.55	**265869**	30-04-2020 21:00	6.58
3	01-01-1990 03:00	6.07	**265870**	30-04-2020 22:00	4.70
4	01-01-1990 04:00	6.29	**265871**	30-04-2020 23:00	4.67

Fig. 1 Wind speed data recoded from the anemometer from 1990 to 2020

Fig. 2 Monthly mean wind speed data

	Month	Speed		Month	Speed
0	1990-01	5.783306	**359**	2019-12	5.063118
1	1990-02	5.739762	**360**	2020-01	5.845403
2	1990-03	6.021290	**361**	2020-02	5.615862
3	1990-04	5.959000	**362**	2020-03	5.958683
4	1990-05	4.398011	**363**	2020-04	5.839861

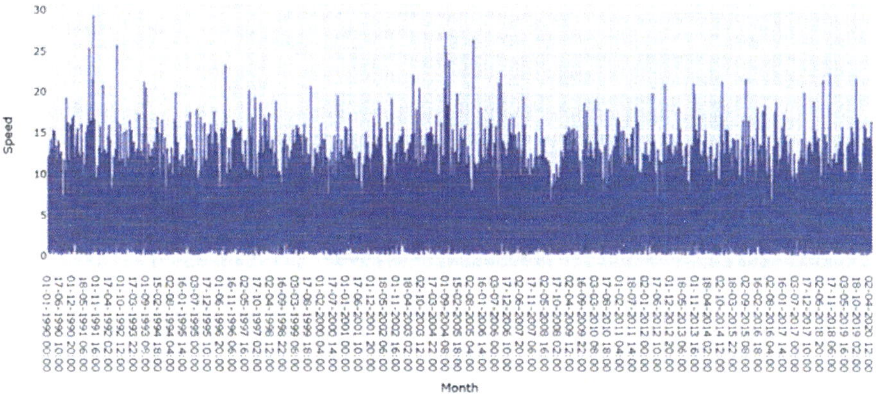

Fig. 3 Wind speed from 1990 to 2020

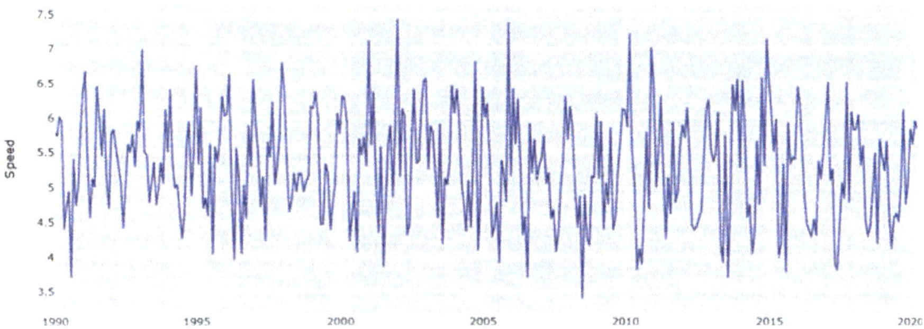

Fig. 4 Monthly mean wind speed from 1990 to 2020

the wind speed recorded. From the plot in Figs. 3 and 4, it can be observed that many dips and rises are seen over the period of time.

2.3 Data Analysis

Correlograms. Correlograms are used to represent the visualization between the errors recorded in a time series from one period to another. It is used to find whether the data are autocorrelated or not [12].

Autocorrelation Function. In time series analysis, autocorrelation describes the connection between lagged values. To check the white noise in the data, at least, 95% of the spikes produced should exist between the limit of "X," where

$$X = \frac{2}{L} \quad (1)$$

Here, "L" denotes the length of the time series. The ACF value is used to find the most appropriate order for the moving average process.

Partial Autocorrelation Function. PACF is the conditional correlation between the recorded period in a time series and its previous observations at lag "l." PACF takes account of responsibility for every one of the intermediate values present in the time series, which is quite different from a regular correlation. The PACF value is used to find the most appropriate for the AR process.

ARIMA Model. ARIMA (autoregressive integrated moving average) is a frequently utilized procedure in fitting data and forecasting it. It is a broader process of the ARMA model (autoregressive moving average). The model is determined by three parameters, p, d, and q. The model which gets generated is of order p, d, and q [13]. Here, p denotes the autoregression(AR) process; d denotes the order of differentiation, and q denotes the moving average (MA) process, respectively. AR and MA are two different methods to fit stationary time series observations, and when both these processes are combined together, they form the ARIMA model, and ARIMA yields a better fit of the model when differencing the data of ARMA is done. It is defined as:

$$\begin{aligned} y_t &= \delta + \left\{\varphi_1 y_{t-1} + \varphi_2 y_{t-2} + \cdots + \varphi_p y_{p-1}\right\} \\ &\quad + \left\{\theta_1 \epsilon_{t-1} + \theta_2 \epsilon_{t-2} + \cdots + \theta_q \epsilon_{q-1}\right\} + \epsilon_t \\ y_t &= \delta + \sum_{i=1}^{p} \varphi_i y_{t-i} + \sum_{i=1}^{q} \theta_i \epsilon_{t-i} \end{aligned} \quad (2)$$

To understand the above mathematical definition, first, we need to understand the AR, MA models. AR Model: The AR model is a linear model where the target variable is forecasted based in the linear combinations of its lagged values, i.e., the model predicts the patterns of the time series based on its previous observations. An AR(n) model will predict the pattern of the "n" numbers of past observation [14]. It is defined as:

$$y_t = \delta + \varphi_1 y_{t-1} + \varphi_2 y_{t-2} + \cdots + \varphi_p y_{p-1} + \epsilon_t \quad (3)$$

where δ is intercept value, y_t is the predicted value at the time "t" which is very similar to linear regression but with lagged observations, φ is the coefficient for the lagged values and ε is the white noise.

MA Model: Unlike the AR model, the MA model forecasts the target variable based on the linear combination of its error values, i.e., the model predicts the patterns of the time series based on its previous observation's error. The moving average is the product of the past error and the coefficients, i.e., the next observation would be the mean of the past observations. A MA(n) model that has an error up to lag "n" is referred to as having an order of n [15]. It is defined as:

$$y_t = \delta + \theta_1 \epsilon_{t-1} + \theta_2 \epsilon_{t-2} + \cdots + \theta_q \epsilon_{q-1} + \epsilon_t \quad (4)$$

where δ is intercept value, y_t is the predicted value at the time "t" which is very similar to linear regression but with lagged observations, θ is the coefficient for the lagged values and ϵ_t ε is the white noise.

Model specification for the ARIMA model can be specified by identifying the p, d, and q values. At first, the d value is determined to identify whether the time series data are stationary or not. The data are stationary when $d = 0$. If not, then, first-order differencing is over the data, and again, checked for its stationarity, this step is iterative until a stationary output is obtained. To find the value of p, PACF can be performed over the data. If the spikes produced go beyond the limit after some lags, then it can be considered as the value of p. To find the value of q, ACF can be performed over the data. If the spikes in the ACF and PACF plot go beyond the limit, then it can be considered as the value of q [16].

LSTM Model. LSTM is abb. as long short-term model which is a deep learning neural network model of a recurrent type used in natural language processing. Before understanding LSTM, one should know about what a neural network is and what a recurrent neural network is. An artificial neural network is a layered construction of associated neurons, similar to the biological neuron. It is the combination of multiple algorithms running simultaneously to perform complex calculations. A recurrent neural network (RNN) is a type of neural network derived from a feed-forward neural network for performing operations temporary data. They produce prescient outcomes based on the sequential information that other algorithms cannot process. RNN has a short-term memory, where it saves all the results from the outputs and include them as a part of the input, which increases the accuracy of the prediction on what's coming next. Due to this type of operation, the output of the function changes since there is a change in input for every step. This is called a gradient. It observes the variation in weight of the model with respect to the change in error. There will be times when the gradient will be small, eventually leading the model to stop learning or take loads of time to generate results. This is a major drawback in using RNN called vanishing gradient. To suppress this, LSTM got introduced. Unlike RNN, LSTM has more memory than RNN which helps them remember the input value for a long period of time. The concept of a "memory cell" is inherited in LSTM, which solves the problem of vanishing gradient. These memory cells also called gated cells have the authority to decide on whether the processed information is to be stored or deleted subjected to its importance [1]. They tend to assign memory cells by learning through algorithms.

As seen in Fig. 5, there are three gates in LSTM, input, forget, and output. At the input gate, it decides whether to allow the data to process or not. The forget gate deletes the information if it is not important, and the resultant is obtained in the

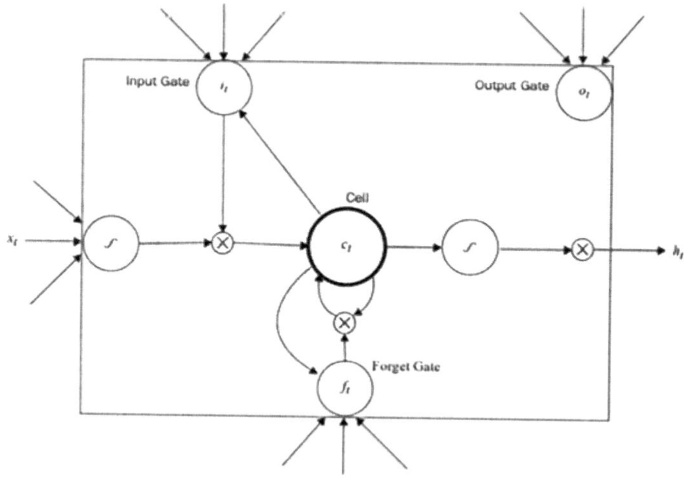

Fig. 5 LSTM architecture

current time step which is received at the output gate. Being in the form of sigmoid, their value ranges from 0 to 1 and are analog in nature.

Prophet Model. Prophet is a model introduced by scientists at Facebook for advanced time series forecasting. Initially, it was used to forecast data with weekly and yearly seasonality, but over the period of time, it has been used in more types of data. It has been observed that a good forecast can be yield when Prophet is applied over time series data that have strong seasonality. Mathematically, it is defined as:

$$y_t = g(t) + s(t) + h(t) + \epsilon_t \tag{5}$$

This shows that the Prophet model is a nonlinear regression model. Here, y_t is the predicted value at the time "t;" $g(t)$ is the growth term (or) linear trend; $h(t)$ is the holiday effect; $s(t)$ defines the seasonal pattern and is white noise.

A Fourier series of observations is present in the seasonal pattern, where the order of annual seasonality is 10 and 3 for weekly seasonality by default. When the change points for the linear trend are not mentioned, then the points are determined by using a logistic function to encapsulate an upper limit for the trend. A Bayesian approach is used in the selection of change points and model characteristics to estimate the model.

Prophet model has been made to forecast univariate time series. It has its own condition of the dataset to satisfy so that it can process. The data frame on which the Prophet model is to be applied should have two columns. One is the date time series named "ds" and the other column where the data recorded at that time interval named as "y".

3 Results

3.1 Correlograms

A correlogram is a good method to figure out the correlation between the data. As the dataset had 364 records, its points will also be the same. Therefore, the spikes were limited at 2/364 = 0.005.

After visualizing the ACF and PACF plot from Figs. 6 and 7, we infer that more than 95% of the AR coefficient were beyond the confidence interval. This gives a clear indication that there is covariance in the data, and white noise is present. To reduce the white noise, a lag of 1 was introduced to the data by first-order differencing.

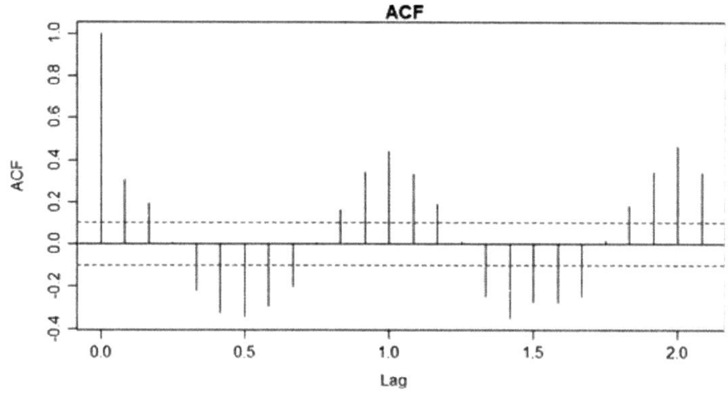

Fig. 6 Resultant ACF plot on the wind speed dataset

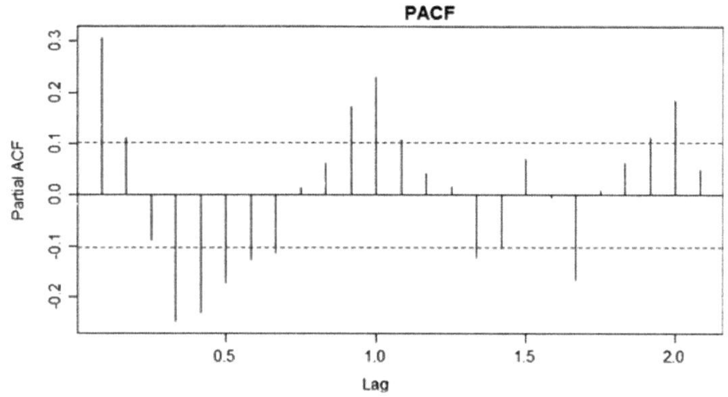

Fig. 7 Resultant PACF plot on the wind speed dataset

3.2 ARIMA Model

From the correlograms, we were able to confirm that white noise exists in our data, and thus, it is non-stationary. After making the data stationary, we did a preliminary seasonal decomposition to check all the components in the time series. Figure 8 is the resultant of the seasonal decomposition. This shows that there is no general trend in the data; therefore, seasonality of the data is also differing with respect to trend. The residual shows the plot when the trend and the stationarity of the data are removed.

Once the preliminary analysis was done via seasonal decomposition, it was time to implement the ARIMA model. To find the most optimal value for the parameter p, q, r for fitting the model, the "auto_arima()" function from the library "pmdarima" was used. It is already observed from figure that there is seasonality in the time series. Thus, the seasonal ARIMA (SARIMA) model can be used rather than ARIMA. SARIMA is an extended version of the ARIMA model which can process univariate time series data with a seasonal component. The model adds the hyperparameter for autoregression, differencing and moving averages for both seasonal components as well as the period of the time series.

The "auto_arima()" function found the best parameter as SARIMAX(1, 0, 0) $\times$ (2, 1, 0, 12). This was found on the basis of the least Akaike information criterion (AIC) and Bayesian information criterion (BIC) value. They are used to measure a statistical model on the basis of its goodness of fit and how simple the model can be for a single statistic [17]. The resultant can be seen Fig. 9.

Later, the data were split into train and test set. The last one of the data was considered for the testing set, while the rest were for training. Using the "SARIMAX()" function, the model was fit to the dataset. The prediction from the ARIMA model was compared against the testing data which can be seen in Fig. 10.

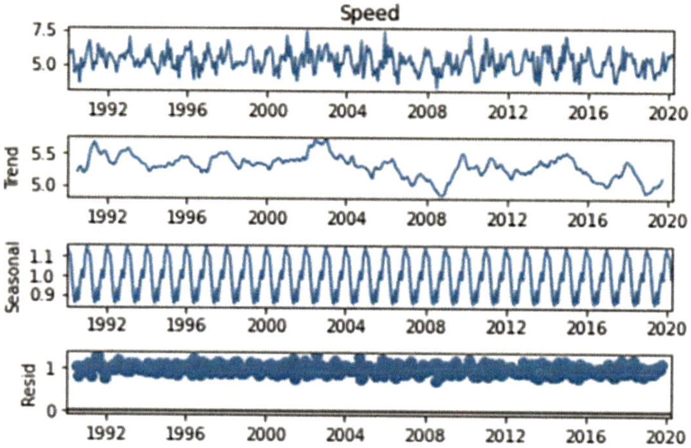

Fig. 8 Seasonal decomposition on the wind speed dataset

```
Best model:  ARIMA(1,0,0)(2,1,0)[12]
Total fit time: 25.404 seconds
                              SARIMAX Results
 Dep. Variable:   y                               No. Observations: 364
        Model:    SARIMAX(1, 0, 0)x(2, 1, 0, 12)  Log Likelihood    -348.142
         Date:    Wed, 19 May 2021                       AIC        704.285
         Time:    06:11:06                               BIC        719.739
       Sample:    0                                      HQIC       710.435
                  - 364
Covariance Type: opg
               coef   std err    z      P>|z|  [0.025  0.975]
   ar.L1    -0.0828  0.052   -1.591   0.112  -0.185  0.019
 ar.S.L12   -0.6944  0.049  -14.139   0.000  -0.791  -0.598
 ar.S.L24   -0.3148  0.053   -5.911   0.000  -0.419  -0.210
  sigma2     0.4156  0.035   11.967   0.000   0.348  0.484
    Ljung-Box (L1) (Q):     0.00  Jarque-Bera (JB):  6.10
              Prob(Q):      1.00         Prob(JB):   0.05
Heteroskedasticity (H):     1.11             Skew:   0.30
  Prob(H) (two-sided):      0.58         Kurtosis:   2.75
```

Fig. 9 Resultant parameter from the model fitting

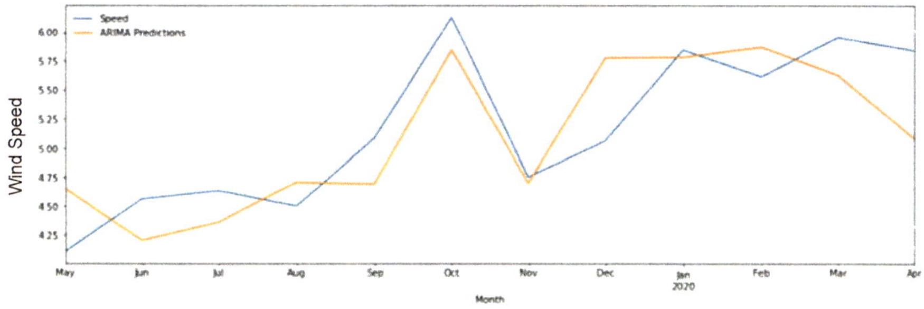

Fig. 10 ARIMA model predictions vs. actual wind speed

3.3 *LSTM Model*

For LSTM, the same training and testing dataset were used. The dataset was using the “MinMaxScaler()” from the Scikit Python library. A time series generator object was created before creating the LSTM model. When it comes to training a supervised learning model like LSTM, time series data require some preparation. The data should be transformed into a series of input and output component. Univariate time series use the prior observation called lag as input and the observation in the current step is the output. To tackle this, the Keras library in Python contains “TimeseriesGenerator” which converts a univariate dataset to that of a suitable supervised learning dataset automatically. Once it gets created, the same Keras library LSTM model could be created, and its predicted value could be obtained. A plot comparing the predicted and the actual wind speed value can be seen in Fig. 11.

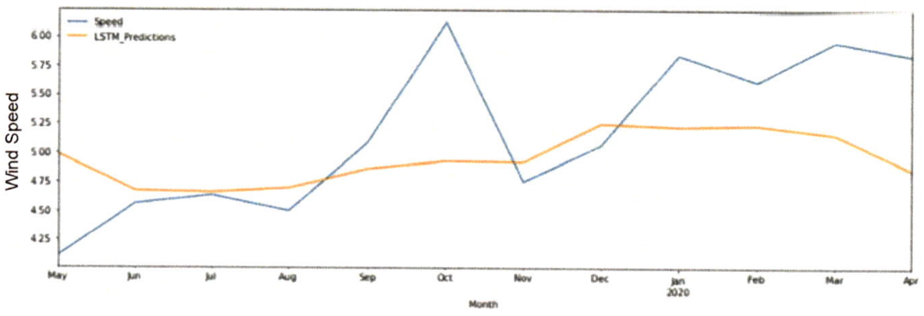

Fig. 11 LSTM model predictions versus actual wind speed

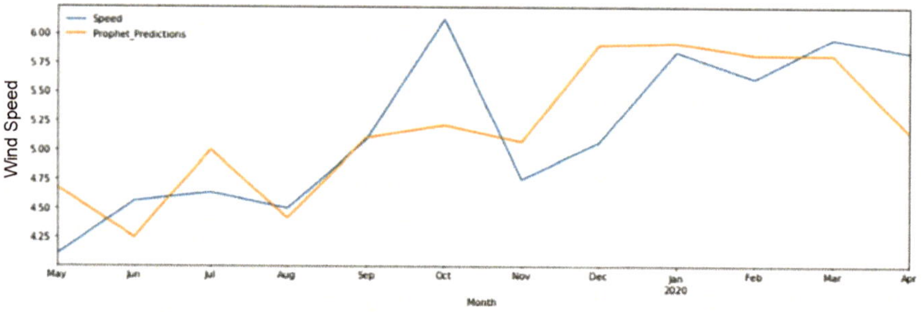

Fig. 12 Prophet model predictions vs. actual wind speed

3.4 Prophet Model

The Prophet model could be fit by invoking the "prophet()" function from the "fbprophet" library in Python. The "make_future_dataframe" function is used to fit the model. A plot comparing the predicted and the actual wind speed value can be seen in Fig. 12.

Figure 13 shows the combination of all comparison between predicted wind speed from fitted models vs. actual wind speed, and Table 1 represents the corresponding predicted model values by the model in contrast to the actual wind speed.

4 Model Estimation

Model evaluation is a fundamental process in the making of model development. It assists in identifying the optimal model for the data and how well the identified model will function later on. We chose MSE, RMSE, and MAPE as the regression metric. The results from the metrics will be obtained from the actual and the predicted values as shown in Table 1.

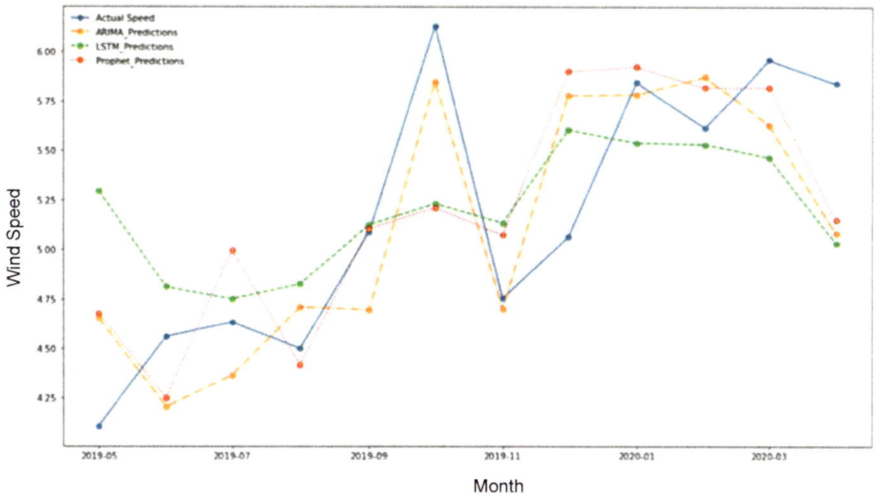

Fig. 13 Comparison of all model predictions with the actual wind speed

Table 1 Date-wise actual speed recorded and its corresponding predicted value from various models

Month	Actual speed	ARIMA model	LSTM model	Prophet model
2019-10-01	6.129543	5.846975	5.232013	5.211597
2019-11-01	4.753222	4.701142	5.134392	5.072160
2019-12-01	5.063118	5.778728	5.605172	5.901023
2020-01-01	5.845403	5.782919	5.539448	5.922132
2020-02-01	5.615862	5.871717	5.530404	5.819033
2020-03-01	5.958683	5.629228	5.464518	5.817778
2020-04-01	5.839861	5.080335	5.032112	5.149565

MSE abb. as mean square error is the most common model evaluator. It is the resultant average of the actual and estimated squared values. The square root of MSE is the RMSE, thus root mean square error. The lower the RMSE value, the better the model. This would indicate that there is very little deviation between the actual and predicted value. MAPE abb. as mean absolute percentage error is a forecast accuracy measurement [18]. It calculates the accuracy in the form of a percentage. It can be measured by taking the average of the difference between absolute percentage error for each time interval by actual values and further dividing the average by the actual value again. A lower MAPE is considered to project a good model. Mathematical notation for the above regression evaluator has been given below.

Table 2 Results of regression metrics for various models

Model	RMSE	MSE	MAPE
ARIMA	0.412699	0.170321	6.937809
LSTM	0.565760	0.320085	8.939733
Prophet	0.479183	0.229616	7.348355

$$\text{MSE} = \frac{1}{N}\sum_{i=1}^{N}(y_i - \hat{y})^2 \tag{6}$$

$$\text{RMSE} = \sqrt{\text{MSE}} = \sqrt{\frac{1}{N}\sum_{i=1}^{N}(y_i - \hat{y})^2} \tag{7}$$

$$\text{MAPE} = \frac{1}{N}\sum_{t=1}^{N}\left|\frac{A_t - F_t}{A_t}\right| \tag{8}$$

Here, N = Number of iterations, y_i = Mean value of i data, $\hat{y}$ = Predicted value, A_t = Actual value, F_t = Forecast value.

The following results in Table 2 were obtained for the models when evaluated with the actual and the predicted wind speed.

From Table 2, ARIMA is projecting the least RMSE and MAPE values followed by the Prophet model and LSTM coming at last. Thus, it is evident that ARIMA is the best model of all where the accuracy is the highest. Forecast for the wind speed for the next 3 years has been plotted and shown in Fig. 14.

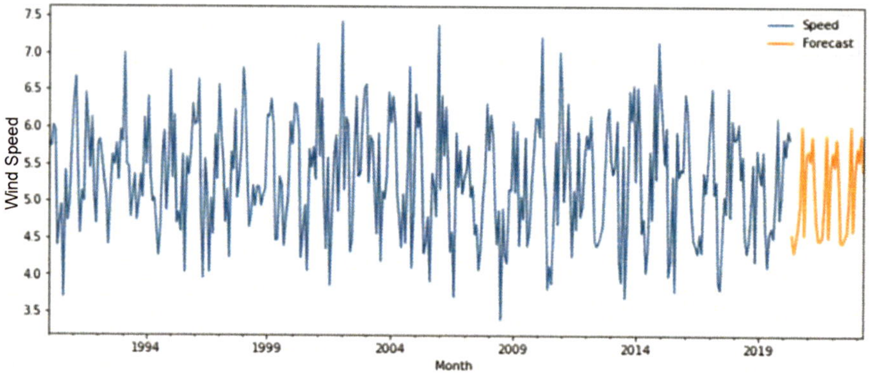

Fig. 14 Forecast plot of wind speed based on ARIMA model

5 Conclusion

As there is an increase in the population and given that it will be a linear increase, the needs of the population will also project the same trend. Nature, being volatile, can't fulfill all the needs for such a large and ever-growing population. We have switched to renewable energy, but still, energy resources like wind, solar, thermal energy are dependent on the acts of nature, which are very existential to predict and forecast. In our case, we took three distinct statistical models and found that ARIMA model is the best and lays out an accuracy of nearly 96%.

Our process to forecast the wind speed might help out wind farmers, energy organization to set up small wind farms by trying a similar statistical, and computational model to forecast wind speed in their locality.

One should not restrict in relying on ARIMA model. Many supercomputers at various organizations can compute the forecast of wind speed at very high accuracy, and there is an inherent sense of developing an algorithm to tune these supercomputers now, and then, but, the quest to increase the accuracy will persist till the dusk of humanity. We complete this research by quoting Antoine de Saint Exupery's views on forecasting—"As for the future, our task is not to foresee it, but to enable it" [1].

References

1. F. Hashmi, *Predicting Stock Prices Using Deep Learning LSTM Model in Python*. https://thinkingneuron.com/predicting-stock-prices-using-deep-learning-lstm-model-in-python/
2. P. Jiang, S. Qin, J. Wu, B. Sun, Time series analysis and forecasting for wind speeds using support vector regression coupled with artificial intelligent algorithms. Math. Prob. Eng. **2015**, 1–14 (2015)
3. R. Burtiev, F. Greenwell, V. Kolivenko, Time series analysis of wind speed and temperature in Tiraspol, Moldova. Environ. Eng. Manage. J. (2013)
4. V. Prema, K.U. Rao, Time series decomposition model for accurate wind speed forecast. SpringerOpen Renew. **2**, 18 (2015).https://doi.org/10.1186/s40807-015-0018-9
5. M. Martín, L. Cremades, J. Santabàrbara, Analysis and modelling of time series of surface wind speed and direction. Int. J. Climatol. **19**, 197–209 (1999)
6. P. Lv, L. Yue, Short-term wind speed forecasting based on non-stationary time series analysis and ARCH model. Int. Conf. Multimed. Technol. **2011**, 2549–2553 (2011). https://doi.org/10.1109/ICMT.2011.6002447
7. Copernicus. https://climate.copernicus.eu/climate-reanalysis
8. R.H. Shumway, D.S. Stoffer, *Time Series Analysis and Its Applications (Springer Texts in Statistics)* (Springer, Berlin, 2005)
9. G. Shmueli, K.C. Lichtendahl Jr., *Practical Time Series Forecasting with R: A Hands-On Guide*, 2nd edn. (Axelrod Schnall Publishers, 2016)
10. J.D. Hamilton, *Time Series Analysis* (Princeton University Press, 2020)
11. R.J. Hyndman, G. Athanasopoulos, *Forecasting: Principles and Practice* (Monash University, Australia). https://otexts.com/fpp2/
12. G. Jain, Time-Series analysis for wind speed forecasting. Malaya J. Matematik (2018)
13. Z. Huang, Z.S. Chalabi, Use of time-series analysis to model and forecast wind speed. J. Wind Eng. Ind. Aerodyn. **56**(2–3), 311–322 (1995)

14. I. Yenidoğan, A. Çayir, O. Kozan, T. Dağ, Ç. Arslan, Bitcoin forecasting using ARIMA and PROPHET, in *2018 3rd International Conference on Computer Science and Engineering (UBMK)* (2018), pp. 621–624
15. Quotes from: Wikiquotes. https://en.wikiquote.org/
16. GeeksforGeeks, *ARIMA Model for Time Series Forecasting*. https://www.geeksforgeeks.org/python-arima-model-for-time-series-forecasting/
17. P.S.P. Cowpertwait, A.V. Metcalfe, *Introductory Time Series with R* (Springer, New York, 2009)
18. M. Fakültesi, A. Endüstri Mühendisli Bölümü, Wind speed forecasting using time series analysis methods. Çukurova Univ. J. Fac. Eng. Archit. **32**(2), 161–172 (2017)

Software-Defined Network-Based Packet Keys to Secure Critical Infrastructures of Internet of Things

Antony Taurshia, Jaspher W. Kathrine, and S. Jebapriya

1 Introduction

The resource-constrained nature of IoT devices limits the use of cryptographic primitives to secure its critical security applications like smart grid, Industry 4.0, health care, etc. Even with a smart home system, the security surveillance system demands high confidentiality, integrity, and availability of data, since if a malicious user is able to eavesdrop on the data, he can easily break into the house when the owners are not around. The devices in the smart home are prone to two kinds of attacks [1]. Passive attack where a malicious user eavesdrops on the network traffic with the aim of collecting the data. Active attacks include Denial of Service (DoS), phishing, manual tampering of device, and collection of user credentials to gain unauthorized access.

Security issues in the smart grid include integrity, availability, and confidentiality of data [2]. Loss in integrity and availability of data will lead to inefficient power management. Lack of confidentiality will hinder the privacy of users. Some of the attacks on the smart grid include SCADA-based man-in-the-middle attack, false data injection on the smart meter, substitution attack to charge Electric Vehicles (EV) using unauthorized tokens, etc. Stuxnet and Duqu worm are examples of attacks on the industrial IoT [3]. The attacks were performed by taking advantage of a vulnerability in the system. Private networks are less prone to attacks as they are isolated and

A. Taurshia (✉) · J. W. Kathrine · S. Jebapriya
Karunya Institute of Technology and Sciences, Coimbatore, India
e-mail: antony18@karunya.edu.in

J. W. Kathrine
e-mail: kathrine@karunya.edu

S. Jebapriya
e-mail: jebapriya@karunya.edu

J. D. Peter et al. (eds.), *Disruptive Technologies for Big Data and Cloud Applications*, Lecture Notes in Electrical Engineering 905,
https://doi.org/10.1007/978-981-19-2177-3_39

avoids unnecessary communication. Isolating critical security applications is one of the ways to ensure security [4].

Security services like firewall, intrusion detection, and deep packet inspection (DPI) are provided as a service by cloud service centers with software-defined networks (SDN) and virtualized functions (VF) as enabling technologies [5, 6]. The security services are built as virtualized functions, and SDN orchestrates the data traffic through the virtualized functions based on the demands of the user. The work in [7] built the virtual authentication, authorization, and accounting functions, and the SDN controller is used for orchestrating the data traffic through the virtual functions. Honey is built as a virtualized function and used for proactive and reactive defense against attacks using SDN in [8]. SDN for intrusion detection in IoT networks is proposed in [9]. The network traffic is captured using Wireshark built on top of SDN which collects the network traffic and Long Short-Term Memory (LSTM)-based deep learning is used for anomaly detection. SDN-centered group management is proposed [10] to ensure horizontal control over the applications. SDN-centered group-based access control is proposed in [11] which uses a Group Management Service Function (GMSF) embedded in fog node to ensure communication only between group members. This scheme isolates a group from another group at network-level using SDN. In the proposed work also an SDN-centered, GMSF is used to secure communication between group members. The group members are distributed with group secrets along with the group key. With the group secret, the users can generate packet keys to communicate. Using packet keys improves the security of the system as the keys are shortlived. It also protects the unreliable UDP traffic from data or packet loss ensuring availability. It also makes the traffic resistant to replay attack since a key is used only once and then discarded. Group-based communication provides resistance to man-in-the-middle attack as the secrets are known only to the authorized group members ensuring confidentiality and integrity of data.

The following work unfolds as follows. Section 1 overviews the existing packet key generation system. Section 2 briefs on the proposed packet key generation system. Section 3 gives the security analysis. Section 4 gives the performance evaluation in terms of computation, storage, and scalability. Section 5 gives the conclusion.

2 Existing Packet Key Generation Scheme

The existing schemes use the Diffie–Hellman technique for generating the packet keys [12, 13]. The group members are distributed with the global parameters q, α. Now the users A and B of the group intend to communicate securely. User A then generates a vector of values $X_{a1}, X_{a2} \ldots\ldots X_{an}$ as secret value vector SVV_a and assign index values $I_1, I_2 \ldots I_n$. The user A then generates $Y_{a1}, Y_{a2} \ldots Y_{an}$ as blind value vectors BVV_a by computing $Y_{ai} = \alpha^{X_{ai}} \text{mod} q$. User A then sends the BVV_a to user B. User B conversely generates its own $X_{b1}, X_{b2} \ldots X_{bn}$ as SVV_b and assign index values. User B then generates $Y_{b1}, Y_{b2} \ldots\ldots Y_{bn}$ as BVV_b by computing $Y_{bi} = \alpha^{X_{bi}} \mathrm{mod} q$. User B then sends BVV_b to user A.

Users A calculate the packet key by computing $K_{pi} = Y_{bi}{}^{X_{ai}} \mathrm{mod} q$. User B then calculates the packet key by computing $K_{pi} = Y_{bi}{}^{X_{ai}} \mathrm{mod} q$. Now users A and B communicate by encrypting each packet with a different packet key exchanged between the users. The existing work uses public-key cryptography, hence has a higher computation cost. The communication cost is higher as n number of BVVs need to be exchanged to generate n packet keys. Moreover, the work is less scalable.

3 Proposed Work

The proposed work uses symmetric key cryptography with lesser computation and communication overhead and is more scalable. The proposed work is an extension of the work in [11]. The service center in the cloud provides group management service to IoT applications with the SDN-centered GMSF embedded in fog nodes depicted in Fig. 1. SDN is used for group-based access control and the group members are distributed with group secret $pp \in$ {primitive polynomials}, and a secret value s for a second level of group member verification using LFSR-based cyclic redundancy check (CRC) and message authentication code (MAC). In the proposed work, the SDN-centered GMSF uses a key management SF to distribute the group secrets

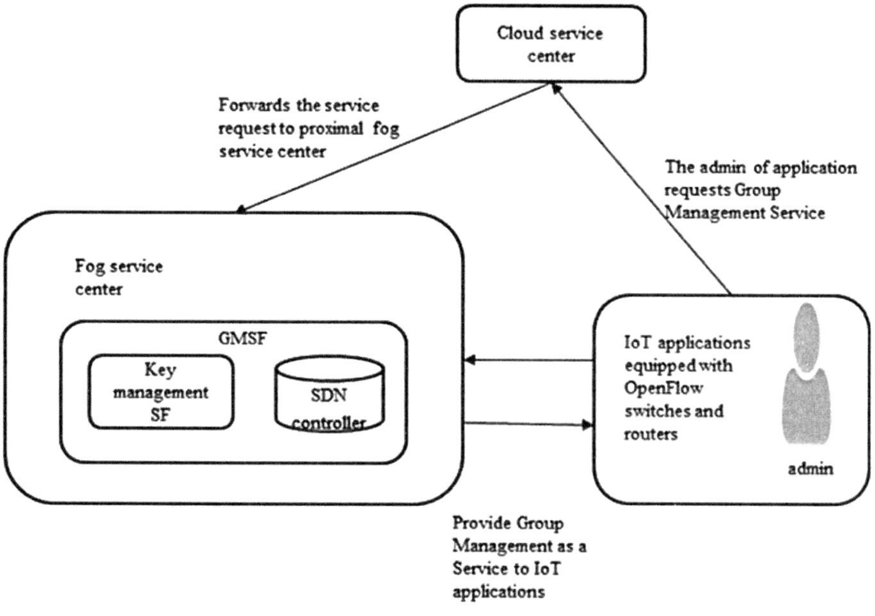

Fig. 1 Group management service to IoT applications

pp, and s only to the authorized group members, and these group secrets are the key materials used to generate a series of packet keys for encryption. The proposed packet key generation scheme is scalable and provides data integrity and availability by preventing packet loss.

The group uses global parameters α and q. User A sends a communication request to user B with its ID encrypted with the group key GK. User B then validates user A as a valid group member as the message is encrypted with the group key. Further verification is done using MAC as the MAC is generated using the polynomial pp, and s value. User B responds with an acceptance message with its ID encrypted using GK. The users A and B generate the packet keys using a pseudorandom generator. Since LFSRs are used for generating CRC and MAC [11, 14] and is compatible with the resource-constrained devices of IoT [15], the proposed work uses LFSR for packet key generation. The key sequences generated using LFSR are flexible [16] and can support both hardware- and software-based implementation [17], making it an appropriate choice for IoT devices.

A binary LFSR of length L consists of L registers. It uses a primitive polynomial of Galois field GF (2) as a feedback function to generate a long non-repetitive key sequence. The output sequence generated using an LFSR is a linear function of its previous state. An LFSR of length l can generate $2^l - 1$ non-repetitive key sequence [13]. For the proposed work, the group secret pp is used as the feedback function and s as input to the LFSR. The key sequence generated is tailored based on the packet key size and index values are assigned to them. The key size is based on the cipher used for encryption. The bits are split into 80-bit key if PRESENT cipher is used, 128-bits in case AES is used. Packet keys of size 16-bit, 32-bit, 64-bit can be paired with RC4-based encryption. The key sequence is split into 80-bit keys as follows

$$b_1, b_2, \ldots b_{80} - K_1 || I_1 \tag{1}$$

$$b_{81}, b_{82}, \ldots b_{160} - K_2 || I_2 \tag{2}$$

$$b_{i+1}, b_{i+2}, \ldots b_{i+80} - K_n || I_n \tag{3}$$

User A calculates the first packet key $K_{p1} = \alpha^{K_1} \text{mod} q$ which encrypts the packet and append its corresponding index value. The second packet is encrypted using the key $K_{p2} = \alpha^{K_2} \text{mod} q$. User B also generates packet keys for encryption similarly. Figure 2 depicts the packet key-based communication between users A and B. User A initiates communication with user B by sending a communication request along with its ID encrypted using the group key. Since the communication request is encrypted using the group key user B identifies user A as an authorized group member and replies with an acceptance message along with its ID encrypted using the group key. User A then starts sending messages by encrypting its first packet using the first packet key generated using the group secrets. Since the group secrets are known only to the group members, encryption using packet keys will act as a second level of verification as an authorized group member. User B decrypts the received first packet

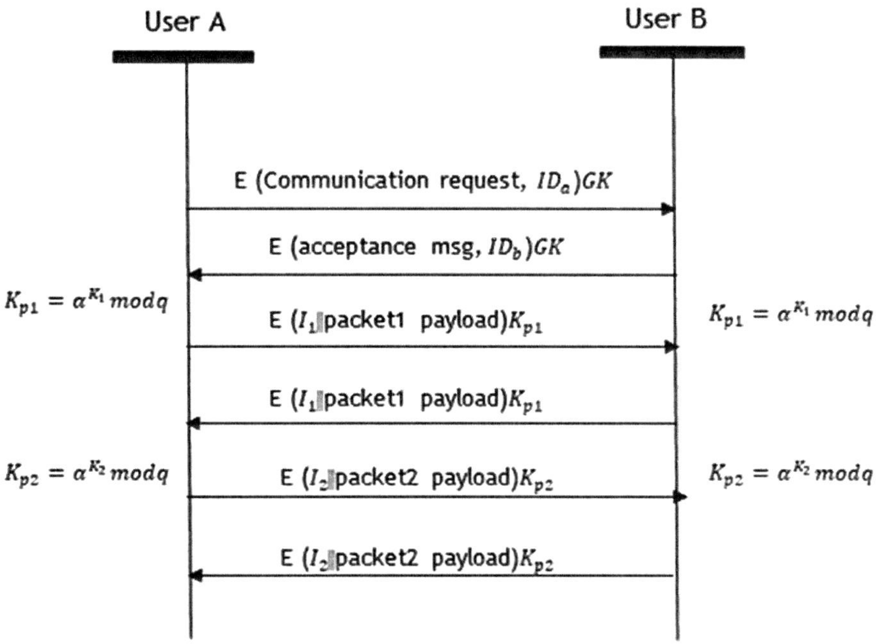

Fig. 2 Users A and B communicate using packet keys

using the first packet key. User B similarly encrypts the first packet it generates using the first packet key and the communication proceeds in a similar fashion. In case if a receiver cannot decrypt a packet using the corresponding packet key, then it indicates a missing packet. The receiver then sends the GMSF a notification of missing packet from the sender. The GMSF will then notify the sender of the missing packet and the packet can be resent.

```
Algorithm for packet based communication
repeat i ← i + 1, j ← j + 1
  for i, j do
    Sender compute packet key K_psi = α^ksi mod q
    Sender encrypt packet_j using K_psi
    Receiver compute K_pri = α^kri mod q
      if K_psi = K_pri
        Receiver decrypt packet_j using K_pri
      else
        Receiver send notification to GMSF
  until i, j < 2^(l-1)
```

Critical security applications like healthcare demand integrity, availability, and confidentiality of data transferred between the communicating devices. The data generated by the monitoring device connected to a critically ill patient should be free of packet loss to ensure the availability and integrity of the medical records. The use of packet keys prevents data loss irrespective of the protocol used for communication ensuring reliability in both Transmission Control Protocol (TCP) and User Datagram Protocol (UDP) transmission.

4 Security Analysis

Theorem *Packet keys make data traffic resistant to replay attack.*

Proof The first packet sent by any user is encrypted with the first packet key generated $K_{p1} = \alpha^{K_1} \bmod q$ and then discarded. The second packet is encrypted with the next packet key generated $K_{p2} = \alpha^{K_2} \bmod q$ using the pseudorandom series and then discarded. The upcoming packets are encrypted conversely. Now a malicious user A captures a packet and replays later. The receiving user can identify it as a replay attack since a key is used only once and discarded. Hence, packet keys make data traffic resistant to replay attack.

Theorem *Packet keys prevent data loss.*

Proof Each packet is encrypted with distinct packet keys generated in serial order, and the index values corresponding to the keys are appended to the packet's payload. Hence, a missing key can signify packet loss.

Theorem *Packet keys increase the complexity of finding the symmetric key.*

Proof Unless the group secrets pp and s are known, the problem of finding K_i is discrete logarithm. This reduces the attack surface to brute-force attack. The probability of finding the key using brute-force attack is $\left(\frac{1}{2}\right)^{sk}$ where sk is the size of key [18]. Since a packet key is used only once and discarded, finding the key of n packets increases the complexity to $n\left(\frac{1}{2}\right)^{sk}$.

5 Performance Evaluation

The proposed work is compared with an existing packet key generation scheme [13] in terms of computation and storage cost. The proposed work is evaluated in terms of communication cost and scalability also.

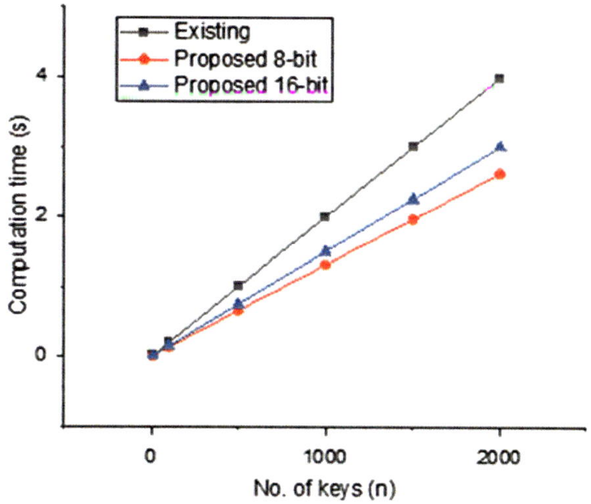

Fig. 3 Comparison of computation cost

5.1 Computation Cost

The evaluation is done for generating 64-bit packet keys in an Intel i5 system with 64-bit operating system with 8 GB RAM. Though the device is not IoT-based, the computation time will remain proportional with any system and also for variable packet key sizes. The average time taken for 8-bit LFSR to generate 255 8-bit key sequence 0.01000 s. To generate a 64-bit key, eight 8-bit sequences can be used hence the time taken is 0.000313 s. The average time taken for 16-bit LFSR to generate 65,535 16-bit key sequences is 8.30000 s. To generate a 64-bit sequence, four 16-bit sequences can be used hence the time taken is 0.00050 s. The average time taken for one exponential and modular operation is 0.00100 s. The existing Diffie–Hellman-based packet key generation scheme uses two exponential and modular operations. The proposed work uses one exponential operation, modular operation, and pseudorandom sequence generation. Figure 3 depicts the comparison of the computation time of the proposed work with the existing work. The proposed scalable packet key generation scheme has lesser computation overhead as two exponential operations are reduced to one and instead uses a simpler pseudorandom sequence generator.

5.2 Communication Cost

The communication cost is measured in terms of number of keys to be transferred. In the proposed work, an 8-bit LFSR can generate 32 64-bit packet keys. The seed for the next set of key generation is sent along with the packet encrypted data by any one of the communicating parties. In the case of 16-LFSR can generate 1023 packet keys. Hence, a seed is exchanged only after 1023 packets.

5.3 Storage Overhead

In the existing work, a user needs to store α,q, its own set of $SVVs$, and the other user's BVVs. However in the proposed work, a user needs to store α,q, pp, and s. Hence, the storage overhead is minimal compared to the existing work.

5.4 Scalability

The proposed work is scalable compared to the existing work as an 8-bit LFSR can generate 32 64-bit packet keys and 16-bit LFSR can generate 1023 64-bit keys. To generate higher bit keys, 128-bit LFSR can be used. Hence, our scheme is scalable.

6 Conclusion

The proposed work is symmetric key-based cryptography and needs a shared secret to establish packet keys for communication. Still, with an SDN-centered group management server, the keys can be securely distributed to the communicating parties. The proposed work has added benefit of resistance to replay attack. The performance analysis shows the efficiency of the proposed work in terms of computation, communication, and storage. The proposed scheme has a lesser computation cost of nearly 34% in case of 8-bit LFSR and 25% in case of 16-bit LFSR. The storage overhead has also reduced by 77%. The proposed work is comparatively more scalable since with a single key exchange, significant number of packet keys can be generated with lesser computation overhead. Building novel applications with SDN-based packet keys to secure network traffic is considered for future work.

References

1. W. Ali, G. Dustgeer, M. Awais, M.A. Shah, IoT based smart home: security challenges, security requirements and solutions, in *ICAC 2017—2017 23rd IEEE International Conference on Automation & Computing, Addressing Global Challenges Through Trade Automation and Computing*, no. September (2017), pp. 7–8
2. B. Khelifa, S. Abla, Security concerns in smart grids: threats, vulnerabilities and countermeasures, in *Proceedings of 2015 IEEE International Renewable and Sustainable Energy Conference*, IRSEC 2015, 2016

3. G. Gualandi, E. Casalicchio, A self-protecting control application for IIoT, in *2019 IEEE 4th International Workshops on Foundations and Applications of Self* Systems, FAS*W 2019*, no. Cl (2019), pp. 152–157
4. Y. Yan, Y. Qian, H. Sharif, D. Tipper, A survey on cyber security for smart grid communications. IEEE Commun. Surv. Tutorials **14**(4), 998–1010 (2012)
5. D. Migault et al., A framework for enabling security services collaboration across multiple domains. Comput. Electr. Eng. **69**(March), 224–239 (2018)
6. K.D. Joshi, K. Kataoka, pSMART: a lightweight, privacy-aware service function chain orchestration in multi-domain NFV/SDN. Comput. Netw. **178**, 107295 (2020)
7. A.M. Zarca, D. Garcia-carrillo, J.B. Bernabe, J. Ortiz, Managing AAA in NFV / SDN-enabled IoT scenarios, in *2018 Global Internet of Things Summit (GIoTS)* (2018), pp. 1–7
8. A.M. Zarca, J.B. Bernabe, A. Skarmeta, J.M. Alcaraz Calero, Virtual IoT HoneyNets to mitigate cyberattacks in SDN/NFV-Enabled IoT networks. IEEE J. Sel. Areas Commun. **38**(6), 1262–1277 (2020)
9. A. Wani, S. Revathi, R. Khaliq, SDN-based intrusion detection system for IoT using deep learning classifier (IDSIoT-SDL). CAAI Trans. Intell. Technol. **6**(3), 281–290 (2021)
10. E. Festijo, Y. Jung, M. Peradilla, Software-defined security controller-based group management and end-to-end security management. J. Ambient Intell. Humaniz. Comput. **10**(9), 1–18 (2018)
11. A. Taurshia, G. Jaspher, Software defined networks centered group based access control service for internet of things applications. Commun. Comput. Inform. Sci. **1347**, 48–60 (2021)
12. Y. Jung, M. Peradilla, R. Agulto, Packet key-based end-to-end security management on a blockchain control plane. Sensors (Switzerland) **19**(10), 1–16 (2019)
13. A. Saini, E.D. Festijo, Y. Jung, Packet key-based security system for the next generation LTE networks. Procedia Comput. Sci. **110**, 297–303 (2017)
14. E. Dubrova, M. Näslund, G. Selander, F. Lindqvist, Message authentication based on cryptographically secure CRC without polynomial irreducibility test. Cryptogr. Commun. **10**(2), 383–399 (2018)
15. M.A. Philip, V. Vaithiyanathan, A survey on lightweight ciphers for IoT devices, in *Proceedings of 2017 IEEE International Conference on Technological Advancements in Power and Energy: Exploring Energy Solutions for an Intelligent Power Grid, TAP Energy 2017* (2018), pp. 1–4
16. A. Ahmad, On design of 8-bit CRC circuits equipped with primitive characteristic polynomials, in *2011 International Conference on Multimedia, Signal Processing and Communication Technologies, IMPACT 2011*, vol. 3 (2011), pp. 264–267
17. R. Muthalagu, S. Jain, Modifying LFSR of ZUC to reduce time for key-stream generation. J. Cyber Secur. Mobil. **5**(4), 257–268 (2016)
18. T.N. Dang, H.M. Vo, Advanced AES algorithm using dynamic key in the internet of things system, in 2019 IEEE 4th International Conference on Computer and Communication Systems, ICCCS 2019 (2019), pp. 682–686.

Automated Face Authentication and Recognition Using Deep Neural Network with SVM Classifier in Cloud Environment

T. Sujatha, N. R. Wilfred Blessing, Sruthi Anand, and Esther Daniel

1 Introduction

Traditionally, data produced by surveillance cameras are stored and processed in private data centers or in the private cloud. The entire process of storing, analysis, computing, communication, etc., was performed in the cloud servers. The dynamic nature of current technology needs, customer requirements, and increasing security issues creating the need for bigger infrastructure, huge storage, high end processing resulting into the distributed environment, and cloud computing [1].

From the discovery of cellular telephones to artificial intelligence, era has come an extended way. And the reality that every person wishes to simply accept is, that we've tailored era with open fingers for alleviating our duties at hand. The feature common to all technologies is artificial intelligence (AI) and, more precisely, deep learning, where a learning happens through data, that too, raw data which is inexpensive and available vast.

Face recognition [2] is a biometric answer to identify and authenticate people without intervening of manual need. The security can be improved using face recognition. Unlike other applications, it identifies the distinctive features of the face image and then makes the contrast based on the stored images in the database. Once match occurs, the face is verified. Because of its adoptability, accuracy, easy applicability,

T. Sujatha (✉)
Sri Krishna College of Engineering and Technology, Coimbatore, India
e-mail: tsujathathankaraj31@gmail.com

N. R. Wilfred Blessing
University of Technology and Applied Sciences, Salalah, Oman

S. Anand
Sri Krishna College of Engineering and Technology, Coimbatore, India

E. Daniel
Karunya Institute of Technology and Sciences, Coimbatore, India

J. D. Peter et al. (eds.), *Disruptive Technologies for Big Data and Cloud Applications*, Lecture Notes in Electrical Engineering 905,
https://doi.org/10.1007/978-981-19-2177-3_40

and automation, the face recognition system has emerged as the master piece of all other recognition systems.

Facial identification system is getting improved and the future in the field of cyber security [3]. Artificial neural network and deep network algorithms are making facial recognition algorithms to be more accurate. The facial recognition system in cloud environment [3] can improve performance in meeting public security missions like the following:

1. Finding missing children and lost adult
2. Finding the exploited children
3. Criminals identification and tracking
4. Providing investigation support
5. Attendance tracking
6. Defense services sector
7. Banking sector, etc.

With precision in results coming from intelligent analysis of data with the help of machine learning techniques in distributed cloud environment, it has helped organizations private and public sectors make quicker and easier decisions and also reduces the dependency on manual security processes. Due to its flexibility and powerful nature, the need of Facial Recognition system is emerging and growing faster.

2 State of Art

In the field of computer vision, face recognition plays a tremendous role. In earlier days before 2000, the performance of face detection algorithms was not good enough until Viola and Jones initiated an algorithm [4]. They have introduced rectangular boxes in the image for detecting the face. But it has lot of drawbacks like feature dimension is massive. Following this proposal, lot of algorithms like HOG, SIFT, SURF, FFN, ACF, NFD, etc., were introduced by researchers to improve the results [4].

The fact of human's capability in face recognition induces the need for machine learning algorithms so that the machines can learn on its own and make accurate predictions. This was the reason behind the lot of research works are being carried out in the use of deep neural networks with popular machine learning algorithms in the task of face recognition. This paper also proposes the dynamic neural network approach with classifier SVM [5] for making accurate predictions on face recognition in distributed cloud environment since all the surveillance cameras are connected with public clouds via private clouds to meet the processing and scalability requirements.

3 Proposed Model

The original photocopy of the image is partitioned into four sub-blocks of 10 × 10 pixels, sixteen sub-blocks of 5 × 5 pixels, and five overlapping sub-blocks of 20 × 6 pixels which makes the neural network architecture to have 25 input nodes for processing. Its purpose is to discover vital face features. Since the basic structure of human faces are alike, it is very important to identify the unique features which differentiate one from the other. So, it has been decided to extract the features of eyes, nose, and mouth. The machine makes use of one hidden layer to symbolize neighborhood functions that signify faces. The sigmoid function is used as activation function with the learning factor value 0.3. The system has series of robust classifiers to make the accurate classification model.

The pattern recognition system of this proposed system shown below in Fig. 1 observe and capture the input data from the real world with sensors. Then, the information is getting preprocessed and noise data will be removed. Before processing the data, the color image is converted into grayscale image. Feature extraction is the important part of data processing. Here, in facial recognition system, other than face part are filtered out. Hence, all the facial parts are given importance. After preparing training dataset, the new test data will be verified with the training set database and if match occurs, notification will be sent.

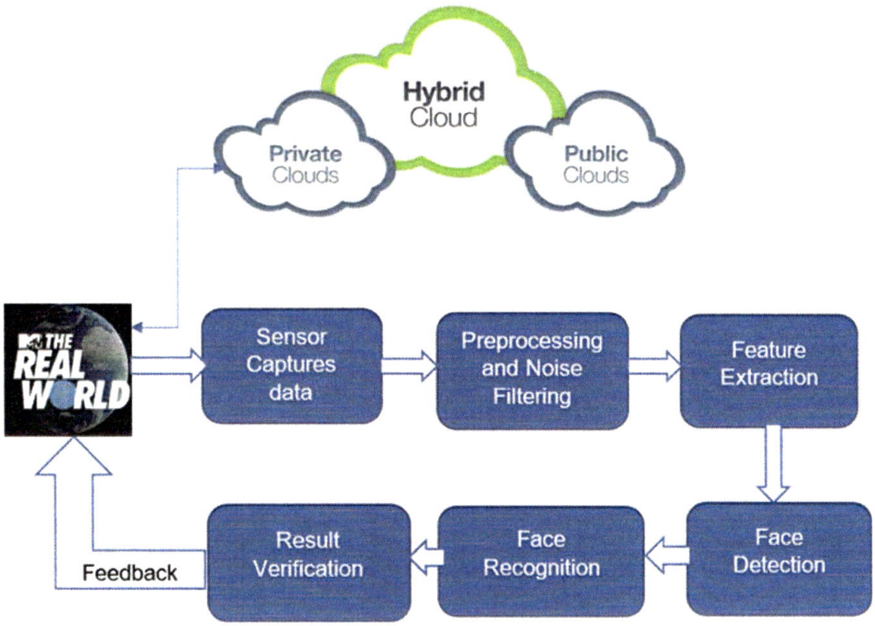

Fig. 1 Proposed work flow diagram

Identification of face shape alignment marks [5] is the most important phase in the face recognition process. The main objective is to extract the feature points of eyes, nose, mouth, cheek, chin, etc., in the face.

Face recognition process involves the following steps which involves face finding, analyzing features, comparison and making prediction as below:

1. On capturing pictures, locate all the face images in it.
2. Ensure every face and be capable of recognize that even supposing a face is growing to become in an unusual course or in terrible lighting, its miles nonetheless the identical person.
3. Select the feature points of the face image which uniquely differentiate the people one from the other.
4. Compare the feature points of the face image with existing database containing set of images to find the match.

3.1 Step 1: Locating All the Face Images in a Picture

The first step in this proposed work is to identify all the face images in the picture. Since the work is about face recognition, the only face part of the image is being focused which is shown in Fig. 2.

The color image is converted into gray scale image first since the color features are not necessary to find the faces. For every single pixel in the image, do the following:

1. Compare the pixel with the neighboring pixels.
2. Find out the darkness of the current pixel with the neighboring pixels.
3. Draw the arrow in the direction of pixel from low intensity to high intensity.

Fig. 2 Identification of face images in the picture (*Source* https://towardsdatascience.com/data-science/)

4. Repeat this all for all the pixels in the image so that the pixels will be replaced by arrows. These arrows are names as gradients which show the flow from low intensity to high intensity all over the image. Simply, identify all the edges.
5. Partition the image into 16 × 16 blocks.
6. Normalize each block by finding the majority of the gradient points in each block replace the entire block with the same.
7. Now, the original face image is getting converted into simple representation which captures the basic skeleton structure of the face.
8. Compare the image with known patterns of face images training set and extract the part of images that are more similar. The matching parts of the images are the faces. Using this methodology, all the face images in the picture can easily be identified.

This is how all the face images in the picture are being located for further processing.

3.2 Step 2: Ensuring Facial Features Consistency

After locating all the face images in the picture, the important thing is to ensure every face and be capable of recognize that even supposing a face is growing to become in an unusual course or in terrible lighting, its miles nonetheless the identical person. That is even after some transformations are being applied, crucial parts of the face like nose, eyes, and lips are always in the same sample location in the image. Hence, it is very important to extract the measurements of feature points in each face. The research works have showed that the most accurate approach is to make the system to figure out those measurements automatically. Deep learning paved a better way to locate which feature points in the face are important to measure. So, it is very important to train the deep neural network to generate 128 feature points for each face. The feature point measurements is shown in Fig. 3.

Thus, the complicated image representation [5] is reduced to the list of computer-generated numbers. The shape of the image can be represented by n points $\{x_i, y_i\}$, $X = (x_1, y_1, \ldots x_n, y_n)^T$ as vector of 2n elements. For simplification, the 2n data space should be reduced. The most popular approach to reduce the dimensions is applying PCA method on the data space as follows:

1. Compute the dataset mean value,

$$\mu = 1/n \sum_{i=1}^{n} x_i \tag{1}$$

2. Compute the covariance matrix

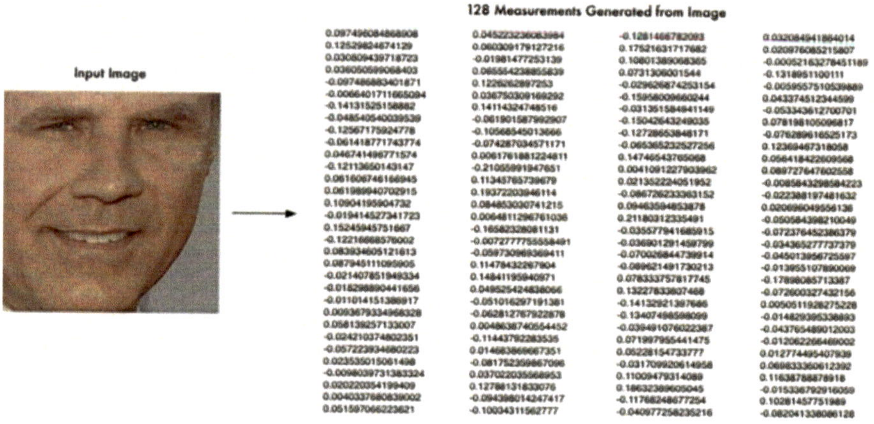

Fig. 3 Representation of face image using 128 feature points

$$\Sigma = \frac{1}{n-1}\sum_{i=1}^{n}(x_i - \mu)(x_i - \mu)^T \tag{2}$$

3. Compute the eigen vectors and eigen values of the dataset.
4. Approximate the value of x from the training dataset.

$$X \approx \mu + P_s b_s \tag{3}$$

where P_s is the variance in the dataset and b_s is the shape parameters.

By applying transformations, the actual shape of the image can be generated as follows:

$$X = T(\mu + P_s b_{s,} x_c, y_c, s_x, s_y, \theta) \tag{4}$$

which includes translation, scaling, and rotation as shown in Fig. 4.

3.3 Step 3: Train the Network Model

The process of training a network model [6] needs lot of computing power and data. But, once the network model is well trained, the model will be able to generate shape feature points of any face. After the network is trained, execute with all the faces to generate the shape model. The important objective here is, the network model should be able to generate the same 128 feature points for different pictures of the same individual.

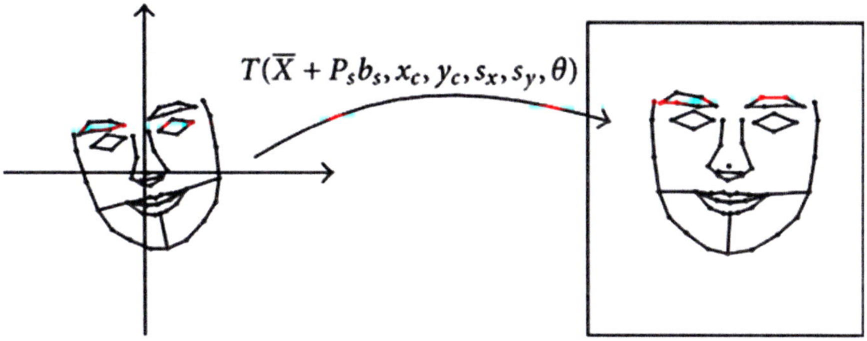

Fig. 4 Generation of face image shape

The training dataset process of the network works by loading different images of the same person and 128 feature points are measures for each image. Then compare the results. The feature point measurements generated for the same person face images will be closer which is shown in Fig. 5.

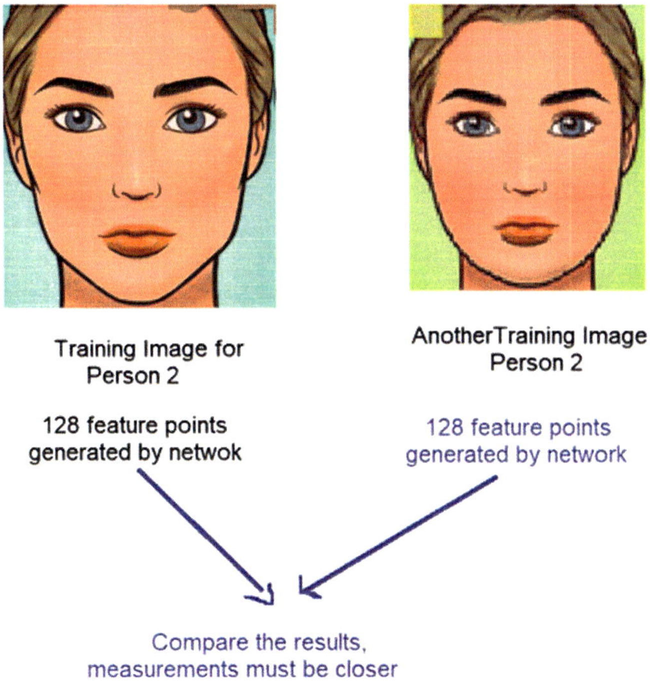

Fig. 5 Illustration of training set dataset

3.4 Step 4: Find the Match

The detected face images are compared with database of crime accused images to find the match. For this purpose, the well-known popular machine learning algorithm is used, that is SVM. Here, the classifier needs to be trained in such a way that it will tell which image in the database is the closest match with the feature points of new test image. The outcome of the classifier is the name of the person. The classes C_1 and C_2 are the inputs to the SVM algorithm [6, 7], which in turn generates the decision, mathematically denoted as follows:

$$C_1 = t_i - t_j | t_i \approx t_j$$
$$C_2 = t_i - t_j | t_i \neq t_j$$

The SVM is the binary classifier, it gives either 'yes' or 'no' as the outcome. C_1 maintains the dissimilarities in facial images of the same person, and C_2 maintains the dissimilarities in facial images of the different persons. Given the difference between facial images, the SVM classifier estimates if the face images in the two images are of the same individual or not. The SVM algorithm also returns the similarity index, $\delta = w.(p1 - p2) + b$ as a measure of matching.

3.5 Step 5: Notification

Once the match occurs, notification will be sent to respective officials to make necessary action. In applications like criminals tracking, the database of criminal's face images will be maintained, compare this with the live video and if match occurs, information will be notified for further action.

4 Performance Analysis

The face recognition module which produces the matching face image in the database along with identity, 'name' of the person. The performance of this system is analyzed with various examples in training set data (from CelebA dataset) against the new samples. The various image transformations were applied to the new samples and

Table 1 Error rate for the samples

Samples	100	200	500	1000
Error rate	0.08	0.074	0.59	0.04

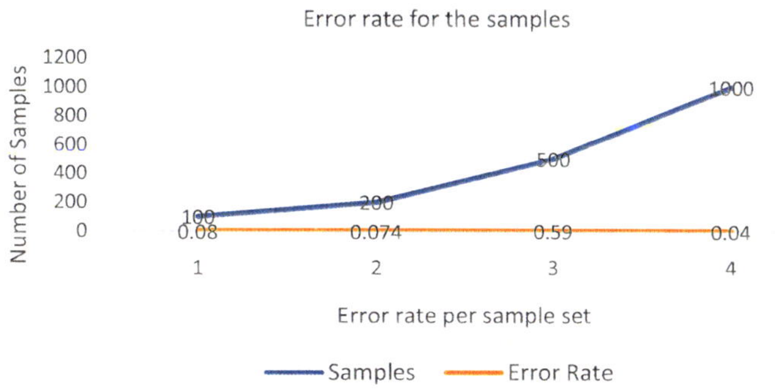

Fig. 6 Plot showing error rate for the samples

tested. The system performs well even with the sample images of side pose, lightening, different resolution, different hairstyle, etc. Table 1 and Fig. 6 show the error rate achieved while testing.

5 Conclusion

The aim of this project is to identify the people on demand like criminals, missing children in any investigation department. Here, the images of the criminals are stored in the database. Since the face is the primary focus of attention in this project, due attention is given to facial changes due to emotion, makeup, hairstyle, etc. The system is aimed to match and recognize the face images even with any sort of transformation applied. With the goal of developing the promising system to recognize the face images with the remarkable capability of human to recognize the face, the artificial intelligence system is developed so that it will learn from data and make predictions like human with the capability of scaling and adoptability in the distributed cloud environment. With the dataset of 1000 samples, an average error rate of this proposed model is 0.04 and it needs to be enhanced to achieve 100% accuracy.

References

1. A. Naik et al., Proposed system for criminal detection and recognition on CCTV data using cloud and machine learning, in *International Conference on Vision Towards Emerging Trends in Communication and Networking* (2019)
2. J. Saragih, S. Lucey, J. Cohn, Deformable model fitting by regularized landmark mean-shifts. Int. J. Comput. Vis. **91**, 200–215 (2010)
3. N. Azma et al., Face recognition for criminal identification: an implementation of principal component analysis for face recognition, in *International Conference on Applied Science and Technology* (2017)

4. P Viola, M. Jones, Rapid object detection using a boosted cascade of simple features, Computer Vision and Pattern Recognition, 2001, in *Proceedings of the 2001 IEEE Computer Society Conference on CVPR 2001*, vol. 1 (2001)
5. W. Jong et al., Multi feature shape regression for face alignment. EURASIP J. Adv. Signal Proces. 51 (2018)
6. T.H. Le, Applying artificial neural networks for face recognition. Adv. Neural Syst. (2011)
7. S. Milborrow, F. Nicolls, Locating facial features with an extended active shape model, in *ECCV* (2018), pp. 504–513

A Data Sharing Protocol to Minimize Security and Privacy Risks of Cloud Storage in Big Data Era

S. Akila and Dhina Suresh

1 Introduction

1.1 Introduction to Cloud Computing

Cloud storage has become great good participant for group associations that experiences the effects of strong point restriction. Cloud computing is a policy that follows worldwide-based computing. The cloud computing system makes use of us to decrease the data board cost and time. What is more, cloud computing is used to store the data that can be gotten to in remote zones. The most testing task in the cloud is to guarantee accessibility, uprightness, and secure document move (Fig. 1). However, according to Zou [1], even though cloud computing is a new methodology concept, it used face a standardized equation. This testing task is used to transmit the data securely in a document or file manner. The most popular meaning of cloud computing is proposed by the USA National Institute of Standards and Technology (NIST).

Cloud computing is a usage of network transmission system of remote users hosted on the network to store, manage, and execute process of data, rather than a local area network or wide area network computer. Hence, following types describe the model of cloud-based system.

- Distributed vs. grid computing
- Fog computing
- Fog and cloud computing
- Evolution of cloud computing
- Challenges in cloud computing.

S. Akila · D. Suresh (✉)
St. Joseph College of Arts and Science for Women, Hosur, India
e-mail: dhinadulcy@gmail.com

J. D. Peter et al. (eds.), *Disruptive Technologies for Big Data and Cloud Applications*, Lecture Notes in Electrical Engineering 905,
https://doi.org/10.1007/978-981-19-2177-3_41

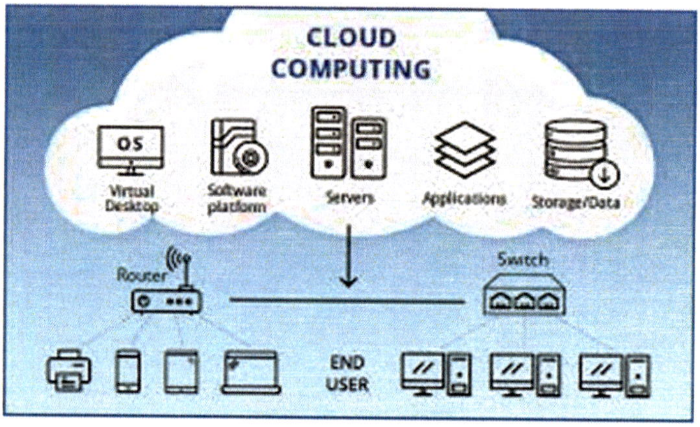

Fig. 1 Cloud computing model

1.2 Cloud Deployment Model

The cloud computing is for the most part dependent on the idea of asset sharing which will in general diminishing the capital cost, support cost and increment the measure of room accessibility. Hence, cloud condition can be sent dependent on the client prerequisites and requirement for security (Fig. 2). Cloud deployment model represents the various types of computing environment architecture in cloud solution.

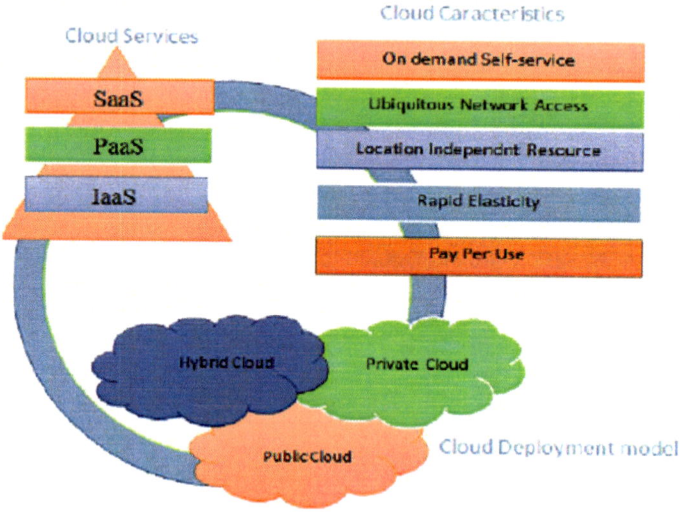

Fig. 2 Cloud deployment model

The cloud organization model gives four sorts of clouds to the remote clients in cloud, and they are as per the following:

- Public cloud
- Private cloud
- Community cloud
- Hybrid cloud.

1.3 Issues in Cloud Computing

The user data can be accessed by the host company with or without permission. And, it as multiple ethical issues in technological criteria effecting issues in cloud server [2]. And, it as set of protocols (set of rules), and it effects on ethics with (T&C) terms and condition based. Figure 3 shows that the cloud security to secure data in transmission file access by cloud computing methodology. The key FBSGK is used in this technological areas.

The challenges of cloud computing are:

- Security issues
- Cost management and containment
- Governance
- Lack of resources

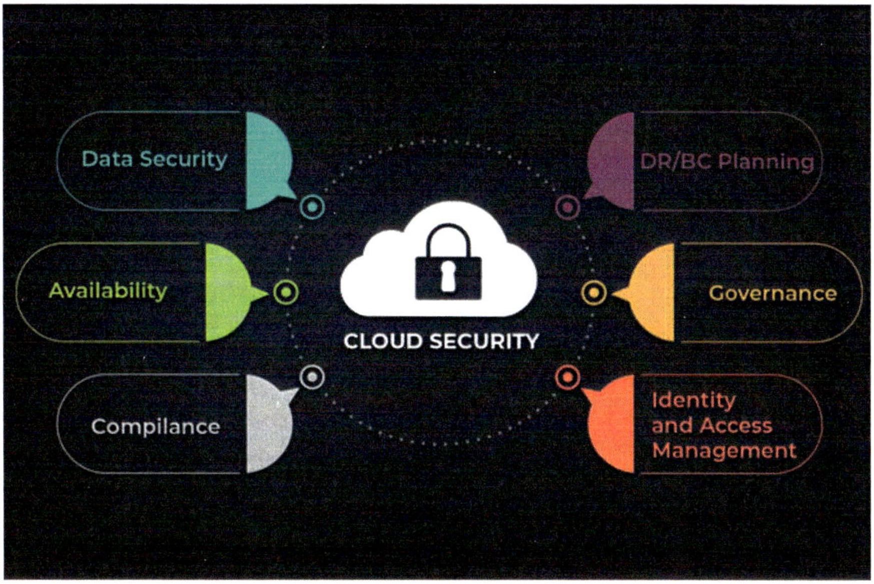

Fig. 3 Cloud security access in secured file transfer

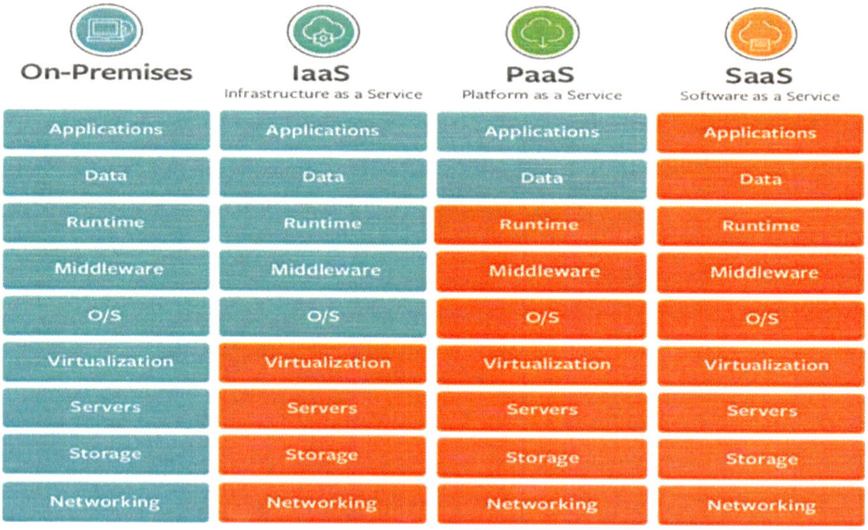

Fig. 4 Cloud computing service models

- Building a private cloud, compliances….etc.

Objective of the Research

- To study the different key management schemes in data sharing protocols
- To study the secret sharing group key management protocol (SSGK)
- To create a forward and backward security in group key management schemes
- A comparative analysis of the existing SSGK and proposed FBSGK.

As per the terms of Fig. 4, it refers to wide range of cloud service models delivered on demand World Wide Web (www). And, it is user-friendly access in cloud deployment model.

The finish of this decade is portrayed by a change in outlook of the mechanical data innovation toward a membership-based assistance plan of action known as cloud computing. It furnishes clients with a not insignificant rundown of benefits, for example,

- Services for assess capabilities
- Worldwide and miscellaneous network access
- Money pooling
- Flexibility with measured services.

1.4 Application of Cloud Computing

- Cloud service provides many applications that are,

- Procedure work,
- Collection of data, data storage,
- Maintenance services, knowledge,
- Recreation,
- Handling,
- Social networking, etc.

2 Literature Overview

The literature overview was completed to locate a fitting procedure that encloses the prerequisites, for example, large room and secured confirmation, clearance record partaking is made sure about condition, offering productive assistance at negligible looking through time and valuation cost. In this point of view, a few quantities of works are finished by the specialists through various strategies and methods which are accounted for in this part [1].

- The cloud computing is a result of many attempts at large-scale computing with faultless access to virtually endless resources.
- There are three kinds of services in this cloud computing:
- Types of services:
  - Platform (PaaS),
  - Software (SaaS),
  - Infrastructure (IaaS),
  - Services-based application programming interface (API).

2.1 *Key Reconstruction and Verification*

(i) Cipher (D) and v from the point to-point public channel
(ii) The following objective remakes K with cooperation adulterated members
(iii) The means of the remaking convention are portrayed as follows.

Stage 1: Every member $P_{i,}$ $I = 1, 2, 3, \ldots n$ aside from O exchanges their sub-shares. P_i gets cipher (s_i) and v from O through open channel and decodes cipher (s_i) utilizing condition 6. P_i gets the public key PK_i of $P_j; j + (I + 1)\text{mod}\, n$; $(\mathrm{I} + 2)$ mod n ...$(\mathrm{I} + n - 1)\text{mod}\, n$ from cloud supplier; if the estimation of j is 0, j would be changed to n. After scrambled s_i, it sends the encoded s_i to $P_j, j = (I + 1)\text{modn}$; $(I + 2)\text{modn}\ldots$ $(I + n - 1)\text{modn}$. On the off chance that the estimation of j is 0, j would be changed to n.

$$\text{Cipher}_j(s_i) = \text{RSA}^{-1}\text{PK}_j(s_i)$$

After these means, each member P_i may get $n - 1$ sub-shares from different members, and the sub-shares got by each member are appeared. Stage 2: Every member $P_i, I = 1, 2, 3 \ldots n$ aside from O computes K_i, a duplicate of K, and V utilizing Lagrange interpolating formula. At that point, it registers V mod K_i, if Vmod $K_i <$ v, it will communicate that there are undermined members; else, it will send its ID scrambled with s_j to the cloud supplier for information access authorization. As P_i figures the bunch key K, it can download the cipher (D) from cloud and decode it by K.

$$D = \text{AES}_K^{-1}(\text{Cipher}(D))$$

$F(0) = a_0$ can be determined as follows utilizing determinant activity.

Where n represents the limit of this sharing gathering. There are m members on the web, and they cloud compute m sub-shares $s_1, s_2, \ldots s_m$ utilizing condition 2. The gathering key $K = F(0) = a_0$ can be determined as follows utilizing determinant activity [3].

Equation 1:

$$\begin{pmatrix} 1^0 & 1^1 & 1^2 & \ldots & 1^n \\ 2^0 & 2^1 & 2^2 & \ldots & 2^n \\ & & \ldots & & \\ m^0 & m^1 & m^2 & \ldots & m^n \end{pmatrix}, |A| = \begin{cases} = 0 \text{ if } m < n \\ \neq 0 \text{ if } m = n \end{cases}$$

Equation 2:

$$A \begin{pmatrix} a_0 \\ a_1 \\ \ldots \\ a_n \end{pmatrix} = \begin{pmatrix} s_0 \\ s_1 \\ \ldots \\ s_m \end{pmatrix}$$

The estimation of K which not be determined as A^{-1} does not exist at the point when $m < n$, with the goal that the accessible m members cannot get any data of K. At the point when $|A| = 0$ and $m = n$, a_i, $i = 1, 2, 3, 4 \ldots n$ cloud be determined by stage 3. Also, every member gets K.

$$\begin{pmatrix} a_0 \\ a_1 \\ \ldots \\ a_n \end{pmatrix} = A^{(-1)} \begin{pmatrix} s_0 \\ s_1 \\ \ldots \\ s_m \end{pmatrix}$$

Table 1 File size comparison

File name	Original file size	SSGK	FBSGK
		Encrypted file size in byte	Encrypted file size in byte
Student_data.mdb (MB)	8	4.8	4
Goodtime.avi (MB)	200	120	80
Mail_collection.doc (KB)	150	90	60
Datacommunication.pptx (KB)	300	180	120

Presently, we accept that P_i chooses to break our convention. P_i sends some unacceptable estimation of Cipher I $- 1(s_i)$ to $P_i - 1$, while it gets the right Cipher I (siC1) from PiC1. $P_i - 1$ misunderstands the K also and figures the irrefutable worth which is not equivalent to (Fig. 3). Cloud security access in secured file transfer is used to transfer the data with key encryption of FBSGK. $P_i - 1$ communicates that the convention is deserted and nobody could get the shared information from cloud. Table 1: File size comparison is used to measure the original file size of SSGK and FBSGK with using of file name by encrypted file size in byte.

Figure 7 and Table 1: average encryption time and decryption time in different computer architecture and that the comparison of encryption and decryption file size. Compared to all file formats, FBSGK has more compression on file compared to SSGK.

Execution Performance

The presentation appraisal of the proposed convention. The accompanying analyzes center around the capacity and calculation overhead of SAPDS and the proposed convention FBSGK. These analyses are running on a worker with Intel Core 2 Deo, 3.01 GHz processor what is more, 4 GB RAM [4].

Table 2 describes the encryption and decryption time of SSGK and FBSGK. In all the file case, FBSGK is quicker than SSGK (Fig. 5). Encryption and decryption file size chart is used to calculate the SSGK and FBSGK encrypted file size in byte using file size KB.

Table 2 Encryption time and decryption time

File name	Original file size	SSGK		FBSGK	
		Encryption time (ms)	Decryption time (ms)	Encryption time (ms)	Decryption time (ms)
Student_data.mdb	8 MB	10.11	10.02	09.18	08.54
Goodtime.avi	200 MB	14.17	14.04	12.01	11.48
Mail_collection.doc	150 KB	0.85	0.81	0.61	0.44
Datacommunication.pptx	300 KB	0.62	0.61	0.49	0.39

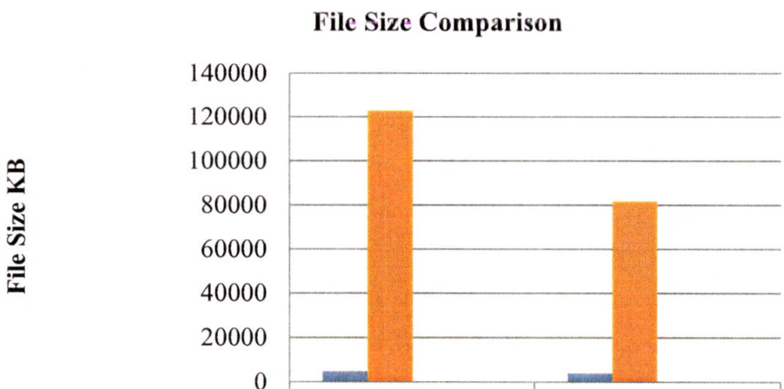

	SSGK-Encrypted File Size in Byte SSGK	FBSGK-Encrypted File Size in Byte FBSGK
Student_data.mdb	4915.2	4096
Goodtime.avi	122880	81920
Mail_collection.doc	90	60
Datacommunication.pptx	180	120

Fig. 5 Encryption and decryption file size

Figure 6 and Table 3 show that the encryption and decryption are performed in different operating systems. In all case, FBSGK is working more efficient compared to SSGK. Ubuntu Linux produces the least time for all the case.

The operating system is used to calculate the file size of time in measurement size (ms) of encryption and decryption time (Fig. 7).

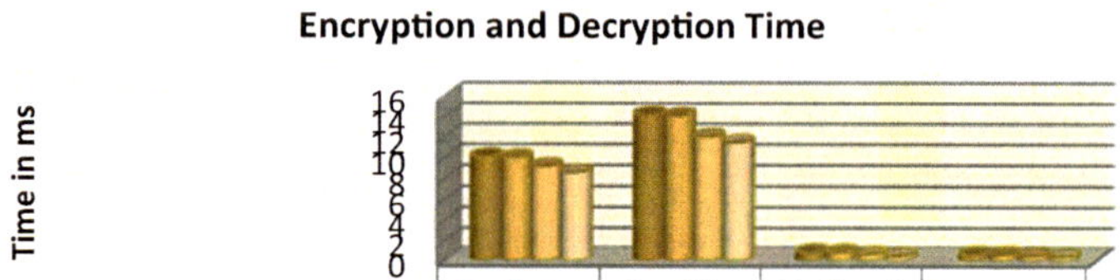

	Student_data.mdb	Goodtime.avi	Mail_collection.doc	Datacommunication.pptx
SSGK-Encryption Time(ms)	10.11	14.17	0.85	0.62
SSGK-Decryption Time(ms)	10.02	14.04	0.81	0.61
FBSGK-Encryption Time(ms)	9.18	12.01	0.61	0.49
FBSGK-Decryption Time(ms)	8.54	11.48	0.44	0.39

Fig. 6 Encryption time and decryption time

Table 3 Average encryption time and decryption time in different computer architecture

Operating system	SSGK		FBSGK	
	Encryption time	Decryption time	Encryption time	Decryption time
Windows 7	10.11	10.02	09.18	08.54
Windows 8	10.04	10.00	09.03	08.04
Windows 10	09.44	09.22	09.01	08.11
Ubuntu Linux	08.67	08.51	08.01	07.54
Red Hat EL	10.08	10.03	09.08	08.34

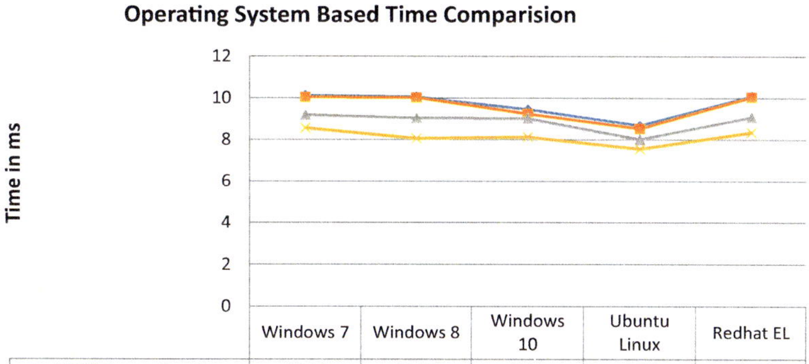

	Windows 7	Windows 8	Windows 10	Ubuntu Linux	Redhat EL
SSGK-Encryption Time	10.11	10.04	9.44	8.67	10.08
SSGK-Decryption Time	10.02	10	9.22	8.51	10.03
FBSGK-Encryption Time	9.18	9.03	9.01	8.01	9.08
FBSGK-Decryption Time	8.54	8.04	8.11	7.54	8.34

Fig. 7 Average encryption time and decryption time in different computer architecture

3 Conclusion and Future Enhancement

This paper shows the basis of cloud computing and its potential of use in the various areas. The concept of cloud computing comes from the networks illustrating the Internet as a cloud, where it is not possible or not important. This paper illustrates the basic area where cloud computing is used, and the criterion of cloud computing its features and the IT industries accepts this calculations. This project is only defined by time and byte. The future enhancement of this project can be more useful in upcoming future as and when needed for the same origination, as it is very user-friendliness in terms of expansion and to expression upcoming future. A cloud-based large information sharing framework uses a storage space from a cloud specialist organization to impart information to authentic clients. This research focused on to analyze the different key management protocols in cloud environment. And also, detailed investigation on the secret sharing group management protocols is performed. Forward and backward securities in group management protocols are created based on the secret sharing group management protocols, named as FBSGK. The proposed method is investigating encryption and decryption time on different files and also different operating system. All the comparisons are performed, and it concludes that all the existing drawbacks on SSGK protocols are solved in proposed FBSGK.

References

1. D. Zou, Y. Xiang, G. Min, Privacy preserving in cloud computing environment. Secur. Commun. Netw. **9**(15), 2752–2753 (2016)
2. V. Casola, A. Castiglione, K.K. Choo, C. Esposito, Healthcarerelated data in the cloud: Challenges and opportunities. IEEE Cloud Comput. **3**(6), 10–14 (2016)
3. L. Wu, Y. Zhang, K.-K.R. Choo, D. He, Efficient and secure identity based encryption scheme with equality test in cloud computing. Future Gener. Comput. Syst. **73**, 22–31 (2017)
4. Y.S. Rao, A secure and efficient ciphertext-policy attribute-based signcryption for personal health records sharing in cloud computing. Future Gener. Comput. Syst. **67**, 133–151 (2017)
5. J. Zhou et al., Securing outsourced data in the multi-authority cloud with fine-grained access control and efficient attribute revocation. Comput. J. **60**(8), 1210–1222 (2017)

An IoT-Based Smart Device to Monitor and Analyse the Performance of Athletes

S. Revathi, V. Muthu Priya, C. A. Bhargavan, and Fatah Mohammed

1 Introduction

In the recent past, modern analytics has found its way into sports including athletics and is being widely used not only in professional, but also in amateur events. The performance of a sportsman or an athlete is done in various ways which include but not limited to video capturing, high-end WSNs, etc. whose cost is very high. Due to this very reason, many players from rural as well as downtrodden backgrounds usually find it difficult to monitor their performance and take necessary steps to improve their performance accordingly. This also makes them struggle against their fellow competitors from urban and developed backgrounds who are very well equipped against them.

To overcome this problem, a wireless sensor network (WSN) based on the Internet of Things (IoT) called Particle Argon is used which makes use of two sensors, namely MAX30100 sensor and MPU6050 sensor to monitor the athletes' heartbeat, blood oxygen percentage and acceleration of the body part(s). The collected data is sent to Particle Cloud via Wi-Fi. This data is stored in a Google Sheets file with the help of an applet created using IFTTT. The data stored in Google Sheets file is processed using a custom-built script, so that the performance of the athlete can be visualised using Google Data Studio and analysed using 'PERFMAX', a desktop application built using Python and Tkinter. This application makes use of a linear regression algorithm to find whether the performance of the athlete has improved overtime or not. It also uses a min-max algorithm to find any use of PEDs.

S. Revathi (✉) · V. Muthu Priya · C. A. Bhargavan · F. Mohammed
Department of Computer Science and Engineering, B. S. Abdur Rahman Crescent Institute of Science and Technology, Vandalur, Chennai 600048, India
e-mail: srevathi@crescent.education

V. Muthu Priya
e-mail: muthupriya@crescent.education

J. D. Peter et al. (eds.), *Disruptive Technologies for Big Data and Cloud Applications*, Lecture Notes in Electrical Engineering 905,
https://doi.org/10.1007/978-981-19-2177-3_42

The reminder of this article is organised as followed: Sect. 2 presents ideas, techniques and background on technologies related to this work. Sections 3, 4 and 5 describe the problem definition, existing systems and proposed system in detail. Section 6 defines system design of the proposed system. Section 7 gives a brief look at the intended result and its analysis. Section 8 provides the conclusion and the future work of the proposed system, respectively.

2 Literature Survey

Mohammad Iman Mokhlespour Esfahani and Maury A. Nussbaum created a smart undershirt (SUS) [1] in 2018, which could detect upper body activities with a 94% accuracy and measure planar joint angles in the lumbar area and shoulder with 1.3 and 9.6° accuracy, respectively. The SUS is also the first textile-based device to detect activities and measure angles in the lumbar and shoulder regions.

Esteban Municio et al. created an athlete monitoring system in cycling situations using an IoT application in 2019 [2]. It recommends that state-of-the-art IoT applications can be used in cycling. The data is collected from the bunch of cyclists who are all connected in a mesh network, and the collected data shows location, heart rate and speed of the cyclist. The goal is to use a local body area network to collect data from the rider and a 6TiSCh network to create a long-range WSN with numerous hops. It is predominantly evidencing its relevance in a cycling context with reliable performance.

In 2019, Kong et al. Wang developed an athlete tracking and action recognition joint framework [3]. In athlete tracking, it uses scaling and occlusion robust tracker (SORT) to localise each athlete's position. It uses compressive tracking (CT). And in action recognition, it uses long-term recurrent region-guided convolutional network (LRRCN), which specifies the tracking results as an input.

In 2018, Lian and Li used the smoothing approach to produce a predictive analysis of e-commerce [4]. Operating costs, exponential smoothing, smoothing constant and predictive analysis are all suggested. The optimal smoothing is achieved here by employing the exact and objective Excel programming tool.

In 2018, Zhang et al. developed a lightweight wearable device [5] which monitors the running speed of the human body in real time and reminds the athlete to distribute the running speed correctly and reasonably in the running process. It also helps the athlete to run in a scientific method. The results showed that the measurements taken by the device were pretty accurate and reliable.

In 2018, Qiu et al. used heterogeneous Internet of Things (HetIot) [6] to build our future. It proposes four-layer architecture comprising networking, cloud computing, sensing and applications, and the discussion is made in the state of HetIot research and application. This paper also provides solutions for major challenges faced in HetIot, like self-organising, big data transmission and implementing HetIot on a large scale. In the future, HetIot will also be applied on sensors, big data analysis and cloud computing.

Shepherd et al. developed a wrist-mounted sensor for analysing kinematic parameters in 2017 [7]. Kinematics were evaluated using three-dimensional retro-reflective motion capture, and retro-reflective motion capture was also performed to assess if a wrist sensor placed in the forearm was a good monitoring strategy.

In 2017, Seshadri et al. developed smart wearable device for sports [8]. The authors proposed hand notation technique which is used in collecting data and their challenges in tracking of athlete position, velocity and displacement. They have also discussed the integration of GPS, accelerometer and heart-rate technology into a single sensor device that makes trainers easy to monitor the performance of athletes. In the future, both biometric and biovital signs will be developed. In 2017, Iervolino et al. examined the potential of a wearable device for the monitoring and the performance analysis of running athletes [9]. They also used tri-axial accelerometers (MEMS) to generate the data to elaborate for assessing the main gait parameters during walking or running road tests.

In 2015, Alfandi et al. discussed the criticality of various physiological parameters in humans from a healthcare perspective [10]. They also defined various categories for physiological data and proposed a routing and integration strategy based on the criticality of physiological parameters. In 2014, He and Bai developed a wearable Zigbee-based wireless sensor network [11] which recognises human activity and integrated the sensors to social media networks to improve usability.

In 2014, Deshpande and Malathi developed an embedded system for data communication between USB and Wi-Fi [12]. The ultimate focus of the work is to transfer data from USB-enabled device to Wi-Fi-enabled device or vice versa and also communication of data between one USB-enabled device and the other. It uses an optimised transmission method, and also Wi-Fi interface has the capacity to operate the data rate of 54 Mbps.

In the year 2013, Nur Hidayah Malek and colleagues created a wireless stethoscope with data logging capabilities that measures and monitors heartbeat signals and saves the information in a database system [13]. They used a Zigbee Pro Series 1 of wireless module to transmit the heartbeat signal across a 100-m range. The wireless heartbeat signal is monitored using a GUI developed using MATLAB software.

In 2013, Chakravorty et al. developed a tool which extracts stroke information during swimming on a real-time basis [14]. Here, two algorithms are used (filtering and signal processing) to extract real-time performance of the athletes in report format. Therefore, it allows quick access to the reports which are used to improve the performance of the athletes in swimming during the training period. On the basis of reusability components, these two algorithms have been developed and it can also be used in other application as well.

In 2010, Dhamdhere et al. developed wireless sensor networks (WSNs) for real-time athlete monitoring [15], in which new application is introduced, i.e. body area sensor network. The challenges are collection of data and describing design choices. Future plans are to relax some assumptions like accurate reports from the base stations back to player nodes.

Presently, the performance of athletes is monitored using high-end motion capture cameras and sensors. Although the proposed system is not a novel idea per se, it is a low-cost alternative to the high-end models already available in the market. The market in recent times has provided many different and disparate systems that cater to deliver solutions for individual aspects of the entire performance monitoring proposal. Apart from such systems, the integrated systems available have not visibly known to present an effective and optimal solution to cater to the problem. Many systems comprising multiple components, involving activity trackers, health trackers, location and movement monitoring, etc., have presented a challenge in successfully delivering cost-effectiveness and optimality in the energy consumption and processing.

A plethora of various technologies have been employed in various different systems in an initiative to successfully realise the portability, feasibility, ease of use and processing of the systems, in regard to the performance monitoring and analysis domain. There are a lot of systems that are being used to monitor and analyse the performance of athletes. A device was developed to monitor and track cyclists cycling in remote areas where it is difficult for them to receive signals. There are also full upper body wearable shirts with multiple sensors embedded in it, and these sensors act as a mesh to track the motion of different joints in the athlete's upper body. Other than these devices, there are also wristbands with sensors in them which measure the health parameters, step count and calories burnt by the athletes.

As the cost of components used in the existing systems is comparatively costly for underprivileged athletes, this system is proposed with the intent of cutting down the costs.

3 Problem Definition

The main idea of this system is to equip athletes who come from underprivileged backgrounds with low-cost performance monitoring devices which helps them in monitoring and analysing their performance, as well as to take the necessary future course of action.

Usually, the cost of performance monitoring devices is very high, which many up-and-coming athletes from underprivileged backgrounds would not be able to afford on their own as many parents in India still do not consider sports to be a lucrative career option. To overcome this problem, an IoT-based low-cost WSN is proposed which collects and stores the data in the cloud which can be used for real-time analysis as well as to predict the future course of action with the help of predictive analysis. The proposed athlete monitoring system usually includes monitoring of the basic health parameters of the patient, such as heart rate, pulse rate and oxygen level as well as acceleration. These parameters may indicate the changes caused due to the effort undertaken by the athlete. All these activities to monitor the patient should be made accessible remotely wherefore the guardian can monitor the progress and be alerted to provide support to the patient.

The proposed system is also used to check if the athlete is using any performance enhancement drugs commonly referred to as PEDs, as the system would show the or coaches) if there is any sudden spike in the athlete's performance. Further- a report could be generated on the basis of the athlete's performance, which used by their coaches to analyse their performance and also to facilitate future of action.

Proposed System

is a developing nation where the majority of the population lives from paycheck ycheck, and sports is not given major attention as it gets quite expensive if you to track and monitor one's performance. Due to this very reason, many bright its give up on a career in sports and settle for some other job which is unrelated ie sports in which they were good at. Hence, a performance monitoring system be implemented to overcome this problem as the athletes can not only track their -time performance, but also use the data to predict their future performance.

The system proposed focuses on providing an efficient, optimal and easy-to-use erface that would facilitate better processing of the functionalities designed. To iver the solution, the technology of IoT is involved to design a wearable device t would allow portability and ease of access. A consolidated, wholesome and igle efficient system, encompassing all the essential components necessary for the lution, enhances the usability and the functionality of the system, to resolve the oblem.

A variety of sensor components are required to process the individual parameters ı the patient, such as heart-rate sensor, pulse oximeter, accelerometer and gyrom- er. All these components are incorporated as digital components in the system. entralised processing of these components is facilitated by incorporating a newgen Vi-Fi-based portable microcontroller, Particle Argon.

A brief about the components used in the proposed system:

- Particle Argon is a powerful Wi-Fi-based IoT development kit that includes the Nordic nRF52840 and Espressif ESP32 processors, as well as a built-in battery charging circuitry that makes it easier to connect a Li-Po battery and 20 mixed signal GPIOs for interacting with sensors, actuators and other electronics.
- MPU6050 sensor is a 6-axis motion tracking device that includes a 3-axis gyroscope, 3-axis accelerometer and a digital motion processor in a tiny package. It also contains an on-chip temperature sensor and an I2C bus interface for communicating with microcontrollers.
- MAX30100 sensor is an integrated pulse oximetry and heart-rate monitor sensor solution that combines two LEDs, a photodetector, optimised optics and low-noise analogue signal processing to detect pulse oximetry and heart-rate signals.

The proposed system is planned in such a way that MAX30100 sensor and MPU6050 sensor collect raw data of the athlete's heartbeat, blood oxygen percentage and acceleration of the body part(s) wherever the sensor is attached to with the help of Particle Argon. The collected data is transferred to the Particle Cloud.

The system also makes use of IFTTT to act as an interface to transfer the data being collected and stored in Particle Cloud to Google Sheets, where a custom script is used to process the data and store it. The processed data is visualised with the help of Google Data Studio. It is also analysed by using a custom desktop application which is built using Tkinter and Python. The parties involved in the runtime execution of the system include the athlete or the wearer and the coach. The concerned parties involved with this system play a noteworthy role in the support of the system, on the basis of receiving alerts and providing timely and appropriate actions and treatment.

A sudden spike in heart rate and acceleration shows that the athlete has made use of performance enhancement drugs commonly referred to as PEDs which are used by the athletes to enhance their performance in any sport. A basic algorithm is formulated to monitor the acceleration of the athlete which is calculated using the data collected from the accelerometer. This data coupled with the data from the heart-rate sensor is used to calculate the progress of the athlete. This algorithm is implemented in WebIDE, an integrated development environment of Particle IoT devices using embedded C to collect the aforementioned parameters and store it in Particle Cloud. The code would be a part of Particle Argon, programmed to sense the values consolidated by the respective sensors.

The collected values/data is further used to analyse the athlete's performance in real time with the help of Python-based data analytics packages—Matplotlib, Sklearn and pandas. Further, linear regression is used by the proposed system to find a relationship between heart rate and acceleration to check whether the performance of the athlete is improving over a period of time or not. The system also implements a min-max algorithm to arrive at conclusive evidence whether the performance of the athlete is improving over a period of time or not.

5 System Design

A variety of sensor components are required to process the individual parameters on the patient, such as heart-rate sensor, pulse oximeter, accelerometer and gyrometer. All these components are incorporated as digital components in the system. Centralised processing of these components is facilitated by incorporating a newgen Wi-Fi-based portable microcontroller, Particle Argon.

The architectural diagram of the proposed system is shown in Fig. 1. The sensors are tracked and processed by the microcontroller, governing the signals to be sent for sensing the athlete's parameter values. These values from the sensors are then stored on a cloud-serviced database, to perform analysis and pattern processing. All these values upon storage, analysis and processing are required to be integrated into a consolidated interface to provide a clear picture of the processing of the system and to promote effortless user interface with the designed system. The proposed system is implemented with the use of the above-mentioned software and hardware components. These components are used to build the five modules of this system.

Figure 2 depicts the data flow of the proposed system.

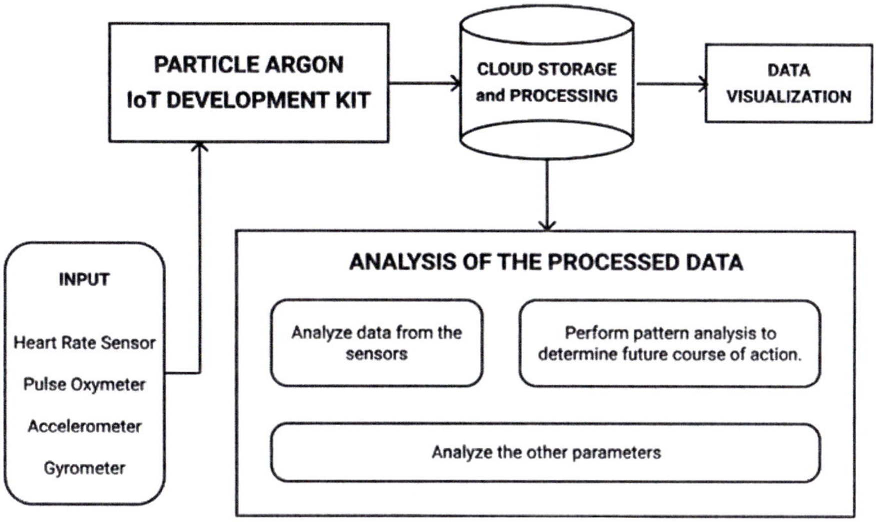

Fig. 1 Architectural diagram of the proposed system

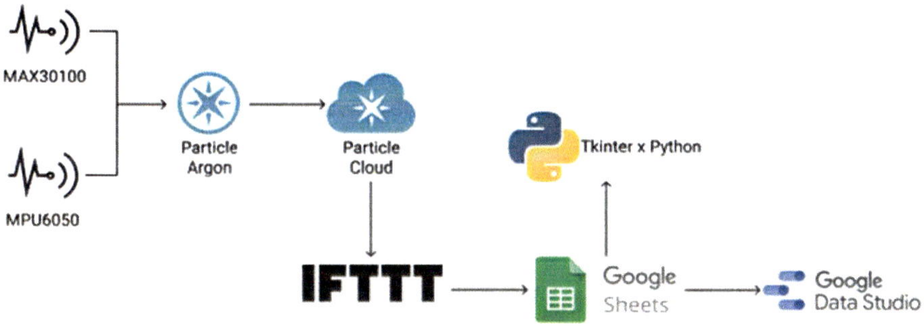

Fig. 2 Data flow of the proposed system

As shown in Fig. 2, the data flow of the proposed system is given. In this, the data is gathered by the two sensors in the system MPU6050 sensor which is used to get the acceleration of the athlete and MAX30100 sensor which is used to measure the heart rate and blood oxygen of the athlete. The collected data is stored in Particle Cloud, and this data is transferred to Google Sheets with the help of an applet created using IFTTT. This file is used by Google Data Sheets and custom-built desktop application 'PERFMAX' to monitor the performance of the athlete and analyse it for future course of action.

This whole system can be implemented in four phases:

5.1 Hardware Assembly and Sending Data to Particle Cloud.

The project setup using the hardware components is given in Fig. 3. In this step, MAX30100 sensor and MPU6050 sensor collect raw data of the athletes' heartbeat, blood oxygen percentage and acceleration of the body part(s) wherever the sensor is attached to with the help of Particle Argon. The collected data is transferred to the Particle Cloud.

Fig. 3 Hardware setup using all the hardware components

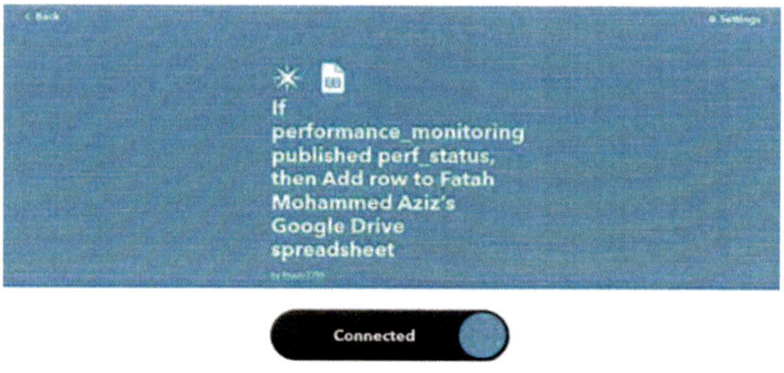

Fig. 4 Applet created on IFTTT

5.2 Create an Applet Using IFTTT to Transfer Data from Particle Cloud into Google Sheets

In this step, an applet is created using IFTTT and is shown in Fig. 4, which transfers collected data from Particle Cloud into Google Sheets whenever the Particle Argon module is turned on. This data is processed and stored in Google Sheets using a custom script.

5.3 Visualise Data on Google Data Studio

In this step, the processed data from the Google Sheets file which was created in the last step is integrated with Google Data Studio to visualise the performance of the athlete and is given in Fig. 5

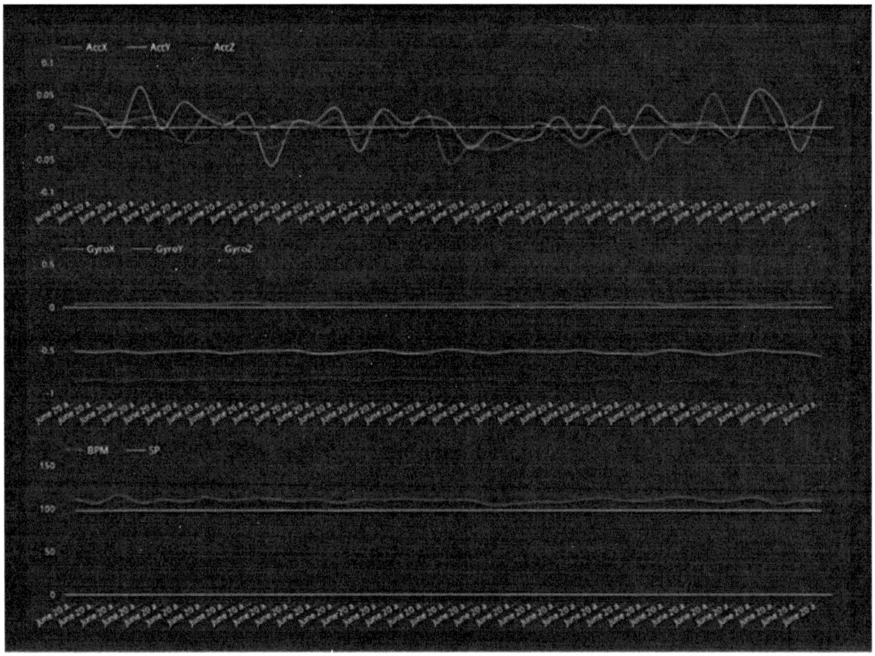

Fig. 5 Data visualised on Google Data studio

5.4 Analyse Data on Custom-Built Desktop Application 'PERFMAX'

In this step, the Google Sheets file which was created in the last step is uploaded into a Python-based GUI application developed with the help of Tkinter where the relationship between Acceleration of the athlete with respect to time is found using linear regression. The application also uses a min-max algorithm to find if the athlete has used any PEDs to enhance their performance. This application is used to determine whether the performance of the athlete has improved overtime or not, as well as to tell the user (coach) if any PEDs have been used by the athlete.

5.5 Algorithm

Step 1: The data collected by the MPU6050 sensor and MAX30100 sensor.

(1a) The sensor MPU6050 is used to get the acceleration of the athlete
(1b) The MAX30100 sensor is used to measure the heart rate and blood oxygen of the athlete

Step 2: The collected data is stored in Particle Cloud.

Step 3: The data from Particle Cloud is now transferred to Google Sheets with the help of an applet created using IFTTT.

Step 4: This file is used by Google Data Sheets and custom-built desktop application 'PERFMAX'.

Step 5: The performance of the athlete is monitored and analysed for future course of action.

Step 6: The report is generated on the basis of the athlete's performance, which can be used by their coaches to analyse their performance and also to facilitate future course of action.

6 Result and Analysis

After the data is sent to the Particle Cloud over Wi-Fi, the applet that was created with the help of IFTTT is used to store it in a Google Sheets file which uses a trigger to process the incoming data for future analysis and visualisation(s). This processed data is downloaded in the form of a CSV file, which is used to analyse the future course of action for the athlete. The system makes use of a linear regression algorithm to find if the performance of the athlete has improved over a period of time or not. Figure 6 showcases the acceleration of the athlete over a period of time, whereas

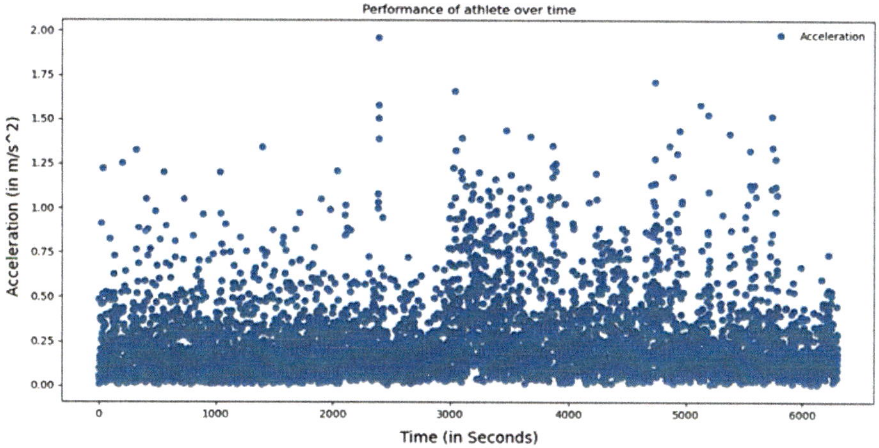

Fig. 6 Scatter plot of sensor data using Matplotlib

Fig. 7 is used to illustrate the average acceleration of the athlete. The performance of the athlete which is collected from the sensor over the period of time is recorded and is used to analyse the upcoming course of action.

Figure 8 is used to show the difference between the actual acceleration and the predicted acceleration, whereas Fig. 9 is used to show whether the performance (acceleration) of the athlete has improved over time or not. It not only analyses the data, but also gives a conclusive outcome whether or not the performance of the athlete has improved over a period of time.

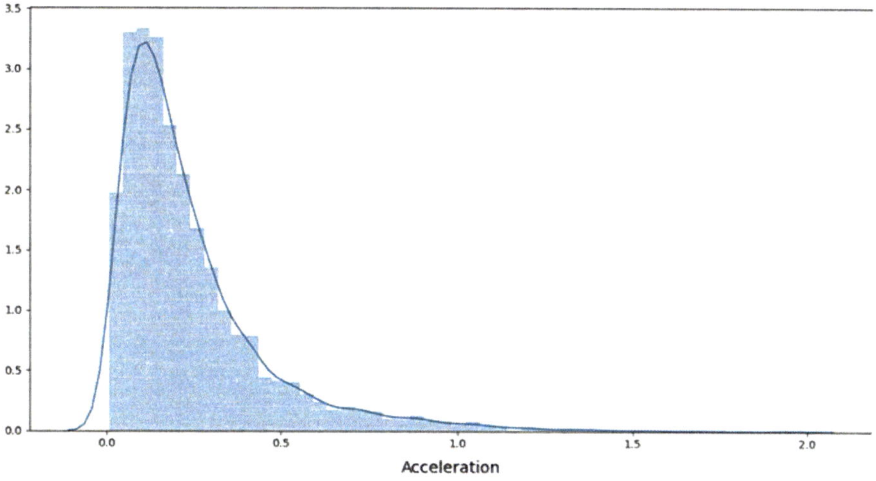

Fig. 7 Average acceleration of sensor data using Seaborn

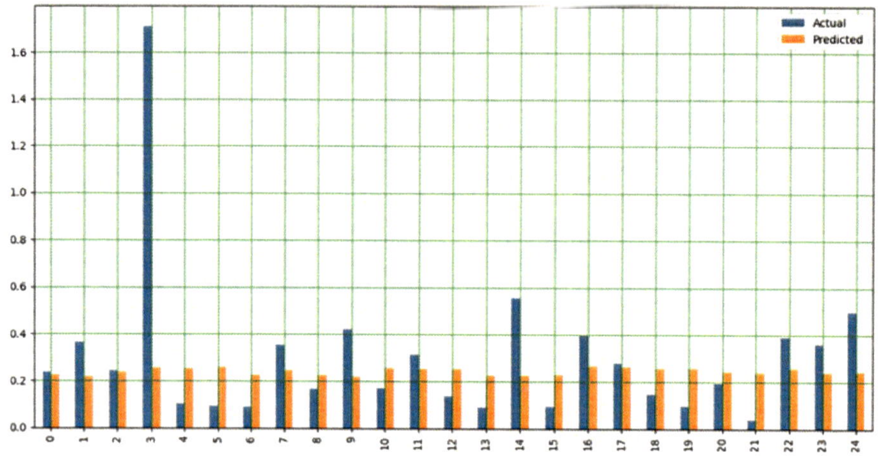

Fig. 8 Difference between actual acceleration and predicted acceleration

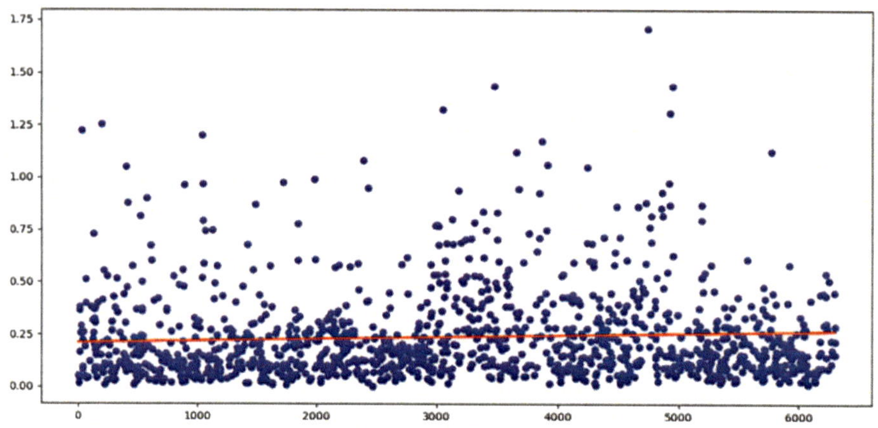

Fig. 9 Acceleration of athletes over a period of time using Matplotlib

Figure 10 shows the Min-Max algorithm to show if the athlete has used any PEDs. This system also makes use of a linear regression algorithm which is used to find whether the athlete has used any PEDs to improve his performance or not using his/her heartbeat.

Fig. 10 Min-Max algorithm to show if the athlete has used any PEDs

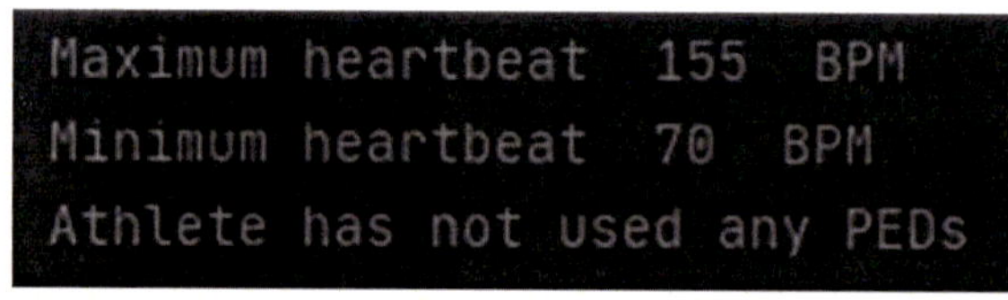

7 Conclusion and Future Enhancement

The aim of this system is to build a low-cost performance (health and motion) monitoring system for athletes who have the potential in them to achieve bigger, but they lack the resources to monitor their performance on a daily basis. This work was successfully implemented with the use of Particle Argon along with MAX30100 sensor and MPU6050 sensor, which collected the health parameters like blood oxygen level and heart rate and motion parameters like acceleration. This collected data was stored and processed in Google Sheets from Particle Cloud with the help of applets created using IFTTT. This data was used for real-time visualisation with the help of Google Data Studio and was also analysed using a desktop application developed by the author of this document.

Going further, parameters like temperature, distance travelled, total active time, etc. can be incorporated in this system. A mesh network between various Particle Argon devices can be created to measure and track movement of arm(s) and/or leg(s). A web application, same as the desktop application, can be developed which could be used by the coach(s) to monitor and analyse the performance of the athlete in their mobile phones, where the users could upload a file with data in it, to see if there is any improvement and predict the future course of action.

References

1. M. Iman Mokhlespour Esfahani, M.A. Nussbaum, A "smart" undershirt for tracking upper body motions: task classification and angle estimation. IEEE Sens. J. **18**(18) (2018)
2. E. Municio, G. Daneels, F. De Turck, S. Latre, Athlete monitoring in cycling environment using Iot application. IEEE Internet of Things J. (2019)
3. L. Kong, D. Huang, J. Qin, Y. Wang, A joint framework for athlete tracking and action recognition in sports videos. IEEE Trans. Circ. Syst. Video Technol. (2019)
4. J. Lian, L. Li, Predictive analysis of e-commerce enterprises soft operating costs based on exponential smoothing technique, in *IEEE 9th International Conference on Information Technology in Medicine and Education* (2018)
5. B. Zhang, J.B. Guo, D.D. Xu, Wearable pace speed monitoring and reminding system, in *International Conference on Intelligent Transportation, Big Data & Smart City (ICITBS)* (2018)
6. T. Qiu, N. Chen, K. Li, M. Atiquzzaman, W. Zhao, How can heterogeneous internet of things build our future. IEEE Commun. Surv. Tutor. 20(3), 2011–2027 (2018)
7. J.B. Shepherd, G. Giblin, G.-J. Pepping, D. Thiel, D. Rowlands, Development and validation of a single wrist mounted sensor for elite netball shot. IEEE Sens. Appl. **1**(5) (2017)
8. D.R. Seshadri, C. Drummond, J. Craker, J.R. Rowbottom, J.E. Voos, New integrated technologies allow coaches, physicians, and trainers to better understand the physical demands of athletes in real time. IEEE Digital Object Identifier (2017)
9. R. Iervolino, F. Bonavolontà, A. Cavallari, A wearable device for sport performance analysis and monitoring, in *IEEE Instrumentation and Measurement Society* (2017)
10. O. Alfandi, S. Ahmed, M. El Barachi, A. Khan, Challenges of leveraging mobile sensing devices in wireless healthcare, in *12th Annual IEEE Consumer Communications and Networking Conference (CCNC)* (2015)
11. Z. He, X. Bai, A wearable wireless body area network for human activity recognition, in *Sixth International Conference on Ubiquitous and Future Networks (ICUFN)* (2014)

12. A.A. Deshpande, P. Malathi, Embedded system for communicating data between "USB" and Wi-Fi, in *IEEE Sixth International Conference on Computational Intelligence and Communication Networks* (2014)
13. S. Nur Hidayah Malek, W. Suhaimizan Wan Zaki, A. Joret, M. Mahadi Abdul Jamil, Design and development of wireless stethoscope with data logging function, in *IEEE International Conference on Control System, Computing and Engineering* (2013)
14. N. Chakravorty, T. Le Sage, S.E. Slawson, P.P. Conway, A. West, Design and implementation of an integrated "performance monitoring tool" for swimming to extract stroke information at real time. IEEE Trans. Hum. Mach. Syst. 43(2) (2013)
15. A. Dhamdhere, H. Chen, A. Kurusingal, V. Sivaraman, A. Burdett, Experiments with wireless sensor networks for real-time athlete monitoring, in *5th IEEE Building Sensor Network Applications* (2010)

Optical Character Recognition-Based Signboard Detection

N. Dinesh and Senthilkumar Mathi

1 Introduction

Traffic sign acknowledgment as use of character recognition has two parts—symbol identification and text identification. This work emphasizes the discovery and identification of text on traffic marks. Text on traffic signs conveys data that can be helpful to drivers like speed limits, options to proceed, alerts, and so on. Acknowledgment of text from street signs can be predominantly troublesome. The difficulties incorporate text style variety, direction, shadows, low image determination, obscuring, and other alterations credited to the movement of the vehicle. Numerous methodologies have been proposed to recognize traffic signs [1]. Customary strategies [2–4] are normally founded on hand-crafted highlights like tone, surface, edge, and other low-level highlights to detect traffic signs in an image. In driving surroundings, because of the variety of the traffic sign presence, the impediment of traffic signs by different items, and the impact of lighting conditions, customary techniques for traffic sign recognition showed meager performance. Machine learning is a solicitation of artificial intelligence that delivers the capability to acquire without the involvement and vibrant programming [5].

The goal of an automatic road symptoms identification system is to locate, classify the road symptoms inside the pictures captured through a camera, and identify the textual content correctly. Text detection and identification machine have several vital software regions, together with progressed driver assistance systems, road surveying, and self-reliant motors. Automatic detection and identification of traffic signs are

N. Dinesh · S. Mathi (✉)
Department of Computer Science and Engineering, Amrita School of Engineering, Amrita Vishwa Vidyapeetham, Coimbatore, India
e-mail: m_senthil@cb.amrita.edu

N. Dinesh
e-mail: cb.en.p2aid19009@cb.students.amrita.edu

J. D. Peter et al. (eds.), *Disruptive Technologies for Big Data and Cloud Applications*, Lecture Notes in Electrical Engineering 905,
https://doi.org/10.1007/978-981-19-2177-3_43

challenging trouble where detection implies the process to detect the traffic sign images from the scene image captured by the camera [6]. The detector checks for the presence of any objects which are most similar to the traffic signs which are given as an input dataset. The next process is the process of recognition which implies the classification of the traffic sign and identifying which sign is present in the signboard. There are several categories of traffic symptoms that can be visible beside the roads [7]. A well-organized driver must notice each of the road symptoms in front of him and need to take measures accordingly. If not, awful things can show up. The preprocessing stage aims to make the photo suitable for dissimilar feature extraction algorithms [8, 9]. Initially, pictures may be affected by noise or blurred so the preprocessing level can make the preliminary pictures more suitable for computation [10, 11].

2 Related Work

Malakar and Chiracharit proposed the recognition of text and non-text signs using a convolutional neural network (CNN) [12]. The work discussed the classifying of the data whether it belongs to text sign or non-text sign using CNN. A dataset by capturing images from public places and downloading from the Internet for classifying the data has been created. A self-designed ten-layered convolutional neural network has been used and trained with a sign image dataset and experimented with in two different ways. First experiment with CNN models and the second experiment with maximally stable extremal regions and optical character reputation (OCR) where CNN accuracy was very low (65–70%) when compared with maximally stable extremal regions and OCR models.

Goodfellow et al. introduced multi-digit number detection from street view imagery using deep CNN. This paper said about the recognition of the number of houses using deep CNN [13] and has used the street view house numbers dataset to evaluate this approach and achieved about 96% accuracy in recognizing complete street numbers and also tried reproducing captcha in this work.

For reproducing the captcha, the captcha puzzles dataset has created where it contains many captchas. This paper implemented a deep neural network to train the large, distributed neural network on high-quality images with eleven hidden layers. Max-pooling and subtractive normalization at alternative layers so half of the layers do not reduce the spatial size of the representation was used. All convolutions use zero paddings on the input to preserve representation size.

Meena investigated information extraction from signboards; this paper discusses the extraction of information from the signboard and arrow detection [14]. Extraction of information is done from nameplates, then upscale images and pass them to the OCR.

In arrow detection, the match for the image against the template to account for the unknown size of the arrow is done. This selects the region of interest to do preprocessing. Secondly, erode and dilate since the arrow is not a single connected component after binarization. At last, edge detection of the arrow is done to find the orientation of the arrow.

Toaha et al. proposed automatic signboard detection from natural scene images in this paper and have identified detection of the name board. The Bangladesh street view signboard objects image dataset was used. This contains the image of some name boards found in Bangladesh [15]. This paper brought a unique approach to observe signboards correctly with the aid of smart picture processing techniques and a statistically determined hyperparameter-based deep learning method. Their proposed classical object segmentation method has unsatisfactory performance on signboard detection and so has used the faster recurrent-CNN method for the segmentation-based learning algorithm.

Lim et al. suggested a robust detection of the area of interest of signboards in a noisy surroundings [16]. In this technique, the Chinese name board dataset was used. Area of interest to acquire the necessary information using pictures or video was used. The proposed preprocessing technique is based on morphology, which means studying the geometric form of a picture that involves erode and dilate operations. The effectiveness of this method is compared with prevailing works for the recognition of the signboard [17] and the experimental outcome shows that the proposed technique has high correctness and suitable processing speed in real-time.

Ren et al. examined real-time object observation. In this paper, PASCAL visual object lessons dataset is used. They have accomplished object detection networks depend on the region proposal algorithm to hypothesize object places [18]. This paper introduced an area proposal network that shares full-picture convolution features with the detection network. The region proposal network is educated, end to end to generate high-quality region proposals, which are used by fast recurrent-CNN for detection. In this proposed method, an accuracy of about 73.2% on the 2007 dataset and 70.4% accuracy on the 2012 dataset was obtained.

3 Proposed Methodology

In this part, the overall architecture of the proposed framework is shown in Fig. 1. The main objective is to find text recognition from the traffic sign board using OCR. To detect and identify the signboard in natural scenes and recognize the text from the signboard. Two sections are there one is the detection and the other one is text recognition. There are numerous traffic signboards present in India which are textual content-based. It is impractical to note all of them manually in an operating vehicle. The possibilities of lacking very crucial signboards had been high. Manual taking pictures of signboards additionally divert the driver's interest. This may also cause an accident on Indian roads.

To prevent accidents superior driver assistance machine plays a key role. Traffic signboard observation and identification are the key issues for studies recently. For traffic signboard detection, an EfficientDet algorithm is been used. This algorithm works better for object detection. The weather also plays a major role in this present

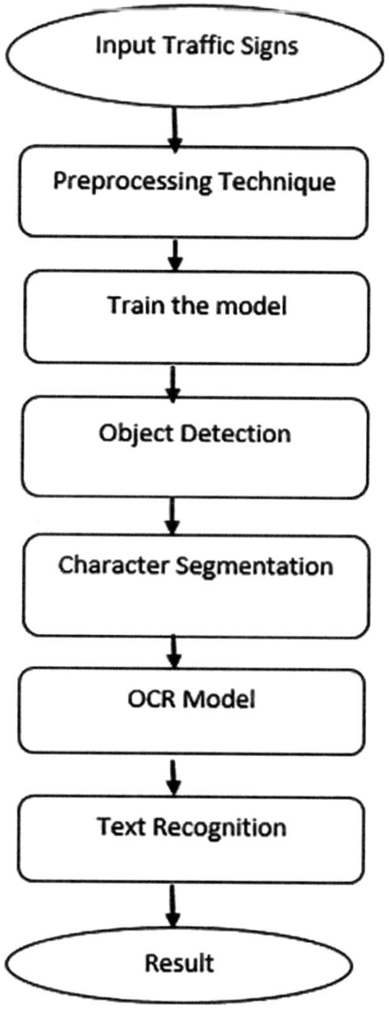

Fig. 1 Flowchart of the proposed work

work because the capture of images during rain, fog, and smoke would not be much clear. The image may be blur or shaky so the capturing of images also be a challenging problem. So, a deep learning algorithm for the detection of text boards is used.

Figure 1 shows the overall flow of the proposed work, in the proposed work the captured images is given as input for signboard detection and recognize the text from the image. It is a process of electronically extracting textual content from pictures. The image is given into the model for object detection and the signboard is been detected. The detected signboard with its region of interest is been segmented and given into the OCR model. OCR is the popularity of published or written textual content characters using a laptop. This involves picture scanning of the textual content person-trough-person, evaluation of the scanned-in picture, and then translation of the individual picture into individual codes, such as inclusive of ASCII, usually used in the statistics processing. OCR is a technology that permits you to transform dissimilar types of documents, such as scanned paper documents, pdf files, or pictures observed by a digital camera into editable and searchable data.

OCR is a significant exploration area in design acknowledgment. The goal of an OCR framework is to perceive alphabetic letters, numbers, or different characters, which are computerized images, without any human interference which is shown in Fig. 2. This is refined via looking through a match between the highlights removed from the given character's image and the library of image models. OCR has been debated significantly and has had great accomplishments lately, for example, in license plate distinguishing proof and text documents. Nevertheless, the comparability with the customary OCR, the content of the normal prospect is stimulating because of the arbitrary designs, night vision, blurry images, foggy images, dissimilarity in the performance, and contextual objects that can force difficulties and cannot be organized. A signboard discovery organization distinguishes the signboards that are available at the edges of the roads, identifies them, and shows them on the vehicle's dashboard. In any case, OCR in systematic settings is still remarkably

Fig. 2 Representation of OCR for text recognition

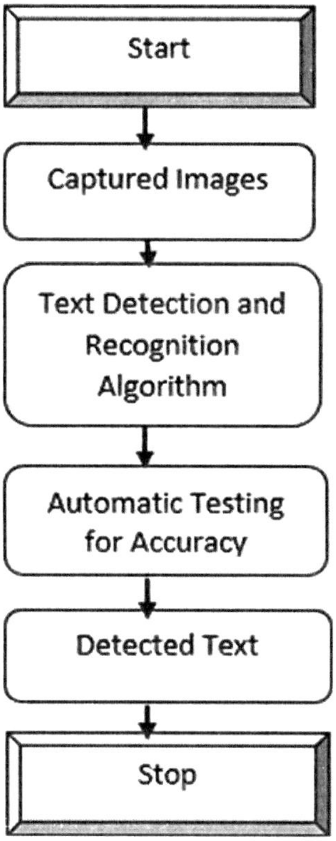

testing because of the many variables that exist, such as a text font, and close by clutter noise.

4 Result Analysis

In the proposed method, OCR for identifying the text from the signboard was used. Text recognition phases are as follows.

1. outlook correction
2. discover textual content characters
3. shape characters into traces
4. OCR of area

The present work shows an overall framework of textual identification from the signboard. The picture first observed from the external surrounding. In the proposed work, the dataset used is the German traffic sign recognition dataset as shown in Fig. 3, which contain images of the various signboards like stop, speed limit, kmph, etc. The proposed work is in the segment of preprocessing; consequently, no finite results can be proven. Here, it gives the design of input and the awaited output. In the training phase of the algorithm, the input picture consists of the training dataset photographs captured. These input photographs incorporate each text in addition to symbols.

All traffic panels from the roadsides are captured. These photographs are given as input to the set of rules. The algorithm is trained to discover and recognize text and logos from traffic signboards. The experimental outcome indicates that the proposed approach provides better performance and resulted in 70 to 85% accuracy as shown in Table 1 for text recognition to signboard. From this work, it is inferred that OCR helps us to convert photocopy into text understandable machine language. In this work, the conversion is performed for the signboard to text which helps the driver to reach the destination on time by avoiding major accidents.

The purpose of this system is to observe and identify signboards and to assess the performance measurement. Table 1 shows the observation of the correctness and the testing sample for recognition of text extraction and detection. The performance was made for the German traffic sign recognition dataset. Testing sample is the number of samples taken for testing purposes and recognition would be how many are correctly

Fig. 3 Sample dataset

Table 1 Analysis of recognition and extraction

Objectives	Number of samples tested	Number of samples recognized	Accuracy (%)
Text detection	500	407	80
Text extraction	500	360	71.2

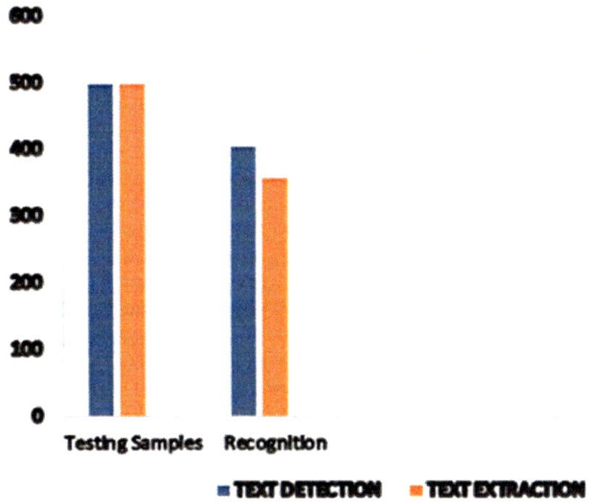

Fig. 4 Number of samples identified

recognized from the proposed system. The plot between the testing sample and recognition is shown in Fig. 4. Accuracy is the closeness of a measurement to a specific value, and in the proposed work, 80% accuracy for text detection on the signboard and 71.2% accuracy for text extraction are obtained and compared with another model as shown in Fig. 5.

5 Conclusion

The traffic sign detection and identification system solve the problem of identifying and interpreting the traffic sign set on either side of the road. Through this current work, it is possible to successfully carry out the process of identifying and interpreting the traffic sign. The picture is captured by the camera, the traffic sign is being extracted, it is detected from the datasets, and if the traffic signboards contain any text, it is been identified using OCR. Perspective recitation and temporal fusion of candidate areas of textual content had been used to enhance OCR results. OCR

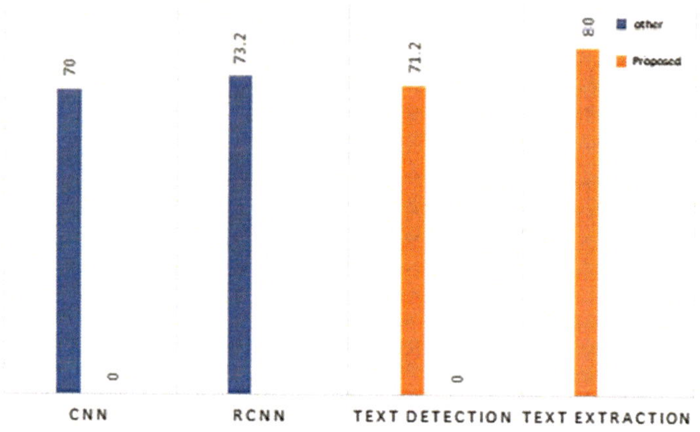

Fig. 5 Accuracy rate

offers better results for textual content recognition. The present study and experimental analysis give us how text is been obtained from the signboard. In future, the aim is to identify the signboard from natural scenes and identify the text from the signboard to do both the detection and recognition of signboards in future work.

References

1. E. Sarku, J. Steele, T. Ruffin, B. Gokaraju, A. Karimodini, Reducing data costs-transfer learning based traffic sign classification approach, in *SoutheastCon 2021* (IEEE, 2021), pp. 1–5
2. W.A. Haque, S. Arefin, A.S. Shihavuddin, M.A. Hasan, DeepThin: a novel lightweight CNN architecture for traffic sign recognition without GPU requirements. Expert Syst. Appl. **15**(168), 114481 (2021)
3. S. Salti, A. Petrelli, F. Tombari, N. Fioraio, L. Di Stefano, Traffic sign detection via interest region extraction. Pattern Recogn. **48**(4), 1039–1049 (2015)
4. S. Uluskan, Automatic detection of regulatory traffic signs via circle detection by post edge detection applied to straight line Hough transform. Int. J. Automot. Sci. Technol. **4**(2), 49–58 (2020)
5. Z. Qin, W.Q. Yan, Traffic-sign recognition using deep learning. Geom. Vis. **1386**, 13 (2021)
6. S. Mahajan, R. Rani, Text detection and localization in scene images: a broad review. Artif. Intell. Rev. **16**, 1–61 (2021)
7. P. Sahare, S.B. Dhok, Review of text extraction algorithms for scene-text and document images. IETE Tech. Rev. **34**(2), 144–164 (2017)
8. S. Mathi, E. Joseph, S. Dharini, V. Mohan Karthik, S. Harishkiran, Design and implementation of message communication to control traffic flow in vehicular networks. Int. J. Eng. Adv. Technol. **9**(1), 848–853 (2019)
9. T. Singh, Foggy image enhancement and object identification by extended maxima algorithm, in *2017 International Conference on Innovations in Control, Communication and Information Systems (ICICCI)* (IEEE, 2017), pp. 1–5
10. N.A. Menon, K.S. Anusree, A. Jerome, K. Sreekumar, An enhanced digital image processing based dehazing techniques for haze removal, in *2020 Fourth International Conference on Computing Methodologies and Communication (ICCMC)* (IEEE, 2020), pp. 789–793

11. P. Ranjitha, K. Rajashekar, A. Shamjith, A review on text detection from multi-oriented text images in different approaches, in *2020 Proceedings of the International Conference on Electronics and Sustainable Communication Systems, ICESC 2020*, art. No. 9156002, pp. 240–245 (2020)
12. S. Malakar, W. Chiracharit, Detection of text and non-text signs using convolutional neural networks, in *International Conference on Engineering and Industrial Technology 2020 May*, vol. 2020, pp. 6–8 (2020)
13. I.J. Goodfellow, Y. Bulatov, J. Ibarz, S. Arnoud, V. Shet, Multi-digit number recognition from street view imagery using deep convolutional neural networks. arXiv preprint arXiv:1312.6082 (2013)
14. A.K. Meena, *Information Extraction from Signboards* (Doctoral dissertation, IIT Delhi India) 2014
15. M. Toaha, S. Islam, S.B. Asad, C.R. Rahman, S.M. Haque, M.A. Proma, M. Shuvo, A. Habib, T. Ahmed, M. Basher, Automatic signboard detection and localization in densely populated developing cities. arXiv preprint arXiv:2003.01936 (2020)
16. D. Lim, Y. Kim, I. Park, J.H. Seung, K. Chong, Robust detection of ROI of signboards in a noisy environment. Int. J. Signal Process. Image Process. Pattern Recogn. (2017)
17. M.A. Panhwar, K.A. Memon, A. Abro, D. Zhongliang, S.A. Khuhro, S. Memon, Signboard detection and text recognition using artificial neural networks, in *2019 IEEE 9th International Conference on Electronics Information and Emergency Communication (ICEIEC)* (IEEE, 2019), pp. 16–19
18. S. Ren, K. He, R. Girshick, J. Sun, Faster R-CNN: towards real-time object detection with region proposal networks. IEEE Trans. Pattern Anal. Mach. Intell. **39**(6), 1137–1149 (2016)

IPL Win Prediction Using Machine Learning

P. V. Kavitha, R. Sai Pavitra, K. Suwetha, and P. Uvashree

1 Introduction

There are various nearby competition which will be going on and they will store the brings about book and it is not secure when something turns out badly or tainted and all the players will not think about their group progress and their triumphant likelihood so we went over the possibility of an application when all is on the web and required no book and pen and can oversee it completely through a pleasant interface and can think about the groups advance and deal with every one of the exercises of the whole period of the competition. The venture named IPL DATABASE gives client advantages to make their own competition of T20 and deal with every one of the exercises, for example, choosing and adding seasons, adding groups, and adding players to the groups and adding matches between groups. IPL win prediction gets into the habit of looking at past and real-time data to predict future demand. It is carried out using the historical data, where the businesses can learn from their past mistakes and avoid making the same mistakes in the future. To predict the outcome of IPL match, machine learning regression algorithms play a vital role.

2 Literature Survey

Sport Analytics uses different machine learning algorithms for predicting the outcome of the games. Wickramasinghe [1] have introduced Longitudinal and Hierarchical Linear methods for predicting the performance of the batsmen in the test

P. V. Kavitha (✉) · R. Sai Pavitra · K. Suwetha · P. Uvashree
Department of Information Technology, Sri Ramakrishna Engineering College, Coimbatore, India
e-mail: kavitha.krishna@srec.ac.in

J. D. Peter et al. (eds.), *Disruptive Technologies for Big Data and Cloud Applications*, Lecture Notes in Electrical Engineering 905,
https://doi.org/10.1007/978-981-19-2177-3_44

series. The limitation of this approach is simple relationship with limited number of variables.

Vistro et al. [2] have proposed techniques like logistic regression to predict the score in different innings and Random Forest algorithm to predict the winner of the match. Both these approaches have the ability to predict the discrete functions.

Jyothsna and Srikanth [3] have introduced simple linear regression to calculate the weight age of a team based on players past performance of IPL. It uses a single variable for the prediction.

Lokhande and Professor [4] have introduced techniques like simple linear regression and Naive Bayes classifier for the prediction of live cricket score. It uses a single variable for the prediction and Naive Bayes assumes that all predictors (or features) are independent, rarely happening in real life.

3 Proposed Methodology

The prediction of IPL match is implemented by multivariable linear regression. Regression is a statistical technique used in Data Mining. In this technique, the dependent variable is continuous, the independent variables can be discrete or continuous, and nature of regression line is linear. Multiple linear regression attempts to model the relationship between two or more explanatory variables and a response variable by fitting a linear. The multivariable linear regression is represented by

$$Y = b_0 + b_1 X_1 + b_2 X_2 + \cdots + b_p X_p$$

where

- Y is the indicator worth of the reliant variable
- X_1 through X_p are p particular autonomous or indicator factors
- b_0 is the worth of Y when the entirety of the free factors are equivalent to 0
- b_1 through b_p is the assessed relapse co-proficient

Multivariable linear regression algorithm is possible to take many factors for prediction. In this proposed work, the five factors that are used for the prediction of win or lose of the game with weightage of each factors are listed below:

- Points—30%
- Win: lose ratio—30%
- Last three matches—15%
- Last match—5%
- Net run rate—20%

The application can be used by users who sign into this application and can be accessible anywhere around the worldwide and can also store all the data related to tournament. This application is not only used for international and national levels but can also be used for local tournaments.

4 Modules of the Proposed System

This application incorporates the modules for the easy access and usage to the user. The modules of the project are as follows:

- Data collection
- Data preprocessing
- Splitting the dataset into training set and test set
- Build the model using multivariable linear regression algorithm
- Training the multiple linear regression model
- Acquisition of prediction graph

Figure 1 depicts the block diagram of the proposed system.

The modules of the proposed system are described as follows:

A. *Data Collection*

Data collection is defined as the procedure of collecting, measuring and analyzing accurate insights for research using standard validated techniques. The dataset is collected from Kaggle consisting of 76014 samples with date, venue, batting team, bowling team, bowler, points, wickets, overs, points, win: lose ratio, last three matches, last match and net run rate.

B. *Data Preprocessing*

It is a machine learning technique that involves transforming raw data into an understandable format. The unwanted columns will be removed and clean rows for missing values.

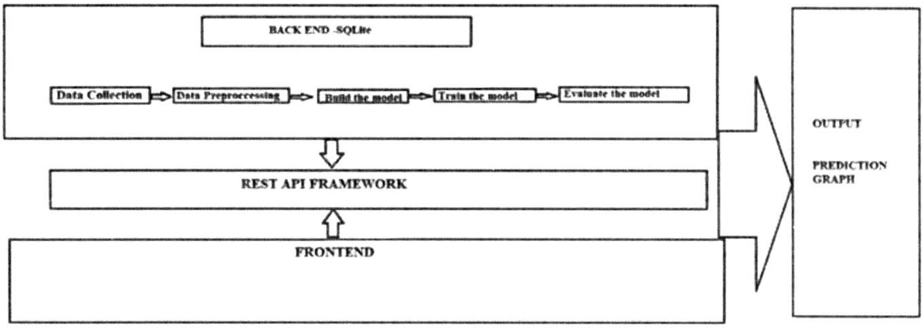

Fig. 1 Block diagram of proposed system

C. *Splitting the dataset into training set and test set*
The next important task is to split our dataset into training set and test set. In order to train our model with one portion of the data called the "training set" and test the prediction results on another set of data called the "test set". The "train_test_split" function is used to split the data in the ratio 80:20. The model is trained on 80% of data and validated against the other 20% of the data.

D. *Build the model*
The prediction of IPL match is implemented using multivariable linear regression. Regression is a statistical technique used in machine learning. In this technique, the dependent variable is continuous, the independent variables can be discrete or continuous, and nature of regression line is linear.

E. *Training the model*
Import the "LinearRegression" class will be applied to training set. This function "regressor.fit" used to fit the training dataset to this LinearRegression class. Training the model, the multiple linear regression algorithm will attempt to "fit" the model to the given data and will be accurate in predicting the win or lose of the match.

F. *Acquisition of Prediction Graph*
Data visualization helps to find out the winning probability of the teams over the years, with the help of a bar plot. Canvas JS is used for plotting graph of data point. Amchart is used for creating vertical bar graph for prediction. Thus the predication graph is shown as output which team which has won or lose.

5 Project Description and Technologies Used

Our application will store all the information you entered in our data set and will show the focuses table of groups dependent on matches added. The application is worked with Django (Python) for backend and PostgreSQL as a data set and for the front end, we have utilized HTML, CSS, and JAVASCRIPT. The application consists of five principles:

A. *Add Player*
In this part, players can be added to a group.

B. *Add Team*
In this segment, group can be added. The group name and group tone can be entered to make a group. The group will be added to the data set once the submit button is clicked.

C. *Matches*
In this segment, the client can choose two groups, and pick who won the throw and for what they selected. Presently the client can foresee the scores and overs

Fig. 2 Login page—IPL

taken by both the groups. Result can be proclaimed and can be submitted. This outcome will show up in the score table including the net run pace of a group.

D. *Create new Season*
In this page, new T20 season can be made and afterward groups furthermore, players can be added into the individual seasons. The client can make different seasons and can see any season's connected information

E. *Points Table*
This is the part where the client can see all the group's exhibitions. This incorporates the logo of the group, matches played, focuses acquired, number of wins sections misfortunes and the net run pace of the group. The victor indicator is utilized in this part and this will anticipate the champ of the competition dependent on the matches set by all groups.

6 Results and Discussion

The output screenshots of the project are depicted as follows (Figs. 2, 3, 4, 5, 6, 7, 8 and 9):

7 Performance Evaluation

The performance metrics used in this proposed work are Mean Square Error (MSE) and Root Mean Square Error (RMSE) and R2 Score. They are described as follows:

Fig. 3 IPL—players list

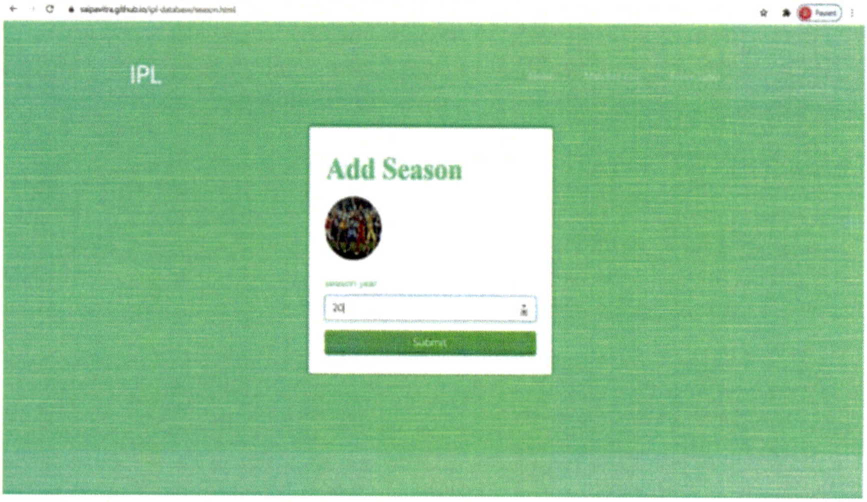

Fig. 4 Add season

A. *Mean Square Error*
It simply refers to the mean of the squared difference between the predicted parameter and the observed parameter.

B. *Root Mean Square Error*
It is the standard deviation of the residuals (prediction errors).

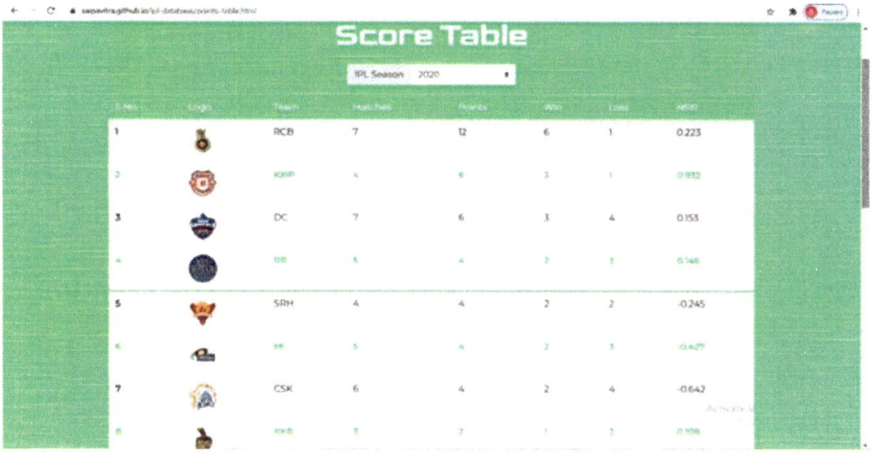

Fig. 5 Score table

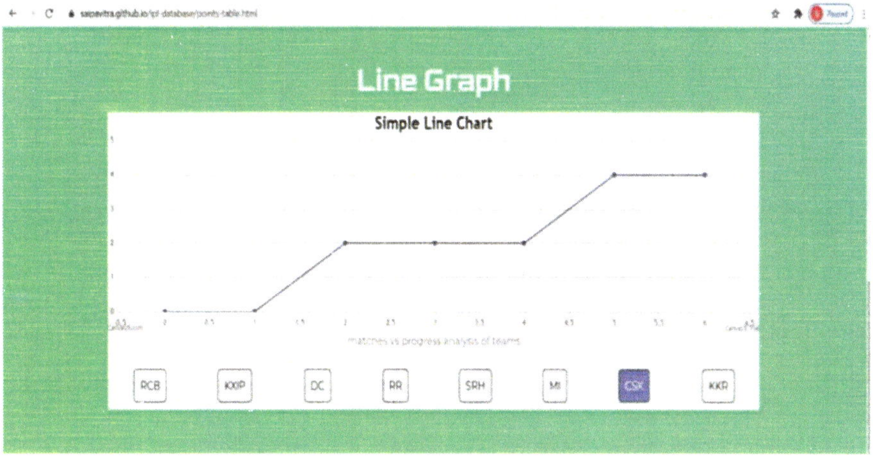

Fig. 6 IPL winner prediction graph - Line chart

C. *R2 Score*

It is the coefficient of determination which is known as regression score function.

Table 1 depicts the performance evaluation of linear regression and multivariable linear regression based on evaluation metrics like Mean Square Error (MSE), Root Mean Square Error (RMSE) and R2 Score.

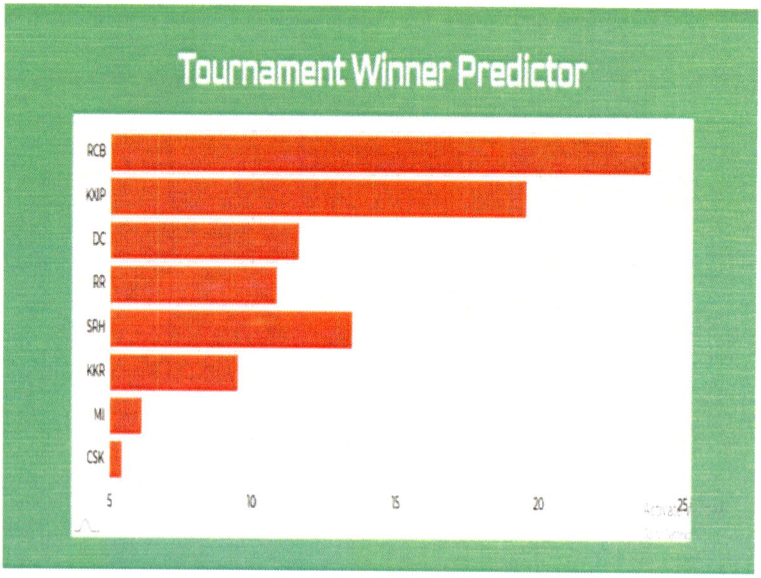

Fig. 7 IPL winner prediction graph

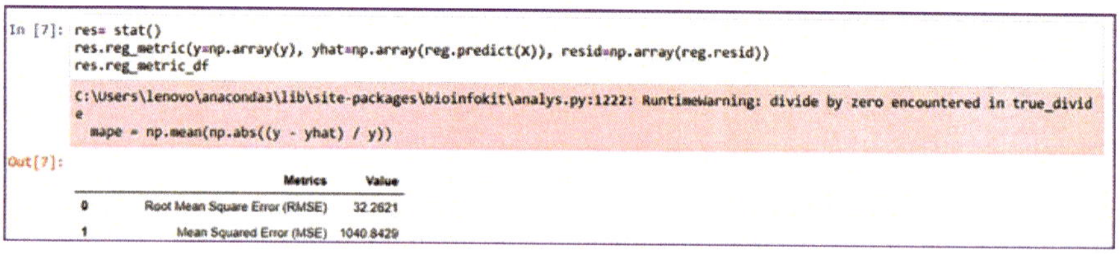

Fig. 8 Linear regression

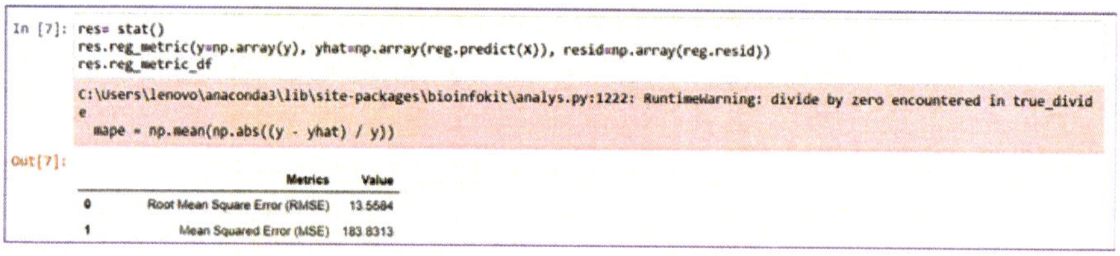

Fig. 9 Multivariable linear regression

Table 1 Performance evaluation

Performance metrics	Linear regression	Multi variable linear regression
MSE	1040.84	183.83
RMSE	32.26	13.55
R2 score	0.56	0.92

8 Conclusion and Future Scope

This framework is more easy to use when contrasted with the current framework. In this framework, client has been given the advantages to play out every one of the exercises of T20 cricket. Also, the client can deal with their own T20 information base and can refresh as they wish, and the project is totally free with no expense inside the data set utilization limit. In this work the prediction of either win or lose of the IPL cricket match is efficiently done with the help of Multi Variable Linear Regression algorithm. Multi Variable Linear Regression algorithm produces better results compared to Linear Regression based on performance metrics like Mean Square Error (MSE), Root Mean Square Error (RMSE) and R2 Score. Future work may be extended by implementing more Machine Learning techniques like Bagging, Boosting, etc., in addition to the techniques used in the project to obtain more accuracy.

References

1. I.P. Wickramasinghe, Predicting the performance of batsmen in test cricket. J. Human Sport Excercise **9**(4), 744–751 (2014)
2. Vistro DM, Rasheed F, David LG, The cricket winner prediction with application of machine learning and data analytics. IJSTR **8**(09) (2019)
3. D. Jyothsna, K. Srikanth, Analyzing and predicting outcome of IPL cricket data. IJIRSET **8**(4) (2019)
4. R. Lokhande, A. Professor, Live cricket score and winning prediction. Int. J. Trend Res. Dev. **5**(1), 2394–9333 (2018)

HEIST DETECTOR: A Secured IOT-Based Real-Time System

R. Abinaya, S. Tharani, A. S. Arunachalam, and K. Suthendran

1 Introduction

In this day and age of availability and brilliant gadgets, there is a critical need to alter our current everyday articles and make them savvy; additionally, it is not the time where we can indiscriminately believe the old and traditional safety efforts, and explicitly talking is our entryway locks. Home security has become a solemn issue in the society. Of course, we can find plethora of technologies available for security but the real problem is it cannot be implemented everywhere this drawback can be of many reasons such as high cost, less efficiency, less safety, high current consumption, and many more. Often people do not know what is happening in their current location so this could be perfect time to implement this security monitoring system that would be helpful for peoples who are seeking for a security within their budgets and the system that should be easily understandable for their knowledge. Also, theft rates were really increased due to the global pandemic (COVID-19) as most of the people were lost their jobs this could be the proper time for implementing our system. It allows the admin to monitor and control their property even they are far apart from the place where they fix the system. It also provides secure entry into home and the system can be installed at convenient locations somewhere in house. This system can work with or without Internet connection. This will turn up silent mode into normal mode when it gets the notification from particular number.

R. Abinaya · S. Tharani · A. S. Arunachalam · K. Suthendran (✉)
Department of Information Technology, Kalasalingam Academy of Research and Education, Srivilliputhur, Tamil Nadu, India
e-mail: k.suthendran@klu.ac.in

J. D. Peter et al. (eds.), *Disruptive Technologies for Big Data and Cloud Applications*, Lecture Notes in Electrical Engineering 905,
https://doi.org/10.1007/978-981-19-2177-3_45

2 Literature Survey

In [1], author proposed an automatic door locking system based on Android mobile application, Bluetooth, and microcontroller. Here, the door is accessed by entering a valid password from the mobile phone application. It enables the users to create a password with digital numbers or alphanumeric. When the password matches, then the door is opened by a motor that rotates the door lock's handle. It also has some additional features such as adding new users and reset the old passwords, among others. However, this requires 24×7 power supply to operate and the password can be easily stolen using social engineering. Moreover, if some theft happens, then it is hard to trace back the wrongdoer. These flaws have made the performance of this application questionable.

The authors in [2] developed a password-based automatic door lock system. This system is based on the idea of a pre-determined password. It raises the security level to deter an attacker from gaining unauthorized access. If a user forgets their passwords, they are given certain privileges to change or reset their passwords. When wrong password attempt takes place, then the security alarm starts. When some tries to open the door using wrong password, then the security alarm will notify the homeowner about such wrong attempt. If no one is at home when such attempt is done, information about such attempt may not reach the homeowner. This system also requires continuous power supply for proper functioning. It is hard to trace the wrongdoer in the absence of homeowner. The door is automated with help of mobile application [3]. The valid login credentials are passed from the mobile application to hardware part through Bluetooth connection, which is in turn makes the hardware part to unlock the door. In [4], authors have proposed a security system based on Bluetooth. This model is based on controlling the door of home and sound alarm in case of unwanted opening. Bluetooth module has been used to give commands to control the door lock by using an application. In this work, the Arduino board plays an important role. However, the alarm itself is not sufficient and it may be failure system when the homeowner is not available at the incident time. This circuit starts once the door sensors are broken. In this way, the circuit saves the power. However, trace back is quite difficult using this model.

In [5], the camera module interface is utilized. This system captures the live face images of the users and it compares it with the database. This database is used to store the captured photos of the authorized users. The access to the person is granted by comparing the face detected with the images in the database. The camera module is used to capture the face images of the person for whom the access is to be validated. The camera in the security alert system in [6] captures the face of the users, and it compares that with the database. If the photo matches, then the person gets the access to the secret locker, and otherwise, notification goes the administrator. Here, uninterrupted power supply is required to operate this device. If the power goes off, when the lock is open, there may be the chance that wrongdoer may threaten the actual user and access to the secret room. Here, the performance of security system is purely based on the backup power.

Android-based smart door locking system was proposed for secret places such as banks with normal mode and multiuser mode operating feature. The user can access the system through mobile application. Multiusers can operate this system but no one gets notification about others who was lastly logged in to the system. Moreover, there is no video surveillance. Therefore again, the security of this system is questionable [7]. By entering valid password, the user can unlock the car door otherwise notification reaches car owner through mobile phone. If any theft attempt happens, that is also sensed by the IR sensor and appropriate notification alerts car owner. The limit switch turns on automatically when the car is towed and car owner gets notified with current location details of the car though GPS module of the security system [8]. When compared to the existing approach, this circuit contains more complexity and thus leads to costly. The authors in [9] proposed a home automation system controlled by smart phone application. Here, the user has provisions to turn on the security system when he/she is far from the home. The home appliances can be controlled being in remote. When some intrusion happens, then appropriate alert reaches the owner. The Raspberry Pi board is configured, and it will enable the relay circuit. The video recordings provide the user with accurate monitoring, and the user checks the status. The authors in [10] proposed a security system with two Android smart phones, ATmega8 microcontroller, a Bluetooth module (HC-05), a PIR sensor (HC-SR501), a security alarm, a cloud-based server. Being in remote, the user can monitor the home appliances.

In [11], the ATM surveillance system was introduced based on deep learning. ATM machines can be secured from theft and robberies using this framework. If an individual approaches the ATM counter with a weapon, such as a knife, the YoloV3 object detection algorithm will detect it. For the connections of different hardware parts, they use a microcontroller. The camera device is used to capture videos of objects that enter the ATM counter when a person enters the counter. Authors in [12] have proposed an IOT security system. When user places an item or object in the safe box, the camera inside the safe box takes a picture, and it is stored in the database as a reference image. When there is a change in the safe box, such as a change in light intensity, the light intensity sensor within the safe box can cause the camera to capture a new image using Raspberry Pi (configured using Python). After that, the new image will be compared with the reference image, and if there is a difference, the user will be notified via a web portal warning and SMS via the Internet. A secured door locking system was proposed in [13]. The sensor detects this and sends a signal to the Raspberry Pi controller when someone enters. The controller then validates the signal, transfers the camera to the area where movement was observed, and transmits the picture over the Internet for the home owner to review [14].

The above survey reports that most of the security system uses sensors on it to sense the environment changes and appropriate actions are taken based on the sensed information. In addition, most of the systems require uninterrupted electricity, and also most of them are out of reach for most people. Until now, no system has sent an image-based notification. Password protected system is prone to social engineering attacks. Many proposed security system fails to pass the notification during the absence of homeowner. Therefore, heist detector, a secured IOT-based real-time

Fig. 1 GSM module

system, is developed to overcome all the issues reported. The experimental results are very effective in preventing theft incidents taking place.

3 Hardware Components

3.1 GSM Module

GSM module collects the data and sends it to the respective mobile communication. The module can also be used for developing IOT and embedded system. It enables communication between microcontroller and GSM module. There are various types in GSM such macro, micro, and umbrella cells. There are some cell sizes in a GSM network macro, micro. The coverage varies with implementation environment. The power supply of GSM module 900 A is between 3.4 and 4.5 V. In this work, GSM module helps us to send the captured image to our mobile phone (Fig. 1).

In order to avail the short message service, the GSM module is to be tuned to text mode first. This is accomplished through AT Command "AT + CMGF = 1." We send this order by composing this in to software serial port. To accomplish this, the serial (), Println () work are utilized.

3.2 Raspberry Pi Module

Raspberry Pi currently available with two types of camera board: an 8MP device and a 12MP high quality (HQ) camera. It is capable of capturing 3280 × 2464 pixel static image. The high-definition 5MP camera delivers outstanding photos. The lightweight camera board allows for it is useful in more practical roles, such as a hidden camera (Fig. 2).

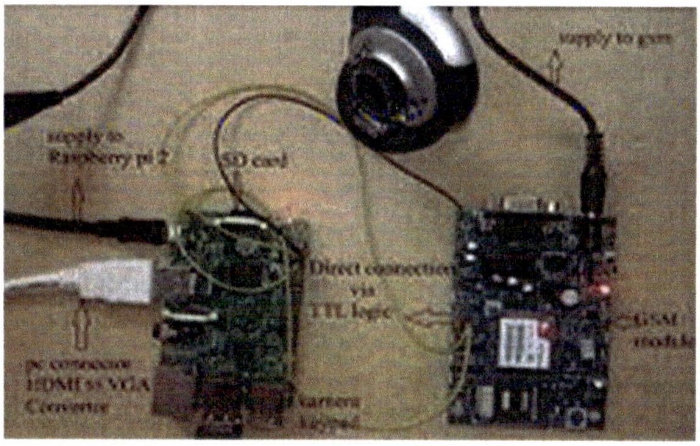

Fig. 2 Raspberry Pi module

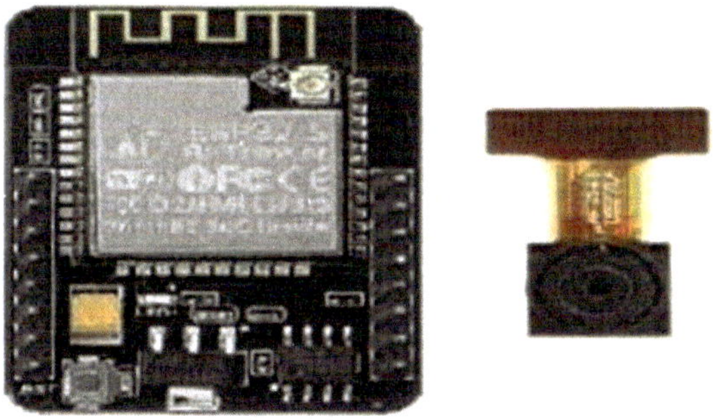

Fig. 3 Esp8266 camera module

3.3 *ESP8266 Wi-Fi Camera Module*

This module captures the images when door sensors are broken and it transfers the captured image to the homeowner via telegram. At the same time, it can also be used for video surveillance whenever it is required by the user. It is available within few hundred ranges in the market (Fig. 3).

3.4 *Magnetic Door Switch*

Door switch set is a small switch assembly specially designed to notify you when the doors, drawers, or any other things get opened. These types of switches are majorly

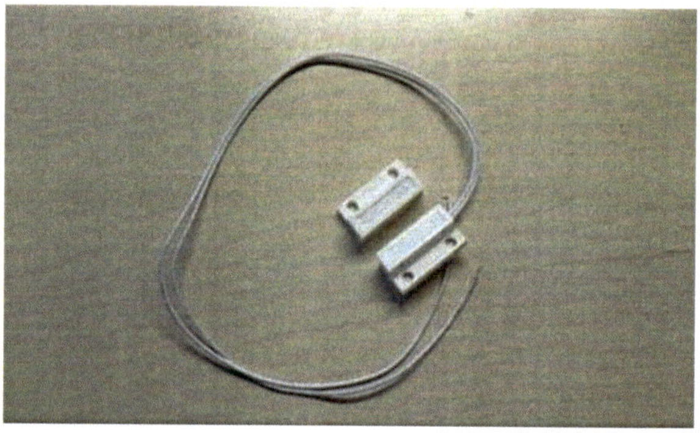

Fig. 4 Magnetic door switch

used in home security systems. When the switches get separated from each other, the contact is released and it triggers an event in the system. When a magnetic field is placed as opposite to the each of the electrical connectors, it joins them together and closes the circuit. It is very less in cost and also very easy in implementation so that it can easily accommodated by all levels of people (Fig. 4).

3.5 Arduino IDE

The Arduino Integrating Development Environment is an open-source programed which is used for programming and compiling the code into the Arduino module, and it is based on our needs. It is a formal Arduino programed that makes extremely simple, allowing someone with no precious knowledge of learning or coding to learn the techniques and make the projects easier. It runs on the Java platform, which includes built-in functions and commands for compiling, debugging and editing code in the environment. It can be easily accommodated by various operating systems such as Linux, Mac, and windows. There are various Arduino Micro, Arduino Uno, Arduino Mega, Arduino Leonardo, and many more. One of the boards of each design is a microcontroller that already has been programmed and accepts data in the form of code. The main code is defined as a sketch, which is generated on the IDE platform and eventually generates a Hex file, which is then transferred and uploaded to a board's controller. The IDE environment is divided into two parts: Editor and compiler. The editor is used to compile and upload the code to the Arduino module. Both C and C++ are supported in this environment.

Instructions to Download Arduino IDE:

The software can be downloaded from the Arduino website. As previously mentioned, the product is available for popular working frameworks such as Linux, Windows, and Max, so it is very important to choose and install the correct programming the code into the given Arduino module. This environment supports both C and C++ languages.

3.6 Power Supply

The Arduino Uno is controlled by the US association. External force comes from Li-ion (2200 mA) battery. The connector can be associated to the board's force jack. Leads from a battery with power connector. The board can work on an outside supply of 6 to 20 V. Whenever it is provided with under 7 V, the 5 V pin may supply fewer than five volts and the board might be unsteady. The voltage controller may overheat and harm the board. The max reach is 7 to 12 V.

4 Existing System

This system uses a motion detection sensor to detect motion and the camera module records the video for the particular time when motion is detected. Then via Internet connection, the e-mail notification is send to the user; if there is any Internet problem, the GSM module automatically sends normal message to the user. The system is explained in a diagrammatic manner as shown in Fig. 5 [6].

Demerits of Existing system:

- Due to usage of PIR sensor, the power supply is required for 24/7 of the time for the system.
- It sends only SMS notification,
- It does not send messages to multiple mobile numbers.
- It does not send captured image to the user

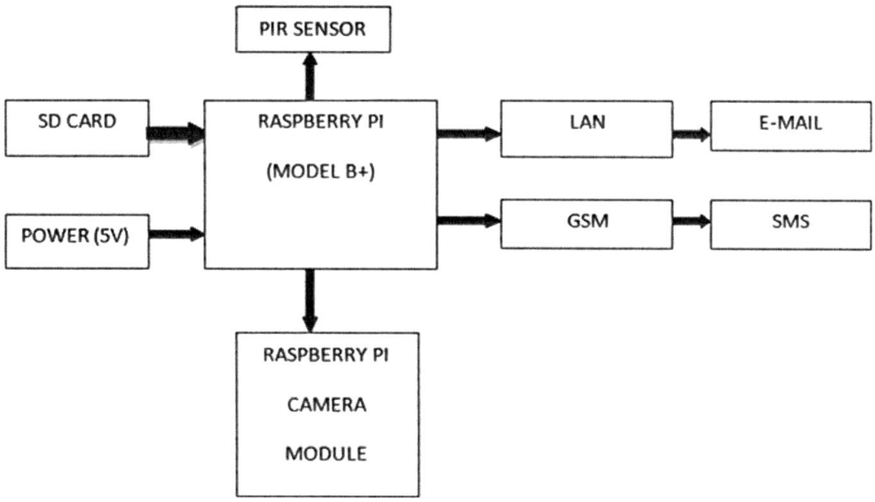

Fig. 5 Existing system

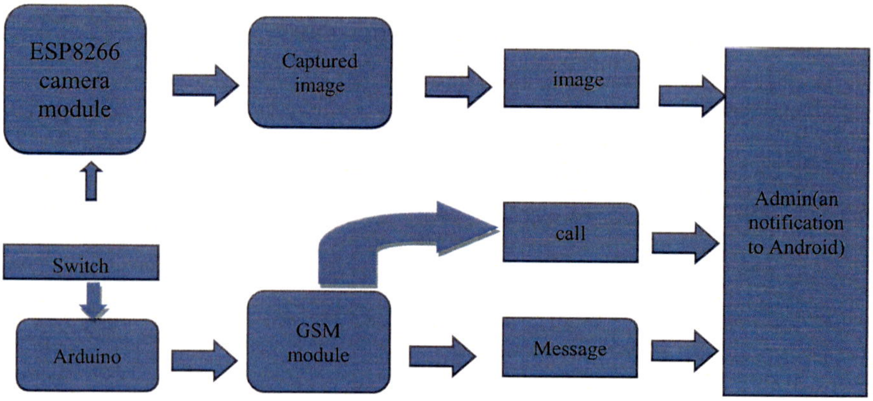

Fig. 6 Proposed smart flask

- It does not make phone alert when it is on the do not disturb mode.

5 Proposed System

When unknown someone opens the door, then the door switch gets detached, the structure gets activated and the connection is passed to the power then the camera gets activated. It starts capturing the image/video. The image is stored on a memory card. GSM module sends the notification to the admin as messages and phone calls to the various mobile numbers. ESP8266 Wi-Fi module is used to capture the image and send the image to the application, and we can also receive the notification by means of ringtone even when phone is in do not disturb mode. The overall organization of the proposed system is shown in Fig. 6.

6 Methodology

During implementation process, the camera is fixed on the secret room doors/cupboard or anywhere; it can also be fixed on the car doors. Camera and GSM start working when the door switch gets detached. Once it turns on, it starts to capture the image having a memory size of 520 KB SRAM and it can be send to the mobile phones. The GSM module which is connected to the switch is programmed to send notifications to the various users via message and phone calls. Notification and image will be sent to the people at the same time. When the door opens, door sensor is activated, and the connection is passed to the power, then the camera gets activated and it captures the image. GSM module gets the information whether door is open or closed then sends notification to admin. If any power failure occurs, 14 V battery which is connected to our system is help us to run our process without any destruction.

The proposed system alerts the homeowner or admin through notifications; he can get call alert even if his/her phone is in silent mode. The developed system has simple connections. Therefore, it is easily implemented. Any user can operate this system easily. The following are the easy steps that are to be followed for smooth functioning of the system.

- For making user phone to ring even the phone is in do not disturb mode, add user GSM mobile as your starred contact in mobile phone, then go to the settings and then call settings next to and there user can find various options select allow incoming calls from starred contacts only.
- Then for video surveillances, just copy and paste the given hotspot IP address
- For receiving images, we would create a chat bot after creating a chat bot we will be receiving an authentication key copy and paste that key in the proper place of the coding to receive the captured image in your telegram account by the simple comment/photo.

7 Results

The development of proposed system is shown in Fig. 7. The magnetic switch as shown in Fig. 4 is actually placed on the secret room door. The remaining system can be placed on the hidden location. So, when someone opens this door, the connection between the magnetic switch is broken that triggers the circuit to turn on.

Once the circuit is turned on, then the GSM module starts functioning. All mobile numbers which are pre-registered in the GSM module are alerted with the theft

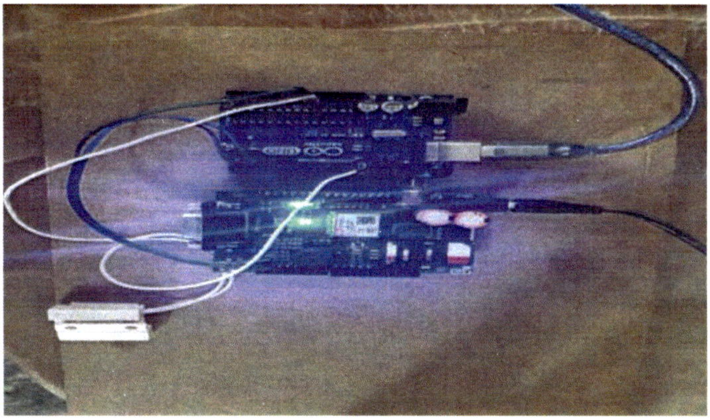

Fig. 7 Implementation of proposed system

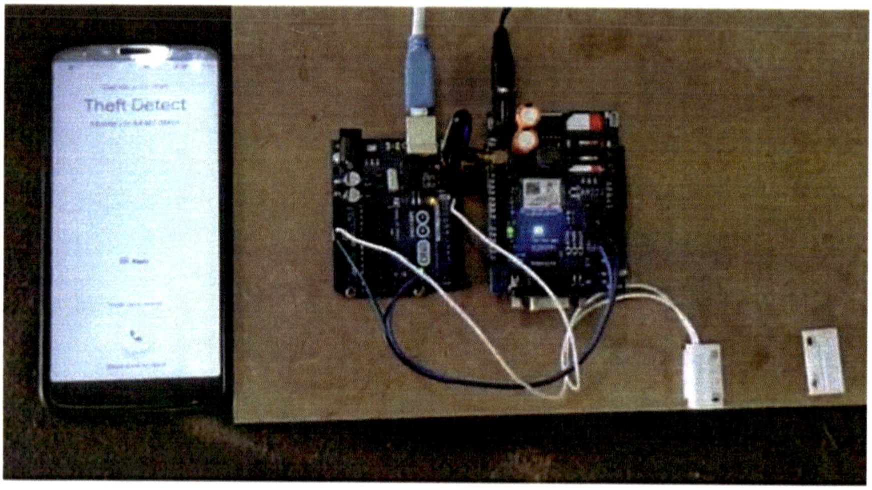

Fig. 8 Functioning of proposed system

detection phone call as shown in Fig. 8. In some existing system, this kind of alert was sent to one mobile system [9].

But, there may be the chance that particular mobile user may miss this alert because of his/her busy schedule and it leads to huge loss. When the alert reaches to multiple people, the probability of decision taking chances is more and some preventing actions may be carried out.

In addition to the call alert, short message service (SMS) alert will also reach the users mobile phone as shown in Fig. 9.

The above services are supported by the proposed system through GSM module even when there is no Internet connection. With Internet facility, the captured photos may be sent to the registered mobile users through telegram application (Fig. 10).

By making use of this photo, the homeowner can see who is accessing the secret door. If it is done by unknown person, then this evidence may be very much useful for the police arrest the thief.

Discussion:

The existing system has video camera facility so it can record everything that takes place in front of it and send the recording to server. Thus, the system works 24 h a day; it consumes larger power. When some movement is detected, then it starts recording. So the amount of recorded video data is very large and requires a server with a lot of memory to store it. For these reasons, the cost of this system is a bit high. The activities that can take place in the secret room can be seen through streaming video being in remote. However this service can be used depending on the available Internet data facility. Sometimes, it may take more time to load. If there is any movement, a short message about it will reach the registered mobile. This existing system will start working even if there is any small interruption in front of the camera which

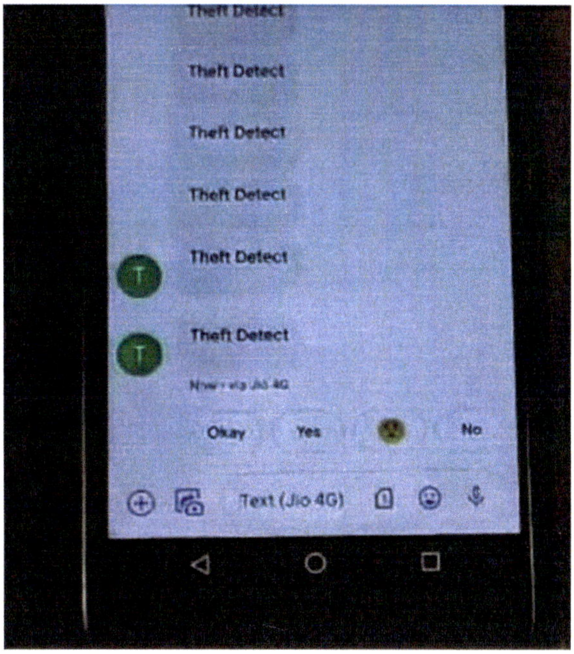

Fig. 9 SMS alert by proposed system

can sometimes cause unnecessary tension to the property owners. In comparison, the proposed heist detector system is very small in size. In addition, when someone opens the door of the secret room, the proposed system starts functioning when the magnet switch attached to it is separated. From that moment onwards, the system will start taking photos and the same will reach the registered mobile phones telegram application via GSM module. It will also share SMS regarding the untoward event.

Advantages

Its size is very small compared to all other existing systems in practice so it can be easily hidden from view. Other systems require 24 × 7 of power to operate. But this system will only start working when somebody enters in to the secret room. Therefore, it consumes less power. With the help of the Li-ion (2200 mA) battery attached to the system, it will function well even in the absence of power (critical time).

Limitations

The proposed heist detector cannot store any data as it only has buffer memory. It is only a few kilo bits in size. Photos taken by the camera mounted on the Arduino UNO circuit are sent to registered mobile phones via GSM module with the help of this buffer memory. The copies of the photos thus taken are nowhere to be found in this circuit. However, in addition to this, there are possibilities for additional memory to be enhanced. This makes the price of this system a bit higher.

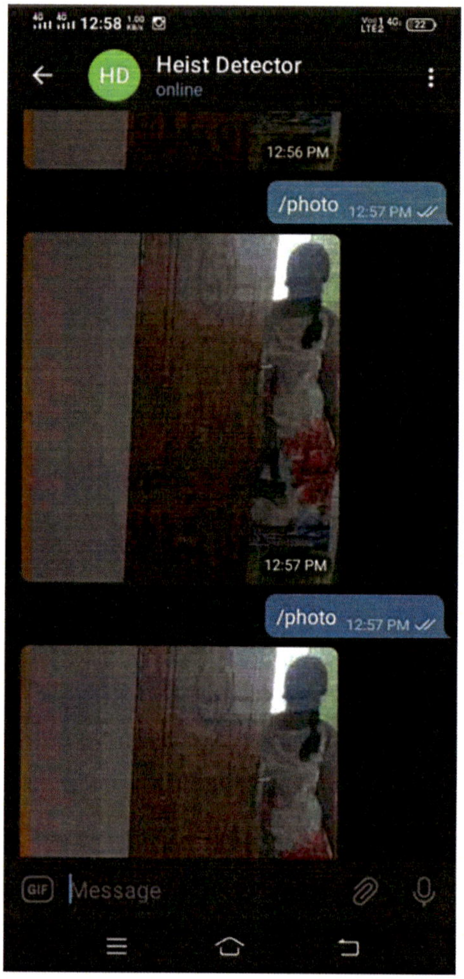

Fig. 10 Image alert through telegram application

8 Conclusion

The Internet of things is a term that links the virtual world of information technology to the physical world of things. RFID and sensor technologies, which are part of the Internet of things, make our lives easier and relaxed. The large number of different technologies needed to support IOT implementation and growth places a focus on interoperability, leading to widespread efforts to establish standards and technical specifications that enable seamless communication between IOT devices and components. The proposed heist detector system was designed and fitted in the secret room. During the test, the door was opened several times, and its notification was checked for accuracy on the mobile phones. The test results confirmed that the implemented heist detector works more accurately then other existing systems at right time with reduced cost and power consumption. In the future, the magnet switch used in this system will be replaced by a wireless switch and an attempt will be made to notify the registered mobile phone in multiple ways in case of any mishap happens in more than one location.

References

1. N. Majgaonkar, R. Hodekar, P. Bandagale, Automatic door locking system. Int. J. Eng. Dev. Res. **4**(1), 495–499 (2013)
2. A. Mishra, S. Sharma, S. Dubey, S.K. Dubey, Password based security lock system. Int. J. Adv. Technol. Eng. Sci. **2**(5), 100–103 (2011)
3. L. Kamelia, S.R. Alfin Noorhassan, M. Sanjaya, W.S., E. Mulyana, Door-automation system using Bluetooth-based android for mobile phone. ARPN J. Eng. Appl. Sci. **9**(10) (2014)
4. A. Mondal, S. Saha, S. Dey, S. Bhattacharya, S. Kundu, S. Nath, R. Islam, M. Banik, Bluetooth controlled security alarm system. Int. J. Res. Appl. Sci. Eng. Technol. (IJRASET) **7**(V), 1–8 (2015)
5. D.B.S. Fernando, A.B. Srivarsha, K. Pirathisha, K. Rubiya, Face recognition for home security. Int. J. Comput. Sci. Eng. (SSRG-IJCSE) **6**(10), 7–12 (2019)
6. A. Arun Rajaa, R. Naveedhab, G. Niranjanadevic, V. Roobinid, An internet of things (IoT) based security alert system using Raspberry Pi. Asia Pac. Int. J. Eng. Sci. **02**(01), 37–41 (2016)
7. A.V. Patil, Android based smart door locking system. Int. J. Eng. Res. Technol. (IJERT) **6**(13), 1–4 (2018)
8. D. Mukhopadhyay, M. Gupta, T. Attar, P. Chavan, V. Patel, An attempt to develop an IoT based vehicle security system, in *2018 IEEE International Symposium on Smart Electronic Systems (iSES) (Formerly iNiS)*, pp. 195–198 (2018)
9. R. Senthilkumar, Vigneshram, Sudharsanbabu, Priyadharsan, Android based home automation and vision surveillance using Raspberry Pi. Int. J. Comput. Sci. Eng. (IJCSE) **4**(2), 29–38 (2015)
10. M.H.A. Sohag, M.A. Ahamed, Smart home security system based on microcontroller using internet and android smartphone, in *International Conference on Materials, Electronics and Information Engineering, ICMEIE-2015* (2015)
11. F.A. Sabu, IoT based ATM surveillance system using deep learning. Int. J. Future Gener. Commun. Netw. 13(3), 1869–1876 (2020)
12. P. Joshi, A. Patel, P. Jaiswal, S. Bajpai, S. Rasal, Smart secure safe box using IoT. Int. J. Innov. Res. Technol. **5**(11), 733–735 (2019)
13. Ohsung, Gyeongsan, Gamasil, A digital door lock system for the internet of things with improved security and usability. Int. J. Secur. Appl. **9**(8), 45–54 (2015)
14. U. Sirisha, D. PoojaSri, N. Gayathri, K. Heshma, G. Raja Sekhar, IoT based anti-theft detection and alerting system using Raspberry Pi. Int. Res. J. Eng. Technol. (IRJET) **07**(3), 1650–1655 (2020)

Blockchain-Based Decentralized E-Voting System Using Smart Contract

R. Priscilla, G. Jaspher Willsie Kathrine, M. Rubavathi, and T. Sruthi Krithika

1 Introduction

Blockchains attract attention as a brand-new era designed to lessen transaction charges through securely allowing direct transactions among a limitless variety of trusted and untrusted users. The main components in a blockchain are the miners and the distributed ledger concept.

Blockchain miners are a newly created block on a blockchain and must be properly mined before it can be integrated to the blockchain [1]. As a peer-to-peer computer system, blockchain mining is used to protect and secure Bitcoin transactions. The mines include miner blockchain operators who enter Bitcoin transaction data into the blockchain's global public Bitcoin ledger regarding past trades. In the registers, the blocks are saved by the miners of the blockchain and linked to form a chain. Providers review new transactions and record them on the blockchain. Block crashes every ten minutes. Miners compete to solve complicated problems based on a cryptographic hash algorithm. The solution found is called Proof of Work. This evidence shows that the miner spent considerable time and resources trying to solve the problem. The block is "dissolved," content transactions are guaranteed, and the Ethereum involved in the transaction can be used.

Miners receive a reward when a transaction takes place on a blockchain. Bitcoin mining is the process of adding transaction records to the Bitcoin public ledger via previous transactions or blockchains. This earlier transaction book is known as a blockchain because it is a series of blocks. Mining requires high computing power. Miners in the blockchain network make their living from mining.

R. Priscilla · M. Rubavathi · T. Sruthi Krithika
St. Joseph's Institute of Technology, Shollinganallur, Chennai, Tamil Nadu 600119, India

G. J. W. Kathrine (✉)
Karunya Institute of Technology and Sciences, Coimbatore, India
e-mail: kathrine@karunya.edu

J. D. Peter et al. (eds.), *Disruptive Technologies for Big Data and Cloud Applications*, Lecture Notes in Electrical Engineering 905,
https://doi.org/10.1007/978-981-19-2177-3_46

Distributed ledger technology (DLT) is a digital process for recording transactions, and all the information is recorded in lots of places simultaneously [2]. Unlike conventional databases, disbursed ledgers no longer have imperative information to keep or control functionality. In a disbursed ledger, every node strategizes and verifies everything, therefore producing a document for every object and developing a real settlement for every object. A disbursed ledger may be used to document each static information and dynamic information. Ledger that is a transaction document and the equal information had been round for heaps of years on paper. They have been made into computer systems through the arrival of computer systems withinside the twentieth century, despite the fact that ledger computer systems frequently mimicked what become on paper. However, during history, the imperative authority has been required to confirm the authenticity of the offerings indexed withinside the ledger. For example, banks want to confirm monetary transactions. A blockchain, which incorporates transactions with blocks tied together, and disbursed to nodes in a network, might be the most famous sort of ledger era delivered.

Electronic vote casting (e-vote casting) is considered, where the citizens are located abroad or in remote areas. India has 31 international locations for casting e-vote. The net result depends primarily on vote casting software which in turn guarantees to lessen election prices and the result evaluation time.

A blockchain is a shared distribution record that is steady among nodes without a link. The blocks withinside the blockchain are linked to every different and saved in a dispensed ledger. Each block withinside the block chain has a timestamp, records, hash, preceding hash, nonce, and issue, non-public and public keys. A timestamp is used to signify the time to block. Hash changed into records the use of a stable hashing algorithm. The preceding hash identifies the block hash right now earlier than the present-day block. Nonce is a counter cost to make sure the distinctiveness of the block withinside the block chain and guard the block from replay attacks. Difficulty is the cost used to calculate zeros than the hash cost that is adjusted whenever duration for the specified mining process. Public key and personal key are used to create and ship virtual signatures in the course of transactions. A blockchain is a distributed system, in that, each one node ought to make sure of the transactions. Verification of transactions is known as a mining process. Adding a block to a blockchain is achieved through all of us withinside the network. Before including a block to a blockchain, it is essential to carry out Proof of Work (PoW). PoW is a mathematical puzzle that makes use of a noncomplicated variety and issue to confirm the complexity of the block mining process. The first block in a blockchain is known as genesis block. Data and hash cost collectively were used to create a virtual signature licensed through miners.

Cryptographic hash function (CHF) is a hash characteristic for use in cryptography. It is a mathematical set of rules that maps the information of the competition to a small length string ("hash value," "hash,") and is a one-manner characteristic, that is, a characteristic that is not possible to convert. The best manner to discover a hash-generating message is to try and forcefully look for inputs which can be probable to look if they are generating matches, or the use of a rainbow comparable matching

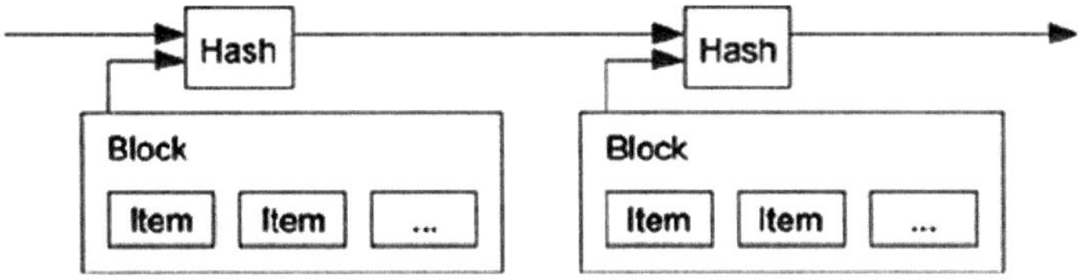

Fig. 1 Linking of a hash with previous hash

table. Cryptographic hash capabilities are a simple device for contemporary-day cryptography. Cryptographic hash capabilities upload protection capabilities to traditional hash capabilities, making it hard to discover message content material or records approximately recipients and senders.

Hashing is a way of writing that converts any sort of information into a completely unique textual content string. Any piece of information may be burned, irrespective of length or type. The hash is designed to paintings as a one-manner characteristic you may input information right into a hashing set of rules and get a completely unique thread; however, while you discover a new hash, you cannot specify the enter information that represents it. Hashing is an easy-to-do math task; however, the toughest job is to change the hash. The extensively used hashing capabilities are MD5, SHA1, and SHA-256 (Fig. 1).

2 Related Work

This phase describes the numerous researches associated with current Ethereum blockchain methods and clever contracts. Ohlin and Jens David [3], highlighted a primary hassle withinside the protection vote, while withinside the 2016 U.S. presidential election, EVM's could not be held and votes might be disrupted. Research has proven that those vintage balloting machines are not simplest liable to screw ups and injuries, however, they are regarded for being smooth to hack and disrupt.

Mahender Kumar, Satish Chand, and CP Katti, end-to-end confirmation (E2E) [4] makes the voter test that his or her ballot has been recorded as she or he supposed and for the general public to make sure that the gadget has nicely counted all recorded votes. On-line balloting structures primarily based totally at the E2E certification precept have many drawbacks; most significantly its safety. There are some E2E balloting applications mentioned during the last decade concerning the evaluation of the balloting gadget and the legalization of its protection requirements.

Rifa and Budi [5] have come to the belief that after we use hash values to report the balloting effects of every polling station, we make this recording gadget greater secure and using virtual signatures make it dependable. The use of the proposed series withinside the blockchain implementation system on this gadget considers that during an unneeded mining election gadget as withinside the Bitcoin gadget due to the fact and balloting numbers are clean and now no longer allowed to vote greater than once, the proposed series guarantees all nodes are legally related and may keep away from visitors collisions.

Somnath Panja, Samiran Bag, Feng Hao, and Bimal Roy [6] have proposed a first-of-its-type balloting regulation with a similarly elected vote gadget via the means of Borda. The protocol no longer requires a dependable putting or score authority to calculate the tally. Voters engage with the general public by way of means of making this rule in a manner that may be publicly verified. Our middle protocol includes rounds. In the primary round, citizens submit their public keys, and withinside the 2D round, they solidify their ballots randomly.

In Xuechao Yang, Xun Yi, Surya Nepal, Andrei Kelarev, and Fengling Han method [7], superior safety features are needed to introduce energetic on-line balloting (e-balloting) worldwide. Elections made on paper price a variety of assets and make a contribution to deforestation, mainly to weather change. Recent on-line balloting reviews in international locations consisting of the United States, India, and Brazil have proven that similarly studies are wanted to enhance protection within the imminent elections, to make sure the integrity of the citizens, and to reinforce their credibility. In this paper, we endorse a particular on-line balloting gadget, which addresses those challenges Shiyao gao, Dong Zheng, Rui Guo, Chunming Jing, and Chenchung Hu [8], as a critical way of creating democratic decisions, balloting has constantly been a reminder of public concern. Compared to traditional, e-balloting is broadly utilized in lots of decision-making conditions due to the fact it is far less difficult, less difficult to take part, and decrease costs. The balloting initiator publishes the balloting content material and the candidate list withinside the blockchain clever contract, detailing the connection among a candidate's call and public key address (name and address).

On the blockchain, the balloting system is a transaction from a voter to a candidate. First, the voter symptoms and symptoms the ballot, after which the signed ballot and its hash fee are broadcast to all nodes withinside the blockchain network. The node seems for hash fee collisions primarily based totally on computing electricity or different consensus methods. After a success verification, the hash fee is mixed with the preceding to shape the Merkle tree that is recorded withinside the new block, in order that every block incorporates all the ballot information. Furthermore, the regulator does now no longer intrude with the election, and our scheme guarantees voter anonymity through using the hoop signature set of rules, while citizens signal the ballots.

When beginning the counting phase, the modern updates the whole variety of ballots for this election, permitting verifiers inquisitive about the election end result to count the ballots through getting access to the records at the block. The transaction variety is the modern overall variety of ballots.

3 Problem Statement

The flow of the system is shown in Fig. 2: The blockchain electronic voting system based on intelligent contracts is written in Python, a free and open source computer language: Jar, Request, Number, Proxy Template, Jsonify, Redirect, and other voters libraries. Candidates are asked to fill out a registration form with their personal details. On the day of the election, declared by the election commissioner, voters cast their votes; after completing the voting process, all details are encrypted, and a hash value is generated. With the permission of the administrator, a different block is defined for each user. Then, the blockchain is formed, and the administrator counts the votes and explains the final reports and results at the end of the election process.

Contestant Module

The election administrators are responsible for registering the candidate phase. When creating an election, election administrators must create a deterministic list of eligible candidates. A government identity verification service component may be required to securely screen and approve qualified individuals. It is important to use such a service to meet the need for secure authentication, which is not guaranteed by default when using a blockchain architecture. A comparable identity wallet would be created for each qualified contender in our endeavors. Each participant receives a unique wallet. For any election, in which, they are entitled to vote.

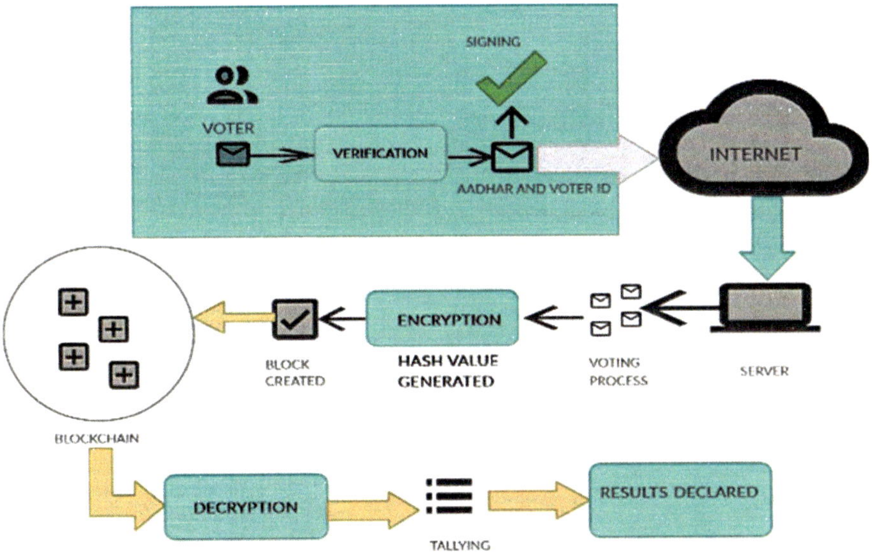

Fig. 2 Blockchain design for E-Voting system

Voter Module

In this phase, the generation of the keys begins in the power of the voters, provided that before this phase, all keys of the committee and the witnesses are saved. This is the phase in which the voters, during the elections start at the entrance to the terminal and end with the exit of the terminal where the elections are taking place. The committee provides the voters with a blank ballot paper available, which must be decrypted with the private voter key. A structured data sequence is generated automatically. At the end of the electoral process, each voter receives a hash that can be used to check the election results. The same hash value is not expected for different voters in each selection terminal.

Election Commission Module

A smart contract consists of three parts: (1) identifying the roles involved in the agreement (in our case, the election agreement), (2) the agreement process (i.e., the election process), and (3) the transactions used in the smart contract (i.e., voting transaction). (1) Elective seats: The roles of a smart contract include the parties that must participate in the agreement. The following roles are performed during the election process: I Election administrator—manages the life cycle of elections. This role can be performed by various reputable institutions and companies. Election administrators create the election, register voters, determine the duration of the election, and assign authorization nodes. Election administrators create ballot papers by defining a candidate list for each constituency. Smart contract district nodes get access to communication with their respective smart contracts after the smart contracts are added to the blockchain.

Smart Contract Creation

A smart contract is a stand-alone computer program that can run automatically when certain conditions are met. A smart contract is a specific account on the Ethereum blockchain that contains data and code with numerous programmable functions. Users can contract smart contracts with Ethereum accounts through binary application interfaces (ABI). A new transaction sent from an account can activate features that are defined in smart contracts. Entities can use this property to implement job functions, including data transfer, request handling, and access management.

A stand-alone script, usually written in Solidity and compiled into binary or JSON files before being deployed to a specific address on the blockchain, is shown in Fig. 3. Similarly, when you run some logic, you can run the smart contract implemented at a

Fig. 3 Smart contract operation

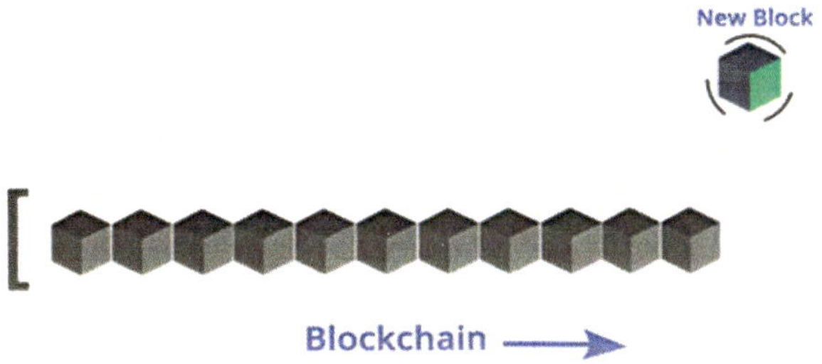

Fig. 4 Block creation

given address by sending the correct data as well as the Ethereum required to invoke the implemented and compiled robustness function. From a business perspective, this means that smart contract works can be monetized inherently. As a collection of code stored on the blockchain network that defines the terms to which all contracting parties agree.

Block Creation

The voter and participants must be approved by the administrator. This administrative review prevents duplicate expenditure attacks. After a successful check, the block builder module in Fig. 4 generates a block with the details. The hash of the block data is stored in each block. A chain of blocks. The key from the key generator module is used to hash the voting system.

The hash of the above block is, in a sense, the chain of block chains. The blocks in the blockchain build on each other because the hash of the previous block is included in the hash of the new block. There would be no connection or chronology between each block if that component does not exist. A hash can contain all transactions in a block. This is the root hash of the Merkle tree.

Hash Value Creation

A cryptographic hash is an algorithm that converts input like the one shown in Fig. 4 into fixed-size output. It seems like a mix of numbers and letters. Crypto hashes come in different forms. For example, the SHA-256 hashing algorithm is used. A hashing algorithm is a computation function that compresses input data to a fixed size and produces an output as a result, known as a hash or hash value. Hashes are used to identify, compare, and perform calculations on files and data strings. Usually, the program calculates a hash before comparing the values with the original files (Fig. 5).

Blockchain technology could and should be the solution for data protection and security. The blocks are connected to one another in thc form of a chain. The first link in the chain is called genesis. Each block consists of one block. Header, a transaction counter and a transaction. The public address also protects the privacy of voters and constants.

Many have advocated blockchain voting as blockchain technology has expanded and evolved over the past few years. Instead of holding elections in person, we can do so online using blockchains. The voters would control all the details of the blockchain instead of a third party, which makes the electoral process less cumbersome. Health

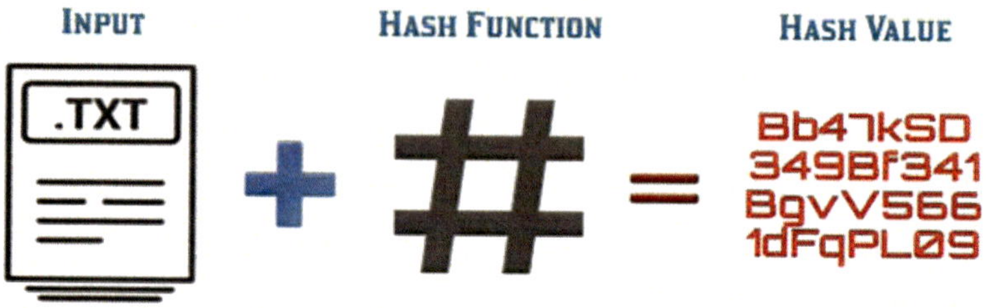

Fig. 5 Hash value creation

information cannot be deleted or tampered with because the blockchain ledgers are immutable. Transactions on the blockchain are accompanied by a timestamp so that those with access can access current information.

4 Result

The home page for both voter and contestants is shown in Fig. 6. Figure 7 shows the page that is for the registration of voters. After successful registration, the user should be approved by the admin. The contestants of an election will also have to register in a similar way like voters. The admin in this case is the Election Commissioner who has access to all the details and can approve or disapprove any contestant or

Fig. 6 Home page

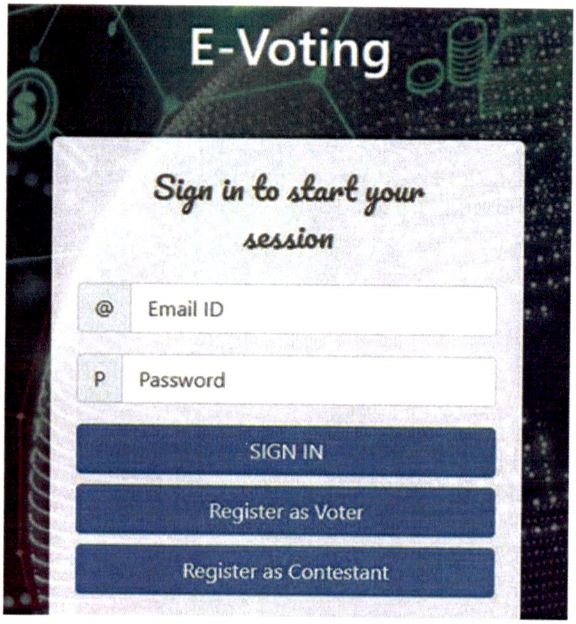

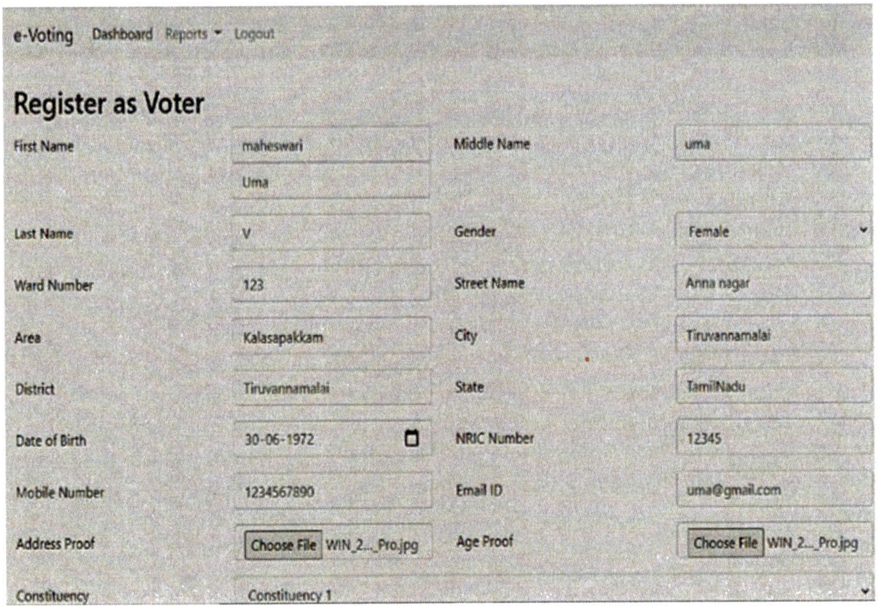

Fig. 7 Registration page for voters

voter. The admin also declares the date of election and can see all the reports such as gender, constituency, results party wise, age wise, and also download the voter list

Figure 8 shows the admin dashboard who creates the blockchain and has all the controls. The display of the election results will also be shown in the work in the form of charts and tables.

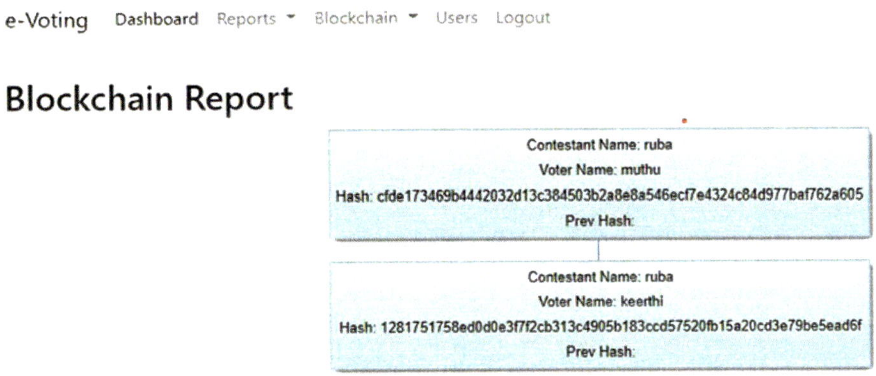

Fig. 8 Admin dashboard

5 Conclusion

As mentioned in the document, electronic voting is a possible solution to the lack of interest in voting among tech-savvy young people. The possible solution to make e-Voting is more open, transparent, and independently tested would be based on blockchain technology. This article examines the power of blockchain technology and its effectiveness in the electronic voting system. This article suggests and implements the voting system. This article describes the performance and related performance metrics as well as the challenges that the blockchain platform poses in implementing complex systems such as electronic voting.

This document identifies certain shortcomings and suggests the next two possible steps to improve the base platform (blockchain technology) to support electronic voting and other similar programs. Blockchain technology shows promise, but it may not realize its full potential as it is. Research into the core blockchain technology requires a concerted effort to improve its capabilities and support complex applications that can run within the blockchain network.

6 Future Enhancement

Based on our analysis of our system, we concluded that the small number of voters has some advantages in terms of security and efficiency, which makes it suitable for small elections. If the number is large, you will achieve greater security by sacrificing efficiency.

References

1. Z. Zheng, S. Xie, H. Dai, X. Chen, H. Wang, An overview of blockchain technology: architecture, consensus, and future trends, in *2017 IEEE 6th International Congress on Big Data* (2017), pp. 557–564. https://doi.org/10.1109/BigDataCongress.2017.85
2. A. Panwar, V. Bhatnagar, Distributed ledger technology (DLT): the beginning of a technological revolution for blockchain, in *2nd International Conference on Data, Engineering and Applications (IDEA)* (2020), pp. 1–5. https://doi.org/10.1109/IDEA49133.2020.9170699
3. J.D. Ohlin, Did Russian cyber interference in the 2016 election violate international law, 95 Texas law review 1579 (2017)
4. M. Kumar, S. Chand, C.P. Katti, A secure end-to-end verifiable internet-voting system using identity-based blind signature. IEEE Sys. J. **14**(2), 2032–2041 (2020). https://doi.org/10.1109/JSYST.2019.2940474
5. R. Hanifatunnisa, B. Rahardjo, Blockchain based e-voting recording system design, in *2017 11th International Conference on Telecommunication Systems Services and Applications (TSSA)* (2017), pp. 1–6. https://doi.org/10.1109/TSSA.2017.8272896
6. S. Panja, S. Bag, F. Hao, B. Roy, A smart contract system for decentralized borda count voting. IEEE Trans. Eng. Manage. **67**(4), 1323–1339 (2020). https://doi.org/10.1109/TEM.2020.2986371

7. X. Yang, X. Yi, S. Nepal, A. Kelarev, F. Han, A secure verifiable ranked choice online voting system based on homomorphic encryption. IEEE Access **6**, 20506–20519 (2018)
8. S. Gao, D. Zheng, R. Guo, C. Jing, C. Hu, An anti-quantum e-voting protocol in blockchain with audit function. IEEE Access **7**, 115304–115316 (2019). https://doi.org/10.1109/ACCESS.2019.2935895

Machine Learning-Based Diagnosis of Diseases Associated with Abnormal and Heavy Menstrual Bleeding: A Literature Review

P. Raji and P. Subha Hency Jose

1 Introduction

Federation of Gynaecology and Obstetrics (FIGO) has identified various diseases associated with abnormal Uterine bleeding and it is included in the formal classification system as PALM CO-EIN (polyp, adenomyosis, leiomyoma, malignancy and hyperplasia and coagulopathy, ovulatory dysfunction, endometrial dysfunction and iatrogenic, Not otherwise classified) [22].

There are various methods adopted so far for the measurement of menstrual blood loss. Various methods are a woman's self-estimate of her menstrual loss, Counting the number of sanitary products, Weighing sanitary products, Full blood count, Menses cup, Chemical analysis of the blood content of used sanitary products, Pictorial blood loss assessment charts, the menstrual pictogram and Counting the number of days of menstruation [19]. It is also determined that measuring menstrual blood loss from menstrual fluid loss can be easily done by using some regression analysis [8]. The method suggested is to measure the weight of the sanitary products before and after use [29].

There is a significant relation between total fluid lost during menstruation and the total blood lost ($r = 0.93$). Menstrual blood loss can be classified into Normal blood flow (<60 mL blood), moderately heavy (60–100 mL) and excessive or heavy menstrual bleeding (>100 mL) [8, 30]. According to Fraser, menstrual fluid consists of 48% of blood for those having a blood loss from moderate to heavy and 50% of blood for women having heavy blood loss.

P. Raji (✉) · P. Subha Hency Jose
Department of Biomedical Engineering, Karunya Institute of Technology and Sciences, Coimbatore, India
e-mail: raji.glitz@gmail.com

J. D. Peter et al. (eds.), *Disruptive Technologies for Big Data and Cloud Applications*, Lecture Notes in Electrical Engineering 905,
https://doi.org/10.1007/978-981-19-2177-3_47

A correspondence of 0.73 between evaluated and measured values was found when diaries were used by women to calculate the blood loss in a single day. Alkaline-haematin method is the standard for measurement of menstrual blood loss [8] and when bleeding diary data along with laboratory parameters and age are taken into account it gives a good correlation with the set standard [29]. A blood volume sensor integrated into the sanitary napkin can be used to measure the total fluid volume. It can be done by calculating the change in the output voltage due to the resistance change when fluid accumulates in the sanitary pad [21]. The regression model for predicting the menstrual blood volume from the total fluid volume was experimentally done. They tested different saline liquids and intravenous human blood in addition to the menstrual blood loss volume.

Since the sensor is placed and tested in a sanitary pad manufactured by a particular brand alone, its performance in other brands or different materials of sanitary pads will require further more testing. The sensor will have to be discarded after use since it is placed inside the pad. The hazard of placing sensor in a sensitive part of human body will require further investigations and testing. Another disadvantage of the proposed method is that since sensor is placed in sanitary pads, the normal physical activities of women like sitting for long hours, running, walking, etc. may result in different output which will require further testing.

2 Abnormal Uterine Bleeding and Diseases Associated with AUB

According to the formal classification system by FIGO, a patient may have more than one entity which may cause AUB [18]. In adolescents if there are endocrine disorders such as hypothyroidism, hyperprolactinemia and polycystic ovary syndrome then it may result in AUB. If the bleeding is so high and the haemoglobin level drops to <10 g/dL then hospitalization is required and medication should be started [5]. In the case of adolescents during the first post menarcheal year normal cycles are considered to be within 19 to 90 days. A cycle outside 19 to 90 days are said to be abnormal and can be an indication of anovulation, which is also indicated by abnormal uterine bleeding. Adolescents having high level of androgens and a menstrual abnormality are having a risk of 50% persistence [27]. Surgical intervention like hysterectomy is available for the treatment of HMB, but it may result in unwanted infertility or various other health issues [30].

It is recommended to start the treatment for abnormal uterine bleeding as early as possible, at least within one year. The symptoms of PCOS can also be analysed by themselves (the external symptoms like inflammatory acne, abnormal menstrual bleeding, etc.) [27]. The main cause for infertility is basically the delay in analysing the symptoms.

If the menstrual blood loss can be quantified then it will be critical turning point for disease prediction, treatment response and further analysis of the development of disease. Furthermore it will also reduce unwanted medical and surgical interventions like hysterectomy [34]. The analysis of the bleeding pattern all through the life of a woman can also lead to a healthy menstrual life.

There is a critical gap in the research on HMB as the methods to detect HMB is not standardised. In different countries across the world different methods are adopted for measuring HMB. Mostly, HMB is taken into consideration only when a woman complaint about heavy bleeding, which may not be accurate also. The lack of agreement between what a woman complaint about HMB and the actual blood loss is making the progress difficult in the research on HMB [12]. At present, there are many health and period tracking apps in IOS and Google. Certain apps like Flo provide insight into reproductive health and care, nutrition tips, workout tips, etc. These apps also provide a feature to measure the HMB by means of pictogram and counting the total pads along with the number of menstrual bleeding days. The data may not be accurate since every woman is analysing the bleeding according to her previous experience only. If menstrual bleeding disorders are diagnosed earlier then proper medical attention can be provided which will prevent various diseases in the uterus. With the help of a menstrual bleeding measurement device and an android app for the analysis of bleeding pattern it will be possible to lead to custom made treatment for every patient as explained in Fig. 1.

The menstrual blood loss can be quantified from the used sanitary pads. The used reusable sanitary pads can be weighed to measure the amount of blood loss and the data collected can be analysed after each period cycle. The weight measurement data collected from reusable pads can be used as an attribute along with several other factors like age, body mass index, interval between each periods, etc. If the blood flow rate is above the normal range then alert can be given to the concerned women. The data collected over a period of one year or more can also be used for diagnosing diseases related to abnormal menstrual bleeding.

3 Machine Learning for Diagnosis of PCOS and Uterine Fibroids

The PCOS is usually seen in women in their reproductive age. A single diagnostic test cannot be done for the diagnosis of PCOD. Various lifestyle interventions like obesity, ovarian dysfunction, diabetes, hypertension, dyslipidaemia, anorexia or bulimia, hypothyroidism, irregular menstrual bleeding, coarse hair growth, acne are some of the diagnostic tools for the detection of PCOS [33]. When a questionnaire was formulated for the diagnosis of PCOS by Pedersen and Corenblum menstrual irregularity, fertility concerns, and hirsutism were taken as the attributes. The accuracy of this diagnostic test was evaluated using the data obtained from 117 patients [24]. Logistic regression modelling was used in the same research paper to examine the prediction of PCOS or no PCOS and every patient having PCOS experienced menstrual irregularities. Therefore, it is clear that if menstrual irregularity is

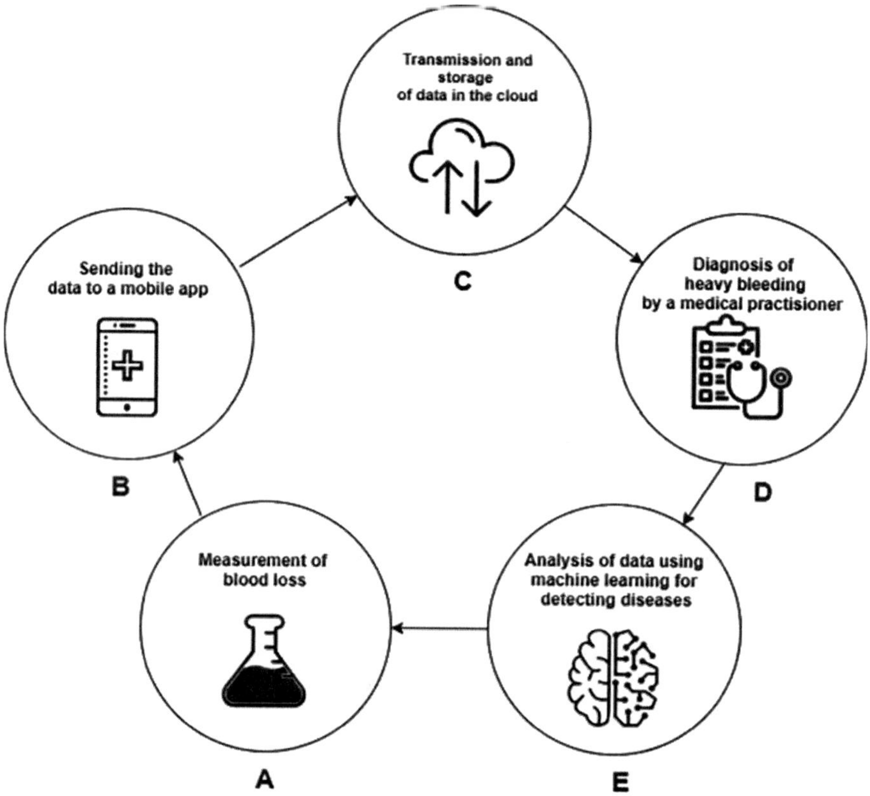

Fig. 1 Schematic representation of the proposed menstrual health assistance system

included as a feature along with several other attributes, physicians will be able to diagnose PCOD efficiently.

Uterine leiomyoma or uterine fibroids are the commonly caught benign myomatous tumours of the uterus, which can be seen in 20 to 40% of reproductive age women and 70 to 80% of perimenopausal women. The symptoms of fibroids are profuse abnormal uterine bleeding, urinary frequency and urgency, constipation, pelvic pressure, urinary retention, low back pain, bowel dysfunction and dyspareunia [36]. Ultrasonography and MRI scanning is used as the initial tools for diagnosis. Anderson had discussed about various multiple Imaging features along with diagnostic algorithms and machine learning to differentiate leiomyoma from leiomyosarcoma (LMS) [32]. Self-reported pictorial blood loss assessment charts did not bring a difference for those women with or without fibroids [26]. When a woman makes a self-assumption on heavy bleeding it seems mostly inaccurate and it cannot be considered for diagnosis purposes.

In an Internet-based survey conducted in 2009, self-reported identification of uterine fibroids were taken from 21,746 women. Heavy bleeding, prolonged bleeding,

bleeding between periods, chronic pelvic pain, frequent periods, painful sexual intercourse, irregular and predictable periods, pressure on the bladder, and pain occurring mid-cycle, after and during menstrual bleeding [36] were taken as the parameters and women with fibroids experienced a combination of any of these symptoms in different proportions.

Prediction of fibroid based on information collected from different women was done successfully by Girija et al. [10, 11]. They prepared a questionnaire with the help of doctors and radiologists and that data was processed using various machine learning algorithms. The Fibroid Disease Prediction System (FDPS) developed by them predicted fibroid with an accuracy of 98% by using various parameters like age of the patient, heavy bleeding, marital status and pelvic pain. A relationship between the menstrual bleeding pattern and the diagnosis of PCOD/fibroids is shown in Table 1.

In another study the relevance of fibroid size and location with the quantity of menstrual blood loss was conducted amongst 50 women with HMB diagnosed with

Table 1 Menstrual bleeding pattern and its relation to PCOD/Fibroids

Reference	Symptoms other than menstrual bleeding pattern	Menstrual bleeding pattern	Chances of PCOD or fibroid
[33]	Obesity, ovarian dysfunction, diabetes, hypertension, dyslipidaemia, anorexia or bulimia, hypothyroidism, coarse hair growth, acne	Irregular menstrual bleeding	31% of women having PCOS were having menstrual dysfunction
[24]	Fertility concerns, acne and hirsutism	Irregular menstrual bleeding	75% of women with PCOS were having irregularities in menstrual bleeding
[26]	Lower haemoglobin and high risk of anaemia	Self-reported pictorial blood loss assessment chart for detecting HMB	Self-reported pictorial blood loss assessment chart brings a *P* value 0.4 for diagnosing fibroid
[36]	Chronic pelvic pain, frequent periods, painful sexual intercourse, irregular and predictable periods, pressure on the bladder, and pain occurring mid-cycle, after and during menstrual bleeding	Heavy bleeding, prolonged bleeding, bleeding between periods	59.8% women diagnosed with Uterine fibroids were having heavy bleeding
[10, 11]	Lower back pain, pain during intercourse, pelvic pain, frequent urination and pressure	Heavy bleeding	98% accuracy in predicting fibroids

fibroid. It was found out that the amount of MBL is not correlated with fibroid size or location whereas the volume of uterus depends on the extent of fibroids [31].

It is important to measure the uterine volume for the treatment follow up of fibroids and MRI images are used for this. The fibroid and uterus can be segmented using image processing from the MRI. The research for fully automatic uterine segmentation with the help of MRI images [6, 15, 17] was successful and provides a good segmentation result when compared with visual evaluation.

Women with HMB has a low risk (0.11%) when compared with those having IMB (0.52%) for endometrial cancer [25]. The initial treatment for fibroid starts with an oral prescription of tranexamic acid, hormonal contraceptives and non-steroidal anti-inflammatory drugs to reduce the heavy bleeding. Surgical treatment includes uterine artery embolization, hysterectomy, myomectomy and magnetic resonance–guided focused ultrasound surgery [1]. Minimally invasive treatment methods like uterine artery embolization (UAE) and laparoscopic/hysteroscopic myomectomy are also some alternatives for open surgery [3]. Cervical cancer can also be diagnosed [7, 16] using machine learning techniques like KNN, SVM, MLP, DT and LR. Various machine learning algorithms used for the detection of fibroid, PCOS and cervical cancer are described in Table 2.

4 Application of Machine Learning in Various Medical Field

In the review article about different machine learning techniques used for diagnosing breast cancer, done by Houssein et al., various ML and DL algorithms were reviewed. There is a huge increase in the number of publications from 2011 to 2020 and mammogram is found as the major tool for detection of breast cancer [13]. Neural network and Naïve Bayes classification algorithm can be used for prediction of breast cancer. The values of Age (years), BMI (kg/m^2) and some routine blood analysis parameters like Glucose, HOMA, Insulin, Leptin, Resistin can be used as the attributes for diagnosis. The algorithms were tested and ANN was tested with 86.95 accuracy whereas Naïve Bayes with 83.54 [28]. Mammography is an imaging technique which is used for the diagnosis of breast cancer. Various algorithms like support vector machine, Naïve Bayes, K-nearest neighbours, decision trees, random forests, artificial neural networks and logistic regression can be used for the diagnosis of breast cancer by analysing the imaging features taken using mammography. ANN was tested with an accuracy of 98.57%, which clearly indicates that machine learning can be helpful for the physicians for assisting them in the prediction of diseases [14].

The performance of various algorithms of ML used for medical diagnosis can be evaluated using kappa error (KE), classification accuracy (ACC) and area under the receiver operating characteristics curve. For medical diagnosis it is found that there is no universal predictor for classification, but various algorithms need to be tried and find the one which gives best accuracy. Rotation forest (RF) classifier algorithm can

Table 2 Machine learning algorithms used for detection of fibroid, PCOS, cervical cancer

Reference	Machine learning algorithm	Diseases	Accuracy (%)	Advantages	Disadvantages
[24]	Multivariate logistic regression	PCOS	85	A true positive result can lead to careful examination of the patient for further diagnosis	The sample size used here was 100. A large sample size can give a more accurate diagnosis
[26]	Linear regression	Fibroid	41% in predicting YES	Bivariate analysis using pictorial blood loss assessment chart (PBLAC) and anaemia or haemoglobin can provide good detection results	The accuracy score is very low. It can be improved if a multi variate analysis is used
[10]	ANN	Fibroid	98	High accuracy. Multi-layer perceptron neural network provides a better accuracy in classification	Extreme care should be taken to avoid over fitting
[11]	Naïve Bayes	Fibroid	96	It can work with even less data number of data	The data was not split into training set and test set
[15]	CNN	Cervical cancer	78 to 86	Accurate in processing MRI images	Requires more computational speed and graphic cards maybe required
[36]	ResNet	Uterine fibroid embolization outcome	84.7	It can predict features of MRI images better than human eye	Computationally expensive. There is not much difference between CNN and ResNet accuracy

be used to improve the accuracy of various ML algorithms [23]. Amongst females, breast cancer is the most commonly diagnosed cancer and the leading cause of cancer death, whereas the cancer in cervix, uterus, ovary, vulva and vagina totally will give 5.5% death rate [2]. HMB or IMB is a common symptom for all the cancer related to the reproductive organs.

Various algorithms are being developed for computer aided diagnosis in medical field and evolutionary dendritic neuron model (EDNM) is one which can be implemented in hardware and gbest-guided artificial bee colony (GABC) algorithm can be used for optimization of the same. Machine learning can be employed in order to understand the complex physical and biological phenomena of human body. The emerging fields in medical area where machine learning is employed in a wide spectrum are personalized medical care, synthetic biology, molecular robotics, clinical trials, drug interaction, pre diagnosis of diseases, regenerative medicine, etc. [35].

Personalized medicine covers disease characteristic identification, drug development and therapeutic effect prediction. Biological data is being analysed using deep learning and reinforcement learning and combining the learning techniques in order to adapt the learning capability for analysing complex biological data is gaining popularity [20]. Disease risk prediction is another area where machine learning has a wide scope. Structured dataset of 706 patients were taken and several learning algorithms were used to predict the risk factor of chronic illness by Chen et al. CNN-based multimodal disease risk prediction (CNN-UDRP) algorithm was employed for medical data and the prediction accuracy was 94.8% [4].

Machine learning to detect ovarian diseases started from early 2000 onwards, types of tissues of ovary were analysed using SVM. Separation of Ovarian cancer tissues from other normal tissues in the body and normal ovarian tissues was discussed by Furey et al. Here 97,802 DNA clones were used for experiments which were either cancerous ovarian tissue, normal ovarian tissue or normal non-ovarian tissue. One normal ovarian tissue sample, N039 and one non-ovarian normal tissue HWBC3 were misclassified as cancerous in all instances [9].

After the analysis of various machine learning algorithms used in detecting diseases for internal organs, it can be summarized that machine learning can improvise the disease diagnosis due to abnormal and heavy menstrual bleeding also. The various improvements expected are performance improvement, reducing the training time of the data set, attribute selection criteria and improving the accuracy of various machine learning algorithms.

5 Conclusion

In this review several papers have been investigated which discuss about the quantification of menstrual blood loss. According to this review, so far the research done in the engineering field for analysing the menstrual blood loss is a few. As a result, it is quite difficult to do any diagnosis accurately, for those diseases in which menstrual blood loss is a symptom (like PCOS, fibroids, uterine cancer, cervical cancer, vaginal cancer or cancer in vulva). The diagnosis can be automated with the help of various machine learning algorithms if there is a dedicated device for the measurement of menstrual blood loss. The scope of engineering research in the area of women sanitation, menstrual hygiene and disease diagnosis of reproductive organs is very large and need more importance. This review concludes the need of a reusable menstrual bleeding measurement device integrated with an android app for the analysis of bleeding pattern of women thereby leading to custom made treatment for every patient.

References

1. D.E. Anderson, Diagnosis and management of uterine torsion. Vet. Tech. Llamas Alpacas. 265–269 (2013). https://doi.org/10.1002/9781118695111.ch58
2. F. Bray et al., Global cancer statistics 2018: GLOBOCAN estimates of incidence and mortality worldwide for 36 cancers in 185 countries. CA. Cancer J. Clin. **68**(6), 394–424 (2018). https://doi.org/10.3322/caac.21492
3. A. Chapman, Thermal ablation of uterine fibroids using MR-guided focused ultrasound-a truly non-invasive treatment modality. 2505–2511 (2007). https://doi.org/10.1007/s00330-007-0644-8.
4. M.I.N. Chen et al., Disease prediction by machine learning over big data from healthcare communities, 8869–8879 (2017)
5. S. Elmaoğulları, Z. Aycan, Abnormal uterine bleeding in adolescents. JCRPE J. Clin. Res. Pediatr. Endocrinol. **10**(3), 191–197 (2018). https://doi.org/10.4274/jcrpe.0014
6. A. Fallahi, M. Pooyan, H. Ghanaati, M.A. Oghabian, H. Khotanlou, M. Shakiba, A.H. Jalali, K. Firouznia, Uterine segmentation and volume measurement in Uterine fibroid patients' MRI using fuzzy C-mean algorithm and morphological operations **8**(3), 150–156 (2011). https://doi.org/10.5812/kmp.iranjradiol.17351065.3142
7. H.K. Fatlawi, Enhanced classification model for cervical cancer dataset based on cost sensitive classifier abstract **4**(4), 115–120 (2017)
8. I.S. Fraser, P. Warner, Estimating blood loss in women with normal and excessive blood loss. Obstet. Gynecol. **98**(5), 806–814 (2001)
9. T.S. Furey et al., Microarray expression data **16**(10), 906–914 (2000)
10. D.K. Girija et al., Data mining approach for prediction of fibroid disease using neural networks, in *Proceedings—2013 International Conference on Emerging Trends in Communication, Control, Signal Processing and Computing Applications (C2SPCA)* (IEEE, 2013), pp. 2–6. https://doi.org/10.1109/C2SPCA.2013.6749370
11. D.K. Girija, M.S. Shashidhara, Data mining techniques used for uterus fibroid diagnosis and prognosis, in *Proceedings—2013 IEEE International Multi-Conference on Automation, Computing, Communication, Control and Compressed Sensing (iMac4s)*, pp. 372–375 (2013). https://doi.org/10.1109/iMac4s.2013.6526439
12. P.K. Gregersen et al., Expert reviews menstruation: science and society, pp. 624–664 (2020). https://doi.org/10.1016/j.ajog.2020.06.004
13. E.H. Houssein et al., Deep and machine learning techniques for medical imaging-based breast cancer: a comprehensive review. Expert Syst. Appl. 114161 (2020). https://doi.org/10.1016/j.eswa.2020.114161
14. M. Islam et al., Breast cancer prediction: a comparative study using machine learning techniques. SN Comput. Sci. 1–14 (2020). https://doi.org/10.1007/s42979-020-00305-w
15. Y. Kurata et al., Automatic segmentation of the uterus on MRI using a convolutional neural network. Comput. Biol. Med. **114**, 103438 (2019). https://doi.org/10.1016/j.compbiomed.2019.103438

16. J. Lu et al., Machine learning for assisting cervical cancer diagnosis: an ensemble approach. Futur. Gener. Comput. Syst. **106**, 199–205 (2020). https://doi.org/10.1016/j.future.2019.12.033
17. Y.H. Luo et al., Deep learning based on MR imaging for predicting outcome of uterine fibroid embolization. J. Vasc. Interv. Radiol. **31**(6), 1010-1017.e3 (2020). https://doi.org/10.1016/j.jvir.2019.11.032
18. M. Madhra et al., Abnormal uterine bleeding: advantages of formal classification to patients, clinicians and researchers. Acta Obstet. Gynecol. Scand. **93**(7), 619–625 (2014). https://doi.org/10.1111/aogs.12390
19. J.L. Magnay et al., A systematic review of methods to measure menstrual blood loss. BMC Women's Health **18**(1), 1–13 (2018). https://doi.org/10.1186/s12905-018-0627-8
20. M. Mahmud et al., Applications of deep learning and reinforcement learning to biological data. arXiv **29**(6), 2063–2079 (2017)
21. M. Mukherjee et al., MenstruLoss: sensor for menstrual blood loss monitoring. Proc. ACM Interact. Mobile Wearable Ubiquit. Technol. **3**(2), 1–21 (2019). https://doi.org/10.1145/3328929
22. M.G. Munro et al., The two FIGO systems for normal and abnormal uterine bleeding symptoms and classification of causes of abnormal uterine bleeding in the reproductive years: 2018 revisions. Int. J. Gynecol. Obstet. **143**(3), 393–408 (2018). https://doi.org/10.1002/ijgo.12666
23. A. Ozcift, A. Gulten, Classifier ensemble construction with rotation forest to improve medical diagnosis performance of machine learning algorithms. Comput. Methods Programs Biomed. **104**(3), 443–451 (2011). https://doi.org/10.1016/j.cmpb.2011.03.018
24. S.D. Pedersen, B. Corenblum, Recherche Syndrome des ovaires polykystiques Polycystic ovary syndrome **53**, 1041–1047 (2007)
25. M.E. Pennant et al., Premenopausal abnormal uterine bleeding and risk of endometrial cancer. BJOG Int. J. Obstet. Gynaecol. **124**(3), 404–411 (2017). https://doi.org/10.1111/1471-0528.14385
26. K. Puri et al., Bleeding and anemia. Am. J. Obstet. Gynecol. **210**(1), 38.e1-38.e7 (2014). https://doi.org/10.1016/j.ajog.2013.09.038
27. R.L. Rosenfield, The diagnosis of polycystic ovary syndrome in adolescents. Pediatrics **136**(6), 1154–1165 (2015). https://doi.org/10.1542/peds.2015-1430
28. M.M. Saritas, A. Yasar, Performance analysis of ANN and Naive Bayes classification algorithm for data classification. Int. J. Intell. Syst. Appl. Eng. **7** (2019). https://doi.org/10.18201//ijisae.2019252786
29. U. Schumacher et al., Estimation of menstrual blood loss volume based on menstrual diary and laboratory data. BMC Women's Health **12** (2012). https://doi.org/10.1186/1472-6874-12-24
30. I. Sriprasert et al., Heavy menstrual bleeding diagnosis and medical management. Contracept. Reprod. Med. **2**(1), 1–8 (2017). https://doi.org/10.1186/s40834-017-0047-4
31. S. Sulaiman et al., Uterine fibroids—do size and location determine menstrual blood loss? Eur. J. Obstet. Gynecol. Reprod. Biol. **115**(1), 85–89 (2004). https://doi.org/10.1016/j.ejogrb.2003.10.027
32. S. Sun et al., How to differentiate uterine leiomyosarcoma from leiomyoma with imaging. Diagn. Interv. Imaging **100**(10), 619–634 (2019). https://doi.org/10.1016/j.diii.2019.07.007
33. H. Teede et al., Polycystic ovary syndrome: a complex condition with psychological, reproductive and metabolic manifestations that impacts on health across the lifespan (2010)
34. G. Warrilow et al., Quantification of menstrual blood loss. Obstet. Gynaecol. **6**(2), 88–92 (2004). https://doi.org/10.1576/toag.6.2.88.26983
35. S. Zhang et al., Learning for personalized medicine: a comprehensive review from a deep learning perspective. IEEE Rev. Biomed. Eng. **12**, 194–208 (2018). https://doi.org/10.1109/RBME.2018.2864254
36. A. Zimmermann et al., Prevalence, symptoms and management of uterine fibroids: an international internet-based survey of 21, 746 women (2012)

Analysing the Resting-State Functional Connectivity of Chronic Pain Patients

V. Rejula, J. Anitha, and R. V. Belfin

1 Introduction

Pain is an alarm system in our body to inform us of the danger, injury, or presence of disease. There are two varieties of pain based on the duration. The first type of pain is acute pain, and it lasts for few days. Similarly, the second type of pain is called chronic pain, and it persists for the long term as three months and more. Chronic pain is a major public health issue across the world. Low back pain causes more impairment than any other ailment on the planet. Low back pain, cervical and thoracic pain, rheumatoid arthritis, osteoarthritis, shoulder pain, headache, cancer, and fibromyalgia are the most common causes of chronic pain. There are three main ways to measure pain. The measures are self-report, observe behaviour, and infer indirect physiology. Self-pain report has been furnished by the patients who suffer pain. This method is not efficient because the intensity of pain reports varies from person to person. Also, infants and unconscious patients cannot report pain by themselves. Likewise, observe behaviour and infer method will have an observer who writes the pain with the help of the patient's expression. Both these methods have their disadvantages of not reporting the right intensity of pain. The indirect physiology method of pain directly takes pain from the brain by using various modalities like EEG, MEG, PET, and MRI. These methods do not involve human reports to identify the intensity of pain. This method uses bio-markers of pain to determine the severity and the duration of pain from the patient.

Brain analysis has become an exciting and essential work nowadays. In brain studies, fMRI-based analysis is one of the critical analyses for neurosurgeons to

V. Rejula · J. Anitha · R. V. Belfin (✉)
Karunya Institute of Technology and Sciences, Coimbatore, India
e-mail: researchbelfin@gmail.com

J. Anitha
e-mail: anitha_j@karunya.edu

J. D. Peter et al. (eds.), *Disruptive Technologies for Big Data and Cloud Applications*, Lecture Notes in Electrical Engineering 905,
https://doi.org/10.1007/978-981-19-2177-3_48

understand various brain-related diseases. The fMRI data analysis can be considered a big data problem because of the vast neural data involved in this analysis. When we think of brain volume, it contains approximately 100,000 voxel measurements, and each experiment contains hundreds of brain volumes. Further, each investigation has to be repeated on multiple subjects to aid population inference. The preprocessing steps of fMRI analysis have been done by image reconstruction, time slice correction, motion correction, co-registration, normalization, and spatial smoothing.

A standard fMRI experiment data contains a massive amount of data inclusive of anatomical and functional data. By detecting variations in blood flow, functional MRI detects brain activity.

Resting-state fMRI in the brain is a marker of normal brain activity. Variations in the resting-state brain might result in bio-markers that represent a distinct pain state or therapeutic responses for a variety of diseases, according to the literature [3, 7–11]. Resting-state fMRI is a fMRI employed to evaluate regional inter-communications that occur in a resting or task-negative situation when a precise task is not performed. It is yet unclear how specific these variations are to the underpinning mechanisms of intractable pain. So, the analysis of resting-state pain signals is critical for understanding any disease conditions.

The steps followed in this manuscript have been given as follows. Section 2 describes the methodologies used to process, clean, and analyse the data. The following Sect. 3 elucidates the experimental setup and result achieved in the investigation. Finally, the manuscript ends with summary given in the Sect. 4.

2 Methodology

This initial work focuses on finding out the variation in the blood flow between the controls and the pain patient's brain. This section describes how each process happens in the flow of brain network analysis from data acquisition to data analysis. The flow of the work has been depicted in Fig. 1.

2.1 fMRI Time Series Data Collection

The data collected during an fMRI experiment consists of a series of individual magnetic resonance images obtained in a way that allows one to study oxygenation patterns in the brain. Therefore, to know the nature of fMRI data and how these images are used to infer neuronal activity, one must first study the acquisition of individual MR images.

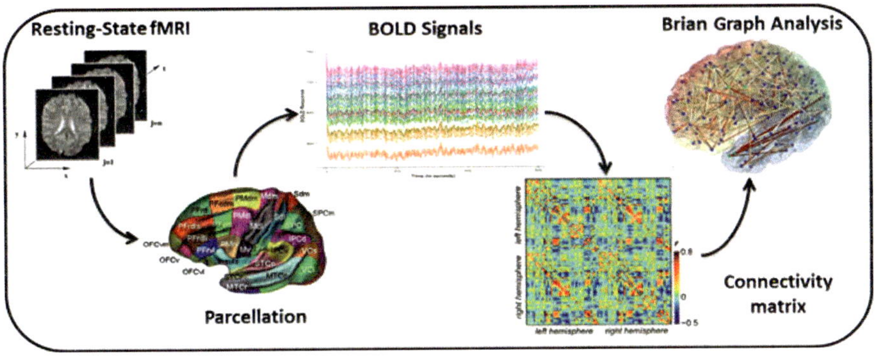

Fig. 1 Flowchart of the analysis steps

2.2 Preprocessing

The preprocessing performed in the dataset for the manuscript has been done using fMRIPrep 20.2.1 [5, 6]. Since this experiment focused on BOLD brain signals, the functional data from the dataset has been used for the investigation. The steps for the functional data preprocessing have been depicted in Fig. 2.

2.3 Cleaning and Confound Regression

The movement of the patient at the time of MRI neuroimaging is a significant problem that neuroimagers face. During the resting-state fMRI collection, the movement can cause a false correlation between the brain region leading to inaccurate conclusions about the brain connectivity. There is a need to splitting the signals into actual brain signals and motion (confounder) signals to solve this problem. Then, the confounds can be removed from the original signal to get the pure BOLD response.

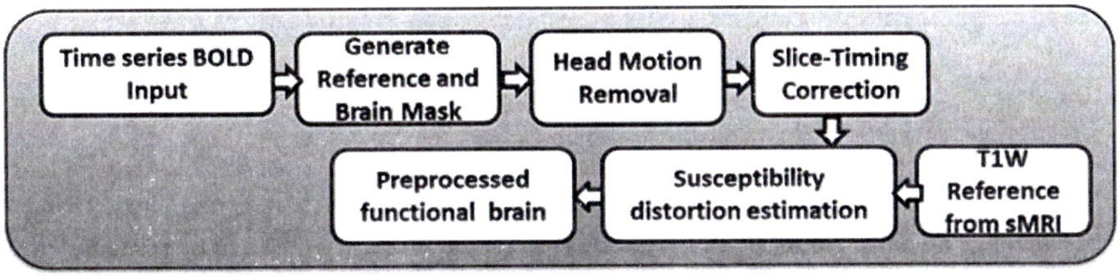

Fig. 2 Flow of functional data preprocessing

2.4 Parcellation

Dividing up the human brain into regions is a longstanding tradition, dating back to Brodmann areas [1]. While these subclasses of the brain originated from a desire to reveal homogeneous neural populations, they have become a powerful communication tool in human neurosciences. As such, a brain parcellation fashions how we think about units of the brain. Atlases are useful in research because they may be used as registration and segmentation objectives to support image analysis in population imaging investigations and, perhaps, in clinical treatment in future. These atlases can also be used as templates for identifying normal brain anatomical variances and, increasingly, for surgical planning. The currently available whole-brain structural MRI atlases have been studied in [4]. An example of the parcelled brain using [2] has been depicted in the Fig. 3.

2.5 Time Series Signal Extraction

Real-time functional magnetic resonance imaging (fMRI) allows researchers to track a subject's brain activity throughout time. In research and clinical trials, as well as neurofeedback applications, real-time information on brain activity is required for constructing and improving interactive fMRI models. Analysis of fMRI time series is often conducted by selecting one or more parameters for the individual voxels. When we do this time series extraction in a parcelled brain, it gives us the relationship between various voxels. The extracted time series from a chronic pain patient has been shown in Fig. 4.

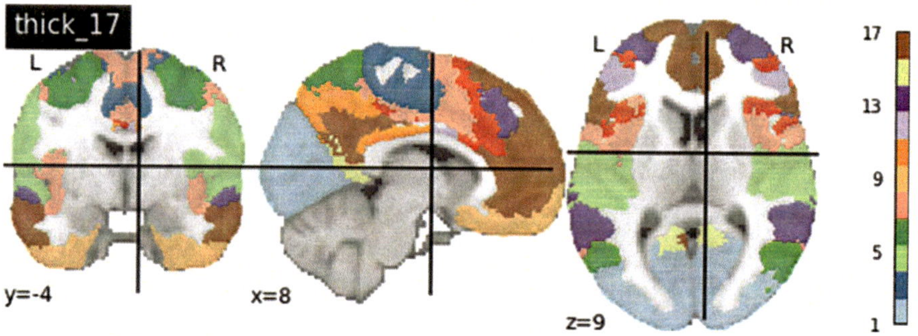

Fig. 3 Sample brain parcellation using Yeo atlas [15]

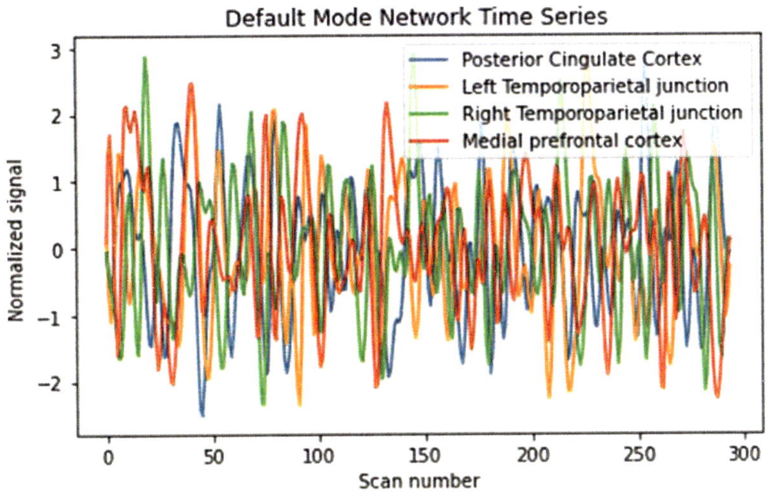

Fig. 4 Time series BOLD signals in various areas of brain during chronic pain

2.6 Brain Region Connectivity Matrix

The patterns of brain connection can be shown as a graph or a matrix. The graph of structural brain connection is sparse and directed. Parameters reflecting connection densities or binary elements indicating the presence or absence of a link can be used to weight the graph. Each component of brain functional connectivity encodes statistical dependency or closeness between two brain regions, resulting in a complete symmetric matrix. The system components can be neurons, recording sites or voxels. The functional correlation matrices will be thresholded to yield binary undirected graphs, with the threshold constraining the degree of sparsity. The matrix created by correlating the brain regions and the signals of two healthy controls and two chronic pain patients is given below in Fig. 5.

2.7 Graph Theory Analysis

Many network analysis tools have been used in parallel efforts to map and depict various biological networks, such as cellular metabolism, gene regulation, or ecology, and many of them are used to investigate and analyse brain connections.

Graph theory is a method for learning directed and undirected graphs. At all levels, we may use graph theory approaches to study structural, functional, and effective brain connection. Graphs are made of vertices (corresponding to neurons or brain regions) and edges (corresponding to synapses or pathways or statistical dependencies within neural elements). A connection matrix or adjacency matrix with

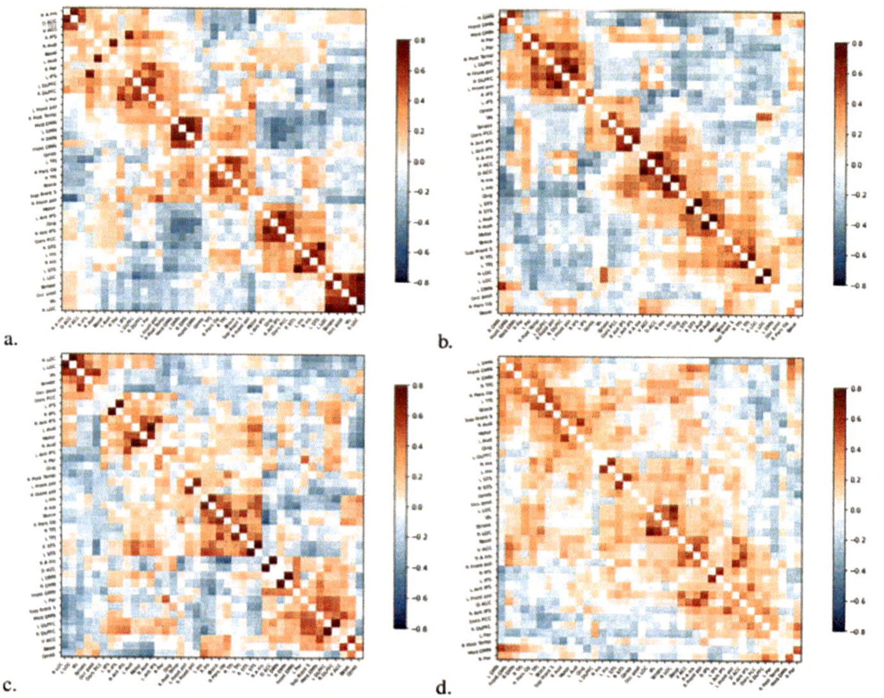

Fig. 5 Correlation between regions of brain during resting state. **a** (Healthy) Sub-01, **b** (Healthy) Sub-02, **c** (Pain) Sub-21, **d** (Pain) Sub-22

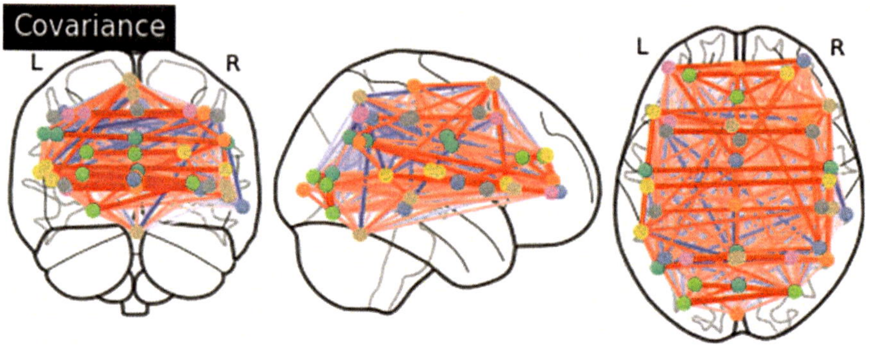

Fig. 6 Brain functional connectivity network using co-variance matrix

binary components reflecting the existence or absence of a directed edge between pairs of vertices can be used to describe graphs in their most basic form. Direct ties between vertices or pathways made up of many edges can be used to interact with them. The network generated from a chronic pain subject is given in Fig. 6.

3 Experimental Setup and Result

The experiment of this work has been conducted with the dataset presented in [13]. This section describes the details about the dataset, the experimental design, and the result.

3.1 Dataset Description

The dataset used for this work is a chronic pain dataset from openneuro.org [13]. This data contains 76 subjects, including 20 healthy controls, 17 osteoarthritis (OA) patients from a 2-week placebo only treatment, and 39 OA patients from a 3-month placebo vs duloxetine treatment. The dataset also contains demographic details such as age, gender, stratification of all the subjects. Precisely for this article, we used only four subjects, two are healthy controls, and two are chronic pain controls.

3.2 Experiment Design

The experiment aims to identify the difference in the brain connectivity between various regions in the brains of the healthy controls and the chronic pain patients. Four of the subjects (Sub-01, Sub-02, Sub-21, Sub-22) from the dataset have been selected for the experiment design. The two healthy controls (Sub-01, Sub-02) and the other two are chronic pain patients (Sub-21, Sub-22).

The experiment from these subjects is to determine the difference in the connectivity between the healthy and chronic pain subjects. The investigation followed the flow given in the methodology section and compared the healthy and chronic pain controls using the network metric called intra-network connectivity. The equation for finding out the intra-network connectivity has been given in the literature [12, 14].

3.3 Result

The subjects (Sub-01, Sub-02, Sub-21, Sub-22) have been selected from the dataset, and the fMRIPrep has preprocessed it. The preprocessed functional image has been overlayed on the atlas and will identify the regions of the brain. Then, the time series signal extraction from the various areas of the brain has been extracted. Once the region-based signal has been extracted, the correlation matrix between the region has been generated. From the correlation matrix generated, the brain connectivity matrix between regions has been generated. The graph has been generated using the connectivity matrix. The graph can be then analysed to find the various network

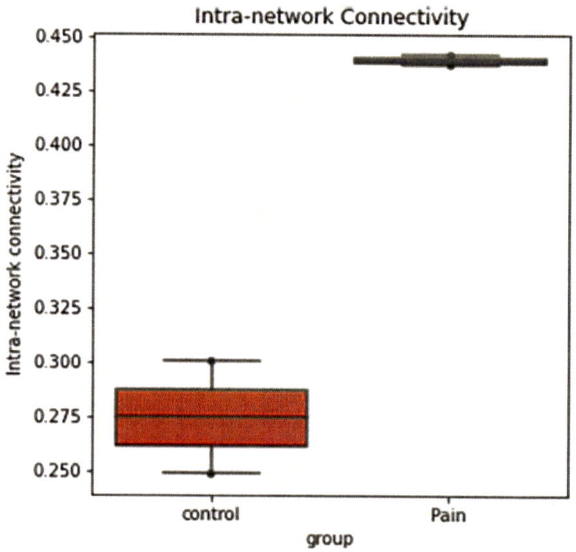

Fig. 7 Comparison of intra-network connectivity between chronic pain and healthy controls

metrics. For this experiment, the difference in the network connectivity between the pain and the healthy controls has been identified using intra-network connectivity. The result shows the difference in inter-network connectivity between pain and healthy controls. The inter-network connectivity for healthy controls has been between approximately 0.265 to 0.280. Likewise, for the chronic pain patients, both the subject's intra-network connectivity has been between 0.425 and 0.450. The box plot showing the result has been given in Fig. 7.

4 Conclusion

The fMRI-based brain analysis has been critical in the current scenario. This study tends to identify the difference in the intra-network connectivity between the pain and healthy controls. The result has been achieved by three significant steps that have been followed. Initially, the data has been cleaned and preprocessed using the fMRIPrep. Then, the motion attributes named confounds have been cleaned and removed to get the pure BOLD signals. Finally, the parcellation and signal extraction has been carried out to relate the brain regions with the BOLD signals. The intra-network connectivity or functional connectivity (FC) results from correlation on the signals extracted between two regions. This study used two healthy controls and two chronic pain subjects as input to do the group analysis of variation in the intra-network connectivity of the brain. The result shows that chronic pain patients show a more denser connection between the regions of the brain.

References

1. K. Brodmann, Vergleichende Lokalisationslehre der Grosshirnrinde in ihren Prinzipien dargestellt auf Grund des Zellenbaues. Barth (1909)
2. R.L. Buckner, F.M. Krienen, A. Castellanos, J.C. Diaz, B.T. Yeo, The organization of the human cerebellum estimated by intrinsic functional connectivity. J. Neurophysiol. **106**(5), 2322–2345 (2011)
3. T. Cavazzi, R. Becerra, Psychophysiological research of borderline personality disorder: review and implications for biosocial theory (2014)
4. D.A. Dickie, S.D. Shenkin, D. Anblagan, J. Lee, M. Blesa Cabez, D. Rodriguez, J.P. Boardman, A. Waldman, D.E. Job, J.M. Wardlaw, Whole brain magnetic resonance image atlases: a systematic review of existing atlases and caveats for use in population imaging. Front. Neuroinf. **11**(1) (2017). https://doi.org/10.3389/fninf.2017.00001, https://www.frontiersin.org/article/10.3389/fninf.2017.00001
5. O. Esteban, R. Blair, C.J. Markiewicz, S.L. Berleant, C. Moodie, F. Ma, A.I. Isik, A. Erramuzpe, M. Kent, D. James, E. Goncalves, DuPre , K.R. Sitek, D.E.P. Gomez, D.J. Lurie, Z. Ye, R.A. Poldrack, K.J. Gorgolewski, fmriprep. Software (2018) . https://doi.org/10.5281/zenodo.852659
6. O. Esteban, C. Markiewicz, R.W. Blair, C. Moodie, A.I. Isik, A. Erramuzpe Aliaga, J. Kent, M. Goncalves, E. DuPre, M. Snyder, H. Oya, S. Ghosh, J. Wright, J. Durnez, R. Drack, K.J. Gorgolewski, fMRIPrep: a robust preprocessing pipeline for functional MRI. Nature Methods (2018). https://doi.org/10.1038/s41592-018-0235-4
7. W. Huang, D. Pach, V. Napadow, K. Park, X. Long, J. Neumann, Y. Maeda, T. Nierhaus, F. Liang, C.M. Witt, Characterizing acupuncture stimuli using brain imaging with FMRI-a systematic review and meta-analysis of the literature. PLoS One **7**(4), e32960 (2012)
8. Huttlin et al.(2017)B. Huttlin, C. Paulo, B. Ting, G. Colby, P. Gygi, E.L. Huttlin, R.J. Bruckner, J.A. Paulo, J.R. Cannon, L. Ting, K. Baltier, G. Colby, F. Gebreab, M.P. Gygi, H. Parzen, et al., Architecture of the human interactome defines protein communities and disease networks. Nature **545**(7655):505–509 (2017)
9. A. Kucyi, K.D. Davis, The dynamic pain connectome. Trends Neurosci. **38**(2), 86–95 (2015)
10. M.L. Loggia, J. Kim, R.L. Gollub, M.G. Vangel, I. Kirsch, J. Kong, A.D. Wasan, V. Napadow, Default mode network connectivity encodes clinical pain: an arterial spin labeling study. PAIN® **154**(1), 24–33 (2013)
11. C. Miniussi, J.A. Harris, M. Ruzzoli, Modelling non-invasive brain stimulation in cognitive neuroscience. Neurosci. Biobehav. Rev. **37**(8), 1702–1712 (2013)
12. S.M. Tashjian, D. Goldenberg, M.M. Monti, A. Galván, Sleep quality and adolescent default mode network connectivity. Social cognitive and affective neuroscience **13**(3), 290–299 (2018)
13. P. Tétreault, A. Mansour, E. Vachon-Presseau, T.J. Schnitzer, A.V. Apkarian, M.N. Baliki, Brain connectivity predicts placebo response across chronic pain clinical trials. PLoS Biol. **14**(10), e1002570 (2016)
14. Wang et al.(2015) P. Wang, B. Zhou, H. Yao, Y. Zhan, Z. Zhang, Y. Cui, K. Xu, J. Ma, L. Wang, N. An, et al., Aberrant intra-and inter-network connectivity architectures in Alzheimer's disease and mild cognitive impairment. Sci. Rep. 5(1):1–12 (2015)
15. B.T. Yeo, M.R. Sabuncu, R. Desikan, B. Fischl, P. Golland, Effects of registration regularization and atlas sharpness on segmentation accuracy. Med. Image Anal. **12**(5), 603–615 (2008)

Detection of Drowsiness Using Artificial Intelligence

J. Sri Nivetha and G. Aswini

1 Introduction

Artificial intelligence is used for many applications like robotics, automobiles and medical science. The important approach of artificial intelligence is to collect the human ideas and to assemble it with the good and correct methods to prove the technology as the best. This paper achieves the ideas of AI to detect the drowsy eyes of the driver and to ping him to be alert. This technique follows many approaches to detect the object without any failures and mistakes.

AI leads with the imagination of human brain that to be learned and executed it with a good plan to solve many issues in technologies too.

2 Literature Survey

1. Mohamad-Hoseyn Sigari, John, Sagar hussain (2016)
   [Drivers face monitoring system In face recognition].
2. IvanGrishchenko, Mike royal, Raison,
   [Tracking the eye using source of OpenCV].
3. Xinjia, Sha zen, Choien (2016)
   [Image Recognition based on deep learning methods].

J. Sri Nivetha · G. Aswini (✉)
Department of Computer Science, St. Joseph's College of Arts and Science for Women, Hosur, Tamil Nadu, India
e-mail: aswini083@gmail.com

J. D. Peter et al. (eds.), *Disruptive Technologies for Big Data and Cloud Applications*, Lecture Notes in Electrical Engineering 905,
https://doi.org/10.1007/978-981-19-2177-3_49

Terms of Index

Methods in AI, Drowsiness detection of a driver, Detection of eye, Positioning eyelids, Detection of eye in different layers, Object recognition, Alert techniques.

Technologies used for Detection Purpose [1]

- Camera for capturing the object
- Object recognition system used in Python code
- Face detection methods
- Eye detection methods
- Eyeballs navigation techniques
- Positioning the eyes.

The above technologies are helpful for entire system to work with good response and enhancing for future predictions.

2.1 *Camera for Object Recognition*

The camera here is placed in front of the driver's view, so that it is very easy for capturing the particular object. The camera speed is very fast in detection to start the initial process of detection (Fig. 1).

For example,

Fig. 1 .

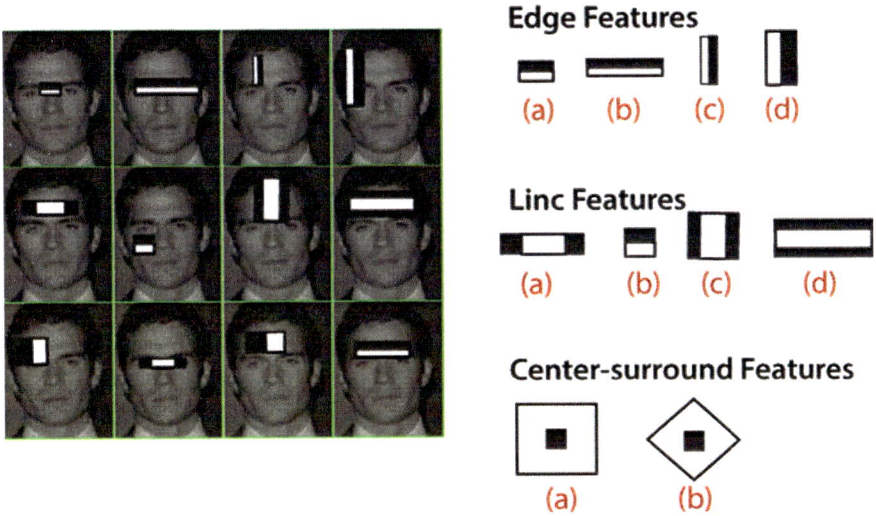

Fig. 2 .

2.2 *Face Detection Methods*

The methods are face detection and are said to be several, but here in this approach, those methods are done for a clear detection. Those are,

- Edge Detection
- Line Detection
- Surrounded Detection
- Centred Detection.

The above detection methods are to detect with face edges, lined pixels in face, surrounding portion of the entire face and mainly the centric position where the particular object of the face is to be detected (Fig. 2).

For example,

2.3 *Eye Detection Methods*

Eye detection is a major role as well the best process for the future detection technology. The security management is mostly based on the eye detection process.

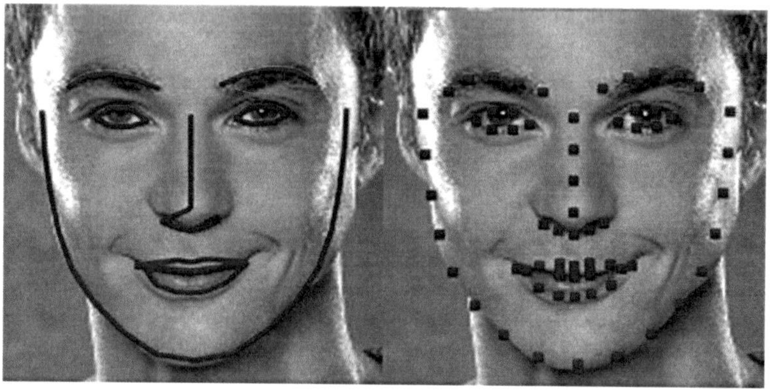

Fig. 3 .

- Knowledge- or rule-based methods
- Feature in-variant method
- Template matching method
- Appearance-based method
- Eye blink detection.

Knowledge-Based Detection

This method is said to be the top-down-based approach where it depends upon the centre portions of the object to be detected. Here the object is said to be the face. The face has two eyes which are symmetric to each other, and it has a nose and mouth. These parts are detected with intensity measures where the intensity is measured with particular portions of the face. This is how the knowledge-based method works (Fig. 3).

Feature in-variant method

The feature in-variant method shows the poses of the object or the image to be recognized. Here in this system, the face gives variety of poses when it is recognized and it starts detected. At that time of process, the poses are considered to be the main feature which is in the detection time. When the driver gives different poses while on driving, the system goes to this kind of method for detection [2]. Analysis of drowsiness system.

- The poses here imported in this system are said to be,
- Head shake pose
- Upper head shake pose
- Lower headshake pose
- Right and left head shake pose
- Off-camera pose
- Rotated head pose.

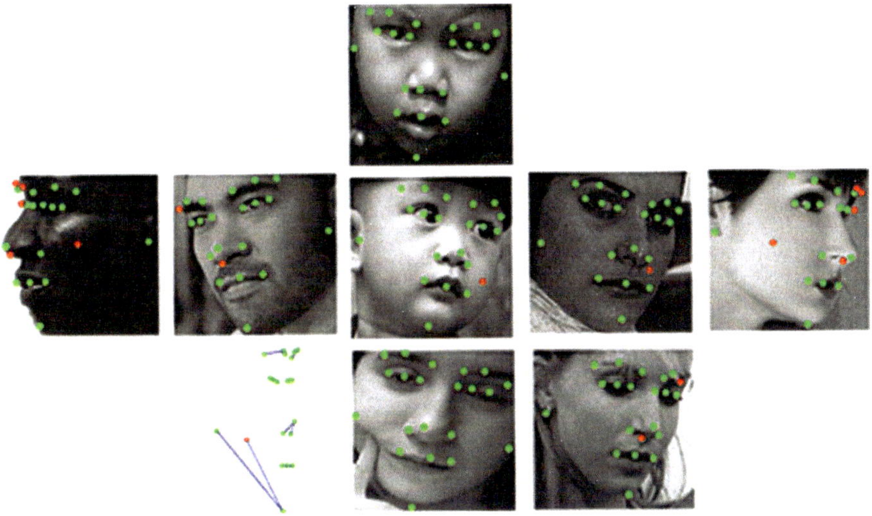

Fig. 4 .

For example,

Figure 4 example poses show how the detection is done using the feature in-variant method.

Template matching method

The template matching mechanism is just recognized, and the face is matching or not with similar qualities. In this system, the template matching is done whether the driver is same or not. It detects the driver which is older one or he/she changes for driving. If the driver gets changed, then the system asks for login to the entire technique to continue driving.

The template for the driver uses his/her eyes to login into the system. This follows eye detection process and continues with the current techniques for eye detection. Then it follows for eye recognition to predict the number of blinks, open and closure timings of the eyes and so on are carried out for eye detection methods (Fig. 5).

For example,

Appearance-based method

Here the images have a set of training images to compare the variability of the other comparing image. The appearance of the training image is said to be the duplicate image or the object to compare the real image or detection. This method has some rules to be followed,

- The image/object must have training set of image
- There contains the other image/object for comparing it with training image
- The condition must satisfy when comparing with training image/object.

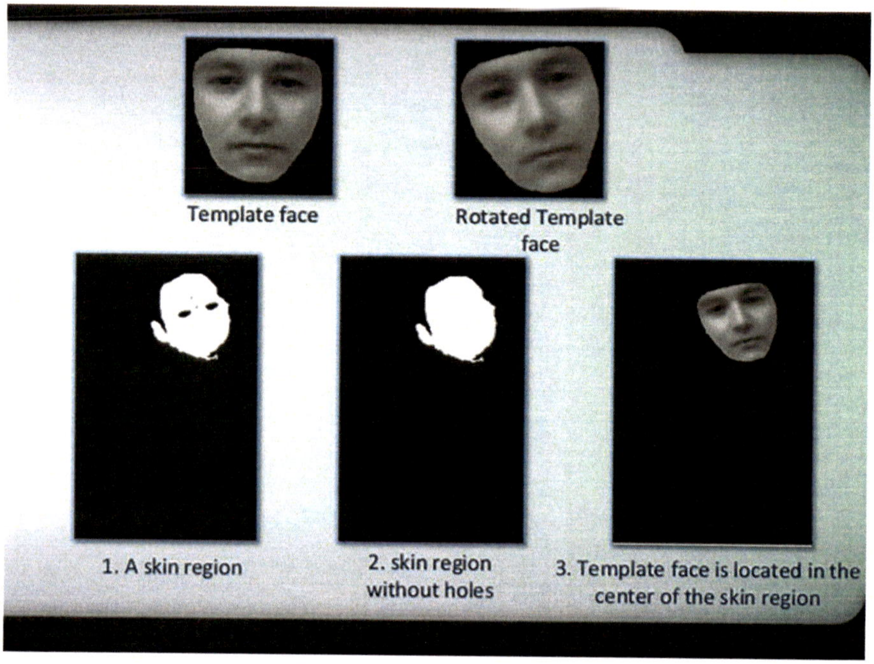

Fig. 5 .

- Now the image satisfies the condition as true or false, then it goes for detection process.
- The detection is done here to predict the real image with correct terms.

Eyeblink method

The eyeblink detection method is used for detecting the number of eyeblinks, motion of the eye, blinking motion, etc.

Here in this system, the driver blinks his/her eyes when driving. This system detects the number of blinks and number of seconds. If the eye's blink increases to the number of seconds, then the alarm is pinged to alert the driver.

For example,

Figure 6 shows the blinks of the eyes and open–close movements of the eyes. The eye's motion is varied with the fraction of seconds as well as in the different poses. At that moment, the poses are recognized using feature variant methods and the blinks are detected using several kinds of algorithms for predicting the time interval for blinks.

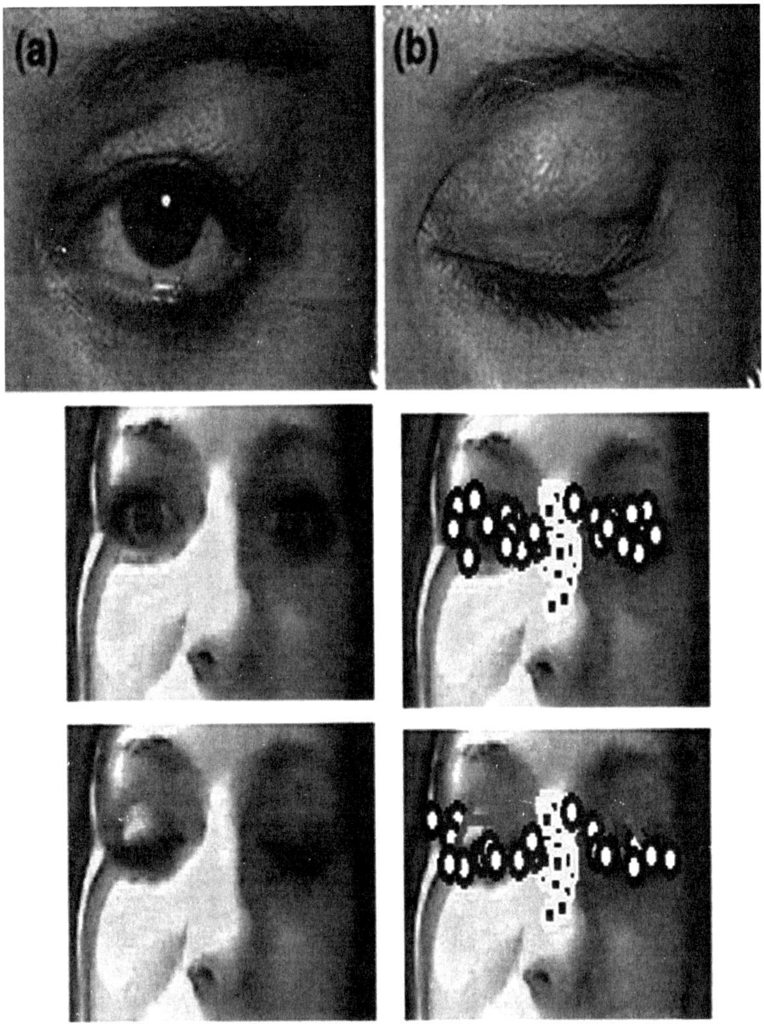

(a) Source images. (b) Displacement of feature points.

Fig. 6 .

2.4 *Eye Ball Navigation Method*

The balls of the eyes are navigated and recognized when the driver is on-road driving. This method completely analyses the movement of the eyeball in basic measures. It calculates with some algorithmic procedures where the movement is detected. This method has algorithmic phenomena by using the navigating system (Fig. 7).

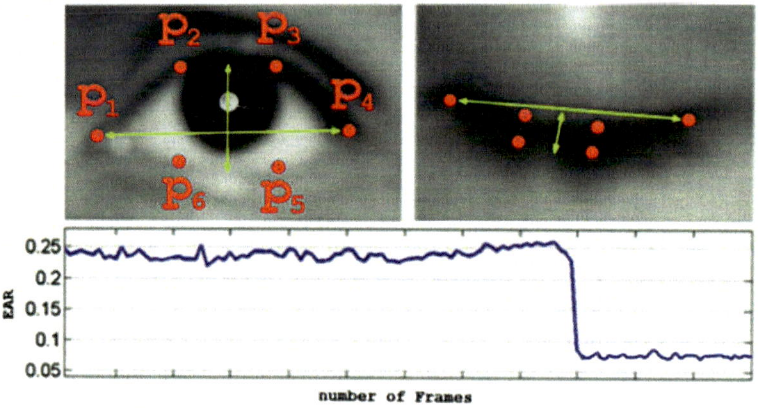

Fig. 7 .

where

EAR is said to be the variable

$P1$ to $p6$ is the pointing positions of the eyeball

2.5 Eye Positioning for Eye Detection

Basically, eye has several positions which changes in fractions. For detecting these positions, we use eye positioning method in the positions which are follows:

- Left positioning—the eye motion is upon left side
- Right positioning—the eye motion is upon right side
- Top positioning—the eye motion is upon upper side
- Bottom positioning—the eye motion is upon down side
- The position which varies recognized by the sensor called camera which detects entire face and eyes. The entire face are detected by the system is helpful for all types of methods which are imported here Batista [3]. *A Drowsiness and point of attention monitoring system* (Vol. volume 4).

Thin Lens Equation

$$\boxed{\frac{1}{f} = \frac{1}{d_o} + \frac{1}{d_i}}$$

d_o = distance of object from lens
d_i = distance of image from lens
f = focal length of the lens

Fig. 8 .

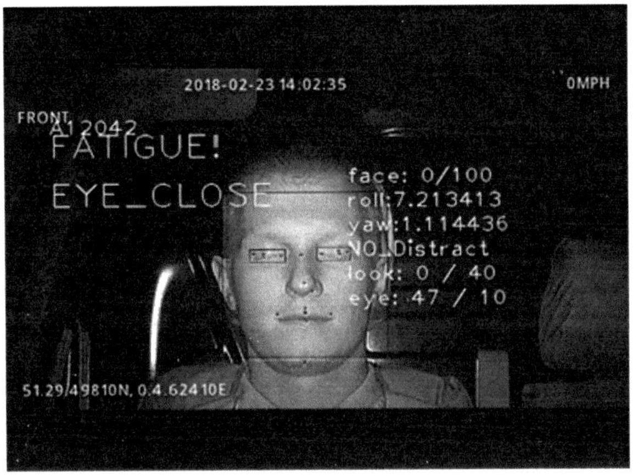

The equation is based on the distance measures. The length of the eye lens is measured with the normal distance of the eye. The eye is predicted with the specific part of an image; that is, for example, the image is said to be the eye. From that, eyeballs are the parts where distance is calculated. Now the distance is calculated with eye and eyeball where the local length is determined. Finally, the centric portion of the image is calculated and measured for detection process.

Detection of Drowsiness using Alarm [4]

The system maintains the alarm for pinging the driver to be more alert. This is the main goal of the entire system. This helps in avoiding the accidents causing on roads. The driver when he feels drowsy, this alerts the driver using the alarm. This alarm is completely designed using the eye detection methods. When the driver closes his/her eyes for more than three seconds, then alarm will alert the driver to be alert; if he opens his eyes, then immediately the alarm will end instead of giving sound messages.

Figure 8 shows the driver feels fatigue and she closes her eyes. Now the system detects the eye for a short time and gives the message that she feels fatigue.

Durations for detecting drowsiness [2]

The duration of this system is detected using the time variants and time intervals. The time here is designed for three seconds interval. The time is assigned as three seconds; if the driver closes his/her eyes for more than three seconds, then the alarm will alert him with sound.

Prediction \ Labels	Alert	Low Vigilant	Drowsy
Alert	0.63	0.25	0.09
Low Vigilant	0.33	0.45	0.26
Drowsy	0.04	0.30	0.65

The above calculation shows the time management in eye detection process.

Algorithm for alarm management [5]

Step 1: Set up the hardware device for alarm.
Step 2: Link the alarm with the eye detection methods.
Step 3: The methods follow the formulations of the timings of the eye close and openings.
Step 4: The technique maintains the number of blinks and it responds with EAR variable with the time of eyeblinks.
Step 5: The seconds are to be set for eye close, so that the alarm will work.
Step 6: Now the three seconds of timings is fixed to process the alarm.

Algorithm for processing video [6]

Step 1: The video is processed with the help of the method called videoStream.
Step 2: The videoStream is processed using some frames where to connect with the user.
Step 3: The number of frames is created for processing the video.
Step 4: The frames are now processed and the user is able to interact with the web camera.
Step 5: Likewise the video is processed using the videoStream method.
Step 6: Here the frames depend upon the particular image of the user.
Step 7: The image may be a face, objects, etc.
Step 8: Finally, the video is processed (Fig. 9).

Source code for Facial and Eye Landmark Index

```
(lStart, lEnd) = face_utils.FACIAL_LANDMARKS_IDXS["left_eye"].
(rStart, rEnd) = face_utils.FACIAL_LANDMARKS_IDXS["right_eye"].
```

Source code for Alarm Setting

```
Counter Seconds
if not ALARM_ON
ALARM_ON = True
```

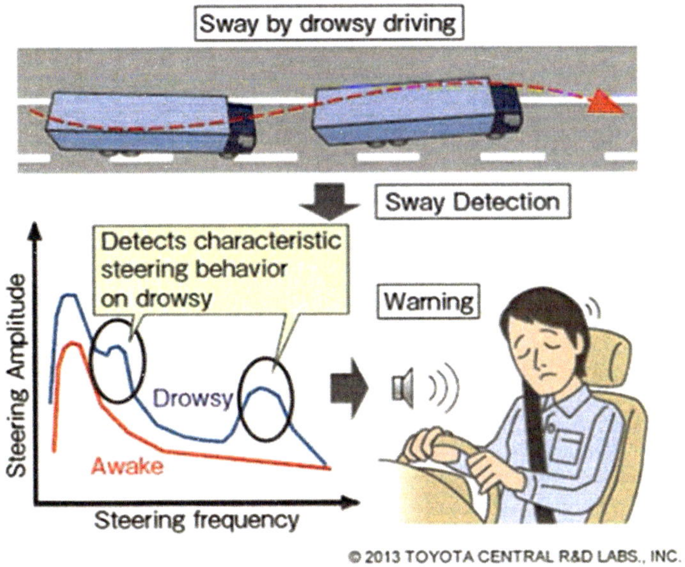

Fig. 9 .

Source code for Destroying Alarm

```
Alarm destroy
COUNTER = 0
ALARM_ON = False
cv2.destroyAllWindows()
vs.stop()
```

Experimental result

Implementation [7]
The implementation of this system can be done by detecting the image of an object in a rapid of seconds. In this system, the image is detected within a fraction of seconds and it needs the ability to detect an object faster than the previous detection processing. The quick detection is done by various methods and algorithms. Those algorithms are followed by this system, and it can be completely processed by implementing the methods as well as algorithms. Thus, this system needs to be implemented with quick detecting methodologies.

3 Conclusion

This system concludes with the detection of drowsiness that has been completed successfully using the help of those techniques and algorithms represented in discussion chapter. Thus, the system with a good model was designed using some techniques called OpenCV and HAAR classifiers. Thus, this system which is developed was neatly tested, and its drawbacks are identified to precede it with the further plans. This system maintained a good process using OpenCV and many methods for drowsiness detection. The eye is detected here when the driver feels drowsy and the alert sound pings the driver to make him/her alert. In this method, we followed the image detection, eye detection as well as the movement of image and eye detection. This is considered to whenever if our eyes are not in correct position; even it detects our eyes in right way. So that the system works in a correct procedure as we had done.

4 Future Study

This system works further with the following aspects,

- We intend to process the more kinds of detection methods with even a minute movement of an object.
- We intend to use the several algorithms to implement the clarity when detection is undergoing.
- We can also implement detection works using the Haar classifier for large videoStream detection.

References

1. P.R. Jones, Detecting bahavioral microsleep. *Argentina* (IEEE, 2010)
2. S. Thomas, V. Ramuste, Analysis of drowsiness system, in *IEEE International Conference* (2013)
3. J. Batista, A drowsiness and point of attention monitoring system **4** (2017)
4. R. Lienhart, J. Maydt, An extended set of haar-like features for rapid object detection, in *IEEE International Conference* (2014)
5. S. Vitabile, Bright pupil detection in drowsiness monitoring system, in *24th IEEE International Conference on Advanced Networking and Application* (2013)
6. T. Akerstedt, *Subjective and Objective Sleepiness in the Active Individual* (International of Data Science, 2016)
7. T. Hong, H. Qin, An improved real time eye state identification, in *IEEE National Conference on Control and Automation*, Vol. 1 (2013)
8. V. Hargutt, H.P. Kruger, Eyelid movements and their predictive value for fatigue stages, in *International Conference of Traffic and Transport Psychology* (2015)
9. Z. Tianet, H. Qin, Real-time drivers detection system, in *IEEE International Conference* (2010)
10. C. Kumar, R. Rahul, Classifying methods of eye detection, in *IEEE on Signal and Image Processing Application* (2012)
11. G. Duncker, Overview of eye detecting methods, in *Deep Learning Methodologies* (2015)
12. F. Friedrichs, B. Yang, Drowsiness monitoring by steering and Lane data based Features. *EUSIPCO* (2015)
13. J.M. Lyznicki, T.C. Doege, Sleepiness driving and motor vehicle crashes. J. Am. Med. Assoc. **2** (2015)
14. J. Maydt, OpenCV functions. *Open Source Computer Vision Library Reference Manual* (2013)
15. L.A. Reyner, J.A. Horne, *Evaluation in Car with Drowsiness Driver*, Vol. 1 (Oxford, 2012)

Classifying Sleep Stages Automatically in Single-channel Against Multi-channel EEG: A Performance Analysis

B. L. Radhakrishnan, E. Kirubakaran, Immanuel Johnraja Jebadurai, and Kummari Gurudev

1 Introduction

Sleep plays a pivotal role in being healthy, but surprisingly, it is often neglected due to misunderstanding, or busy lifestyle [17]. Sleep quality influences many essential tasks such as learning, concentration, and memorisation [5]. Prolonged lack of sleep could influence the growing danger of sleep apnoea, diabetes mellitus, hypertension, cardiovascular disease, obesity, Parkinson's, and Alzheimer's diseases, and even reduce the immunity system's efficiency [2, 9]. Accurately identifying sleep stages and sleep cycles is essential for understanding the aforementioned sleep-related diseases and disorders.

The gold standard tool for sleep evaluation in clinical diagnosis and treatment of sleep problems is polysomnography (PSG). In a standard PSG recording, many physiological signals are recorded, including electroencephalogram (EEG), electrooculogram (EOG), electromyography (EMG), electrocardiogram (ECG), respiration, and pulse oximetry. An epoch in PSG recordings is 30 s long. Domain experts allocate each epoch to a particular sleep stage using the American Academy of Sleep Medicine (AASM) [3] guidelines. Based on the AASM sleep scoring method, the PSG recordings can be divided into five sleep stages: wake state(w), non-rapid eye movement (NREM) sleep stage 1 (N1), NREM sleep stage 2 (N2), slow-wave sleep (SWS) that mixes NREM stages (N3 and N4), and rapid eye movement (REM).

B. L. Radhakrishnan (✉) · I. J. Jebadurai · K. Gurudev
Karunya Institute of Technology and Sciences, Coimbatore 641 114, India
e-mail: blrk.research@gmail.com

K. Gurudev
e-mail: kummarigurudev@karunya.edu.in

E. Kirubakaran
Grace College of Engineering,Thoothukudi 628005, India

J. D. Peter et al. (eds.), *Disruptive Technologies for Big Data and Cloud Applications*, Lecture Notes in Electrical Engineering 905,
https://doi.org/10.1007/978-981-19-2177-3_50

The traditional in-clinic PSG recording has drawbacks such as adhesive sensors, change in the regular sleep environment, and multiple wires that cause inconvenience to the person taking EEG and affect sleep quality. Single-channel EEG is a competing way of recording EEG, mostly use dry electrodes [10]. In-home sleep monitoring systems use single-channel EEG for improved accuracy.

Traditional sleep scoring is time-bound and dependent on an expert's experience; therefore, efficient sleep scoring can provide better clinical information about patients. Automatic machine learning schemes have been developed to solve this issue with promising accuracy [4]. These studies fall into the multi-channel and single-channel EEG category. Multi-channel uses EEG, ECG, EMG, EOG, respiration, and pulse oximetry signals; however, only a single EEG channel is used by single-channel EEG systems.

There is an increasing demand for in-home sleep monitoring systems. Devices such as Neurosky uses a single dry electrode [8]. This work aims to compare a machine learning algorithm's performance in multi-channel and single-channel to validate the application of single-channel-based EEG for remote sleep monitoring. This work aims to compare the performance of sleep-stage classification in single-channel with multi-channel EEG.

The remainder of this article is organised as follows. Section 2 describes the methodology. In Sect. 3, results and discussion are presented. Section 4 concludes the work.

2 Methodology

This work evaluated the performance of sleep-stage classification of single-channel EEG with multi-channel EEG using machine learning algorithms. Figure 1 shows the steps of our scheme to classify multi-channel and single-channel EEG. The preprocessing step reads the EEG signal and then performs filtering, converting the signal to a 30s epoch and signal decomposition to identify α, β, γ, σ, and δ waves. Feature extraction extracts time, frequency, and time-frequency domain features from the signal. The EEG signal samples were trained and tested using k-fold validation. Finally, the performance of the sleep-stage classification in multi-channel and single-channel was compared.

2.1 Dataset Used

This study used the Sleep-EDFx expanded, a public dataset by Physionet Kemp et al. [12]. In the Sleep-EDFx database, there are 197 whole-night EEG sleep recordings exist that comprise EEG, EOG, chin EMG, and event markers. Respiration and body temperature are also recorded for some subjects. The sleep experts used R&K rule to

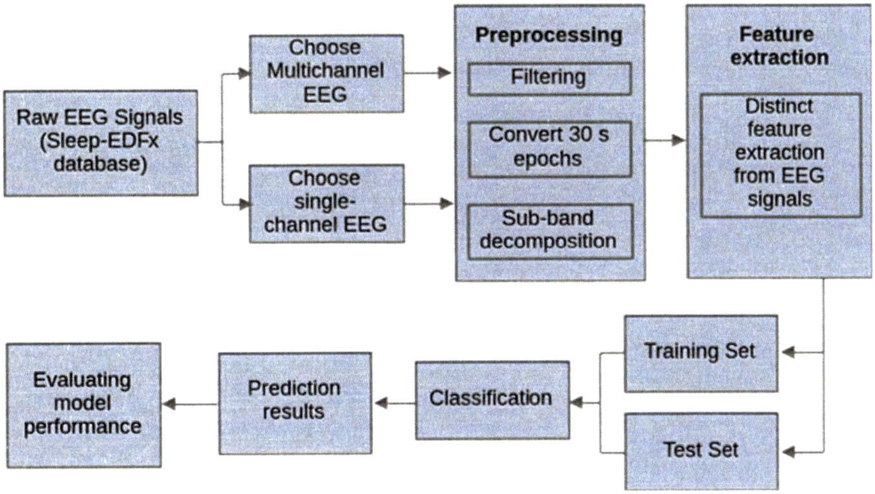

Fig. 1 Block diagram of multi-channel and single-channel EEG sleep-stage classification

manually score related sleep patterns (hypnograms) using Fpz-Cz/Pz-Oz EEGs [12, 16]. The database is sampled 100 Hz. In this study, single-channel uses only EEG Pz-Oz channel and multi-channel used both the EEg channels.

2.2 Preprocessing

The existence of γ waves plays a notable role on sleep-stage classification [15]. This study applied an FIR filter from the MNE Python library [7].

2.3 Feature Extraction

Feature extraction remains a vital step in automated sleep-stage detection because it extracts unique characteristic patterns of EEG signals related to different sleep stages. The sampling rate of the dataset 100 Hz, $s(n)$ represents a characteristic wave of a 30 s epoch, which comprises 3000 samples. The machine learning algorithms used the following features.x

Mean: An epoch's mean electric potential was determined. An epoch's data points were added together and then divided by 3000, where s_i denotes the ith sample.

$$\bar{s} = \frac{\sum_{i=1}^{i=N} s_i}{N} \tag{1}$$

Standard deviation (SD): This feature was used to measure the amount of variation or distribution EEG sub-bands data, where s_i denotes the ith sample of $s(n)$, $\bar{s}$ denotes the mean of $s(n)$, and N is 3000.

$$\sigma = \sqrt{\frac{1}{N}\sum_{i=1}^{N}(s_i - \bar{s})^2} \tag{2}$$

Skewness: Skewness is a way of expressing higher-order statistics (third moment). It was used to determine the degree of asymmetry of a distribution's measure. Skewness is zero in a normal distribution, while positive and negative skewness indicate skewed right or left data, respectively.

$$\text{skewness} = \frac{\sum_{1=1}^{N}(s_i - \bar{s})^3}{\sigma^3} \tag{3}$$

Kurtosis: Kurtosis is a higher-order statistics test (fourth moment). In terms of the concentration of values near the mean, it was used to quantify the peakedness or flatness of a frequency distribution graph, especially compared to the normal distribution.

$$\text{kurtosis} = \frac{\frac{\sum_{1=1}^{N}(s_i - \bar{s})^4}{N}}{\sigma^4} \tag{4}$$

Entropy: It was used to calculate the epoch's data points randomness. The lack of order or predictability described by entropy. The stochastic process denoted by a high entropy that does not make a particular pattern. The p is the probabilities of s 1 to N.

$$\text{entropy} = -\sum_{1=1}^{N} p_i * log_2(p_i) \tag{5}$$

Frequency bands EEG: The EEG signal was converted from the time domain to the frequency domain using the Fourier transform, which extracted the band information. EEG signals were split into five bands based on their frequency. These bands are part of the overall EEG signal collected by a single electrode. Many researchers used frequency bands as a feature for many machine learning techniques to classify the sleep stages [6]. Once the signal was converted to the frequency domain, the mean power of every band (δ, θ, α, β, γ) is extracted. Frequency band data can be obtained using mathematical algorithms such as Fast Fourier transform (FFT). The list of different frequency bands and its frequencies are given in Table 1.

$$\text{average power} = \frac{1}{N}\sum_{i=1}^{N}|s(i)|^2 \tag{6}$$

Table 1 List of EEG bands

Wave	Bandwidth (Hz)
δ	0–4
θ	4–7
α	8–12
β	12–30
γ	30–100+

Power spectral density (PSD): FFT can be used to calculate PSD directly. It was used to show how a signal's power distribution changed with frequency. The squared value of the EEG signal is referred to as power. PSD stands for "energy per frequency".
Petrosian Fractal Dimension (PFD): It was used to find the pattern changes. This feature is taken from the PyEEG library [1]. All of the preceding features were used in the machine learning models.

2.4 *Classification*

Various machine learning methods for classifying sleep stages have been suggested over the years. Extreme gradient boosting (XGBoost), light gradient boosting machine (LGBM), and voting classifiers are chosen and used to boost the classification efficiency of sleep stages [4].

XGBoost is widely used to classify sleep stages. It has a high level of reliability and the capacity to solve real-world problems and a high level of adaptability to various class classification issues. It does not need any multiplication operations, making it very hardware-friendly. Furthermore, it is easy to migrate learning with larger samples and fine-tune with smaller samples, ideal for producing various wearable devices and in-home sleep monitoring [13].

LGBM is a tree-based learning algorithm and used as a boosting framework. LGBM uses a method of gradient boosting on the decision tree algorithm utilised for classification. In LGBM, the leaf-wise approach lowers more loss than the level-wise approach when increasing on the same leaf, ending in much higher accuracy [14].

A voting classifier learns from a set of models and predicts an outcome (class) based on the output of the chosen class with the highest probability. It adds the outcomes of each classifier passed into the voting classifier and determines the output class that has won the most votes. Our model determines the performance using the majority of cumulative votes for every output class [18].

2.5 Evaluation Metrics

Evaluating the performance of the algorithm is very important in any machine learning study. This study uses the $F1$-score ($F1$), precision (P), and sensitivity (SEN) to estimate the performance of sleep-stage classification [11]. Besides, this study employs accuracy (ACC), kappa (k), and weighted $F1$-score (WF1) to estimate the performance.

$$\text{SEN} = \frac{\text{TP}}{\text{TP} + \text{FN}} \tag{7}$$

$$P = \frac{\text{TP}}{\text{TP} + \text{FP}} \tag{8}$$

$$F1 = \frac{2 \times R \times P}{R + P} \tag{9}$$

$$\text{accuracy} = \frac{\text{TP} + \text{TN}}{\text{TP} + \text{FP} + \text{TN} + \text{FN}} \tag{10}$$

TP, FP, TN, and FP denote the number of true positive, false positive, true negative, and false positives.

$$\text{WF1} = \sum_{i=1}^{N} w_i \times F - 1i \tag{11}$$

$$\kappa = \frac{ACC - P_e}{1 - P_e} \tag{12}$$

w_i and F_{1i} describe ith class' weight and $F1$-score. P_e expresses the agreement by chance using hypothetical probability. All the evaluation mentioned above metrics were obtained on the Sleep-EDFx dataset using multi-channel and single-channel EEG signals. This study compares the performance of sleep-stage detection using multi-channel and single-channel. Equations 7–12 explain the calculation techniques for evaluation metrics.

3 Results and Discussion

A Dell Power Edge R730 server with 32 CPUs and 128GB RAM was used to perform the experiments. The preprocessing and feature extraction were done as mentioned in Sect. 2. This study used five-stage sleep classification according to AASM manual. Data matrix of 7917000 samples from five subjects is used in the experiment.

3.1 XGBoost

This experiment used a tenfold cross-validation technique for training and testing. According to the findings, the overall accuracy obtained in multi-channel and single-channel (87.96, 86.53%) from XGBoost classifiers. Table 2 shows the output of the XGBoost classifier with three parameters (precision, recall, $F1$-score, and support) for each sleep-stage level. Besides, individual sleep-stage misclassification in multi-channel and single-channel was analysed and presented in Tables 3 and 4.

3.2 LGBM

For training and testing, a tenfold cross-validation methodology was used. The overall accuracy achieved in the multi-channel and single-channel was (86.24, 86.81%). Table 5 shows the output of the LGBM classifier with four parameters (precision, recall, $F1$-score, and support) for each sleep-stage level. Additionally, individual sleep-stage misclassification in multi-channel and single-channel was analysed and presented in Tables 6, 7, and 8.

Table 2 XGBoost performance in multi-channel and single-channel EEG

Sleep stage	Precision		Recall		$F1$-score		Support	
	MC	SC	MC	SC	MC	SC	MC	SC
$N1$	86	88	67	52	75	65	27	27
$N2$	87	86	93	93	90	89	153	153
$N3$	93	93	87	86	90	89	77	77
R	85	83	87	85	86	84	53	53
W	87	83	87	90	87	86	39	39

Table 3 XGBoost multi-channel EEG confusion matrix

	Actual class					
Predicted class	Sleep stages	$N1$	$N2$	$N3$	R	W
	$N1$	18	4	0	1	4
	$N2$	0	142	5	6	0
	$N3$	0	10	67	0	0
	R	0	6	0	46	1
	W	3	1	0	1	34

Table 4 XGBoost single-channel EEG confusion matrix

	Actual class					
Predicted class	Sleep stages	$N1$	$N2$	$N3$	R	W
	$N1$	14	5	0	4	4
	$N2$	0	142	5	3	3
	$N3$	0	11	66	0	0
	R	0	8	0	45	0
	W	2	0	0	2	35

Table 5 XGBoost single-channel EEG confusion matrix

Sleep stage	Precision		Recall		$F1$-score		Support	
	MC	SC	MC	SC	MC	SC	MC	SC
$N1$	87	86	52	44	62	59	27	27
$N2$	86	86	93	93	89	89	153	153
$N3$	92	94	86	87	89	91	77	77
R	82	87	87	85	84	86	53	53
W	87	80	85	92	86	86	39	39

Table 6 LGBM multi-channel EEG confusion matrix

	Actual class					
Predicted class	Sleep stages	$N1$	$N2$	$N3$	R	W
	$N1$	14	5	0	4	4
	$N2$	0	142	6	5	0
	$N3$	0	11	66	0	0
	R	0	6	0	46	1
	W	4	1	0	1	33

3.3 Voting Classifier

For training and testing, tenfold cross-validation was used by the voting classifier. The voting classifier was found to have an overall high accuracy among other classifiers used in the multi-channel and single-channel (88.25, 87.1%). Moreover, individual sleep-stage misclassification in the multi-channel and single-channel was analysed and presented in Tables 9 and 10.

Table 7 LGBM single-channel EEG confusion matrix

	Actual class					
Predicted class	Sleep stages	$N1$	$N2$	$N3$	R	W
	$N1$	12	6	0	3	6
	$N2$	0	143	4	3	3
	$N3$	0	10	67	0	0
	R	0	8	0	45	0
	W	2	0	0	1	36

Table 8 LGBM performance in multi-channel and single-channel EEG

Sleep stage	Precision		Recall		$F1$-score		Support	
	MC	SC	MC	SC	MC	SC	MC	SC
$N1$	84	80	59	59	70	68	27	27
$N2$	87	86	94	93	90	89	153	153
$N3$	96	93	87	86	91	89	77	77
R	85	88	89	85	87	87	53	53
W	87	85	87	90	87	88	39	39

Table 9 Voting classifier multi-channel EEG confusion matrix

	Actual class					
Predicted class	Sleep stages	$N1$	$N2$	$N3$	R	W
	$N1$	16	5	0	2	4
	$N2$	0	144	3	5	1
	$N3$	0	10	67	0	0
	R	0	6	0	47	0
	W	3	1	0	1	34

3.4 Comparison of XGBoost, LGBM, and Voting Classifier

To summarise the classification performance, it was observed that the voting classifier had the highest accuracy rate in both multi-channel and single-channel (88.25, 87.1%). XGBoost, on the other hand, was ranked second and reached an overall accuracy score of (87.96, 86.53%) in both multi-channel and single-channel. Finally, LGBM came in third place for classifying sleep levels, with an overall lower accuracy rate (86.24, 86.81%) in both multi-channel and single-channel. Table 11 shows the accuracy, $F1$-score, precision, recall, and kappa values among the three classifiers on multi-channel and single-channel. Also, it reveals the percentage of overall classification accuracy between XGBoost, LGBM, and voting classifiers.

Table 10 Voting classifier single-channel EEG confusion matrix

		Actual class				
Predicted class	Sleep stages	*N*1	*N*2	*N*3	*R*	*W*
	*N*1	16	5	0	2	4
	*N*2	1	142	5	3	2
	*N*3	0	11	66	0	0
	R	0	8	0	45	0
	W	3	0	0	1	35

Table 11 Overall accuracy, F1-scroe, precision, recall, and kappa

Classifier/metric	XGBoost		LGBM		Voting	
	MC	SC	MC	SC	MC	SC
Accuracy	87.96	86.53	86.24	86.81	88.25	87.1
*F*1-score	85.59	82.77	82.04	81.97	85.04	84.06
Precision	87.65	86.53	84.89	86.44	87.86	86.42
Recall	84.09	81	80.35	80.42	83.24	82.48
Kappa	83.02	80.92	80.53	81.3	83.37	81.75

4 Conclusion

Researchers widely used machine learning approaches to classify the sleep stages. The application of in-home sleep monitoring systems is on the rise due to technological advancements. This increases the adoption of single-channel-based sleep monitoring systems for better accuracy. The reliability of single-channel EEG sleep monitoring systems needs to be standardised, similar to multi-channel EEG systems. This study evaluated the sleep-stage classification in single-channel against multi-channel. Both multi-channel and single-channel EEG signals were classified using XGBoost, LGBM, and voting algorithms. Our work demonstrated that the accuracy of single-channel EEG systems on par with multi-channel EEG systems. Hence, the single-channel-based EEG systems can be used as an alternative to traditional multi-channel EEG systems in-home sleep monitoring. However, proper validation studies across healthy, unhealthy in different age groups will improve its reliability.

References

1. F.S. Bao, X. Liu, C. Zhang, Pyeeg: an open source python module for EEG/MEG feature extraction. Comput. Intell. Neurosci. **2011** (2011)
2. D.A. Barone, S. Chokroverty, Neurologic diseases and sleep. Sleep Med. Clin. **12**(1), 73–85 (2017)

3. Berry et al.(2017) R.B. Berry, R. Brooks, C. Gamaldo, S.M. Harding ,R.M. Lloyd, S.F. Quan, M.t. Troester, B.V. Vaughn, ASAM scoring manual updates for 2017 (version 2.4) (2017)
4. K. Boostani, R. Nami Boostani, F. Karimzadeh, M. Nami, A comparative review on sleep stage classification methods in patients and healthy individuals. Comput. Methods Programs Biomed. **140**, 77–91 (2017)
5. P. Ellenbogen, J.M. Stickgold Ellenbogen, J.D. Payne, R. Stickgold, The role of sleep in declarative memory consolidation: passive, permissive, active or none? Curr. Opinion Neurobiol. **16**(6), 716–722 (2006)
6. N. Estrada, B. Behbehani, E. Lucas Estrada, H. Nazeran, P. Nava, K. Behbehani, J. Burk, E. Lucas, EEG feature extraction for classification of sleep stages, in The 26th Annual International Conference of the IEEE Engineering in Medicine and Biology Society, vol. 1 (IEEE, 2004), pp. 196–199
7. A. Gramfort, M. Luessi, E. Larson, D.A. Engemann, D. Strohmeier, C. Brodbeck, R. Goj, M. Jas, T. Brooks, L. Parkkonen, M.S. Hämäläinen, MEG and EEG data analysis with MNE-Python. Front. Neurosci. **7**(267), 1–13 (2013). https://doi.org/10.3389/fnins.2013.00267
8. F. Grosselin, F. Navarro-Sune, A. Vozzi, K. Pandremmenou, F. De Vico Fallani, Y. Attal, M. Chavez, Quality assessment of single-channel EEG for wearable devices. Sensors **19**(3), 601 (2019)
9. O. Itani, M. Jike, N. Watanabe, Y. Kaneita, Short sleep duration and health outcomes: a systematic review, meta-analysis, and meta-regression. Sleep Med. **32**, 246–256 (2017)
10. C.J. James, O. Gibson, M. Davies, On the analysis of single versus multiple channels of electromagnetic brain signals. Artif. Intell. Med. **37**(2), 131–143 (2006)
11. D. Jiang, Y. M. Lu Yn, W. Yuanyuan, Robust sleep stage classification with single-channel EEG signals using multimodal decomposition and hmm-based refinement. Exp. Syst. Appl. **121**, 188–203 (2019)
12. B. Kemp, A.H. Zwinderman, B. Tuk, H.A. Kamphuisen, J.J. Oberye, Analysis of a sleep-dependent neuronal feedback loop: the slow-wave microcontinuity of the EEG. IEEE Trans. Biomed. Eng. **47**(9), 1185–1194 (2000)
13. Y. Liao, M. Zhang, Z. Wang, X. Xie, Design and FPGA implementation of an high efficient xgboost based sleep staging algorithm using single channel EEG, in *International Conference on Cognitive Systems and Signal Processing* (Springer, 2018), pp. 294–303
14. J. Liu, Q. Li, Y. Xin, X. Lu, Obstructive sleep apnea detection using sleep architecture, in *2020 IEEE International Conference on Mechatronics and Automation (ICMA)* (IEEE, 2020), pp. 255–260
15. P. Memar, F. Faradji, A novel multi-class EEG-based sleep stage classification system. IEEE Trans. Neural Syst. Rehabil. Eng. Publ. IEEE Eng. Med. Biol. Soc. **26**(1), 84–95 (2018)
16. M. Mourtazaev, B. Kemp, A. Zwinderman, H. Kamphuisen, Age and gender affect different characteristics of slow waves in the sleep EEG. Sleep **18**(7), 557–564 (1995)
17. X. Tan, L. van Egmond, C.D. Chapman, J. Cedernaes, C. Benedict, Aiding sleep in type 2 diabetes: therapeutic considerations. Lancet Diabetes Endocrinol. **6**(1), 60–68 (2018) (2018)
18. C. Chouvarda Timplalexis, K. Diamantaras, I. Chouvarda, Classification of sleep stages for healthy subjects and patients with minor sleep disorders, in *2019 IEEE 19th International Conference on Bioinformatics and Bioengineering (BIBE)* (IEEE, 2019), pp 344-351

Design and Development of a Weather Forecasting Android Application

Nandini S. Hinduja, D. Jayashree, O. Pandithurai, A. R. Monica, and N. V. Keerthana

1 Introduction

Weather forecasting has been in practice since time immemorial. In olden days, people used to predict the weather based on cloud patterns, astrology, lunar phases, electric telegraphs, and so on. These methods of weather prediction were not always reliable. But nowadays, we can get reliable weather predictions with the help of satellites and weather stations. But, these forecasts will fetch best results only when they are communicated to the public in a convenient fashion. This convenient fashion is the mobile phone because mobile phones have now become the main source of our day-to-day information.

Android technology is extremely accessible these days, and therefore, making a project for the android platform will prove to be relevant and fruitful [1]. Also, a weather forecasting application is a real-world application and can be helpful to almost everyone. Moreover, weather forecasting can be applied to enhance the day-to-day experiences of traveling and touring. We used Android Studio, Java, and XML to implement this application. The weather forecasting android application will be connected to the Internet, and it will communicate with Web APIs to get the weather data which will then be displayed in the application. The application shows the weather forecast along a route while traveling from one place to another. Furthermore, the application will also display the weather forecast for the next five days with an emphasis on the current day's weather. By scrolling through the screen, the users can look at the long-range forecast, the hourly forecast for the current

N. S. Hinduja · D. Jayashree · O. Pandithurai (✉) · A. R. Monica · N. V. Keerthana
Department of Computer Science and Engineering, Rajalakshmi Institute of Technology, Chennai, India
e-mail: pandics@gmail.com

N. S. Hinduja
e-mail: nandhinishinduja.2018.cse@ritchennai.edu.in

J. D. Peter et al. (eds.), *Disruptive Technologies for Big Data and Cloud Applications*, Lecture Notes in Electrical Engineering 905,
https://doi.org/10.1007/978-981-19-2177-3_51

day, the concentration of air pollutants in the air currently, the daily forecast for the next sixteen days, the weekly forecast for the current week and the sunrise, sunset, moonrise, and moonset information. By clicking on one of the forecasts, the users can get a detailed view of the weather forecast which has weather details such as temperature, pressure, humidity, and wind. The users can also share the weather information with their friends by clicking on the share button. The application can even post system wide notifications to let the users know about the weather conditions without even entering the application. Also, the application gives recommendations to the users about how they can prepare for the day according to the day's weather. The users can go to the settings screen to change the temperature units and the location. The users can also look at the weather news, weather facts, and weather for specific tourist spots. We used Maps SDK, routing, networking, and Web APIs to show the weather along a route while traveling. We also used multiple Web APIs to show the weather for tourist spots, weather for cities, weather news, and so on. The four pillars of android—activity, broadcast receivers, content providers, and services were utilized in this application. Several Java classes such as SimpleDateFormat class, MediaPlayer class, AlarmManager class, date class were used in the making of this application. Firebase cloud messaging was also used to implement the weather alerts feature.

2 Related Work

We studied various android weather applications, namely AccuWeather, The Weather Channel, Weather and Radar, Highway Weather, Travel Weather, Windy.com, and FarmWeather [1]. The AccuWeather android application gives the current weather, hourly forecast, and daily forecasts along with a weather radar. The main disadvantage of this application is that it does not give the three-hourly forecasts, weekly forecasts and sunrise, sunset, moonrise, and moonset information and air pollution information [2].

The next application which we studied was The Weather Channel android application which displays the hourly forecast, current forecast, daily forecast, sunrise, sunset, moonrise, and moonset information, air pollution information, seasonal information along with a weather radar and a COVID tracker. The prime limitation of this application is that it does not give the user any recommendation about how they can prepare for the day according to the weather conditions of the day [3].

The third application which we analyzed was the FarmWeather application which allows the user to create a farm and also to delete and edit an existing farm. The user can enter several details of the farm such as crops grown in the farm, location of the farm, and number of acres of land of the farm. Based on the details entered by the user, this application shows the weather of the farm [4].

The fourth application which we researched was Highway Weather. This application allows the user to enter the starting location, destination location, and mode of travel. Based on these details, the application shows the weather along the route selected by the user. The major downside of this application is that it does not show the weather forecast for popular tourist spots [5].

The fifth application which we looked into was Weather & Radar. This application shows the current forecast and daily forecast along with weather videos, weather news, weather camera, weather map, and a weather Web site. The leading hitch in this application is that it does not show weather facts [6].

The sixth application which we explored was Travel Weather. This application shows the weather along a route. The primary fault in this application is that it takes enormous time to load the weather, and the user interface is extremely complicated [7].

The seventh application which we came across was Windy.com. This application exhibits a weather map which can be zoomed, and several weather features such as temperature, wind, wind gusts, rain, thunder, waves, clouds, and air quality can be seen on the map. The main defect in this application is that it does not show the forecast in a text format which is more readable and easily interpretable [8].

3 Implementation

This paper explains in detail about the process of developing a weather forecasting android application. The integrated development environment used is Android Studio. The programming language used is Java. The markup language used is XML. First, a new project has to be created in Android Studio. Then, we are supposed to select the project template [9] (Figs. 1 and 2).

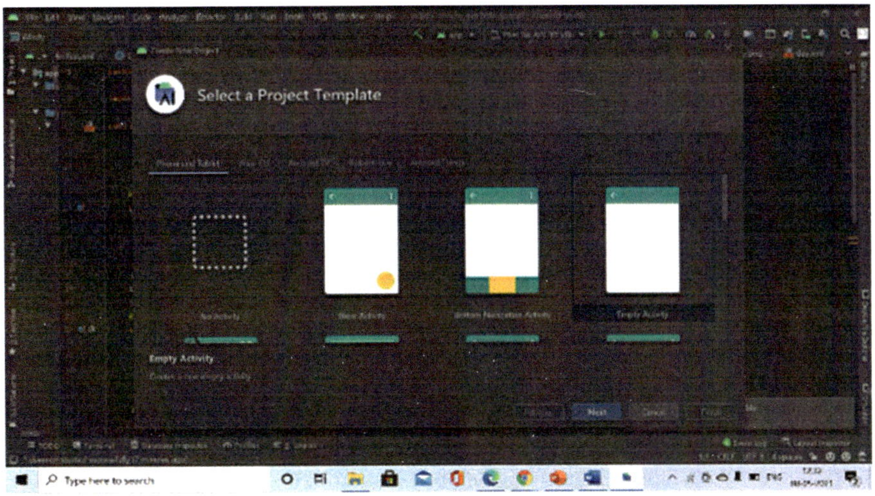

Fig. 1 Selecting a project template

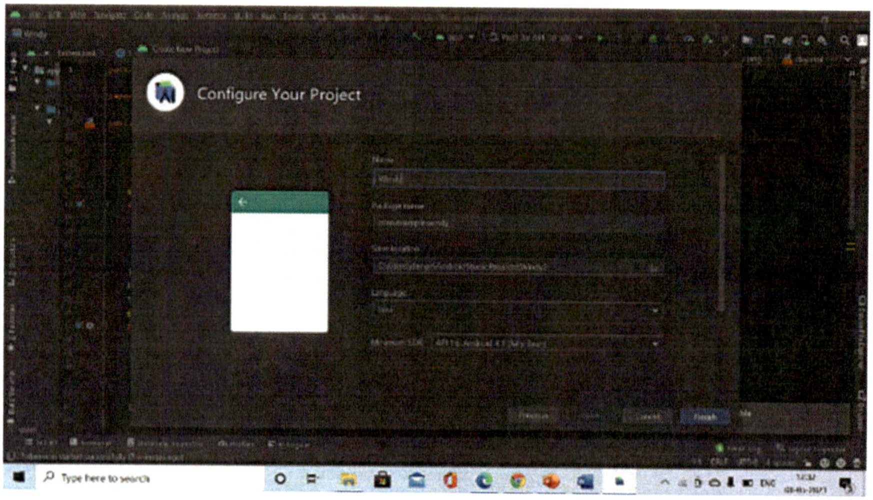

Fig. 2 Creating a new project

While selecting the project template, we also have to select the device for which we are creating our application. In this case, we selected **Phone and Tablet** as our devices since we are creating this application for android mobile devices. Next, we can select the project template from the given options which are as follows: basic activity, bottom navigation activity, map activity, empty activity, etc. Here, we selected the **Empty Activity** project template for our application. Then, we have to give our project a name, and we have to select a programming language for our project. We can select Java or Kotlin as the programming language. We named our project **Windy,** and we selected **Java** as the programming language of our project (Figs. 3 and 4).

While creating a weather forecasting application, the weather data need to be collected from an application programming interface (API). Some APIs provide the weather data for free, and some other APIs provide the weather data for a small fee, and still, some other APIs provide both free and paid weather data. We used four different Web APIs for this project. We used the OpenWeatherMap API for getting the three-hourly forecast for the next five days, detailed weather forecast and current weather forecast [10]. We used the Weatherbit API to get the daily forecasts for the next 16 days and weekly forecasts for the current week [11]. Further, we used the WeatherAPI to get the air pollution information, sunrise, sunset, moonrise and moonset information, and the hourly forecasts for the current day [12]. Furthermore, we used the GNews API to get the weather news [13]. In order to get the data from these APIs, first, we need to create an account on their Web site, and then, we should choose a plan to get the weather data. Most APIs provide several plans such as free, startup, developer, professional, and enterprise. We selected the **Free** plan in all four APIs for the development of this project. After selecting a plan, we need to retrieve

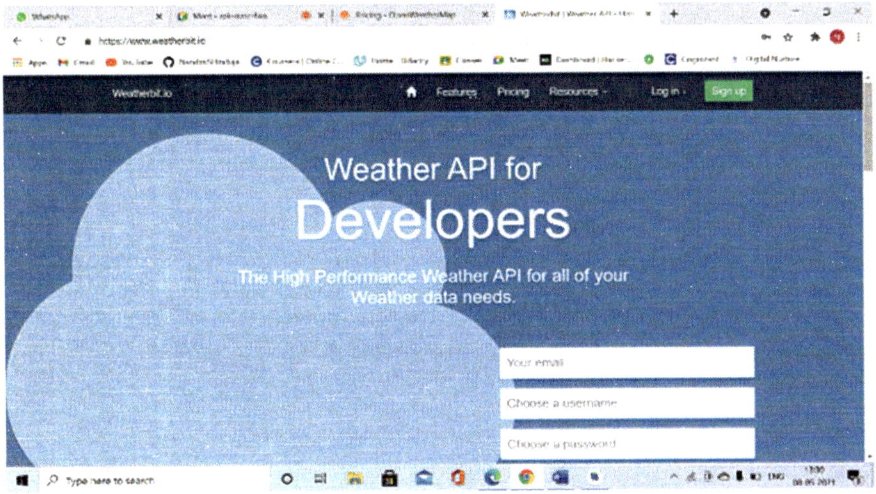

Fig. 3 Creating an account in a weather API Web site

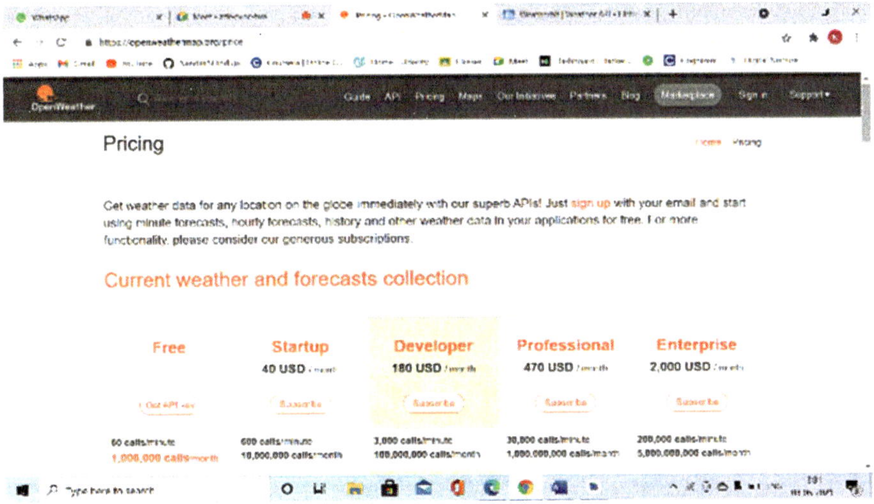

Fig. 4 Choosing a plan to get the weather data

our API key from the Web site, and we can then use our API key along with the API URL to retrieve the weather data (Figs. 5 and 6).

The weather data will be retrieved in JSON format. We can read the documentation provided by each weather API Web site in order to understand the type and format of data that the API provides. The JSON file obtained from an API URL is almost always unreadable by humans, and so, a JSON formatter Web site can be used to format the JSON content in a way which is more readable [14]. This project is a multiscreen

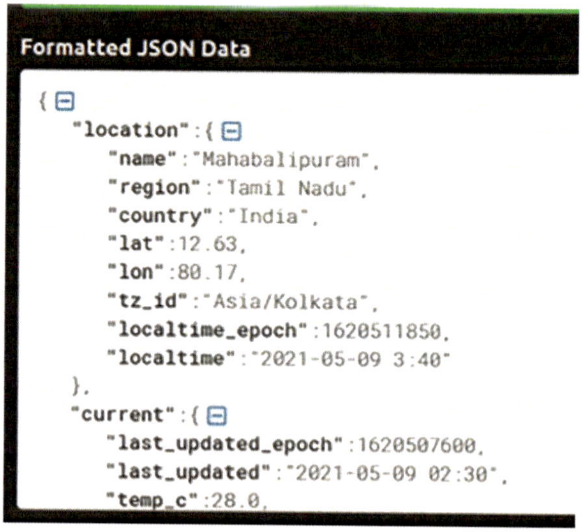

Fig. 5 Formatted JSON data in a JSON formatter

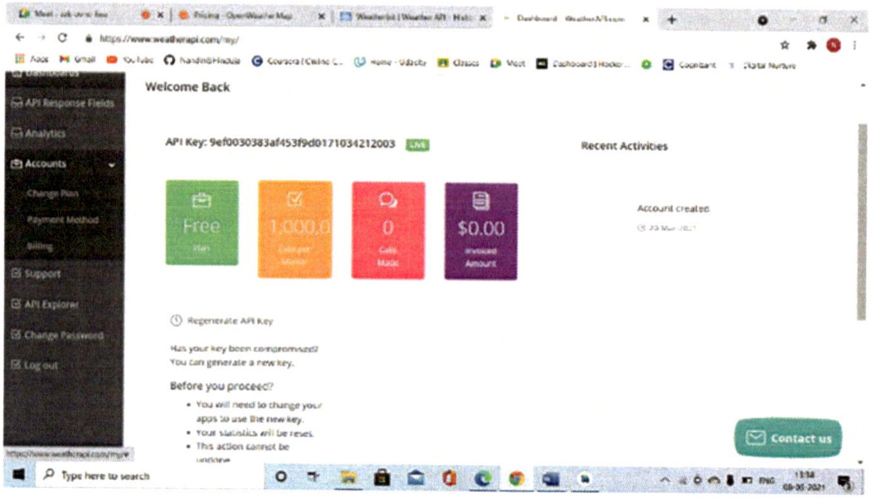

Fig. 6 Process of obtaining an API key

application and has several screens and thus several activities [15]. Some of the most significant activities of this application are route weather activity, tourist weather activity, live weather activity, weather news activity, weather facts activity, and day preparation activity (Figs. 7 and 8).

The main screen displays all the features of the application. This application can show live weather, route weather, tourist weather, day preparation, weather facts, and weather news. The live weather feature displays the current forecast, the three-hourly

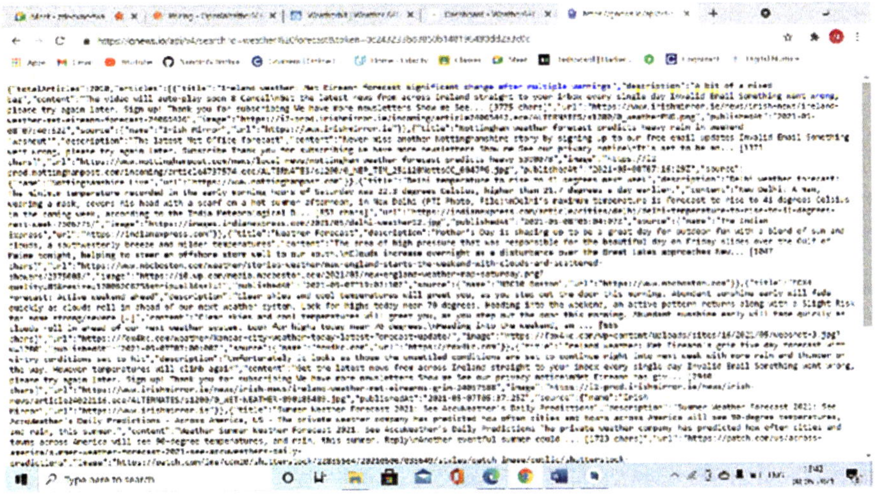

Fig. 7 JSON file which is retrieved from an API URL

Fig. 8 Main screen of the application

forecast for the next five days, the hourly forecast for the current day, the daily forecast for the next sixteen days, the current forecast, the detailed forecasts, the weekly forecast for the current week, the air pollution information and the sunrise, sunset, moonrise, and moonset information. The weather can also be shared through other messaging applications. Users can also turn on notifications to get regular, periodic notifications regarding the weather. The users can also select the temperature units such as Fahrenheit or Celsius [16]. Further, the users can also select the location for which they wish to view the weather. The top part of the live weather screen has a video background. This video background was implemented using a **VideoView** in the XML and several methods such as onCreate(), onStart(), onPrepared(), onDestroy() and onResume() in Java. A different video plays in the background during the day, and a different video plays during the night. This was implemented using the **Date class** is Java. Furthermore, the video is able to play in the background due to the **MediaPlayer** class is Java. The users are able to share the weather through other messaging applications due to the **Share Intent** which is being used in our application. The notifications feature is enabled by using **AlarmManager** class in Java. The user is able to select the location and temperature units because **Shared Preferences** were implemented in our application. **JSON parsing** was done using JSON arrays, JSON objects, and strings to get the weather data from the JSON file and display the data in our application. The **Glide library** was used to display images from the API in our application. These images give a visual representation of the weather forecast. The **Volley library** was used for the networking required to get the data from the API [17]. An **ImageView** was used to display the image representing a weather forecast. **TextViews** were used to display the temperature, date, day, time, pressure, and weather conditions [18] (Figs. 9 and 10).

CardViews were used to display the hourly forecast, daily forecast, weekly forecast, and air pollution information. By clicking on a particular forecast on the screen, the users would enter into a detailed view of the forecast where they can see the pressure, temperature, humidity, day, date, weather conditions, and an icon which represents the weather conditions. The application uses **horizontal ScrollViews** and **vertical ScrollViews** in several activities. **Switch layout** was used to get a toggle button for displaying the sunrise, sunset moonrise, and moonset information. **Firebase Cloud Messaging** from **Firebase** was used to send weather alerts to the users during natural weather calamities such as storms, cyclones, tsunamis, and so on [19]. **Vertical linear layouts, relative layouts, frame layouts,** and **horizontal linear layouts** were used throughout the application. **EditTextPreference** and **ListPreference** were used for the preference screen. Our application uses cyan, white, and black as the theme (Figs. 11 and 12).

If the users click on the air pollution information which is on the main screen of live weather, they will enter into a detailed view of the air pollution information which displays several air pollutants such as carbon monoxide (CO), nitrogen dioxide (NO2), Ozone (O3), sulfur dioxide (SO2), particulate matter less than 2.5 microns (PM2.5), and particulate matter less than 10 microns (PM10). The concentration of these air pollutants present in the air currently is also displayed in the application.

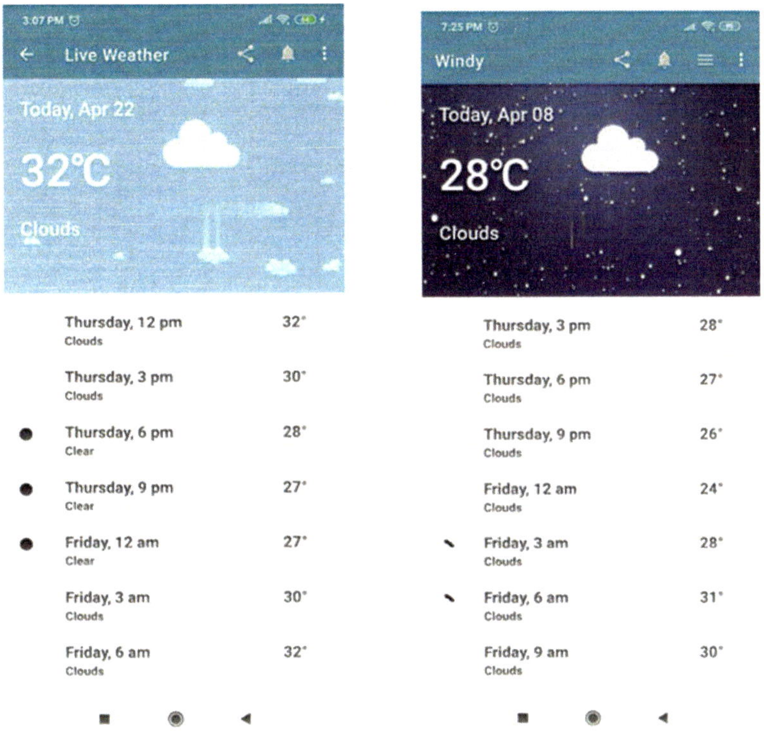

Fig. 9 Live weather screen during the day and night

The **SimpleDateFormat** Java class was used for parsing date and time in several places in our application. Several firebase dependencies such as **firebase-analytics, firebase-messaging,** and **firebase-bom** were added in our application to enable the weather alerts feature. Next, we implemented the tourist weather feature. This feature was implemented for tourist spots in Chennai and Mumbai. This feature shows the weather for several parks, beaches, temples, resorts, and other tourist spots in Chennai and Mumbai. We also implemented the route weather feature. Here, the user has to enter the starting location and destination location. Once the user enters the starting and destination location, they will get the weather for the starting location, destination location, and locations in between. The users will be able to view all these details on a map. The map is included in this application by using **Maps SDK** from the **Google Cloud Platform** [20]. The route from the starting location to the destination location is displayed using **polylines** and **routing**. The **latitude and longitude coordinates** and **markers** were used to identify the locations between the starting location and destination location. Next, we implemented the weather news feature which shows the latest weather news. The headline, an image representing the news, publication date, source, and publication time of the news are displayed in the application. Once the user clicks on a particular news article, a Web site which contains the entire

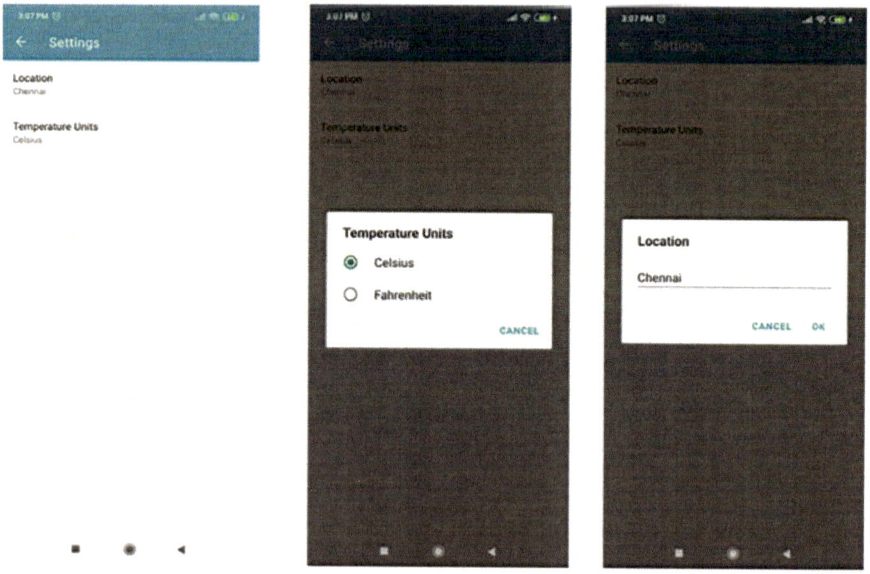

Fig. 10 Settings screen

news article opens, and then, the user can read the entire news article. When the users click the back button, they will be navigated back to the application. The Web site containing the entire news article is opening in a Web browser due to the **CustomTabsIntent** [21] (Figs. 13, 14, and 15).

Next, we implemented the weather facts feature. Here, several weather facts are displayed. Each fact has an audio recording, a title, an image, a share button, and some text associated to it. The **MediaPlayer class** in Java was used to implement the audio recording of the facts. The audio recording can be stopped, played, paused, and resumed. The weather fact can also be shared by clicking on the share button. **CardViews** were used for the layout of each weather fact. The share feature was implemented by using **Share Intents.** Finally, we implemented the day preparation feature. This feature tells the user how to prepare for the day based on the temperature outside. If the temperature is above 30 °C, the application tells the users to stay at home; if the temperature is above 20 °C and less than or equal to 30 °C—the application tells the users to wear sunglasses and apply sunscreen; if the temperature is between 10 and 20 °C—the application tells the users to take an umbrella or wear a raincoat if they go outside, and finally, if the temperature is below 10 °C, the application tells the user to wear gloves (Figs. 16 and 17).

Fig. 11 Detailed forecast screen

4 Results

4.1 Results for Storage

We collected the storage consumption data of several other weather forecasting android applications along with the storage consumption data of our weather forecasting application Windy. Table 1 shows the storage data for various applications, namely: FarmWeather, Windy.com, Windy, AccuWeather, and The Weather Channel.

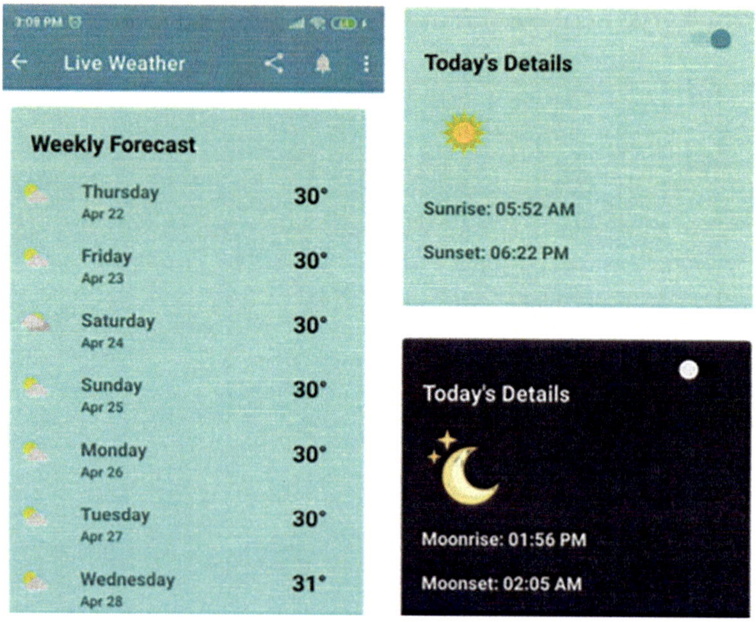

Fig. 12 Weekly forecast, sunrise, sunset, moonrise, and moonset information

4.2 Results for Data Usage

We collected the data usage information of several other weather forecasting android applications along with the data usage information of our weather forecasting application Windy. Table 2 shows the data usage information for various applications, namely: FarmWeather, Windy.com, Windy, AccuWeather, and The Weather Channel.

4.3 Results for Response Time

The different activities of our application are tested by measuring their response time. Table 3 shows the response time of different activities of our application Windy. The response time test has been repeated three times for accuracy.

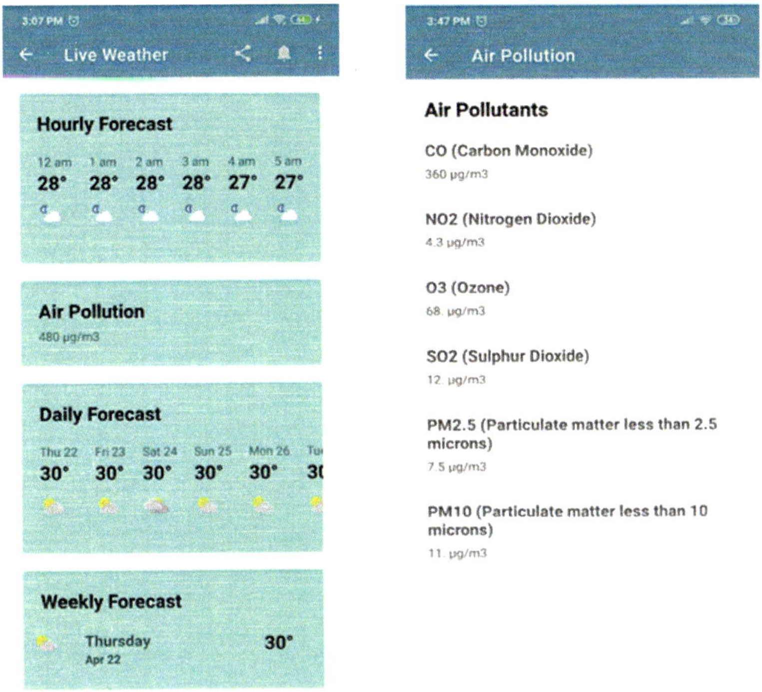

Fig. 13 Hourly forecast, air pollution information, and daily forecast of live weather screen and air pollution screen

5 Discussion of Results

5.1 Analysis of the Results of Storage Consumption

Our weather forecasting application—Windy consumes less storage than The Weather Channel application and the AccuWeather application. Since Windy consumes less storage space than these applications, it will work faster than these applications.

Also, Windy consumes more storage than FarmWeather, Highway weather, Windy.com, Weather and Radar and Travel weather applications. This is because our application Windy provides more features than these applications, and therefore, it requires more storage.

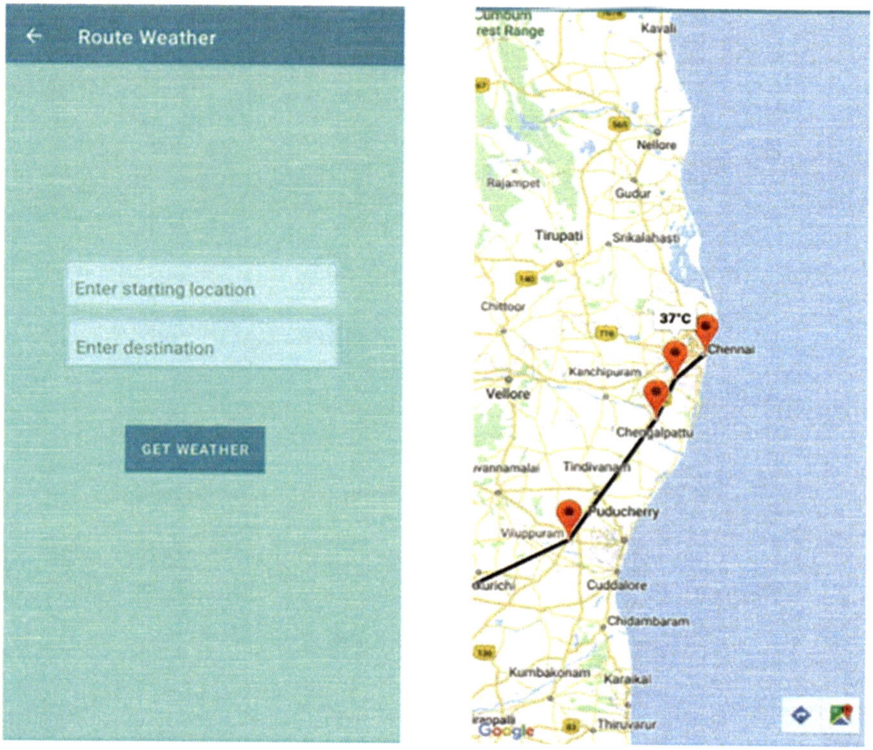

Fig. 14 Route weather screens

5.2 Analysis of the Results of Data Usage

Windy uses less data than AccuWeather and Weather and Radar applications. This reduces the cost which the user spends on Internet.

Also, Windy consumes more data than The Weather Channel application. This is because Windy provides more weather information than The Weather Channel application.

The other applications used in this experiment did not provide the data usage information, and so, they are excluded from this analysis.

5.3 Analysis of the Results of Response Time

All the activities, more or less, have the same response time. This is because all of them are fetching data from the Internet through Web APIs.

In test 1, the tourist weather activity has the highest response time. This is because the weather for specific tourist spots is fetched in this activity.

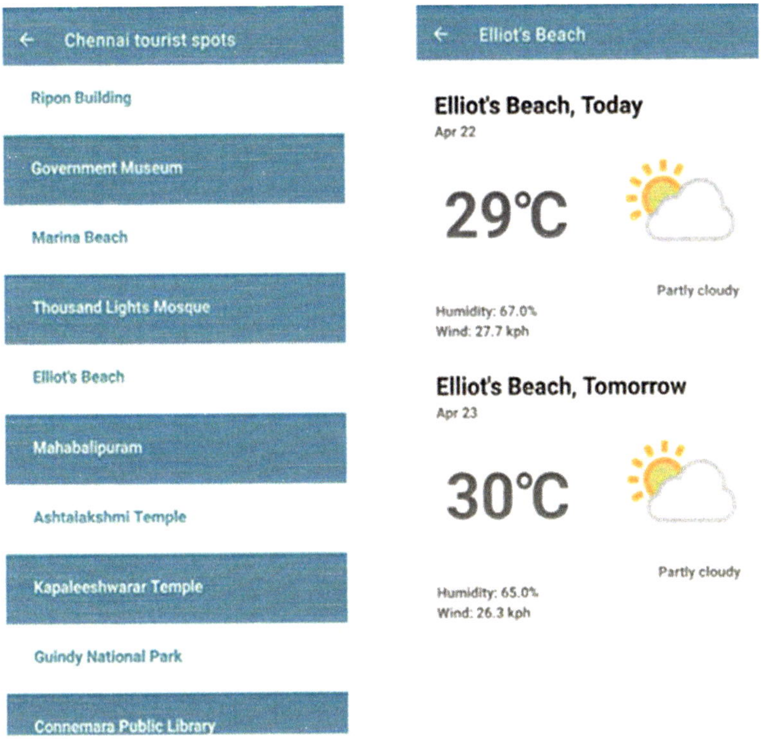

Fig. 15 Tourist weather screens

In test 2, the weather news activity has the highest response time. This is because the latest news have to be fetched from the API frequently.

In test 3, the live weather activity has the highest response time. This is because the location coordinates and temperature units have to be fetched in this activity from shared preferences.

On an average, the tourist weather activity has the highest response time (Figs. 18, 19, and 20).

6 Conclusion

In this research paper, we discussed the implementation details of the weather forecasting application Windy. To sum it all up, our application provides the hourly forecast for the current day, the three-hourly forecast for the next five days, the daily forecast for the next sixteen days, detailed weather forecasts, current weather forecast, weekly weather forecasts for the next seven days, air pollution information, sunrise, sunset, moonrise, and moonset information. Our application also provides

Fig. 16 Weather facts screen and weather news screen

weather forecast for tourist spots, weather along a route, weather facts, and recommendations of ways to prepare for the day based on the weather of the day. Moreover, the application allows the user to select the location and temperature units. Furthermore, the application sends weather alerts and weather notifications to the users. This application will prove to be useful to tourists, hustlers, travelers, and common people. This application will also be of great assistance during weather calamities. We have also compared the storage consumption and data usage of Windy with other weather applications. In the future, we plan to incorporate several other features in our application such as farm weather, weather chat, weather journal, weather videos, weather maps, and a camera which shows the weather for the location which is clicked through it, and a camera which has a wide variety of weather filters. The merits of this application are the accuracy, adaptability, reusability, easy-to-understand user interface, and so on. And the main disadvantage of this application is that it cannot be used in iOS phones, and it cannot be accessed through the Web.

Fig. 17 Day preparation screen

Table 1 Storage consumption of different weather applications

Name of the application	Storage (MB)
FarmWeather	29.29
Windy.com	40.37
The weather channel	199
AccuWeather	299
Weather and radar	80.48
Highway weather	19.49
Travel weather	29.29
Windy (our application)	106

Table 2 Data usage information of different weather applications

Name of the application	Data usage
FarmWeather	0B (no data available)
Windy.com	0B (no data available)
The weather channel	232.2 KB
AccuWeather	433.3 KB
Weather and radar	861.1 KB
Highway weather	0B (no data available)
Travel weather	0B (no data available)
Windy (our application)	290.5 KB

Table 3 Response time information of our weather forecasting application—Windy

Activity	Response time test 1 (ms)	Response time test 2 (ms)	Response time test 3 (ms)
Live weather activity	2260	3390	4030
Tourist weather activity	8430	2810	1610
Route weather activity	3220	1610	1450
Weather news activity	4750	5310	2580
Weather facts activity	2980	2660	2740
Day preparation activity	2650	2170	2500

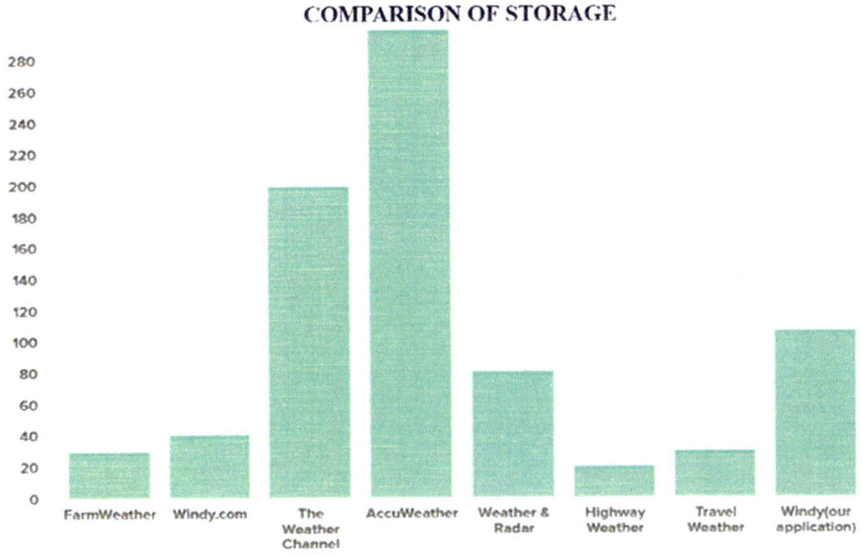

Fig. 18 Comparison of storage

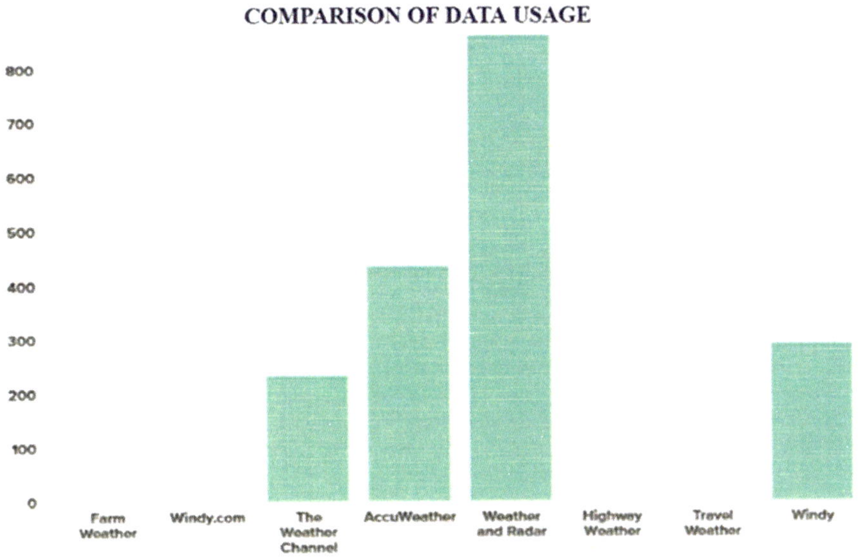

Fig. 19 Comparison of data usage

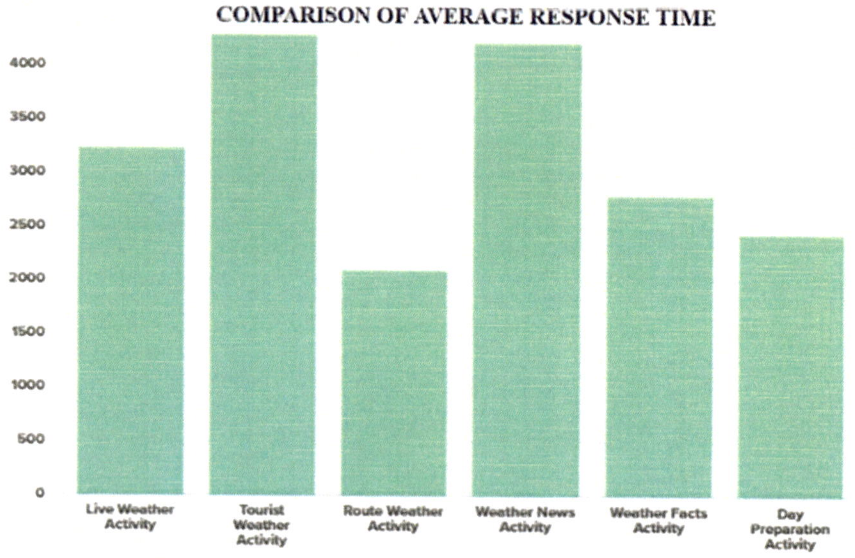

Fig. 20 Comparison of average response time in ms

References

1. B.G. Patel, Design and implementation of location based weather application on android platform. Int. J. Res. Appl. Sci. Eng. Technol. (IJRASET) (2017)
2. AccuWeather website: https://www.accuweather.com/
3. Weather Channel website: https://weather.com/en-IN/?Goto=Redirected
4. https://play.google.com/store/apps/details?id=com.yarafarmweather.farmweather&hl=en_IN&gl=US
5. https://play.google.com/store/apps/details?id=com.voyagegames.weatherroute&hl=en_IN&gl=US
6. https://play.google.com/store/apps/dev?id=5635094786271580047&hl=en_IN&gl=US
7. https://play.google.com/store/apps/details?id=pozzo.apps.travelweather&hl=en_IN&gl=US
8. https://play.google.com/store/apps/details?id=com.windyty.android&hl=en_IN&gl=US
9. S. Aleksandrov, S. Vulović, Creating an android weather forecast application in the android studio, in *7th International Scientific Conference Technics and Informatics in Education, Faculty of Technical Sciences* (Čačak, Serbia, 2018)
10. OpenWeatherMap website: https://openweathermap.org/api
11. Weatherbit website: https://www.weatherbit.io/
12. WeatherAPI website: https://www.weatherapi.com/
13. GNews website: https://gnews.io/
14. JSON Formatter: https://jsonformatter.curiousconcept.com/
15. https://www.udacity.com/course/android-basics-multiscreen-apps--ud839
16. https://classroom.udacity.com/courses/ud851
17. Android official documentation: https://developer.android.com/
18. https://classroom.udacity.com/courses/ud834
19. https://classroom.udacity.com/courses/ud0352
20. Google Cloud Platform: https://console.cloud.google.com/getting-started
21. Stack Overflow website: https://stackoverflow.com/

Enhanced Monotonic Activation Function in Convolutional Neural Network for Multiclass EEG Signal Classification

M. Bhuvaneshwari, E. Grace Mary Kanaga, and J. Anitha

1 Introduction

Brain–computer interface (BCI) is one of the emerging technologies that enable communication between the human brain and computer. People affected with paralysis experience impaired communication which might increase their mental illness despite physical illness. Recent technologies make effort to balance the situation; one such emerging technology is brain–computer interface (BCI). BCI technology attempts to make unimaginable innovations in future healthcare industries. This technology uses the computer to decode the brain signals pinned with the stimuli for EEG signal classification applications. Electroencephalogram (EEG) is one of the safest and non-invasive techniques for capturing human brain signals. In EEG signal acquisition, multiple electrodes are placed over the specific scalp regions to record the activities of the brain. EEG records the brain signals with a high temporal resolution which makes it adaptive for medical applications such as seizure detection [1], motor-imagery classification [2], and rehabilitation [3]. Apart from these applications, EEG data has also been applied recently to design a communicative device for paralytic people [4]. As EEG data has a high temporal resolution, low signal-to-noise ratio [5], and extremely high dynamic characteristics, extraction of robust features becomes a highly difficult and important step in communicative tool designing. The recent evolution in deep learning techniques boosted up the researchers to adopt them for analyzing large datasets [6]. Also, the automated feature extraction of deep

M. Bhuvaneshwari (✉) · E. Grace Mary Kanaga · J. Anitha
Karunya Institute of Technology and Sciences, Coimbatore, India
e-mail: bhvnshwari@gmail.com

E. Grace Mary Kanaga
e-mail: grace@karunya.edu

J. Anitha
e-mail: anitha_j@karunya.edu

J. D. Peter et al. (eds.), *Disruptive Technologies for Big Data and Cloud Applications*, Lecture Notes in Electrical Engineering 905,
https://doi.org/10.1007/978-981-19-2177-3_52

learning techniques [7] proved to be better than the hand-engineered method of feature extraction and prevents information loss [8].

1.1 Deep Learning

Deep learning has a wide variety of state-of-the-art architectures convolutional neural network (CNN), recurrent neural network (RNN), long short-term memory (LSTM), and autoencoder. Among other architectures, CNN is one of the popular and widely used architecture models in the recent decade. A survey by [9] states that CNN is widely employed in supervised feature learning-based applications.

CNN architecture has four important layers: convolutional layer (C), pooling layer, fully connected layer, and flattened layer. In the CNN model, convolutional layers are employed to extract the features from the raw data. In addition to the convolutional layer, activation functions are used to enhance the feature extraction. Convolutional layers at the initial step learn the local features and learn global features at the end. CNN with different activation functions enhances the feature learning mechanism. Hence, the proposed study aims to develop new activation functions that perform well than the existing activation functions.

The general architecture consists of various types of layers in the CNN model. Each layer contributes differently to classify the data. A convolutional layer is otherwise a filter that extracts specific features by multiplying the input values with the filter values which is called a feature map. This feature map diminishes the input dimensions, and the reduced feature map is passed on to the next layer to generate the output. The convolutional layer can also be stacked together to learn the most appropriate features in the early layer. This helps to identify the global features more precisely in the later layers. The pooling layer helps to reduce the dimensionality of the feature map. This pooling operation is applied separately to each feature map.

Deep learning techniques process the objective of the problem in the form of layers and the network as a whole. These interconnected layers are called hidden layers, and the final layer is called the output layer. The parameters like weight, bias, activation functions, and learning rate are applied to represent the values while configuring the network. The changes made in the parameter like weight tune the network to learn any complex function. These tuned parameters get applied to a function called activation function which provides nonlinearity to the network.

The classic activation functions experience saturation and vanishing gradient problem. Recent researches [10–12] focused on tuning the activation functions to overcome the issues in the existing activation function. Hence, designing an efficient activation function has become the recent research area among the deep learning fraternity. The proposed method aims to tune the activation function of the output layer. The proposed activation function ALU introduces algebraic functions to normalize the probability distribution of the output. The results are tested and evaluated against the classical activation functions and different standard datasets.

The remaining sections of this paper are organized as follows: Sect. 2 mentions some classical monotonic activation functions. Sect. 3 entails the proposed work, and the experimental results are discussed in Sect. 4. Section 5 concludes the paper.

2 Related Work

Activation functions are mathematical models attached to every neuron and determine its activation to predict the output of a neural network based on its input. Linear and nonlinear activation functions are two types of activation functions that are applied in recent researches. In the linear activation function, no transformation will take place, whereas in the nonlinear activation function the input value is transformed to a value that ranges between 0.0 and 1.0. According to [13], activation functions have a significant impact on the learning process of the neural network. The standard activation function does not depend on any learning parameters, and the changes in weights and biases are made only during the training phase. On the other hand, [14] introduced a learnable parameter to the activation function which has become more suitable for larger datasets.

The activation function tunes the output of the neuron by applying varying weights and bias of the neuron. The choice of the activation function greatly improves the performance of the network. Different monotonic activation functions like ReLU [15], Leaky ReLU [16], Softplus [17], Exponential Linear Unit [18], and Scaled Exponential Linear Unit [19] have been proposed to improve the classification performance.

ReLU is one of the widely applied activation functions in CNN. ReLU is defined as

$$y = \max(0, x) \begin{cases} x \text{ if } x > 0 \\ 0 \text{ if } x < 0 \end{cases} \tag{1}$$

where y is the output and x is the data. This generates the output in the range of 0–1 which infers that there is no flow of negative values in the network. In ReLU, the positive values are mapped to 1 and negative values to 0 which makes the network sparse and eases the computation process. Though ReLU overcomes the vanishing gradient problem, it experiences dead neuron proliferation which leads to the evolution of Leaky ReLU [20].

Leaky ReLU does this by restoring the negative part to a minimal value of 0.01 instead of 0 in ReLU. Despite improves the classification task it is found to be not stable hence exponential-based activation functions like ELU and SELU have been proposed.

ELU is a form of combinational activation function given by

$$g(x) = \begin{cases} a(e^x - 1 & \text{if } x < 0 \\ x & \text{if } x \geq 0 \end{cases} \tag{2}$$

which behaves like sigmoid and ReLU for negative and positive inputs, respectively. This combination makes the mean of the activation function reach toward zero that brings the steepest descent of the gradient close to the natural gradient which in turn accelerates the learning speed.

SELU is an improvization from ELU by introducing two fixed parameters α and λ whose values are derived from the input. These fixed parameters scale the output function of the negative input and are given by

$$g(x) = \lambda \begin{cases} x & \text{if } x > 0 \\ \alpha e^x - \alpha & \text{if } x \leq 0 \end{cases} \tag{3}$$

3 Methodology

A function g in domain Z is said to be monotonic if any pair $x, y \in Z$ with $x \leq y$ should eventually evolve $f(x) \leq f(y)$. Monotonic functions affect the cost function on tuning the weight of the network. This reduces the error in the loss function and performs the classification task well. The architecture of the proposed system is shown in Fig. 1.

3.1 Algebraic Linear Unit (ALU)

On reviewing other monotonic activation functions, most of the meta-units of the activation functions are expressed in exponential form. The proposed method is designed in a new way whose meta-units are expressed in algebraic functions form.

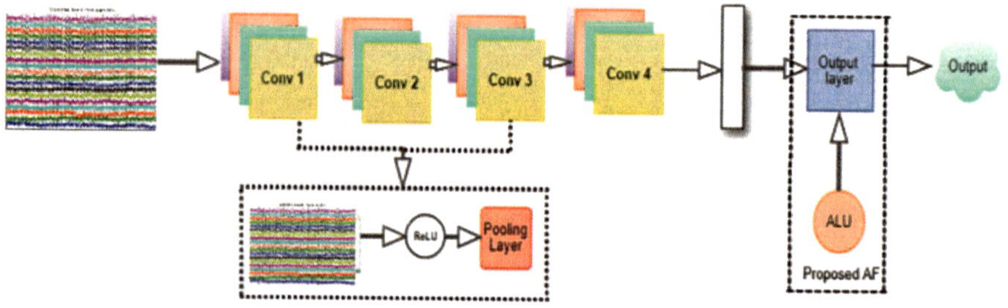

Fig. 1 Architecture of the proposed system

ALU is defined by

$$g(x) = \frac{1 + x}{\sqrt{1 + x^2}} \tag{4}$$

Its first derivative is defined by

$$g'(x) = \sqrt{x^2 + 1} + \sin h^{-1}(x) \tag{5}$$

Algorithm 1: Algorithm for CNN-ALU

Input: x and y are the features and labels of the dataset
Output: Accuracy Score
Required Parameters:
C as No of Convolutional Layers, f as Filter Size, e as no of epochs, O as Optimizer, L Loss Function
Begin

```
Initialize the f, C, e, O and L
for i=1 to C do
    Conv_i= Convolution(x_i-1, f)
    x_i=maxpooling(Conv_i)
end for
Repeat
    For each j in e do
        Conv_c= Convolution(x_i, f, activation_function=ALU())
        e= Max(Conv_c)
        Score=average(e[1],e[2]...e[n])
    End for
Until e_j < len(e)
function ALU():
    x=(1+x)/sqrt(1+pow(x,2))
    return x
```

Some properties of ALU are as follows:

- $g(x)$ is derivable $\forall x \in R$
- if $x = 0$, $g(x) = 1 g'(x) = 1$
- when $x > 0$, $g(x) > 0$ $g'(x) > 0$

Proof

$$\forall x \in R \frac{1 + x}{\sqrt{1 + x^2}} \in (0, \infty)$$

$$\text{When } x > 0, \quad g(x) > 0$$

$$x < 0, \quad g(x) < 0$$

The algebraic meta-units in the proposed activation functions allow the negative input quadrant and zero derivative values inside the network. Thus, ALU enhances the probability distribution in the output layer much better than other activation functions. The algorithm of the proposed activation function is shown in Algorithm 1.

4 Experimental Design

4.1 Data Acquisition

4.1.1 Stimuli

There are six different images, each representing the generic needs of the subjects, flickering at various frequencies ranging from 5 to 20 Hz. Each image is displayed for 10 s followed by 10 s of interval. The images are selected and designed by the author who is a non-subject.

4.1.2 Participants

Seven right-handed healthy subjects aging from 20 to 24 participated voluntarily in this experiment. None of them have visual impairments. All the subjects were instructed with the general guidelines and overall objective of the application. Few such instructions are avoiding body and facial movements, sleeping, and deficit attention.

4.1.3 Experimental Protocol

The subjects strictly followed the guidelines during the experiment. The entire experiment includes displaying six flickering images; each flickered for 10 s in a specific frequency with 10 s of interval between each image. Before recording the stimuli response, one minute of the preparatory period is allotted to prompt the start of the experiment. The subjects are instructed to concentrate only on the screen during the recording of brain signals.

4.1.4 Equipment and Environment

The stimuli response was recorded by a g-tec recorder powered by a g-USB amplifier. The experiment used 16 electrodes that are placed following the international 10–20 electrode placement system with a sampling rate of 256 Hz. The experiment was carried out in a well-ventilated and noiseless environment.

4.2 Standard Dataset

4.2.1 EEG-SSVEP Dataset

This dataset contains SSVEP signals for the visual stimulation of 11 subjects. The EEG signals have been acquired from 256 channels for five different frequencies of 6.66, 7.50, 8.57, 10.00, and 12.00 Hz with a sampling frequency of 250 Hz.

4.2.2 BCI-SSVEP Dataset

The database contains the visual stimuli recording of 30 subjects who were subjected to focus visual stimuli. EEG signal has been acquired from 16 channels with a sampling frequency of 128 Hz. Each subject performed different tests and is grouped. These tests include three different five-box tests and two visual image searches.

4.2.3 BCI Competition IV 2a

This dataset contains EEG signals of nine subjects with a sampling frequency of 250 Hz. Signals have been recorded from 22 EEG channels and 3 EOG channels. Subjects performed four different motor-imagery tasks for two sessions, each session recorded on different days. Each session includes 6 runs with a short interval and comprises 48 trials per run.

The proposed activation function is embedded with CNN and is evaluated with three standard datasets, such as EEG-SSVEP dataset, BCI-SSVEP dataset, and BCI Competition IV 2a, and acquired dataset. In order to reduce the time and space complexities, the data is stored in the cloud. The summary of the experimented datasets is shown in Table 1. The performance of the proposed method is also compared with classical activation functions.

Table 1 Summary of EEG datasets

Datasets	Type	Subjects	Class	Channels
EEG-SSVEP dataset (Exp 1)	Visual stimuli	11	5	256
BCI-SSVEP dataset (Exp 2)	Visual stimuli	30	5	16
BCI Competition IV 2a (Exp 3)	Motor imagery	9	4	22
Acquired dataset (Exp 4)	Visual stimuli	7	6	16

Table 2 Parameters of the implemented CNN architecture

Parameter	Values
Convolutional layers	4
Dimension of convolutional filters	2 × 2
Pooling layers	2
Optimizer	Adadelta optimizer
Hidden layer neurons	3000
Fully connected layers	1
Pooling block size	2 × 2
Dropout probability	0.5
Training rate	10^{-5}
Activation functions	Proposed and classical activation functions

4.3 CNN Parameter Configuration

This paper uses CNN architecture to implement the proposed activation function, and the performance of the model is compared with classical activation function with the same CNN architecture. CNN architecture of this paper has four convolutional layers, with a standard window size of 2 * 2. This paper splits the training and testing data in the ratio of 70:30 using train_test_split of sklearn package in Python. The architecture uses the mean squared logarithmic error loss function and Adadelta optimizer throughout the experimental analysis. Classification accuracy is evaluated by changing the activation function of the output layer. For the training process of all four datasets, the number of epochs is fixed to 100. The different parameter used in the implemented CNN is tabulated in Table 2.

5 Result and Analysis

The proposed activation function can model nonlinear data more robust with minimal data loss. The proposed monotonic function directly influences the cost function when the input of the network changes. This in turn reduces the data loss and boosts the classification performance of the network.

5.1 Performance Metrics

The evaluation of the proposed activation function with other classical activation functions is mandatory to explore its performance. In this paper, performance of the proposed functions is evaluated by adopting the following metrics:

$$\text{Accuracy} = \frac{\text{TP} + \text{TN}}{\text{TP} + \text{TN} + \text{FP} + \text{FN}} * 100 \tag{6}$$

$$\text{Precision} = \frac{\text{TP}}{\text{TP} + \text{FP}} \tag{7}$$

$$\text{Recall} = \frac{\text{TP}}{\text{TP} + \text{FN}} \tag{8}$$

$$F1\text{-Score} = 2 * \frac{\text{Precision} * \text{Recall}}{\text{Precision} + \text{Recall}} \tag{9}$$

Performance evaluation results of ALU with classical activation function and different datasets are summarized in Table 3. All the datasets employed in this paper are multiclass classification data. The classification results across different datasets for different activation functions are determined and tabulated. The shape of the proposed activation function and other classical activation functions is shown in Fig. 2a, b, respectively. The shape proves that the proposed method reveals that both the positive and negative inputs increase monotonically and makes the data distribution more robust. Both the proposed activation function and its derivative are monotonic.

Table 3 shows the classification results of the proposed activation function and classical activation function considered in this paper with CNN architecture across

Table 3 Classification accuracies of different activation functions

Activation function	Metrics	Sigmoid	Tanh	Leaky ReLU	Softplus	ELU	SELU	Proposed AF
Exp 1	Accuracy	82.44	83.14	81.46	84.44	85.16	84.48	86.34
	Precision	0.78	0.81	0.81	0.81	0.81	0.81	0.81
	Recall	0.73	0.84	0.84	0.82	0.87	0.87	0.84
	*F*1 score	0.75	0.82	0.82	0.81	0.84	0.82	0.82
Exp 2	Accuracy	86.39	84.59	83.96	87.98	84.09	84.71	88.91
	Precision	0.84	0.81	0.8	0.86	0.82	0.83	0.80
	Recall	0.81	0.80	0.82	0.88	0.80	0.85	0.83
	*F*1 score	0.82	0.80	0.81	0.87	0.81	0.84	0.81
Exp 3	Accuracy	86.91	87.36	86.33	89.4	88.31	86.81	89.45
	Precision	0.85	0.86	0.84	0.86	0.85	0.82	0.88
	Recall	0.87	0.83	0.85	0.89	0.86	0.81	0.89
	*F*1 score	0.86	0.84	0.85	0.87	0.85	0.81	0.88
Exp 4	Accuracy	80.32	78.4	79.93	82.14	80.38	81.67	83.56
	Precision	0.78	0.79	0.78	0.81	0.80	0.80	0.80
	Recall	0.80	0.77	0.80	0.80	0.77	0.81	0.81
	*F*1 score	0.79	0.78	0.79	0.80	0.78	0.80	0.80

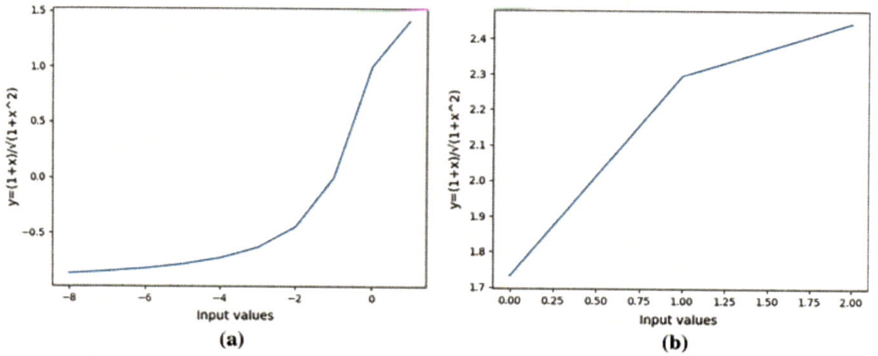

Fig. 2 Shape of **a** ALU, **b** derivative of ALU

multiple datasets. Based on the experimental results, it is made clear that ALU classifies better than other classical monotonic activation functions. The proposed ALU performs better than other monotonic functions like Leaky ReLU, Softplus, ELU, and SELU.

Exp 1, Exp 2, Exp 3, and Exp 4 have been performed with the classical activation function and proposed Activation function, and the results are tabulated in Table 3. The performance of the system is measured with the metrics such as accuracy, precision, recall, and $F1$ score, and it has been shown in Fig. 3.

Figure 3a depicts the classification performance on different datasets graphically. In Exp 1, the classification performance of sigmoid, tanh, Leaky ReLU, Softplus, ELU, SELU, and ALU is 82.44%, 83.14%, 81.46%, 84.44%, 85.16%, 84.48%, and 86.34%, respectively. Compared to Leaky ReLU, one of the widely adapted monotonic activation functions, the ALU classifies 4% higher. In Exp 2, the classification results of sigmoid, tanh, Leaky ReLU, Softplus, ELU, SELU, and ALU are 86.39%, 84.59%, 83.96%, 87.98%, 84.09%, 84.71%, and 88.91%, respectively. Here, ALU performs 5% better than Leaky ReLU. Similarly for Exp 3 and Exp 4, the ALU outperforms 3% and 4%, respectively, than Leaky ReLU.

6 Conclusion

The activation function is the key ingredient of the deep learning architectures. The paper tunes the existing linear units by introducing some algebraic functions into them. Using those algebraic functions, ALU gains the flexibility to train the neural network for dynamic inputs. An extensive experiment has been conducted to evaluate the performance of the proposed method. The comprehensive analysis concluded that the performance of the ALU has a positive influence on the classification result of all the datasets. On comparing with other monotonic activation functions, the performance of the ALU improves from 1.5 to 4% accurately than other activation

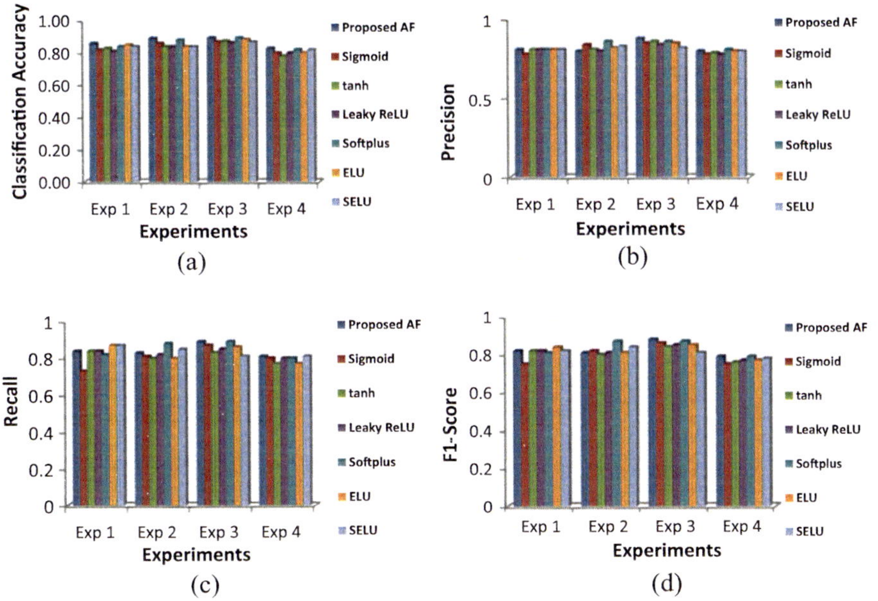

Fig. 3 Performance comparison of different experiments with various activation functions **a** classification accuracy, **b** precision, **c** recall, **d** $F1$ score

functions. ALU performs better than other monotonic activation functions for multi-class classification problems. This monotonic activation function will be applied to a communication tool that applies deep learning-based classification techniques to identify the needs of the paralytic people based on stimuli classification results. Also, handling these types of data in cloud environment is effective in increasing the processing speed. Future work of this research aims at extending this type of monotonic activation functions to other deep learning architectures.

Acknowledgements This work is supported by the Department of Science and Technology—ICPS, Ministry of Science and Technology, Government of India [grant number: DST/ICPS/CLUSTER/Data Science/2018/General].

Conflict of Interest The authors declare no conflict of interest.

References

1. H. Niknazar, S.R. Mousavi, M. Niknazar, V. Mardanlou, B.N. Coelho, Performance analysis of EEG seizure detection features. Epilepsy Res. **167**(October), 106483 (2020). https://doi.org/10.1016/j.eplepsyres.2020.106483
2. M. Rajabioun, Motor imagery classification by active source dynamics. Biomed. Signal Process. Control **61**, 102028 (2020). https://doi.org/10.1016/j.bspc.2020.102028

3. S. Lou, X. Xue, Application of electroencephalographic (EEG) biofeedback therapy in the rehabilitation of patients with chronic diseases. Psychiatry Res. **293**(August), 113371 (2020). https://doi.org/10.1016/j.psychres.2020.113371
4. A. Kübler, Brain-computer interfaces for communication in paralysed patients and implications for disorders of consciousness. Neurol. Conscious. 217–233 (2009). https://doi.org/10.1016/B978-0-12-374168-4.00017-4
5. B. Burle, L. Spieser, C. Roger, L. Casini, T. Hasbroucq, F. Vidal, Spatial and temporal resolutions of EEG: Is it really black and white? A scalp current density view. Int. J. Psychophysiol. **97**(3), 210–220 (2015). https://doi.org/10.1016/j.ijpsycho.2015.05.004
6. N. Padfield, J. Zabalza, H. Zhao, V. Masero, J. Ren, EEG-based brain-computer interfaces using motor-imagery: techniques and challenges. Sensors (Switzerland) **19**(6), 1–34 (2019). https://doi.org/10.3390/s19061423
7. S.U. Amin, M. Alsulaiman, G. Muhammad, M.A. Mekhtiche, M. Shamim Hossain, Deep learning for EEG motor imagery classification based on multi-layer CNNs feature fusion. Futur. Gener. Comput. Syst. **101**, 542–554 (2019). https://doi.org/10.1016/j.future.2019.06.027
8. W. Qiao, & X. Bi, Deep spatial-temporal neural network for classification of EEG-based motor imagery, in *ACM International Conference Proceeding Series* (2019), pp. 265–272 . https://doi.org/10.1145/3349341.3349414
9. Y. Roy, H. Banville, I. Albuquerque, A. Gramfort, T.H. Falk, J. Faubert, Deep learning-based electroencephalography analysis: a systematic review (2019). *ArXiv*
10. A. Apicella, F. Isgrò, R. Prevete, A simple and efficient architecture for trainable activation functions. Neurocomputing (2019). https://doi.org/10.1016/j.neucom.2019.08.065
11. M. Tanaka, Weighted sigmoid gate unit for an activation function of deep neural network. Pattern Recogn. Lett. **135**, 354–359 (2020). https://doi.org/10.1016/j.patrec.2020.05.017
12. K. Vijayaprabakaran, K. Sathiyamurthy, Towards activation function search for long short-term model network : a differential evolution based approach. J. King Saud Univ. Comput. Inf. Sci. (2020). https://doi.org/10.1016/j.jksuci.2020.04.015
13. Y. Guo, L. Sun, Z. Zhang, H. He, Algorithm research on improving activation function of convolutional neural networks, in *Proceedings of the 31st Chinese Control and Decision Conference, CCDC 2019* (2019), pp. 3582–3586. https://doi.org/10.1109/CCDC.2019.8833156
14. K. He, X. Zhang, S. Ren, J. Sun, Delving deep into rectifiers: Surpassing human-level performance on imagenet classification, in *Proceedings of the IEEE International Conference on Computer Vision, 2015 Inter* (2015), 1026–1034. https://doi.org/10.1109/ICCV.2015.123
15. K. Eckle, J. Schmidt-hieber, K. Eckle, J. Schmidt-hieber, A comparison of deep networks with ReLU activation function and linear spline-type methods. Neural Netw. (2018). https://doi.org/10.1016/j.neunet.2018.11.005
16. F. Agostinelli, M. Hoffman, P. Sadowski, P. Baldi, Learning activation functions to improve deep neural networks, in *3rd International Conference on Learning Representations, ICLR 2015—Workshop Track Proceedings, 2013* (2015), pp. 1–9
17. C. Dugas, Y. Bengio, F. Bélisle, C. Nadeau, R. Garcia, Incorporating second-order functional knowledge for better option pricing, in *Advances in Neural Information Processing Systems* (2001)
18. D.A. Clevert, T. Unterthiner, S. Hochreiter, Fast and accurate deep network learning by exponential linear units (ELUs), in *4th International Conference on Learning Representations, ICLR 2016—Conference Track Proceedings* (2016), pp. 1–14
19. A. Madasu, V.A. Rao, *Effectiveness of Self Normalizing Neural* (n.d.)
20. J. Xu, Z. Li, B. Du, M. Zhang, J. Liu, Reluplex made more practical: Leaky ReLU, in *Proceedings—IEEE Symposium on Computers and Communications, 2020-July* (2020). https://doi.org/10.1109/ISCC50000.2020.9219587

BPSO-PSO-SVM: An Integrated Approach for Cancer Diagnosis

Amrutanshu Panigrahi, Santosini Bhutia, Bibhuprasad Sahu, Mohammad Gouse Galety, and Sachi Nandan Mohanty

1 Introduction

Cancer is caused due to some of the irregular changes in genes that play a vital role in the development of the human cell. These modifications in genes incorporate the transformation of the DNA structure which is responsible for developing the genes. For the most part, cancer cells display fundamentally more hereditary modification as compared to ordinary cells, albeit tumors indicate some distinctive explicit combination of genetic modifications in various individuals [1]. A couple of changes in the genetic behavior may cause the cancer disease. But the main challenge behind the issue is to detect cancer in its early stage by means of which the appropriate remedies can be carried out. Hence, the early detection of cancer may increase the recovery rate. Machine learning suggests various models by using the patient test results. As per the WHO and NCI in 2018, approximately 18 million people have been detected for cancer out of which around 10 million patients have died due to the unavailability of a proper diagnosis system. So, cancer classification has emerged as a major area of research. Many researchers have proposed various cancer classification models by using machine learning. Machine learning is now day provides

A. Panigrahi · S. Bhutia
Department of Computer Science and Engineering, Siksha'O' Anusandhan (Deemed to Be University), Bhubaneswar, India

B. Sahu (✉)
Department of Artificial Intelligence and Data Science, Vardhaman College of Engineering (Autonomous), Hyderabad, India
e-mail: prasadnikhil176@gmail.com

M. G. Galety
Department of Information Technology, Catholic University in Erbil, Erbil, Iraq

S. N. Mohanty
Department of CSE, College of Engineering, Pune, India

J. D. Peter et al. (eds.), *Disruptive Technologies for Big Data and Cloud Applications*, Lecture Notes in Electrical Engineering 905,
https://doi.org/10.1007/978-981-19-2177-3_53

various AI based models to detect and classify types of cancer [2]. The model for classifying the disease can be constructed by using either the biopsy dataset or the microarray dataset. The biopsy data contain a very limited number of information which puts a constraint in taking it for model development. To construct an accurate model, more genetic information is being required by the model, and the microarray data give that scope to the developer [3]. The microarray data will have a huge number of genetic information based on which the model can be trained properly to provide high accuracy. But the main concern in applying the microarray data is the high-genetic information with the lower number of samples which is also known as high dimensionality.

The high dimension of the microarray data causes high-computational cost and high-computational complexity. Along with these, if any ML algorithm is applied directly, then the model will suffer from the small sample size (SSS). The solution for the SSS is to reduce the dimension of the dataset. IT can be done in the following two approaches [4].

- Feature Selection—This will remove some irrelevant gene expression data and only keeps the useful information for classification purpose.
- Feature Extraction—It will add some features to the missing values of the microarray dataset.

The primary objective of the microarray data analysis is to define a model from the provided data which will predict the class of the unknown sample. The accuracy of the ML model will depend upon the outlier present in the microarray dataset. The irrelevant data not only create the noise but also include some unnecessary noise to gene expression analysis [5]. So the dimensionality reduction technique plays a vital role whenever the ML model is deployed to handle the microarray dataset. There are several approaches such as principal component analysis (PCA), linear discriminant analysis (LDA), independent component analysis (ICA), and random forest (RF) can be used for dimensionality reduction purposes. In addition, the feature selection techniques based on searching strategies can be listed as follows. The filter feature selection techniques include correlation-based feature selection (CFS), fast correlation-based filter (FCBF), INTERACT algorithm, information gain, relief, minimum redundancy maximum relevance (mRMR), and consistency-based filter (CBF).

The wrapper feature selection techniques include distance sensitive rival penalized competitive learning—support vector machine (ADSRPCL-SVM), genetic algorithm—support vector machine (GA-SVM), particle swarm optimization—genetic programming (PSO-GP), genetic algorithm—kernel density estimation—Bayes (GA-KDE-Bayes), and successive feature selection (SFS). The embedded feature selection techniques include SVM based on recursive feature elimination (SVM-RFE), kernel penalized SVM (KP-SVM), probably approximately correct—Bayes (PAC-Bayes), iterative perturbation method (IFP), random forests (RF), and first order inductive learner (FOIL) rule-based feature subset selection (FRFS).

The hybrid feature selection techniques include correlation-based feature selection, mRMR-GA, Chi-GA, PCA-ACO, multiple filter multiple wrapper (MFMW), stratified random forest (SRF), and relief—mRMR—GA (R-m-GA) [6–9].

In the current study, BPSO has been considered as the feature selection algorithm, and to the featured dataset, the PSO algorithm has been implemented for the optimization purpose. And finally to the optimized and featured dataset, the SVM has been implemented as the classifier.

The structure of the paper is as follows. Section 2 deals with the associated cancer classification work. Section 3 will focus on the methodology that has been used during the research work. Section 4 heads the experimental analysis part, and finally, the conclusion has been drawn in Sect. 5.

2 Related Work

In [10], the author has used the SVM and NN classification technique to the lung cancer microarray data and found the classification accuracy for the proposed model is about 98%. The author in [11] CNN has been implemented for cervical cancer classification purposes, and the accuracy obtained is 99.5%. Ragab et al. in [12] had used DCNN for classification of breast cancer microarray data and have compared the result with the SVM-PCA model and found that the classification accuracy becomes 98% for DCNN, and it is quite high as compared to the SVM-PCA model.

The author in [13] had implemented ANN for lung cancer microarray dataset and had compared the results with SVM, KNN state of the machine learning algorithms and found that for the proposed model, the CA becomes 98.6%. The author in [14] has implemented LR, SVM with stochastic gradient descent optimization, and MPN for classification using voting mechanism and found classification accuracy for the ensemble method for breast cancer dataset becomes 99.42%. In [15], KNN has been taken as the classifier for breast cancer dataset tetrolet feature extraction algorithm, and the author has obtained 95% classification accuracy. In [16], the author had used SVM as the classification algorithm and GWO as the optimization technique with breast and colon cancer datasets and obtained the classification accuracy of 97%. In [17], the author has used the SCF-SVM model for gene expression data and obtained the classification accuracy of 96%. In [18], mRMR has been used as the feature selection algorithm, and the BBHA algorithm has been taken as the optimization algorithm. SVM is being used as the classification algorithm and the breast cancer, and colon cancer dataset has been taken for experimental purpose. The author has obtained the classification accuracy for the abovesaid model as 97%.

Table 1 Dimension of the used dataset

Dataset	Size
Breast cancer	33 × 2198
Prostate cancer	55 × 2796
Cervical cancer	36 × 1794
Colon cancer	64 × 2250
Lung cancer	182 × 1628
Brain cancer	27 × 1072

3 Dataset and Methodology

3.1 Dataset Description

Different cancer microarray dataset has been considered for the analysis of proposed model such as the breast cancer, prostate cancer, cervical cancer, colon cancer, lung cancer, and brain cancer having different samples and attributes [19]. The attributes matrix is specified in Table 1.

3.2 SVM

Support vector machine (SVM) is a supervised machine learning algorithm that draws an optimal hyperplane so that the class present in the dataset can be separated. The hyperplane focuses on maximizing the margin in between the two classes of the corresponding dataset, and the margin is the distance of support vectors or SVs from the obtained hyperplane. Initially, the SVM was being developed for classifying the linear data. But the introducing the hyperparameter such as the kernel trick, the SVM can be successfully implemented in a huge dataset. Excluding the kernel trick, the other hyperparameters such as the cost and gamma function can be tripped for better performance. The available kernels of SVM are RBF (R), linear (L), polynomial (P), and sigmoid (S). In the current research work, we have included all kinds of the kernel for better understanding.

3.3 PSO

Particle swarm optimization (PSO) is an evolutionary algorithm that stimulates the behavior of various natural aspects such as bird flocking. PSO starts with a random population of swarms. Each swarm moves in the selected search space with the dimension D. During the movement, every particle keeps updating its corresponding position. The current position of different vectors is represented as the vector $Z =$

$\{zi_1, zi_2, zi_3, \ldots zi_D\}$with a velocity limited to VLmax. Every swarm is allowed to move inside the search space with a velocity range [-VLmax, VLmax]. From the position vector, the local best for every particle has been found out from which the global best is obtained as the optimal solution.

3.4 Binary PSO

MI or mutual information is being used as the feature selection algorithm for finding out the relevance present in between two different attributes. The main objective is to choose n number of attributes with minimum redundancy. So BPSO considers both relevancy and redundancy as the criteria for the feature selection. The fitness function F is being calculated by using Eq. (1).

$$F = \sum_{x \in X} \text{IG}(x; \text{cl}) - \sum_{xi,xj \in X} \text{IG}(xi; xj) \quad (1)$$

IG is the information gain, X is the selected features, and cl is the class level of the dataset. The following are the steps for BPSO.

Step 1. Find out the distribution factor for the dataset and determine the training, testing dataset.
Step 2. Initialize the position and velocity for every particle.
Step 3. If maximum iteration does not exceed, then

- Find out the local best.
- Find out the global best.
- Detect the position.

Step 4. Return the best features based on the global best.

3.5 Methodology

In the current research work, we have considered BPSO, PSO, and SVM for six different kinds of cancer datasets. Initially, the microarray dataset has been considered for the feature selection purpose to reduce the dimensionality, and BPSO does the same. After applying the BPSO, a featured dataset is being obtained as the output on which the PSO has been implemented as the optimizer resulting in an optimized and featured dataset. To the obtained dataset, the SVM classification algorithm with different kernel tricks has been implemented to classify all of the datasets linearly as the number of the class present in all of the datasets is 2. Finally, the performance of the proposed model is measured on the basis of key parameters such as accuracy, specificity, sensitivity, and F1 score. Figure 1 shows the flow of the proposed model.

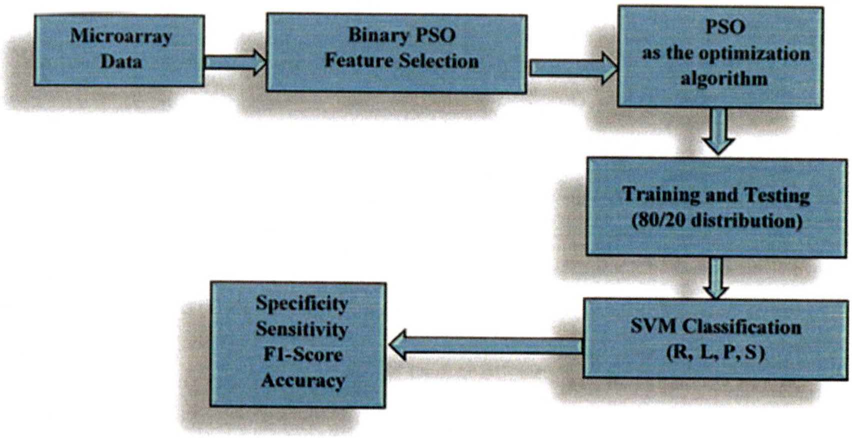

Fig. 1 Block diagram of BPSO-PSO-SVM model

4 Empirical Analysis

To implement the proposed work, Python 3.0.3 has been taken as the primary environment with Anaconda IDE with a system having i5 processor, 8 GB RAM, and Windows 10 OS. BPSO and PSO have been implemented as the feature selection and optimization algorithm, respectively, while the SVM is applied to the featured matrix with its kernel trick. Table 2 shows the results of the BPSO-PSO-SVM model. Figure 2 indicates the performance comparison concerning 4 key parameters with six different microarray datasets.

5 Conclusion

We have briefed the concepts of the machine learning microarray data and its importance on cancer classification. Simultaneously, we have summarized the basic difficulties faced during the microarray data analysis. To overcome the well-known issues of the microarray data, the feature selection and feature extract are the possible solutions present. In our research, we have focused on the feature selection approach. We have proposed a model based on BPSO-PSO-SVM, where BPSO has been considered as the feature selection algorithm, PSO as the optimization technique, and VM as the classification technique. The experimental analysis shows that the SVM with RBF kernel shows the highest CA as 97.7%.

Table 2 Accuracy of SVM with different kernel

	Kernel	Specificity	Sensitivity	$F1$ score	Accuracy
Breast cancer	R	0.979	0.936	0.955	0.977
	L	0.962	0.958	0.944	0.966
	P	0.965	0.954	0.943	0.958
	S	0.923	0.939	0.952	0.923
Cervical cancer	R	0.955	0.954	0.956	0.956
	L	0.954	0.953	0.949	0.945
	P	0.934	0.938	0.937	0.935
	S	0.92	0.93	0.923	0.918
Colon cancer	R	0.935	0.934	0.934	0.942
	L	0.944	0.942	0.952	0.937
	P	0.964	0.963	0.962	0.953
	S	0.952	0.953	0.952	0.95
Prostate cancer	R	0.964	0.963	0.966	0.963
	L	0.945	0.943	0.945	0.951
	P	0.944	0.943	0.944	0.945
	S	0.952	0.95	0.955	0.954
Lung cancer	R	0.923	0.933	0.93	0.938
	L	0.955	0.945	0.948	0.958
	P	0.941	0.942	0.95	0.953
	S	0.96	0.94	0.956	0.954
Brain cancer	R	0.935	0.943	0.933	0.939
	L	0.944	0.965	0.94	0.94
	P	0.935	0.942	0.946	0.943
	S	0.953	0.952	0.951	0.956

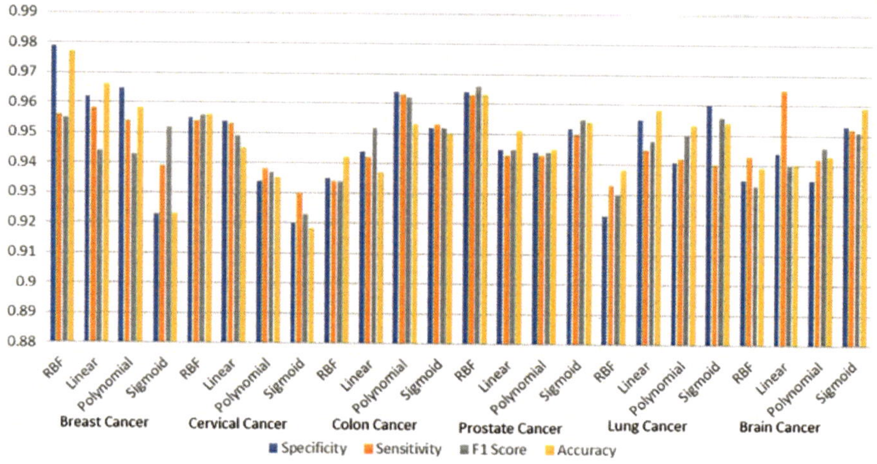

Fig. 2 Specificity, sensitivity, $F1$ score, and accuracy BPSO-PSO-SVM

References

1. M. Amrane, S. Oukid, I. Gagaoua, T. Ensari, Breast cancer classification using machine learning, in *2018 Electric Electronics, Computer Science, Biomedical Engineerings' Meeting (EBBT)* (IEEE, 2018), pp. 1–4
2. A.K. Dwivedi, Artificial neural network model for effective cancer classification using microarray gene expression data. Neural Comput. Appl. **29**(12), 1545–1554 (2018)
3. T.J. Brinker, A. Hekler, J.S. Utikal, N. Grabe, D. Schadendorf, J. Klode, C. Berking, T. Steeb, A.H. Enk, C. Von Kalle, Skin cancer classification using convolutional neural networks: systematic review. J. Med. Internet Res. **20**(10), e11936 (2018)
4. S.M. Ayyad, A.I. Saleh, L.M. Labib, Gene expression cancer classification using modified K-nearest neighbors technique. Biosystems **176**, 41–51 (2019)
5. L. Dora, S. Agrawal, R. Panda, A. Abraham, Optimal breast cancer classification using Gauss-Newton representation based algorithm. Expert Syst. Appl. **85**, 134–145 (2017)
6. D. Kashyap, A. Somani, J. Shekhar, A. Bhan, M.K. Dutta, R. Burget, K. Riha, Cervical cancer detection and classification using Independent Level sets and multi SVMs, in *2016 39th International Conference on Telecommunications and Signal Processing (TSP)* (IEEE, 2016), pp. 523–528
7. S.K. Majumder, N. Ghosh, P.K. Gupta, Support vector machine for optical diagnosis of cancer. J. Biomed. Opt. **10**(2), 024034 (2005)
8. A.V.D. Sánchez, Advanced support vector machines and kernel methods. Neurocomputing **55**(1–2), 5–20 (2003)
9. X. Liu, A. Krishnan, A. Mondry, An entropy-based gene selection method for cancer classification using microarray data. BMC Bioinf. **6**(1), 1–14 (2005)
10. P. Nanglia, S. Kumar, A.N. Mahajan, P. Singh, D. Rathee, A hybrid algorithm for lung cancer classification using SVM and neural networks. ICT Express (2020)
11. A. Ghoneim, G. Muhammad, M.S. Hossain, Cervical cancer classification using convolutional neural networks and extreme learning machines. Futur. Gener. Comput. Syst. **102**, 643–649 (2020)
12. D.A. Ragab, O. Attallah, M. Sharkas, J. Ren, S. Marshall, A framework for breast cancer classification using multi-DCNNs. Comput. Biol. Med. **131**, 104245 (2021)

13. V. Patel, S. Shah, H. Trivedi, U. Naik, An analysis of lung tumor classification using SVM and ANN with GLCM features, in *Proceedings of 1st International Conference on Computing, Communications, and Cyber-Security (IC4S 2019)* (Springer, Singapore, 2020), pp. 273–284
14. S.K. Baliarsingh, S. Vipsita, B. Dash, A new optimal gene selection approach for cancer classification using enhanced Jaya-based forest optimization algorithm. Neural Comput. Appl. **32**(12), 8599–8616 (2020)
15. A.A. Ali, S. Mishra, B. Dappuri, Breast cancer classification using tetrolet transform based energy features and K-nearest neighbor classifier, in *Recent Trends and Advances in Artificial Intelligence and Internet of Things* (Springer, Cham, 2020), pp. 39–46
16. M. Loey, M.W. Jasim, H.M. El-Bakry, M.H.N. Taha, N.E.M. Khalifa, Breast and colon cancer classification from gene expression profiles using data mining techniques. Symmetry **12**(3), 408 (2020)
17. D. Nagajyothi, R. Addagudi, T. Gunda, S. Santhoshi Logitla, Detection of lung cancer using SVM classifier. Int. J. **8**(5) (2020)
18. E. Pashaei, E. Pashaei, Gene selection for cancer classification using a new hybrid of binary black hole algorithm, in *2020 28th Signal Processing and Communications Applications Conference (SIU)* (IEEE, 2020), pp. 1–4
19. https://archive.ics.uci.edu/ml/index.php

Detecting the Lateral Movement in Cyberattack at the Early Stage Using Machine Learning Techniques

Ashwathy Anda Chacko, Bijolin Edwin, and M. Roshni Thanka

1 Introduction

Advanced persistent threat (APT) is a sophisticated, sustained, elongated, and an aimed cyberattack. The intruder gets a hold to a network and remains unnoticed in the long run. The goal of APT is data theft. Advanced persistent threats have caused several large, costly data breaches [1]. Some examples of advanced persistent threats are GhosNet, Stuxnet, DeepPanda, APT28, etc. Early detection of APT [2] is a crucial factor to mitigate its damage. Starting with reconnaissance, moving on into delivering an attack which causes the initial intrusion detection more accurate [3, 4]. Once inside, the threat tries to contact a command-and-control server which later is used to move laterally inside the network and finally fulfilling the goal by extracting the sensitive data which the attacker tries to steal. Arguably, the most critical stage in this chain of events is the lateral movement (LM) phase. Here, the attacker tries to map the inside of the network and gain access to as much as possible of the network. Stopping the attack before the goal is achieved is the only way to eliminate the expensive costs of these advanced cyberattacks.

As the cyber-threats are continuing, the traditional security measures that we use are not enough to stop these cyberattacks. Now itself, some companies have started combining big data analytics and machine learning to provide a more powerful approach to cybersecurity. When we use both big data analytics and machine learning, big data analytics help to aggregate and manage the huge volume of data. And

A. A. Chacko · B. Edwin (✉) · M. R. Thanka
Karunya Institute of Technology and Science, Coimbatore, Tamil Nadu, India
e-mail: bijolin@karunya.edu

A. A. Chacko
e-mail: ashwathy@karunya.edu.in

M. R. Thanka
e-mail: roshni@karunya.edu

J. D. Peter et al. (eds.), *Disruptive Technologies for Big Data and Cloud Applications*, Lecture Notes in Electrical Engineering 905,
https://doi.org/10.1007/978-981-19-2177-3_54

Table 1 Comparison between different machine learning algorithms

S. No	Classifier	Accuracy	Merits	Demerits
1	ANN	98.66	Ability to work with incomplete knowledge	Unexplained behavior of the network
2	RF	99.67	Flexible to both classification and regression	Requires much time for training
3	Data mining approach	97	Supports incremental learning	Does not perform model selection and does not use tunable parameters
4	RF and AODE	90.51	Easier to tune	Not easily interpretable
5	KNN	99.96	Very simple and intuitive	Choosing *k* may be tricky

machine learning algorithm helps to analyze this data to uncover hidden patterns and detect threats. For machine learning, the main component is big data. Machine learning will analyze this data and find the patterns, correlations, and anomalies within the data. In the field of cybersecurity, machine learning means processing massive amount of security data using an algorithm and helping to detect the security threats [5]. With big data analytics and machine learning approach, several malicious activities can be caught at the early stage itself.

2 Literature Survey

In [6], the authors adapted an ANN algorithm to classify the attacks. It recorded an accuracy of 98.66%. In [7], the authors are using random forest classifier which recorded an accuracy of 99.67%. In [8], the authors proposed a NIDS for the knowledge discovery in data. The authors used average one dependence estimator (AODE) classification algorithm for detecting the attacks. In [9], the detection of cyberattack is done using the classifier which is a fusion of random forest and AODE algorithm. In [10], the authors developed an IDS using KNN algorithm which recorded an accuracy of 99.96% (Table 1).

3 Machine Learning in Detecting the Cyberattacks

With the help of machine learning, cybersecurity systems can easily analyze patterns. They will learn from these patterns and will help to prevent similar attacks and respond to the switching behavior. It helps the cybersecurity people to be more ardent in preventing threats and reacting to active attacks in real time. It can decrease the amount of time spent on routine tasks and helps the organizations to use their resource more deliberately.

Machine learning is all about developing patterns. Those patterns are manipulated with algorithms. For developing patterns, we need a lot of rich data. Rich data mean, the data which we gather have to contain all details about everything that happened, not just the menace. It should provide details about machines, tools, appliances, applications, protocols, and network sensors. It needs to connect about what has happened between what you see on the network and what you see on the destination point. In machine learning, both quantity and quality of the data are important. The data that are collected should be cleaned, organized, and structured, i.e., the data should make some sense, before giving to an algorithm for predicting the correct outcomes. Then, only, the algorithms and model can realize the data and constructively apply the machine learning capacity. In short, there should be a blended approach between machine learning and data acquisition, organization, and structuring.

As the cyberattacks are increasing day by day, the best solution to control these attacks is early detection. We know that a cyberattack will pass through several stages before its termination (Fig. 1).

Stages in Cyberattack

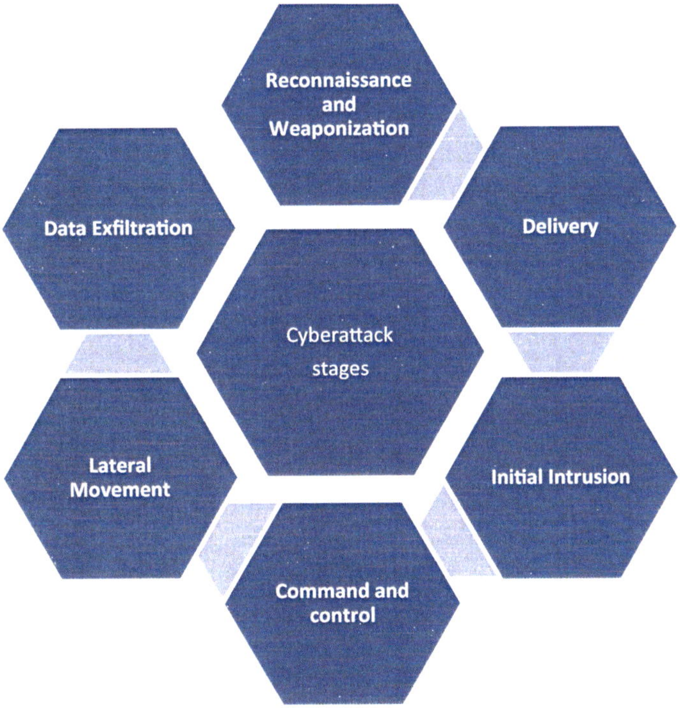

Fig. 1 Stages of cyberattack

1. **Reconnaissance and Weaponization**—gathering information about the infrastructure and data about the employees before the attack.
2. **Delivery**—in this stage, the attacker delivers the exploits used to gain entry to the system.
3. **Initial Intrusion**—the attacker gains the first unauthorized access to the system.
4. **Command and control**—here, the intruder gains control over the infected assets. A remote access tool (RAT) is installed. It allows an attacker to access and control the asset remotely.
5. **Lateral Movement**—once the attacker has remote access to the network, the attacker would move inside the network to attempt to gain access to as much as possible and occurs over a long period to avoid detection.
6. **Data Exfiltration**—transfer the sensitive data to be in control of the attacker.

If we are able to detect this lateral movement at the early, it is possible to minimize or stop the future cybercrimes.

4 Proposed Method

4.1 *Stacked Autoencoder (SAE)*

Foe predicting the cyberattack more effectively and more efficiently, it is better to use stacked autoencoders (SAEs) [11]. Because through hierarchical feature learning, it can determine nonlinear patterns. Autoencoders belong to neural network family which consist of input, hidden, and an output layer. It tries to minimize the reconstruction error. At the same time, optimizes the parameters of the network [12]. The hidden layer can be written using the following equation:

$$h_i = f(x_i) = \sigma(W_1 x_i + b_1)$$

where

- x_i input sample in a sample set X
- W weight matrix of neurons
- b bias vector of all neurons between the input and hidden layers
- σ function of hidden layers

To minimize the loss functions, an AE will use a gradient descent algorithm [12]. Using greedy layer-wise training, the networks define their parameters during the training process. After training each layer, backpropagation algorithm [13] is used for finely tuning the AE. Stacked autoencoder gives better accuracy (Fig. 2).

For detecting the lateral movement using stacked autoencoder, there are different steps like:

- Data collection

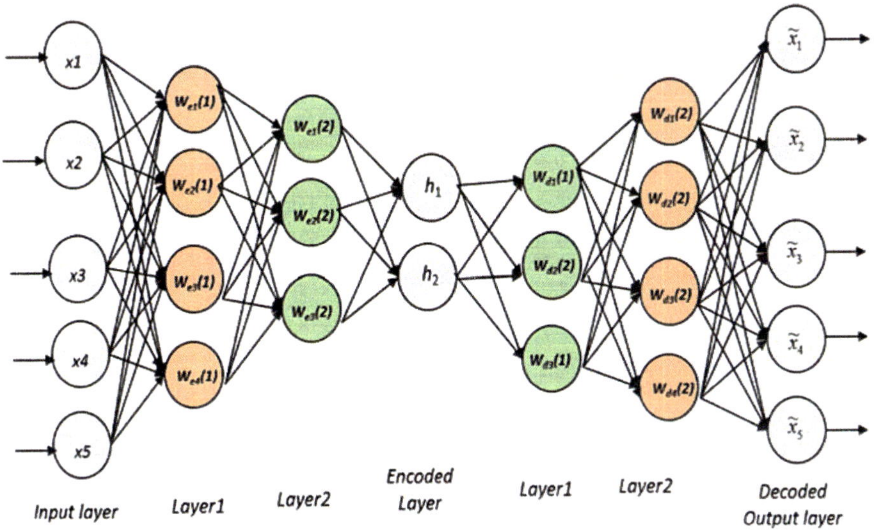

Fig. 2 Stacked autoencoder

- Data preprocessing
- Training and evaluation
- Application.

4.2 Data Collection

The quality of the data that we have collected will be reflected in the final performance of the machine learning model. Creating a dataset containing active directory (AD) [14] logs with the traces of LM is very challenging [15]. For creating the dataset from a live environment, we are using threat cloud. From this, we can extract the log file which will give the basic format of the dataset. The fields in the dataset will be as in Table 2.

The activities related to LM are pass-the-hash (PTH), pass-the-ticket (PTT), tool breaking user policy, and AD enumeration by both a user and admin account (Table 3).

Active directory is a directory service that helps the administrators to control access to network resources. AD enumeration is the process of getting critical information from active directory. Sharphound [16] is a tool to find the shortest path for doing the attack. Pass-the-hash and pass-the-ticket are the lateral movement attacks done using the tool, Mimi Katz [17]. PS Exec is another tool used for stealing the user's credentials.

When we are executing the LM activities that are given in the above table, in a lab environment, it will produce the AD log entries. With the use of the software tool

Table 2 Data fields of the dataset created using live environment

Dataset information	Value
Format	JSON
Log source	AD domain controller
Log type	Windows security event log
Size	418 MB
Normal attacks	1,460,702
Attack events	0
Total events	1,460,702
Attack ratio	0.0
Time span	01:10:40

Table 3 Five indications of LM that are chosen to be investigated

Label	Attack	Tool
1	AD enumeration by admin account	Sharphound
2	Tool breaking user policy	PS Exec
3	Pass-the-hash (PTH)	Mimi Katz
4	Pass-the-ticket (PTT)	Mimi Katz
5	AD enumeration by user account	Sharphound

Splunk [18, 19], it is possible to isolate the LM traces from the AD logs. Splunk is a software tool used for monitoring and analyzing the machine-generated data in real time.

4.3 Data Preprocessing

Preprocessing means the data collected have to be translated into a dataset that machine learning algorithms can understand. Here, different features are to be extracted.

4.4 Training and Evaluation

In this step, the dataset is often divided into two subsets, one for training and one for evaluation. During the evaluation step, the labels are hidden from the machine learning models [20]. It calculates metrics like accuracy, recall, precision, $F1$-score, and computational cost to indicate how well the model can predict unseen data.

Algorithm

1. Split the dataset (normal and attack samples) training-80%, testing-20%

2. Normalize the data
3. Train the SAE model by extracting different balanced sets.
4. For creating new representations, pass the balanced set to different SAEs.
5. From each SAEs, take the new representation and pass it to a deep belief network (DBN) [21] attack detection model.
6. Train the DBN attack detection model.
7. For testing,
   (i) Normalize the test sample
   (ii) Pass all test sets through the SAEs.
   (iii) Pass all new representations from SAEs to the DBN.
   (iv) Result of each DBN is combined with a ReLU [22] activation function for achieving best performance measures.
   (v) Then, it is passed to a random forest classifier to detect attacks.

4.5 Application

The last step is the application. This is where the machine learning model is checked whether it has reached the desired performance in when evaluating.

Since stacked autoencoders are training different hidden layers with so many features extracted in different layers, it gives better accuracy while predicting the cyberattacks and also helps to detect the suspicious user movements. For the better detection of lateral movement in cyberattacks, it will be good to use real-time data [23]. While using real-time data, it is better to use machine learning algorithms because they can analyze the real-time data immediately and suggest the methods the cybersecurity has to follow.

5 Conclusion

In the beginning, advanced persistent threats targeted high prominent organizations or companies with critical data. But now, they are becoming more familiar among smaller and less eminent companies. Due to the expensive nature of APTs, it is crucial to detect them as early as possible and to provide ways to identify them in all of its stages. In this paper, the different stages of cyberattack and stacked autoencoder approach for detecting the lateral movement and which data are relevant for finding the lateral movement was discussed. In future, by finding an approach to discover an APT in the LM stage practically, the security professionals can reduce the risk of a successful attack.

References

1. Pone Mon Institute LLC, Cost of a data breach study: global overview. Technical report (2018)
2. P. Chen, L. Desmet, C. Huygens, A study on advanced persistent threats. **8735** (2014)
3. R. Bace, Intrusion detection. Macmillan Technical Publishing, 201 West 103rd Street, Indianapolis, IN 46290, 2000
4. K. Scarfone, P. Mell, Guide to intrusion detection and prevention systems (IDPS). Technical report 800–94, National Institute of Standards and Technology, Gaithersburg, MD 20899–8930, Feb 2007
5. S. Wagh, V. Pachghare, S. Kolhe, Survey on intrusion detection system using machine learning techniques. Int. J. Comput. Appl. **78**(16) (2013)
6. L. haripriya, M.A. Jabbar, A novel intrusion detection system using ANN and feature subset selection. Int. J. Eng. Technol. (2018)
7. N. Farnaaz, M.A. Jabbar, in *Random Forest Modeling for Network Intrusion Detection System* (Elsevier, Science Direct, 2016)
8. A. Sultana, M.A. Jabbar, Intelligent network intrusion detection system using data mining techniques, in *IEEE Explore* 2017
9. M.A. Jabbar, R. Aluvalu, S. Sai Satyanarayana Reddy, in *RFAODE: A Novel Ensemble Intrusion Detection System* (Elsevier, ICACC-2017)
10. M. Nikhitha, M.A. Jabbar, K nearest neighbor based model for intrusion detection system. Int. J. Recent Technol. Eng. (IJRTE) **8**(2) (2019). ISSN: 2277-3878
11. M. Ranzato, Y.L. Boureau, Y. LeCun, Sparse feature learning for deep belief networks, in *Advances in Neural Information Processing Systems* (MIT Press, Cambridge, MA, USA, 2008), pp. 1185–1192
12. F. Lv, M. Han, T. Qiu, Remote sensing image classification based on ensemble extreme learning machine with stacked autoencoder. IEEE Access **5**, 9021–9031 (2017)
13. H. Karimipour, V. Dinavahi, On false data injection attack against dynamic state estimation on smart power grids, in *2017 5th IEEE International Conference on Smart Energy Grid Engineering* (SEGE, 2017)
14. https://en.wikipedia.org/wiki/Active_Directory
15. I. Ullah, Detecting lateral movement attacks through SMB using bro. Master's thesis, University of Twente, November 2016
16. Extending BloodHound, *Red team Adventures*, 6 February 2020. Available: https://riccardoancarani.github.io/2020-02-06-extending-bloodhound-pt1/
17. https://github.com/gentilkiwi/mimikatz/wiki
18. G.K. Befekadu, V. Gupta, P.J. Antsaklis, Risk-sensitive control under markov modulated denial-of-service (DoS) attack strategies. IEEE Trans. Autom. Control **60**(12), 3299–3304 (2015)
19. H. Bourlard, Y. Kamp, Auto-association by multilayer perceptrons and singular value decomposition. Biol. Cybern. **59**(4–5), 291–294 (1988)
20. G.E. Hinton, G.E. Hinton, R.S. Zemel, in *Autoencoders, Minimum Description Length and Helmholtz Free Energy* (1994)
21. C.S. Wickramasinghe, D.L. Marino, K. Amarasinghe, M. Manic, Generalization of deep learning for cyber-physical system security: a survey, in *Proceedings of the IECON 2018–44th Annual Conference of the IEEE Industrial Electronics Society* (Washington, DC, USA, 2018), pp. 745–751
22. K. Fukushima, Cognitron: a self-organizing multilayered neural network. Biol. Cybern. **20**, 121–136 (1975)
23. A.N. Jahromi, J. Sakhnini, A. Dehghantanha, A deep unsupervised representation learning approach for effective cyber-physical attack detection and identification on highly imbalanced data, pp. 2–11

Deep Learning-Based Big Data Analytics Model for Activity Monitoring of Elderly People

M. Roshni Thanka, Sujitha Juliet, E. Bijolin Edwin, and R. Raahul John

1 Introduction

Senior citizens are the greatest assets to our mankind, so treating them with respect and caring them with more attention are more essential. As the age progress, falling down is one of the threats faced by the elderly people. Every year, 650,000 people die because of fall of which 80% of them from many of the countries ranges with low or medium income [1]. Caring the elderly people seems to be difficult in many of the situations like increased movement and shifting of workplace environment of younger generations, which reduced the timely help by their family members.

The current technologies ensure the elderly people to lead an independent and confident life by continuous monitoring of them by their loved ones. Timely assistance of fall detection is very essential to avoid severe health problem such as immobility and loss of quality of life. Falls can lead to significant wounds, like broken bones, dangerous conditions like hip cracks, head injury, and other severe injuries. Moreover, a fall, even without a physical issue, can vigorously impact a senior's personal satisfaction. This can prompt actual decay and sorrow. Quick detection and remedy after the incident will reduce the harmful effects. Although lot of existing techniques based on wearable devices and senor techniques are available for fall detection, the outcome of these systems is not satisfying because of its false alarms. As a result, elective methodologies, for example, vision-based strategies have emerged.

M. Roshni Thanka · S. Juliet (✉) · E. Bijolin Edwin · R. Raahul John
Karunya Institute of Technology and Sciences, Coimbatore, India
e-mail: sujitha@karunya.edu

M. Roshni Thanka
e-mail: roshni@karunya.edu

E. Bijolin Edwin
e-mail: bijolin@karunya.edu

J. D. Peter et al. (eds.), *Disruptive Technologies for Big Data and Cloud Applications*,
Lecture Notes in Electrical Engineering 905,
https://doi.org/10.1007/978-981-19-2177-3_55

Deep learning's recent effect on computer vision has changed the landscape, enhancing performance in a variety of related area such as computer vision, object recognition, segmentation, and improvising visual effects. For the fall detection, convolutional neural networks (CNNs) are used here to detect falls in optical flow images. CNNs' ability to sequentially train various datasets is taken into consideration to train the limited size of traditional fall datasets which computes the optical flow images of successive frames and train the model to detect different actions. Optical flow images are used to make the model independent of the environmental features. These images depict only the transit of successive video frames, which does not include the details such as color information, intensity, or contract.

Elderly people can wear a fall detection device based on an accelerometer and gyroscope which is used for motion sensors, senses the body positioning and movement. Accelerometer and gyroscope sensors are used to analyze signals to distinguish between activities of daily living (ADL) and falls. A wearable device with these sensors with less restriction can be used in both indoor and outdoor. These monitored data can be transmitted to cloud-based data center which will reduce the size of storage required.

2 Related Works

In the recent past, the field of human activity recognition has gained extensive consideration and significant development. Systems are required to detect the unusual but dependable monitoring of behavior of older adults when they are alone at home. As a result, detecting and recognizing the human behavior using surveillance films or sensors embedded in mobile phones, smart watches, and other devices have been a hot area of current research. There are several systems that have been developed with competitive performance. Few of the recent research works are highlighted here.

Yu et al. [2] proposed a fall detection system based on a camera-based technique that performs posture identification. Background subtraction technique was used to segregate the foreground region of a person's posture, along with additional post-processing to improve the findings. The experiment's results showed a significantly high rate of fall detection with little false positives. However, this method failed to identify the fall when there are several moving objects and occlusions. Autonomous monitoring system with a facial recognition capability is also studied by researchers. The Webcams collect the information such as the pace of a person's movement, their position, and the distance between them and the camera. Image processing tools were utilized on these data to identify whether the person was experiencing a fall [3]. Also, in [4], omni-directional camera and a dedicated computer server were used for fall detection. The advantage of employing an omni-camera is that it may concurrently collect images in 360° in one frame. The traditional camera's problem of blind spots is no longer an issue. However, this system requires the user to provide personal information such as height and BMI index, which will increase the cost of implementation.

Winkley et al. [5] developed verity—a unique system for monitoring the elderly and others requiring passive care. The system makes use of a cutting-edge system on chip (SoC), as well as an ultra-low-power sensor interface and radio frequency communication. For real-time monitoring, the wearer's ambient/skin temperatures, accelerations, and heart rate are all measured. Experiments show that the suggested classifier outperforms traditional classifiers in one-pass classification.

Wang et al. [6] offered an accurate fall detection approach that uses a single device in a top-view configuration to investigate the depth frames of the human body, with the subjects positioned beneath the device inside a room. A classifier based on a binary support vector machine learning algorithm is trained using features retrieved from depth frames. Ziaeefard et al. [7] also developed a system with SVM where the Kinect sensor and analysis of three-dimensional skeletal joint information are used to offer a fall detection approach for indoor environments. However, the Kinect sensor has a rather small field of view when compared to a monocular camera.

Luo et al. [8] have suggested a deep learning approach based on a CNN that has been trained to maximize a sensitivity-based metric. In recent years, many researchers have contributed for the development of ANN model for fall detection and human activity recognition. Westhoven et al. [9] have developed a model where the Softmax function is implemented in the last layers of the ANN. The output of this function is standardized to a probability distribution over the multiple output classes since it is employed as the activation function of the nodes of the last layer of a neural network. Softmax is also employed in the final layers of ANN, in the models developed by [10, 11].

The multilayer perceptron (MLP) is a multilayered ANN with hidden layers between the entrance and exit layers that can sort out nonlinearly separable classes. Each node of this network is a neuron with a nonlinear activation function, which is used for classification in [12, 13]. Radial basis function neural networks (RBFNNs) are utilized to classify the feature vectors from previous CNN layers [14]. This ANN is distinguished by the use of radial basis functions as activation functions, and it outperforms other architectures like as Softmax in terms of generalization because it is trained by minimizing the generalized error calculated by a localized-generalization error model (L-GEM).

ANN architectures' final layers are frequently fully connected, as in [8, 15], where all nodes in one layer are connected to all nodes in the next. The input layer in these structures is used to flatten and transform outputs from prior layers into a single vector, while succeeding layers employ weights to determine proper tagging and, as a result, properly classify events. The autoencoder structure is another suitable ANN structure for classification [16]. This type of ANNs can be used for classification purposes by decreasing the number of output dimensions to the number of final predicted classes, as the number of output dimensions can be lowered. In recent years, ANNs are becoming more popular as their capacity to learn gives them an unrivaled edge. CNN has shown to perform well in the field of artificial vision, by extracting feature maps from images that describe what can be seen.

Though a variety of strategies have been investigated for human activity recognition, the search for a reliable fall detection algorithm for the elderly continues.

The ability to recognize objects and abstract their movement shows promise in this area. Therefore, this paper proposes deep learning-based big data analytics model for activity recognition, especially on detecting the unexpected and fall-related behaviors.

3 Proposed Methodology

The general flow of proposed model for activity recognition is shown in Fig. 1. The raw and big data acquired from the sensors (accelerometer and gyroscope) and video data are processed for normalization during the preprocessing stage. Then, the normalized data are given to deep feature extraction and model building phase where CNN and ANN models are trained and tested to classify the activities such as: standing, sitting, laying, walking, climbing stairs, and going down the stairs. Finally, the activities are recognized and classified based on the model training.

The vision-dependent data have been obtained from Web-based camera or DepthVision camera or digital camera. Fall detection and 3D-based information can be acquired through motion capture cameras. Wearable devices based on accelerometer and gyroscope sensor, where activities of daily living (ADL) data are acquired by placing them in different places of human body, were cost-effective and can be used both in indoor as well as outdoor environment for the purpose of fall detection. These data can be transmitted to cloud for processing by microcontroller. Wearable devices solve the problem of operation range. Human activities can be classified as static postures and dynamic postures. Static postures can be standing, sitting, laying, walking, walking_downstairs, and walking_upstairs, whereas dynamic posture is the gesture between the static postures. Human static postures can be read by placing two tri-axial accelerometers on their chest and thigh. The data collected through the sensors are then transmitted to cloud to train the model, and the validation of the model can be performed through validation dataset.

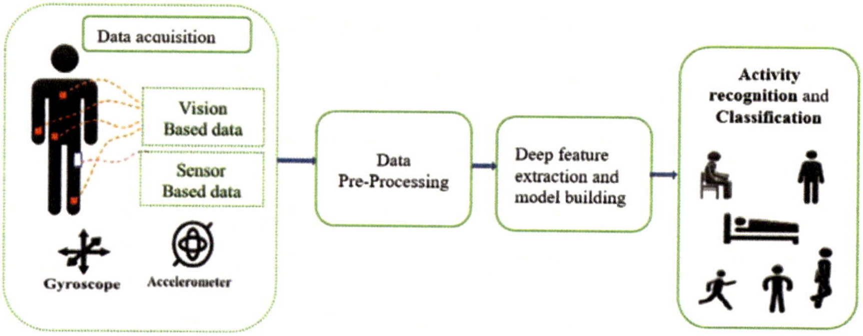

Fig. 1 General flow of the proposed model for fall detection and classification

3.1 Proposed CNN Architecture

The proposed CNN model which processes the input images and predicts the objects is shown in Table 1. The model starts with a convolution layer takes image matrix and kernel as two inputs and computes functions such as sharpening, box blur, Gaussian blur, and blur by implementing kernels; here, the input image shape is (128, 128, 3) color image (RGB) with 3 × 3 kernels to extract 16 feature maps which preserves the correlation among pixels by understanding the image properties and learns 444 parameters. These features are further passed into activation layer which transforms the input to output layer. This is followed by a maxpooling layer for dimensionality reduction with a pooling size of 2 × 2. The features remain here are passed to subsequent convolution layer with 3 × 3 kernel which extracts 4640 features followed by activation layer and maxpooling with 2 × 2 pooling size.

The extracted features are further passed through convolution layer, activation, and maxpooling layer with 18,496 features. The output is further flattened and connected to fully connected layer with 128 nodes followed by activation layer and a dropout layer to normalize the network with a probability of 0.5. The output is connected to the output dense layer with sigmoid as activation layer which maps the resulting value between fall and not fall prediction.

Table 1 Model summary of CNN architecture based on camera data

Layer	Type	Output shape	# Parameter	Connected to
input_1	(InputLayer)	(None, 128, 128,3)	0	
conv2d_9	(Conv2D)	(None, 126, 126,16)	448	input_1
activation_5	(Activation)	(None, 126, 126, 16)	0	conv2d_9
max_pooling2d_9	(MaxPooling2D	(None, 63, 63, 16)	0	activation_5
conv2d_10	(Conv2D)	(None, 61, 61, 32)	4640	max_pooling2d_9
activation_6	(Activation)	(None, 61, 61, 32)	0	conv2d_10
max_pooling2d_10	(MaxPooling2D	(None, 30, 30, 32)	0	activation_6
conv2d_11	(Conv2D)	(None, 28, 28, 64)	18,496	max_pooling2d_10
activation_7	(Activation)	(None, 28, 28, 64)	0	conv2d_11
max_pooling2d_11	(MaxPooling2D	(None, 14, 14, 64)	0	activation_7
flatten_3	(Flatten)	(None, 12,544)	0	max_pooling2d_11
dense_6	(Dense)	(None, 128)	1,605,760	flatten_3
activation_8	(Activation)	(None, 128)	0	dense_6
dropout_1	(Dropout)	(None, 128)	0	activation_8
dense_7	(Dense)	(None, 1)	129	dropout_1
activation_9	(Activation)	(None, 1)	0	dense_7

Total params: 1,629,473
Trainable params: 1,629,473
Non-trainable params: 0

Table 2 Model summary of ANN architecture based on sensor data

Layer	Type	Output shape	# Parameters	Connected to
input_1	(InputLayer)	(None, 561)	0	
dense_65	(Dense)	(None, 1024)	575,488	input_1
dense_66	(Dense)	(None, 512)	524,800	dense_65
dense_67	(Dense)	(None, 256)	131,328	dense_66
dense_68	(Dense)	(None, 128)	32,896	dense_67
dense_69	(Dense)	(None, 64)	8256	dense_68
dense_70	(Dense)	(None, 32)	2080	dense_69
dense_71	(Dense)	(None, 6)	198	dense_70

Total params: 1,275,046
Trainable params: 1,275,046
Non-trainable params: 0

3.2 Proposed ANN Architecture

Artificial neural network with neuron as basic unit is inspired from human brain process which brought lot of revolution in machine learning. The proposed network has 563 input neurons followed by seven fully connected layers with node 1024, 512, 256, 128, 64, 32, and 6, respectively, that have learned 575,488, 524,800, 131,328, 32,896, 8256, 2080, 198 features, respectively. The output layer has six neurons to depict the outputs such as 'STANDING', 'SITTING', 'LAYING', 'WALKING', 'WALKING_DOWNSTAIRS', and 'WALKING_UPSTAIRS'. Table 2 shows the model summary of ANN architecture based on sensor data.

The proposed model aims at creating a vision-based model employing a CNN to classify and predict the fall activity. Also, the sensor-based model using ANN is utilized to classify and prediction of fall activities from activities of lifestyle (ADL). The parameters like 'Triaxial acceleration', 'Triaxial Angular Velocity' from accelerometer and gyroscope, respectively, help in determining the fall of the patient.

4 Results and Discussion

4.1 Dataset and Performance Measures

For the evaluation of the proposed methodology, we use the publicly available UR fall dataset [17]. This dataset contains 70 sequences (30 falls + 40 activities of daily life, image format of 1920 × 1080) captured with two Microsoft Kinect cameras and accelerometer data. The Kinect cameras were positioned on the ceiling, and the accelerometer was worn near the subject's spine on the lower back. Only, one device

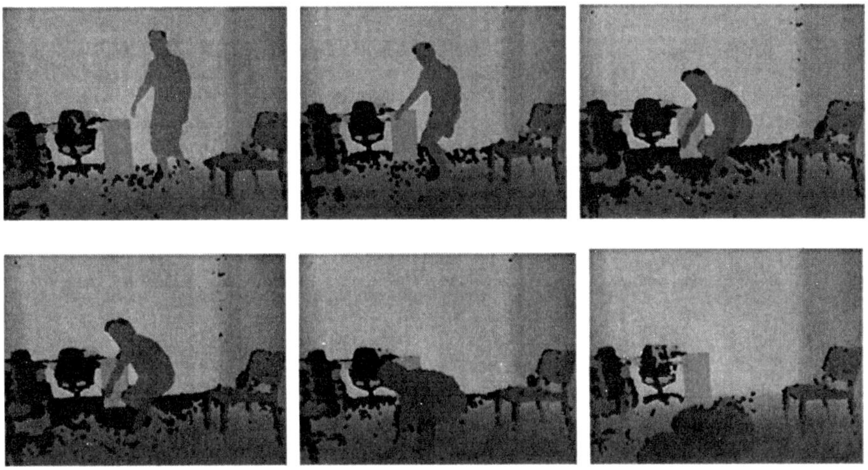

Fig. 2 Sample sequence of optical flow image used from the UR fall detection dataset [17]

(camera 0) and accelerometer were used to record ADL events. Sensor data were collected using PS Move (60 Hz) and x-IMU (256 Hz) devices.

In UR dataset, each row provides a succession of depth and RGB images for two cameras (camera 0 and 1 installed on the floor and ceiling, respectively), as well as synchronization data and raw accelerometer data. Each video stream is saved as a png picture sequence in a separate zip package. PNG16 format is used to hold depth data. Figure 2 shows the sample sequence of optical flow image used from the UR fall detection (URFD) dataset.

The performances of the proposed CNN and ANN models are computed through the metrics accuracy, sensitivity, specificity, precision, and F1-score.

4.2 *Performance on Fall Detection and Activity Recognition*

The CNN model is trained to classify the fall ('fall' and 'not fall') which produces 99.8% accuracy by considering the validation set, and the ANN model is with six output classes such as standing, sitting, laying, walking, walking_downstairs, and walking_upstairs which produces 94.6% accuracy. The confusion matrix generated for camera-based CNN model for fall detection and sensor-based ANN model for activity recognition is shown in Figs. 3 and 4.

Table 3 shows the empirical results of the proposed model with four other methods with URFD. From Table 3, it is inferred that our model has significant improvements in terms of accuracy (Figs. 5 and 6).

The acquired results suggest that human activity detection may be done successfully utilizing sensor data, with some activities having an accuracy of up to 94.6%.

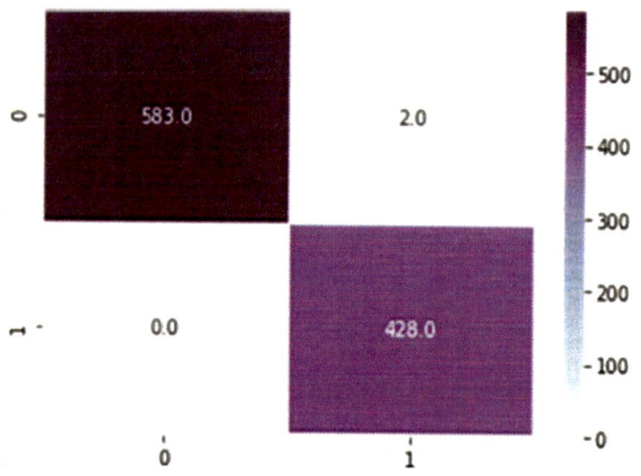

Fig. 3 Confusion matrix of CNN model

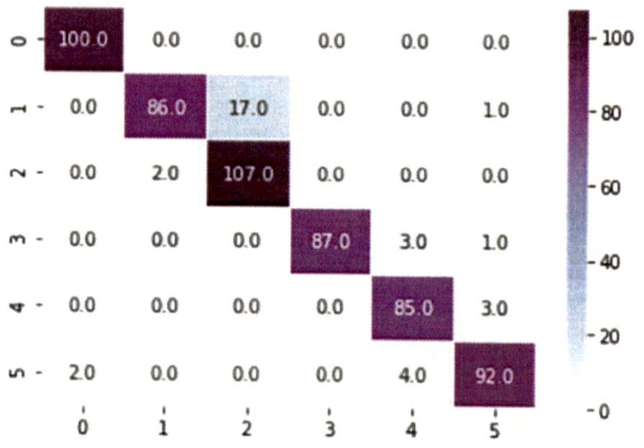

Fig. 4 Confusion matrix of ANN model

Table 3 Performance comparison of our approach on fall detection with literature on the URFD database

Methodology	Accuracy (%)
Dai et al. [18]	94
Harrou et al. [19]	96.60
Zerrouki and Houacine [20]	96.80
Marcos et al. [21]	95
Proposed method	99.8

Because of their similarities, some activities are more difficult to distinguish. This is the case with climbing and descending stairs, which are frequently confused with each other or even walking. The proposed method, on the other hand, accurately identifies these actions, albeit with a somewhat greater error rate. The average accuracy achieved by the proposed method declares the robustness of the proposed model.

Classification Report

	precision	recall	f1-score	support
fall	1.00	1.00	1.00	585
nf	1.00	1.00	1.00	428
accuracy			1.00	1013
macro avg	1.00	1.00	1.00	1013
weighted avg	1.00	1.00	1.00	1013

Fig. 5 Performance of the CNN model

	precision	recall	f1-score	support
0	1.00	1.00	1.00	100
1	0.98	0.83	0.90	104
2	0.86	0.98	0.92	109
3	1.00	0.96	0.98	91
4	0.92	0.97	0.94	88
5	0.95	0.94	0.94	98
micro avg	0.95	0.94	0.95	590
macro avg	0.95	0.94	0.95	590
weighted avg	0.95	0.94	0.95	590
samples avg	0.94	0.94	0.94	590

Fig. 6 Performance of the ANN model

5 Conclusion

Elderly people care is a very essential responsibility of mankind, but often, these responsibilities are ignored by the younger generation because of their life style and working environment. This situation can be overcome by the technological advancement such as remote monitoring, IoT, cloud, and big data. With the help of suitable cameras and by sensor-based data analysis, elderly people can be closely monitored for one their elderly disable activity that is 'fall'. In this proposed work on deep learning-based big data analysis for activity recognition, the vision-based and sensor-based data are processed by the CNN and ANN-based models. The proposed CNN model for classification of vision-based data outperforms the conventional models with 99.8% accuracy by considering the validation set. The activities of daily life provided by sensor-based data are also monitored using ANN model which provides

94.6% accuracy. This system can be further enhanced by applying in real-life situation by fixing camera in different positions and training the model with different transfer learning models.

References

1. D. Mrozek, A. Koczu, B.M. Mrozek, Fall detection in older adults with mobile IoT devices and machine learning in the cloud and on the edge. J. Inf. Sci. 132–147 (2020)
2. M. Yu, A. Rhuma, S. Naqvi, L. Wang, J. Chambers, A posture recognition-based fall detection system for monitoring an elderly person in a smart home environment. IEEE Trans. Inf. Technol. Biomed. **16**(6), 1274–1286 (2012)
3. S. Khawandi, B. Daya, P. Chauvet, Implementation of a monitoring system for fall detection in elderly healthcare. Procedia Comput. Sci. **3**, 216–220 (2011)
4. S. Miaou, P.-H. Sung, C.-Y. Huang, A customized human fall detection system using omni-camera images and personal information, in *1st Trans disciplinary Conference on Distributed Diagnosis and Home Healthcare* (Arlington, USA, 2006)
5. J. Winkley, P. Jiang, W. Jiang, Verity: an ambient assisted living platform. IEEE Trans. Consum. Electron. **58**(2), 364–373 (2012)
6. P. Wang, W. Li, P. Ogunbona, J. Wan, S. Escalera, RGB-D-basedhuman motion recognition with deep learning: a survey. Comput. Vis. Image Underst. **171**, 118–139 (2018)
7. M. Ziaeefard, R. Bergevin, Semantic human activity recognition: a literature review. Pattern Recognit. **48**(8), 2329–2345 (2015)
8. X. Luo, Q. Guan, H. Tan, L. Gao, Z. Wang, X. Luo, Simultaneous indoor tracking and activity recognition using pyroelectric infrared sensors. Sensors **17**(8), 1738 (2017)
9. M. Westhoven, C. Plegge, T. Henrich, T. Alexander, Posture based recognition of the visual focus of attention for adaptive mobile information systems, in *International Conference on Augmented Cognition* (Springer, 2016), pp. 416–427
10. W. Wang, A.X. Liu, M. Shahzad, K. Ling, S. Lu, Device-free human activity recognition using commercial WiFi devices. IEEE J. Sel. Areas Commun. **35**(5), 1118–1131 (2017)
11. R.L.S. Torres, D.C. Ranasinghe, Q. Shi, A.P. Sample, Sensor enabled wearable RFID technology for mitigating the risk of falls near beds, in 2013 *IEEE International Conference on RFID (RFID)* (IEEE, 2013), pp. 191–198
12. Q. Zhang, L.T. Yang, Z. Chen, P. Li, A survey on deep learning for big data. Inform. Fusion **42**, 146–157 (2018)
13. A.M. Wempe, R.F. Keefe, S.M. Newman, T.B. Paveglio, Intent to adopt location sharing for logging safety applications. Safety **5**(1), 7 (2019)
14. Y. Gu, F. Ren, J. Li, PAWS: Passive human activity recognition based on WiFi ambient signals. IEEE Internet Things J. **3**(5), 796–805 (2016)
15. W. Ruan, Q.Z. Sheng, L. Yao, X. Li, N.J. Falkner, L. Yang, Device-free human localization and tracking with UHF passive RFID tags: a data-driven approach. J. Netw. Comput. Appl. **104**, 78–96 (2018)
16. A. Ignatov, Real-time human activity recognition from accelerometer data using convolutional neural networks. Appl. Soft Comput. **62**, 915–922 (2018)
17. B. Kwolek, M. Kepski, Human fall detection on embedded platform using depth maps and wireless accelerometer. Comput. Methods Programs Biomed. **117**, 489–501 (2014)
18. B. Dai, D. Yang, L. Ai, and P. Zhang, A novel video-surveillance- based algorithm of fall detection, in *Proceedings of 11th International Congress on Image and Signal Processing, BioMedical Engineering and Informatics (CISP-BMEI)* (Beijing, China, 2018), pp. 1–6
19. F. Harrou, N. Zerrouki, Y. Sun, A. Houacine, An integrated visionbased approach for efficient human fall detection in a home environment. IEEE Access **7**, 114966–114974 (2019)

20. N. Zerrouki, A. Houacine, Combined curvelets and hidden Markov models for human fall detection, Multimedia Tools Appl, 1–20 (2017)
21. A. Núñez-Marcos, G. Azkune, I. Arganda-Carreras, Vision-based fall detection with convolutional neural networks. Wirel. Commun. Mob. Comput. **9474806**, 1–16 (2017)

Customized Internet of Things-Based Bus Tracking and Management System

K. Akilan, Anusha Chandrasekaran, Arunesh Kumar, S. B. Jashwaanth, A. Joshua, S. Rajalakshmi, and S. Angel Deborah

1 Introduction

As cities develop, their transportation network expands vastly. Buses connect various parts of the city and make travelling easier. Especially for colleges and universities, availability of college buses is a good alternative over staying at the college hostel. Private contractors lease buses to colleges and monitor them. However, an obvious problem with this arrangement is that the college cannot keep track of the buses as they are all leased from different contractors. While each contractor may have a way to track his/her buses, the college bus coordinator does not. Since the college bus coordinator is unable to track the buses, they do not have access to any statistics regarding the buses such as the departure time and arrival time of the bus, the route the bus travels in, any deviations in its route as compared to the predestined route,

K. Akilan · A. Chandrasekaran · A. Kumar (✉) · S. B. Jashwaanth · A. Joshua · S. Rajalakshmi · S. Angel Deborah
Department of Computer Science and Engineering, Sri Sivasubramaniya Nadar College of Engineering, Chennai, Tamil Nadu, India
e-mail: arunesh18024@cse.ssn.edu.in

K. Akilan
e-mail: akilan18016@cse.ssn.edu.in

A. Chandrasekaran
e-mail: anusha18020@cse.ssn.edu.in

S. B. Jashwaanth
e-mail: joshua18065@cse.ssn.edu.in

A. Joshua
e-mail: jashwaanth18063@cse.ssn.edu.in

S. Rajalakshmi
e-mail: rajalakshmis@ssn.edu.in

S. Angel Deborah
e-mail: angeldeborahs@ssn.edu.in

J. D. Peter et al. (eds.), *Disruptive Technologies for Big Data and Cloud Applications*, Lecture Notes in Electrical Engineering 905,
https://doi.org/10.1007/978-981-19-2177-3_56

and a few basic derived information such as the fuel consumption of the bus as well as the distance travelled by it. Another problem that arises from this arrangement is that the students who take the college bus do not have a way to track their buses or receive any information if there are any changes in the route.

The proposed system aims to solve these problems by using a GPS module integrated with an Arduino placed in the bus, which will send the location data to a web server in a periodic manner. The collected data is then processed and analysed using Haversine formulae [1]. The results are used to calculate the distance travelled and fuel consumption of each bus. The novelty of this system lies in the fact that it is sculpted to suit the specific needs of colleges and college students. Specific features such as updating of bus routes, live tracking of buses as well as alerts in case of deviation. Our system also provides administrative privileges over the overall working of the system, insights on the data collected by the system as well as features to add and remove drivers and buses. The paper is organized as follows: Sect. 2 comprises of a literature survey of the related existing works. Section 3 describes the working of the proposed system. Section 4 discusses the features of the proposed system. Data analysis is described in Sect. 5, and 6 shows the results and discussions. The conclusion and future work are noted in Sect. 7. Section 8 consists of the acknowledgements.

2 Literature Survey

The most commonly used method to obtain the location of a vehicle is the Global Positioning System (GPS) signal. The GPS system has 24 satellites revolving around the earth to provide location with accuracy upto 30 cms [2]. The GPS has gained popularity amongst researchers as well as service providers due to its unmatched accuracy, ability to send other data such as speed and altitude and it is low cost price to make a GPS receiver. Model proposed by Singla and Bhatia [3] uses the GPS and artificial neural networks (ANN) for predicting the location of the bus. The GPS signal is transmitted by a local device in the bus to a server which uses Kalman filtering model, K-means clustering, ANN and support vector machines to predict the time of arrival of the bus. The details regarding the live location of the bus are then presented to the user by means of an Android application. The model does not discuss the hardware configurations. Saad et al. [4] have produced real-time location tracking by using faster databases such as MariaDB. Their model uses ESPresso lite WiFi chipset to transfer the GPS signal to a database server on the cloud. From the cloud, the model then uses a web server to transfer the data to the web client in real time in a Google Maps Application Programming Interface (API) using Asynchronous JavaScript And XML (AJAX) and JavaScript Object Notation (JSON) to reduce delay. No further calculations regarding speed or distance estimate are done. With the arrival of many commercial GPS modules, models such as [5] tend to use them for better performance and lower maintenance costs at a higher price. The model then proceeds to provide the data to users using a web server.

The authors Priya et al. [6] and Mane et al. [7] have used the GPS receiver and transmitter installed inside a smartphone to collect the GPS location information. Though they are easy to install and configure, using smartphones or GPS enabled phones are not scalable enough to be practically implemented. Jisha et al. [8] have come up with an IoT-based solution to the problem. The model proposes the use of IoT to provide real-time data instead of using a server. The model then provides bus arrival time prediction using Haversine formula and Kalman filtering. In the novel solution provided by Sugumar et al. [9], the usage of GPS is avoided due to claimed inconsistency of signal. Instead, the model uses an active Radio-Frequency IDentification (RFID)-based solution where each area where the bus is travelling is categorized into wireless sensor network (WSN) and then further divided into subzones, and each subzone has an active RFID tracker that consistently monitors the location of the bus. The WSN then uses IoT protocols (Raspberry Pi 3) to send the real-time data into the IoT cloud. The data is then provided to the client through a web server. Thus, yielding a cost-effective and low bandwidth solution. This idea can most certainly be implemented in a small scale to enhance performance and data security. However, the usage of RFID in cities, towns and districts is quite unlikely. In the field of prediction and estimation of arrival time of a bus, Han et al. [10] use GPS data calibration and three separate long short-term memory (LSTM) algorithm models to predict the current and future location of the bus to very high accuracies. The model has been rigorously tested and proved of its efficiency. Lui et al. [11] use spatial and temporal feature vectors along with LSTM and artificial neural networks to provide accurate bus arrival time prediction. And to predict the different traffic anomalies in order to predict the future location of the bus, Sai et al. [1] have used Hadoop MapReduce-based solution whence a more cost-effective solution was proposed by Prathilothamai et al. [12] by using Apache spark.

3 Working of the Proposed System

The proposed system shown in Fig. 1 has three major components, namely hardware, server, and user interface (UI). The hardware sends the real-time location of the bus to the server where it stores the location in a database. This location information is then processed and is presented on the UI in a more digestible way.

1. **Hardware:** The hardware consists of a single SIM808 module and an Arduino Uno. Each bus is equipped with these modules.

   - **SIM808**: A SIM card is inserted into the SIM808 module. SIM808 uses Global Navigation Satellite System (GNSS) for best-in-class signal acquisition, tracking sensitivity, time to first fix (TTFF) and accuracy. A Global System for Mobile (GSM) supported SIM card is inserted into the SIM808 module, and the connections are made as shown in the circuit diagram depicted in Fig. 2. The SIM808 comes equipped with a GSM module, and we use this GSM module to send a HTTP request with the coordinates to the server that is hosted

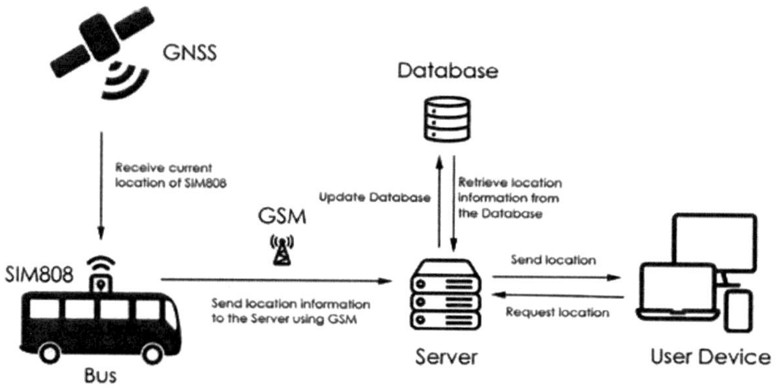

Fig. 1 Architecture of the system

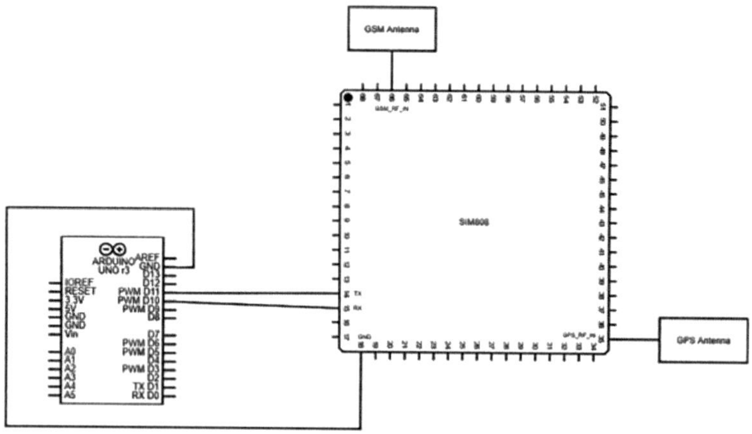

Fig. 2 Circuit diagram

online using HTTP GET method to store the coordinates. It has Quad-band 850/900/1800/1900MHz, GPRS multi-slot class 12/10, GPRS mobile station class B, Control via AT commands, Supply voltage range 3.4–4.4 V [13].

- **Microcontroller**: The Arduino ATmega328p microcontroller serves as the core part of the tracking module, controlling the school/college bus tracking system. A python language program, saved in the microcontroller's memory, controls the modules. The Arduino ATmega328p microcontroller board has 14 digital I/O pins out of which six are capable of pulse-width modulation (PWM) output, 6 analog I/O pins and is programmable with the Arduino Integrated Development Environment (IDE), via a type B USB cable. It can be powered by the USB cable or by an external 9-volt battery, though it accepts voltages between 7 and 20 V [14].

2. **Server** The server being used is a free web hosting website, "000webhost.com". The entire project is deployed using the website. The database that is used is MySQL with phpMyAdmin as the administrative tool. The languages used are HTML, CSS, PHP, JavaScript and Ajax. PHP and MySQL are used to log the values of the coordinate into the database. The SIM808 module sends a HTTP GET request to the server over GSM. A PHP file in the server processes the data received and inserts it into the database. This inserted data is later displayed by the UI.
3. **User Interface**: In any application, the layer in direct contact with the user is called the user interface or UI. This layer is designed to provide the user to access the product or service with ease. The model discussed in this paper uses the commonly used web client designing languages HTML5, CSS and JavaScript. Bootstrap was used for responsive web design and a themed look. The Google Maps API was integrated with HTML5 to invoke the real-time maps feature. The UI also has an area to upload and download current bus routes.

4 Features of the Proposed System

1. **A Single Entity**: The bus management system is designed to empower the bus in-charge representative of an organization (school/college/company) to manage all the buses being used for an organization from different contractors through a single system.
2. **Independent GPS Device**: The main feature of this system is that the GPS device is privately owned. The system has the right to privacy of the GPS data, the database and the server because of the devices being non-commercial. The commercially available devices are more expensive and data privacy may be compromised.
3. **Geo fencing**: A Geo-fence is a virtual perimeter of a real-world geographic area. It could be generated as a radius around a specific location or be predefined by a set of boundaries [15]. The given system defines the Geo-fence of the buses that are under the organization.
4. **Tracking of bus routes**: One of the main features of the system is to help the administrator keep track of the active bus routes. In case of deviation from the predefined route by any bus, the administrator is notified of the same.
5. **Admin/Student modules**: The system provides separate modules for administrator and student. Both users have a set of basic functionality while the administrator has extra functionality for in-depth analysis. Students can check and track their buses, while the admin can track and monitor all the buses and the activities.
6. **Fuel Consumption**: Fuel consumption is a major consideration in the logistic division as the administration needs to know how much fuel is consumed as well as the fuel cost per day to manage their expenses. The proposed system implements a mileage-based fuel consumption analysis which depicts the amount of fuel consumed based on the route and the average mileage of the bus.

7. **Bus stop tracking**: For the students or clients of this model, it is important to know whether their bus has arrived at their designated stop based on the routes assigned. This information is provided by the system using real-time bus tracking. Moreover, it alerts them when the bus has crossed their stop using a pop-up notification. It even alerts the administrator, if any bus misses the stop or goes off-route.
8. **Data Analytics**: The data received by the server is used to calculate number of students on the bus per day, number of buses deployed on any day, fuel consumed per day, as well as the average mileage of each bus. Based on this data, graphs are plotted and displayed using the user interface.
9. **Route History Analysis**: For every route travelled by the bus, a table is maintained in the server which contains the number of bus stops the given bus has reached at the predefined time. This table helps analyse the accuracy of the arrival of the buses at their stops for a given route with respect to their predefined time.

5 Data Analysis

In our system, the data was collected by placing GPS modules in college buses. The traced route was between Kalavakkam and Sholinganallur, which spans 18Km. The GPS data consisting of the latitude, longitude, speed and timestamp was sent to the server every 10 s. The data received is processed to give the mileage and fuel consumption of the bus. The distance travelled is calculated using Haversine formulae shown in equation below, which computes the shortest distance over the surface of the earth. It accomplishes this by assuming the earth to be spherical [2].

$$d = 2r \arcsin\left(\sqrt{\sin^2\left(\frac{\phi_2 - \phi_1}{2}\right) + \cos(\phi_1)\cos(\phi_2)\sin^2\left(\frac{\lambda_2 - \lambda_1}{2}\right)}\right)$$

$r =$ Radius of the Earth
$d =$ Distance between the 2 points
$(\lambda_1, \phi_1) =$ Latitude and Longitude of Point 1
$(\lambda_2, \phi_2) =$ Latitude and Longitude of Point 2

6 Results and Discussions

The system was built from scratch by following the specifications described in the working of the proposed system section. The model uses the signals from the GPS module to transfer real-time location of the bus to the main server. The server then uploads this information to the MySQL DB. Now, any client who can connect to the server is able to view the real-time location of the bus.

- **User Interface**: The user interface is designed using HTML5, CSS and JS with the help of the Bootstrap framework. The dashboard of admin user is shown in Fig. 3a. It consists of important features such as number of buses, number of students and visual representation of some data analytics on fuel consumption. The buttons navigate the user to the students bus page where the location information about all buses are available.
- **Hardware Connection**: The hardware setup to be kept in the bus is shown in Fig. 3b. The connection to the hardware is done as per the circuit diagram as shown in Fig. 2. The module receives the GPS signal from the satellite and transmits the data thorough the Internet to the database. The database is shown in Fig. 4a.
- **Server**:
  - **Plotting Bus Location**: Once the server database updates the location of the bus, it is displayed to the user through the Google Maps API rendered in the UI as shown in Fig. 4b.
  - **Deviation in Path alert**: When the bus proceeds to move in a direction that is not intended, it immediately alerts the admin user about the change in direction along with the bus number as shown in Fig. 5a. The left side of the upper half of the figure shows the correct route, and the right side shows the deviated route. The lower half of the figure shows the alert notification in admin user login.
  - **Plotting Entire Path**: When the bus begins to move in the desired direction, its location is captured by the GPS module which updates the database and consequently is plotted as a line in the Google Maps API rendered through the UI. The predefined stops through which the bus was supposed to travel are also displayed as shown in Fig. 5b.
  - **Data analysis**: The data about the location of the bus is used to find the distance it travelled using the Haversine formula as shown in Fig. 6. This distance is further used to predict the fuel consumption rate as shown in Fig. 7. This data can be analysed and used to predict the information regarding the trend of the buses.

- **Performance measures**:

The SIM808 GSM module used in the proposed system was tested by placing in two different locations. The number of successful attempts to connect to the server as well as GPS data for both indoor and outdoor placements have been measured and tabulated as shown in Table 1.

The tracking performance of the proposed system depends on the GPS signal strength and frequency of successful server pings, which ranges from an average of 25 pings for every 30 attempts. Overall, the tracking performance is reliable. A few existing models such as that proposed by Sugumar et al. [9] are better in tracking due to the high performance of active RFID modules but require the installation of RFID chips in multiple locations throughout the route and is hence difficult to implement.

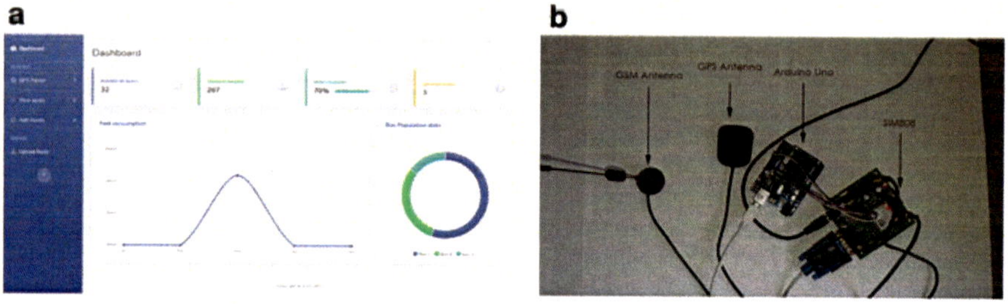

Fig. 3 **a** Dashboard of the system. **b** Hardware model

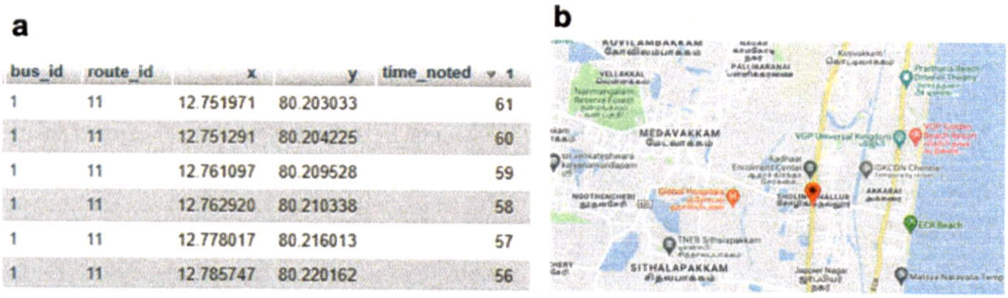

bus_id	route_id	x	y	time_noted
1	11	12.751971	80.203033	61
1	11	12.751291	80.204225	60
1	11	12.761097	80.209528	59
1	11	12.762920	80.210338	58
1	11	12.778017	80.216013	57
1	11	12.785747	80.220162	56

Fig. 4 **a** Current location database. **b** Plot of the bus location

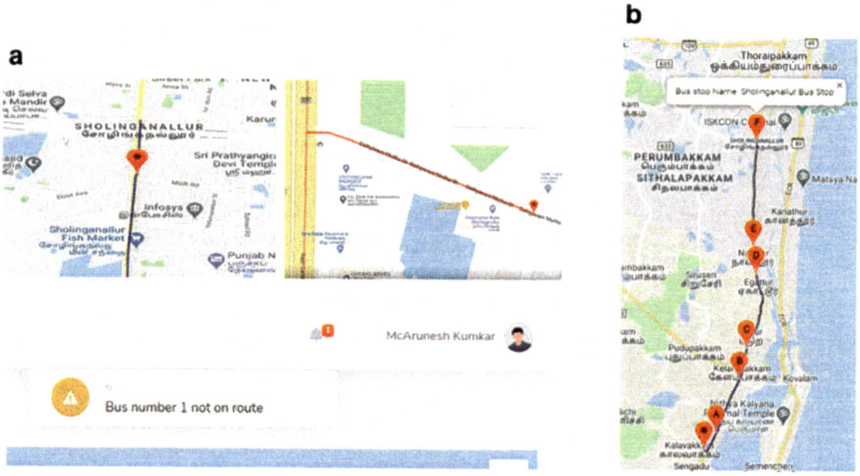

Fig. 5 **a** Alert sent when there is deviation in predefined bus route. **b** Bus pickup locations in a predefined route

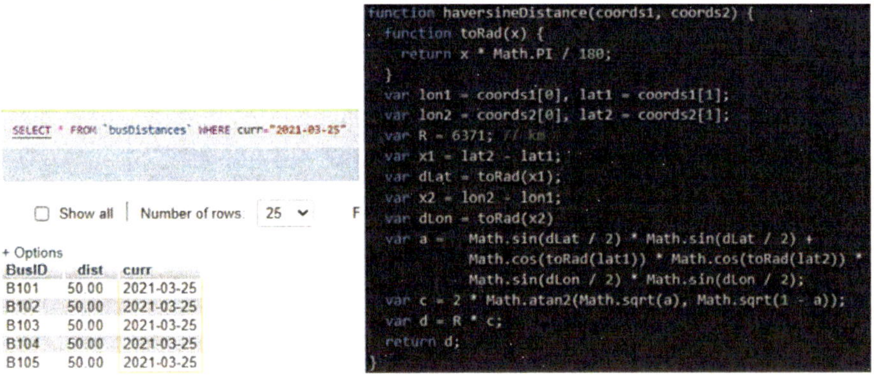

Fig. 6 Haversine formula implementation and distance database

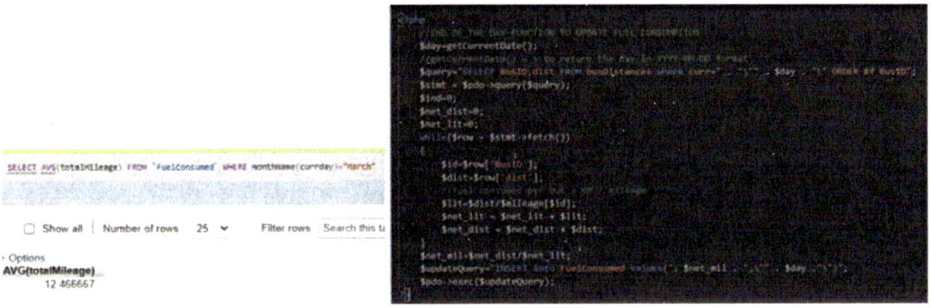

Fig. 7 Fuel implementation and calculation

Table 1 GPS signal strength comparison

Location	Accuracy (m)	Average SNR	Successful Pings (30)
Location A (indoor)	32	21.7	15
Location A (outdoor)	3	31.5	29
Location B (indoor)	24	28.9	19
Location B (outdoor)	8	32.4	28

On the other hand, modules such as those suggested by Priya et al. [6] and Pradip Suresh et al. [7] require the use of a smartphone strapped to the inside of a bus, which is comparatively unstable due to the large amount of power that gets drained while using the smartphone's GPS connectivity. Furthermore, the smartphone requires to be powered regularly and needs to be reinstalled after a few years of usage, hence leading to low feasibility.

7 Conclusion and Future Work

The proposed system helps the user to track their college/school/company bus from a website using any device. This system solves the various problems faced by students/employees stated in the problem statement by providing them with real-time information about their designated buses and if there are any changes regarding the same. The system provides a separate module for the administrator. The administrator is able to monitor all the buses in the system and acquire information such as the mileage and time details. The proposed system provides complete control over the fleet of buses to the bus in-charge, i.e. administrator of the organization, and hence solves issues faced due to the different contractors for the buses. The system's server can be deployed in any hosting service which supports HTML, CSS, JavaScript, PHP, MySQL and allows web scraping to upload the GPS information to the server. Both the SIM808 and Arduino Uno can be powered using a lithium ion battery which will be recharged using the 12 V battery that will be already available in the bus. The total cost of hardware for one bus will be Rs. 4500. The system can be further enhanced by placing RFID tags in bus ID to help in monitoring count of on-board people, and by doing further data analysis to predict various parameters such as fuel efficiency, percentage possibility of driving in wrong route, possibility of breakdown based on previous breakdown events, possibility of delay in reaching destination. Furthermore, the system can be enhanced by adding a driver's interface where they can find their allotted route for the day.

8 Acknowledgements

We would like to convey our heartfelt gratitude to the management of Sri Sivasubramaniya Nadar College of Engineering, Kalavakkam, Chennai, for funding our project on customized bus tracking and monitoring system.

References

1. A. Sai, S. Salim, P.K. Binu, R.C. Jisha, A hadoop based architecture using recursive expectation maximization algorithm for effective and foolproof traffic anomaly detection and reporting. Int. J. Appl. Eng. Res. **10**(55), 2101–2106 (2015)
2. C.C. Robusto, The cosine-haversine formula. Am. Math. Mthly. 64(1), 38-40 (1957). JSTOR, www.jstor.org/stable/2309088. Accessed 13 Apr. 2021
3. L. Singla, P. Bhatia, GPS based bus tracking system, *2015 International Conference on Computer, Communication and Control (IC4), Indore, India* (2015), pp. 1-6. https://doi.org/10.1109/IC4.2015.7375712.
4. S.A. Saad, A.A.B. Hisham, M.H.I. Ishak, M.H.M. Fauzi, M.A. Baharudin, N.H. Idris, Real-time on-campus public transportation monitoring system, in *2018 IEEE 14th International*

Colloquium on Signal Processing and Its Applications (CSPA) (2018). https://doi.org/10.1109/cspa.2018.8368715
5. M. Kumbhar et al., Real time web based bus tracking system. Int. Res. J. Eng. Technol. **3**(2), 632–635 (2016)
6. S. Priya, B. Prabhavathi, P.S. Priya, B. Shanthini, An android application for tracking college bus using google map. Int. J. Comput. Sci. Eng. Commun. **3**(3), 1057–1061 (2015)
7. P.S. Mane, V. Khairnar, Analysis of bus tracking system using GPS on smartphones. IOSR J. Comput. Eng. **16**, 80–82 (2014)
8. R.C. Jisha, A. Jyothindranath, L. S. Kumary, Iot based school bus tracking and arrival time prediction, *2017 International Conference on Advances in Computing, Communications and Informatics (ICACCI), Udupi, India* (2017), pp. 509-514. https://doi.org/10.1109/ICACCI.2017.8125890.
9. D. Sugumar, T. Anita Jones, K.S. Senthilkumar, R.J.S. Jeba Kumar, G. Thennarasi, "Chapter 4-Smart Vehicle Monitoring and Tracking System" Powered by Active Radio Frequency Identification and Internet of Things, The Cognitive Approach in Cloud Computing and Internet of Things Technologies for Surveillance Tracking Systems. Academic Press (2020), pp. 51–64. ISBN 9780128163856, doi.org/10.1016/B978-0-12-816385-6.00004-0
10. Q. Han, K. Liu, L. Zeng, G. He, L. Ye, F. Li, A bus arrival time prediction method based on position calibration and LSTM. IEEE Access **8**, 42372–42383 (2020). https://doi.org/10.1109/ACCESS.2020.2976574
11. H. Liu, H. Xu, Y. Yan, Z. Cai, T. Sun, W. Li, Bus arrival time prediction based on LSTM and spatial-temporal feature vector. IEEE Access **8**, 11917–11929 (2020). https://doi.org/10.1109/ACCESS.2020.2965094
12. M. Prathilothamai, A.M. Sree Lakshmi, D. Viswanthan, Cost effective road traffic prediction model using Apache spark. Indian J. Sci. Technol. 9(17) (2016)
13. https://www.simcom.com/product/SIM808.html [White paper]
14. https://en.wikipedia.org/wiki/ArduinoUno
15. https://en.wikipedia.org/wiki/Global_Positioning_System

An Emerging Paradigm in IoT-Based Indoor Positioning System

Shilpa Shyam, Sujitha Juliet, and Kirubakarn Ezra

1 Introduction

Internet of things is considered to be an architecture which stands out as an empowering factor for the evolution and implementation of several applications which promise to eradicate insecurity, lack of reliability and competence [1]. The drift towards IoT is in relation to compact devices, energy efficiency and extensive variety of smart devices [2, 3]. Positioning has become an emerging field of application in IoT from management and logistics [4, 5] to monitoring staffs and patients in a hospital [6, 7]. With recent developments in indoor positioning, it claims to have an accuracy of less than a metre in determining a particular position [8].

1.1 Challenges in IoT-Based Indoor Positioning System

When it comes to IoT-based indoor positioning system, localization accuracy is regarded as highly necessary for the efficient running of such devices but a few more IoT specific factors need to be considered and taken in account. IoT devices at times require extensive phases and additional complex procedure which makes it difficult for implementations. They extend from very simple devices to extensive use of sensors and processors equipped with multiple Radio Access Technologies (RATs) Also, in Cloud computing, IoT is highly dependent on Internet connection and the usage of cloud which allows it to handle transfer and collection of data without local processing, Indoor positioning systems based on IoT fail to look beyond accuracy and error and does not take into account the specificity of IoT devices and applications.

S. Shyam · S. Juliet (✉) · K. Ezra
Karunya Institute of Technology and Sciences, Coimbatore, India
e-mail: sujitha@karunya.edu

J. D. Peter et al. (eds.), *Disruptive Technologies for Big Data and Cloud Applications*, Lecture Notes in Electrical Engineering 905,
https://doi.org/10.1007/978-981-19-2177-3_57

A detailed review on the IoT localization methods is presented in [9]. It is one of the first reviews to have made a comprehensive study on error sources and mitigation. Review [10] aims to provide a brief study on the recent advancements of IoT-based positioning system focusing only on the wireless networks and intelligent applications of IoT. Review [11] is a coherent study on selective technologies including Wi-Fi, BLE and LoRa meanwhile ignoring other technologies. Review [12] focuses on basic technologies and their advancements till this date. A review having a detailed study on both wireless methods and emerging intelligent methods was lacking. This paper fulfils the purpose.

The intricate study presented in this paper has three sections in all. The introduction Sect. 1 details the role and importance of IoT-based indoor positioning in our lives, its challenges and related works done in this domain. Section 2 deals with the various kinds of IoT-based indoor localization technologies along with the significance and drawbacks. Section 3 deals with the emerging technologies in this field.

2 IoT Localization Technologies

This section would deal with the significant localization technologies used in indoor habitats for tracking and monitoring purposes. The various technologies are detailed below.

Wireless Local Area Network (WLAN)—Wi-Fi technology is much sought after as it does not demand additional hardware requirements, however in certain cases chances of increase in hardware complexities when techniques including TDoA, ToA [13–16] and Angle of Arrival are been implemented. Wi-Fi-based fingerprinting systems proved to acquire better results than triangulation technology [17]. The only drawback was it been very tedious, time consuming and very much susceptible to environmental changes. Considering the affinity of RSS signals to interference Paulose et al. [18] proposed a hybrid deep learning model based on indoor positioning, where RSSI heat maps were used instead of raw RSSI signals. The work proved to provide smaller localization errors than conventional approaches. Ijaz et al. [19] introduced a fusion method of technologies including Wi-Fi for IoT application and proved that a fusion of Wi-Fi with Bluetooth and PDR provided better results in average error and accuracy [20] for localization compared to other fusion of technologies.

Radio Frequency Identification (RFID) is of mainly two types, Active and Passive RFID. Zafari et al. [16] stated that Passive RFID are commonly not suitable for positioning services because of its limited range which makes it only applicable for communication purposes. RFID is preferred by many users for IPS devices since the tags can be operated at a higher range that of the reader. As seen in [21] location of moving objects could be tracked using active RFID by implementing an extended Kalman filter. Even though passive RFID tags are less used, they were used for localization in 2-D and 3-D indoor environments in [22]. The primary disadvantage of RFID was shown to be its short range and to overcome this disadvantage a huge number of tags are needed.

The most recent and advanced version funded by Apple is Bluetooth Low Energy (BLE), found in the form of beacons. Similar to the former cases, RSSI is the most commonly used technique along with others. Ranging-based (RSSI, ToA, AoA, TDoA) and non-ranging-based (fingerprinting) are the two basic kinds of algorithms in BLE-based IPS systems [23]. Moreover, the advanced iBeacons are also not preferred much due to its delay in values reaching the user device.

Unlike other technologies, Ultra-Wide Band technology (UWB) is not hindered by interference and path loss which made researchers to study this technology more and extend its capabilities to the next level. The two dominant techniques used in this case are ToA or TDoA [24]. When localization accuracy is considered as a main metric, it is been chosen over other technologies as it can provide higher range of accuracy up to centimetre level. UWB implementing ToF is capable of achieving an accuracy up to 10 cm, whereas in [24], Yassin et al. proposed that an accuracy of 15 cm could be reached if ToA or TDoA is been used. This technology though seems to overpower others, its short range, expensive nature for additional hardware is still an issue [25].

ToF is the most prominent techniques used in Ultrasound IPS along with multiliterate algorithms like trilateration which is capable of detecting nodes with cm accuracy. However, this technology requires additional anchor nodes and compulsory synchronization between nodes [26]. LoS is an issue commonly faced in UIPS for which De Angels et al. in [27] proposed a model to overcome it. However, to acquire high precision as an IPS system it demanded huge number of beacons. Another disadvantage of this technology is that it gets highly disturbed by Doppler Effect and noise [28].

Accuracy of Visible Light Communication (VLC) technology could be amplified by using multiple LEDs in an indoor environment but the result will not be satisfying if the emitted light overlaps each other. The main technique followed by this technology RSS, AoA, TDoA, and ToA. The authors in [29] claim that the most apt techniques to be used would be TDoA and ToA; but would result in an expensive system. In the course of finding exact coordinates algorithms including trilateration and triangulation are been used. However, flexibility in this technology is not always up to the mark. Ghimire et al. in [30] proved that an IPS using multiple LEDs would be able to transmit synchronized OFDM signals and each AP produces same symbols in order to convey data symbols.

Zigbee is primarily used to design Personal area networks and deal issues with short range communication [31]. Having the advantage of low-cost implementation, high scalability, high availability, and its ability to support dynamic routing topology it is at times preferred over other technologies [32]. The main technique used for indoor localization is the fingerprinting approach in [33] a system based on Zigbee was implemented for indoor localization. It proved the efficiency of using multiple algorithms instead of one. A higher accuracy was obtained in this case. However multipath fading limits its functionally to an extent. Moreover, zigbee is not much used for IoT applications. A comparative analysis of the performance metrics used

Table 1 Comparison of metrics used for the evaluation of various technologies in indoor positioning system

Criteria/technology	Wi-Fi	BLE	Active RFID	Semi passive RFID	ZigBee	UWB	NFER
Positioning accuracy	Low	Medium	High	High	Very high	Very high	Very high
Communication range	Very high	Very high	High	Medium	Medium	Medium	Medium
Deployment complexity	Very high	Very high	Medium	Low	High	Low	Low
Energy efficiency	Low	Very high	Low	High	Very high	High	Medium
Compactness	Yes	Yes	Yes	Yes	Yes	Yes	Yes
Communication channel	Very high	Medium	Very high	Very high	Low	Very high	Medium
Noise-Tolerance	Medium	Medium	Medium	Medium	Medium	Very high	Medium
Identification of an object of tracking	Yes	Yes	Yes	Yes	Yes	Yes	Yes
Communication channel security	High	High	Medium	Medium	High	Very high	Medium

Table 2 Different positioning algorithms with their characteristics

Positioning algorithm	Accuracy	Time consumption	Deployment cost	Energy consumption	Communication utilization
Fingerprint	Elevated	Low	Elevated	Stunted	Standard
Reference tags	Standard	Elevated	Standard	Standard	Standard
AOA	Elevated	Standard	Standard	Elevated	Elevated
TOA/TDOA	Stunted	Stunted	Stunted	Stunted	Elevated

for the evaluation of various technologies in indoor positioning system is depicted in Table 1. The different positioning algorithms with their characteristics are covered in Table 2.

3 Emerging Intelligent IoT-Based IPS System

Optimization methods have taken the lead at present and has made numerous irreplaceable contributions in case of IoT applications regarding indoor positioning. Notable advancements in this field are been discussed in Table 3.

Table 3 Notable works in intelligent-based IoT systems

Reference	Optimization method	Proposed methodology	Results
Varma et al. [35]	Random forest-based ML algorithm	Enhancing location accuracy using IoT for smart buildings	Higher accuracy and better test prediction
Boudani et al. [39]	Deep Neural Network	Asset tracking in complex indoor environment	DNN performed well with lowest classification rate
Wang et al. [43]	Deep Learning	Recognizing location, activity and gesture in an indoor laboratory	Proved to be less time consuming and less requirement of labour
Alhammadi et al. [45]	Bayesian graphical model with fingerprinting	Indoor localization for a testbed with a dimension of 51 × 22 m^2	On comparison with Madigan model accuracy shot up to 25.65%

Standing out first is the machine learning which learns by itself from different patterns and regularities in a data set. The random forest-based ML algorithm discussed in [34, 35] enhanced the location accuracy in indoor localization for IoT-based smart buildings. The success rate was showed up to a rate of 14% proving to be better than other categorizations. Georgiadis et al. [36] developed a hydroponics monitoring system with IoT. The parameters of the solution were monitored and observed within the greenhouse so as to bring about an increase in the production of the crops. Considering cost as a metric Jain et al. [37] proposed a low-cost indoor localization system using machine learning.

A neural network happens to imitate the functioning of human brain and adhere to a sequence of algorithms to develop relationships in a data set. It is looked forward by various IPS systems for amplifying estimation accuracy [38]. 5G IoT networks have also begun to play an irreplaceable role for numerous smart application and devices. One such application was developed by Boudani et al. [39]. The deep neural network was implemented for 3-D positioning to track assets and location in a complex indoor environment. It was also demonstrated against other ML algorithms.

Extraction of both supervised and unsupervised features, analysis and study of patterns and their categorization is been imparted by a group of ML algorithms collectively known as deep learning algorithms [40]. Similar to a neural network, deep learning algorithms works in enhancing the accuracy along with reduced errors. Numerous algorithms based on deep learning [41–43] showed an improvement in location accuracy for indoors. In the case of deep learning K-nearest neighbour is commonly implemented algorithm [44].

The Bayesian network was introduced in indoor positioning considering the improvement of estimation accuracy fulfilling the bays' theorem [13]. Very few IPS systems based on IoT have inculcated this method and many more advancements are to be seen in this domain. Usage of additional hardware at times limits the functionality of an IoT devices. Alhammadi et al. [45] came up with a 3-D indoor positioning system for complex indoor IoT applications. It greatly depends on Wi-Fi signal strength and was initially tested for results inside a building.

Bio inspired algorithms [46–48] imitate the behaviour of organisms and have also proved to provide better results. A Cuckoo search is a bio inspired algorithm proposed by Khan et al. [11] to increase accuracy and reduce computational speed by using K-nearest neighbour (KNN) and support vector machine (SVM). However, the algorithm showed a result of 99.87% in accuracy.

4 Conclusion

Keeping the huge demand and scope of IoT and IPS system in mind, a comprehensive study on its challenges, technologies, applications and drawbacks is been done. This review also illustrates the emerging technologies and their future applications. Future work would include in considering other metrics and measurements as well.

References

1. D. Macagnano, G. Destino, G. Abreu, Indoor positioning: a key enabling technology for IoT applications, in *2014 IEEE World Forum on Internet of Things (WF-IoT)* (IEEE, 2014), pp. 117–118
2. L. Atzori, A. Iera, G. Morabito, The internet of things: a survey. Comput. Netw. **54**(15), 2787–2805 (2010)
3. L. Bittencourt, R. Immich, R. Sakellariou, N. Fonseca, E. Madeira, M. Curado, L. Villas, L. DaSilva, C. Lee, O. Rana, The internet of things, fog and cloud continuum: Integration and challenges. Internet of Things **3**, 134–155 (2018)
4. B. Khoo, RFID-from tracking to the Internet of Things: a review of developments, in *2010 IEEE/ACM Int'l Conference on Green Computing and Communications & Int'l Conference on Cyber, Physical and Social Computing* (IEEE, 2010), pp. 533–538
5. S. Amendola, R. Lodato, S. Manzari, C. Occhiuzzi, G. Marrocco, RFID technology for IoT-based personal healthcare in smart spaces. IEEE Internet Things J. **1**(2), 144–152 (2014)
6. A.M. Sazdar, N. Alikhani, S.A. Ghorashi, A. Khonsari, Privacy preserving in indoor fingerprint localization and radio map expansion. Peer-to-Peer Netw. Appl. **14**(1), 121–134 (2021)
7. S.R. Pokhrel, H.L. Vu, A.L. Cricenti, Adaptive admission control for IoT applications in home WiFi networks. IEEE Trans. Mob. Comput. **19**(12), 2731–2742 (2019)
8. H. Liu, H. Darabi, P. Banerjee, J. Liu, Survey of wireless indoor positioning techniques and systems. IEEE Trans. Syst., Man, Cybernet., Part C (Appl. Rev.) **37**(6), 1067–1080 (2007)
9. Y. Li, Y. Zhuang, X. Hu, Z. Gao, J. Hu, L. Chen, Z. He, L. Pei, K. Chen, M. Wang, X. Niu, Location-enabled IoT (LE-IoT): a survey of positioning techniques, error sources, and mitigation (2020). *arXiv preprint* arXiv:2004.03738
10. K.A. Kordi, A. Alhammadi, M. Roslee, M.Y. Alias, Q. Abdullah, A review on wireless emerging IoT indoor localization, in *2020 IEEE 5th International Symposium on Telecommunication Technologies (ISTT)* (IEEE, 2020), pp. 82–87
11. F.U. Khan, M. Awais, M.B. Rasheed, B. Masood, Y. Ghadi, A comparison of wireless standards in IoT for indoor localization using LoPy. IEEE Access (2021)

12. A. Billa, I. Shayea, A. Alhammadi, Q. Abdullah, M. Roslee, An overview of indoor localization technologies: toward IoT navigation services, in *2020 IEEE 5th International Symposium on Telecommunication Technologies (ISTT)* (IEEE, 2020), pp. 76–81
13. A. Alhammadi, F. Hashim, M.F.A. Rasid, S. Alraih, A new indoor localization system based on Bayesian graphical model, in *2017 International Conference on Wireless Communications, Signal Processing and Networking (WiSPNET)* (IEEE, 2017), pp. 1960–1964
14. S. Alraih, A. Alhammadi, I. Shayea, A.M. Al-Samman, Improving accuracy in indoor localization system using fingerprinting technique, in *2017 International Conference on Information and Communication Technology Convergence (ICTC)* (IEEE, 2017), pp. 274–277
15. A. Alhammadi, M.Y. Alias, S.W. Tan, C. Sapumohotti, An enhanced localisation system for indoor environment using clustering technique. Int. J. Comput. Vis. Robot. **7**(1–2), 83–98 (2017)
16. F. Zafari, A. Gkelias, K.K. Leung, A survey of indoor localization systems and technologies. IEEE Commun. Surv. Tutor. **21**(3), 2568–2599 (2019)
17. B. Jang, H. Kim, Indoor positioning technologies without offline fingerprinting map: a survey. IEEE Commun. Surv. Tutor. **21**(1), 508–525 (2018)
18. A. Poulose, D.S. Han, Hybrid deep learning model based indoor positioning using Wi-Fi RSSI heat maps for autonomous applications. Electronics **10**(1), 2 (2021)
19. R. Ijaz, M.A. Pasha, N.U. Hassan, C. Yuen, A novel fusion methodology for indoor positioning in IoT-based mobile applications, in *2018 IEEE 4th World Forum on Internet of Things (WF-IoT)* (IEEE, 2018), pp. 742–747
20. M.U. Ali, S. Hur, Y. Park, Wi-Fi-based effortless indoor positioning system using IoT sensors. Sensors **19**(7), 1496 (2019)
21. J. Wang, R.K. Dhanapal, P. Ramakrishnan, B. Balasingam, T. Souza, R. Maev, Active RFID based indoor localization, in *2019 22th International Conference on Information Fusion (FUSION)* (IEEE, 2019), pp. 1–7
22. M. Liu, H. Wang, Y. Yang, Y. Zhang, L. Ma, N. Wang, Rfid 3-d indoor localization for tag and tag-free target based on interference. IEEE Trans. Instrum. Meas. **68**(10), 3718–3732 (2018)
23. Y. You, C. Wu, Indoor positioning system with cellular network assistance based on received signal strength indication of beacon. IEEE Access **8**, 6691–6703 (2019)
24. A. Yassin, Y. Nasser, M. Awad, A. Al-Dubai, R. Liu, C. Yuen, R. Raulefs, E. Aboutanios, Recent advances in indoor localization: a survey on theoretical approaches and applications. IEEE Commun. Surv. Tutor. **19**(2), 1327–1346 (2016)
25. H. Perakis, V. Gikas, Evaluation of range error calibration models for indoor UWB positioning applications, in *2018 International Conference on Indoor Positioning and Indoor Navigation (IPIN)* (IEEE, 2018), pp. 206–212
26. J. Qi, G.P. Liu, A robust high-accuracy ultrasound indoor positioning system based on a wireless sensor network. Sensors **17**(11), 2554 (2017)
27. A. De Angelis, A. Moschitta, P. Carbone, M. Calderini, S. Neri, R. Borgna, M. Peppucci, Design and characterization of a portable ultrasonic indoor 3-D positioning system. IEEE Trans. Instrum. Meas. **64**(10), 2616–2625 (2015)
28. A. Lindo, E. Garcia, J. Urena, M. del Carmen Perez, A. Hernandez, Multiband waveform design for an ultrasonic indoor positioning system. IEEE Sens. J. **15**(12), 7190–7199 (2015)
29. Q. Wang, H. Luo, A. Men, F. Zhao, X. Gao, J. Wei, Y. Zhang, Y. Huang, Light positioning: a high-accuracy visible light indoor positioning system based on attitude identification and propagation model. Int. J. Distrib. Sens. Netw. **14**(2), 1550147718758263 (2018)
30. B. Ghimire, J. Seitz, C. Mutschler, Indoor positioning using OFDM-based visible light communication system, in *2018 International Conference on Indoor Positioning and Indoor Navigation (IPIN)* (IEEE, 2018), pp. 1–8
31. M. Uradzinski, H. Guo, X. Liu, M. Yu, Advanced indoor positioning using zigbee wireless technology. Wireless Pers. Commun. **97**(4), 6509–6518 (2017)
32. C.N. Huang, C.T. Chan, ZigBee-based indoor location system by k-nearest neighbor algorithm with weighted RSSI. Procedia Comput. Sci. **5**, 58–65 (2011)

33. S.H. Fang, C.H. Wang, T.Y. Huang, C.H. Yang, Y.S. Chen, An enhanced ZigBee indoor positioning system with an ensemble approach. IEEE Commun. Lett. **16**(4), 564–567 (2012)
34. A.H. Salamah, M. Tamazin, M.A. Sharkas, M. Khedr, An enhanced WiFi indoor localization system based on machine learning, in *2016 International Conference on Indoor Positioning and Indoor Navigation (IPIN)* (IEEE, 2016), pp. 1–8
35. P.S. Varma, V. Anand, Random forest learning based indoor localization as an IoT service for smart buildings. Wireless Pers. Commun. **117**(4), 3209–3227 (2021)
36. G. Georgiadis, A. Komninos, A. Koskeris, J. Garofalakis, Implementing an integrated Internet of Things system (IoT) for hydroponic agriculture, in *Data Science and Internet of Things: Research and Applications at the Intersection of DS and IoT* (2021), pp. 83–102
37. C. Jain, G.V.S. Sashank, S. Markkandan, Low-cost BLE based indoor localization using RSSI fingerprinting and machine learning, in *2021 Sixth International Conference on Wireless Communications, Signal Processing and Networking (WiSPNET)* (IEEE, 2021), pp. 363–367
38. M.R. Hassan, M.S.M. Haque, M.I. Hossain, M.M. Hassan, A. Alelaiwi, A novel cascaded deep neural network for analyzing smart phone data for indoor localization. Futur. Gener. Comput. Syst. **101**, 760–769 (2019)
39. B. El Boudani, L. Kanaris, A. Kokkinis, C. Chrysoulas, A. Dagiuklas, S. Stavrou, Positioning as service for 5G IoT networks, in *2021 Telecoms Conference (ConfTELE)* (Institute of Electrical and Electronics Engineers (IEEE), 2021)
40. F. De Vita, D. Bruneo, A deep learning approach for indoor user localization in smart environments, in *2018 IEEE International Conference on Smart Computing (SMARTCOMP)* (IEEE, 2018), pp. 89–96
41. Y. Zhu, R. Mottaghi, E. Kolve, J.J. Lim, A. Gupta, L. Fei-Fei, A. Farhadi, Target-driven visual navigation in indoor scenes using deep reinforcement learning, in *2017 IEEE International Conference on Robotics and Automation (ICRA)* (IEEE, 2017), pp. 3357–3364
42. L. Li, Y. Lv, F.Y. Wang, Traffic signal timing via deep reinforcement learning. IEEE/CAA J. Automat. Sin. **3**(3), 247–254 (2016)
43. J. Wang, X. Zhang, Q. Gao, H. Yue, H. Wang, Device-free wireless localization and activity recognition: a deep learning approach. IEEE Trans. Veh. Technol. **66**(7), 6258–6267 (2016)
44. J. Hu, D. Liu, Z. Yan, H. Liu, Experimental analysis on weight K-nearest neighbor indoor fingerprint positioning. IEEE Internet Things J. **6**(1), 891–897 (2018)
45. A. Alhammadi, S. Alraih, F. Hashim, M.F.A. Rasid, Robust 3D indoor positioning system based on radio map using Bayesian network, in *2019 IEEE 5th world forum on Internet of Things (WF-IoT)* (IEEE, 2019), pp. 107–110
46. X. Chen, S. Zou, Improved Wi-Fi indoor positioning based on particle swarm optimization. IEEE Sens. J. **17**(21), 7143–7148 (2017)
47. A. Alhammadi, F. Hashim, M. Fadlee, T.M. Shami, An adaptive localization system using particle swarm optimization in a circular distribution form. Jurnal Teknologi 105–110 (2016)
48. D.V. Le, N. Meratnia, P.J. Havinga, Unsupervised deep feature learning to reduce the collection of fingerprints for indoor localization using deep belief networks, in *2018 IEEE International Conference on Indoor Positioning and Indoor Navigation (IPIN)* (2018), pp. 1–7

SMART CARAFE: An IoT-Based Real Time System

K. Suthendran and T. Subburaj

1 Introduction

In early 100 BC'S the kettle was used in ancient Mesopotamia for cooking and it was made up of iron, heated using firewood. Later, electric kettle came in the year of 1893 with some limitations and then in 1922 it was demonstrated by the engineer Birmingham although it has the features of automatic heating of the water it is heavy in weight and it cannot maintain the same temperature for a long. On the other hand for cooling water in ancient days they had used the sand pot and in later dispenser came with the gas inside the dispenser in which it makes the water as cool. Even though it produces the hot and cold water, it is big in size and high in cost [1].

Smart flask enables us to have both the hot and cold water in a single bottle. The user requirement may be communicated through appropriate switch. For heating, we have used copper wire at the bottom of the bottle and for cooling, a Peltier is attached in aluminium rectangular tube with heat sink fan. In addition, the message notification reaches the user when the water attains its correct heating temperature of 55 °C or cooling temperature of 22 °C and automatically the lights and system gets off without any manual support.

K. Suthendran (✉) · T. Subburaj
Department of Information Technology, Kalasalingam Academy of Research and Education, Srivilliputhur, Tamil Nadu, India
e-mail: k.suthendran@klu.ac.in

Department of Information Science and Engineering, RajaRajeswari College of Engineering, Bangalore, India

J. D. Peter et al. (eds.), *Disruptive Technologies for Big Data and Cloud Applications*, Lecture Notes in Electrical Engineering 905,
https://doi.org/10.1007/978-981-19-2177-3_58

2 Literature Survey

When we look back, the people in North America have used tea kettles for hot water and tea. The kettle was made up of material known as copper (cu) which heats quickly. In downstream, Chinese have used the porcelain to make kettles. When time passes by, in western Asia, people used bronze made of kettles. Before nineteenth century, kettles were made from iron.

In 1906 [2], HW Taylor and L Haws have introduced the water cooler. The main purpose of this was to chill the water as well as to provide the purified water. About ages ago, the water cooler contents were chilled out based on glaciers. But it involves huge human resource to arrange it. L Haws did register first Patent for water spout. As the time passes by, it took different shapes and dimensions and these days many compact models are in use.

In early nineteenth century, the electricity based coolers introduced. In 1981 [3], the US based carpenter electric Inc. coined first the electrical kettle. After a decade, the swan industry came in to play with instant heating kettles in which the heating component was kept inside the metal tube and positioned at the water compartment. In 1955 [4], Russell Hobbs managed to develop an automatic electrical stainless steel kettle. Here, the steam was generated by thermostat boils the water. The process is paused once the water is heated. In [5], the researchers have deliberated the Peltier effects and its role in both heating and cooling. In 2015, the authors have popped up new water heater system based on solar. It is coiled with insulated collector box in which inside having copper reflector to absorb heat and produce hot water [6]. In [7, 8], the authors have demonstrated a possible heating and cooling systems in public area which covers 10 cm insulation padding to hold hotness with matching vapour barrier and plaster boards. In 2013, African authors have considered electric water heater competence and principles in practice. They have examined the standing loss and energy and provided effective solution for the step up. In H2OPAL hydration tracker system is reminds the user about the amount of water to be consumed. This system works with the help of mobile phone [9]. Next gululu is one of the most interactive water bottles mainly useful for kids to make them drink more water in which the adventures levels are unlocked when the consumed water is high and also the messages (consumption of water) are sent to their parents. Then, ozmo smart cup and app is one of the Bluetooth smart cup hydration trackers in which it is made for having both water as well as coffee. This shows the coffee level consumption and level of hydration via Bluetooth in mobile app [10]. Finally moikit gene is a water bottle which is mainly used by the sports person in which it records the water intake automatically made up of tritan [11]. It is water resistant and easy to wash. In 2015, Thakkal Mohit introduced an idea on "Peltier cooling module" it works with thermoelectric refrigeration, aims to provide cool water [12]. In [13], the cost analysis was extensively done for solar-based system.

Therefore by analysing the previous system and survey of the existing system on heating and cooling process we had come up with an idea of designing a water bottle of 1 L by the material of stainless steel which is free from corrosion and weight less product to carry everywhere by having both the heating and cooling system attached to it. The drawbacks and issues of the existing systems like heavy weight and long process are avoided in this proposed system by having both the heating system with insulation, suitable copper wire and cooling system. By using Peltier in aluminium rectangle tube fixed on one side with heat sink fan to produce cool water is a cooling system. This water bottle will have two switches on the holder indicating hot and cold by processing either of the switches it starts processing the respective system and automatically turns off when it reaches the certain temperature and sends a message to the user simultaneously. This helps in the time management and weightless product to use. The same system can be fitted in to the solar-based technology in future with some additional cost.

3 Hardware Components

3.1 Heat Sink

It is an inactive heat interchanger as shown in Fig. 1. It is also called as heat spreader. The heat generated from electronic or mechanical device will be transferred based on heat sink. Cellular materials are used for the construction of very small heat sinks. It is very desirable for a wide range of applications where size, weight and efficiency. Heat sinks are used to cool the CPUs, GPUs and RAM modules. There are different kinds of heat sinks are available based on the material like copper, aluminium and graphite heat sink. The aluminium or copper is mostly preferred to design a heat sink. It is commonly used in laptops, computers, etc. In electronic circuit heat is generated by every element or component which must be dissipated to improve its reliability and to prevent from the premature component failure. In this work, we have used

Fig. 1 Heat sink

Fig. 2 Copper pipe

heat sink 3.8 cm length, 3.8 cm width and 1.7 cm height pin fin model with fan to absorb the heat from the water and produce the cool air to make the water cool.

3.2 *Copper*

It is one of the best suited materials to develop water pipes as shown in Fig. 2. Its characteristics is durable, recyclable, light weight, malleable and its durability is one of the term to make it a long term value. When compared to PVC the Copper pipes are environment friendly. It is possible to have copper pipes with anti-corrosion coating. Further, copper is a best conductor because it has more free electrons at outer orbit. Nichrome is other option for copper, has a higher resistance and heat faster, but it has a lower melting temperature. So nichrome is easy to burn. Copper has less voltage drop and no burning process. In this work, we have used copper wire (0.25 in. diameter) for heating the water by keeping it all directions around the inside layer of the flask.

3.3 *Peltier*

A two terminal electronic piece of equipment or module which creates the temperature difference at the junction when an external electrical energy is given as shown in Fig. 3. The applied voltage causes current flow in the device which makes one side of the module as hot and other side as cool. In many real world applications this module deployed as a cooling component. Depending up on the requirement the heat can be removed from one side by connecting heat sink with fan or water. In thermoelectric effect, the Peltier cooler works on the Peltier effect [5]. In this work, we have used 40 mm x 40 mm × 3.6 mm dimension Peltier (TEC1-12,706) for cooling the water by fixing heat sink with fan to absorb the heat and provide cold air and water.

Fig. 3 Peltiar

3.4 Thermostats

A suitable device for retaining a required temperature on a selected system is called thermostat. Depending upon the sensor detecting the value it will off the heating or cooling system. There are three types of thermostats in practice. Firstly, bimetallic thermostat is a strip with two odd metals when it attains the desired temperature the connection between them goes off. Secondly, the wax type of thermostat, when excess heat is sensed the wax gets expanded that pushes the stick to move out which breaks the connection and allows the system to be cooled. Finally, the electronic thermostat, the change in temperature will result changes in resistance. In this work, we have used thermostat to sense the temperature and makes it ON or OFF when it reaches the attainable set temperature.

3.5 Battery

A battery powers any circuit or system which is made up of one or many electrochemical cells as shown in Fig. 4. It is also called as secondary battery or secondary cells. These days, rechargeable batteries are in practice. It can hold the charge and discharge across the load. It is available in different shapes and size. The cost of the battery is higher than the disposable batteries and it is having an environmental impact. Automobile starters, light vehicles, golf carts and wheel chairs are the few real time examples used the rechargeable batteries. It contains two terminals made

Fig. 4 Battery

of different materials one is anode and the other is cathode. These batteries are easy and convenient to recharge with a simple battery charger.

3.6 *Block Diagram*

The block diagram of proposed smart flask system is shown in Fig. 5. The flask is equipped with Arduino controller circuit which makes our job easier.

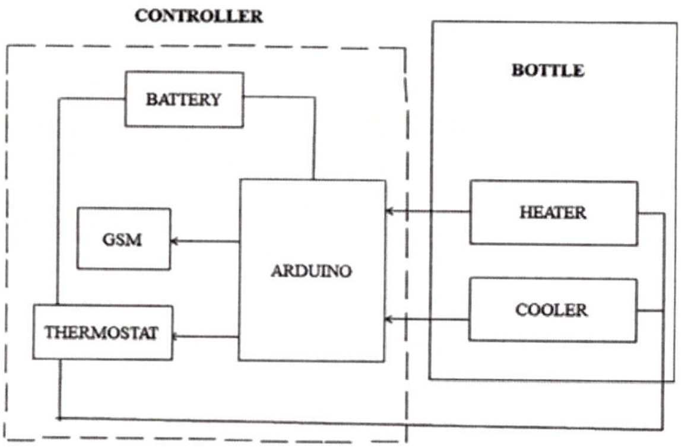

Fig. 5 The block diagram of proposed system

3.7 *Working Principle*

The kit consists of stainless steel flask (1 L) in which it is resist to corrosion along with Arduino and GSM. GSM is used for sending the message to the customer once the water is ready with the temperature fixed. In controller side GSM and thermostat is connected with Arduino and power supply is insisted the bottle fixed with heating and cooling system, for heating, the copper wire is kept in all directions around the inner layer of the bottle and for cooling, Peltier along with heat sink (with fan) is fixed both connections are given in Arduino and thermostat to attain temperature.

4 Existing System

There are many water bottles systems in practice for example the hydration level checking and temperature displaying via the Bluetooth technology provided with USB socket for charging and some with coin cell battery is one type as shown in Fig. 6. The smart lid bottle system is made up of co polyester (plastic type) and it is a washable bottle attached with USB port in the top of the bottle cap. The intake of the water is continually monitored. The water level intake and temperature is shown in the connected device (smart phone) via Bluetooth technology [14]. This bottle

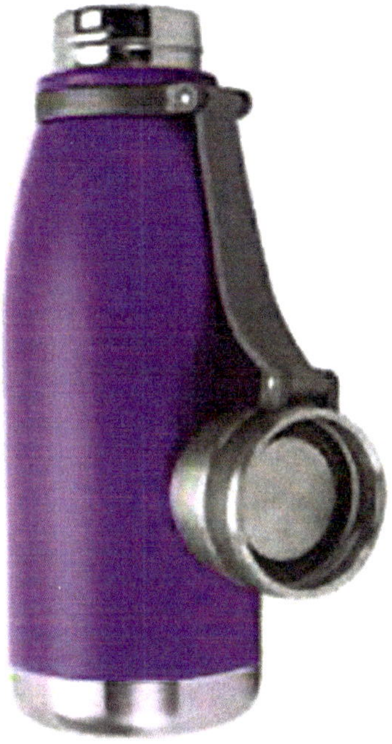

Fig. 6 Existing system

system tells how much water we have consumed and reminding the hydration level as per the data we had stored in the application.

Also, in a stainless steel flask system vacuum keeps cold things as cold and hot things as hot. It consists of cork inside a metal resting with two silver layered glass vessels. It maintains the hotness and coldness for a long time inside a bottle.

5 Proposed System

A new smart flask with heating and cooling system includes IoT and stands for the name of smart in the system as shown in Fig. 7. In existing system there is a hydration level tracker and temperature display system. The sophistication and the favour of our choice in the water are tracking in which it is to be filled either hot or cold water only. To overcome this we have designed the bottle made up of stainless steel and structured as the flask by having both heating and cooling system in a single bottle in which it takes to the next level of smartness by having the hydration level with the favour of our choice in the water. In this system the water will starts heating and in a heated state when the hot button is pressed in the bottle which is placed in the flask holder and similarly for the cooling system also by stepping into the smartness the system gets automatically off when it affairs the correct temperature. Once the water is in a correct state either hot or cold the message will also send to the user by reminding the water status in the format of "The water is ready to drink". This system is highly durable, safe and water proof and it is easy to carry everywhere and helps in the time management.

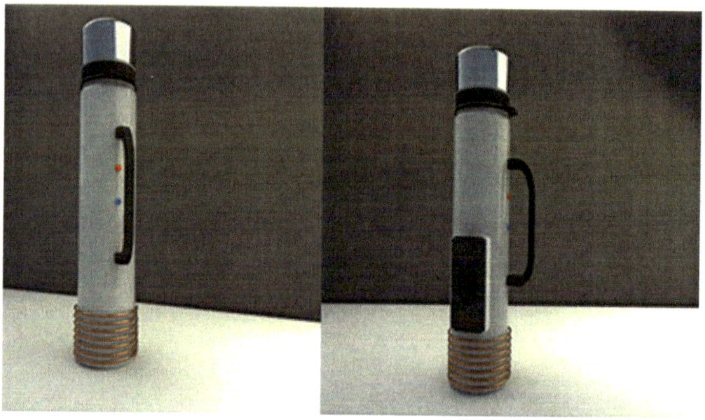

Fig. 7 Proposed smart flask

Operating Procedure

Before using the bottle, the user who is going to use this bottle has to register his/her mobile number in the GSM module. After connecting the circuit with Peltier and heating element in the bottle, the required drink must first be filled into the bottle. There is a button to turn on and off the bottle. The bottle is given with the facility to select the desired temperature. Once the temperature is selected using it, the bottle will start to function. As soon as the drink reaches the selected temperature in the bottle, the corresponding information will be sent as an SMS to the mobile registered in the GSM. As in the existing system, with this facility the user does not have the job of monitoring its temperature from near the bottle. Its specialty is that the temperature of the drink thus prepared remains constant for more than two days as shown in Fig. 9. The bottle can be turned off as soon as the drink is ready.

6 Results

The Peltier operates at 15.4 V is used here to prepare the cooling water. It is actually placed on inner the wall of the container that is by cutting the inner layer, one side of the Peltier is placed and other side comes outside the inner layer where the heat sink is placed. The heat generated on the other side of the Peltier is dissipated with help of heat sink. In order to prepare the hot water, the copper tubes which are covered at the bottom of the bottle are used.

It operates with the support of rechargeable battery. The user is given hot and cold options to select and further has the provision to set the desired temperature. Figure 8 shows how long the temperature of the drink in the existing system is maintained. In Fig. 9, the result of the test comparison is shown that the smart carafe kept the temperature of the drink constant for about 2 days. After that the temperature starts to change. This is because of the metals used in it and the design of this bottle. Figure 10 witness that the proposed bottle takes approximately 10–15 min to increase/decrease the temperature of the drink.

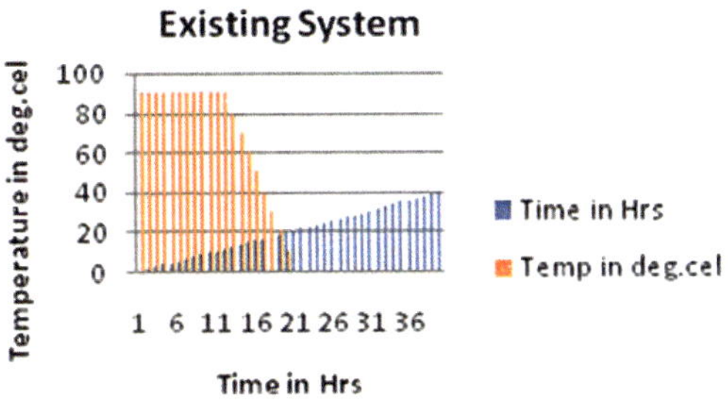

Fig. 8 Existing system performance

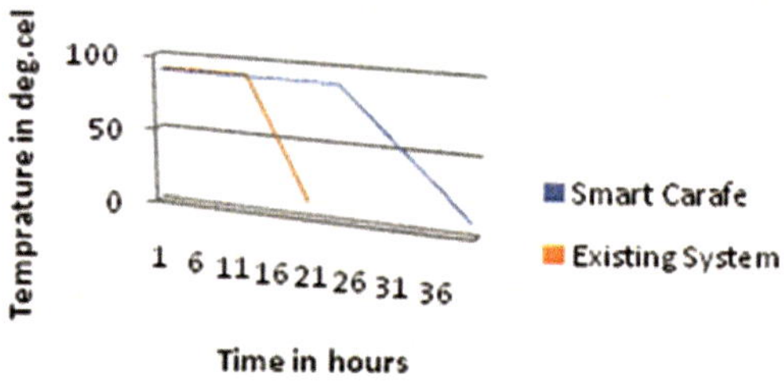

Fig. 9 Comparison of smart carafe and existing system

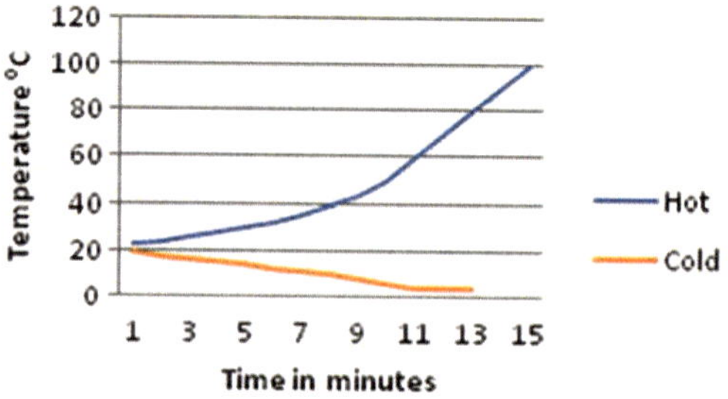

Fig. 10 Performance of smart carafe

7 Conclusion

The proposed system is designed in a 1 L water bottle (stainless steel material) with both heating and cooling system. This will be useful for everyone, which is in light weight and easily rechargeable water bottle flask to drink water in the favour of their choice. The Peltier used in this bottle supports quick cooling. The proposed smart carafe performs better than the existing system which means that it can maintain the hot/cold for an extended time period as shown in Figs. 8 and 9. The performance of smart carafe is shown in Fig. 10. From the results, the proposed system prepares the hot and cold water in couple of minute's time. Once the set temperature is reached user gets notification through mobile phone which makes user no need to spend time to monitor the process. The cost of exiting flask and proposed carafe is comparable. Since the electronic items used in the proposed system has better life and results in cost effective with additional features. The future work covers experimenting with DHT11 or LM35 sensor for predicting the temperature of the drink.

References

1. C. Hommalee, S. Wiriyasart, P. Naphon, Development of cold & hot water dispenser with thermoelectric module systems. Heat Transfer Asian Res. **48**(3), 854–863 (2019)
2. https://www.aquaidwatercoolers.co.uk/10180#:~:text=The%20water%20cooler%20as%20we,but%20to%20provide%20purified%20water
3. K. Subahan, E. Siva Reddy, R. Meenakshi Reddy, CFD analysis of pin-fin heat sink used in electronic devices. Int. Res. Sci. Technol. **8**(9), 562–569 (2019)

4. https://web.archive.org/web/20070927020152/, http://www.appliancemagazine.com/news_print.php?article=9241&zone=0&first=1
5. A. Nikam, J. Hole, A review on use of Peltier effects. Int. J. Sci. Spirit. Bus. Technol. **2**(2), 6–12 (2014)
6. V.G. Shelke, C.V. Patil, K.R. Sontakke, Solar water heating system. Int. J. Sci. Eng. Res. **3**(4), 13–17 (2015)
7. N. Khamkar, A. Waghmode, A. Joshi, P. Supekar, A. Kumar, K. Londhe, V.R. Kaushik, Heat sink design for optimal performance of compact electronic appliances. J. Adv. Res. Appl. Sci. **4**(5), 13–21 (2017)
8. V. Geetha, V. Siva chidambaranathan, An overview of designing an induction heating system for domestic applications. Int. J. Power Electron. Drive Syst. **10**(1), 351–356 (2019)
9. Rajendra, P. Patil, P. Suryawanshi, A. Pawar, A. Pawar, Thermoelectric refridgeration using peltier effect. Int. J. Eng. Sci. Res. Technol. **5**(5), 614–618 (2017)
10. https://www.cnet.com/reviews/ozmo-active-smart-cup-review/
11. M. Kosir, A. Krainer, M. Dovjak, R. Perdan, Kristl, Alternative to the conventional heating and cooling systems in public buildings. J. Mech. Eng. **56**(9), 575–583 (2010)
12. S.O. Tan, H. Demirel, Performance and cooling efficiency of thermoelectric modules on server central processing unit and Northbridge. Comput. Electr. Eng. **46**, 46–55 (2015)
13. S. Lv, W. He, Q. Jiang, Study of different heat exchange technologies influence on the performance of thermoelectric generators. Energy Convers. Manage. **156**, 167–177 (2018)
14. S. Vishwakarma, A. Goud, A literature review on: smart bottle using IOT. Int. J. Comput. Sci. Eng. **7**(6), 300–302 (2019)

Deep Learning-Based Lung Cancer Detection

S. Mahima, S. Kezia, and E. Grace Mary Kanaga

1 Introduction

The kind of cancer that tends to affect the lungs is referred to as lung cancer. According to studies, non-small cell lung cancer (NSCLC) and small cell lung cancer (SCLLC) are the two different forms of lung cancer (SCLC).

Among the patients, 80–85% of them suffer from the first type. The SCLC is more like a recurrent type, and it is usually spread across the lungs even before the first diagnosis. According to statistics, lung cancer is the second-highest frequent type of cancer that humans acquire where it is observed that the patients are mostly 65 years and above. The mortality rate of lung cancer is more than colon, breast, and colon cancers combined. Researchers across the world are currently more involved in the prevention and early detection of this cancer. Many different types of models by applying artificial intelligence techniques are being utilized for the early diagnosis of lung cancer in which the deep learning techniques outweigh them all. Numerous works have been proposed by researchers with regard to the early prediction of lung cancer using many machine learning and deep learning techniques. To the best of our knowledge in this field, the deep learning techniques LSTM and autoencoders have never been offered as a combination strategy for lung cancer diagnosis.

This will allow patients and their doctors to take the required steps more quickly and effectively. Furthermore, the use of deep learning for the identification of this

S. Mahima (✉) · S. Kezia · E. Grace Mary Kanaga
Department of Computer Science and Engineering, Karunya Institute of Technology and Sciences, Coimbatore, India
e-mail: smahima20@karunya.edu.in

S. Kezia
e-mail: kezias20@karunya.edu.in

E. Grace Mary Kanaga
e-mail: grace@karunya.edu

J. D. Peter et al. (eds.), *Disruptive Technologies for Big Data and Cloud Applications*, Lecture Notes in Electrical Engineering 905,
https://doi.org/10.1007/978-981-19-2177-3_59

cancer is a boon, so hence, more research in this field will yield a slew of useful and much-needed novel ideas. In this regard, in our paper, we have proposed an approach by combining these two methods for an early detection of lung cancer.

2 Technical Concepts

2.1 *Long Short-Term Memory (LSTM)*

The concept of long short-term memory (LSTM) was initially introduced in the year 1997, and its efficiency has been widely observed and applied in different fields of research. The primary goal of LSTM is to overcome backflow of errors. It can adapt to integrate time spans higher than 1000 stages with even noisy, axisymmetric input patterns without degrading small time lag skills. The long short-term memory structure was prompted via a study of erroneous flows in conventional RNNs, which revealed how prior techniques could not handle long time lags when backpropagated error either collapses or decays exponentially. An LSTM system is composed of memory chunks that are connected consecutively. There are one or much more repeatedly linked memory cells for each one, as well as three multiplicative units (input, output, and forget gates) that enable continuous analogues of write, read, and reset operations for the cells. The net can only interface with the cells through the gates.

2.2 *Autoencoders*

An autoencoder is a type of artificial neural network that may be used to acquire a compressed representation of original data. An encoder and a decoder are the two main that make up an autoencoder. The encoder reduces the data, while the decoder attempts to recreate the inputs from the reduced version of the encoder. The encoder model is maintained after training, whereas the decoder is deleted.

After training, the encoder model is saved and the decoder is discarded. Their main goal is to learn an "informative" description of the information in an unsupervised way that can be used for various purposes such as clustering. When expanded further, it consists of four sub-models, namely encoder, bottleneck, and decoder and reconstruction loss. Backpropagation is then used in the training to reduce the network's reconstruction failure. Autoencoders tend to be data-oriented. As a result, they are capable of compressing the data which is close or related to what they have been made to learn. They differ with respect to dimensionality reduction techniques such as gzip in that they learn features that are specific to the training samples. Autoencoders are a fantastic way to reduce dimensionality. Because of their simplicity, they are a common teaching tool in introductory deep learning courses.

3 Related Works

3.1 Clustering Algorithm Based Works

A CAD system, also known as computer-aided designing, is a computer program that allows you to design parts or products before they are made. A computer-aided design (CAD) system is used to create vehicle parts and other items. Originally used in the field of mechanical engineering and construction, CAD is now being used in other areas like health care also. Radiologists use CAD in the detection of cancer cells. One main disadvantage is that these systems require huge amounts of data for detection. The authors of [1] have suggested a generative adversarial network (GAN) as a potential data augmentation approach for improving CAD systems by generating extra training data. They have also suggested a convolutional autoencoder deep learning framework for learning unsupervised image characteristics from non-labeled data for lung nodules. They have come up with clustering augmented learning method (CALM) for improving the accuracy and precision. The results obtained utilizing the lung image database consortium (LIDC) dataset from the National Cancer Institute (NCI) indicated an overall accuracy of 95.3% and a precision of 94.9%.

In the image capturing process of lung cancer detection, noise signal gets in the way of producing a quality image. To avoid this, the major essential stages for improving the nature of the input image are edge detection, lung image resampling, lung image upgrade, and image denoising. The authors of this paper [2] have dealt with improving the quality of lung image and diagnosis of lung cancer by reducing misclassification. The scientists employed an improved profuse clustering method (IPCT) to improve picture quality and weighted mean histogram equalization to remove noise from the Cancer Imaging Archive (CIA) dataset. For predicting, deep learning instantaneously trained neural network has been used. The MATLAB-based simulation results show 98.42% of accuracy with minimum classification error of 0.038.

3.2 Neural Network-Based Works

A computerized tomography scan (CT or CAT scan) creates cross-sectional images of the body using computers and rotating X-ray devices. These images contain more information than standard X-ray images. Soft tissues, blood arteries, and bones in various regions of the body can all be seen. The authors of [3] have used three different deep neural network types, namely convolutional neural network (CNN), deep neural network (DNN), and sparse autoencoder (SAE) for the purpose of lung cancer calcification. The CT images of benign and malignant lung nodules are fed into these neural networks, the dataset being obtained from LIDC-IDRI database. The results obtained indicate that CNN bags the honor of being the best-performing network among the three with an accuracy of 84.15%, sensitivity of 83.96%, and specificity of 84.32%.

When dealing with the detection of lymph nodes using histologic slides, the problem of false prediction occurs as a part and parcel. The authors of [4] have created a revolutionary two-step deep learning system to handle the problem of false-positive prediction while still detecting cancer accurately. The authors gathered 349 whole-slide lung cancer lymph node pictures for this investigation. Errors were decreased by 36.4% on average and up to 89% in slides using this two-step strategy, resulting in 100% sensitivity in situations of mega metastases, micro-metastases, and isolated tumor cells. Immune checkpoint inhibitors are used in checkpoint blockade immunotherapy, which is a novel treatment that targets a variety of cancers. These drugs, in particular, can aid the body's immune system in recognizing and attacking malignant cells. The authors of [5] looked at the possibility of using deep learning to predict anti-PD-1 response directly from H&E images. The authors achieved a state-of-the-art model with an area under the curve (AUC) of 0.778 for 54 melanoma testing samples and 55 lung cancer samples by segmenting 190 H&E slides of melanoma.

When a patient has acquired cancer, the surviving period of the patient is usually diagnosed by the treating physician. The authors of [6] have proposed numerous survival prediction models based on deep learning techniques to address both cancer survival classification and regression issues. The authors examined the performance of the most common deep learning architectures—artificial neural networks (ANN), convolutional neural networks (CNN), and recurrent neural networks (RNN)—versus classical machine learning models. The lung cancer portion of the Surveillance, Epidemiology, and End Results (SEER) cancer registry provided the dataset. The results show that the deep learning models outperform any other conventional models with a sweeping 71.18% accuracy. So far, reducing the processing time and increasing the accuracy of CAD systems have been a difficult undertaking. The authors of this research [7] have offered a useful deep learning (DL) method for lung cancer detection based on the Targeted Weighted Elman DL Neural Network (TWEDLNN) and the Mask Unit (MU)-based 3FCM algorithm. For the implementation, the scientists used CT images from the LIDC-IDRI database. The findings show that the TWEDLNN is more effective at detecting lung cancer, with a 96% accuracy rate.

3.3 U-Net Architecture-Based Works

The authors of [8] have studied the images from CT scans of lung cancer patients with the motive to detect and classify the lung nodules based on their malignancy level. U-Net architecture is used by the authors to study these images. The authors have proposed a multipath VGG-like network with the data being taken from LIDC-IDRI database. The predictions from the U-Net architecture and 3D VGG-like network make up the end outcome. An accuracy of 95.60% is obtained for the classification of the ling nodules based on their malignancy levels.

The authors of [9] have created a modified three-dimensional (3D) U-Net deep learning model for the automated detection of lung nodules on chest CT scans, with the goal of calculating the model's accuracy. By training the deep learning model with the dataset from Lung Image Database Consortium and Image Database Resource Initiative, the results were obtained for internal and external model validation. The authors have come up with a high-performing model with an internal validation CPM of 94.7% and an external validation CPM of 83.3%.

3.4 *ProNet Model*

The authors of [10] have taken CT images of 920 patients with lung cancer, including 554 patients with ADC (lung adenocarcinomas), 175 patients with lung SCC (squamous cell carcinomas), and 191 patients with SCLC (small cell lung cancers) as the database for their work. According to histopathologic data, they employed 3D deep learning (called ProNet) and radiomics (named com radNet) to classify lung tumors into the three kinds stated above. The weighted average F1-score for the ProNet model was 73.2%, whereas the weighted average F1-score for the com radNet was 72.2%. The results had an accuracy of 71.6% and 74.7%, respectively.

3.5 *DeepLR Method*

Lung nodule screening is a mandatory step in the detection of lung cancer. To determine the timing of the next CT scan, the screening guidelines consider the mean diameter, volume, or density of the largest lung nodule on the previous CT scan or the presence of a new nodule. The authors of [11] have used a deep learning algorithm named DeepLR to estimate the lung cancer risk with 2 CT scans of a particular patient. A dataset of 25,097 participants in National Lung Screening Trial (NLST) across a two-year span is taken into account. The findings demonstrate that the DeepLR model correctly identified 94%, 85%, and 71% of incident and interval lung cancers diagnosed within one year, two years, and three years following the second screening CT scan, respectively.

3.6 *Other Deep Learning Methods*

For clinical adoption of any proposed model, it is necessary that the model has a strong generalization and stability and has been tested using large datasets. The authors of [12] compared a deep learning framework to state-of-the-art models while examining larger and heterogeneous datasets. The authors employed three low-dose CT lung cancer screening datasets: the National Lung Screening Trial (NLST), data from

Lahey Hospital and Medical Center (LHMC), data from the Kaggle competition, and data from the University of Chicago. The results show that the suggested model outperformed the widely used PanCan Risk Model, with an AUC of 86–94%.

The deep learning techniques employed for the detection of lung cancer show improvement in performance in a predictive fashion. The ultimate aim is to classify an input sample into predefined categories. The authors of this paper [13] have investigated how deep learning algorithms create cancer diagnosis predictions based on clinical, radiological, and pathological image data, as well as the creation and use of clinical deep learning algorithms. The authors have also discussed the future development and scope in the detection of lung cancer, a much-needed area.

4 Deep Learning-Based Lung Cancer Detection

4.1 Input Data Collection

For this proposed work, we would make use of the publicly available dataset for lung cancer from the Kaggle website using [14] that consists of photographs of lung tissues. Every photograph is stored as jpeg files at a dimension of 768 by 768 pixels. The photographs were produced using a collection of 750 lung tissue photographs obtained from HIPAA-compliant and authorized sources. In the dataset, there are 250 photographs of benign lung tissue, 250 images of lung adenocarcinomas, and 250 images of lung squamous cell carcinomas in all of these 750 images.

4.2 Methodology

The detailed architecture diagram for the proposed work is represented in Fig. 1.

In the obtained input image dataset, many images are observed to be highly blurred which makes it difficult to identify. Initially, in the proposed method, we would load these input images to the autoencoders. Autoencoders are applicable for a good denoising process which will help us to get a clear image. Once the denoising is performed by the autoencoders, they are clustered by the technique into three clusters. These images are then passed as the input to the long short-term memory (LSTM) which would classify them as benign and malignant. The images that are under the categories: "Lung adenocarcinoma" and "Lung squamous cell carcinoma" as malignant and the others benign. Also, to evaluate the performance of the proposed method, metrics such as accuracy score would be applied.

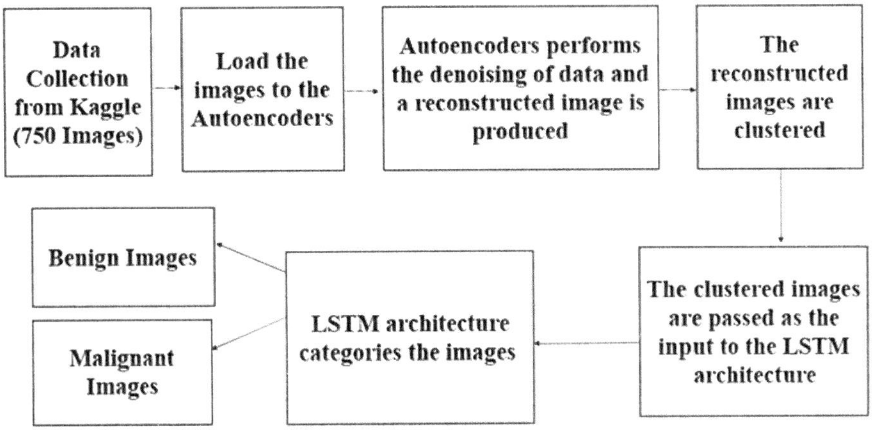

Fig. 1 Architecture diagram for the lung cancer detection system

4.3 Key Benefits of Our Proposed Method

The most popular methods for detecting the presence of cancer cells in an individual are CT scans and X-rays. However, adequate information is not always gathered in this manner, and as a result, an effective remedy for a lung cancer patient is not always available. As per our analysis, our proposed method can aid doctors to get proper images the process of denoising the images could produce much clearer images. This may help in better diagnosis, and as a result, better treatment plan could be given to the patient. In addition, we are proposing the detection as malignant and benign using accuracy score as a metric. The model would be trained until the best accuracy is obtained. Hence, manual errors such as incorrect diagnosis can be prevented.

5 Conclusion

As previously stated, lung cancer is the second most prevalent cancer, and it is critical to forecast and diagnose nodules at an early point. Researchers have proposed a variety of experiments to identify lung cancer early using a multitude of artificial intelligence methodologies. According to our research in this area, the deep learning techniques LSTM and autoencoders have never been proposed as a combined strategy for lung cancer detection. Keeping this factor in mind, we have proposed a methodology that is quite cost-efficient for a good early detection of lung cancer. This will help the patients as well as their physicians to take the necessary actions at a faster and effective pace. Also, the use of deep learning in the detection of this cancer is a boon and further research in this area will bring up many useful and much-needed novel ideas.

References

1. S.S. Ghosal, et al., Lung nodule classification using convolutional autoencoder and clustering augmented learning method (CALM). HSDM@WSDM (2020)
2. P. Shakeel, Mohamed, M.A. Burhanuddin, M. Desa, Lung cancer detection from CT image using improved profuse clustering and deep learning instantaneously trained neural networks. Measurement 145 (2019). https://doi.org/10.1016/j.measurement.2019.05.027
3. Q.Z. Song, L. Zhao, X.K. Luo, X.C. Dou, Using deep learning for classification of lung nodules on computed tomography images. J. Healthc. Eng. **2017**, 1–7 (2017). https://doi.org/10.1155/2017/8314740
4. H.H.N. Pham, M. Futakuchi, A. Bychkov, T. Furukawa, K. Kuroda, J. Fukuoka, Detection of lung cancer lymph node metastases from whole-slide histopathologic images using a two-step deep learning approach. Am. J. Pathol. **189**(12), 2428–2439 (2019). ISSN 0002-9440. https://doi.org/10.1016/j.ajpath.2019.08.014. https://www.sciencedirect.com/science/article/pii/S0002944019307187
5. J. Hu, C. Cui, W. Yang, L. Huang, R. Yu, S. Liu, Y. Kong, Using deep learning to predict anti-PD-1 response in melanoma and lung cancer patients from histopathology images. Transl. Oncol. **14**, 100921 (2021). https://doi.org/10.1016/j.tranon.2020.100921
6. S. Doppalapudi, R.G. Qiu, Y. Badr, Lung cancer survival period prediction and understanding: deep learning approaches. Int. J. Med. Inform. 148, 104371 (2021). ISSN 1386-5056. https://doi.org/10.1016/j.ijmedinf.2020.104371. https://www.sciencedirect.com/science/article/pii/S1386505620319079
7. L. Tiwari, R. Raja, V. Awasthi, R. Miri, G.R. Sinha, M.H. Alkinani, K. Polat, Detection of lung nodule and cancer using novel Mask-3 FCM and TWEDLNN algorithms. Measurement **172**, 108882 (2021). ISSN 0263-2241. https://doi.org/10.1016/j.measurement.2020.108882. https://www.sciencedirect.com/science/article/pii/S0263224120313701
8. R. Tekade, K. Rajeswari, Lung cancer detection and classification using deep learning, in *2018 Fourth International Conference on Computing Communication Control and Automation (ICCUBEA)*, Pune, India, 2018, pp. 1–5. https://doi.org/10.1109/ICCUBEA.2018.8697352
9. K. Suzuki, Y. Otsuka, Y. Nomura, K. Kumamaru, R. Kuwatsuru, S. Aoki, Development and validation of a modified three-dimensional U-Net deep-learning model for automated detection of lung nodules on chest CT images from the lung image database consortium and Japanese datasets. Acad. Radiol. (2020). https://doi.org/10.1016/j.acra.2020.07.030
10. Y. Guo, Q. Song, M. Jiang, Y. Guo, P. Xu, Y. Zhang, C.-C. Fu, Q. Fang, M. Zeng, X. Yao, Histological subtypes classification of lung cancers on CT images using 3D deep learning and radiomics. Acad. Radiol. (2020). ISSN 1076-6332. https://doi.org/10.1016/j.acra.2020.06.010. https://www.sciencedirect.com/science/article/pii/S1076633220303603
11. P. Huang, C. Lin, Y. Li, M. Tammemagi, M. Brock, S. Atkar-Khattra, Y. Xu, P. Hu, J. Mayo, H. Schmidt, M.D. Gingras, S. Pasian, L. Stewart, S. Tsai, J. Seely, D. Manos, P. Burrowes, R. Bhatia, M.-S. Tsao, S. Lam, Prediction of lung cancer risk at follow-up screening with low-dose CT: a training and validation study of a deep learning method. The Lancet Digital Health (2019). https://doi.org/10.1016/S2589-7500(19)30159-1

12. S. Trajanovski, D. Mavroeidis, C.L. Swisher, B.G. Gebre, B.S. Veeling, R. Wiemker, T. Klinder, A. Tahmasebi, S.M. Regis, C. Wald, B.J. McKee, S. Flacke, H. MacMahon, H. Pien, Towards radiologist-level cancer risk assessment in CT lung screening using deep learning. Comput. Med. Imaging Graph. 101883 (2021). ISSN 0895-6111. https://doi.org/10.1016/j.compmedimag.2021.101883. https://www.sciencedirect.com/science/article/pii/S0895611121000318
13. R. Daneshjou, B. He, D. Ouyang, J.Y. Zou, How to evaluate deep learning for cancer diagnostics—factors and recommendations. Biochim. Biophys. Acta (BBA)—Rev. Cancer **1875**(2), 188515 (2021). ISSN 0304-419X. https://doi.org/10.1016/j.bbcan.2021.188515. https://www.sciencedirect.com/science/article/pii/S0304419X21000147
14. Dataset, https://www.kaggle.com/andrewmvd/lung-and-colon-cancer-histopathological-images

Pneumonia Detection from Chest X-Ray Images Using Deep Learning Methods

C. Lenny, A. Ajitha Margharet, B. Shiny, Sabnam Tigga, and S. Thomas George

1 Introduction

Pneumonia is an infection that happens in the lungs because of microorganisms, viruses, and fungi. Diagnosis of the respiratory disorder may be a difficult task exploitation chest X-rays pictures which require identification accuracy must be improved. In this work, an associate degree economic model for the detection of respiratory disorder that square measure trained on digital chest X-ray pictures is planned, which might aid the radiologists in their method of deciding and conjointly avoid errors which occur due to misdiagnosis. This paper focuses on comparison of the detection and measurement of respiratory disorders exploiting different computer-aided techniques based on deep learning techniques and recommends an effective model for respiratory disorder detection. Deep learning belongs to a part of a larger family of machine learning method strategies supported by artificial neural networks with illustration learning. Learning is often supervised, semi-supervised, or unsupervised. Deep learning uses multiple layers to progressively extract higher-level options from the raw input that comes under the category of machine learning algorithms. For instance, in the image process, lower layers might determine edges, whereas high layers determine the ideas relevant to a personality's like digits or letters, or faces. CNN (short for convolutional neural networks) became the algorithms for a large style of tasks in deep learning, particularly for computer vision and medical images. Recently, researchers are experimenting and attempting to come up with higher-order architectures to boost model's accuracy on different tasks. CNN is a square measure regularized versions of multilayer perceptron whose full connectivity of these deep

C. Lenny · A. A. Margharet · B. Shiny · S. Tigga · S. T. George (✉)
Department of Biomedical Engineering, Karunya Institute of Technology and Sciences, Coimbatore, India
e-mail: thomasgeorge@karunya.edu

J. D. Peter et al. (eds.), *Disruptive Technologies for Big Data and Cloud Applications*, Lecture Notes in Electrical Engineering 905,
https://doi.org/10.1007/978-981-19-2177-3_60

neural networks creates them vulnerable to overfitting information which should be avoided.

2 Literature Survey

In general for respiratory disease, symptom dysfunction, respiratory organ scarring, and others on chest X-ray examination mislead, and they can resemble respiratory illness [1]. This is frequently the most common cause of X-ray image misclassification in the dataset. As a result, the implementation of effective algorithms for detecting pectoral diseases such as respiratory illness will improve clinical settings in remote areas. X-ray image analysis is a tedious and crucial task for radiology specialists. Therefore, researchers have proposed many computer-based algorithms to analyze X-ray pictures. Also, many computer-assisted diagnosing tools are developed to supply an insight into X-ray pictures. However, these tools are not able to give adequate data to support doctors in creating choices [2]. A model proposed, supported by deep learning and convolutional neural networks, is capable of mechanically classifying whether a patient has respiratory illness or not. The proposed technique employs a deep transfer learning formula to derive options from an X-ray picture that mechanically explains the existence of illness and reports whether it is a case of respiratory illness [3]. Training CNNs to identify images from these datasets is often accomplished by dividing the dataset into three subsets: train (data directly used to learn model parameters), tune (data used to choose hyper-parameter settings, often known as "validation"), and test as well as take a look at (data used completely for performance analysis of models learned victimization train and tune data). With the first two, CNNs are trained for completion, and the ultimate set is used to estimate the model's expected output on current from previously unseen data [4]. Throughout the training, models such as evolutionary-based algorithms and reinforcement learning are used to find optimal network hyper-parameters. However, these methods are computationally expensive, requiring a lot of processing power. As an alternative, research works developed later propose a conceptually simple but cost-effective network model to address the problem of respiratory illness classification [5]. Chest X-rays square measure the common technique accustomed to diagnose respiratory illness, and it wants a doctor to judge the results of the X-ray. With the advancement of technology, it is now possible to design an automatic system to detect pneumonia, respiratory illness, or respiratory condition, and treat the disease, particularly if the patient is in a distant place and medical services are required. Convolutional neural network is optimized to perform the sophisticated task of identifying illnesses like respiratory illness to assist doctors in diagnosing and treatment of the disease. The authors of this work developed many models to figure out the only model in identifying respiratory illness with the foremost correct results [6]. Rahman et al. [7] aim to use digital X-ray images to mechanically detect microorganisms and virus infection. It begins with a detailed study on advancements in the proper diagnosis of respiratory illness, followed by the authors' methodology.

For transfer learning, four separate deep convolutional neural networks (CNNs) were used: AlexNet, ResNet18, DenseNet201, and SqueezeNet. Moreover, several factors, such as the appearance of sickness, which may be obscure in chest X-ray pictures or is often confused with other illnesses, diagnostic tests are normally subjective. As a result, clinicians would need to be driven by computer-aided diagnosis systems [8]. It is troublesome to identify respiratory illness simply by observing chest X-rays. This study aims to alter the respiratory illness detection method for specialists moreover as for novices. A very distinctive deep learning framework for the detection of respiratory illness victimization the construct of transfer learning is proposed [9]. As a feature extractor, a convolutional neural network was used, and several existing convolutional neural network models were used. As a result of the findings, deep options offered stable and reliable options for respiratory illness identification, and the minimum redundancy of most connectedness techniques was discovered to be a valuable tool for reducing the feature set's size [10]. Current advancements in the area of computer science would be extremely beneficial as convolutional neural networks would be used to interpret chest X-ray images to determine whether or not the respiratory disease is present. Ninety-six percent accuracy was achieved using a pixel-based technique, and eighty-eight percent accuracy was achieved using a victimization feature-based technique [11]. With computer-assisted diagnosing of pneumonia or respiratory illness or respiratory disorders, specifically, physicians will accurately and expeditiously diagnose the disease [12]. Another goal of this research was to create a deep learning system that could mechanically diagnose respiratory illness using chest X-ray images and identify the results as typical cases of respiratory illness cases, making it easier to diagnose the illness rapidly and easily. Their architectures are highly adaptable, so they can learn right away from the data and improve their predictive accuracy by feeding it a large amount of data [13]. A predicted model is in the works is enforced by convolutional deep neural networks (CNNs) with Python artificial language. When preprocessing data, the application uses a variety of machine learning methods that have been trained to match the performance of CNN with popular and common used classifiers. Promising results have been achieved, once the results of normal classifiers like SVM, random forest, AdaBoost. [14]. We tend to use deep transfer learning ideas to create a framework for diagnosing respiratory illness in this paper. To this end, effort to investigate the efficiency of various deep neural network architectures as feature extractors, which were previously used in tandem with ancient classification methods, is carried out. Another solution that addresses these issues is deep transfer learning, in which a model that has been previously trained to perform one particular task is tailored and reused for another similar task [15].

3 Methodology

The details of the overall methodology are explained in Fig. 1 where using four different models for pneumonia detection are used which are namely VGG16, ResNet 50, InceptionV3, and EfficientNet50.

Initially, the dataset is preprocessed, and data augmentation methods are done. All the preprocessed data are passed through the architectures, namely InceptionV3, ResNet50, EfficientNet50, and VGG16 for the classification of pneumonia. Now, the model classifies whether an input X-ray image is pneumonia or otherwise initially for binary classification. In the case of multiclass classification as a next part, the input image that is fed will be categorized into three classes, namely bacterial pneumonia, viral pneumonia, and normal. In this paper, we tried to compare the efficiency of different models trained on the pneumonia dataset. One new model that we trained to compare was EfficientNet50. The other models were being used as feature extractors in several proposals.

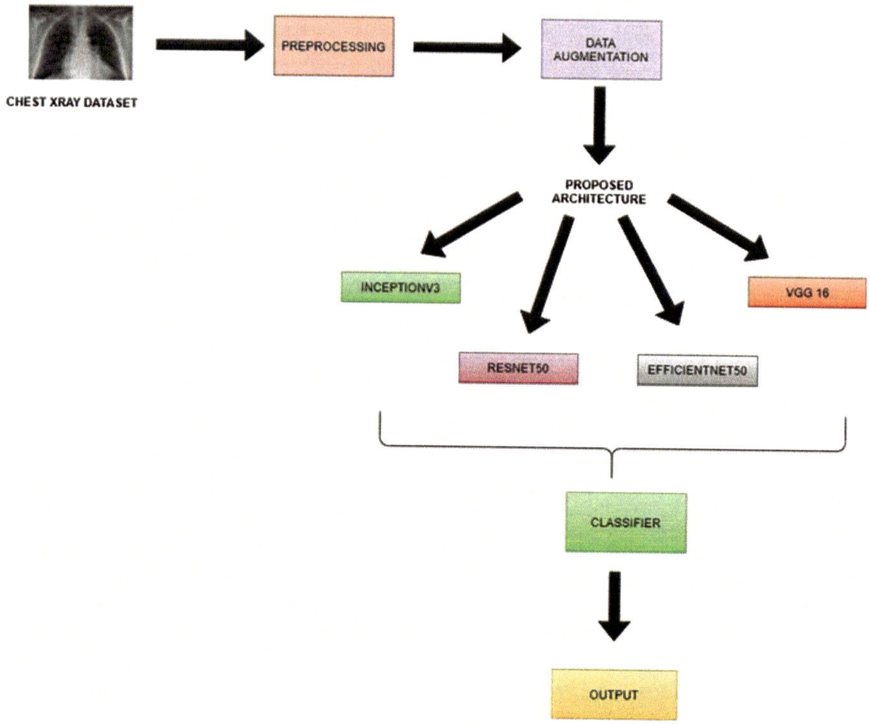

Fig. 1 Overall block diagram

4 Dataset

4.1 Flow of Dataset

The dataset used here has a chiliad of approved optical coherence tomography (OCT) and digital X- rays of the chest specified as well as tested. These X-rays would be separated into two, namely training and testing dataset. The X-rays would be labeled: (illness)—(random ID of the patient)—(patient's image no.), and separated and categorized within the four directories, namely: DRUSEN, DME, CNV, and normal.

4.1.1 Binary Classification—Normal, Pneumonia (Bacterial, Viral)

In binary classification, the dataset is split into train and test. For this classification, we have separated 80% for training and 20% for testing. In each train and test set, two classes, namely normal and pneumonia, as shown in Fig. 2 are considered. The number of images which we have split for normal and pneumonia is depicted in Table 1

For training, we have used a total of 5233 images, out of which normal images are 1349 and pneumonia affected are 3884. For testing, we have used a total of 624 images, out of which normal images are 234 and pneumonia affected is 390.

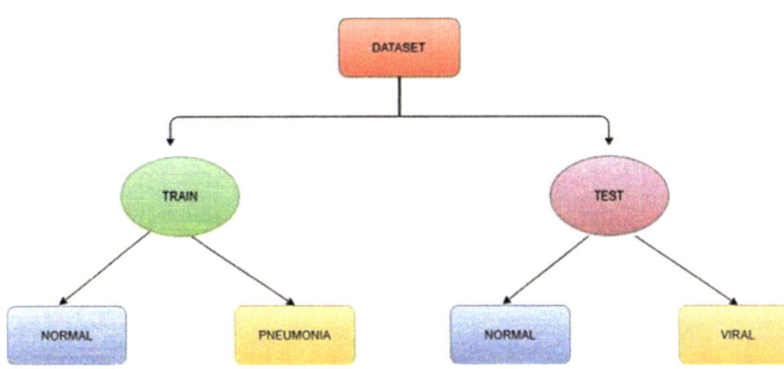

Fig. 2 Dataset—binary classification

Table 1 Dataset—binary classification

Dataset	Classes	No. of images
Train	Normal	1349
	Pneumonia	3884
Test	Normal	234
	Pneumonia	390

4.1.2 Multiclass Classification—Normal, Bacterial Pneumonia, Viral Pneumonia

In multiclass classification, we have split the dataset into train and test. For this classification, we have taken 70% for training and 30% for testing. In each train and test, we have three classes, namely bacterial, normal, and viral as shown in Fig. 3. The number of images that we have split for bacterial, normal, and viral is depicted in Table 2

For training, we have used a total of 4097 images, out of which normal images are 1106 and pneumonia affected due to bacteria is 1946 and viral is 1045. For testing, we have used a total of 1757 images, out of which normal images are 475 and pneumonia affected due to bacteria is 834 and viral is 448 (Figs. 4, 5, and 6).

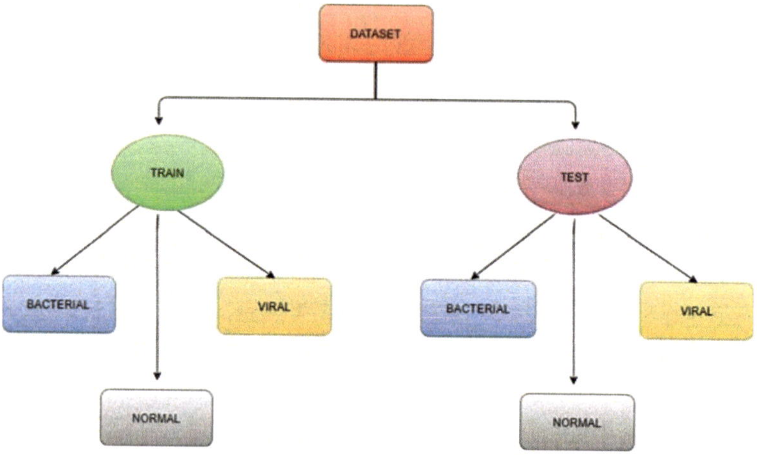

Fig. 3 Dataset—multiclass classification

Table 2 Dataset—multiclass classification

Dataset	Classes	No. of images
Train	Bacteria	1946
	Normal	1106
	Viral	1045
Test	Bacteria	834
	Normal	475
	Viral	448

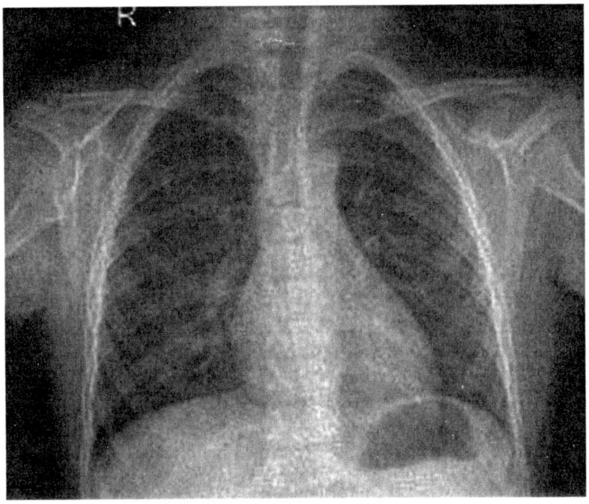

Fig. 4 Normal [16]

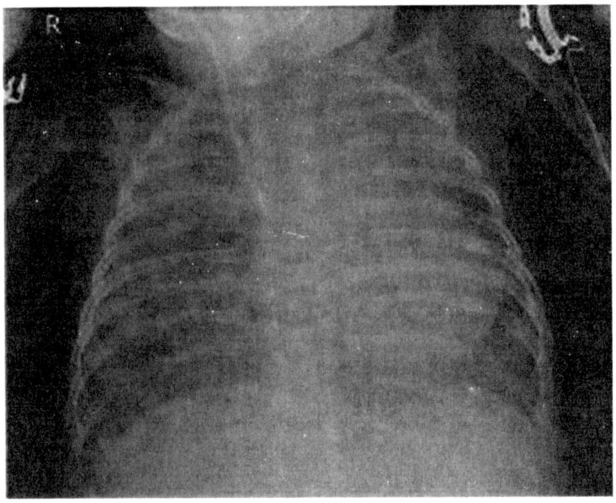

Fig. 5 Bacterial [16]

4.2 *EfficientNet 50*

EfficientNet focuses on the efficiency of models and not just on improving the accuracy. To bring up the efficiency of the architecture on a specific work, scaling is done (Fig. 7).

Fig. 6 Viral [16]

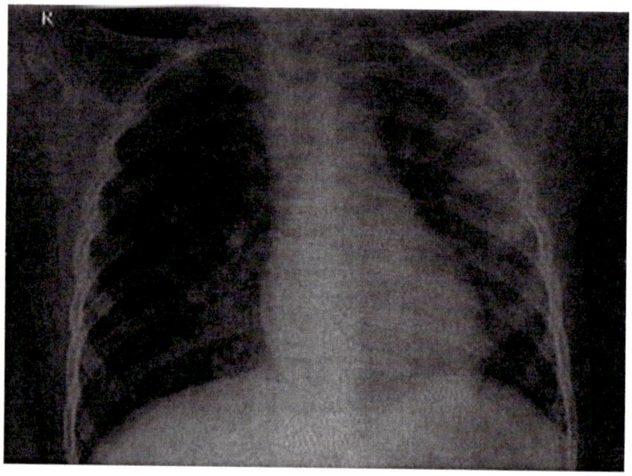

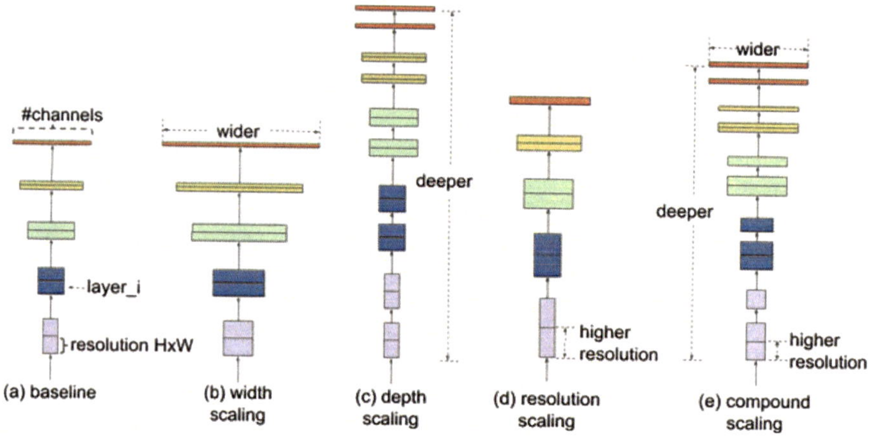

Fig. 7 EfficientNet architecture [17]

4.3 ResNet 50

A residual neural network (ResNet) is an artificial neural network (ANN) of a kind that builds on constructs known from pyramidal cells in the cerebral cortex. Residual neural networks do this by utilizing *skip* connections, or shortcuts to jump over some layers (Fig. 8).

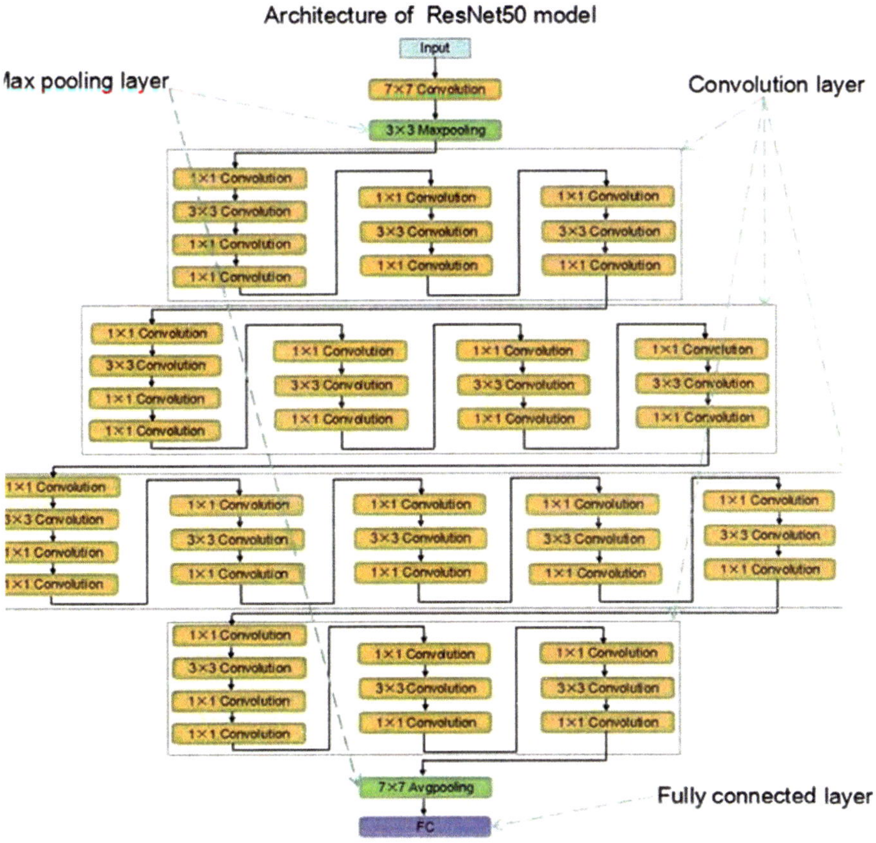

Fig. 8 ResNet 50 architecture [18]

4.4 *Inception V3*

The inception network is a crucial milestone within the development of CNN classifiers. Almost, all of the famous CNNs are just stacking more convolution network layers over it again and longing for better results (Fig. 9).

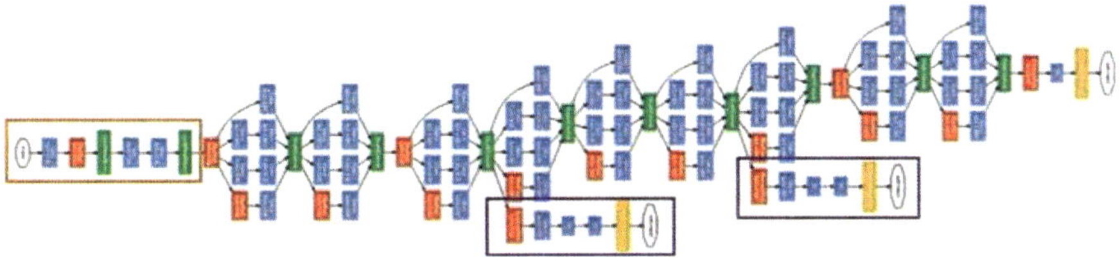

Fig. 9 InceptionV3 architecture [19]

4.5 *VGG16*

The model shown in Fig. 10 is VGG16.

The input to the cov1 layer is a 224 × 224 RGB image with a fixed size. The image is composed of a stack of convolutional (Conv.) layers, with the filters applied to a very narrow receptive field: 33 (the smallest size that can capture the concepts of left/right, up/down, and center). It also uses a 1 * 1 convolution filter in one of the setups, which can be thought of as a linear transformation of the input channels (followed by nonlinearity). The convolution stride is set to one pixel, and the abstraction cushioning of the convolution layer input is set. When convolution is used, the abstraction resolution is maintained, i.e., the cushioning is 1-pixel for 3 * 3 conv. layers. Five max-pooling layers match a number of the conv. layers and provide abstraction pooling (not all the conv. layers are followed by max-pooling). With a stride, a pair of max-pooling is done over a 2 * 2-pixel frame.

5 Comparison

Tables 3 and 4 show the comparison of different models with varied classifications.

5.1 *Comparison—Binary Classification—Normal, Pneumonia (Bacterial, Viral)*

See Table 3.

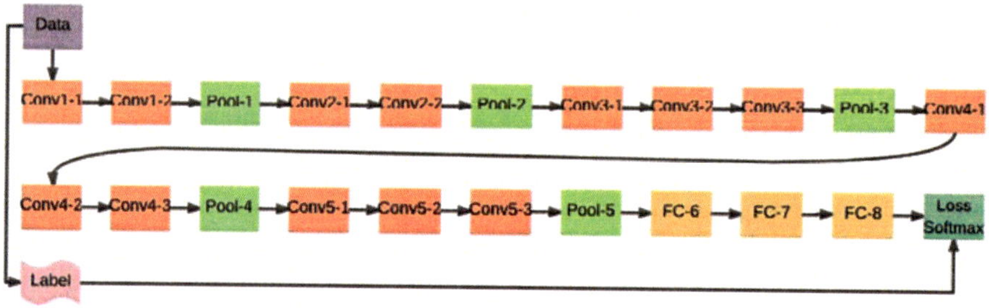

Fig. 10 VGG16 architecture [13]

Table 3 Comparison—binary classification

Network	VGG16	RESNET50	INCEPTIONV3	EfficientNET50
Accuracy (%)	95	94	96	94
Loss	–	56–27	10–8	1.5–0.1
Accuracy graph	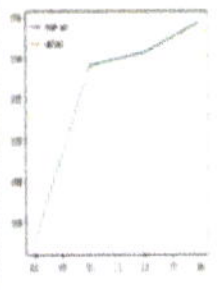	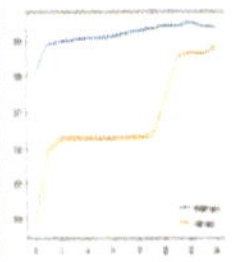	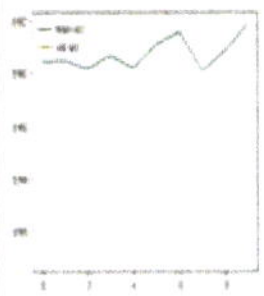	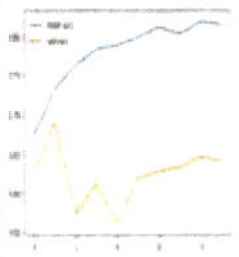
Loss graph	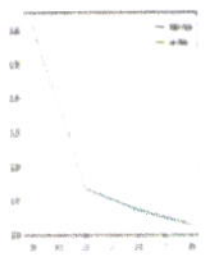	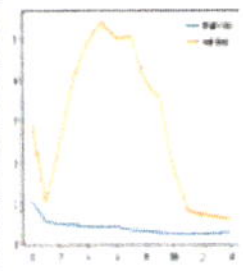	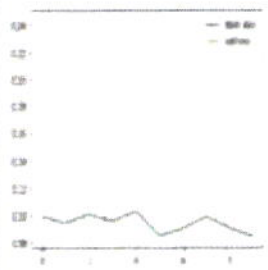	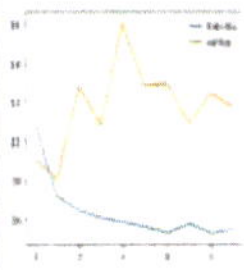

Table 4 Comparison—multiclass classification

Networks	VGG16	RESNET50	INCEPTIO N V3	EfficientNET 50
Accuracy (%)	90	87	86	
Loss	95–25	53–42	41–35	1.8–0.5
Accuracy graph	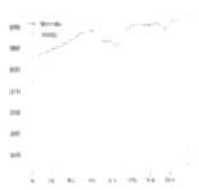	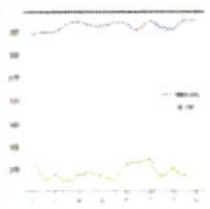	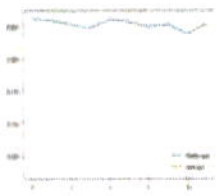	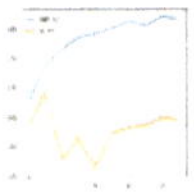
Loss graph	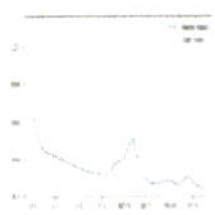	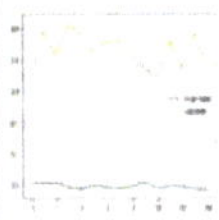	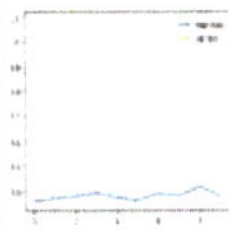	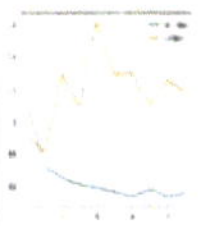

5.2 *Comparison—Multiclass Classification—Normal, Bacterial Pneumonia, Viral Pneumonia*

Table 4 shows the accuracy metric and the corresponding loss for each neural network. The performance of the model is determined with the help of accuracy metric.

6 Results and Discussions

This work shows the comparative study of various neural networks on the chest X-ray dataset. The accuracy metric is the parameter that compares the neural network's efficiency. Based on the accuracy, we calculate the performance and can determine the loss. It is found out that VGG16 outperforms several other networks in multiclass classification, but in the case of binary classification, InceptionV3 shows outstanding results. The number of trainable parameters is more in VGG16. The larger the number of parameters with larger the features can be extracted, and new architecture can be worked out in the future.

References

1. D. Varshni, K. Thakral, L. Agarwal, R. Nijhawan, A. Mittal, Pneumonia detection using CNN based feature extraction, in *Proceedings of 2019 3rd IEEE International Conference on Electronics, Computing and Communication Technologies, ICC CT 2019* (2019), pp. 1–7. https://doi.org/10.1109/ICECCT.2019.8869364
2. A.K. Jaiswal, P. Tiwari, S. Kumar, D. Gupta, A. Khanna, J.J.P.C. Rodrigues, Identifying pneumonia in chest X-rays: a deep learning approach. Meas. J. Int. Meas. Confed. **145**, 511–518 (2019). https://doi.org/10.1016/j.measurement.2019.05.076
3. M.F. Hashmi, S. Katiyar, A.G. Keskar, N.D. Bokde, Z.W. Geem, Efficient pneumonia detection in chest X-ray images using deep transfer learning. Diagnostics **10**(6), 1–23 (2020). https://doi.org/10.3390/diagnostics10060417
4. J.R. Zech, M.A. Badgeley, M. Liu, A.B. Costa, J.J. Titano, E.K. Oermann, Variable generalization performance of a deep learning model to detect pneumonia in chest radiographs: a cross-sectional study. PLoS Med. **15**(11), 1–17 (2018). https://doi.org/10.1371/journal.pmed.1002683
5. O. Stephen, M. Sain, U.J. Maduh, D.U. Jeong, An efficient deep learning approach to pneumonia classification in healthcare. J. Healthc. Eng. **2019** (2019). https://doi.org/10.1155/2019/4180949
6. S. Shah, H. Mehta, P. Sonawane, Pneumonia detection using convolutional neural networks, in *International Conference on Smart Systems and Inventive Technology, ICSSIT 2020* , vol. 9, no. 04 (2020), pp. 933–939. https://doi.org/10.1109/ICSSIT48917.2020.9214289
7. T. Rahman, et al., Transfer learning with deep convolutional neural network (CNN) for pneumonia detection using chest X-ray. Appl. Sci. **10**(9) (2020). https://doi.org/10.3390/app10093233
8. E. Ayan, H.M. Ünver, Diagnosis of pneumonia from chest X-ray images using deep learning, in *2019 Scientific Meeting on Electrical-Electronics and Biomedical Engineering and Computer Science, EBBT 2019* (2019), pp. 1–4. https://doi.org/10.1109/EBBT.2019.8741582
9. V. Chouhan, S.K. Singh, A. Khamparia, D. Gupta, V.H.C. De Albuquerque, Applied sciences a novel transfer learning based approach for pneumonia detection in chest X-ray images
10. M. Toğaçar, B. Ergen, Z. Cömert, F. Özyurt, A deep feature learning model for pneumonia detection applying a combination of mRMR feature selection and machine learning models. IRBM **41**(4), 212–222 (2020). https://doi.org/10.1016/j.irbm.2019.10.006
11. A. Tilve, S. Nayak, S. Vernekar, D. Turi, P.R. Shetgaonkar, S. Aswale, Pneumonia detection using deep learning approaches, in *International Conference on Emerging Trends in Engineering and Technology, ic-ETITE 2020* (2020), pp. 1–8. https://doi.org/10.1109/ic-ETITE47903.2020.152
12. B. Antin, J. Kravitz, E. Martayan, Detecting pneumonia in chest X-rays with supervised learning, *Semant. org* (2017), pp. 1–5
13. N.M. Elshennawy, D.M. Ibrahim, Deep-pneumonia framework using deep learning models based on chest X-ray images. Diagnostics **10**(9), 1–16 (2020). https://doi.org/10.3390/diagnostics10090649
14. P. Rudraraju, K. Sri Harsha, K. Usha, B.V.S.S. Madhuri, Pneumonia detection by X-ray images using deep learning through CNN. J. Xi'an Univ. Archit. Technol. **22**(1), 8–11 (2020). https://doi.org/10.9790/0661-2201040811

15. K.T. Islam, S. Wijewickrema, A. Collins, S. O'Leary, A deep transfer learning framework for pneumonia detection from chest X-ray images, in *VISIGRAPP 2020—Proceedings of 15th International Joint Conference on Computer Vision, Imaging and Computer Graphics Theory and Applications*, vol. 5, no. April, pp. 286–293 (2020). https://doi.org/10.5220/0008927002860293
16. D. S. Kermany et al., Identifying medical diagnoses and treatable diseases by image-based deep learning. Cell **172**(5), 1122–1131.e9 (Feb. 2018). https://doi.org/10.1016/j.cell.2018.02.010
17. M. Tan, Q. v. Le, EfficientNet: Rethinking Model Scaling for Convolutional Neural Networks. May 2019, [Online]. Available: arXiv: 1905.11946
18. J. Peng et al., Residual convolutional neural network for predicting response of trans arterial chemoembolization in hepatocellular carcinoma from CT imaging. Eur. Radiol. **30**(1), 413–424 (Jan. 2020). https://doi.org/10.1007/s00330-019-06318-1
19. C. Szegedy et al., Going Deeper with Convolutions, Sep. 2014, [Online]. Available: arXiv: 1409.4842

A Meta-Analysis on the Algorithms for Virtual Machine Consolidation

Rose Rani John and E. Grace Mary Kanaga

1 Introduction

Cloud computing is defined as a way of providing massively scalable IT-enabled capabilities "as a service" to customers around the globe using internet technologies [1]. Services over cloud are provided using the revolutionary "pay-as-you-go" paradigm model [2]. According to NIST [3], the architecture of cloud computing is a versatile model for delegating undisturbed, suitable network access to shared computing resources. And the backbone of cloud computing is the thousands and thousands of data centers spread across the various geographical locations around the globe. And the essential components of a basic data center consist of network infrastructure, storage infrastructure, computing resources, and support system. The energy efficiency of the data center plays a major role in the management and maintenance of the data center. Many approaches are explored by various researchers on this objective. Energy-efficient virtual machine (VM) consolidation is a popularly thought-after area of research.

This paper focuses on surveying the existing VM Consolidation approaches and analyzing the performance of the existing basic algorithms. Section 2 discusses the prominent approaches in the literature for virtual machine consolidation, methods for detecting overloaded hosts, and algorithms for virtual machine placement. Section 3 focuses on the analysis of the simulation results on the existing algorithmic combinations. Section 4 gives the conclusive inferences and narrates the future scope in this area.

R. R. John (✉) · E. Grace Mary Kanaga
Department of Computer Science and Engineering, Karunya Institute of Technology and Sciences, Coimbatore, India
e-mail: roseranijohn@gmail.com

E. Grace Mary Kanaga
e-mail: grace@karunya.edu

J. D. Peter et al. (eds.), *Disruptive Technologies for Big Data and Cloud Applications*, Lecture Notes in Electrical Engineering 905,
https://doi.org/10.1007/978-981-19-2177-3_61

2 Related Work

2.1 Introduction to Related Work

The main focus of researchers in cloud computing is optimizing the energy consumption of data centers. VM Consolidation can be brought down to three processes: Host overload/underload detection, VM selection and VM placement problem [5]. This paper provides a meta-analysis of the existing algorithms in the VM Consolidation process.

2.2 VM Consolidation Approaches

Consolidation over the cloud is mentioned as server consolidation and VM Consolidation alternatively. In this paper, the name VM Consolidation will be used for both. Beloglazov [4–6] gave better clarity on the various steps involved in VM Consolidation. So the research involved in VM Consolidation is split up into three sub-areas such as host overload/underload detection, VM selection, and VM placement [5]. So in this section, researchers who worked with a clear awareness of these basic steps are considered, studied, and consolidated.

Ranjbari and Akbari Torkestani in [7] proposed a reinforced learning-based approach for VMC. The work of Witanto et al. [9] stands unique as he focuses on a neural network-based approach for selecting the best VMC approach based on the requirements. A generic framework for VMC is provided by Khattar et al. in [10]. A predictive approach for VMC is proposed by Moghaddam et al. Unsworth in [11] using the various machine learning models.

2.3 VM Selection Approaches

The effective basic approaches used for VM selection included the work of A. Beloglazov and R. Buyya, in which he has used the VM selection policies such as maximum correlation (MC), minimum utilization (MU), and random selection (RS) which are used by CloudSim [6, 13] as a standard approach for VM selection. Duong-Ba et al. in [14] have proposed a multilevel join VM placement and migration (MJPM) based on the relaxed convex optimization framework. Farahnakian et al. [12] have used maximum load and minimum load VM selection policy. Khattar et al. [10] have used two different VM selection policies such as MRCPMB and MPCBM, which are depending on CPU utilization, memory, and bandwidth. [8] has used MCAMS, which is based on migration and capacity.

2.4 Host Overload/Underload Detection Algorithms

Zhou et al. [15] proposed an algorithm for host selection based on network topology. They choose a subnet with just enough resources and look out for minimal data transfer traffic and use of network resources. [5] Belogalzov has proposed interquartile range (IQR), local regression (LR), local regression robust (LRR), median absolute deviation (MAD), static threshold (ST)-based host selection policies which are used in CloudSim as standard algorithms. Li et al. in [8] use an estimation of the overload probability (EOP) algorithm for host overload detection.

2.5 VM Placement Approaches

Virtual machine placement problem is considered an NP-hard problem. Researchers have attempted from time to time to come up with optimal solutions for this through various approaches. The mapping of the problem is significant while considering the solution approaches. Many researchers have considered VMPP as a bin packing problem. Kaaouache and Bouamama in [16] have mapped VMPP as a bin packing problem and proposed a hybrid genetic algorithm as a solution.

Considering the VMPP, the prominent objectives by the researchers are around minimizing energy/power consumption, the number of active hosts, resource wastage SLA violations, and maximizing resource utilization. The basic approaches were approximation algorithms such as first fit, best fit, average fit, and worst fit, and the several variants of the same. Beloglazov proposed modified versions of best fit decreasing such as PABFD [5] and MBFD [17], which are used with CloudSim. Dubey et al. [18] have come up with the efficiency of service best fit decreasing policy which is a modified version of the basic BFD policy which gives better results than FF and BF policies. Zhao et al. [19] proposed a power-aware performance-guaranteed VMP(PPVMP). They have formulated VMP as a bi-objective optimization problem and used ant colony optimization to solve it.

3 Experimental Analysis

3.1 Simulation Settings

CloudSim toolkit is used for simulation of VM Consolidation, and the basic settings used are given below. Two HPProLiant server combinations are used, and the configuration of the same is given in Table 1 [5]. The power consumption benchmark of the servers is given in Table 2 [20]. Four different types of VMs are used in the simulation, and the configuration of the VMs is given in Table 3 [20]. The combinations of five basic statistical models available for host selection and four different algorithms

Table 1 Configuration of the servers

1	HP ProLiant ML110 G4 (Intel Xeon 3040, 2 Cores x 1860MHz, 4GB)
2	HP ProLiant ML110 G5 (Intel Xeon 3075, 2 Cores x 2660MHz, 4GB)

Table 2 Power consumption benchmark provided by spec for the hosts used

	0%	10%	20%	30%	40%	50%	60%	70%	80%	90%	100%
G4	86	89.4	92.6	96	99.5	102	106	108	112	114	117
G5	93.7	97	101	105	110	116	121	125	129	133	135

Table 3 VM types

Type	CPU (MIPS) (GHz)	Cores	Memory
I	2.5	1	0.85 GB
II	2.0	1	3.75 GB
III	1.0	1	1.7 GB
IV	0.5	1	613MB

for VM selection have experimented with the power-aware BFD algorithm for VM placement. The standard workload of PlanetLab is used for studying the performance of existing algorithmic combinations.

3.2 Performance Metrics

The most significant parameters used to measure the performance are discussed below.

1. Resource utilization(UR): Most of the researchers focus on CPU utilization in their works. Some of them look into memory utilization, bandwidth utilization as well. Some approaches focus on host utilization, while others consider the VM CPU utilization as well. In this paper, the host CPU utilization is considered.
2. Resource wastage (WR): This parameter measures the percentage of unused CPU. The main objective of most of the works is to reduce resource wastage and thereby maximize resource utilization.

$$W_R = \frac{\mathrm{CPU_T} - \mathrm{CPU_U}}{\mathrm{CPU_T}} \times 100 \tag{1}$$

where $\mathrm{CPU_T}$ is the total CPU available, and ($\mathrm{CPU_U}$) is the utilized CPU.

3. The number of VM migrations ((VMM_n)): This is a very straightforward parameter that measures the number of VM migrations during a VM Consolidation simulation.
4. Average SLA violations (($\mathrm{SLAV}_{\mathrm{AVG}}$)): This parameter gives the average percentage of service level agreement violations with VMC algorithmic combinations. Average SLA violation can be calculated by finding the mean of the SLA violations [21].

$$\text{SLA Violation\%} = \frac{\text{totalRequestedMIPS} - \text{totalAllocatedMIPS}}{\text{totalRequestedMIPS}} \times 100 \quad (2)$$

5. Energy Consumption(P): Each server has a SPEC power consumption benchmark as per Table 2. Following the standard benchmark, the CPU utilization of each host and the corresponding energy consumption are calculated.
6. The number of active Hosts ((H_n): This is a straightforward parameter that showcases the total number of hosts which are actively in use and not idle in the data center.
7. Host overload percentage (($H_{\mathrm{OL}}\%$)): In this paper, a static upper resource utilization threshold of 80% is assumed for convenience in this paper for the result evaluation purpose.

$$H_{\mathrm{OL}}\% = \frac{H_{\mathrm{OL}}}{\mathrm{AH}_n} \times 100 \quad (3)$$

where (($H_{\mathrm{OL}}\%$)) is the number of hosts with UR >=80%.

8. Host underload percentage (($H_{\mathrm{UL}}\%$)): In this paper, a static lower resource utilization threshold of 20% is assumed for convenience in this paper for the result evaluation.

$$H_{\mathrm{UL}}\% = \frac{H_{\mathrm{UL}}}{\mathrm{AH}_n} \times 100 \quad (4)$$

where (($H_{\mathrm{UL}}\%$)) is the number of hosts with (U_{R}) <=20%

3.3 *Result Analysis and Inferences*

The possible combinations of considered existing VM selection and host selection algorithms along with the VM placement algorithms are simulated, and the results are analyzed. The simulation is run for 800 hosts in which the resource utilization of each host is monitored for each algorithmic combination. For getting a better understanding, the average resource utilization of the active hosts under each algorithmic combination is calculated. As per the results plotted in Fig. 1, it is evident that the LrrRs combination is having the maximum average resource utilization and MadMu with the minimum utilization. Considering an algorithmic average, it is seen that LRR-based combinations maximize resource utilization, whereas THR-based algorithms give minimum resource utilization. LRR-based algorithms give approximately

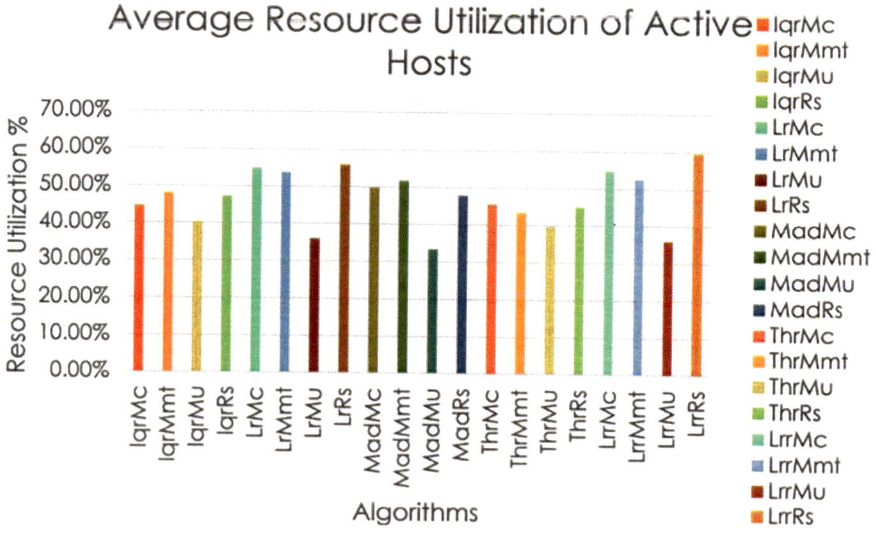

Fig. 1 Resource utilization with each algorithmic combination

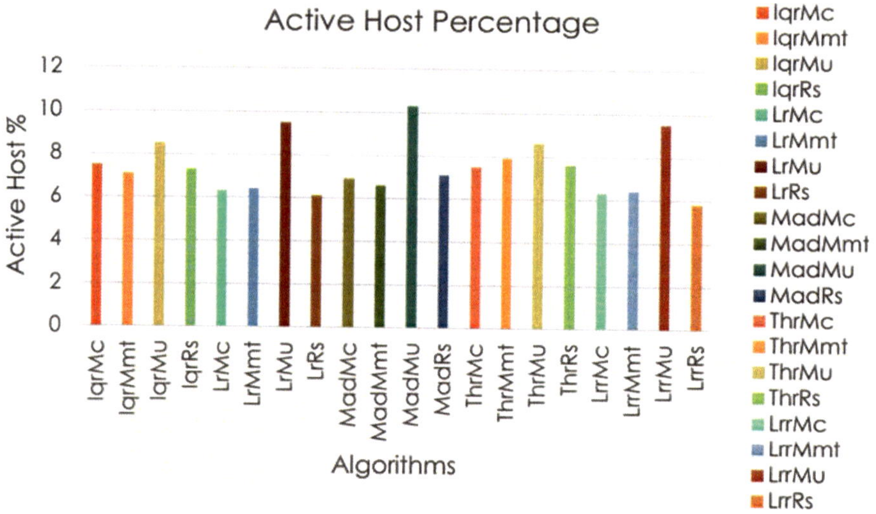

Fig. 2 Active host percentage with each algorithmic combination

7% better resource utilization than THR-based algorithms. RS-based VM selection policy maximizes resource utilization, whereas MU-based algorithms give minimum resource utilization. RS-based algorithms give approximately 14% better results than MU-based algorithms.

The main objective of any algorithmic combination would be to decrease the number of hosts which are in use as it linearly influences the power consumption. The

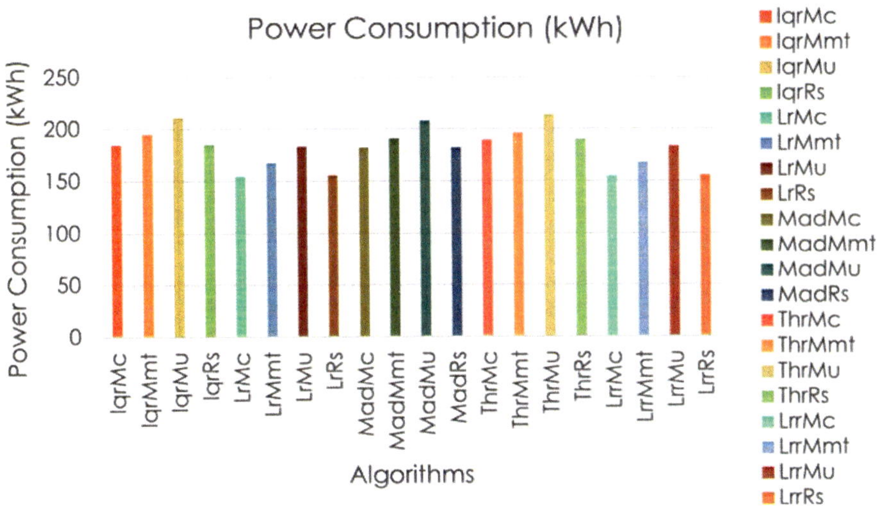

Fig. 3 Power consumption by each algorithmic combination

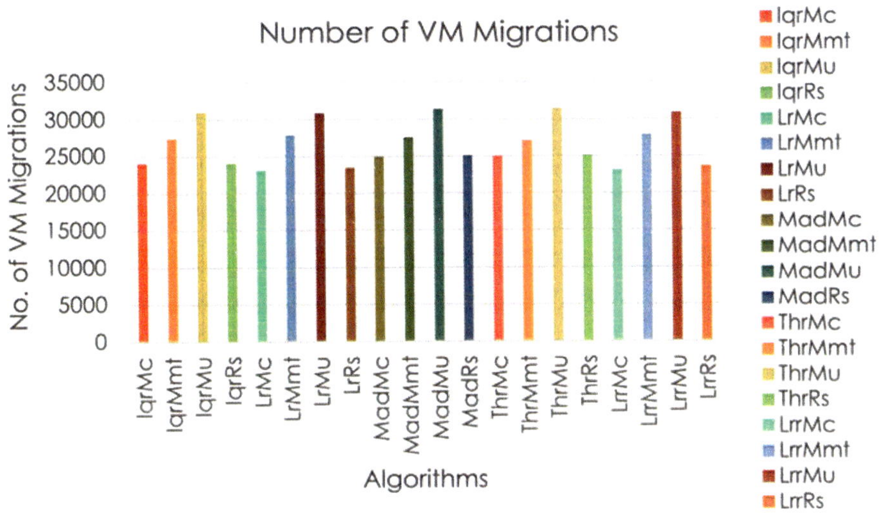

Fig. 4 Number of VM migrations with each algorithmic combination

performance of the algorithms is plotted in Fig. 2, from which it is quite evident that LrRs and LrrRs give the minimum number of active hosts, whereas MadMu algorithmic combination gives the maximum number of hosts active. While considering the performance of the host selection algorithms, it is observed that the performance variance between the combinations is minor. LRR-based algorithms give approximately 1% better results than THR-based algorithms. Comparing in terms of the

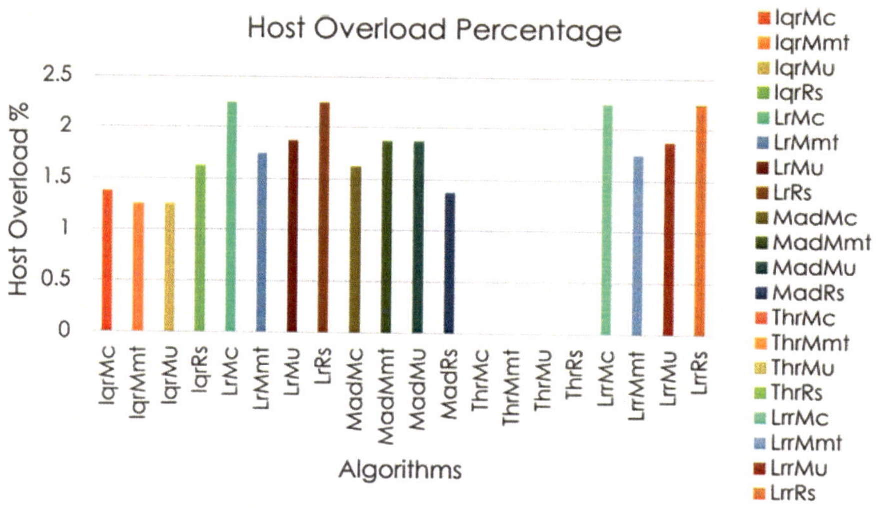

Fig. 5 Host overload percentage with each algorithmic combination

VM selection policy, RS-based algorithms give 2.5% better results than MU-based algorithms.

As per Fig. 3, LrMc and LrrMc give better performance considering the given circumstances. LRR-based algorithms give approximately 9% better results than THR-based algorithms. MC-based algorithms give approximately 7% better results than MU-based algorithms. VM migrations have a linear dependence on power consumption.

From Fig. 4, it is seen that the ThrMu combination has the maximum number of VM migrations, and LrMc has the minimum number of migrations. LR-based algorithms give approximately 2% better results than MAD-based algorithms. MC-based algorithms give approximately 13% better results than MU-based algorithms (Fig. 5).

Most of the algorithms are considering a safety parameter to detect overloaded hosts. In this paper, all the active hosts in each algorithmic simulation are considered and have tried to figure out how many hosts are still at the risk of being overloaded and would still need a migration requirement. An upper utilization threshold of 80% is used for detecting overloaded hosts, and the host overload percentage is calculated. Active hosts simulated based on THR algorithms seem to be safe from host overload in the considered scenario. MU-based algorithms give approximately 9% better results than RS-based algorithms.

A consolidation of the results is given in Table 4. An ideal combination of host selection and VM selection algorithms for each parameter is given based on the result analysis. The most commonly sought-after objective of many researchers is to reduce the energy/power consumption, which has a direct dependence on maximizing resource utilization, minimizing the number of active hosts and number of VM

Table 4 Consolidated results

S. No.	Parameter	Best host selection algorithm	Best VM selection algorithm	Ideal combination
1	Resource utilization	LRR	RS	LRR-RS
2	Host overload %	THR	MU	THR-MU
3	Host underload %	LRR	MMT	LRR-MMT
4	Active hosts	LRR	RS	LRR-RS
5	VM selection time	THR	RS	THR-RS
6	Host selection time	THR	MC	THR-MC
7	VM reallocation time	THR	MC	THR-MC
8	VM migrations	LR	MC	LR-MC
	Power consumption	LRR	MC	LRR-MC
10	Average SLA violations	LRR	MC	LRR-MC

migrations. Another serious consideration is to improve the quality of service and reduce SLA violations.

4 Conclusion and Future Scope

Algorithms based on linear regression in combination with random selection and maximum correlation use relatively minimum number of hosts, gives minimum hosts underload percentage, average SLA violations, number of VM migrations, and power consumption and thus seem to provide a better result considering the experimented scenario. The static threshold-based approach causes a very minimal number of overloaded hosts. The standard outputs from these basic algorithms based on the Planet Lab workload can be used to leverage a better predictive mechanism for VM Consolidation, inculcating an optimal VM placement strategy.

References

1. D.C. Plummer, T.J. Bittman, T. Austin, D.W. Cearley, D.M. Smith, Cloud computing: defining and describing an emerging phenomenon (2008)
2. S. Ibrahim, B. He, H. Jin, Towards pay-as-you-consume cloud computing, in *Proceedings—2011 IEEE International Conference on Services Computing* (SCC 2011, 2011), pp. 370–377
3. P. Mell, T. Grance, The NIST definition of cloud computing, Recommendations of the National Institute of Standards and Technology (n.d.)
4. A. Beloglazov, J. Abawajy, R. Buyya, Energy-aware resource allocation heuristics for efficient management of data centers for cloud computing. Future Gener. Comput. Syst. **28**, 755–768 (2012)

5. A. Beloglazov, R. Buyya, Optimal online deterministic algorithms and adaptive heuristics for energy and performance efficient dynamic consolidation of virtual machines in Cloud data centers. Concurrency Comput. Pract. Experience **24**, 1397–1420 (2012)
6. A. Beloglazov, R. Buyya, *Managing Overloaded Hosts for Dynamic Consolidation of Virtual Machines in Cloud Data Centers Under Quality of Service Constraints* (IEEE Trans. Parallel Distrib, Syst, 2013)
7. M. Ranjbari, J. Akbari Torkestani, A learning automata-based algorithm for energy and SLA efficient consolidation of virtual machines in cloud data centers. J. Parallel Distrib. Comput. (2018)
8. Z. Li, C. Yan, X. Yu, N. Yu, Bayesian network-based virtual machines consolidation method. Future Gener. Comput. Syst. (2017)
9. J.N. Witanto, H. Lim, M. Atiquzzaman, Adaptive selection of dynamic VM consolidation algorithm using neural network for cloud resource management. Future Gener. Comput. Syst. (2018)
10. N. Khattar, J. Singh, J. Sidhu, An energy efficient and adaptive threshold VM consolidation framework for cloud environment. Wirel. Pers. Commun. **113**, 349–367 (2020)
11. S. Mashhadi Moghaddam, M. O'Sullivan, C. Walker, S. Fotuhi Piraghaj, C.P. Unsworth, Embedding individualized machine learning prediction models for energy efficient VM consolidation within cloud data centers. Future Gener. Comput. Syst. **106**, 221–233 (2020)
12. F. Farahnakian, T. Pahikkala, P. Liljeberg, J. Plosila, N.T. Hieu, H. Tenhunen, Energy-aware VM consolidation in cloud data centers using utilization prediction model. IEEE Trans. Cloud Comput. (2019)
13. A. Beloglazov, R. Buyya, Optimal online deterministic algorithms and adaptive heuristics for energy and performance efficient dynamic consolidation of virtual machines in cloud data centers. Concurrency Comput. Pract. Experience. (2012)
14. T.H. Duong-Ba, T. Nguyen, B. Bose, T.T. Tran, A dynamic virtual machine placement and migration scheme for data centers. IEEE Trans. Serv. Comput. **1374**, 1–14 (2018)
15. A. Zhou, S. Wang, B. Cheng, Z. Zheng, F. Yang, R.N. Chang, M.R. Lyu, R. Buyya, Cloud service reliability enhancement via virtual machine placement optimization. IEEE Trans. Serv. Comput. (2017)
16. M.A. Kaaouache, S. Bouamama, ScienceDirect solving bin packing problem with a hybrid genetic algorithm for VM placement in cloud-review under responsibility of KES International. Procedia Comput. Sci. **60**, 1061–1069 (2015)
17. A. Beloglazov, R. Buyya, OpenStack Neat: a framework for dynamic and energy-efficient consolidation of virtual machines in OpenStack clouds. Concurrency Comput. **27**, 1310–1333 (2015)
18. K. Dubey, A.A. Nasr, S.C. Sharma, N. El-Bahnasawy, G. Attiya, A. El-Sayed, Efficient VM placement policy for data centre in cloud environment. Adv. Intell. Syst. Comput. **1053**, 301–309 (2020)
19. H. Zhao, J. Wang, F. Liu, Q. Wang, W. Zhang, Q. Zheng, Power-aware and performance-guaranteed virtual machine placement in the cloud. IEEE Trans. Parallel Distrib. Syst. **29**, 1385–1400 (2018)
20. T. Chaabouni, M. Khemakhem, J. Supercomput. Energy management strategy in cloud computing: a perspective study **74**, 6569–6597 (2018)
21. A. Mosa, N.W. Paton, Optimizing virtual machine placement for energy and SLA in clouds using utility functions. J. Cloud Comput. **5**, 1–17 (2016)

Implementation of Compression Technique for Endoscopy Video Using Intra-Coding HEVC

Suvarna Nandyal and Heena Kouser Gogi

1 Introduction

The real-time access to medical data such as images and video in the medical fraternity has increased in the past few decades. With the availability of cloud-based healthcare service offers several advantages, not only in case of electronic record but also in the case of storage and access of medical data. The main factor of concern in the medical is size and the quality of data.

With the help of a cloud-based health centre, access to medical imagery will reduce the cost of distribution, reduce duplication of medical imagery and reduce the cost of medical image storage infrastructure. Medical images must be stored at very high quality and efficient compression algorithms will have a direct impact on lowering costs.

In general, four sets of PET-CT medical images of one patient may require 4+ GB of storage space. For example, 1 h of HD video needs 500 GB of space. Due to the large amount of memory required for HD video, compression is almost inevitable used during recording.

Endoscopy is a nonsurgical procedure used to examine a person's digestive tract. Using an endoscope, a flexible tube with a light and camera attached to it, doctor can view pictures of digestive tract on a colour TV monitor. For an average patient, this procedure takes 20–30 min of monitoring and video recording. Picture and video quality is defined by a variety of factors, including the performance resolution (e.g. standard definition [SD] vs. high definition [HD]), data transfer rate and compression

S. Nandyal
Department of CSE, PDA College of Engineering, Kalaburgi, Karnataka, India

H. Kouser Gogi (✉)
Department of ISE, HKBK College of Engineering, Kalaburgi, Karnataka, India
e-mail: heena.shakir@gmail.com

J. D. Peter et al. (eds.), *Disruptive Technologies for Big Data and Cloud Applications*, Lecture Notes in Electrical Engineering 905,
https://doi.org/10.1007/978-981-19-2177-3_62

process. By extracting all non essential files, compression shrinks file sizes. Uncompressed HD video takes up a lot of room on your hard drive. One hour of HD video, for example, requires 500 GB of storage. Compression is almost always used during recording because of the large amount of memory needed for HD video.

A codec is a programme that compresses and decompresses images and video into different image formats.

Digital Imaging and Communications in Medicine (DICOM) is a common protocol for handling and transmitting medical images and related data that is commonly used in hospitals or cloud-based health centre [1]. DICOM aim is to ensure that systems used to build, store, exchange, view, send, query, process, retrieve and print medical images, as well as related workflows, are interoperable. Endoscopy-specific DICOM formats for photographs and videos, visible light endoscopic image and video endoscopic image, are provided by DICOM (Storage) [2]. DICOM images are much superior to thermal paper and videotape images in terms of efficiency. As a result, the endoscopy video had to be shared in the DICOM while retaining video quality.

High Efficiency Video Coding (HEVC), also known as H.265 and MPEG-H Part 2, is a video encoding format that was created as a substitute for the widely used Advanced Video Coding standard (AVC, H.264, or MPEG-4 Part 10). In contrast to AVC, HEVC provides 25–50% better data compression at the same video quality level, or dramatically enhanced video quality at the same bit rate.

The ITU-T Video Coding Experts Group and the ISO/IEC Moving Picture Experts Group are currently working on a new video coding standard called High-Performance Video Coding (HEVC). The HEVC format outperforms other compression standards such as H.264/AVC which JPEG/J2K and can save substantial amounts of space in medical image applications [3–5]. The main aim of the HEVC standardisation initiative is to achieve substantially better compression efficiency than current standards, in the region of 50% bit-rate reduction for comparable perceptual video quality.

Both still image and video coding require intraframe coding [6]. Although lossless compression for medical images is preferable, lossy compression that does not compromise diagnostic accuracy may be appropriate. H.265 can be used for the compression of the video in DICOM. Several image compression specifications have been proposed by international standardisation organisations during the last few decades [4, 5, 7]. Our work's key contribution is the concept of an appropriate PSNR value and compression ratio range for both HEVC Intra and HEVC Intra-encoding of medical images.

As a result, for medical images, an efficient compression algorithm is needed, which has a direct effect on lowering costs and improving access to ubiquitous and heterogeneous mobile devices.

By extracting all non-essential files, compression shrinks file sizes. Uncompressed HD video uses a significant amount of storage space. Compression is almost always used during recording because of the large amount of memory needed for HD video.

2 Related Work

The compression techniques used in medical imaging are discussed in this section. Since a decade, several algorithms and techniques have been proposed. However, there are major variations in the datasets used and the measurement parameters. Below is a list of the numerous compression methods and techniques that have been used.

Ramachandran et al. [8] discuss medical image compression and image consistency. The authors address how current methods can still be improved in terms of compression ratio. As a result, a new image matrix technique for efficient image compression in the medical pitch is proposed. Effective hybrid technique (EHT) is a recently proposed medical image compression technique. The image quality (IQ) and compression ratio (CR) of medical images are determined by EHT. The proposed mechanism is new and operates on the storage parameter of medical images while preserving consistency.

Mohamed et al. [9] discuss predictive image coding method. Even after compression, the consistency of the medical image in the diagnostically relevant region (DIR) is maintained using the proposed process. Using a graph-based segmentation technique, the image is initially segmented into two sections, namely DIR and non-DIR portions. The gravitational search method is found to be successful. As opposed to particle swarm and back-propagation methods, the gravitational search method achieves higher PSNR values.

Manel et al. [10] propose an artificial neural network-based medical image compression algorithm (ANN). It is a straightforward algorithm that leaves all of the image data intact. The PSNR values in the experimental results are in the 30–50 db range.

Senthilkumaran et al. [11], for lossless image compression, propose an enhanced back-propagation neural network technique. By comparing X-ray images to three metrics such as compression ratio, transmission time and compression efficiency, the method shows that the enhanced back-propagation neural network technique outperforms the current Huffman coding.

Technique used for lossless image compression, so now the main compression parameters used here for the measurements are compression ratio, compression time and PSNR. Now in our proposed work of HEVC implementation of endoscopic videos, we take into consideration the same parameters to evaluate the results.

The motivation behind the work is the requirement in improvement of compression technique for the medical dataset [2], which is the driving factor for the storage and transmission in cloud-based health centre. Tele-endoscopy endoscopy can be real time (synchronous), in which the video is streamed live and analysed by a remote observer during the event, or store-and-forward (asynchronous), in which the video is displayed after the event has occurred. Synchronous tele-endoscopy necessitates more technological assistance than asynchronous tele-endoscopy (e.g. high-bandwidth Internet, logistical support).

More than endoscopies were performed by local endoscopists and recorded with both high-quality digital video compressed video (25 megabits/second [Mbps], 720–480 pixels) and highly compressed MPEG-1 video (2.0 Mbps, 352–240 pixels) simultaneously. In 85% of cases, endoscopist reviewers graded the videos as diagnostic, but just 18% of videos yielded the same diagnoses. According to our research, the image management system must evolve in terms of storage and archiving, which can be done by compression.

2.1 Video Coding Standard

The picture (a medical image or a frame) from a video series is represented in the DICOM format by three components: Luma (y-component), which represents the brightness of the medical material at each pixel position, and U- and V-components, which reflect the colour difference at each pixel in the picture. Each frame is represented by a rectangular array which consists of Y-, U- and V-components.

In most of the cases, medical photographs are depicted in DICOM format, and the images are in sequence. A frame (a medical image) from a video sequence (a medical image series) is defined by three components of digital video coding: The luma (Y-component) represents the brightness of the medical material at each pixel position, while the other two chroma components (commonly referred to as U- and V-components) reflect the colour difference at each pixel in the image. Each file (medical image) is made up of three rectangular arrays of Y-, U- and V-components.

The block-based coding framework of H.264, a widely used hybrid video coding standard, splits each medical image into multiple macro-blocks of size 16 × 16 pixels, 32 × 32 pixels and 64 × 64 pixels. The coding efficiency of a greater pixel block size is typically improved. H.265 requires various block sizes to be coded for projected blocks. Each marco-block is made up of two 16 × 16 chroma component arrays and 16 × 16 Luma component arrays For a 4:4:4 YUV scheme, each marco-block has two 16 × 16 Luma component arrays and two 16 × 16 chroma component arrays. In this case, each component array is of equal size. Other YUV systems are 4:2:2, of which chroma samples are half the width but the same height as the Y-component, and 4:2:0, in which all chroma components are half the width and height of the Y-component. The YUV format of 4:0:0 is generally recommended for medical pictures, which are mainly grayscale images. Macro-blocks are further split into smaller blocks ranging in size from 16 × 16 to 4 × 4 [12].

3 Implementation of Compression Technique for Endoscopy Videos Using Intra-Coding HEVC

The medical endoscopy videos are used for the study purpose and checking the various parameters of compression [13]. DICOM is a medical imaging standard which enable heterogeneous medical applications and equipment to interact with one another [14]. The architecture of the HEVC standard is used here to achieve high coding efficiency. The ability to encrypt video at the lowest possible bit rate while retaining a certain level of video quality is how reliability is calculated in video coding and filter for quality [15]. There are two common approaches for assessing the coding performance of a video coding standard: using an empirical criterion such as peak signal-to-noise ratio (PSNR) or subjective video quality appraisal. HEVC intra-coding [6, 7, 12, 16], which comes with a higher computational cost [17], achieves higher compression performance.

HEVC is standard which potential to deliver the better performance than earlier standard such H.264 [18]. Here for our work purpose endoscopy videos are considered. The format of the video is AVI, which is raw video, and direct sample is applied for the study of compression.

The compression method comprises of the following process at the ends, one at sender called as encoding and other at the receiver end called as decoding.

Endoscopy videos consist of a sequence of video frames which are given as inputs to the pre-processing phase, where the colour conversion takes place as shown in Fig. 1. After this phase, the greyscale frames are forward for block generation of 16 × 16 to 64 × 64 bits. Then the transformation and quantisation take place, and finally, the encoder encodes and generates compressed bits. The compressed bitstream is stored or transmitted. A video decoder decompresses the bitstream to create a sequence of decoded frames as shown in Fig. 2. In our experiment, 0–300 frames are generated per endoscopy video of 200–250 MB [19].

This compressed bit stream is transmitted over the network or stored at the server. The implementation is done in MATLAB, and therefore, the experimental results proved the effectiveness of proposed video compression technique in terms

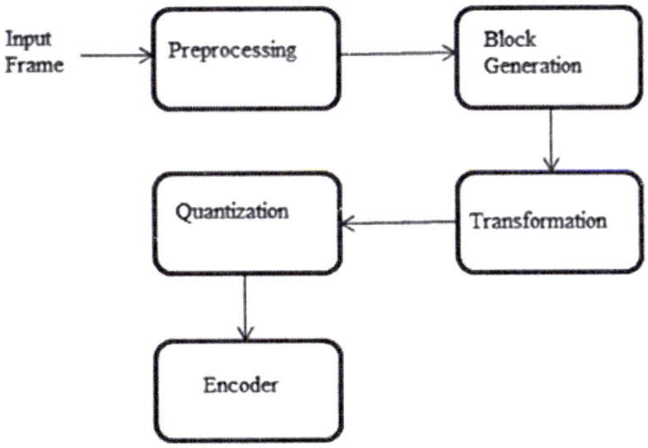

Fig. 1 Video encoding in HEVC

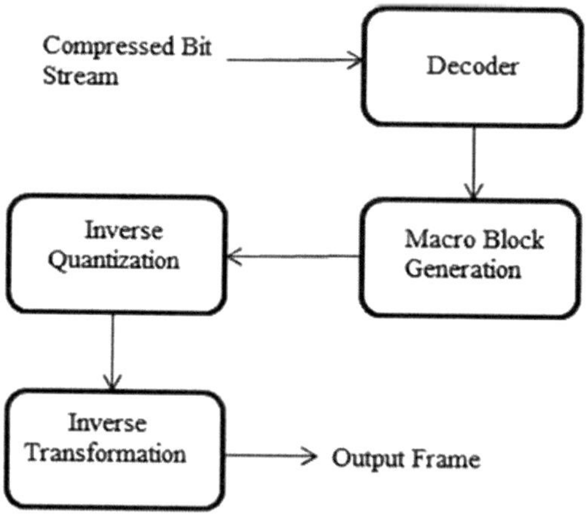

Fig. 2 Video Decoding in HEVC

of compression ratio, PSNR and compression time. Now let us consider the definition aspect of the three parameters utilised for our experiment.

In PSNR, p and q are the size of the image, $f\,(x,\,y)$ is the original input images and $g\,(x,\,y)$ is the reconstructed image. The PSNR is measured in terms of decibels.

$$\text{PSN} = 10\log_{10}\left[\frac{\frac{1}{p\times q}\sum_{i=1}^{p}\sum_{j=1}^{q}(f(x,y))^2}{\frac{1}{p\times q}\sum_{i=1}^{p}\sum_{j=1}^{q}(f(x,y)-g(x,y))^2}\right]$$

Compression ratio is defined as the ratio of the original image and compressed images.

$$\text{CR} = \frac{\text{Uncompressed}_{\text{image}}(\text{Bytes})}{\text{Compressed}_{\text{image}}(\text{Bytes})}$$

Compression time is defined as the time difference between the last frame and the first frame during the compression process.

$$\text{Compression time} = \text{Tok} - \text{Tik}.$$

where Tok is the time recorded at which the last frame is compressed out. Tik is time recorded at which the first frame taken into consideration for recording. Compression time is measured in turn of seconds.

Algorithmic Steps:

Encoding Process

Step 1: Read frames from medical video. MF: Read Medical frame.

Step 2: Conversion of image take place.

Step 3: Generation of Macro-block.

MF: Now the MF will be in range of 16X16 to 64X64 blocks.

Step 4: Encoding process starts.

4.1 Define the size of Frame as 64 × 64.

4.2 Encode each frame.

Step 5: Then perform Transformation on each frame using DCT.

Step 6: Apply matrix Quantisation.

Step 7: Apply Intra-Prediction control on the macro-block.

Step 8: Out Put CF = Compressed Frame.

Decoding Process

Step 1: Input: CG = Compressed Frame.

Step 2: Decoding process.

Decoding of frames: MF = Medical Frame.

Step 3: Generation of macro-block of size 64 × 64.

Step 4: Apply Inverse Quantisation.

Step 5: Apply Inverse transformation.

4 Results and Discussion

Here for experiment purpose, we have considered five samples of endoscopic videos out of which three pictorial output is shown as Frames 1, 2 and 3 uncompressed and compressed. The result is shown in Table 1. For experiment purpose, the parameter

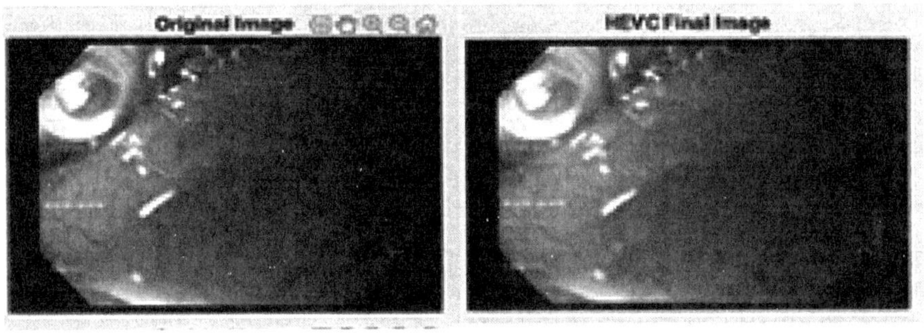

Frame 1 First Image

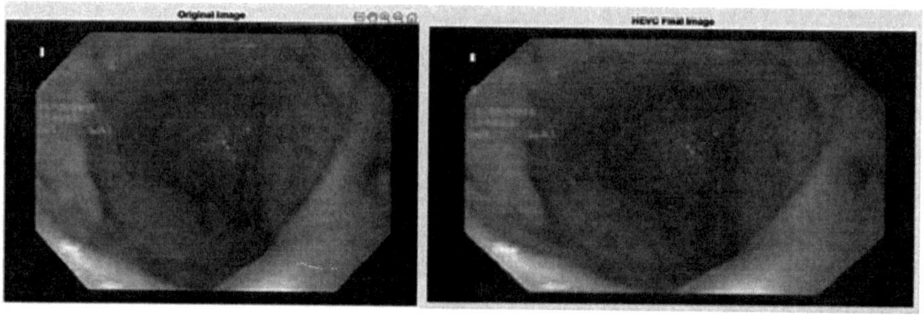

Frame 2 Second Image

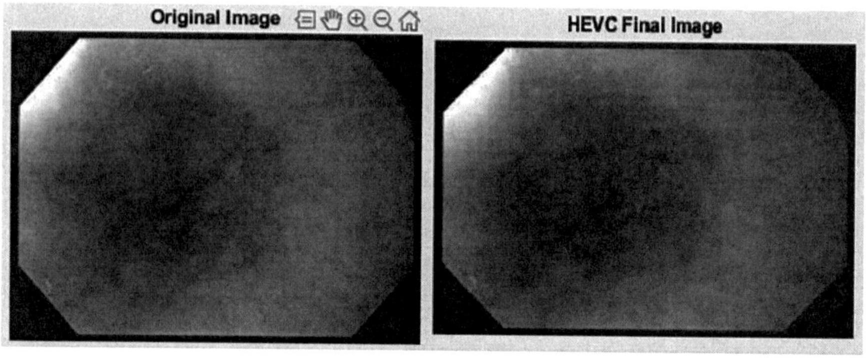

Frame 3 Third Image

Table 1 Result of five different videos in intra-coding HEVC

Serial No.	PSNR value (db)	Compression time (s)
1	44.58	175.84
2	43.47	175.76
3	44.68	154.00
4	49.35	182.68
5	45.03	157.04

such as PSNR, compression time and compression ratio. The compression ratio achieved in our experiment is 1:5.

Table 1 shows the compression result of five consecutive endoscopic videos.

Compression ratio shows that five pixels of video frame are compressed to 1 pixel in compressed video frame [20].

The graph shown in Fig. 3 is a plot of PSNR of five endoscopic videos with their corresponding PSNR value.

Figure 4 shows the compression time with respect to its corresponding endoscopy video compression in seconds.

Fig. 3 PSNR in intra-coding HEVC

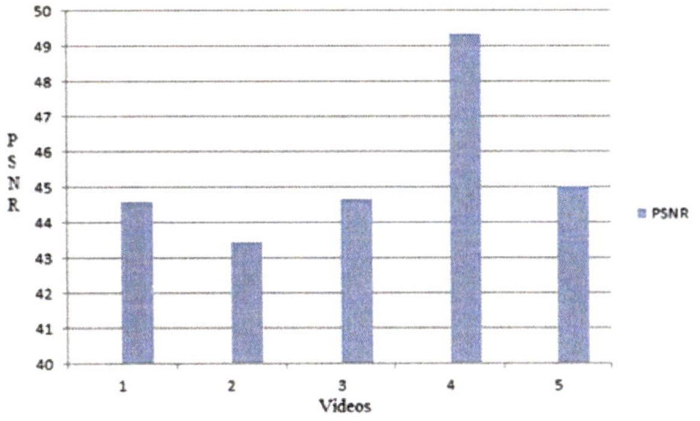

Fig. 4 Compression Time in Intra Coding HEVC

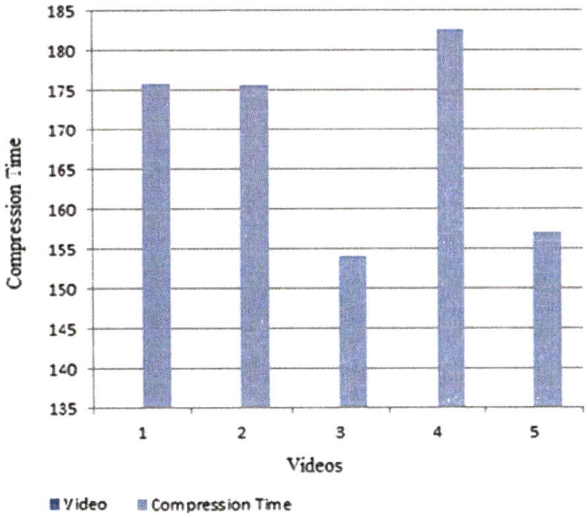

Comparison of the PSNR is shown in Fig. 5, with our five sample videos with achieving a PSNR in the range of 43–50 db, compared to that of 40–46 db [8].

With the implementation of HEVC, it provides a step change improvement in video compression compared with earlier standards [21]. HEVC was developed to address the following trends:

- Relaxing in terms of network parameters, with help better compression standard.
- Reduction in terms of storage and applicable use of video resolution beyond HD, which w.
- Using the computation capabilities of mobile handset or tablet which helps in continuing improvements in processing capacity. More than desktops.
- HEVC techniques can transfer and storage the videos in the more efficient ways compared to H.264.

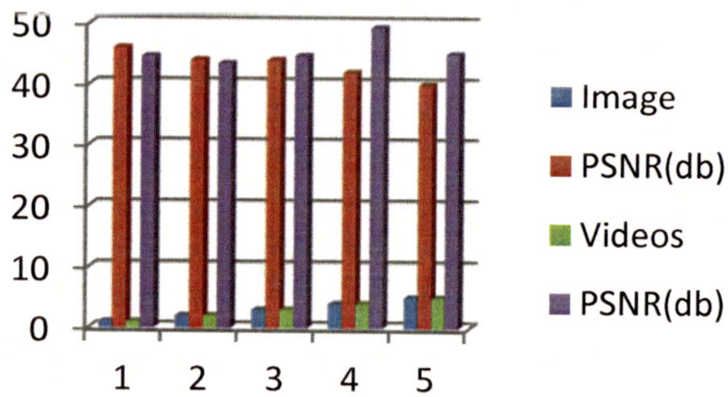

Fig. 5 Comparison of PSNR of X-ray image and endoscopy Video in Intra-Coding HEVC

- At the same picture size and quality, an HEVC video sequence should occupy less storage or transmission capacity than the equivalent H.264 video sequence.

HEVC uses the subsequent twenty division matrices that depend upon the dimensions and kind of the remodel block [22].

5 Conclusions

The issue with video compression is that the image quality, compression ratio and speed are all poor. The HEVC video compression method is proposed in this paper to address and resolve these problems. For the parameters compression time, compression ratio and PSNR, comparison results are obtained. The efficiency of the proposed was demonstrated by comparison tables and graphs. As a result, the current compression issues are minimised to a larger degree with the assistance of this proposed technique.

References

1. NEMA PS3/ISO 12052, Digital Imaging and Communications in Medicine (DICOM) Standard, Rosslyn, VA, USA: National Electrical Manufacturers Association. USA
2. F.M. Murad, S. Banerjee, Image management systems. Technology Status Evaluation Report Asge Technology Committee. Gastrointest. Endosc. **79**(1) (2014)
3. G.J. Sullivan, J.R. Ohm, W.J. Han, T. Wiegand, Overview of the high efficiency video coding (HEVC) standard. IEEE Trans. Circuits Syst. Video Technol. **22**(12), 1649–1668. http://ieeexplore.ieee.org/stamp/stamp.jsp?tp=&arnumber=6316136&isnumber=6403920
4. B. Bross, J.W. Han, J.R. Ohm, S. J, T. Wiegand, High Efficiency Video Coding (HEVC) Text Specification Draft 8, JCT-VC document, JCTVC-J1003," Stockholm, Sweden, 2012
5. ITU-T H265 Recommendation, 4 2015. Available: https://www.itu.int/rec/T-REC-H.265-201504-I/en. Accessed 3, 2016
6. M. Zhou, W. Gao, M. Jiang, H. Yu, HEVC lossless coding and improvements. IEEE Trans. Circuits Syst. Video Technol. **22**(12), 1839–1843 (2012). Available: http://ieeexplore.ieee.org/stamp/stamp.jsp?tp=&arnumber=6317159&isnumber=6403920

7. S. Parikh, H. Kalva, V. Adzic,Evaluation of HEVC compression for high bit depth medical images, in *2016 IEEE International Conference on Consumer Electronics (ICCE)*, Las Vegas, NV, USA, 2016, pp. 311–314
8. A. Ramachandran, E.K. Subramanian, J. Jegan, An effective image matrix compression technique for medical pitch. Int. J. Recent Technol. Eng. (IJRTE) **7**(6s4) (2019). ISSN: 2277-3878
9. M.U.A. Ayoobkhan, E. Chikkannan, K. Ramakrishnan, Feed-forward neural network-based predictive image coding for medical image compression. Arab J Sci Eng **43**(8), 4239–4247 (2018)
10. M. Dridi, B. Bouallegue, M.A. Hajjaji, A. Mtibaa, An enhancement medical image compression algorithm based on neural network. Int. J. Adv. Comput. Sci. Appl. **7**(5) (2016)
11. N. Senthilkumaran, J. Suguna, Neural network technique for lossless image compression using X-ray images. Int. J. Comput. Electr. Eng. **3**(2), 1793–8163 (2011)
12. V. Sanchez, J. Bartrina-Rapesta, Lossless compression of medical images based on HEVC intra Coding, in *IEEE International Conference on Acoustic, Speech and signal Processing*, 2014, pp. 6622–6626
13. A. Panayides, Z. Antoniou, M.S. Pattichis, C.S. Pattichis, A.G. Constantinides, High efficiency video coding for ultrasound video communication in m-health systems, in *Annual International Conference of the IEEE Engineering in Medicine and Biology Society (EMBC)*, 2012, pp. 2170–2173. https://doi.org/10.1109/EMBC.2012.6346391
14. A.S. Panayides, M.S. Pattichis, A.G. Constantinides, C.S. Pattichis, M-health medical video communication systems: an overview of design approaches and recent advances, in *Proceedings of 2013 35th Annual International Conference of the IEEE Engineering in Medicine and Biology Society (EMBC)*, 2013, pp. 72537256. https://doi.org/10.1109/EMBC.2013.6611232
15. A.S. Panayides, M.S. Pattichis, C.P. Loizou, M. Pantziaris, A.G. Constantinides, C.S. Pattichis, An effective ultrasound video communication system using Despeckle filtering and HEVC. IEEE J. Biomed. Health Inform. **19**, 668–676 (2015). pmid:24951708
16. A.S. Panayides, M.S. Pattichis, C.P. Loizou, M. Pantziaris, A.G. Constantinides, C.S. Pattichis, An effective ultrasound video communication system using despeckle filtering and HEVC. IEEE J. Biomed. Health Inform. **19**(2), 668–676 (2015)
17. S. Parikh, D. Ruiz, H. Kalva, G. Fernandez-Escribano,Content dependent intra mode selection for medical image compression using HEVC, in *2016 IEEE International Conference on Consumer Electronics (ICCE)*, Las Vegas, NV, USA, 2016, pp. 561–564
18. P. Hanhart, M. Rerabek, P. Korshunov, Ebrahimi, Subjective evaluation of HEVC intra coding for still image compression. Joint Collaborative Team on Video Coding (JCT-VC) of ITU-T SG, 2013
19. K.K.R. Shenthil, L. Nithyanandan, Medical video communication using modified HEVC over WiMAX network, in *Proceedings of 2014 International conference on Communications and Signal Processing (ICCSP)* (IEEE, 2014). pp. 808–812
20. A. Panayides, M.S. Pattichis, C.S. Pattichis, HEVC encoding for reproducible medical ultrasound video diagnosis, in *Proceedings of 2013 Asilomar Conference on Signals, Systems and Computers* (IEEE, 2013), pp. 1117–1121
21. JCTVC-P1006, Common test conditions and software reference configurations for HEVC range extensions, 2014
22. G.J. Sullivan, J. Ohm, W. Han, T. Wiegand, Overview of the high efficiency video coding (HEVC) standard. IEEE Trans. Circuits Syst. Video Technol. **22**, 1649–1668 (2012)

Suvarna Nandyal received the B.E. degree in Computer Science and Engineering, Gulbarga University, Karnataka, and M.Tech. degree in Computer Science and Engineering, Visvesvaraya Technological University, Belagavi, India, in 1993 and 2003, respectively, and the Ph.D. degree in Computer Science and Engineering from Jawaharlal Nehru Technological University, Hyderabad, India, in 2013. She is currently a Professor and HOD with the Department of Computer Science and Engineering, PDA College of Engineering, Gulbarga. With over 22 years of teaching experience, she is guiding eight Ph.D. students and guided many undergraduate and postgraduate projects. She has authored or co-authored over 49 articles in refereed international journals and conferences. Her current research interests include the image processing, video and sound processing, information retrieval, mobile applications and cloud computing.

Heena Kouser Gogi received master degree in Computer Science and Engineering from VTU, Belagavi, India. She is currently the Ph.D. scholar of the PDA College of Engineering, Affiliated with VTU, Kalaburgi, India. Her research interests include image processing, cloud computing, video processing and artificial intelligence.

Hybrid Multi-filter and Harmony Search Algorithm-Based Gene Selection Method for Cancer Classification

Bibhuprasad Sahu and Mohammad Gouse

1 Introduction

Micro-array technology is an emerging approach adopted by various researcher, biologists for identification of biomarker genes from the high-dimensional gene expression dataset where thousand numbers of genes are available in a single sample for experiment. Coarse of dimensionality is a serious concern as the machine learning algorithms are suitable to work with in small no. of genes for accurate classification performance. To provide a solution for this, particular's identification and removal of redundant and irrelevant genes are achieved by the concept called feature selection(FS). The main disadvantage of univariate filter approach is not reliable as gene interaction is very less and gene evaluation is done individually. So, the multivariate filter is the best option adopted by the researchers for accurate prediction of cancer diagnosis. Wrapper method implemented machine learning approach as a fitness function for the feature subset evaluation. In this paper, we have proposed a cancer prognosis model(PM) combination with multi-filter harmony search method. The model consists of two phases: first phase, the feature subset is identified using different well-recognized filter methods such as ReliefF, Fisher score, SU and CFS, and 15% high-ranked features will be selected from the individual filter and by removing the duplication of the features. And in the second phase, a union feature subset is generated, and the selected gene featured subset is to provide as a input to the meta-heuristic algorithm called harmony search to identify the optimal subset of genes with less noise and more relevant genes. This performance of two-stage hybrid approach

B. Sahu (✉)
Vardhaman College of Engineering (Autonomous), Kacharam, Shamshabad, Hyderabad 501218, Telangana, India
e-mail: prasadnikhil176@gmail.com
URL: https://vardhaman.org

M. Gouse
Department of Information Technology, Catholic University, Erbil, Kurdistan Region 44001, Iraq

J. D. Peter et al. (eds.), *Disruptive Technologies for Big Data and Cloud Applications*, Lecture Notes in Electrical Engineering 905,
https://doi.org/10.1007/978-981-19-2177-3_63

is evaluated using various well-known machine learning classifiers such as support vector machine (SVM).

The flow of this paper is organized as follows: Sect. 2 presents the survey of related published papers related to our proposed model. The detail study of the datasets and methodology is mentioned in Sect. 3 and 4. Section 5 renders the experiment details with discussion of the result. Finally, Sect. 6 conclude the study.

2 Background Study

Table 1 presents the performance evaluation of different machine learning algorithms with micro-array datasets. In most of the reviewed articles, filter-wrapper technique is taken into consideration.

Loey Ramadan et al. [1] presented a novel IG-GWO model for breast and colon cancer gene expression data classification. The classification performance of the proposed methods achieves 71.795% in 75 iterations in case breast cancer dataset, whereas 85.484% in colon cancer. However, Gao et al. [2] adopted multi-filter techniques to identify the featured genes using IG, GR, ReliefF and correlation. In both studies, the authors used well-known classifier called support vector machine (SVM) for evaluation of the proposed model. Out of four filter techniques, IG performs better with accuracy of 100% for lung cancer and DLBCL and 90.32%,96.08%, 98.61%, respectively, for Colon cancer, prostate cancer and leukemia. Al-Rajab [3] presented a two-stage multi-filter known as IG-GA-MRMR. In the first phase of the model, IG-GA used as filter, and then MRMR is used to recognize and rank the feature subset before classification with well-known classifiers such as DT, KNN and NB and then concluded that the proposed model achieves with an accuracy of 94.0% for colon cancer and 100% for colorectal dataset. Dagnew et al. [4] developed a model

Table 1 Hybrid multi-filter meta-heuristic methods literature review for classification accuracy.

No.	References	Method adopted	Dataset
1	[1]	IG-GWO	Micro-array dataset
2	[2]	IG-SVM	Micro-array dataset
3	[3]	IG-GA-MRMR	Micro-array dataset
4	[4]	RF-MRMR-DT/KNN/NB	Micro-array dataset
5	[5]	SVM-RFE-BDF	Micro-array dataset
6	[6]	SU-HS	Micro-array dataset
7	[7]	Multi-filter CSO-KNN	Micro-array dataset
8	[8]	MRMR-PSO-SVM	Micro-array dataset
9	[9]	Ensemble filter-GA-KNN/MLP/KNN	Micro-array dataset
10	[10]	MRMR-PSO/CS/SBC-KNN	Micro-array dataset

with RF-based ensemble classifiers for cancer diagnosis. LR, MLP, SVM, RF, DT and KNN are used as classifier in training set, whereas ensemble classifiers such as hard voting and soft voting are used for test dataset to evaluate the proposed model and concluded that RF ensemble approach performs better as compared to others. A combination of embedded wrapper approach SVM-RFE and BDF was proposed by [5] by taking the average of C-SVM, υ -SVM and LS-SVM as objective function and achieve 100% accuracy in colon, leukemia, ovarian and DLBCL, whereas 89.47% in breast cancer. A hybrid model is proposed by [6] using symmetrical uncertainty (SU) as filter and harmony search algorithm as filter to evaluate the performance micro-array dataset. A three-stage classification model [7] was proposed using six numbers of filter approaches, and 5% best-ranked featured genes are considered, then CSO is applied to identify the optimal featured genes. In this, FIS concept is used to balance between exploration and exploitation. But [9, 11] used the same multi-filter approach wrapper for classification of micro-array datasets. Ghosh et al. [9] proposed a hybrid model using MRMR as filter and discrete PSO algorithm as wrapper which performs better in terms of accuracy and computational performance with respect to its counterparts. Mohamed [10] presented a hybrid approach with MRMR as filter and three meta-heuristic approaches such as PSO, CS, ABC and KNN as classifier for accuracy prediction. Various authors differently [12–28] adopted filter-wrapper approaches for cancer accuracy prediction (Fig. 1).

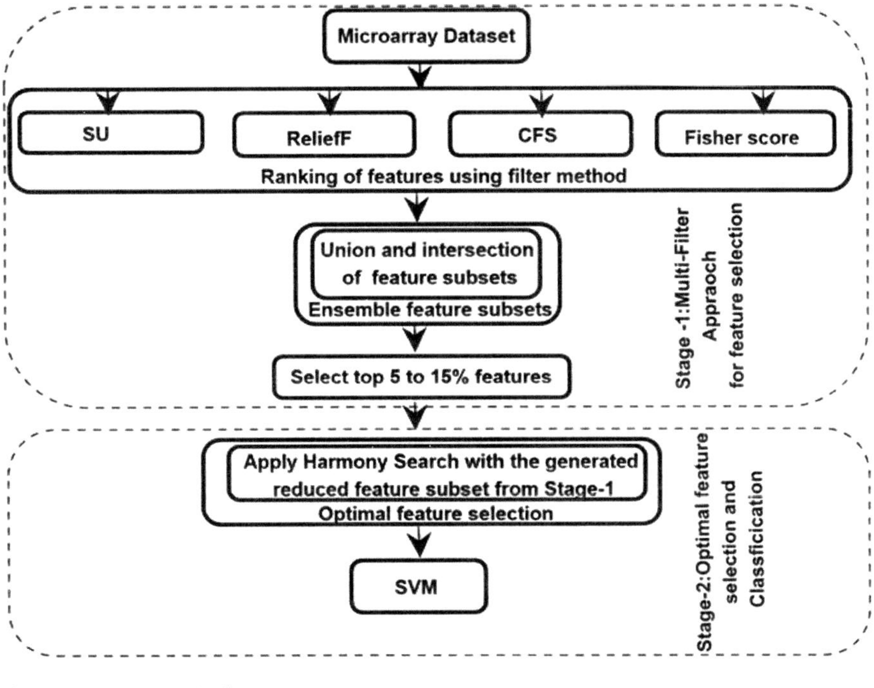

Fig. 1 Proposed block diagram

2.1 Limitation of the Existing Approaches

After reviewing different existing article, we conclude the following limitations.

- Most of the researchers used different filter approaches to tackle with the coarse of dimensionality issues. So it results in the problem of under-fitting issue when it results less no. of feature subset to enhance the classification accuracy. Selecting less no. of features by ignoring other important features is not advisable.
- Most of the researchers concluded that less of gene selected by filter helps to enhance the performance, but it is not accurate to remove the important genes from the classification process.
- After a rigorous survey, it is found that only few researchers applied ensemble filter to achieve better result. But none of the researchers used harmony search as wrapper techniques so far. So we have considered HS in this study for improvisation of the classification accuracy. Following qualities of HS inspired us to implement this for our study.

The main motto of this study is to present a hybrid framework for selecting biomarker genes from micro-array cancer datasets in two stages: The first ensemble multi-filter is used to identify the featured high-ranked feature subset. From out of individual filter approach, we will consider the best 5, 10, 15% of features as input for the hybrid stage (HS) to optimize the quality data from the dataset. Both stages will perform as selection algorithms along with SVM to predict the micro-array datasets. These hybridizations of algorithms are proposed to obtain high-ranked gene subsets with a minimal number of relevant genes. Then we have to test the accuracy using SVM classifier to ensure the investigation is conducted rigorously.

1. HS is proved to be a good meta-heuristic algorithm for micro-array datasets.
2. Hybrid HS is good option for classification as per the literature review.
3. From different study, we have concluded that HS is able to identify optimal feature subset from high-dimensional dataset.
4. Combining the HS and multi-filter approach (stage 1) of this framework model will generate top-ranked features subset.
5. Utilizing the (IG, CFS, MI, Fisher) as a multi-filter (stage 1) will refine the subset generated from original micro-array repository datasets, and we can identify the highly correlated and relevant genes with the class that has the least correlation between the features.

3 Dataset Description

For our study, we have considered six well-recognized databases generated from the micro-array experimental study that are ALL-AML leukemia, breast cancer, colon, ovarian and lung cancer. The abovementioned datasets are taken from the Kent Ridge Public Bio-medical Data Repository [21].

4 Preview of the Proposed Methodology

This study reports two phases of multi-filters harmony search wrapper approach. In the filter first phase, ReliefF(R), Fisher score(f), SU and IG are applied to the original micro-array dataset to evaluate the best-ranked feature subsets individually. Then on the basis of ranking, we consider the best-ranked features (15%) from individual filter algorithms, and feature pool is prepared using first rank features of the pool in order to reduce the variability of the ranked features and generate a more robust filter algorithm. Then the identified features are features(genes) reduced by applying the harmony search(HS) meta-heuristic algorithm. Finally, the methodology employs a support vector machine (SVM) classifier for cancer type classification. Six well-recognized micro-array datasets are used in this study for evaluation of the proposed method. It is concluded that the performance of our proposed method has achieved a higher accuracy in comparison with the existing methods.

In this section, the basic concepts different filter approaches(ReliefF(R), Fisher score(F), SU and CFS are described briefly along with the basic knowledge about harmony search meta-heuristic algorithm also reviewed.

4.1 Study of Different Filter Methods

ReliefF
ReliefF is the well-known inherited version of proposed ReliefA and ReliefD. It is a weight-based approach which provides statistical solution to the FS problems. A sample is selected from the training samples, and from individual sample feature based on Euclidean criteria, the 'nearest hit' and 'nearest miss' are calculated, which encourages to evaluate the ability of features to separate all pairs of classes regardless of which two classes are closest to one another. ReliefF is defined using Eq. (1)

$$\begin{aligned}\text{ReliefF\_Score} = \frac{1}{c}\sum_{j=1}^{l}\Big(-\frac{1}{m_j}\sum_{x_r\in NH(j)} d(X(j,i)-X(r,i)) \\ +\sum_{y=y_j}\frac{1}{h_{jy}}\cdot\frac{p(y)}{1-p(y)}\sum_{x_r\in NM(j,y)} d(X(j,i)-X(r,i))\end{aligned} \quad (1)$$

Here NH(j) presents nearest hit of x_j within the same class, where as $NM(j, y)$ presents nearest miss in class y.

Fisher score
Fisher score is one of the supervised feature selection algorithms used by many researchers for filter purpose to achieve reduced no. of feature subsets. The features are selected in the following concept. Fisher score is the ratio between the feature sample within same similar class and feature values of samples from different dis-

similar classes. Fisher score of individual feature will be produced in a descending order.

$$\text{Fisher\_Score} = \frac{\sum_{j=1}^{c} n_j (\mu_{ij} - \mu_i)^2}{\sum_{j=1}^{c} n_j \sigma_{ij}^2} \quad (2)$$

Symmetric Uncertainty

The statistical independence distribution difference between two features and between features with respective target class can be evaluated using mutual information (MI). Correlation presents the nonlinear measurement between. Symmetric Uncertainty can be produced by normalizing the MI. It is used to identify the best feature subset. SU is presented as follows:

$$\text{SU}(X, Y) = \frac{2 \times \text{Gain}(X|Y)}{H(X) + H(Y)}$$

Correlation Feature Selection

The ranking of features is based on the heuristic evaluation function in correlation. Heuristic function evaluates feature subsets made of attribute vectors, which are correlated with respective class label, and independent of each other. The features having low correlation with the class label should be ignored by the algorithm. The main aim of CFS is to identify strongly correlated features with one or more of the other attributes. The criterion used to assess a subset of features can be expressed as follows:

$$M_{\text{s}} = \frac{\overline{l t_{\text{cf}}}}{\sqrt{l + l(l - t)\overline{t_{\text{ff}}}}} \quad (3)$$

where M_{s} is the evaluation of a subset of S consisting of one feature, t_{cf} is the average correlation value between features and class labels, and $\overline{t_{\text{ff}}}$ is the average correlation value between two features. The main disadvantage of CFS is heavily dependent on the model, so they can fail to fit the data well (Fig. 2). **Pseudo-code of Harmony Search**

4.2 Proposed Methodology

Algorithm:
Multi-filter feature ranking model

1. Let f_i be the feature subset in the micro-array dataset, where $f_i : f_1, f_2, f_3 \ldots F_n$ and cl_i presents the class level, where $cl_i : cl_1, cl_2 \ldots$.
2. Evaluate the featured genes as per the rank and arrange the features with respect to the class cl_i.
3. Top 5–15% features of each filter method are selected.
4. Prepare a complete feature subset using union and intersection method.
5. Remove the rest low rank feature subset.

```
Harmony Search
begin
    Objective function f(x), x=(x1,x2, ...,xd)^T
    Generate initial harmonics (real number arrays)
    Define pitch adjusting rate (r_pa), pitch limits and bandwidth
    Define harmony memory accepting rate (r_accept)
    while ( t<Max number of iterations )
        Generate new harmonics by accepting best harmonics
        Adjust pitch to get new harmonics (solutions)
        if (rand>r_accept), choose an existing harmonic randomly
        else if (rand>r_pa), adjust the pitch randomly within limits
        else generate new harmonics via randomization
        end if
        Accept the new harmonics (solutions) if better
    end while
    Find the current best solutions
end
```

Fig. 2 Pseudo-code of harmony search

Optimal feature selection and classification

1. The output features are considered as the input for harmony search meta-heuristic algorithm to achieve the optimal feature.
2. The performance evaluation of the model can be tested using three well-known classifiers such as SVM.

Dataset	F_nos	IG-SVM	R-SVM	F-SVM	SU-SVM
SRBCT	2308	97.79	98.21	96.72	93.41
Tumors_9	5726	93.77	92.68	95.22	94.11
Leukaemia3	7129	96.76	96.69	99.12	91.43
Colon	10937	94.24	93.22	95.26	95.13
Lung	12601	94.56	96.48	97.46	97.11
Breast	24482	97.89	97.19	98.87	96.94

Table 2 Accuracy evaluation (without wrapper)

Dataset	F_nos	IG-DT	R-DT	F-DT	SU-DT	IG-NB	R-NB	F NB	SU-NB
SRBCT	2308	97.79	98.21	96.72	93.41	95.14	97.21	97.11	94.78
Tumors_9	5726	93.77	92.68	95.22	94.11	97.23	94.59	95.91	93.29
Leukaemia3	7129	96.76	96.69	99.12	91.43	96.11	97.47	96.01	95.96
Colon	10937	94.24	93.22	95.26	95.13	96.75	95.41	95.11	97.29
Lung	12601	94.56	96.48	97.46	97.11	96.37	94.11	96.75	96.14
Breast	24482	97.89	97.19	98.87	96.94	96.45	97.21	98.45	97.34

Table 3 Accuracy evaluation (with wrapper and classifier)

Dataset	IH-SVM	RH-SVM	FH-SVM	SH-SVM	PM-SVM	IH-DT	RH-DT
SRBCT	97.21	96.82	96.54	91.59	98.65	97.23	97.56
Tumors_9	95.46	93.43	97.11	94.72	99.21	96.75	97.11
Leukaemia3	97.75	96.84	97.21	93.47	99.76	96.97	98.34
Colon	90.13	96.72	96.31	96.01	98.49	97.11	97.78
Lung	93.26	96.91	97.96	94.28	98.43	98.02	96.79
Breast	96.98	99.11	98.76	97.04	99.76	96.95	97.78

Dataset	FH-DT	SH-DT	PM-DT	IH-NB	RH-NB	FH-NB	SH-NB	PM-NB
SRBCT	97.11	95.42	97.54	98.59	97.65	98.96	96.38	99.05
Tumors_9	96.43	96.71	97.89	96.72	98.11	97.94	96.01	98.17
Leukaemia3	99.23	91.84	97.82	98.72	97.86	97.19	97.56	97.46
Colon	96.83	96.12	97.31	96.81	97.09	96.78	96.43	97.91
Lung	98.85	98.91	96.56	94.98	97.81	97.95	97.07	98.01
Breast	97.88	97.41	98.16	98.14	98.86	98.89	97.83	97.76

The proposed model is an enhanced version of two-stage filter-wrapper approach. As each filter method uses its own statistical method to identify the high ranked ones, and each of having it own demerits of considering the same as discussed in Sect. 4.1. Instead of using one filter approach, we have used four different approaches and to select the top-n features by making ensemble of four different filter methods. As we have implemented the ensemble approach, if any important feature is missed by any filter approach, then there is a chance that any rest of filter method may consider that. So ensemble of filter may help to achieve best selection of feature subset as compared to a single filter. The details about the proposed model are mention in Sect. 4.

5 Experimental Setup and Results Study

We have deployed the proposed multi-filter harmony search. Our analysis was carried using WEKA which is a well-known machine learning tool consisting of various machine learning algorithms for data mining need. The parameters of all method were set to default as per the WEKA package. We have divided the performance comparison in three ways. First, we have applied simple SVM classifier (no feature selection) in 2, and then we have applied wrapper method based on harmony search with same datasets. Second we have applied various feature selection approaches with wrapper HS with SVM in 3. From the result, it is clear that the multi-filter with harmony search achieves better accuracy as compared to the other studies done in literature survey. Abbreviation of IG-HS, ReliefF-HS, Fisher-HS and SU-HS is denoted as IH, RH, FH and SH, respectively.

6 Conclusion

In this study, a multi-filter-wrapper framework is proposed for optimal feature selection. This framework is embedded withe meta-heuristic algorithm HS into a wrapper method. Instead of single filter, to achieve the best feature subsets, we have adopted four different filter approaches and the resultant features are considered as input for HS to identify optimal feature subset. As per the result achieved, our proposed framework performs better as compared to various hybrid filter-wrapper methods. To evaluate the performance of the hybrid approach, we have used well-known classifier called SVM. It is noteworthy to state that multi-filter-HS-SVM achieves better accuracy. In future, this method may be applied for real-world issues.

References

1. M. Loey, et al., Breast and colon cancer classification from gene expression profiles using data mining techniques. Symmetry **12**(3), 420 (2020)
2. L. Gao et al., Hybrid method based on information gain and support vector machine for gene selection in cancer classification. Genomics Proteomics Bioinf. **12**(3), 389–395 (2017)
3. M. Al-Rajab, J. Lu, Q. Xu, A framework model using multifilter feature selection to enhance colon cancer classification. Plos one **16**(4), 0249094 (2021)
4. G. Dagnew, B.H. Shekar, Ensemble learning-based classification of microarray cancer data on tree-based features. Cogn. Comput. Syst. **3**(1), 48–60 (2021)
5. S.A. Medjahed et al., Kernel-based learning and feature selection analysis for cancer diagnosis. Appl. Soft Comput. **51**, 39–48 (2017)
6. S.S. Shreem, S. Abdullah, M.Z.A. Nazri, Hybrid feature selection algorithm using symmetrical uncertainty and a harmony search algorithm. Int. J. Syst. Sci. **47**(6), 1312–1329 (2016)
7. M.B. Dowlatshahi, V. Derhami, H. Nezamabadi-pour, A novel three-stage filter-wrapper framework for miRNA subset selection in cancer classification, in *Informatics*, vol. 5. no. 1 (Multidisciplinary Digital Publishing Institute, 2018)
8. A. Unler, A. Murat, R.B. Chinnam, mr2PSO: a maximum relevance minimum redundancy feature selection method based on swarm intelligence for support vector machine classification. Inf. Sci. **181**(20), 4625–4641 (2011)
9. M. Ghosh et al., mr2PSO: genetic algorithm based cancerous gene identification from microarray data using ensemble of filter methods. Med. Biol. Eng. Comput. **57**(1), 159–176 (2019)
10. N.S. Mohamed, S. Zainudin, Z.A. Othman, Metaheuristic approach for an enhanced mRMR filter method for classification using drug response microarray data. Expert Syst. Appl. **90**, 224–231 (2017)
11. B. Sahu et al., A combo feature selection method (filter+ wrapper) for microarray gene classification. Int. J. Pure Appl. Math. **118**(6), 389–401 (2018)
12. B. Sahu, Multi-tier hybrid feature selection by combining filter and wrapper for subset feature selection in cancer classification. Indian J. Sci. Technol. **12**(3), 1–11 (2019)
13. B. Sahu, et al., MRMR-BAT-HS: a clinical decision support system for cancer diagnosis. Leukemia 7129, **73**, 48 (2020)
14. V.R. Elgin Christo, et al., Correlation-based ensemble feature selection using bioinspired algorithms and classification using backpropagation neural network. Comput. Math. Methods Med (2019)
15. B. Sahu, et al., 7 An intelligence-based health biomarker identification system using microarray analysis, in *Applied intelligent decision making in machine learning*. (CRC Press, 2020), pp. 137–161

16. B. Sahu et al., A hybrid cancer classification based on SVM optimized by PSO and reverse firefly algorithm. Int. J. Control Autom. **13**(4), 506–517 (2020)
17. B. Sahu, et al., MMFA-SVM: new bio-marker gene discovery algorithms for cancer gene expression, in *Materials Today: Proceedings* (2021)
18. B. Sahu, et al., CDA-SVM: a chaotic dragonfly enriched support vector machine for micro array data sets, in *2021 International Conference on Emerging Smart Computing and Informatics (ESCI)* India (2021)
19. B. Sahu et al., CMBA-SVM: a clinical approach for Parkinson disease diagnosis. Int. J. Inf. Technol. **13**(2), 647–655 (2021)
20. B. Sahu et al., A hybrid approach for breast cancer classification and diagnosis. EAI Endorsed Trans. Scalable Inf. Syst. **6**(20), 5 (2019)
21. J. Li, H. Liu, A hybrid filter-wrapper gene selection method for cancer classification. Kent ridge bio-medical data set repository. Institute for Infocomm Research. http://sdmc.lit.org.sg/GEDatasets/Datasets.html (2002)
22. F. Moslehi, A. Haeri, A novel hybrid wrapper-filter approach based on genetic algorithm, particle swarm optimization for feature subset selection. J. Ambient Intell. Hum. Comput. **11**(3), 1105–1127 (2020)
23. A. Chaudhuri, T.P. Sahu, GA novel hybrid PROMETHEE-based hybrid feature selection technique for high-dimensional biomedical data: application to Parkinson's disease classification. Electronics Letters **56**(25), 1403–1406 (2020)
24. S.S. Hameed, et al., A comparative study of nature-inspired metaheuristic algorithms using a three-phase hybrid approach for gene selection and classification in high-dimensional cancer datasets. Soft Comput. 1–19 (2021)
25. A. SMoayedikia, et al., Feature selection for high dimensional imbalanced class data using harmony search. Eng. Appl. Artif. Intell. **57**, 38–49 (2017)
26. A.K. Shukla, *Feature selection inspired by human intelligence for improving classification accuracy of cancer types* (Comput, Intell, 2020)
27. A.K. Shukla, P. Singh, M. Vardhan, A new hybrid wrapper TLBO and SA with SVM approach for gene expression data. Inf. Sci. **503**, 238–254 (2019)
28. K. Das et al., A metaheuristic optimization framework for informative gene selection. Inf. Med. Unlocked **4**, 10–20 (2016)

COVID-19 Cases Prediction Using Different LSTM Models and Comparison of Effectiveness of Different Models

Essmily Simon, Swapna Sasi, and Aswathy Wilson

1 Introduction

The COVID-19 pandemic is first happened on china and it has spread to different 213 countries across the world. It is a very transmittable disease that has been declared a pandemic by the World Health Organisation (WHO). According to WHO, there have been 17 crore confirmed cases and 32 lakh fatalities across the world during the last of May 2021. COVID-19 is an infectious disease which is badly effect on human respiratory system. It was first reported in December 2019 in Wuhan, China, and quickly spread throughout the world, resulting as pandemic. One of the artificial neural network model is recurrent neural network used for time series prediction. RNN used for the time series data which contain dynamic information. One of the special property in RNN, the previous step's output is used as input for the current step. RNN has a hidden state that stores the sequence information. RNN has reduce the complexity of parameters compared to other neural networks. But the training of RNN is difficult and it has faced the gradient vanishing problem. It cannot process very long sequences using tan h activation function. To solve this problem, Hochreiter et al. and Schmidhuber et al. has suggested LSTM model with input, forget and output gate to capture the relation between long-term sequence.

E. Simon (✉) · S. Sasi · A. Wilson
Jyothi Engineering College, Cheruthuruthy po, Thrissur, Kerala, India
e-mail: essmilysimon@gmail.com

J. D. Peter et al. (eds.), *Disruptive Technologies for Big Data and Cloud Applications*, Lecture Notes in Electrical Engineering 905,
https://doi.org/10.1007/978-981-19-2177-3_64

The LSTM model is employed to analyse and forecasting the growth of pandemic. It overcomes the problems faced during using the previous learning algorithm recurrent neural network (RNN). LSTM is mainly used for time series data prediction and it is also used for handwriting recognition, anomaly detection in network traffic, speech recognition, or in intrusion detection system.

There are four types of LSTM models. They are multivariate, uni-variate and multi-step with multivariate LSTM models. For uni-variate LSTM models, it is used to forecast uni-variate time series problem which consisting of a single series of observation and it is learn from previous observation to get the next value of sequence. For uni-variate LSTM models, there are different models like simple LSTM or Vanila LSTM which has one hidden layer and have output layer used to make prediction. For the case of stacked LSTM, multiple hidden layers can be stacked on top of each other and it has input in three dimensions and output in two dimensions. The another type of LSTM is bidirectional LSTM or Bi-LSTM which is used to learn both forward and backward input sequences and concatenate the interpretation. It wrap the first hidden layer unit and used for uni-variate time series. Here the uni-variate LSTM models used to determine the number of various parameters like number of confirmed cases, death cases and the recovered cases.

For multivariate LSTM model, they have each time step contains multiple observations. There are two multivariate LSTM models such as several series of inputs and several parallel series. The main property of multiple input series is that two or more time series input data streams that are parallel can be applied. It means that each series contains observations made at the same stages. Multi-step LSTM models are used to avoid the problem where the forecast horizon on interval is divided into many time steps. There are two kinds of multi-step LSTM models such as model of vector output and model of encoder-decoder. In the vector output model, each output time series' time step has been anticipated as the vector. Encoder-decoder model is used to forecast variable length output sequences. These models widely used in language translation. This paper provides a case study which has the ability to train COVID-19 stages over time. We collect the data from each country and split into the testing and training data. The hyper parameters are searched heuristically and the best LSTM model can be evaluated during the validation process. At the phase of evaluation, we also compare the different previous LSTM model and RNN.

This paper introduces a framework for predicting the confirmed cases, recovered cases and death cases per day and it helps to understand and examine the future trends in these three classifications. This paper provides a comparison for different time series data prediction models by measuring their performance.

2 Related Work

In December 2019, a series of pneumonia cases with clinical symptoms similar to viral pneumonia were reported in Wuhan, Hubei, China. Deep sequencing analysis of lower respiratory tract samples revealed a novel corona virus, named 2019 novel corona virus (2019-nCoV). Airborne protection, such as a fit-tested N95 respirator and other personal protective equipment, is strongly advised. Corona virus disease (COVID-19) spread predicting is an important task in tracking the pandemic's progression [1]. Long short-term memory (LSTM) methods can be used to discover the relationship between COVID-19 growth and time. Until the best validation score is achieved. Heuristic search is used to find the structure of an LSTM layer. This can be trained using training data from all over the world with confirmed cases. Favourable performance can be achieved with the recurrent neural network (RNN) method, which has a comparable low validation error. The evaluation is carried out using visualisation of graph and the root mean squared error (RMSE). LSTM provides a resemblance between actual cases and prediction [2].

There are two types of recurrent networks: relaxation and standard. Relaxation networks begin in a given state and progress to a fixed point, which typically denotes a class. Convergence can be guaranteed by imposing constraints on the feedback connections. Rather than reaching a fixed point, recurrent networks model a trajectory. The changing input in time series prediction results in a recurrent neural network that rarely reaches a fixed point. Recurrent networks can be used that can forecast a continuous range of values, as opposed to classification networks that use relaxation networks. In RNN, hidden unit patterns feedback to themselves: the internal representations that emerge reflect tasks in the context of previous internal states. This involves the use of recurrent links as well as supply of networks with a dynamic memory and time is represented rather than explicitly, its effects on processing are conveyed implicitly.

The processing of sequential inputs has been accomplished in parallel distributed processing models by providing it with representation of space. Rather than explicitly representing time, it would be preferable to represent it implicitly. That is, rather than representing time as an additional dimension of the input, represent it as an effect on processing. One method for dealing with patterns that have a temporal scope, it is necessary to explicitly represent time by associating the pattern's serial order with time with the dimensionality of the pattern vector. The first temporal event is represented by the pattern vector's first element, the second temporal event by the pattern vector's second position, and so on. The model processes the entire pattern vector in parallel. If there is a sequential input to be processed, as well as a clock that regulates the input's presentation to the network. The following sequence of events would then constitute processing. The input units receive the first input in the sequence at time t. Depending on the nature of the problem, each unit could be a single scalar value or a vector. The hidden units are activated by both the input and context units. The hidden units then feed forward to activate the output units. The hidden units also feedback to activate the context units. This is referred to as forward activation. This time cycle may or may not include a learning phase, depending on the task. If this is the case, the output is compared to a previous input, and error propagation is performed. This time, the context units contain values that match the hidden unit values at time t exactly. As a result, these context units serve as memory for the network [3].

3 Methodology

3.1 Data Collection

We used 100 countries as training data and Indonesia COVID-19 cases as validation data to forecast the number of cases around the country. The dataset have the parameters like country, state/province, latitude, longitude, confirmed cases, recovered cases and death cases. Each sample's sequence is divided into two parts to provide the input and label, with the first part running as the input training from 22 January 2020 to 29 March 2020 and the second part running between 30 March 2020 and 1 May 2020 as the output label. We make use of state wise data to predict the number of COVID-19 cases using W-LSTM, which includes parameters such as recovered, confirmed and decreased cases for different states across the country. The link contains a dataset of novel corona viruses which is used for Bi-LSTM, GRU machine learning models [4]. Column wise .csv files of confirmed cases, death cases and recovered cases from all countries are provided. From 22 January 2020 to 27 June 2020, each of these three categories is assigned to its own file. The COVID-19 dataset contains 158 samples with confirmed cases, deaths, and recovered cases from 22/1/2020 to 10/5/2020. The COVID-19 dataset contains 158 samples with confirmed cases, deaths and recovered cases, and we used cases from 22/1/2020 to 10/5/2020 for training and to predict cases from 11/5/2020 to 27/6/2020. Data for each country includes given cases for 110 days and must be predicted for the within the next 48 days. Before it is fed into ML models for training, the data is preprocessed using mean square error.

3.2 Data Preprocessing

Because the LSTM is extremely susceptible to normalisation, particularly when taking time series data, we use a MinMaxScaler to normalise the data. At first, we convert the data to similar scale to keep bias out of training and validation. The scaling function has been established as follows:

$$\text{Min max scaler}(X\text{scaled}) = 1/n \sum_{i=1}^{n} \frac{X - X\min}{X\max - X\min}$$

where X is the dataset for training input and Xscaled is the normalised output dataset for training.

3.3 Model

3.3.1 LSTM

The memory cells ct and working cell ht. in the LSTM are created when a cell is divided. Memory cells are in charge of retaining the sequence controlled by the forgetting gate ft. Each memory cell's output is the working memory ht, and the output gate ot manages the portion of ct will be remembered. The input gate regulates the former state ht-1 as well as the current input xt. Our architecture has a single to four hidden layers, each with a hidden unit ranging from 1 to 30. The final layer is a layer of linear that predicts 100 sequences. The linear layer's output is routed through the sigmoid activation function to ensure a range of 0–1. To avoid overfitting, we employ a 0.1 dropout. The assessment and learning consists of an input provided to the model and an output is generated. The input and output are both in the form of everyday cases. The outcome of daily cases are then aggregated to depict the time-lapse growth curve.

During the training phase, an input of 100 sequence is divided as the first to 67th days as the input graph and the 68th to 100th days as indicated by the label. The normalisation function is used to normalise input before processing it. Before feeding them into the trained model, the testing and validation data are normalised utilises the scaling factors derived from the training data.

$$f_t = \sigma\left(w_f \times \left[h_{t-1}, x_t\right] + b\right) \quad (1)$$

$$i_t = \sigma((w_f \times \left[h_{t-1}, x_t\right] + b_i \quad (2)$$

$$C_t = \tan h(w_c \times \left[h_{t-1,} x_t\right] + b_c) \quad (3)$$

$$c_t = f_t \times c_{t-1} + i_t \times C_t \quad (4)$$

$$o_t = \sigma\left(w_c \times \left[h_{t-1}, x_t\right] + b\right) \quad (5)$$

$$h_t = o_t \times \tan h(c_t) \quad (6)$$

3.3.2 W-LSTM Model

Weibull-LSTM (W-LSTM) is a time series model that predicts appropriate parameter sets by capturing the model parameters like k, α, β, γ. To analyse and forecast the best GIW parameters, we employ LSTM model. The number of new infections and deaths occurred on a daily basis was modelled for several countries using "Robust Weibull" approach used in machine learning that corresponds to the Generalised Weibull Distribution (GIW) as shown in Eq. 7.

$$f(x) = k \times \gamma \times \beta \times \alpha^{\beta} \times x^{-1-\beta} \times \exp\left(-\gamma\left(\frac{\alpha}{x}\right)\beta\right) \quad (7)$$

Algorithm : W-LSTM training

- *Countries:* All countries with data are listed here.
- *Data*:Each country's data
- $\epsilon_{training}$:Threshold parameter for training
- m: Sequence length for training

```
Procedure TRAIN LSTM MODEL
    t0 ← 0
    Initialize new LSTM model
    while True do
        h0 ← 0
        C0 ← 0
        loss_t ← 0
        MAPE←[0]*m
        lossvect← [0] * m
        for countries in countries do
            if ∑ data[countries] < 2000 then
            continue
            for i from 1 to m do
                output ← ROBUSTCURVEFITTING(data[country][0 : n− m − 1 + i])
                hi, ci ← LSTM(output, ht−1, ci−1)
                LOSS_t ← MSE(output, hi)
                k, α, β, γ ← hk
                pred ← k· γ· β · α^β · x^(−1−β) · exp (-γ (α/x)β ) ∀ x∈ data[country]
            end for
            loss_t ← loss_t + (1 − MAPE/100)
            BACKPROP(LSTM, loss_t) ⊙loss_t
            if t> 0 and |loss_t − loss_(t−1)| < ϵ_training then
                Break
            t ← t + 1
        end for
    end while
end procedure
```

3.3.3 Bidirectional LSTM (Bi-LSTM)

Memory cells are self-contained and that store the network's current state and are three gates control; the input gate, the output gate and the forget gate. The flow of memory cell inputs and outputs into the rest of the network is controlled by input and output gates. When you use bidirectional, your inputs will be routed in both directions, one from past to future and one from future to past and differs this approach from unidirectional is that in the LSTM that runs backward, information from the future is preserved. Whereas we can save information from the past and future at any point in time by combining these two hidden states.

3.3.4 Gated Recurrent Unit (GRU)

There is no explicit memory unit in GRU. The memory unit is combined with the network. In GRU, there is no forget or update gate. They are combined, and thus the number of parameters is reduced. When compared to LSTM, GRU performs well but has a slight drop in accuracy. However, because there are fewer trainable parameters, it is more convenient to use.

4 Prediction Accuracy Measurement

4.1 Root Mean Square Error (RMSE)

For this purpose, the root mean square error (RMSE) is calculated for prediction accuracy for the LSTM model. The RMSE is calculated as

$$\text{RMSE} = \sqrt{\frac{1}{N}\sum_{I=1}^{N}(P - A)^2} \quad (8)$$

It is worth noting that P represents the sequence of prediction and A represents the actual sequence.

4.2 Mean Absolute Percentage Error (MAPE)

The mean absolute percentage error (MAPE) is a measure of the precision of the forecast system. It calculates the average absolute percent error for each time period minus actual values and dividing them by actual values to calculate this accuracy as a percentage. Where M, n, A_t represents the mean absolute percentage error, no. of iterations, actual and forecast value F_t respectively.

$$M = \frac{1}{n}\sum_{t=1}^{n}\left|\frac{A_{t-F_t}}{A_t}\right| \tag{9}$$

4.3 r2-Score

r2 score is the difference in the variance of the dependent and independent parameters.

$$r2\,\text{score} = 1 - \frac{\sum\left|c - c^i\right|}{c - \hat{c}} \tag{10}$$

5 Experimental Results and Discussion

In the case of LSTM, begin with training and testing are carried out only once to forecast a long growth curve effect of 22 January 2020 to 1 May 2020. Second, to reduce the bias caused by random initialisation, training and testing are repeated five times. We made use of them with a learning rate of 0.001, Adam optimizer, and 10,000 iterations. Additionally, we can obtain the mean and range of several curves predictors in this manner. Meanwhile, RMSE is used for quantitative evaluation. For Indonesia, RMSE value is 1111.52 and Sweden has RMSE value is 1756.58. The prediction curve follows an exponential pattern similar to the actual growth. The prediction is several days prior to the actual. The RMSE decreases as the number of

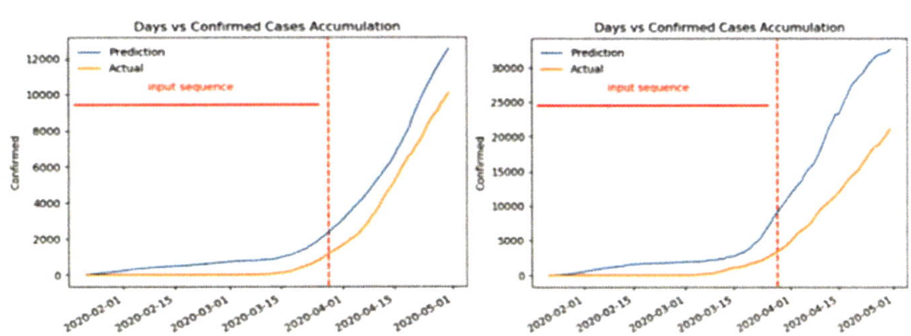

Fig. 1 **a** Confirmed COVID-19 cases in Indonesia, **b** confirmed COVID-19 cases in Sweden from 2 January 2020 to 1 May 2020

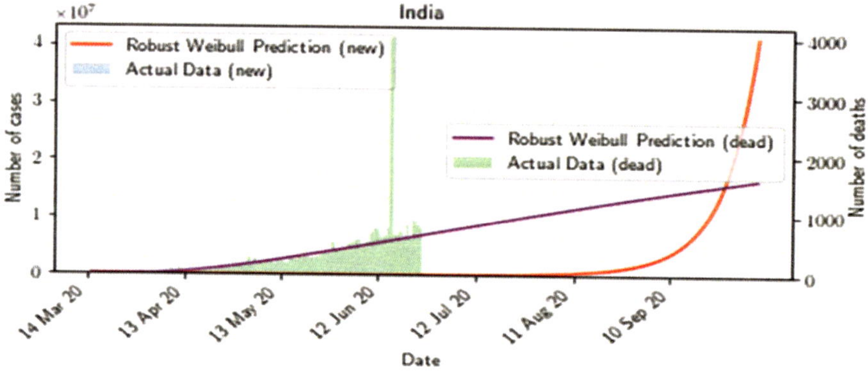

Fig. 2 New daily cases and deaths cases curves for India

hidden states increases. Figure 1 shows the confirmed COVID-19 cases in Indonesia and Sweden from January 22 to May 1, 2020 (Fig. 1).

Robust Weibull approach based on LSTM (W-LSTM) examined findings based on COVID-19 infection data on India. When compared to Simple LSTM models using infection data from COVID-19, it produces noticeably better results. W-LSTM-based approach outperforms others when it comes to MSE, R2-score and MAPE as performance measures. For India, the MSE value is 8.2158, the r2 score is 0.9111, and the MAPE value is 20.083 (Fig. 2).

For GRU, China had MAE values of 2.855, 0.032 and 7.049, respectively, and RMSE values of 3.316, 0.0402 and 8.4009 for confirmed cases, deaths and recoveries. Bi-LSTM forecast the best three cases for China with the greatest precision of all methods for all countries, with the lowest MAE and RMSE values for deaths in China of 0.0070 and 0.0077, respectively.

6 Conclusion

Indonesia has the lowest RMSE value for the LSTM-based prediction model. The RMSE decreases as the number of hidden states grows. Thus, even have 30 hidden states, the LSTM model achieves a high level of accuracy. The pattern of the prediction curve is exponentially similar to that of the actual development of COVID-19 cases. When used on infection data from COVID-19, the W-LSTM model performs better than the simple LSTM model. When compared to the previous model, this study's W-LSTM-based approach outperforms others in terms of MSE, r2 and MAPE. GRU performs good for the country China, with MAE values of 2.8553, 0.0321 and 7.04867 and RMSE values of 3.316, 0.040 and 8.4001 for confirmed cases, deaths and recoveries, respectively. Amongst all countries, Bi-LSTM predicts the best three cases for China, with the highest accuracy of any method and the lowest MAE and RMSE values for China for the parameter death cases of 0.007 and 0.008, respectively. The best r2 score value for recovered cases in China is 0.999.W-LSTM produces better results compared to another models compared to simple LSTM, Bi-LSTM and GRU.

References

1. C.Y. Huang, X. Wang Li, L. Ren, J. Zhao, Y. Zhao, L. Hu. G. Zhang, J. Fan Xu, X.Z. Gu, T. Cheng, J. Yu, Y. Xia, W.X. Wu. W. Xie's Xie Xie, X.H. Yin, H. Yin, H. Yin, H.Y.M. Li, Y. Liu, H. Xiao, H. Xiao, H. Xiao, H.L. Gao, J. Guo, G. Xie, X.R. Wang, Z. Jiang, Q. Gao, J. Jin, Clinical features of patients infected with 2019 novel coronavirus in Wuhan, China, Wang and B. Cao. The Lancet **395**(10223), 497–506 (2020). Available http://www.sciencedirect.com/science/article/pii/S0140673620301835
2. N. Yudistira, Covid-19 growth prediction using multivariate long short term memory, 2020. Available https://www.researchgate.net/publication/340030744
3. Jiang, Wei & Schotten, Hans. (2020). Deep Learning for Fading Channel Prediction. IEEE Open Journal of the Communications Society. PP. 1-1. 10.1109/OJCOMS.2020.2982513.
4. S. Rajkumar, Novelcoronavirus2019dataset. Available https://www.kaggle.com/sudalairajkumar/novel-corona-virus-2019-dataset, 2020
5. L. Wang, Y. Wang, D. Ye, Q. Liu, Review of the 2019 novel coronavirus (sars-cov-2) based on current evidence. Int. J. Antimicrobial Agents **55**(6), 105948 (2020). Available http://www.sciencedirect.com/science/article/pii/S0924857920300984
6. J. Brownlee, Deep learning for time series, 2018. Available https://machinelearningmastery.com/how-to-develop-lstm-models-for-time-series-forecasting
7. S. Tuli, S. Tuli, R. Tuli, and S. S. Gill, Predicting the growth and trend of covid-19pandemic using machine learning and cloud computing. Internet of Things **11**, 100222 (2020)
8. A. Tomar, N. Gupta, Prediction for the spread of covid-19 in India and effectiveness of preventive measures. Sci. Total Environ. **728**, 138762 (2020)
9. F. Shahid, A. Zameer, M. Muneeb, Predictions for covid-19 with deeplearning models of lstm, gru and bi-lstm. Chaos, Solitons Fractals **140**, 110212 (2020)
10. Covid-19 data repository by the center for systems science and engineering (CSSE) at Johns Hopkins University, 2020. Available https://raw.githubusercontent.com/CSSEGISandData/COVID-19/master/csse_covid_19_data/csse_covid_19_time_series/time_series_covid19_confirmed_global.csv
11. S. Hochreiter, J. Schmidhuber, Long short-term memory. Neural Comput. **9**, 1735–1780 (1997). https://doi.org/10.1162/neco.1997.9.8.1735
12. G. Van Houdt, C. Mosquera, G. Nápoles, A review on the long short-term memory model. Artif. Intell. Rev. **53**, 5929–5955 (2020). https://doi.org/10.1007/s10462-020-093838-1
13. S. Siami-Namini, N. Tavakoli, A.S. Namin, The performance of LSTM and BiLSTM in forecasting time series, in *2019 IEEE International Conference on Big Data (Big Data)*, 2019, pp. 3285–3292. https://doi.org/10.1109/BigData47090.2019.9005997
14. A. Aziz Sharfuddin, M. Nafis Tihami, M. Saiful Islam, A deep recurrent neural network with BiLSTM model for sentiment classification, *2018 International Conference on Bangla Speech and Language Processing (ICBSLP)*, 2018, pp. 1–4. https://doi.org/10.1109/ICBSLP.2018.8554396
15. M.S. Islam, A deep recurrent neural network with BiLSTM model for sentiment classification (2018)
16. R. Dey, F.M. Salem, Gate-variants of gated recurrent unit (GRU) neural networks, in *2017 IEEE 60th International Midwest Symposium on Circuits and Systems (MWSCAS)*, 2017, pp. 1597–1600. https://doi.org/10.1109/MWSCAS.2017.8053243
17. P. Simard, Y. Bengio, P. Frasconi, Learn long-term dependencies using gradient descent. IEEE Trans. Neural Netw. **5**(2), 157–166 (2002). https://doi.org/10.1109/72.279181

18. J. Chung, et al., Empirical evaluation of gated recurrent neural networks on sequence modeling. arXiv preprint: 1412.3555 (2014)
19. R. Rana, Gated recurrent unit (GRU) for emotion classification from noisy speech. arXiv preprint: 1612.07778 (2016)
20. D. Parbat, M. Chakraborty, A python based support vector regression model for prediction of COVID19 cases in India. Chaos, Solitons Fractals **138**, 109942 (2020)

EEG-Based Home Automation System Using Brain Sense Device

Christina Saju, Samson T. Anil, and S. Thomas George

1 Introduction

The recent years have seen great advancements in the field of home automation making the daily life more comfortable. Several smart home applications have been created to execute various tasks in the house, for example, voice-activated controlling of the temperature at home, switching appliances on or off, controlling the lighting, etc. They also possess the ability for self-learning. There have been increased efforts to make it user-friendly. However, the usage of such advancements by people with disabilities like paralysis, quadriplegia is very difficult. Quadriplegia is a condition in which paralysis is caused by sickness or injury that results in the partial or total usage loss of all limbs. This often results in the loss of sensory and motor controls. Wafa Alrajhi et al. observed that imagination of doing physical activities produced brain signals [1]. The user's stress, excitement were detected using Emotic Epoch + EEG sensor. Clemens Holzner et al. observed an EEG-based home system using a VR system; it was found that the user was able to control TV commands, opening and closing of windows and doors [2, 3].

The brain–computer interface technology is considered as a communication system that assists human beings to control and utilize the devices by using their thoughts [2]. This method uses EEG signals which are the outcome of various neural interactions. EEG signals are easy to use, portable, non-invasive, and have a high temporal resolution. It also gives waves of varying amplitude and frequencies. EEG waves include delta, theta, alpha, beta, and gamma corresponding to different frequency bands. Mainly, there exist two methods of obtaining EEG signals for brain–computer interface: invasive and non-invasive [4]. The invasive method involves the

C. Saju · S. T. Anil · S. Thomas George (✉)
Department of Biomedical Engineering, Karunya Institute of Technology and Sciences, Coimbatore, India
e-mail: thomasgeorge@karunya.edu

J. D. Peter et al. (eds.), *Disruptive Technologies for Big Data and Cloud Applications*, Lecture Notes in Electrical Engineering 905,
https://doi.org/10.1007/978-981-19-2177-3_65

Waves	Frequency Bands (Hz)	Behaviour Trait	Signal Waveform
Delta	0.3-4	Deep Sleep	
Theta	4-8	Deep Meditation	
Alpha	8-13	Eyes Closed, Awake	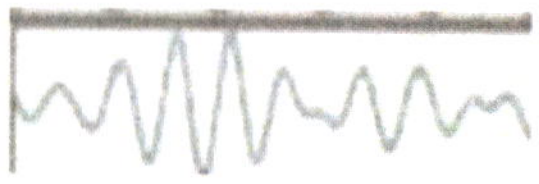
Beta	13-30	Eyes Opened, Thinking	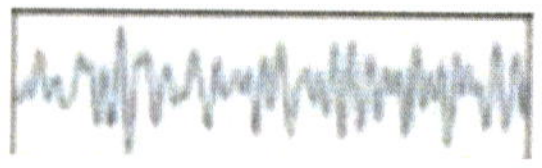
Gamma	30 and above	Unifying Consciousness	

Fig. 1 Different types of brain waves

recording of brain signals by electrodes that are surgically implanted within the depths of the brain. On the contrary, the non-invasive method involves signal acquisition with the help of EEG sensors which are placed on the surface of the scalp [4]. This technique is non-invasive approach. Eye blink is one of the mechanisms through which we can collect signals; this is done via BCI.

Eye blinks can be divided into three types: voluntary, involuntary, and reflexive [5]. Reflex blinks occur due to the external stimulus such as when objects appear quickly in front of the eye. Involuntary blinks are done without conscious efforts or in response to any external stimuli. Voluntary blinks are pre-determined courses of action that involve the usage of various areas of the brain [5]. Voluntary blinks produce evoked potential that is generated from specific parts of the brain. Signals can be recorded from the brain stem, spinal cord, peripheral nerves, and muscles. Evoked potential are responses that are recorded from central nervous system structures [5–7]. Various frequency bands of EEG are given in Fig. 1.

2 Brain–Computer Interface

A brain–computer interface also termed a neural control interface is a direct transmission channel between the brain and an external system. BCIs play a major role in researching, assisting, mapping, or repairing human cognitive or sensory-motor functions [8, 9]. The main purpose of BCIs is to compute the brain activity, extract

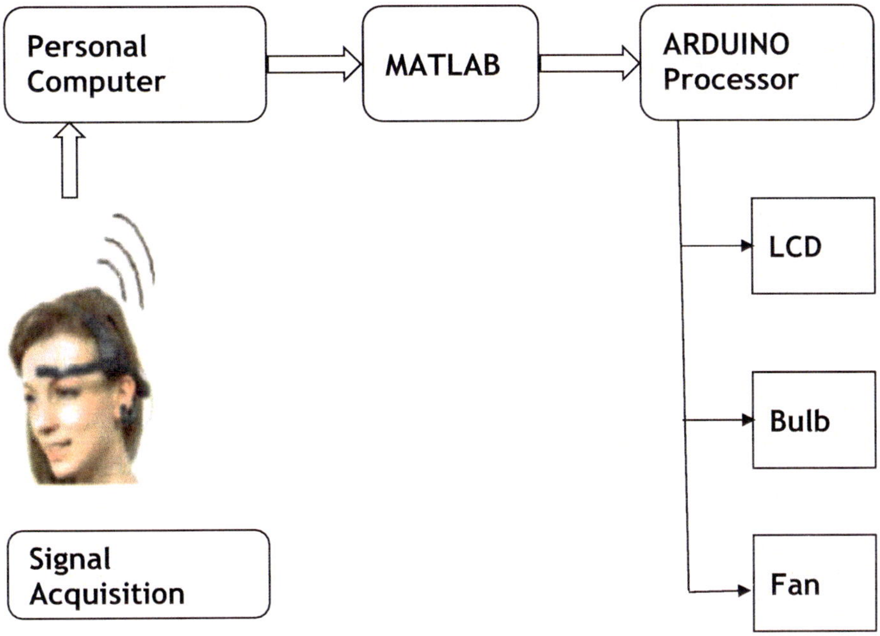

Fig. 2 Parts of a BCI system

features from the brain waves generated and change those features into outputs that replace, restore, improve, assist, or improve human-able functions. BCIs also play a vital role in regaining the ability to control the body, by stimulating the nervous system or muscles that control the limbs. Several BCIs use the method of obtaining electrical signals that are detected using electrodes placed invasively in the surface of the cortex or noninvasively on the scalp [4]. A customized block diagram of a BCI system is given in Fig. 2.

2.1 Importance of Brain–Computer Interfaces

According to a report that was made by the Social Statistics Division of the Ministry of Statistics and Program Implementation, Government of India, it was stated that a certain 2.21% of the Indian population were experiencing one or the other kind of disability. This means that an approx. 2.68 crore (26.8 million) people in India are disabled and find it difficult to lead an independent lifestyle. It was found that about 69% of the overall disabled Indian population resided in rural or village areas. Only a small percentage of disabled people were found to reside in urban or developed areas (Fig. 3) [10].

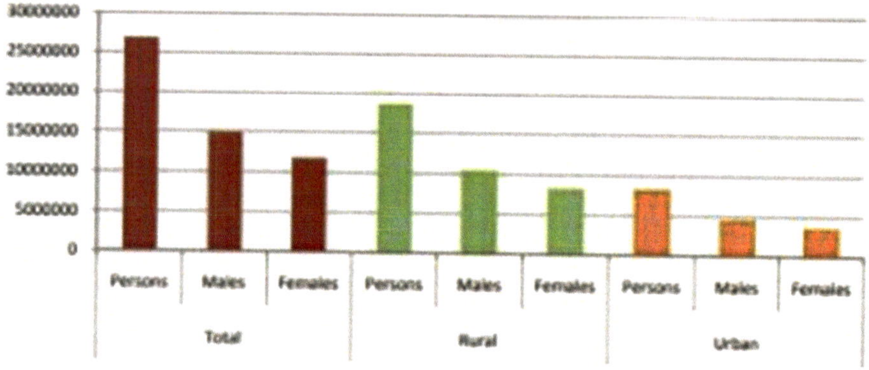

Fig. 3 Disabled population chart

2.2 *Types of Brain–Computer Interfaces*

The brain–computer interface (BCI) in its technical environment incorporates several hardware and software components which help its user in establishing a successful communication with the computer by intentionally altering his or her brain waves. The most predominant types of brain-computer interface approaches are [11]:

- Invasive BCI.
- Minimally invasive BCI.
- Non-invasive BCI.

In the invasive BCI method, the brainwave sensor chip would be implanted surgically into the gray matter of the human brain. One of the main advantages of this method was that it produced quality signals. However, it also caused the issue of leaving a scar behind. The minimally invasive approach makes use of the brainwave sensor chip, which is surgically placed inside the skull but would rest outside the gray matter of the brain. This approach resulted in good resolution signals without requiring any rigorous training to facilitate the output. A steady rise in the performance is observed. Implantation of the microelectrode follows a surgical procedure which eventually results in a selection of neurons. The recordings that consisted of larger samples of neurons are preferred to enhance the accuracy and reliability of the system. In the non-invasive method, the sensor chip is placed on the surface of the scalp. In this case, a dry electrode is used to sense the brain signals, and an ear clip is used as the reference electrode. This method is widely used, since it is simple to handle and wear without any surgical involvement as compared to the other two methods (Fig. 4).

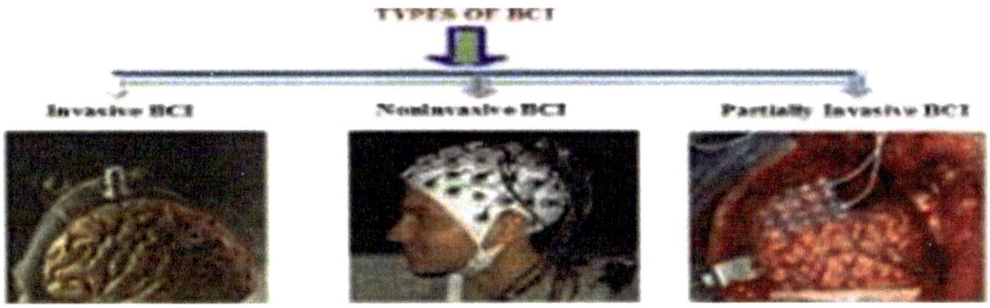

Fig. 4 Types of BCI

3 Materials and Methods

In this work, we used brain sense which is a commercial device to measure the brain activity occurring in the brain. Brain sense is a brain–computer interface (BCI) device used by a wide variety of audiences including students, researchers, and various wellness communities. It is a compact, single-channel, wireless headset that keeps track of your brain activity and translates EEG into meaningful data one can understand. It uses Neurosky technology. The value that will be detected using brain sense is blink detection. The signal collected by the brain sensor is divided into packets, and the packet data is transmitted via Bluetooth which is the wireless medium. The characterization of the eye blinks is generally carried out by peaks and relatively strong voltages. Based on the threshold values set, the eye blinks can be identified. The amplitude of the peaks in the signals varies with subjects of testing.

On the selection of the required action, the program sends a unique signal to the Arduino Uno board. Arduino Uno board is an open-source electronics platform, which has both hardware and software components that are easy to use. Moreover, it is equipped with several (13 digital and 6 analog) pins that can be configured as input or output based on the requirement of the user. This facility makes it possible to connect too many devices. The Arduino receives signals from the home management software according to the command selected by the user; it can be either switching 'on' or switching 'off' the appliance. Several sensor values are obtained by the input pins, which allows checking the device's operations and the actions that are selected by the user. The accomplished objectives of this works involve:

1. Switching the fan on and off.
2. Switching the bulb on and off.

The number of subjects taken for the experiment was four. Each individual was given two trials which consisted of switching on the fan and the bulb, respectively. It was observed that the average time taken by each subject to switch on the bulb and the fan was 22 s. The subjects had to be trained properly to blink to detect the brain wave. The brain sense has two electrodes, the reference electrode which was attached to the ear and the measuring electrode which was situated above the eyebrow. The subject had to sit straight during the measurement (Figs. 5, 6 and 7).

Fig. 5 Experimental setup

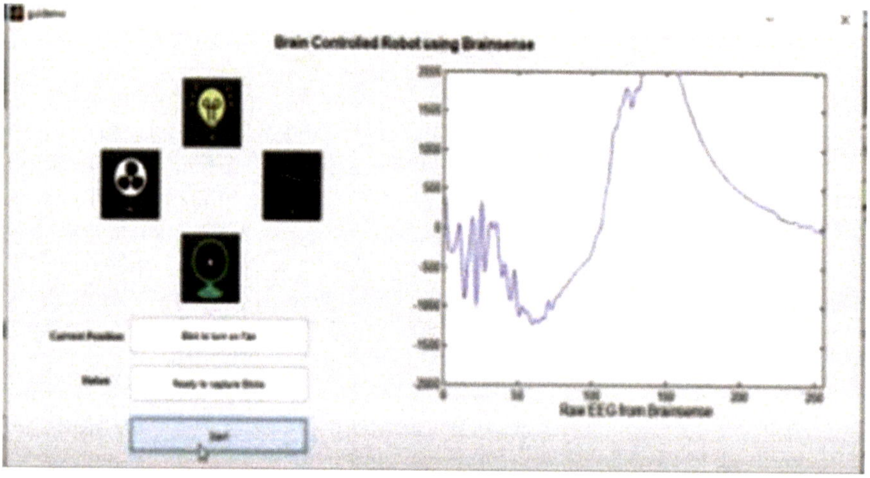

Fig. 6 Peak detection during eye blink

4 Information Transfer Rate

ITR stands for information transfer rate which is the standard method for calculating and determining the performance accuracy of communication for control systems, specifically for brain–computer interfaces [12]. It is defined as the quantity of information transferred per unit time and is usually calculated in bits/min. The standard formula to calculate ITR is given below:

$$B\left(\frac{\text{Bit}}{\text{Trial}}\right) = \log_2 N + P \times \log_2 P + (1 - P) \times \log_2\left(\frac{1 - P}{N - 1}\right)$$

$$Q\left(\frac{\text{Trails}}{\text{Min}}\right) = \frac{S}{T}$$

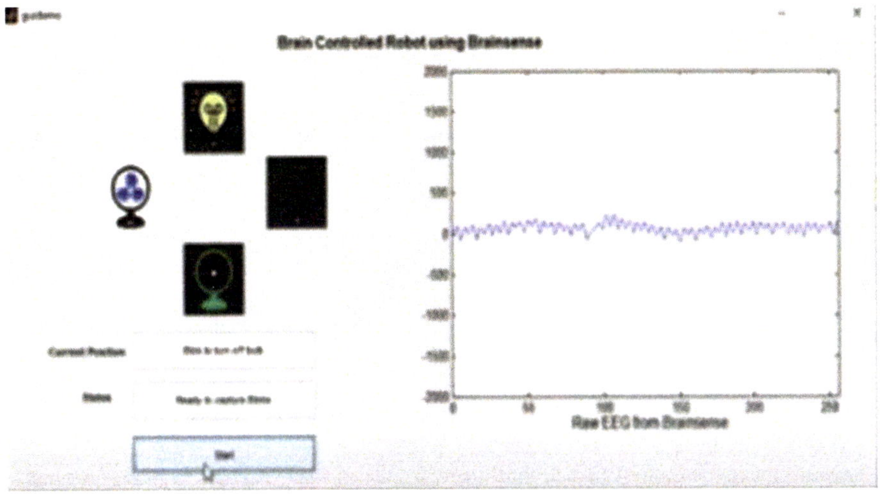

Fig. 7 Brain wave when eye is closed

5 Clinical Tests and Further Researches

Recent years have seen unprecedented growth and technological developments that are coming up with efficacious smart devices that can be managed by the home automation system [13]. The proposed model is used to control various home appliances, with main focus on the individuals with dysfunctional ailments. Systems have been developed that involve the usage of different audible pitches to facilitate and control home appliances. It was observed that the system can be developed further to include a much-expanded range of audible pitches which would allow users for controlling a greater number of appliances like garage door openers, door locks, thermostats (controlling the temperature of the home), and security systems among the many devices that can be controlled and operated.

6 Conclusions

An EEG controlled home automation system is proficient to perform numerous household functions such as switching on lights and fans. The controlling of such devices is accomplished purely on the criteria of eye blinking signals that is obtained from the user end. A couple of frequently used home appliances were operated in the system execution: an electric bulb and a ceiling fan. The experimental testing of the system was carried out on four subjects. Furthermore, each state was accompanied by two trials of these two applications. Accuracy in the completion of the objective and the response time taken by the system to complete each task were the two main parameters which were observed. These tasks were completed successfully.

References

1. W. Alrajhi, D. Alaoola, A. Albarqaw, Smart home: towards daily use of BCI systems (2017)
2. C. Holzner, C. Guger, G. Edlinger, C. Gronegress, M. Slater, Virtual smart controlled by thoughts (2009)
3. M. Shaifulrizal b Abd Rani, Wahidah bt. Manso, Detection of eye blinks from EEG signals for home lighting system activation (2009)
4. S. Sneha Pushpa, N.S. Chandrashekar, EEG based smart home automation using non-invasive brain computer interface approach, vol. 4, pp. 1–4 (2016)
5. S. Kharchenko, Y. Turovsky, R. Mescheryakov, A. Iskhakova, Restrictions of the measurement system and a patient when using visually evoked potentials (2019)
6. V.K.K. Shivappa, B. Lu, M. Solis, K. George, Home automation system using brain computer interface paradigm based on auditory selection attention (2018)
7. M. Agarwal, R. Sivakumar, Blink: a fully automated unsupervised algorithm for eye-blink detection in EEG signals (2019)
8. B.S. Sourab, H.S. Ranganatha Chakravarthy, S. D'Souza, Implementation of home automation using eye blink sensor (2014)
9. M.S. Munna, J. Shetara, S. Bhattacharjee, M. Zahed Hossain Bhuiyan, Design & implementation of a mind controlled home automation system (2017)
10. S.N. Abdulkader, A. Atia, M.-S.M. Mostafa, Brain computer interfacing: applications and challenges (2015)
11. K.J. Panoulas, L.J. Hadjileontiadis, S.M. Panas, Brain-computer interface (BCI): types, processing perspectives and applications (2010)
12. P.P. Roy, P. Kumar, V. Chang, A hybrid classifier combination for home automation using EEG signals (2020)
13. V. Ricquebourg, D. Menga, D. Durand, B. Marhic, L. Delahoche, C. Loge, The smart home concept: our immediate future (2006)

Fertilizer Recommendation System Using Machine Learning

D. Jayashree, O. Pandithurai, L. Paul Jasmin Rani, Praveena K. Menon, Mahek V. Beria, and S. Nithyalakshmi

1 Introduction

India is an immensely populated country and frequent changes in climatic conditions are required to stabilize the world's food resources. During drought conditions, the farmers face large difficulties. The type of fertilizer used by the farmers is the deciding factor in the production of the crop. Recommending the type of fertilizer may help the farmers to make better decisions to increase the growth of the crop. To improve the crop yield, we need to consider the type of soil, soil fertility, and also one of the major factors such as rainfall and groundwater availability. 15 agro-climatic regions are there in India which are divided based on the type of land.

The system is classified into different phases. The first phase is to collect the input data, i.e., soil type, crop name, moisture, temperature, and nitrogen, potassium, and phosphorus levels. The input dataset is pre-processed for missing values in specified

D. Jayashree · O. Pandithurai · L. Paul Jasmin Rani · P. K. Menon · M. V. Beria (✉) ·
S. Nithyalakshmi
Department of Computer Science and Engineering, Rajalakshmi Institute of Technology, Chembarambakkam, India
e-mail: mahekvberia.2018.cse@ritchennai.edu.in

D. Jayashree
e-mail: jayashree.d@ritchennai.edu.in

O. Pandithurai
e-mail: pandics@ritchennai.edu.in

L. Paul Jasmin Rani
e-mail: pauljasminrani.1@ritchennai.edu.in

P. K. Menon
e-mail: praveenakmenon.2018.cse@ritchennai.edu.in

S. Nithyalakshmi
e-mail: nithyalakshmi.s.2018.cse@ritchennai.edu.in

J. D. Peter et al. (eds.), *Disruptive Technologies for Big Data and Cloud Applications*, Lecture Notes in Electrical Engineering 905,
https://doi.org/10.1007/978-981-19-2177-3_66

range. The pre processed data is divided into testing and training dataset [4]. The dataset is further classified using three ML Algorithms which are Decision Tree, Naïve Bayes, and Random Forest algorithms. The algorithm with highest accuracy predicts the fertilizer.

Machine learning comprehends the algorithm based on three different learning algorithms. They are Supervised, Unsupervised, and Reinforcement learning algorithms. Each of these algorithms has its advantages and disadvantages.

Supervised machine learning is the algorithm that builds a mathematical model from the inputs and the desired outputs that form a dataset. Unsupervised machine learning is the algorithm that constructs a mathematical model from inputs that forms dataset but it contains no desired output labels. Semi-supervised machine learning is an algorithm that develops mathematical models from incomplete training dataset containing sample input that does not have labels.

2 Related Works

Dr. S. UshaKiruthika et al. [1]: This paper provides a technique for estimating nutrient content in the soil and suggesting a fertilizer. The system involves four stages, i.e., soil analysis, data pre-processing, data analysis, and recommendation of fertilizer. The IoT-based device utilizing NPK sensor is used to calculate the NPK ratio of the soil sample. The data is formulated into a dataset. The classification of soil quality is done using machine learning technique—clustering and classification. The fertilizer is recommended using the crop and fertilizer datasets. The system provides smart farming and precision farming using NPK value and also provides new opportunities in the Robotics field.

Jianwen Wang et al. [2]: The main focus of this paper is the recommendation of wheat-based fertilizer. A top-N recommendation is used for the large collection of items. The Nitrogen recommendation model is combined with the Nitrogen fertilizer effect function, and normalized difference vegetation index (NVDI) is established. This paper proves that Neural Network Interface is a useful method to determine the nitrogen top-dressing rate at critical growth stages.

Komal Bodake et al. [3]: In this paper, the system will analyze soil using sensors to increase the field. Fertilize recommendations will be based on the soil analysis report. Statistical predictive modeling is used for predicting and recommending fertilizers. For enhancing efficiency and smart farming, the Internet of Things is utilized. It results in an increase in profitability and productivity. This system evaluates soil conditions and performs accordingly to help farmers.

Nidhi H Kulkarni et al. [4]: This paper uses an ensembling machine learning technique for crop recommendation based on the soil type. This model uses Linear SVM, Random Forest, and Naïve Bayes as base learners. The dataset contains climatic conditions and soil-specific characteristics. This system classifies the dataset and recommend crop type, i.e., Kharif and Rabi. The average accuracy obtained is 99.91%. The system helps the farmer to select the crop accurately.

Vinod Sharma et al. [5]: This paper approaches various ML techniques for prediction of crop yield on the basis of macronutrient and micronutrient content in a dataset. This dataset is used for the crop yield prediction and the criteria present in the dataset are macronutrients, such as hydrogen level, organic carbon, emulsifiable concentrate, nitrogen, phosphorous, potassium, and sulfur, and micronutrients, such as zinc, iron, manganese, and copper that are present in the samples. ML algorithms are used to calculate the types of yield. This type will further evaluate the crop earnings. A classification is obtained from different classifier algorithms to estimate the yield of the crop. The system helps in prediction of yield based on soil composition.

Ankit Pawar et al. [6]: This paper proposes a procedure that uses a soil database, crop provided, and parameters such as soil obtained through soil testing. The soil testing provides a dataset which is given to the predicting system. The data collected will be used to create an ensemble model using Support Vector Machine and Artificial Neural Network as learners to commend a crop for a specific parameter with high efficiency and accuracy. This system prevents soil degradation and reduces the usage of chemicals on the cultivated land.

Qingshan Liu et al. [7]: The paper proposes an algorithm for estimating the functional properties of soil samples from some important properties of these soil samples. The algorithm used is Savitzky–Golay filter for pre-processing the present evolution of a single-layer perceptron feed forward network learning procedures for prediction. This technique is also called extreme learning machine.

Xiaonan Lu et al. [8]: This paper proposes a model to calculate the estimated amount of the variety of nutrients that are needed during the crop growth, recommending a suitable fertilizer and arranging fertilizer time decomposition of the model by method of object-oriented to comply with the requirements of C++ programming. A specific computer program to control local entities and the rules form introduced to the system under a certain strategy will produce an applicable recommendation. The system has provided with a convenient solution for the flexible integration between knowledge and system.

3 Proposed Work

The proposed system aims to help farmers in cultivating the crop for a better production. This system predicts fertilizer based on the following parameters:

- Crop Type
- Soil Type
- Temperature
- Moisture

- Nitrogen
- Potassium
- Phosphorus.

Three machine learning algorithms are used for predicting the fertilizer based on the parameters. The above algorithms is used to classify and exhibit confusion matrix, Precision, Recall, f1-score, average values, and accuracy percentage as output at the end.

The input from the user is advanced to Naïve Bayes, Random Forest, and Decision Tree machine learning algorithms [4]. This system will predict the fertilizer on the basis of the present input.

The system will also be provided in three different languages, i.e., English, Hindi, and Tamil.

The user will be able to browse fertilizer shops near their area using the built-in maps in the system.

Random Forest Algorithm: Random Forest is a user-oriented machine learning algorithm. It produces output, even without hyper-parameter tuning. The simple and diverse nature of this algorithm makes it the most commonly used algorithm. The random forest algorithm combines different algorithms that are of a similar type. This algorithm can be used for different problems such as classification and regression problems.

Decision Tree Algorithm: This is a non-parametric machine learning algorithm, used for regression and classification tasks. This is a tree-structured classifier. It is a graphical representation for getting all the possible ways to solve a problem/decision based on given conditions. The decision tree is a modeling technique for prediction that is mostly used in statistics, data mining, and machine learning.

Naïve Bayes Algorithm: **Naive Bayes** is a heuristic machine learning **algorithm** that can be applied in a comprehensive variety of classification tasks. It helps to build machine learning models that make fast predictions. It is a group of algorithms that share a general principle, i.e., every pair of features that are generally classified as unrelated to one another (Fig. 1).

4 Implementation

This is the sample dataset that is being used in this project. The data used to predict the fertilizer is based on 7 factors which are crop type, soil type, temperature, moisture, nitrogen, potassium, and phosphorous. Using these data, a machine learning model is created and the model is trained. This is later used to predict the fertilizer name. Table. 1 represents the sample dataset.

The data is cleaned by removing the NULL values so it is trained easily. The data is being divided into test and train by importing train_test_split. This is used to divide the pre-processed data into train and test based on the weight of the code

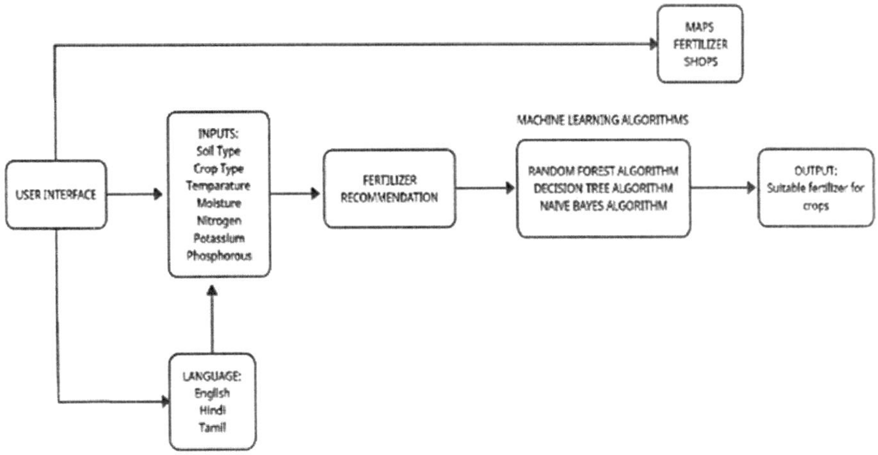

Fig. 1 System architecture

Table 1 Sample dataset

Temperature	Moisture	Crop type	Soil type	N	K	P	Fertilizer name
26	38	Maize	Sandy	37	0	0	Urea
34	62	Cotton	Black	7	9	30	14–35–14
32	41	Paddy	Clayey	24	0	22	28–28
34	53	Sugarcane	Loamy	12	14	12	17–17–17
30	27	Tobacco	Red	4	17	17	10–26–26

given. The data splitting into train and test is done in 0.2 and 0.8 which is 20 and 80%, respectively (Fig. 2).

The given data goes to the three machine learning algorithms (Decision Tree, Naïve Bayes, and Random Forest) [4].

The most accurate result is produced by the random forest algorithm as seen from the graph (Fig. 3), the accuracy of random forest is higher than Decision Tree and Naïve Bayes, and thus, the given data in the project will use the random forest algorithm for the fertilizer prediction (Fig. 4).

The Random Forest Algorithm suits for all types of data either huge or small for efficient prediction. This algorithm forms various decision trees and the number of trees which produce same prediction is checked. The count will be based on the number of votes and the trees which give same output. After which, tree with maximum count will be concluded as the output (Fig. 5).

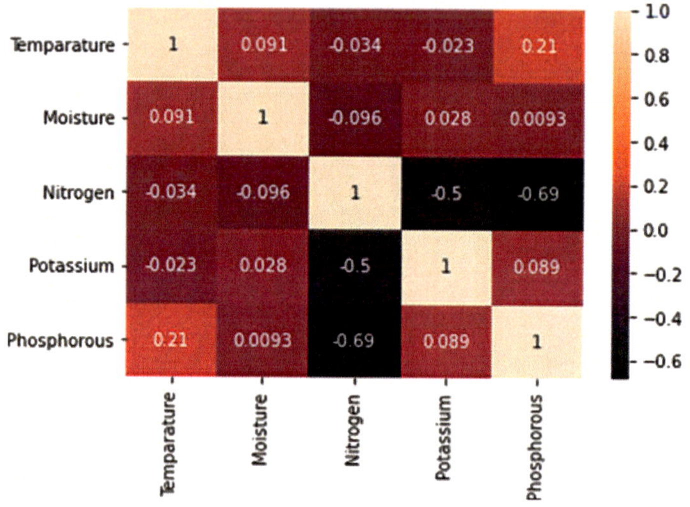

Fig. 2 Heatmap of the dataset

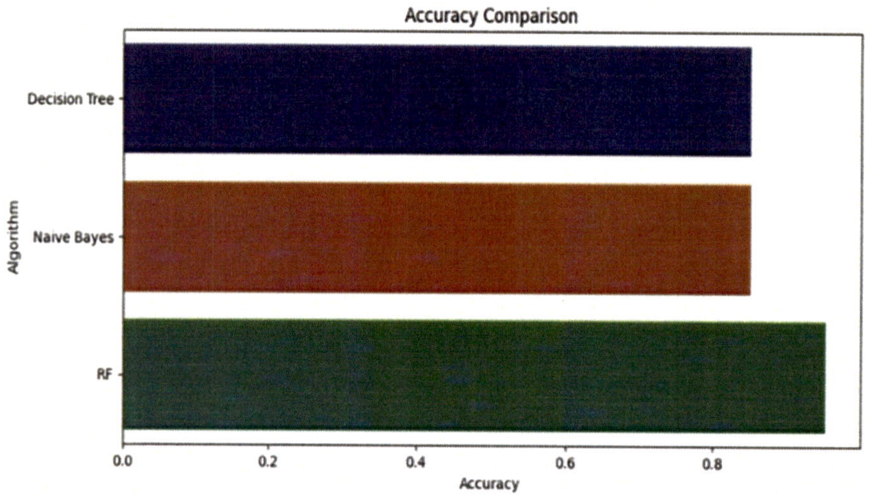

Fig. 3 Accuracy comparison

5 Conclusion

The paper proposes the use of machine learning algorithms to provide appropriate fertilizer. The results from the recommendation systems are optimized with respect to parameter consideration. This model will help the farmer to get details regarding the required fertilizer from different parameters. The proposed system will provide fast

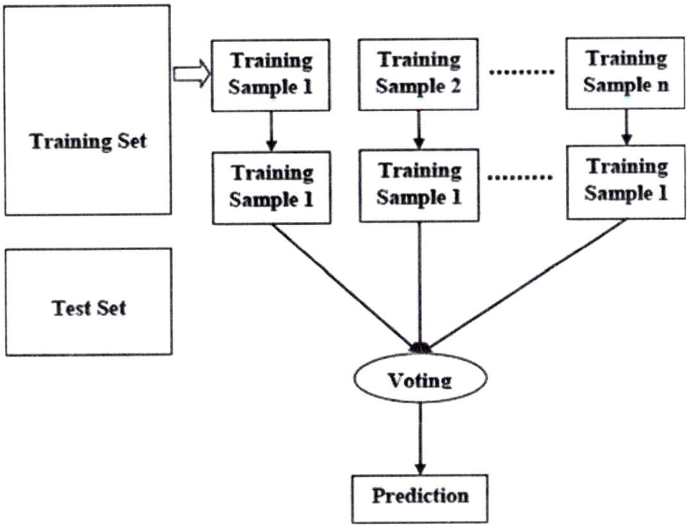

Fig. 4 Flowchart for random forest algorithm

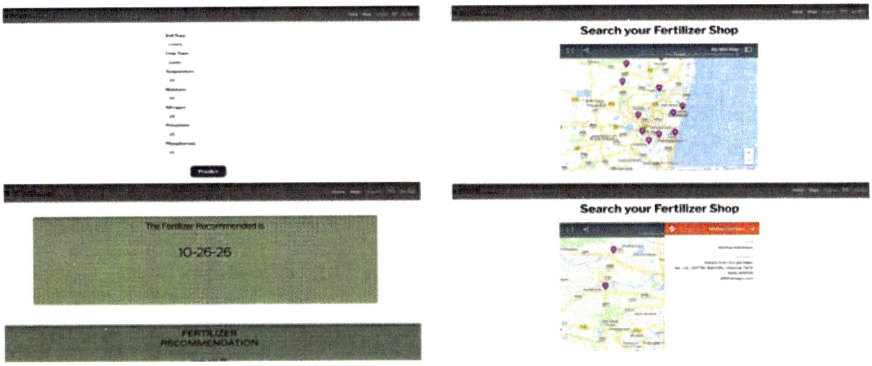

Fig. 5 Fertilizer recommendation system

and accurate results. Our system makes appropriate fertilizer recommendations for the crop in the most economical way. The work can be further extended by building a mobile application and help the farmers with crop diseases using image processing.

References

1. S. UshaKiruthika, S. Kanaga Suba Raja, S.R. Ronak, S. Rengarajen, P. Ravindran, Design and implementation of fertilizer recommendation system for farmers. Test Eng. Manage. **83**, 8840–8849 (2020)

2. J. Wang, Y. Zhao, L. Zhenhai, W. Yang, D. Duan, X. Xu, Recommendations for nitrogen fertilizer in winter wheat based on nitrogen nutrition index, in *2019 8th International Conference on Agro-Geoinformatics (Geoinformatics)*
3. K. Bodake, R. Ghate, H. Doshi, P. Jadhav, B. Tarle, Soil Based recommendation system using Internet of Things. MVP J. Eng. Sci. **1** (2018)
4. N.H. Kulkarni, G.N. Srinivasan, B.M. Sagar, N.K. Cauvery, Improving crop productivity through a crop recommendation system using ensembling technique in *2018 3rd IEEE International Conference on Computational Systems and Information Technology for Sustainable Solutions*
5. A. Sarwar, V. Singh, V. Sharma, Analysis of soil and prediction of crop yield (Rice) using Machine Learning approach. Int. J. Adv. Res. Comput. Sci. **8** (2017)
6. R.K. Rajak, A. Pawar, M. Pendke, P. Shinde, S. Rathod, A. Devare, Crop recommendation system to maximize crop yield using machine learning technique. Int. Res. J. Eng. Technol. (IRJET) (2017)
7. S. Arik, T. Huang, W.K. Lai, Q. Liu, Soil property prediction: an extreme learning machine approach, in *2015 Proceedings of International Conference*
8. Z. Ren, X. Lu, Design of fertilization recommendation knowledge base and application, in *2012 First International Conference on Agro-Geoinformatics (Geoinformatics)*

Land Use and Land Cover Mapping of Landsat Image using Segmentation Techniques

M. Mohith and R. Karthi

1 Introduction

Land cover data refer to surface cover on the ground for a region like forests, wetlands, grass cover, bare land, and water types. Land use refers to how humans use the land for different purposes agriculture, wildlife habitat, built-up area, water reservoir, etc. Land use land cover (LULC) can be determined by analyzing satellite and aerial images. The Landsat program provides us access medium resolution images with a resolution of 30 m per pixel. In this work, Landsat 8 satellite images are used to determine land use land cover using deep learning-based segmentation techniques.

Deep learning is a neural network-based machine learning technique that is used for image recognition [1], classification, and image segmentation [2]. Machine learning and deep learning models are used in different areas of research like biological cell analysis, movement analysis [3], scene detection, remote sensing [4], speech recognition, and many more applications. Recent studies indicate that image segmentation performed by convolutional neural networks is effective compared to the legacy image segmentation techniques [5]. Rithin et al. [6] proposed machine learning algorithms using decision trees, random forest, SVM, neural network algorithms to generate LULC maps from satellite images using pixel-based methods. The objectives of the work are to build a deep neural network models that generate a semantically segmented image for generating LULC map for the given image.

Heryadi et al. [7] used Sentinel-2 data covering region in Indonesia and built a CNN model to perform land cover classification. Image segmentation is performed to

M. Mohith · R. Karthi (✉)
Department of Computer Science and Engineering, Amrita School of Engineering, Amrita Vishwa Vidyapeetham, Coimbatore, India
e-mail: r_karthi@cb.amrita.edu

M. Mohith
e-mail: cb.en.p2cse19014@cb.students.amrita.edu

J. D. Peter et al. (eds.), *Disruptive Technologies for Big Data and Cloud Applications*, Lecture Notes in Electrical Engineering 905,
https://doi.org/10.1007/978-981-19-2177-3_67

identify different regions like the forest, grassland, water body, bare land, and settle ment. Agapiou et al. [8] performed image segmentation on Landsat 8 and Sentinel-2 datasets by using the Gram-Schmidt algorithm where the images are combined and used to identify objects like visible monuments, observe excavated regions, and detecting buried archaeological and historical objects. Potlapally et al. [9] performed instance-image segmentation using the mask R-CNN to segment remote sensing dataset covering rural area of India. Segmentation was done to classify cultivable land, roads, water regions, and areas containing built-up structures. Wu et al. [10] used PolSAR images to perform semantic segmentation based on FCN network with ResNet-50 backbone and U-Net with frozen layers. FCN with ResNet-50 backbone was able to outperform the U-Net model. Xu et al. [11] used high-resolution image and created a custom Res-U-Net deep learning model to perform image segmentation on the data. The model was able to extract buildings from diversified objects in each urban area. Hand-crafted features and artificially labeled ground truth data were used to train and test the deep learning model. Mehra et al. [12] compared many models like encoder-decoder network, feature pyramid network, U-Net, Deeplab V3+ , path aggregation network for image segmentation. The pyramid parsing network gave better results on synthetic aperture radar image with four classes-Urban, agriculture, water, and forest. Chantharaj et al. [5] used Landsat 8 image and performed image segmentation where SegNet model performed well compared to FCN, SegNet, GSN models. The authors classified the dataset into agriculture, forest, urban area. Khryashchev et al. [13] performed a comparative study on SegNet, TlinkNet, U-Net for satellite data and found that U-Net to perform better compared to other models. Many research papers for land use land cover mapping have used widely accepted deep learning models like U-Net, PSP-Net, SegNet, or modified versions of U-Net and SegNet for image segmentation.

The objective of the work is to build a deep neural network model that generates a semantically segmented image for generating LULC map for Indian terrains using Landsat 8 images. The challenges in adapting existing models for the study are many image segmentation methods which are applied to segment objects in high-resolution satellite images such as ships, cars, trees, buildings, and road networks. Landsat images are medium resolution images, and due to the heterogeneous landscape conditions represented in the image, the development of segmentation algorithms is a challenge. Three deep learning-based image segmentation algorithms are applied and tested for LULC classification. The paper is organized as follows: Sect. 2 describes the satellite images selected for the study, and Sect. 3 discusses the processing chain that is used for satellite image segmentation. This section discusses the pre-processing and deep learning algorithms with their architectures used for segmentation. The experimental setup and the results obtained when applying the DL-based segmentation techniques to the datasets are reported in Sect. 4. Section 5 summaries the conclusions from this study and presents further directions for research.

2 Study Area and Data Collection

The Landsat 8 scene used in this study covers the south-west region of India. The Landsat 8 scene LC08_L1TP_144053_20200127_20200210_01_T1 captured on 2020-January-27 is downloaded from the USGS Website and used for the study. The image data acquired are geo-referenced to "WGS84" ellipsoid with UTM projection. The indigenously created dataset is available in Kaggle [14].

3 Deep Learning-Based Segmentation Algorithm for LULC Classification

LULC classification process consists of several stages: data pre-processing, model construction, testing, and validation. Figure 1 shows the steps involved in building the LULC segmentation model.

3.1 Pre-processing

Pre-processing is done on the original image to obtain the input images and segmented images required to build the ML segmentation models. For performing the image segmentation, SWIR-1, NIR-1, and blue bands (bands 6, 5, 2) are considered in the study as these bands discriminate moisture content of soil and vegetation. For pre-processing, the bands are then loaded in the QGIS application where image is enhanced by setting the contrast enhancement parameter to minmax with mean ± standard deviation. Contrast and saturation are set to 50, and the brightness is

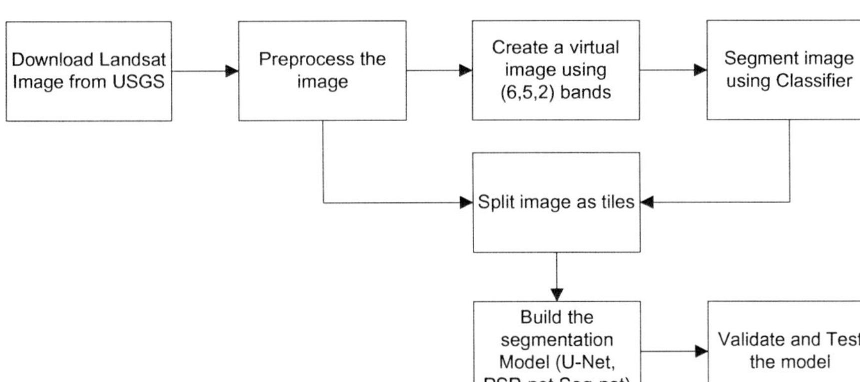

Fig. 1 Architecture for building land cover segmentation model

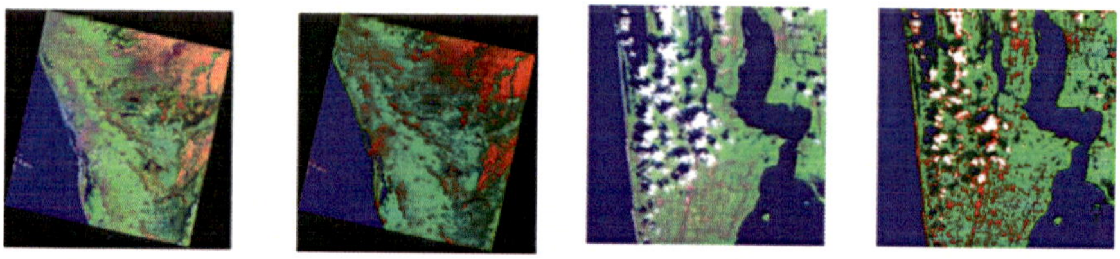

(a) Virtual Image **(b) Segmented Image** **(c) Sample Image slice** **(d) Segmented Image slice**

Fig. 2 Virtual image and image slices

reduced and set to −50. After performing the color correction on the image, the three bands are combined to create a new virtual layer. The combined image is exported as a TIFF image which is used as the input data for performing classification. Segmentation of input image is performed using semi-automatic classification plugin. In the ROI, points and polygons were created manually to specify the classes. The ROIs were labeled into six classes as water, forest, cloud, cultivable land, bare land, and unclassified. After the classes are set, the classification algorithm is run by choosing the minimum distance algorithm. The classified image is generated as a new layer which is the ground truth layer, segmented output for the input image. This segmented ground truth data are exported as a TIFF file. Figure 2a, b shows the original image and segmented results generated by the classification algorithm for the downloaded Landsat 8 image. Figure 2c, d shows a sample image slice and segmented image slice generated from the virtual image 2a and 2b.

3.2 Segmentation Algorithm

Deep learning algorithms used for segmentation with their architectures applied for LULC classification are discussed in this section. Three popular segmentation model such as U-net, PSP-Net, and SegNet are adapted for Landsat 8 image segmentation is summarized below.

U-Net architecture which was proposed by Ronneberger et al. [15] is used to perform segmentation on satellite images. The architecture of U-Net has a "U-shape" which contains two symmetric paths, such as encoder and decoder path. The network accepts images of any size and does not contain dense layers. In the contracting path, each process has two convolutional layers and the number of channels changes from 1 to 64. The number of convolution layers increases with respect to the depth of the model. In the expansive path, the image is upsized in further steps to reach its original size as given by the user. The architecture diagram is given in Fig. 3 [15].

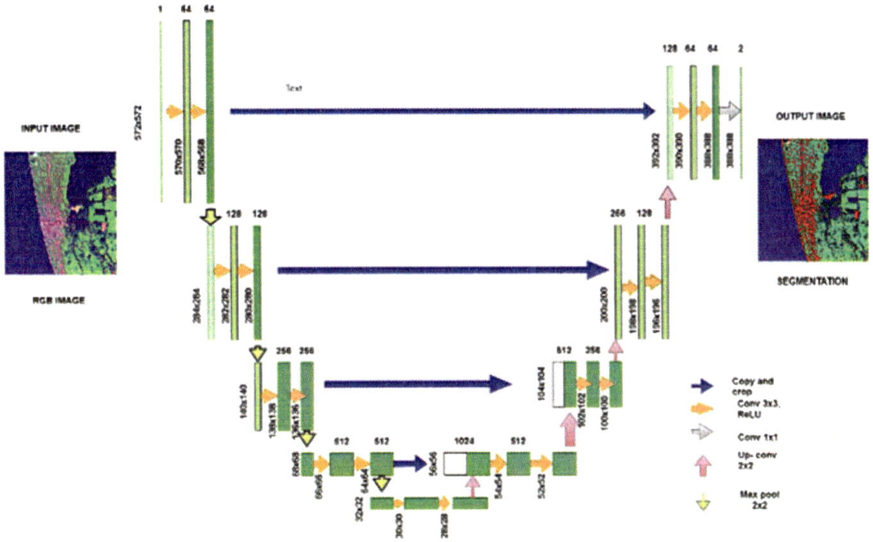

Fig. 3 U-Net architecture

PSP-Net which was put forward by Zhao et al. [16] used to perform segmentation of satellite images as shown in Fig. 4. A pre-trained ResNet model is used for PSP-Net. Dilated convolutions are used to extract the feature map. The feature map is one-eighth of the input image.

The pyramid pooling module is used to gather the context from the image input. Using the four-level pyramid, the pooling kernel in the model covers the entirety, half portions of, and small parts of the image. The pooling kernels are fused to form a global prior, and the global prior is connected with the original feature map in the final layer, and this is followed by a convolution layer, which is used to make the

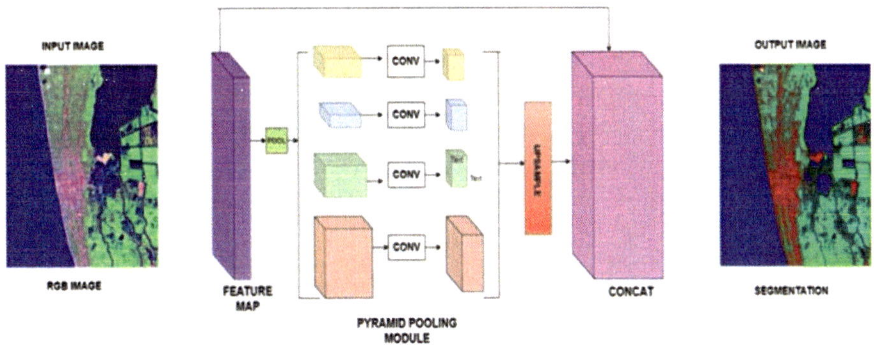

Fig. 4 PSP-Net architecture

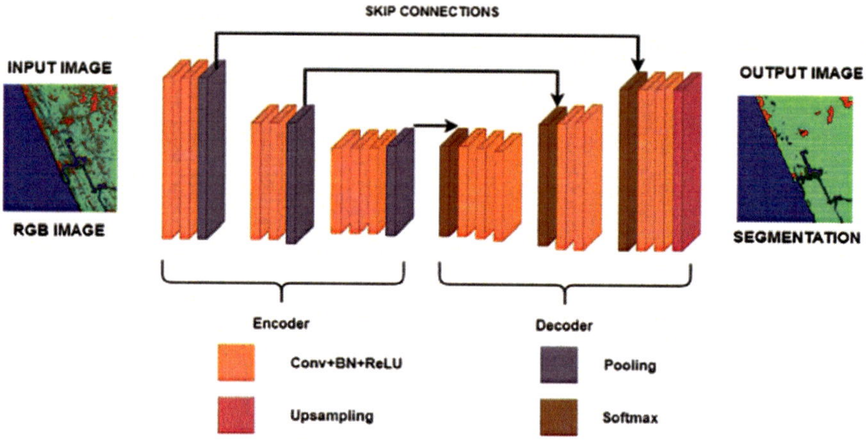

Fig. 5 SegNet architecture with the encoder-decoder network

final prediction of the image. PSP-Net architecture provides a highly effective global contextual prior for pixel level image segmentation.

SegNet which was brought forward by Badrinarayanan et al. proposed a deep learning network, with encoder-decoder architecture. SegNet has an encoder and decoder framework, followed by a final pixel-wise classification layer. Figure 5 [17] shows the SegNet model used for satellite image segmentation. The encoder network consists of 13 convolutional layers, and VGG16 network is used for classification of objects. Each encoder layer has a corresponding decoder layer, and hence, the decoder network also has 13 layers.

4 Experimental Setup

Semantic segmentation algorithm requires segmented ground truth data with precise borders for each entity/object in an image to build the model efficiently. The virtual image generated after combining bands 6, 5, 2 is used as the input image data for segmentation. The classified image generated by the minimum distance algorithm in QGIS is taken as the output or ground truth image for segmentation. The input image is of size 576×576 with 3 bands, and the classified image has dimensions of 576×576.

The input and the classified image are used to generate the train and test data. The input image and output images are sliced into 156 tiles to create the dataset for DL segmentation algorithm. Input image slices and the segmented output sliced image have pixel association as these images are geo-referenced. The images are sliced using a slice-image package in Python, and all the input tiles with their corresponding segmented ground truth tiles are stored in image slice and annotation folders. The resolution of all the generated tiles is 576×576 pixels. Figure 6a–d shows the input

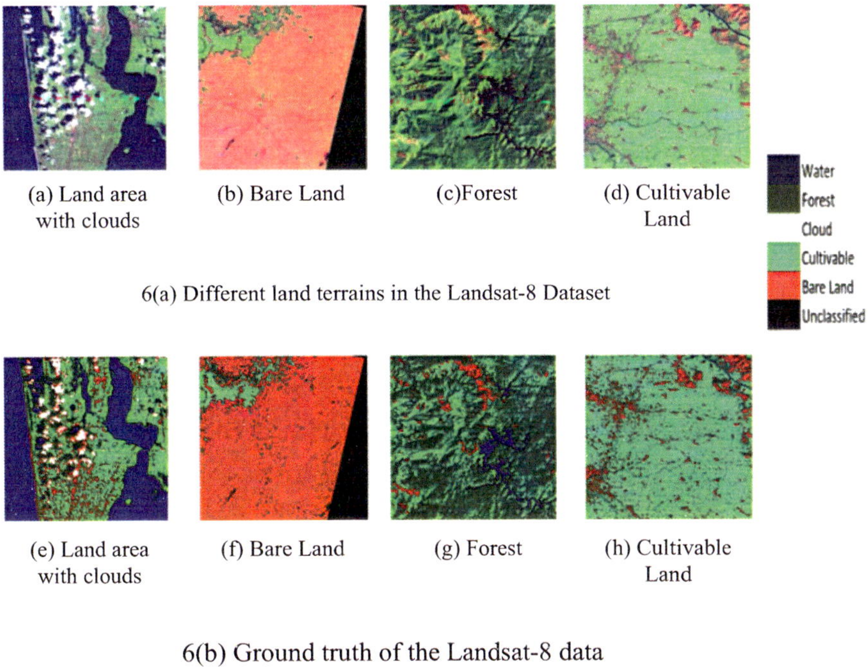

Fig. 6 **a** Different land terrains in the Landsat 8 Dataset. **b** Ground truth of the Landsat 8 data

image slices and the corresponding ground truth image slices with the majority class labels are shown in Fig. 6e–h. The newly created dataset was trained and tested using the deep learning models U-Net, PSP-Net, SegNet with GPU mode enabled in Google Colab Environment.

5 Results and Discussion

Land cover classification is a semantic segmentation problem where the algorithm identified the land cover class for each pixel in the satellite image. This is a multi-class segmentation problem. The datasets are labeled to belong to six classes as water, forest, cloud, cultivable land, bare land, and unclassified. During the training, 90% of samples were used for training, and the others are used for validation. For training, the batch size is set as 2, and step per epoch is set to 512 with Adam optimizer. The models used categorical cross entropy as loss function during training. The models were implemented using the Keras framework and tested in Google Colab Environment with GPU mode enabled. The input image size is 576 * 576, and around 140 image tiles were used to train the deep learning models. U-Net model took significantly

Table 1 Comparison of accuracy and m-IOU of deep learning models

Segmentation model	Accuracy (%)	m-IOU
PSP-Net	78.25	0.5948638
VGG-SegNet	82.33	0.5623949
U-Net	**92.35**	**0.696896**

Bold indicates the best values obtained by the algorithm

more time compared to SegNet and PSP-Net models. The following metrics were used to evaluate the performance of the 3 segmentation models.

5.1 Accuracy

Accuracy is the ratio of the number of correct predictions to the total number of predictions. Upon performing the image segmentation using different deep learning models, the U-Net model was able to give a better accuracy value as shown in Table 1. U-Net has the highest accuracy value of 92.35% compared to the other models SegNet and PSP-Net. The U-Net model was able to detect bare lands and agricultural lands correctly. However, the other models performed averagely when compared with the performance of the U-Net segmentation model. The tiny details of the dataset were preserved by the U-Net model. Many points were misclassified by SegNet and PSP-Net models.

5.2 Mean Intersection-Over-Union (m-IOU)

m-IOU is used for evaluating the performance of semantic segmentation algorithms. To compute the mean IOU, IOU for each semantic class is computed and then averaged over classes. IOU is defined from the confusion matrix as IOU = true_positive/(true_positive + false_positive + false_negative). From the results, we infer that U-Net model reached a value of 0.696896 and are able to distinguish other classes compared to PSP-Net and VGG-SegNet.

For qualitative comparisons, segmentation results produced by the U-Net, VGG-SegNet and PSP-net is shown in Fig. 7. The results show the ability of U-Net to segment smaller class variations in the image. VGG-SegNet and PSP-Net are not able to delineate and retain the smaller classes in the image. PSP-Net was used to perform segmentation on cityscape's data, which has objects of larger size. The SegNet model was trained and tested for road scene and indoor scene segmentation. PSP-Net and SegNet model were not able to provide more reliable results in the satellite image segmentation such as detection of the minor regions like clouds as this model was also designed to segment the indoor, road, and high-resolution images and scene. In Fig. 8, it shows the superimposed images of ground truth image and the

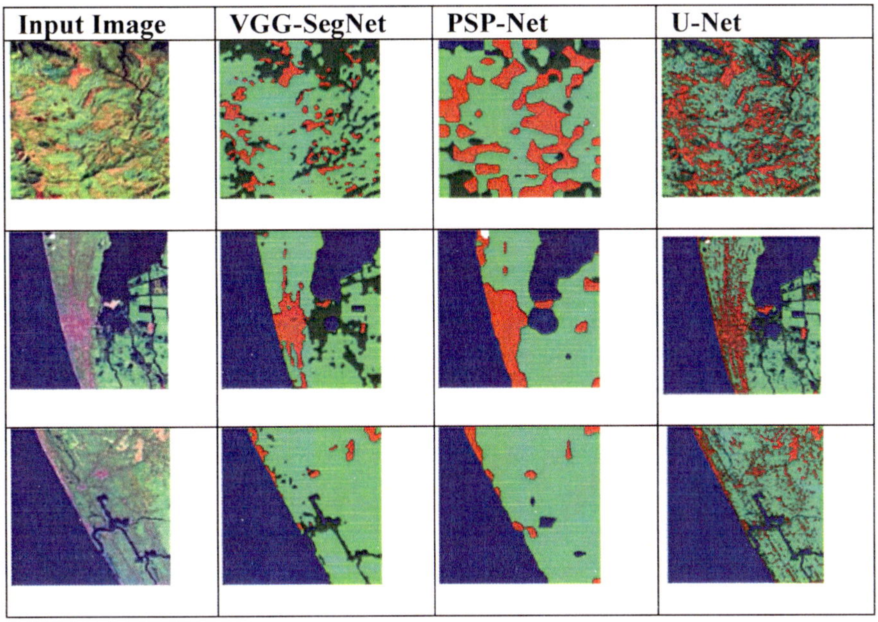

Fig. 7 Input image, segmented output of VGG-SegNet, PSP-Net, U-Net model

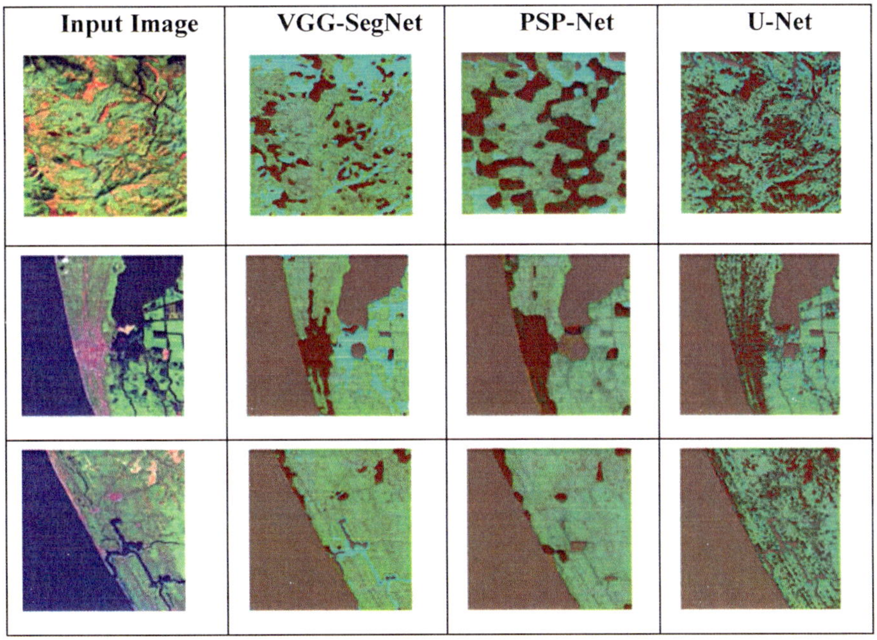

Fig. 8 Input image, superimposed images of ground truth and the image outputs by the different segmentation models (VGG-SegNet, PSP-Net, U-Net)

Table 2 Testing accuracy of different classes

Class	VGG-SegNet	PSP Net	U-Nct
Water	65.35736	80.2728	**91.4127**
Bare Land	28.1327	39.483	**58.9207**
Forest	**40.8832**	7.8210	39.4336
Cultivable Land	**90.5846**	86.1948	75.0207
Cloud	NA	**83.6601**	**83.6601**

Bold indicates the best values obtained by the algorithm

image generated by the segmentation model. The top layer is the ground truth with 20% transparency, and the bottom image layer is the image output generated by the models. From the images, we infer that the some of the portions were not classified properly by the models, especially the VGG-SegNet and PSP-Net model.

5.3 *Class Wise Accuracy of the Models*

Table 2 shows that the accuracy of the models in segmenting different classes. The images in the testing dataset were loaded, and the output was generated using the deep learning models. To analyze the models by class, accuracy metric was used. Initially, the classes were recorded from the annotations data, and the positions were recorded for a specific class. These positions are then again looked upon in the output generated by the segmentation model. The number of actual predictions for a specific class in the ground truth is noted. Then, the number of correct predictions by the segmentation model is found. Once this data are record, class wise accuracy is calculated as the ratio of the number of correct predictions to the total number of predictions.

From the results, we infer that U-Net model was able to perform well in majority of the classes. In the detection of water, U-Net performed well. However, in detecting the forest region and cultivable land, VGG-SegNet model has higher accuracy. PSP-Net was giving poor results in the detection of the forest regions. In the detection of cultivable land, VGG-SegNet was not able to detect the cloud regions in the test images correctly. Overall, U-Net model gave better results in the distinction of different classes present in the image.

6 Conclusion

In this paper, semantic segmentation of Landsat image is done using various deep learning models for the generation of LULC maps. Given a set of images and the corresponding land cover classes associated with each image, deep learning segmentation models are trained to segment the land cover. Three deep learning-based segmentation models such as U-Net, PSP-Net, and SegNet models are tested for segmentation. The accuracy of U-Net model is high compared to other models considered in this study. U-Net segmentation models use spatial information from data and provide accuracy of 92.35% during segmentation. The improved performance is due to the ability of U-Net model to understand the minute details in the dataset. U-Net model was designed for the segmentation of biomedical images. The intricate feature present in biomedical images has patterns similar to land cover in satellite images. These learned patterns have helped the U-Net model to perform image segmentation for satellite images with better accuracy. PSP-Net and SegNet model were not able to provide better results in the satellite image segmentation as the models were designed for scene segmentation applications. The amount of work done in segmentation of land use land cover classification in the Indian Landscape is very less. The Indian dataset has varying conditions of topography and higher heterogeneity, and there is a need to understand every infinitesimal detail in the dataset during ground truth mapping and segmentation. The experiment results are promising and deeper analysis with different models, and hyperparameter tuning can be tested for segmentation of Landsat data.

References

1. R. Sourav, V. Dayanand, R. Gowtham, P. Sini Raj, M. Senthilkumar, A layered approach to detect elephants in live surveillance video streams using convolution neural networks. J. Intel. Fuzzy Syst. (Preprint) 1–8 (2020)
2. M. Midhila, K.R. Krishnan, R. Sudhakar, A study of the phases of classification of liver diseases from ultrasound images and gray level difference weights based segmentation, in *2017 International Conference on Communication and Signal Processing (ICCSP)*. IEEE (2017), pp. 1582–1587
3. B.A. Sabarish, R. Karthi, String reduced dimensional representation of spatial trajectory for clustering. J. Eng. Sci. Technol. Rev. **13**(3), 115–125 (2020)
4. H. Song, Y. Kim, Y. Kim, A patch-based light convolutional neural network for land-cover mapping using Landsat-8 images. Remote Sens. **11**(2), 114 (2019)
5. S. Chantharaj, et al., Semantic segmentation on medium-resolution satellite images using deep convolutional networks with remote sensing derived indices, in *2018 15th International Joint Conference on Computer Science and Software Engineering (JCSSE)*. IEEE (2018)
6. K. Rithin Paul Reddy, S.S. Srija, R. Karthi, P. Geetha, Evaluation of water body extraction from satellite images using open-source tools, in *Intelligent Systems, Technologies and Applications. Advances in Intelligent Systems and Computing*, vol. 910, ed. by S. Thampi et al. (Springer, Singapore, 2020). https://doi.org/10.1007/978-981-13-6095-4_10
7. Y. Heryadi, E. Miranda, Land cover classification based on Sentinel-2 satellite imagery using convolutional neural network model: a case study in Semarang Area, Indonesia, in *Asian Conference on Intelligent Information and Database Systems* (Springer, Cham, 2019)
8. A. Agapiou, Evaluation of Landsat 8 OLI/TIRS Level-2 and Sentinel 2 Level-1C fusion techniques intended for image segmentation of archaeological landscapes and proxies. Remote Sens. **12**(3), 579 (2020)
9. A. Potlapally, et al.Instance segmentation in remote sensing imagery using deep convolutional neural networks, in *2019 International Conference on Contemporary Computing and Informatics (IC3I)*. IEEE (2019)

10. W. Wu, et al., PolSAR image semantic segmentation based on deep transfer learning—Realizing smooth classification with small training sets. IEEE Geosci. Remote Sens. Lett. **16**(6), 977–981 (2019)
11. Y. Xu, et al., Building extraction in very high resolution remote sensing imagery using deep learning and guided filters. Remote Sens. **10**(1), 144 (2018)
12. A. Mehra, N. Jain, H. Shanker Srivastava, A novel approach to use semantic segmentation based deep learning networks to classify multi-temporal SAR data. Geocarto Int. **37**, 1–16 (2020)
13. V. Khryashchev et al.,Comparison of different convolutional neural network architectures for satellite image segmentation, in *2018 23rd Conference of Open Innovations Association (FRUCT)*. IEEE (2018)
14. Landsat-8 South India Dataset, Accessed on February 1, 2021 [Online]. Available: https://www.kaggle.com/mohith4397/landsat8-south-india
15. O. Ronneberger, P. Fischer, T. Brox, U-net: Convolutional networks for biomedical image segmentation, in *International Conference on Medical Image Computing and Computer-Assisted Intervention* (Springer, Cham, 2015)
16. H. Zhao et al., Pyramid scene parsing network, in *Proceedings of the IEEE Conference on Computer Vision and Pattern Recognition* (2017)
17. V. Badrinarayanan, A. Handa, R. Cipolla,*Segnet: A Deep Convolutional Encoder-Decoder Architecture for Robust Semantic Pixel-Wise Labelling* (2015). arXiv preprint arXiv:1505.07293

Analyzing the Financial Soundness and Resilience of Select Small Finance Banks with RBI's Big Data

T. Augustus Immanuel Pauldurai, J. Anitha, and M. Vijila

1 Introduction

Digital media such as mobile and social technologies are becoming a standard platform for money transactions both domestic and international, where the quantum of data the banks consuming and producing is a stunning volume. Such big data throws challenges of both business and IT before the banks and at the same time; they create opportunities for growth, efficient risk and compliance management, keeping financial crimes at check thereby improving operational efficiency. Given the staggering volume of data, it is highly difficult to store them and process those using traditional database and software techniques. Mckinsey and Company's report dated September 19 2018 says banks have a strong foundation in using data. The study among more than 20 banks in Europe, Middle East, and Africa revealed that banks can improve returns using big data analytics by aligning analytics priorities to strategic vision, embedding analytics into decision making and workflows [1].

Banks by virtue of providing financial services and settlement services to the public generate a huge volume of valuable data being stored for many years. Reserve Bank of India has provided a communication backbone in the form of satellite-based INFINET using VSAT technology to the Banking and Financial Sector for upgrading

T. Augustus Immanuel Pauldurai
Department of Commerce and International Trade, Karunya Institute of Technology and Sciences, Coimbatore, India
e-mail: augustusimmanuel@karunya.edu

J. Anitha
Department of Computer Sciences, Karunya Institute of Technology and Sciences, Coimbatore, India
e-mail: anitha_j@karunya.edu

M. Vijila (✉)
Department of Mathematics, Karunya Institute of Technology and Sciences, Coimbatore, India
e-mail: vijilamoses@karunya.edu

J. D. Peter et al. (eds.), *Disruptive Technologies for Big Data and Cloud Applications*, Lecture Notes in Electrical Engineering 905,
https://doi.org/10.1007/978-981-19-2177-3_68

payment and settlement systems. The communication network has been designed and developed by the Institute for Development and Research in Banking Technology (IDRBT). In this process, RBI captures the data related to all intrabank and interbank digital transactions [2].

RBI, being the Central Bank as well as the financial regulator of India, shoulders diverse responsibilities such as monetary management, debt management, inflation management, financial market intelligence, banking regulation and supervision, financial inclusion, and overall financial stability and it has to collect and analyze relevant data for its surveillance and forecasting that will aid policy formulation and implementation. With the explosion of information gathering, computing capability, and analytical toolkits, in addition to the data collected through regulatory returns, large volumes of data, both structured and unstructured, called big data, sourced from consumer interactions in the digital world, also aids policymaking. Accordingly, RBI has set up a Banks' Data Warehouse under its Centralized Information Management System (CIMS) to store these data which are available to users for big data analytics. While some sensitive and confidential data can be accessed only by power users, other data can be accessed by general users for analytical and research purposes. RBI has named this as "Data Base on the Indian Economy (DBIE)" which can be accessed through https://dbie.rbi.org.in [3].

Various data on Indian economy are classified under the following seven major heads such as Real Sector, Corporate Sector, Financial Sector, Financial Market, External Sector, Public Finance and Socio-Economic Indicators.

The agricultural sector and micro, small and medium enterprise (MSME) sector are key sectors that drive the growth of the Indian economy. While the agricultural sector contributes to 20% of Gross domestic product (GDP), the MSME sector contributes to around 32% of GDP. Inadequacies of credit available and lesser attention to the poor and weaker section are significant problems faced by these sectors. In order to have more penetration into the unbanked areas and to provide these sectors easy access to credit and investment products, RBI came up with newer institutional variants such as Small Finance Bank (SFB) for promoting financial inclusion [4].

The journey of SFBs in India started in 2017 with just two SFBs and it has grown to ten as of 31.03.2020. At this juncture, a performance analysis of the SFBs is imperative to understand how they have performed so far and how far they are healthy, stable, and sound, so that they can take forward and implement the agenda of financial inclusion successfully. There are 10 SFBs as on the date of relevant data. In this work, four banks (AU SFB, Equitas SFB, Utkarsh SFB, and Ujjivan SFB) are selected on the basis of the following.

1. The banks having the total business of more than Rs.10000 Crores.
2. The banks which are in the business for a minimum of three years.

This paper aims at processing relevant data available from the cloud database maintained by the Reserve Bank of India and analyzing the performance and soundness of the SFBs in order to find how far they are resilient to achieve their objectives. The objective of this work aims with the following aspects:

1. To study the performance as well as the soundness of the select SFBs by analyzing six important parameters viz, business growth, total income, profitability, liquidity, capital adequacy, and asset quality.
2. To fit the trend line for the business growth and income.

2 Literature Review

Since the banking sector generates an enormous amount of unstructured and structured customer data while processing financial transactions, this huge amount of data needs to be stored in order to generate useful insights from it. Big data and cloud computing services [5] allow the bank to utilize the resources in an efficient manner with the help of data analytics. The data is utilized to derive the various functionalities ranging from sentiment analysis, feedback analysis, transactional analysis, security and fraud analysis, product cross selling, channel usages, customer segmentation, and profiling, reputational risk management, regulatory compliances management, financial crime management, and much more [6].

Sharma et al. [7] explores the importance of big data analysis in the banking sector. The working principles, advantages of applying big data in banking, the challenges behind it, the tools and technologies used, and the various application areas are discussed. Further, Huttunen et al. [8] discusses the advantages of big data, cloud computing, and advanced analytical tools like data science and machine learning in the field of the financial sector to obtain greater insights into their businesses.

Prantik Ray [9] has observed that small banks have the potential for financial inclusion but the performance of existing localized banks like Local Area Banks and Urban Co-operative Banks has not been satisfactory. So any preference for SFBs will require addressing critical issues relating to their size, numbers, capital requirements, exposure norms, regulatory prescriptions, and corporate governance. Kittu et al. [10] have observed that there is unmet potential demand lying in the rural areas and other unbanked centers for financial services and this must be tapped. It is worth trying on new institutional variants to tap the unmet demand. However, given the existence of differentiated markets and consumer groups, we may have to contextualize the concept according to the needs of the customers. With regard to the health of the differentiated banks, a balance between financial inclusion goals and long-term sustainability must be created.

Bhaskar Arora et al. [11] have concluded that financial literacy will define the success of financial inclusion. The SFBs need to support the Government in organizing workshops to impart financial literacy to people. As there is a positive correlation between financial inclusion and economic growth, a sturdy and financially inclusive economy in India can be built with the assistance of small banks.

3 Methodology

a. **Data Set**

The data for this work is extracted from the secondary data on Small Finance Banks by accessing the database of RBI through https://dbie.rbi.org.in and also from websites of the respective SFBs [12].

b. **Financial performance and soundness of the select SFBs**

The data obtained from the database are statistically analyzed with various parameters to show the financial performance and soundness of the select SFBs. Figure 1 shows the flow diagram of analyzing the financial soundness and resilience of SFBs.

iii. **Statistical analysis and inference**

In this part, we used Origin Pro statistical software to analyze a trend line for business and income growth rates, as well as statistical systems that allow for easy comparison of SFB parameters.

Figure 2 shows the trend line of the business growth of the select SFBS. Deposit and Advances constitute the business of a banking company. The business performance of any bank is measured by the bank's ability to mobilize deposits and efficiency in deploying the same. The business of the select SFBs has witnessed striking growth. The trend line predicts further growth. From the past trend, we observe that the growth is a sustainable one since the trend line never went flat or sloping downward.

Figure 3 displays the trend line for income growth. The earning capacity of the bank is measured by total income. The income growth of the SFBs over the period of

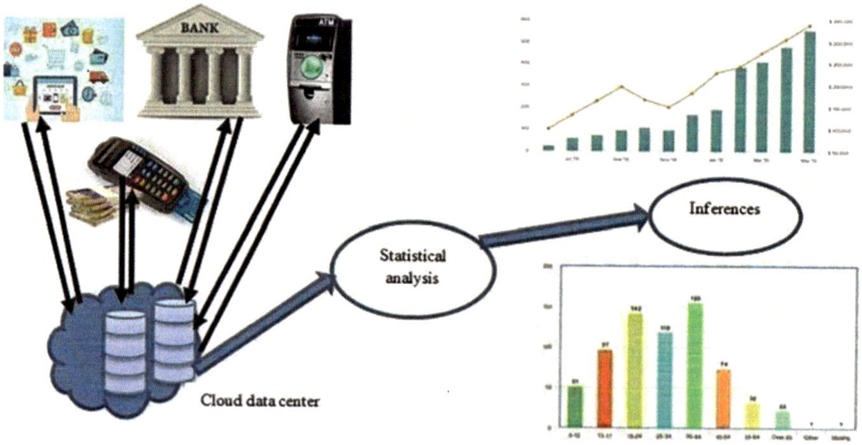

Fig. 1 Architecture for analyzing the financial soundness and resilience of banking system

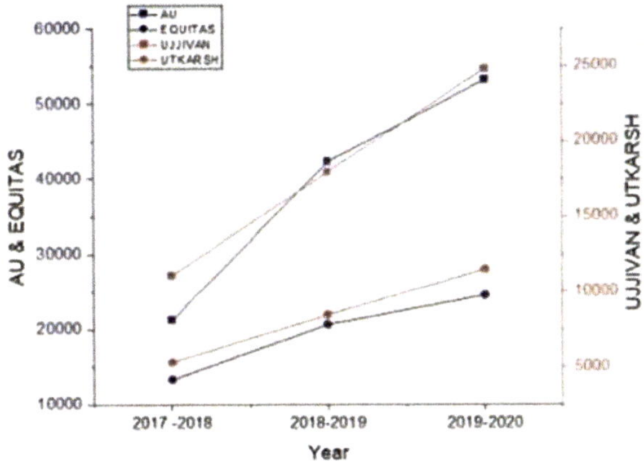

Fig. 2 Business growth for SFBs

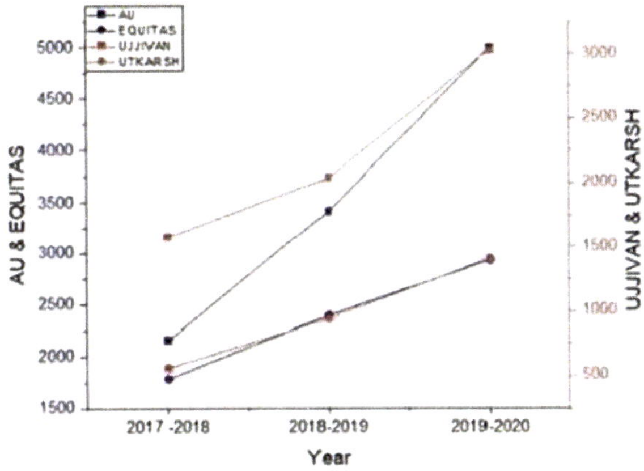

Fig. 3 Income growth for SFBs

three years is more than satisfactory. The trend line here also predicts further growth in the future and the past trend depicts that the growth shall be a sustainable one.

Figure 4a, b exhibit the profitability in terms of net profit and return on total assets, respectively. We have measured the profitability of the banks in terms of net profit in absolute terms and Return on Assets as a percentage of profit to assets. All the SFBs have reported a very handsome return on assets which indicate the efficient deployment of funds as well as sound credit management. Among the four banks, the RoA is the highest for Utkarsh SFB, followed by Ujjivan SFB, AU Small SFB, and Equitas SFB.

Figure 5 reveals the liquidity position of the banks as measured by the Liquidity Coverage Ratio. Liquidity means the bank's ability to meet its short-term obligations at ease. Liquidity Coverage Ratio (LCR) means the bank should maintain High-Quality Liquid Asset (HQLAs) to meet 30 days net cash outflows. It has been mandated for Indian Banks by RBI. Following is the road map laid by RBI for maintaining LCR by banks.

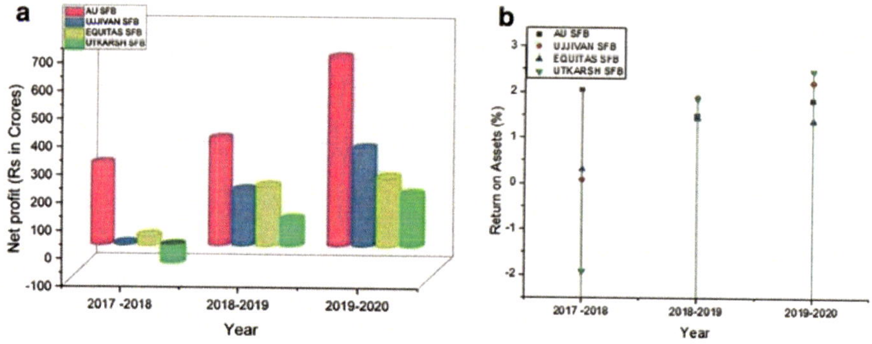

Fig. 4 **a** Business growth for SFBs. **b** Income growth for SFBs

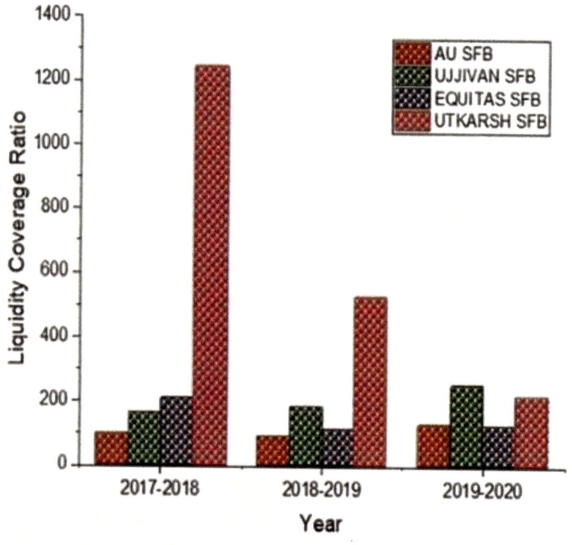

Fig. 5 Liquidity coverage ratio

Jan 1 2018	Jan 1 2019	April 2020	Oct 1 2020
90%	100%	90%	100%

The liquidity Coverage Ratio of all the four select SFBs is far above the stipulated level of 90% as of 31.03.2020, meaning these banks are capable of meeting their short-term obligations with very ease. However as far as Ujjivan SFB and Utkarsh SFB are concerned, if they reduce the ratio a little, it will improve their return on income.

Figure 6 exhibits the Capital Adequacy Ratio of the banks. Capital Adequacy Ratio (CAR) is a measure of a bank's financial strength by comparing its capital to risk-weighted assets. A higher capital ratio means the bank is less likely to become insolvent when an unexpected loss occurs. Capital structures of all the banks are sound as indicated by CAR of more than 20% against the minimum stipulated ratio of 8%.

Figure 7a, b visualize the level of Non-performing assets in terms of Gross NPA and Net NPA. Non-performing assets (NPAs) affect the profitability which in turn affects the internal accretion of the capital of any financial institution. The NPA ratio as measured by Net NPA and Gross NPA is very low which underlines healthy asset

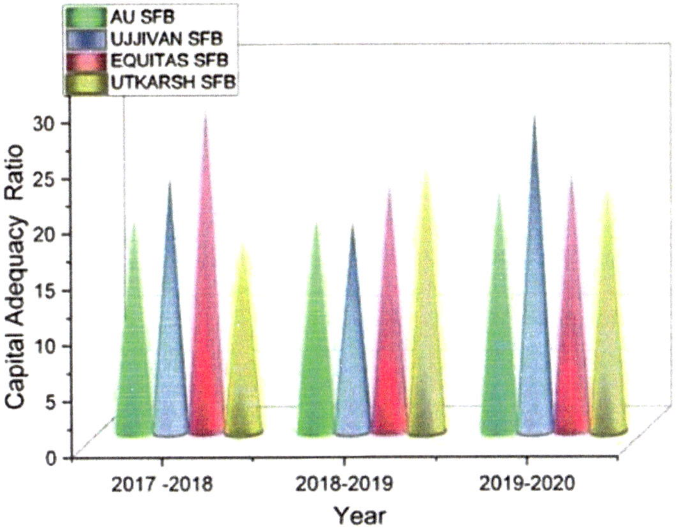

Fig. 6 Capital adequacy ratio

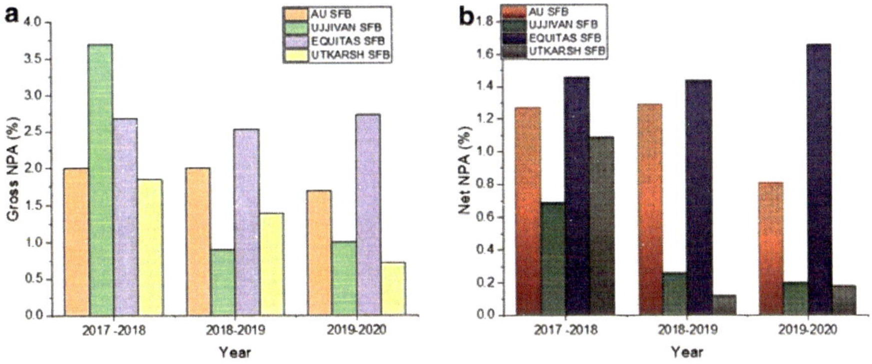

Fig. 7 **a** Gross NPA. **b** Net NPA

quality and lower loan defaults. The select SFBs are able to report very low NPA ratios because of their better management and supervision of the credit portfolio through the employment of social collateral of self-help groups. Their low NPA ratio also reflects better management of credit risk despite serving a small-sized clientele.

4 Conclusion

The huge volume of structured and unstructured data collected through regulatory returns, captured from consumer interactions in the digital world by RBI under CIMS and stored in data warehouse located in cloud. On statistical analysis of the data to study the profitability, liquidity coverage ratio, asset quality, capital adequacy, and the growth rate of business and income, it is found that all the select banks are performing well with regard to business and income growth. They are also financially sound as indicated by higher Liquidity Coverage Ratio and Capital Adequacy Ratio and keeping better asset quality. Within the banks, AU Small Finance Bank is leading in business and income growth, Ujjivan Small Finance Bank tops in profitability, liquidity and capital adequacy, and Utkarsh Small Finance Bank is the most efficient in management of credit portfolio. Trend lines were set for business and income and it is observed that the business and income growth of all the banks are not just positive but sustainable too.

It can be concluded that the select SFBs have performed financially well they are resilient enough to carry on the mission of providing banking services to the unreached and underserved section of the society and fulfill their main objective of financial inclusion. Further research can be made on more objectives like, how far they have achieved the social objectives like coverage of unbanked areas, making banking services, particularly credit, available to the un-served and underserved, etc.

References

1. Mckinsey & Company's Report Dated 19th September 2018
2. RBI Press Release—Apr 2018
3. RBI Report. https://dbie.rbi.org.in. Accessed on 30 May 2021
4. Small Finance Banks: Balancing Financial Inclusion and Viability—RBI Bulletin January 2021
5. N. Niazmand, The impact of cloud computing in the banking industry resources. IJISSM **4**(2), 436–440 (2015)
6. U. Srivastava, S. Gopalkrishnan, Impact of big data analytics on banking sector: learning for Indian banks. Procedia Comput. Sci. **50**, 643–652 (2015)
7. S. Sharma, T. Sharma, B. Kotak, A. Hasotkar, Big data analysis in banking sector. Int. J. New Technol. Res. (IJNTR) **5**(10), 69–72 (2019)
8. J. Huttunen, J. Jauhiainen, L. Lehti, A. Nylund, M. Martikainen, O.M. Lehner, ACRN J. Finan. Risk Persp. **8**, 16–30 (2019)
9. P. Ray, Small banking in India—issues and challenges, in *Proceeding of the First American Academic Research Conference on Global Business, Economics, Finance and Social Sciences*

(AAR16 New York Conference), New York, USA. 25–28 May 2016. Paper ID: N624 (2016). ISBN: 978-1-943579-50-1
10. R.S. Kittu, Smt. Mahananda, B. Chittawadagi, Role of small finance banks in the inclusive growth of our economy. Int. J. Appl. Res. Manage. Econ. **1**(4), 1–12 (2018). ISSN 2538-8053-INTL
11. B. Arora et al., Study on role of small finance banks for achieving financial inclusion in India. UNNAYAN: Int. Bullet. Manage. Econ. **IX** (2018) (special issue)
12. Annual Reports of AU Small Finance Bank, Ujjivan Small Finance Bank, Equita Small Finance Bank, Utkarsh Small Finance Bank

Enhancing Data Security for Sharing Personalized Data in Mobile Cloud Environments

G. Kavitha, P. Latchoumy, and A. Sonya

1 Introduction

Mobile cloud computing is a technology that allows mobile devices to act as clients and utilizes the infrastructure of cloud through the services of Cloud Service Providers [1]. The amount of data stored and shared by the mobile clients are increasing day by day. The users share their photos, videos, documents, and personal files to family and friends through the handheld mobile devices. Due to an increase in the number of mobile devices and the number of mobile applications available, storing and processing data on the mobile devices is very expensive and it also introduces excessive overhead if the compute and storage resources are consumed on local device. Further, the mobile device has limited resources in terms of processing power, battery lifetime, and memory.

The resources of the mobile device can be efficiently utilized by offloading the compute-intensive tasks to the cloud for execution. The offloading process saves the power consumed by the mobile device and also speed up the processing of application. As data is distributed in virtualized cloud storage, some of the key challenges faced in mobile cloud environments are data access control and data integrity in addition to the security issues due to user mobility. Therefore, the security issues in mobile cloud computing are categorized from two perspectives: the handheld mobile devices and virtualized cloud resources. The mobile device can be infected through

G. Kavitha (✉) · P. Latchoumy · A. Sonya
Department of Information Technology, B.S.A. Crescent Institute of Science & Technology, Vandalur, Chennai, India
e-mail: gkavitha.78@gmail.com

P. Latchoumy
e-mail: latchoumy@crescent.education

A. Sonya
e-mail: sonya@crescent.education

J. D. Peter et al. (eds.), *Disruptive Technologies for Big Data and Cloud Applications*, Lecture Notes in Electrical Engineering 905,
https://doi.org/10.1007/978-981-19-2177-3_69

malicious codes or viruses and cause data leakage or data corruption [2]. The privacy of the user's data is an important concern, and it is necessary to standardize mechanisms for ensuring data integrity during data sharing, data archival, and destruction of data. The existing cloud security methods cannot be directly applied to mobile cloud computing environments due to limited storage space and energy consumption of the mobile devices.

The cloud storage should ensure that users store and access their sensitive and personal data in a secure way through controlled accessibility of data [3]. The different security issues that needs to be addressed are authentication, authorization, and confidentiality of data [4]. To improve the data security and confidentiality, Advanced Encryption Standard (AES) is used for encrypting the data that is to be stored in cloud and then the encrypted data is uploaded in cloud. AES algorithm is the widely adopted symmetric encryption algorithm for performing encrypting and decrypting user's data. The shared information can be decrypted only by authorized persons with the key [5, 6].

When there is an increase in the number of users, the size of the data uploaded in cloud also increases and this causes congestion and delay. Hence, there is a necessity to reduce the computational overhead without compromising the security. The threats to shared user data are further reduced and to ensure maximum protection for the data stored in cloud, the proposed work implements enhanced AES through one-time download, share time expire, and self-destruction of shared files. In this proposed work, a lightweight data sharing technique is developed to process the data at a faster pace and achieve the performance comparable with local execution on the mobile device.

The organization of the paper is as follows: the related literature work is discussed in Sect. 2. The design of secure data sharing technique for Mobile Cloud Computing is presented in Sect. 3. The performance of the proposed enhanced AES algorithm is discussed in Sect. 4. The conclusions and future work are presented in Sect. 5.

2 Literature Survey

Data security in cloud systems has been studied by many researchers, and there are different approaches to securely store and retrieve the user's sensitive data on the cloud. Some prominent challenges with mobile cloud security are the mobility and location of the users, transfer of data over unencrypted channels, insider attacks, and limited resources of the mobile device. This section discusses about various cryptographic methods to ensure data security in cloud.

The unauthorized access to data that are offloaded in cloud and the possibility of insider attacks are discussed by Reza and Sonawane [7]. The authors have proposed the mechanism of text steganography to hide the information, thereby improving the privacy of data maintained on cloud. The user's information is embedded in images along with a key to provide additional security and maintain data confidentiality. An additional level of protection can be given by encrypting the hidden message, and this avoids any malicious user trapping the user's data. The user of the mobile device uses a mobile application to embed the image and key to create a stego-image and the data is retrieved from the image if the user enters a matched key. The disadvantage of this work is that this method is not more robust to malicious attackers and if the user wants to share a large volume of data, the processor load increases and it degrades the performance of the mobile device.

A managed system for deploying and running modern applications on the cloud is presented by Lee et al. [5]. In their proposed HEROKU cloud platform, the data security is implemented using Advanced Encryption Standard (AES). The plain text is encrypted to obtain the ciphertext, and the ciphertext is again encrypted using 128 key bits. The customer's apps encrypt the data for storing in cloud storage to make data more secure. The files are downloaded by users with the key, and the data files are decrypted and read by the user. The major drawback of AES in cloud is that it is vulnerable to attacks like brute-force and linear crypt analysis attacks. Further, the key exchange is done manually and the decrypted files are sent manually through e-mail for the recipients. The delay incurred for the encrypting process also increases linearly with the size of the data.

In [8], the authors present a mechanism for security of mobile device environments in mobile cloud using trusted platforms. The trusted execution environment is essential for maintaining the integrity of data while transferred over the Internet. The proposed work requires a crypto processor with trust storage to implement the cryptographic algorithms like AES. The trusted platform module uses an endorsement private key and hashing sequence for user authentication and remote attestation. To avoid illegal access to user's personal information, the remote attestation mechanism is not effective in cloud storage environment. Further, to avoid illegal distribution of digital contents, each user need to have a decryption key shared between the user's mobile device and the cloud and it has to be maintained in the cloud.

The security breaches on the data that is being shared among multiple users and illegitimate access to stored data are discussed by Vohra and Dave [9]. As the amount of data shared over the network is huge, a single authority approach will not be sufficient. Hence, the authors have proposed offline and online encryption and a partial decryption using a proxy server. In the resource-constrained environment, the concept of partial decryption saves the resources like battery life time and energy of the mobile device. The privacy of the data is still maintained as the decrypted file from the proxy server is still in the form of ciphertext and the final decryption done by mobile nodes. The attributes of the user are subject to change over a time period, and thus, a user granted access earlier may not be able to access the data after the changed attributes. It is found that the encryption time increases during offline encryption process of ciphertext generation which makes the proposed method less efficient.

Waseem et al. has discussed about the possible security threats, when user access the data from mobile device [10]. The threats can be minimized if the data is handled properly. Various encryption methods like AES, DES, RSA, and Blowfish are studied to ensure data security in cloud. To improve the mobile device security, the authors proposed a method to implement personal security token on the mobile device. Some of the attacks like DDOS and XML signature wrapping would prevent the clients from receiving the data from storage provider. The data encryption methods presented in this work enable the user to securely share the information but this is done at the cost of increased time.

An enhanced data security mechanism using AES is proposed in this work to resolve the issues stated in the above survey. The effectiveness of encryption depends on the type of encryption algorithm adopted and also the key strength. Therefore, it is necessary to consider the time required for encrypting and decrypting data and also the complexity involved in data handling and storage for designing an efficient data security approach.

3 Secure Data Sharing Using Enhanced AES

The security of data stored in the cloud storage is a critical challenge in Mobile Cloud environments. The existing methods of encryption use the same key size usually 128 bits for all type of data without considering the level of confidentiality required for the user's data. Sharing personal information among friends and family is possible because of cloud storage providers and access to this private information has to be granted only to authorized persons. In this paper, we have designed and developed an enhanced AES-256 encryption algorithm to maintain high confidentiality of user's data and prevent unauthorized access. The mobile cloud data sharing scheme presented in this work includes OTP verification, AES-256 encryption, one-time download, and automatic file destruction after time expiry. The architecture diagram of the proposed Secure Data Sharing Scheme using Advanced AES (SDSS-AA) is given in Fig. 1.

The proposed work is designed as three modules; the first module performs the user registration process. The user gets registered by providing all his personal information details like user name, e-mail ID, password, mobile number, gender, and date of birth. This data gets stored in the database for further authentication and authorization purposes. The database used in this work is RDBMS MySQL, and it is managed by SQLyog which is connected to the CloudMe file storage service. When the registered user needs to access the data, the user performs login by using his user name and password credentials to get access to cloud storage. To further confirm that only the authorized user is accessing the system, a one-time password (OTP) is sent to the respective registered e-mail ID. The OTP is entered, and the user can get access to the storage system.

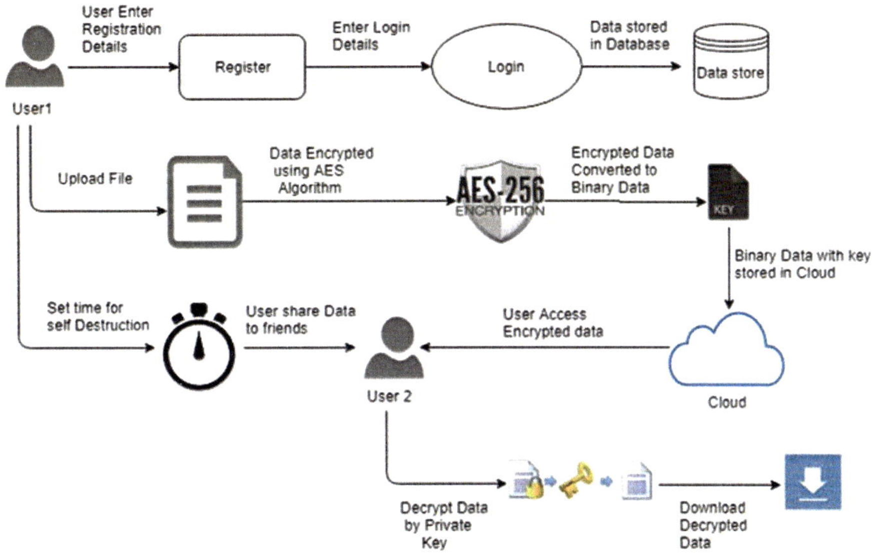

Fig. 1 Secure data sharing scheme using advanced AES

The second module performs data encryption using AES-256 and uploads the data in the CloudMe storage. AES is a symmetric key algorithm which uses the same key for encrypting and decrypting data. The user information which is in plain text is converted to encrypted ciphertext and the encryption key is the secret key for retrieving the stored information from the cloud. The SDSS-AA uses AES-256 to sustain maximum computation hardness for a much longer amount of time than AES-128. The data encrypted using a 256 bit private key requires 2^{256} possible combinations of keys for anyone to identify the data and hence cannot be cracked at any time. The encrypted data is then converted as binary data and uploaded in cloud, thus preventing unauthorized access to personal data. Now, the recipients can download the file once the sender shared the private key with them through e-mail for the corresponding file. The SDSS-AA scheme supports only one-time download. Hence, once the file is downloaded from the cloud, the decrypted file will be deleted at the recipient end. This one-time download option reduces the risk of insider attacks and also minimizes the chance for compromising data from the user side.

Apart from one-time download, this scheme also implements the concept of share time expire. The file that is shared has a time frame within which it has to be downloaded by the recipient. If the time expires, then the user who has shared the file will be notified that the file sent has not been decrypted and downloaded at the other end. The data owner has to resend the file again with a new time stamp. Similarly, the shared data can be accessed by the user's friends using the private key received through e-mail and get access to the data. The maximum number of attempts given to provide the private key for data decryption is three. If the recipient is not able to provide the correct private key for consecutive three times, he will not be able to access the file and this will be notified to the data owner via e-mail.

In the third module, the file sharing and file autolysis is performed. The user has all the files that are to be uploaded to the cloud, and he can select one or multiple files that has to be shared to friends. The user can choose one or multiple friends from the registered user's list and share the file to them. On choosing the file, he can set the time and the file will be available to the shared users only for that duration. If there is a time-out, then the file is self-destructed and the access for the file will be denied. If the recipient of the shared file has not downloaded the file within the time limit, it will be notified to the sender via a notification mail. Hence, file autolysis is enabled if the file is not downloaded within the specified time for self-destruction. The file autolysis is disabled once the file is downloaded and after one-time download, the decrypted file is not accessible to that user. If the file gets deleted by file autolysis, the intimation e-mail will be sent to data owner. Hence, the SDSS-AA scheme provides enhanced data security for the personal data shared among friends and avoids the possible security breaches.

4 Results and Discussion

This section presents the performance evaluation of the proposed SDSS-AA scheme with the existing AES-128 algorithm for data security. The performance metrics evaluated in this work are computational time for encryption and decryption and computational complexity in terms of the amount of resources required for secured data sharing. The resources considered in this work are battery lifetime and mobile storage capacity. In this work, we have used CloudMe for cloud storage. All the data files are encrypted and stored in CloudMe file storage service. The MySQL database is used to store the registration details of the users, and it is managed by SQLyog software. The databases are hosted in CloudMe and managed by SQLyog.

The users Key data items, name, e-mail ID, password, mobile number, gender, and date of birth, are received from the user from the registration page. Now this data gets stored in the cloud through SQLyog which is connected to CloudMe. When a user logs in, the system retrieves the data from the database as shown in Fig. 2.

	fileid	username	filepath	filename
	1	manimaran	C:/SeDaSC/manimaran/Original Files/aj.txt	raj.txt
	2	manimaran	C:/SeDaSC/manimaran/Original Files/hh.txt	hh.txt
	3	ismail	C:/SeDaSC/ismail/Original Files/hh.txt	hh.txt
	4	vignesh	C:/SeDaSC/vignesh/Original Files/CONFIDENTIAL.txt	CONFIDENTIAL.txt
*	(NULL)	(NULL)	(NULL)	(NULL)

Fig. 2 User information retrieved from SQLyog

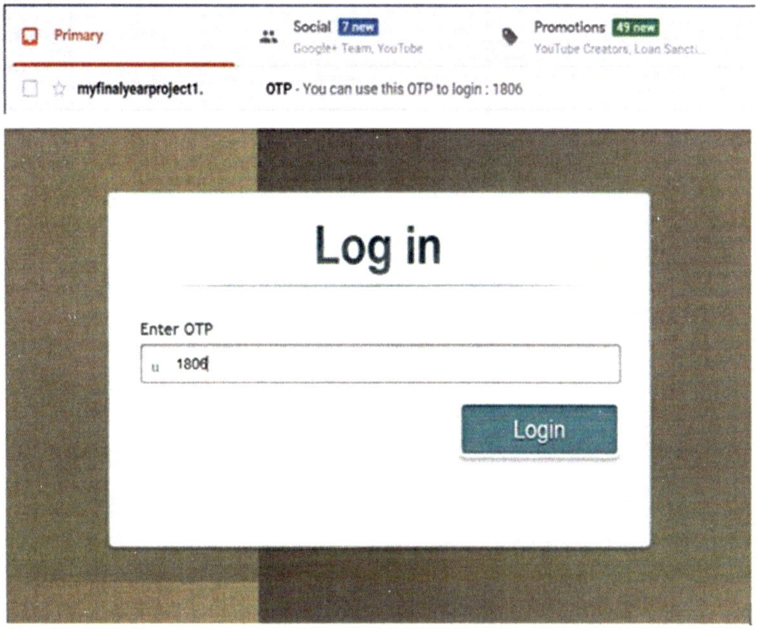

Fig. 3 One-time password generated and sent to user

To ensure the authenticated user is accessing the data, the system will send a one-time password to the respective registered e-mail ID of the user. The user enters the OTP and gets access to the data stored in cloud as shown in Fig. 3.

The user chooses the file and uploads the file to the cloud after encryption using AES encryption algorithm. The file will be stored in the connected CloudMe account in the respective folder as shown in Fig. 4. In SQLyog, the credentials of the file like owner name, time stamp, and file name are stored and maintained.

The process of sharing files to another user other than the file owner is done with a constraint imposed on time limit. The users requesting the files from cloud are viewed, and the file owner can select and grant access to other user. The file owner can set the time limit before which the requesting user has to download the file or else the file will be deleted. After this, the private key of the shared data will be sent to registered e-mail ID of requesting user as shown in Fig. 5.

After receiving the private key, the file can be downloaded and stored in local file system. If a wrong private key is attempted for more than thrice, the file will be deleted stating the reason as unauthorized access as shown in Fig. 6. After download, the file will be deleted and the decrypted file will no longer be available.

The file will be automatically deleted if the requesting user does not download the file successfully within the time limit given by the file owner and this is known as file autolysis implemented in this work.

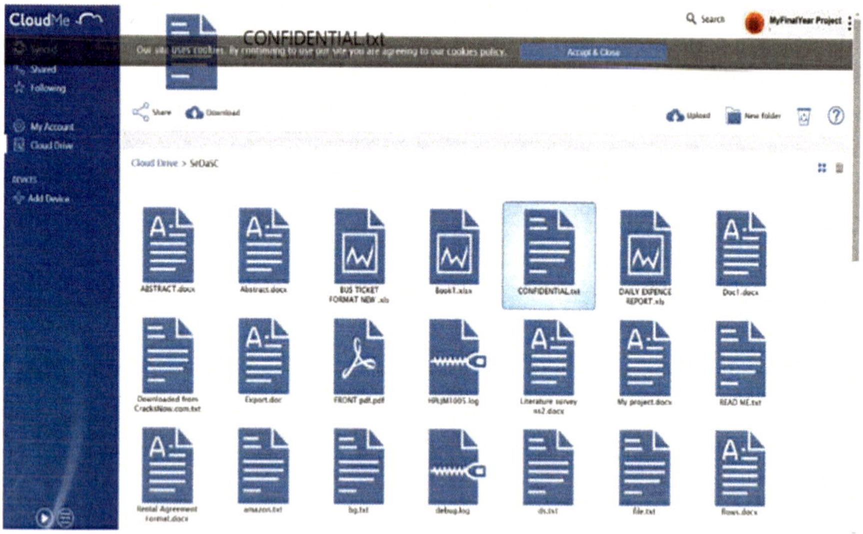

Fig. 4 Files uploaded in CloudMe

Fig. 5 Private key generated and sent to requesting user

Further, we have evaluated the performance of proposed secure data sharing scheme for mobile cloud devices in terms of Authorization and encryption/decryption overheads, respectively. The computational overhead of the proposed scheme is compared with AES, and it is clear that the computational complexity depends on the number of attributes. The proposed method has shown about 20% improvement in terms of computation time for authorization of user requests for secure file sharing and depicted in Fig. 7.

It is shown from Fig. 8 that there is an improvement in the computational time of encryption and decryption with respect to the number of attributes in the control policy. It is observed that the proposed method has provided a moderate reduction in the time of encrypting data while uploading files and decrypting data after the file has been downloaded by authorized user.

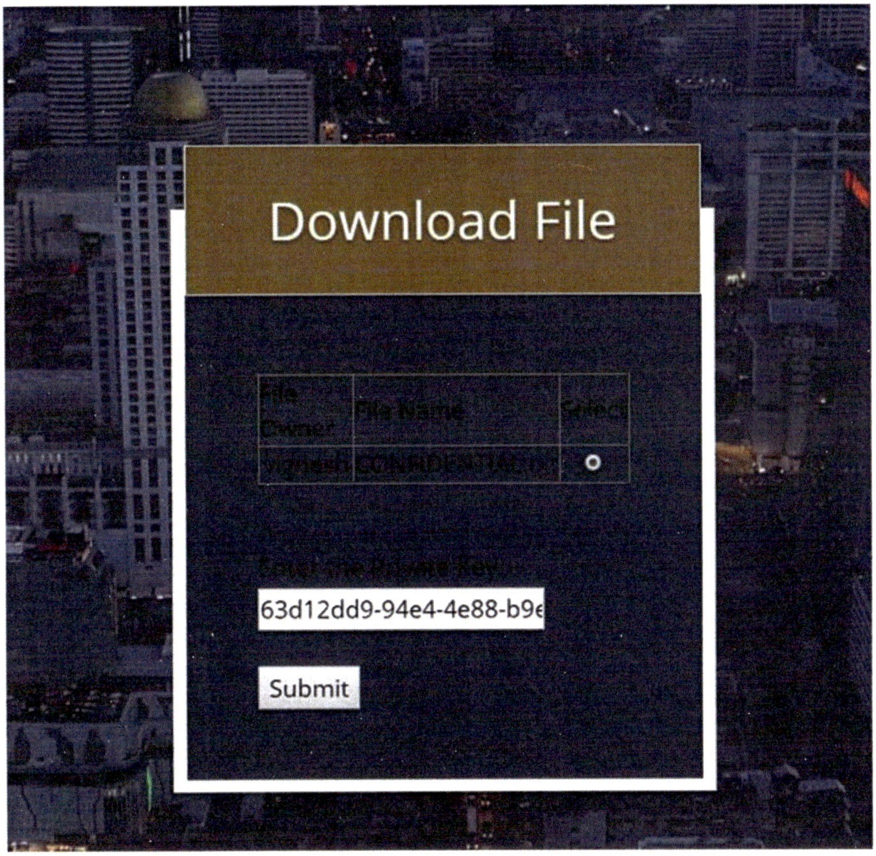

Fig. 6 File downloaded with private key

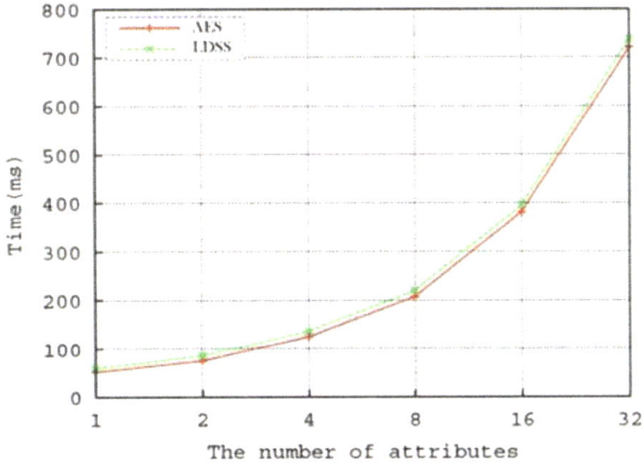

Fig. 7 Computation complexity for user authorization

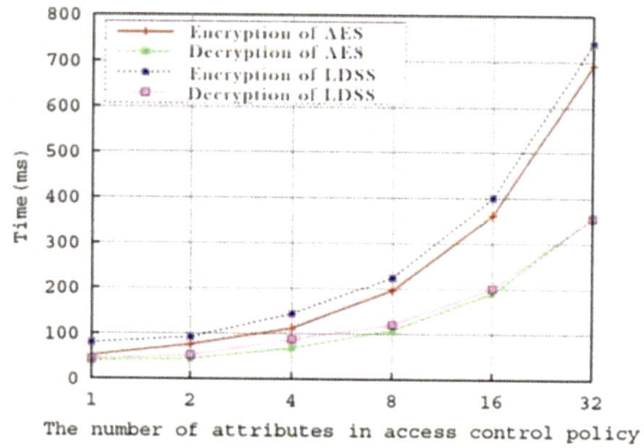

Fig. 8 File encryption and decryption time

5 Conclusion

Adopting cloud for data storage is gaining attention as it can be applied to small, medium, or large organization for efficient storage and retrieval with minimum latency and cost. When the data is moved to cloud, there are vulnerabilities and risks associated with data leakage, incomplete data deletion, stolen credentials, complex key management, and encryption services. This work is aimed to provide enhanced data security by sharing files to users with right credentials and deleting the file from cloud after it has been used once by requesting user. The encryption time and decryption time of the proposed method is comparatively less than existing AES algorithm. Multi-level authentication is used for granting access to users requesting files that are stored in cloud. The one-time password generated for every access securely provides data sharing and avoids data leakage. The work implements file autolysis to destruct the file after it has successfully downloaded one time by requesting user. The work can be further extended to allow multiple clients to share files simultaneously without compromising security.

References

1. G. Huerta-Canepa, D. Lee, A virtual cloud computing provider for mobile devices, in *Proceedings of the 1st ACM Workshop on Mobile Cloud Computing & Services: Social Networks and Beyond*, pp. 1–5 (2010). https://doi.org/10.1145/1810931.1810937
2. M. Baqer Mollah, Md. Abul Kalam Azad, A. Vasilakos, Security and privacy challenges in mobile cloud computing: survey and way ahead. J. Netw. Comput. Appl. **84**, 38–54 (2017)
3. N. Singh, A.K. Singh, Privacy protection mechanisms in cloud. Data Sci. Eng. (Springer) **3**, 24–39 (2018)

4. T. Bhatia, A.K. Verma, Data security in mobile cloud computing paradigm: a survey, taxonomy and open research issues. J. Supercomput. **73**, 2558–2631 (2017). https://doi.org/10.1007/s11227-016-1945-y
5. B. Lee, E.K. Dewi, M.F. Wajdi, Data security in cloud computing using AES under HEROKU cloud, in *2018 27th Wireless and Optical Communication Conference (WOCC)*, pp. 1–5 (2018). https://doi.org/10.1109/WOCC.2018.8372705
6. H. Ma, Z. Wang, Z. Guan, Efficient Ciphertext-policy attribute based online/offline encryption with user revocation. Secur. Commun. Netw. **2019**, 1–12 (2019). https://doi.org/10.1155/2019/8093578
7. R. Hassan, S. Madhuri, Enhancing mobile cloud computing security using steganography. J. Inform. Secur. **7**(4), 249–259 (2016)
8. S. Bouzefrane, L.V. Thinh, Trusted platforms to secure mobile cloud computing, in *2014 IEEE International Conference on High Performance Computing and Communications*, pp. 1068–1075 (2014)
9. K. Vohra, M. Dave, Multi-authority attribute based data access control in fog computing. Procedia Comput. Sci. **132**, 1449–1457 (2018). https://doi.org/10.1016/j.procs.2018.05.078
10. M. Waseem, A. Lakhan, I.A. Jamali, Data security of mobile cloud computing on cloud server. Open Access Libr. J. **3**(4), 1–11 (2016). https://doi.org/10.4236/oalib.1102377

A Decision Support System for Scheduling Lockdown in COVID-19 Pandemic

Abirami Sreerenganathan and R. Joshua Samuel Raj

1 Introduction

A lockdown is a limitation approach for people to stay in their home, especially on an account of high risks to themselves or other people if they can move and interact freely. When a new disease enters the populace with no immunity and advances at an exponential rate, resulting in a large number of deaths and imposing a serious burden on the front-line workers, carrying out a lockdown, where a physical intervention forces the populace to seclusion curbs the growth of the disease, hence allowing hospitals to confront patients given their capacity restraints and thereby reducing the number of deaths [1].

The cost of lockdown includes the lost services and merchandise from closing down economic activity. It also includes the lost utility from impeding an individual's freedom [1]. Other costs involve

- A surge in the number of deaths expected from unemployment.
- A surge in the tally of deaths from overdose and other deaths of despair.
- Lost educational opportunities.
- Further effects of school closure.
- Increased domestic violence.
- Lost non-COVID-19 medical assistance.

The lockdown has even caused the power sectors across the world to encounter ripple effects. Lockdown has recessed the utilization of electricity in the commercial and industrial sector, whereas the electricity demand has increased in the residential sector. An appropriate change in the daily load profiles needs to be adjusted to accommodate this change.

A. Sreerenganathan · R. Joshua Samuel Raj (✉)
Department of ISE, CMR Institute of Technology, VTU, Bengaluru, India
e-mail: joshuasamuelraj@gmail.com

J. D. Peter et al. (eds.), *Disruptive Technologies for Big Data and Cloud Applications*, Lecture Notes in Electrical Engineering 905,
https://doi.org/10.1007/978-981-19-2177-3_70

The label "stay home, stay safe" is usually meant for people to restrict themselves from moving out of their home excluding the case of emergency. The government has implemented lockdown, which allows the people to step up for isolation of themselves in their homes which can at least control a bit of spread of COVID-19 pandemic which is very busy in taking the lives of all our dearest ones. Precautionary measures like social distancing, wearing a mask, washing hands frequently are followed which ensures the prevention of the pandemic.

The paper can be outlined as follows; the COVID's impact on livelihood is discussed in Sect. 2, and round robin scheduling algorithm in Sect. 3. Section 4 elaborates on the methods and analysis done in our work. Conclusions and recommendations for future work are found in Sect. 5.

2 COVID's Impact on Livelihood

The economic and collective disruption caused by the pandemic is very destructive. Millions of mankind are at high risk of falling into extreme indigence [2]. When it comes to senior citizens, they are not just struggling with greater health risks but are also less capable to prop up themselves in isolation. General individual self-care and other safety valves against the COVID-19 pandemic can demand challenges for people with disabilities. For instance, some people with disabilities may have trouble in implementing measures to keep the virus at bay, which includes personal hygiene and persistent cleaning of surfaces and homes. Indigenous peoples are especially unsafe at this time owing to a notably higher percentage of diseases, absence of approach to necessary services.

3 Round Robin Scheduling Algorithm

It is a pre-emptive scheduling algorithm. It does not rely on any process characteristics but on a fixed slice of time (a short period). This small unit of time is known as time quantum. Time quantum commonly ranges from 10 to 100 ms.

Processes are arranged in the READY queue by adopting the FCFS policy. Every process in the queue is given a 1-time slice. The dispatcher dispatches, the process, with the help of a timer (for interrupting) if the process CPU burst exceeds 1-time quantum. If the CPU burst time of the presently running process exceeds the one-time quantum, the timer goes off, interrupting the process. The process is pre-empted by the scheduler, a context switch is executed, and the process is placed at the end of the READY queue. The process scheduler then selects the next process in the queue [3].

4 Methods and Analysis

In our proposal of deciding when to enforce a lockdown and organization of the same, we adopt a dynamic time quantum in our proposed dynamic time quantum round robin algorithm (DTQRR) as a part of the intelligence added to the RR for better outcome [4]. The processes represent the percentage of new cases in this dynamic time quantum is equal to the lockdown percentage of the respective week. Here, the processes do not arrive at the same time and hence have varying arrival times [5].

Here processes are the percentage of the population with active cases.

Steps in our dynamic time quantum round robin scheduling algorithm:

1. Start
2. Form the READY queue.
3. Schedule the processes using the round robin (RR) scheduling algorithm. Every process has a dynamic time quantum whose magnitude is the same as the process's execution time.
4. The process is then removed from the READY queue after its execution.
5. End.

We have considered the lockdown implementation from April 2021 in Karnataka [6]. The percentage of positive cases is tallied out of population strength of 50,000. The percentage of lockdown duration is computed based on an hourly fashion for a month, i.e., 1 month $= 24 * 30 = 720$ h.

The extent of a lockdown to be implemented is based on the following Table 1.

We can infer that an effective lockdown could have been put on immediately when there was an observed significant rise in the number of positive cases, i.e., when there was an occurrence of a steep slope as shown in Fig. 1.

The average of new cases in a week is calculated [7]. The respective percentage is calculated. Table 2 shows the corresponding statistics for April 2021.

Table 3 depicts the random approach that is followed for lockdown, and the Fig. 2 portrays the impact of the same.

Here, the lockdown has been in effect since April 27 [8].

Table 4 depicts the DTQRR approach that is followed for lockdown, and Fig. 3 portrays the impact of the same.

Table 1 Lockdown guidelines

Positive cases%	Hours per day	Number of days	Hours * days	Lockdown %
> 45	24	14	336	46.67
35–45	20	14	280	38.88
25–35	19	14	266	36.94
15–25	16	14	224	31.11
5–15	12	14	168	23.33
< 5	0	0	0	0

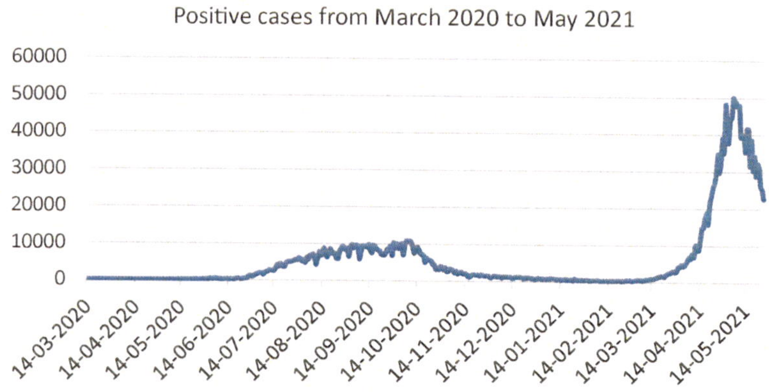

Fig. 1 Positive COVID cases analysis

Table 2 Statistics summary

Week	Average of confirmed cases	Ratio of population (50,000)	Percentage of population (50,000)	Approximate value of percentage
Week 1	5222.29	0.11	10.45	10
Week 2	8764.57	0.18	17.53	18
Week 3	18,184.29	0.36	36.37	36
Week 4	33,437.78	0.67	66.88	66

Table 3 Random approach for lockdown

Weeks	Positive cases (%)	Lockdown (%)	Arrival time (%)
W1	10	0	0
W2	18	0	23.33
W3	36	0	46.66
W4	66	10	70

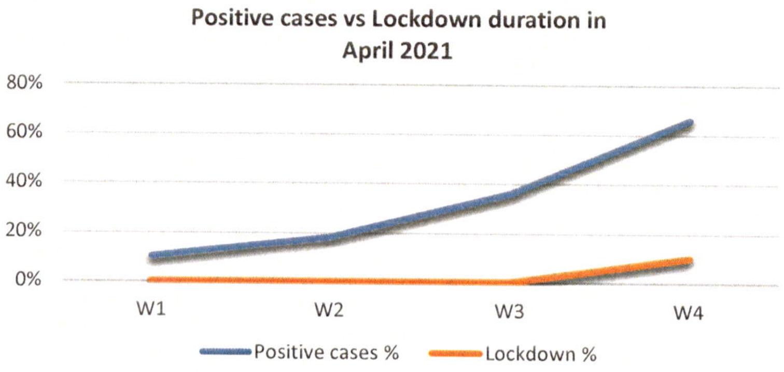

Fig. 2 Positive COVID cases versus lockdown

Table 4 DTQRR approach for lockdown

Weeks (processes)	Positive cases (%)	Lockdown (%)	Arrival time (%)
W1	10	23.33	0
W2	18	31.11	23.33
W3	36	38.88	46.66
W4	66	46.67	70

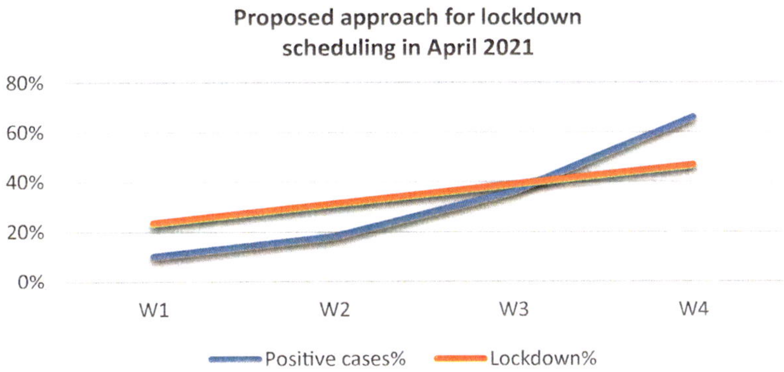

Fig. 3 Positive COVID cases versus lockdown as an impact of DTQRR approach

By scheduling the average cases of four weeks by adopting the dynamic round robin scheduling, the average waiting time is estimated.

The respective Gantt chart is prepared and shown in Fig. 4.

$$\begin{aligned}\text{Avg. wt. time} &= \{0 + (23.33 - 23.33) + (54.44 - 46.66) + (93.32 - 70)\}/4 \\ &= 7.775\%\end{aligned}$$

Figure 5 contrasts the approach for lockdown with respect to the COVID cases when considered as in randomness and also with our proposed dynamic time quantum round robin approach.

Scheduling the lockdown by employing DTQRR helps is in better management as compared to the random approach. In the random approach, the lockdown has been implemented at the very near end of the month, while the state has been witnessing a very high count of positive cases since the beginning of the month [9]. Whereas in

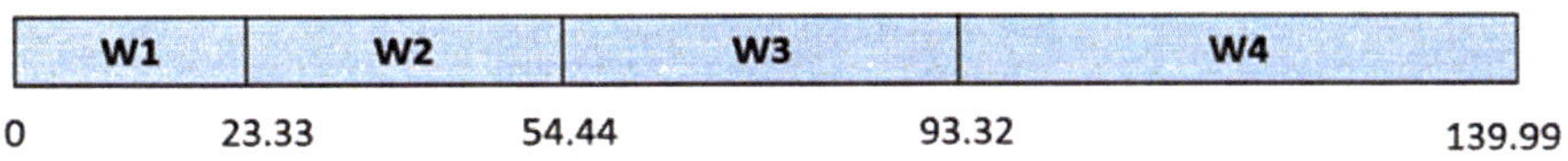

Fig. 4 Gantt chart for dynamic round robin scheduling

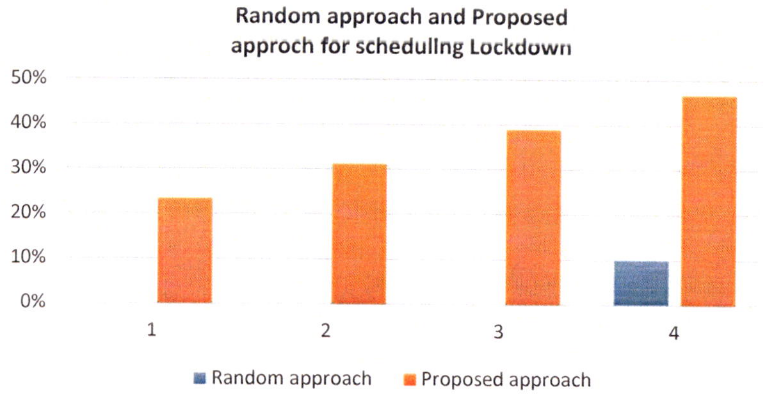

Fig. 5 Random approach versus DTQRR approach

our proposition, the lockdown is scheduled appropriately with regards to the count of cases.

5 Conclusion

We conclude that from the analysis that scheduling lockdown based on the DTQRR algorithm results in better management of patients and in handling the pandemic as can be seen from the comparative graphs. Our approach also ensures in preventing overcrowding in hospitals (additional patients will not get treatment when hospitals are running at full capacity, and ensuring critical patients can get urgent treatment in hospitals, whereas the mild cases who do not need immediate medical attention can get treated at homes. We also observe that the strategic lockdown and bring about dramatic decrease in cases also. In the near future, we plan to extend the work with artificial intelligence for detecting the after effects of COVID and their preventive measures.

References

1. Douglas W. Allen, *Covid Lockdown Cost/Benefits: A Critical Assessment of the Literature* (2021)
2. D. Howard, *Genetic Programming Visitation Scheduling Solution Can Deliver a Less Austere COVID-19 Pandemic Population Lockdown* (2020). arXiv preprint arXiv:2006.10748
3. A. Silberschatz, P.B. Galvin, G. Gagne, *Operating System Concepts*, 9th edn. (Wiley, Hoboken)
4. J.S. Raj, V. Vasudevan, Smart bacterial foraging optimization algorithm for scheduling in grid. Eur. J. Sci. Res. **94**(2), 253–260 (2013)
5. D. Barman, Dynamic time quantum in round robin algorithm (DTQRR) depending on burst and arrival time of the processes. Int. J. Innov. Technol. Explor. Eng. (IJITEE) ISSN: 2278-3075
6. https://en.wikipedia.org/wiki/COVID-19_pandemic_in_Karnataka

7. https://api.covid19india.org/
8. https://www.india.com/news/india/karnataka-lockdown-for-14-days-from-27-april-2021-live-total-shutdown-in-karnataka-bangalore-for-2-weeks-whats-allowed-whats-shut-full-lockdown-guidelines-4616088/
9. J. Panovska-Griffiths et al., Exploring overcrowding trends in an inner city emergence department in the UK before and during COVID-19 epidemic. BMC Emerg. Med. **21**(1), 1–10 (2021)

Simulation and Analysis of Intrusion Resilient Smart Metering System

Abhishek Patil, Adeesh Acharya, B. Y. Manthan, S. Nagasundari, and Prasad B. Honnavalli

1 Introduction

The smart meter is one of the most important electrical appliances in modern days. It is an electrical device that is set up at consumer's premises for measuring the consumption of various commodities such as electricity, water, and gas. Smart meters, when compared to traditional analog meters, provide lots of benefits. The smart meter provides a very detailed bill, and users can quickly analyze and check the readings. Bill can be generated without involving any middleman, unlike traditional meters. The transparent billing system [1] is one of the major sources of complaints and major demands of the customers, and as customers become proactive in making their energy consumption decisions, this demand will only increase. According to science direct, it is expected to have nearly 400 million installed smart meters by 2020.

The smart meter contains an LCD that displays the readings, and some user interfaces so that customers can interact with it. This setup also includes a current sensor (ACS712) for measuring the current consumed by the electrical appliance and a voltage sensor for measuring the voltage between the electrical appliances. By using these two values, power can be derived. The smart meter provides up-to-date usage information. In future, energy consumption charges may vary based on several different factors, such as time of consumption, source of electricity from which energy is being consumed, and many more. Hence, consumers will need to know about the time when the consumption will have a low tariff so that they can use their heavy load appliances at that time.

A. Patil · A. Acharya · B. Y. Manthan · S. Nagasundari (✉) · P. B. Honnavalli
Department of CSE, C-ISFCR, PES University, Bangalore, India
e-mail: nagasundarisuresh@gmail.com

A. Patil
e-mail: princeap173@gmail.com

J. D. Peter et al. (eds.), *Disruptive Technologies for Big Data and Cloud Applications*, Lecture Notes in Electrical Engineering 905,
https://doi.org/10.1007/978-981-19-2177-3_71

One of the major differences between the conventional electric grid and the smart grid is communication [2]. The components of any electric grid are the generation of power, the transmission of power, distribution of power with the help of substations, distribution centers, and finally, end customers. In the conventional electric grid, communication is from generation to substations, and in a few cases, it may also include distribution centers. In smart grids, there is end-to-end bidirectional communication between generation and consumption. Hence, this provides many benefits like monitoring the status of the different components present in the entire grid, quick identification of the source of problems that might occur due to a component failure.

Several problems can be caused if the communication between the central server and smart meter is not secure. The privacy of consumers, like their electricity consumption patterns, might be compromised if the attackers can modify the consumption details in the packets that are being sent by the smart meter to the central server. Then, the attackers might be able to cause significant losses to the consumers or the company providing the service. The idea of a smart grid is not new, but the full-scale implementation is not carried out since it is still facing many challenges that are yet to be overcome [3]. The main challenge is in the way of implementing smart grid security. Security is one of the major concerns of a full-scale smart grid.

The key contributions of this paper are mainly divided into four phases:

- Building the testbed of smart meters using Arduino mega, ESP8266, ACS712 current sensor, Raspberry Pi as a neighborhood data collector and sending the real-time data to a central server.
- Developing the simulation of smart meters using Docker containers.
- Perform DDoS attack and man-in-the-middle attack on the simulation.
- Implement prevention mechanisms for the MITM attack and suggest the prevention mechanism for the DDoS attack.

The rest of the paper is organized as follows. Section 2 outlines the detailed problem definition of this paper. Section 3 contains the literature review, to understand the relevant existing work. Section 4 outlines the detailed design of the system comprising a physical smart meter testbed with a simulation environment, the attack methodologies, and the prevention mechanisms for the performed attacks (DDoS, MITM). The results are depicted and discussed in Sect. 5. Section 6 provides the conclusion and the future work to be accomplished.

2 Problem Definition

A smart grid consists of several components which must be implemented properly to function efficiently. One component among them is the advanced metering infrastructure (AMI), which contains a smart meter as one of its components, which is the device that will be installed in every household. A smart meter is an apparatus that will be installed in households to measure the consumption of vendible like gas, electricity, water, etc., and communicate the details of consumption to the central server which manages the billing details for each customer.

There are many security challenges in implementing the smart grid on a large scale. The major challenge in implementing the smart grid is securing the data transfer between the smart meter and the central server [4]. If the data transfer is not secure, then this might lead to several problems like compromising consumer privacy, economic losses to the consumer or to the company. Large-scale security attacks can lead to the blackout of the entire grid. Not all smart meter-related attacks and their prevention can be done by building a testbed. It is necessary to have an addon simulation environment. Thus, it can be inferred that different kinds of attacks can be performed, and the prevention mechanisms can be implemented as well.

The simulation architecture is built with the same flow as that of the smart meter testbed. Users can spawn any number of smart meters and send data to the data collection center. The simulation is created in such a way as to accommodate the addition of any number of physical smart meters. A user interface is developed in the simulation environment so that the users can toggle the switch of any bulb or fan or any devices that are present in the house and monitor the energy consumption. The energy consumption is visualized as a real-time line graph for each smart meter according to the number of devices that are switched on.

Smart meters are easy targets for several reasons: They are deployed in large numbers and are installed in every household. Usually, there is a data collector in the neighborhood which collects the consumption data between various smart meters and communicates the same to the central server. These smart meters are usually connected to the smart meter through Wi-Fi, which is susceptible to many security attacks. For facilitating the large-scale implementation of the smart grid, these issues must be sorted out. An intrusion resilient smart meter must be developed. Henceforth, this work focuses on analyzing the resiliency of the smart meter.

The main drawback to the research of security on smart meters is the non-availability of low-cost affordable smart meters for doing security tests [5]. To overcome this, we intend to create a testbed for analyzing the effects of security attacks on a smart meter, then create a simulation environment of smart meter and test the effects of DDoS and MITM attacks.

3 Literature Review

The purpose of this literature survey is to provide a brief overview or a summary of the current research of our topic of interest, i.e., analysis of intrusion resilient smart meter. A brief outline of the smart grid architecture, the components constituents of the smart grid, the vulnerabilities in implementing it on a large scale, and the challenges involved in the development of an intrusion resilient smart meter are the area of focus. An overview of one of the existing methods of implementing a smart meter is also elaborated.

3.1 Overview of Smart Meter Architecture

According to Aloul et al. [6], smart meters are built with a long-life expectation. While designing the smart meter, the designer should ensure that it accommodates the support for software upgrades, updation of sensors, or the addition of new sensors.

Two main security functionalities that must be incorporated in the smart grid are the continuous availability of power supply and the protection of information. The main two components of the smart grid are the system component and the network component. System component includes devices like a smart meter, data collection center, service provider, electrical appliance, etc. The network component includes all the components that are related to communication. The network components consist of the home area network (HAN) and the wide area network (WAN). HAN consists of the network of all the smart meters from every house. This network of smart meters is in turn connected to the data collection centers or servers through WAN.

Vulnerabilities that are inherent in this architecture are physical security, low-cost equipment which does not last long, use of common protocols like Internet protocol (IP) for communication, commercial off-the-shelf hardware and software, and customer data security. The possible attacks are injecting false information, firewall breakdown, eavesdropping, malware spreading, DDoS, and MITM attack. A smart grid should provide two-way communication most efficiently. The server analyzes the user data and performs the required operations and provides useful information to the user.

The methods involved in mitigating the security issues are

(i) perform encryption and decryption operations on the user data.
(ii) ensure that the data transmitted by the smart meter are not modified.
(iii) use a new protocol for the transmission of data.
(iv) update the meter software regularly.
(v) ensure regular checks at the hardware level.
(vi) ensure that the user himself is not allowed to modify the readings.
(vii) provide an alert for sudden variation in consumption data.

3.2 Impact of DDoS Attack in Smart Grid

Mahjabin et al. [7] discuss the DDoS attack which causes a threat to a lot of large server stations which can be performed very easily by any novice user. There are many powerful tools available using which DDoS attacks can be performed. The most popular tools are LOIC and HOIC. The impact of the DDoS attack can be analyzed by performing it in a simulation environment. One of the most popular simulation environments is "NESSI" which provides various programmable APIs. NESSI is a java-based simulation environment. The environment contains three main components, i.e., client, server, and bot. The client acts as a smart meter and performs all the smart meter-related functions, the server acts as a data collection center that receives the data from the client, and bot here is a system fully loaded with DDoS malware which when triggered creates havoc. The DDoS attack is initiated by sending series of UDP packets to the simulation environment continuously by employing multiple bots. Initially, the application checks for the incoming requests on the open port and replies with "destination unreachable" packets. Since there are several incoming requests to the application server, the application server will spend most of the time sending ICMP replies.

4 Smart Grid Design

Smart meters have a processing unit, a controller node, and storage. There are different protocols through which this data communication takes place between servers and various other smart meters (Fig. 1).

The entire smart grid architecture and the data are depicted pictorially in Fig. 2.

The contributions of the paper are organized into four stages:

- Build a physical smart meter testbed.
- Develop AMI simulation environment.
- Perform DDoS and MITM attacks on the simulation environment and physical smart meter testbed.
- Implement prevention mechanisms for the MITM attack and suggest the prevention mechanism for the DDoS attack.

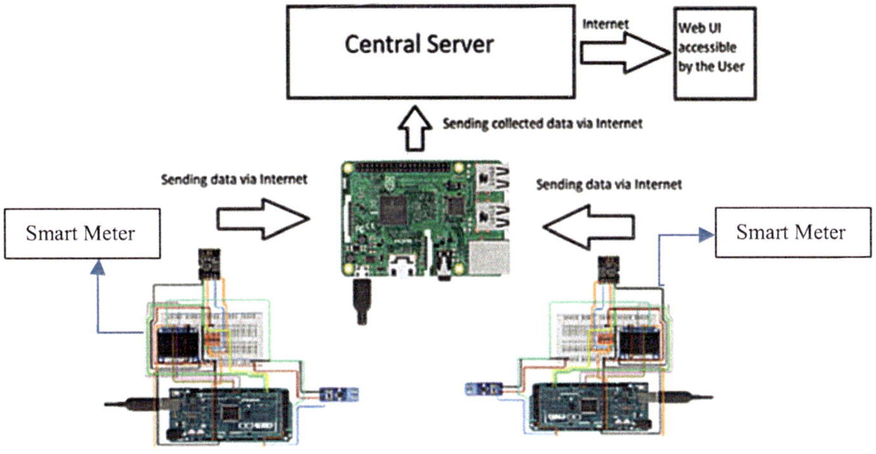

Fig. 1 AMI architecture

Fig. 2 Smart meter testbed

4.1 Building Smart Meter Testbed

The smart meter testbed is constructed with components such as Raspberry Pi 2, ACS 712 current sensor, voltage sensor (ZMPT101B), Arduino mega 2560, ESP8266 Wi-Fi module. The testbed also involves an electrical setup to measure load, attack tools, and operating system with security testing facilities like Kali Linux and Parrot. Arduino is used here as a processing unit, where power calculations are computed. The current sensor, ACS 712, is used to measure the current consumed from the electrical appliances, and LCD is used to display the usage readings.

Arduino mega with a built-in ESP8266 Wi-Fi module is used for the construction of a smart meter testbed. The smart meter (Arduino) records the readings and transmits them to the ESP8266 module which in turn forwards it to a data collection center (Flask server). The power consumption readings can be visualized using a Web app that uses the chart.js framework. The data are displayed in the form of a graph where the x-axis corresponds to time and the y-axis corresponds to usage data.

The experiment makes use of a high-wattage bulb and a hairdryer for the load. The load is measured using ACS712 current sensor, and total power usage is calculated. The computed data are sent to the data collection center using Wi-Fi and also displayed in LCD as shown in Fig. 2.

The Arduino code in Fig. 3 calculates the total power usage by reading the current from the ACS 712 sensor. It takes a maximum value out of 150 readings to get better accuracy. This maximum value corresponds to peak current, and it is then converted to RMS value. The power is calculated using this RMS value and voltage and updated to the total consumption.

```
int peakAdc = 0;
for (int i = 0; i < 150; ++i)
{
  adcVal = analogRead(analogPin);          // take reading from analog pin of smart meter
  peakAdc = max(adcVal, peakAdc);          // take the maximum current value
}
double vout = (peakAdc / 1023.0) * 5000.0;
double peakCurrent = (vout - 2500) / scaleFactor;
double effCurrent = peakCurrent * 0.707;
effCurrent = effCurrent < 0 ? 0 : effCurrent;

usage += ((230.0 * effCurrent) / 3600000.0);    // add to total usage
```

Fig. 3 Arduino code

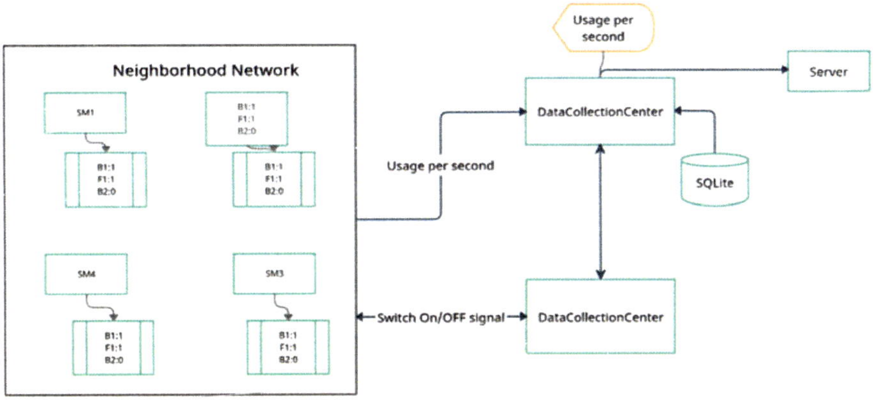

Fig. 4 AMI simulation environment

4.2 *Simulation Environment*

The simulation environment (Fig. 4) is built using the **Docker** container. Here, each Docker container represents a home with a smart meter. Each container has several electric appliances like bulbs, fans, televisions which periodically update the total power usage value. An orchestrator (neighborhood server) is situated between the smart meter and data collection center so that all the user requests for updating the status of the smart meter will go through the orchestrator. The orchestrator forwards the request to the specific smart meter. In addition to the above update request message, each smart meter sends the consumption data to the server every second. For visual analysis, a real-time line graph is provided in the user interface.

Flask API is used to provide endpoint requests and handle cross-origin requests for transmitting consumption data to the data collection center. SqlLite3 is used for persistent data storage.

Advantages of the Simulation Environment

The simulation environment is easy to execute because the simulation of the AMI is done using the Docker containers. It is highly scalable as the users can spawn any number of Docker containers, thereby emulating the same number of smart meters. For visualization purposes, front-end support is provided so that users can update the status of any device easily.

Assumptions

A static value is assigned to the electric current consumption because electrical appliances usually consume some specific amount of electric current. This electric current is almost constant. For example, any bulb would use around 0.4 A of current, and the fan would use on average 0.5–1 A of current.

The "send usage()" function (in Fig. 5) is used in every Docker container spawned. This function is used to send the consumption data to the server every second. This is achieved by using a threading library that calls the function "send usage ()" at regular intervals. The consumption data can be visualized in the UI in the form of a line graph. All the Docker containers can be differentiated using an ID which will be set as an environment variable while creating the container. All the requests to update the status of an electrical appliance will first reach the orchestrator. The orchestrator will validate the smart meter ID and send the request to a specific container. If the ID does not correspond to any valid container, the request is discarded.

The container code in Fig. 6 is used for simulation purposes. This piece of code is copied in every container. The "update_readings()" function will be called when the user changes the status of any electrical appliance in the simulation environment.

```
def sendUsage():
    threading.Timer(1, sendUsage).start()
    curVal = 0
    global usage
    global glob_current
    for key in status:
        if status[key]:
            curVal += current[key[0]]

    glob_current = curVal
    usage += ((230.0*curVal)*curVal)/3600000.0
    requests.post(serverUrl, data={
            "meter_id": meter_id, "current": curVal, "usage": usage})
```

Fig. 5 Container code_2

Fig. 6 Container code_1

```
# updates the given device
# format : {'B1':1, B2: 0, ..}
@app.route("/updateReadings")
@cross_origin()
def updateReadings():
    data = request.get_json()

    for device in data:
        status[device] = data[device]

    return "Updated Successfully", 200
```

4.3 Implementation of DDoS and MITM Attack

The attacks are implemented with the aid of tools like LOIC, Wireshark and OS, Kali Linux with security attack capabilities.

DDoS Attack

The DDoS attack is performed using LOIC. In this tool, the IP address of the target (neighborhood data collector or the smart meter) must be provided. This tool will start flooding the victim with data packets. Due to this, the smart meter spends most of the time replying to the data packets sent by the attacker. This prevents the smart meter from handling genuine requests.

Prevention Mechanism

In the current attack scenario, a tool called LOIC—low orbit ion cannon has been used to perform a DDoS attack (by running the tool on two different laptops). The DDoS attack can be performed easily using LOIC, but preventing the same is complicated. In the current implementation, the DDoS attack has been prevented by disabling the ping feature in the neighborhood data collector. But given that the DDoS attack has three variants, namely ping attack, land attack, and SYN flooding attack, a more robust attack prevention mechanism is recommended. The following prevention mechanism is suggested to prevent DDoS attack.

Intrusion Detection Algorithm

The frequency of data communication between the components of the AMI (e.g., smart meter, neighborhood data collector) needs to be at the minimum interval of 1 min or more. This prevents the components from getting flooded from requests from the neighboring components. It would be better if the communication from various devices is spanned over different time intervals rather than having all the devices sending the requests at the same time. By having the components communicate in a manner defined above, intrusion detection system (IDS) should have a feature of tracking the number of requests from each IP address/MAC address.

Since all the devices communicate at an interval of a minimum of 1 min, it means that each device will send a request only once within 1 min. Hence, ideally, only one request must be obtained from a non-suspicious IP address every 1 min. But if there are two requests from an IP address within 30 s, it means that the IP address is suspicious (because in the event of a DDoS attack, there will be a large number of requests from a specific IP address targeted toward the device), and all requests from the IP address can be blocked, thereby performing the packet filtering at the gateway device itself. All the other devices in the AMI must be configured to communicate with only the gateway device itself, and not directly to each other. The algorithm is detailed below:

Algorithm :*DDOS Attack Prevention Algorithm for use in Intruction Detection System*

```
For every request obtained :
        Ifreq_IP is not in the list of IP addresses :
                Add IP address as the key in the table with the timestamp being   the value
        Else :
                Prev_timestamp = get_timestamp(req_IP )
                if( current_timestamp - prev_timestamp< 30 seconds) :
                        Block all requests from the current IP address
```

Man-in-the-Middle Attack

The attack is performed using the ARP spoofing technique. It is assumed that the attacker has gained access to the network. Once the attacker has access to the network used by the devices within the AMI infrastructure to communicate with each other, ARP spoofing is performed to carry out the attack.

Prevention Mechanism

Given that it is very difficult to identify whether man-in-the-middle attack is carried out or not, the only way to prevent the sniffing attack is to have end-to-end encryption of the consumption information as it travels through the network (Fig. 7).

In the current implementation, we use an encryption mechanism, where the data are encrypted in the smart meter, and it is only decrypted in the central server where the data are stored in the database.

Due to hardware limitations, the current implementation does not have a robust encryption mechanism like RSA encryption. To make this more robust and to prevent the attacker from identifying the key, the use of different keys for different smart meters is recommended and tried and tested encryption mechanisms like RSA or TLS encryption. This would ensure that even if the attacker can sniff the contents of the data packets exchanged between the smart meter and the neighborhood data collector, the attacker would not be able to make any sense of the data as it is encrypted (Fig. 8).

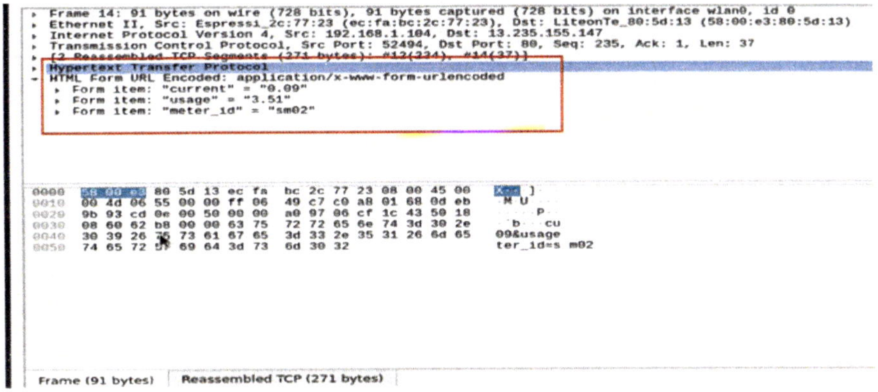

Fig. 7 Usage data clearly visible (without encryption)

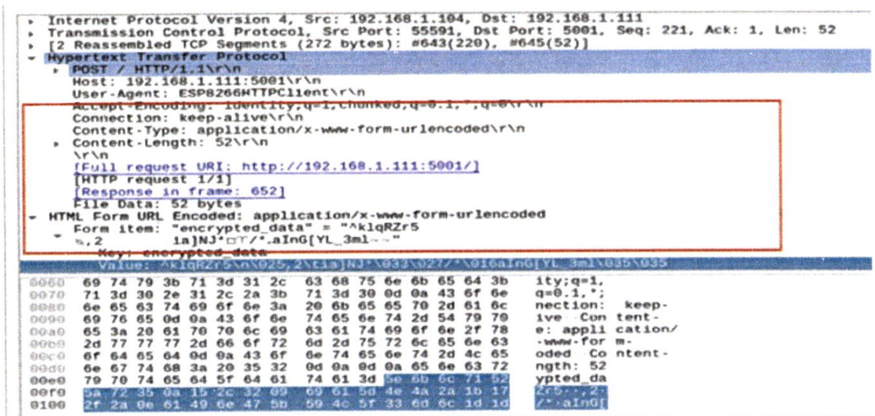

Fig. 8 Consumption info is encrypted

5 Results and Analysis

The Fig. 9 displays the real-time line graph for depicting the total power consumption until a given moment. The attribute "device" contains the list of electrical devices, like the bulb, fan, etc., that are present in the corresponding home. The current consumption represents the usage of current for a particular device.

Tabular values of number of packets versus response time of DDoS attack

No of packets	Response time (in ms)
100	8.413
500	1232.5

(continued)

(continued)

No of packets	Response time (in ms)
1000	2934.5
5000	3056.44
> 15,000	No response

Number of packets sent during DDoS attack versus percentage loss

Number of packets	Percentage loss
100	45.8
500	46.45
1000	47.05
5000	65.45
10,000	70.6

Man-in-the-Middle Attack

The smart meter collects the consumption information and communicates it with the neighborhood data collector on a schedule (for e.g., every 15 s or 30 s). In the attack scenario, the attacker spoofs the smart meter and the neighborhood data collector. Due to this, the MAC address of the attacker's device is mapped to the neighborhood data collector's IP address in the smart meter's ARP table, and the MAC address of the attacker's device is mapped to the IP address of the smart meter in the neighborhood data collector's ARP table (Figs. 12 and 13).

The attacker enables the IP forwarding feature in his laptop. So, all the data from the smart meter and the neighborhood data collector now flow through the attacker before reaching either source or destination. The attacker can use software that reads the contents of the packet (e.g., Wireshark) when the data flow through his machine and the attacker can read the contents of the packet and electricity consumption information. Figure 14 shows a comparison of the time taken for a ping reply before the MITM attack is performed, and after, the MITM attack is performed. Before the MITM attack, the ping request from a neighborhood data collector to the smart meter took 52 ms because the data packets were sent to the smart meter directly. But after the MITM attack, the time taken for the ping reply was increased because the packets sent by the neighborhood data collector to the smart meter (and smart meter to the neighborhood data collector) were routed through the attacker's device.

6 Conclusion and Future Work

The existing infrastructure for advanced metering of commodities like water and electricity is susceptible to various security attacks. The major ones being the distributed

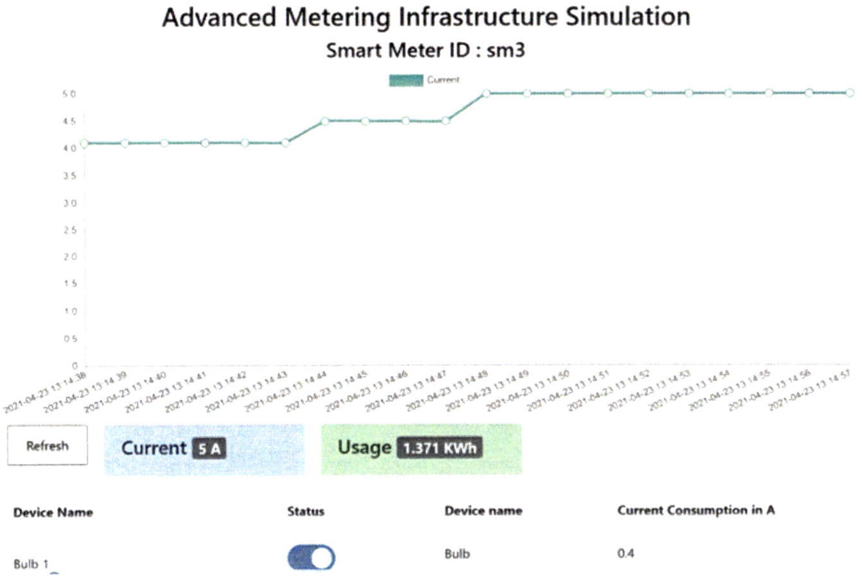

Fig. 9 Real-time graph for smart meter readings

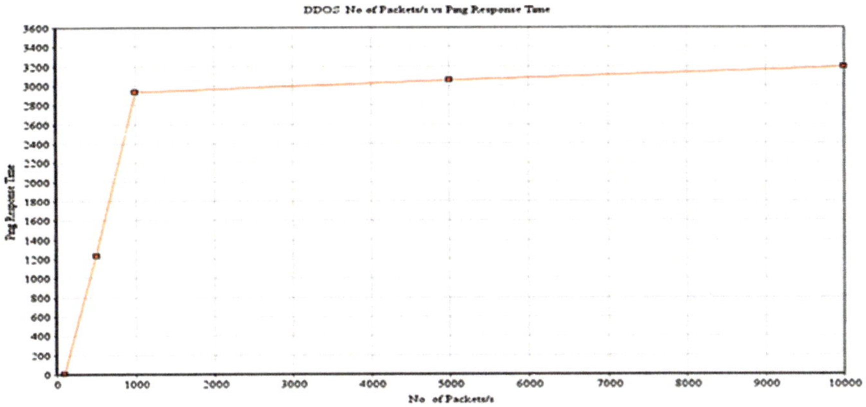

Fig. 10 No. of packets versus response time of DDOS attack

denial of service (DDoS) attack and the man-in-the-middle (MITM) attack, the paper focused on these attacks. The paper includes an overview of the smart grid, smart meter, simulation environment, and vulnerabilities. The various attack surfaces on the smart meter were studied and analyzed the effects of these attacks. A physical testbed and simulation environment were developed. This testbed and simulation environment have all the major functionalities that a commercial AMI infrastructure

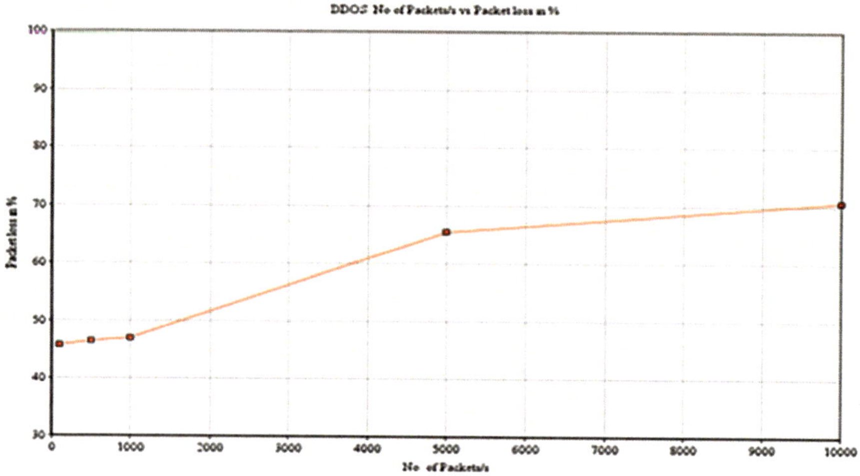

Fig. 11 No. of packets versus percentage loss during DDOS attack

contains. DDoS and MITM attacks were successfully carried out, exposing vulnerabilities in the existing AMI architecture. Prevention mechanisms for the performed attack (DDoS and MITM) have been outlined in this paper. Although, there have not been many published papers regarding the implementation of the effectiveness of these security measures against the security attacks when implemented in smart meters.

The following can be considered for future work:

- Analyzing and implementing different attacks on the physical smart meter testbed and the simulation environment and coming up with the prevention mechanism for the same.
- Enhancing the physical smart meter testbed and simulation environment architecture to include other components of the smart grid.

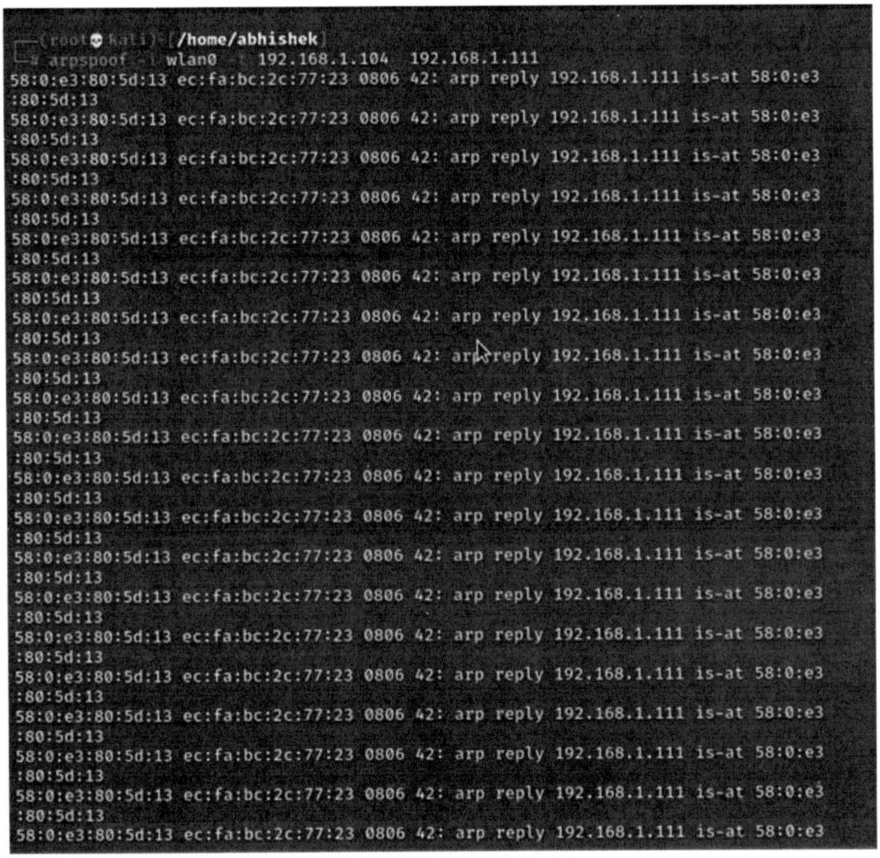

```
┌──(root💀kali)-[/home/abhishek]
└─# arpspoof -i wlan0 -t 192.168.1.104 192.168.1.111
58:0:e3:80:5d:13 ec:fa:bc:2c:77:23 0806 42: arp reply 192.168.1.111 is-at 58:0:e3
:80:5d:13
58:0:e3:80:5d:13 ec:fa:bc:2c:77:23 0806 42: arp reply 192.168.1.111 is-at 58:0:e3
:80:5d:13
58:0:e3:80:5d:13 ec:fa:bc:2c:77:23 0806 42: arp reply 192.168.1.111 is-at 58:0:e3
:80:5d:13
58:0:e3:80:5d:13 ec:fa:bc:2c:77:23 0806 42: arp reply 192.168.1.111 is-at 58:0:e3
:80:5d:13
58:0:e3:80:5d:13 ec:fa:bc:2c:77:23 0806 42: arp reply 192.168.1.111 is-at 58:0:e3
:80:5d:13
58:0:e3:80:5d:13 ec:fa:bc:2c:77:23 0806 42: arp reply 192.168.1.111 is-at 58:0:e3
:80:5d:13
58:0:e3:80:5d:13 ec:fa:bc:2c:77:23 0806 42: arp reply 192.168.1.111 is-at 58:0:e3
:80:5d:13
58:0:e3:80:5d:13 ec:fa:bc:2c:77:23 0806 42: arp reply 192.168.1.111 is-at 58:0:e3
:80:5d:13
58:0:e3:80:5d:13 ec:fa:bc:2c:77:23 0806 42: arp reply 192.168.1.111 is-at 58:0:e3
:80:5d:13
58:0:e3:80:5d:13 ec:fa:bc:2c:77:23 0806 42: arp reply 192.168.1.111 is-at 58:0:e3
:80:5d:13
58:0:e3:80:5d:13 ec:fa:bc:2c:77:23 0806 42: arp reply 192.168.1.111 is-at 58:0:e3
:80:5d:13
58:0:e3:80:5d:13 ec:fa:bc:2c:77:23 0806 42: arp reply 192.168.1.111 is-at 58:0:e3
:80:5d:13
58:0:e3:80:5d:13 ec:fa:bc:2c:77:23 0806 42: arp reply 192.168.1.111 is-at 58:0:e3
:80:5d:13
58:0:e3:80:5d:13 ec:fa:bc:2c:77:23 0806 42: arp reply 192.168.1.111 is-at 58:0:e3
:80:5d:13
58:0:e3:80:5d:13 ec:fa:bc:2c:77:23 0806 42: arp reply 192.168.1.111 is-at 58:0:e3
:80:5d:13
58:0:e3:80:5d:13 ec:fa:bc:2c:77:23 0806 42: arp reply 192.168.1.111 is-at 58:0:e3
:80:5d:13
58:0:e3:80:5d:13 ec:fa:bc:2c:77:23 0806 42: arp reply 192.168.1.111 is-at 58:0:e3
:80:5d:13
58:0:e3:80:5d:13 ec:fa:bc:2c:77:23 0806 42: arp reply 192.168.1.111 is-at 58:0:e3
:80:5d:13
58:0:e3:80:5d:13 ec:fa:bc:2c:77:23 0806 42: arp reply 192.168.1.111 is-at 58:0:e3
:80:5d:13
58:0:e3:80:5d:13 ec:fa:bc:2c:77:23 0806 42: arp reply 192.168.1.111 is-at 58:0:e3
```

Fig. 12 ARP spoof attack on smart meter in MITM attack

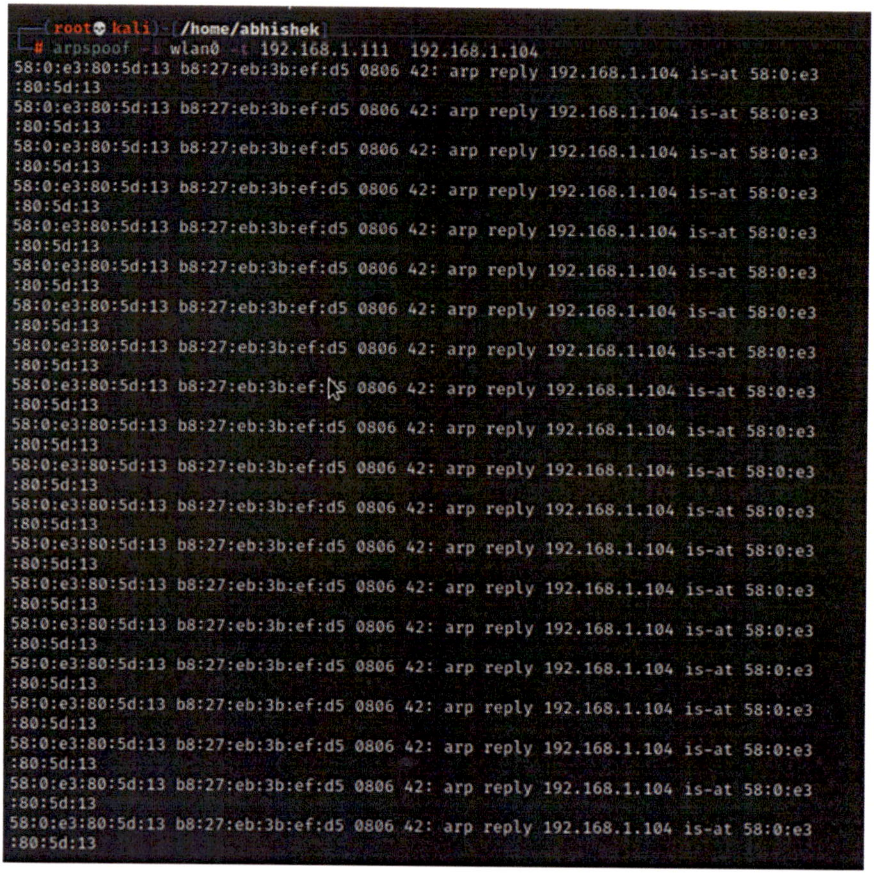

Fig. 13 ARP spoof attack on neighborhood data collector

Fig. 14 Roundtrip time before versus after attack

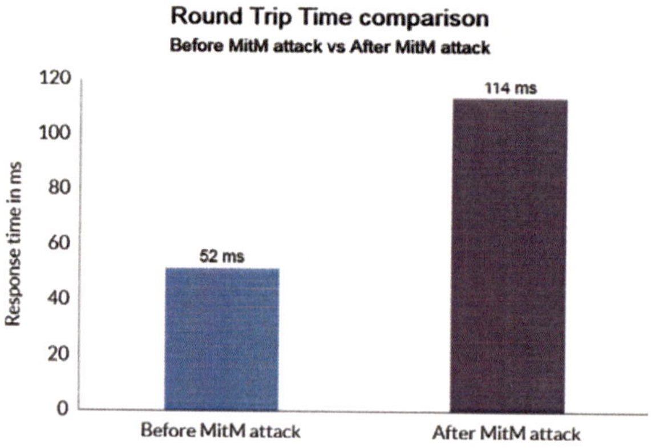

Acknowledgements We would like to express our gratitude to the Department of Computer Science of PES University for the support as this work is a part of the Capstone Project of the course.

References

1. H. Gharavi, R. Ghafurian (eds.), *Smart Grid: The Electric Energy System of the Future*, vol. 99. IEEE (2011)
2. F. Skopik, Z. Ma, T. Bleier, H. Grüneis, A survey on threats and vulnerabilities in smart metering infrastructures. Int. J. Smart Grid Clean Energy **1**(1), 22–28 (2012)
3. M. EL Brak, S. EL Brak, M. Essaaidi, D. Benhaddou, Wireless sensor network applications in smart grid, in *2014 International Renewable and Sustainable Energy Conference (IRSEC)*, pp. 587–592. IEEE (2014)
4. L.I.N. Hsiao-Chung, W.A.N.G. Ping, Implementation of an SDN-based security defense mechanism against DDoS attacks. DEStech Trans. Econ. Bus. Manage. iceme-ebm (2016)
5. A. Singh, J. Bapat, D. Das,Two tier communication architecture for smart meter, in *2013 Fifth International Conference on Communication Systems and Networks (COMSNETS)*, pp. 1–2. IEEE (2013)
6. F. Aloul, A.R. Al-Ali, R. Al-Dalky, M. Al-Mardini, W. El-Hajj, Smart grid security: threats, vulnerabilities and solutions. Int. J. Smart Grid Clean Energy **1**(1), 1–6 (2012)
7. S. Mahjabin, Implementation of DoS and DDoS attacks on cloud servers. Periodicals Eng. Nat. Sci. (PEN) **6**(2), 148–215 (2018)
8. W. Hlaing, S. Thepphaeng, V. Nontaboot, N. Tangsunantham, T. Sangsuwan, C. Pira,Implementation of WiFi-based single-phase smart meter for Internet of Things (IoT), in *2017 International Electrical Engineering Congress (iEECON)*, pp. 1–4. IEEE (2017)
9. K. Shuaib, Z. Trabelsi, M. Abed-Hafez, A.M. Gaouda, M. Alahmad, Resiliency of smart power meters to common security attacks, in *ANT/SEIT*, pp. 145–152 (2015)
10. A. Triantafyllou, J.A.P. Jimenez, A. Del Real, T.L. Torres, K. Rantos, P. Sarigiannidis, The challenges of privacy and access control as key perspectives for the future electric smart grid. IEEE Open J. Commun. Soc. **1**, 1934–1960 (2020)
11. Y. Xiao (ed.), *Security and Privacy in Smart Grids* (CRC Press, Boca Raton, 2013)
12. S. Kulkarni, R.K. Rahul, R. Shreyas, S. Nagasundari, P.B. Honnavalli, MITM intrusion analysis for advanced metering infrastructure communication in a smart grid environment, in *International Conference on Computational Intelligence, Security and Internet of Things*, pp. 256–267 (Springer, Cham, 2020)
13. R. Van Gerwen, S. Jaarsma, R. Wilhite,Smart metering, in *Leonardo-Energy*. org 9 (2006)
14. S. Garg, K. Kaur, G. Kaddoum, J.J.P.C. Rodrigues, M. Guizani, Secure and lightweight authentication scheme for smart metering infrastructure in smart grid. IEEE Trans. Industr. Inform. **16**(5), 3548–3557 (2019)
15. E. Rodriguez-Diaz, E.J. Palacios-García, M. Savaghebi, J.C. Vasquez, J.M. Guerrero, A. Moreno-Munoz, Advanced smart metering infrastructure for future smart homes, in *2015 IEEE 5th International Conference on Consumer Electronics-Berlin (ICCE-Berlin)*, pp. 29–31. IEEE (2015)

Design and Development of Big Data Framework Using NoSQL–MongoDB and Descriptive Analytics of Indian Green Coffee Export Demand Modeling

Saivijayalakshmi Janakiraman and N. Ayyanathan

1 Introduction

In Asia, India is the third-biggest maker and exporter of coffee, and the sixth biggest maker and fifth-biggest exporter of coffee in the world. Though we have the potential to grow and remain as leading exporter, still we could not gain this position, and even if we do, it is not sustainable. While we attempted to find out the reason, the requirement of analysis of huge loads of data consisting of numerous variables as well as its variability depending on seasons, places, people, economy, pricing, logistics, etc., literally renders inability to analyze and conclude. An amazing supply chain brings about better wealth management, more revenue generation, improved customer loyalty and expanded organizational productivity. In order to arrive the results of supply chain management, this needs to be managed properly [1]. Big data analytics is a blended form of web analytics and data analytics [2]. Huge data and its pop up advancements including big data analytics not only make enormous changes in the way the e-commerce and e-services operate but also make conventional data analyzation and business analytics bring new big chances for institutional community and enterprises [2]. Fig. 1 which describes the general framework for big data anlytics. We are planning to implement our descriptive analysis for unstructured information by using MongoDB in our future processes.

S. Janakiraman (✉)
B. S. Abdur Rahman Crescent Institute of Science and Technology, Chennai, India
e-mail: saivijaya_ca@crescent.education

N. Ayyanathan
Department of Computer Applications, B. S. Abdur Rahman Crescent Institute of Science and Technology, Chennai, India
e-mail: n.ayyanathan@crescent.education

J. D. Peter et al. (eds.), *Disruptive Technologies for Big Data and Cloud Applications*, Lecture Notes in Electrical Engineering 905,
https://doi.org/10.1007/978-981-19-2177-3_72

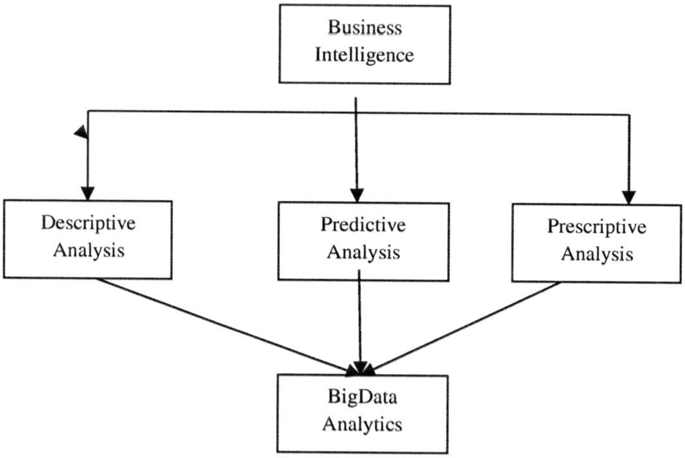

Fig. 1 Framework for big data analytics

1.1 Problem Definition

The supply chain model is amazingly similar to the cost minimizing model which deals with an organization of different suppliers, supplying through a distribution center which in turn distributes to an enormous number of identical retailers design in the Indian green coffee [3]. In this research, we defined a framework for Indian green coffee export demand modeling. This framework also covers descriptive analytics phase for Indian green coffee export demand modeling with the data set that we have gathered from the Indian coffee board, Bengaluru [4]. The insights generated from the big data framework would give the necessary business decision making capacity for the higher level management of the coffee industries.

This paper is categorized into five sections. Section 2 gives a review on related works. Section 3 depicts about the data. Section 4 discusses about the techniques for descriptive analytics across various processes of big data. The last section describes the conclusion of this paper.

2 Literature Review

Business examination implies to the wide usage of information obtained from different sources, factual and quantitative investigation, illustrative and prescient models and truth-based administration to drive choices and activities to legitimate partners [5]. To achieve this, business analytics implemented different methods from data science, machine learning, information system domains and operational research [6]. Business analytics uses engaging models for giving crucial bits of insights and relevant decisions for business execution. Finally, business investigation has progressed past a basic crude information examination on big data sets with a mean to give affiliations an upper hand [7].

Business analytics is categorized to three principle stages described by different levels of difficulty, worth, and insight (i) descriptive analytics, addressing the inquiries "What has occurred?", "For what reason did it happen?", "What is going on now?"; (ii) predictive analytics, responding the inquiries "What will happen?" and "Why will it happen?" in the future; (iii) prescriptive analytics, answering the inquiries "How would it be a good idea for me to respond?" and "For what reason ought to I get it done?" [8–10]. Soltanpoor & Sellis described that it helps organizations to hold the reasons of the events that happened in earlier and comprehend relationships among different kinds of information [11]. However, comparatively to other research works, we considered it as part of descriptive analytics. The justification is to ensure uniformity among three phases of analytics so that each one answers the questions "What?" and "Why?" [9, 10].

The fundamentals of big data analytics consists of mathematics, statistics, engineering, human interface, computer science, and information technology [2, 12]. There is a developing attention to analysis of consumption behavior and preferences using forecasts acquired from client data and transaction records in order to manage product supply chain [7, 13]. Yang and Sutrisno applied and computed regression analysis and neural organization methods to determine request figures for short-lived products. They figured out that daily forecasts are possible with knowledge of sales numbers in the first few hours of the day using either of the above methods with greater accuracy [14].

Lu and Swaminathan uttered that the inventory management concentrates around stream of products, administrations, and data from starting places to end users places through a chain of substances and exercises that are associated with each other [15]. According to Hofmann and Rutschmann review, the key questions to answer are why, what, and how big data analytics/machine learning algorithms could magnify forecast accuracy in similarity to conventional statistical forecasting techniques [16].

Mekuria et al., in their investigation stated that " The situation with Coffee Production and the Potential for Organic Conversion in Ethiopia," features in the conference that the fluctuation in coffee prices is contributing to a socio-financial decline affecting an estimate of 125 million people worldwide. At last, they found that Ethiopia has the capability to deliver high-quality coffee due to the suitable growing conditions and the good variety of genetic resources in coffee Arabica [13].

Gopalsamy and Arul Kumar analyzed in their paper that India's export performance of coffee with respect to USA till 2018 was showing positive performance in terms of quantity and it was expected to increase [17]. The increasing as well as decreasing value on both quantity and cost denotes the quality and quantity of demand imposed by USA toward India [18]. Their examination emphasizes big data analytics identified unstructured data at least 95% of the big data. They also assessed analytics technique for sound, video, text, and web-based media information, and they suggested new tools and techniques for predictive analytics for structured data. Since, big data is noisy, unreadable, and interrelated, there is a need to emerge out a new statistical model for the data come up with social media text, sound, and video [19, 20].

2.1 Review Summary

S. No.	Author	Title/year of publication	Methodology	Inference
1	Punam K., Pamula R., Jain P.K. GUCON	A two-level statistical model for big mart sales prediction (GUCON) Conference: 2018	KNN, support vector regression, two-level statistical model	Two-level model outperformed the single model approaches
2	Burney S.M.A., Ali S.M., Burney S.	A survey of soft computing applications for decision making in supply chain management. ICETSS 2017, 2018	Fuzzy logic, artificial neural network, genetic algorithm	All algorithm widely used for data mining in various areas within inventory management
3	Xindong Wu, Xingquan Zhu, Gong-Qing Wu and Wei Ding	Data mining with big data *IEEE transactions on Knowledge and Data Engineering*, January 2014	Mining uncertain, and incomplete data spare, HACE theorem	Big data mining framework needs to consider complex connections between variables, models, and data sources, along with their open out changes with time and other possible factors
4	R. Soltanpoor, T. Sellis M.A. Cheema, W. Zhang, L. Chang (Eds.)	Prescriptive analytics for big data. ADC 2016	ARIMA, logistic regression	dynamic and cycle viability by assisting examiners with getting binds results to explicit circumstances
5	N. Ayyanathan, A. Kannammal	Combined forecasting and cognitive decision support system for Indian green coffee supply chain predictive analytics IEEE 2015 International Conference on Cognitive Computing and Information Processing (CCIP)	ARIMA, support vector machine	The support vector machine(LS)-expected values and conventional SVM predicted values can be joined by the examination configuration to deliver the new tracking signal spectrum and measure the precision

2.2 Research Gap

Further analysis of data is expected to look deeper in such a way that it should strengthen the export quantity of Indian green coffee system. To extend the operationalization of the conceptual framework, further research is needed. Most of the researchers used various models for statistical analysis. The need is to determine the statistical inference accurately by using charts and to study the relationship between the variables.

3 Data Description

Last ten years data, i.e., from 2009 to 2020, have been considered for analyzing the exporting trend of coffee, and this was gathered from Coffee Board of India, Bengaluru. Indian green coffee export to the forty countries and the research data set for the period 2009–2020 was gathered from distributed sources of the Indian Coffee Board web portal [4]. Coffee Board of India is exporting Indian green coffee every year to other countries. Demand quantity in metric ton is the principle indicator of the total supply chain network [21]. Data set comprises various fields like country-wise coffee exports quantity, percentage to total exports quantity, unit rate, etc. With the collected data sets, we have done the descriptive analytics using NoSQL–MongoDB and visualized the data using charts available in the MongoDB. MongoDB enables the data to build easily, get used quickly, and change friendly design charts, powerful queries.

4 Methodology

MongoDB is the notable NoSQL database. MongoDB is an open-source and cross-stage document-oriented database. The term "NoSQL" characterizes "non-relational." It derives that MongoDB did not depend upon the table-like design; however, it gives an inside and out various systems to capacity and recuperation of data [22]. Relational database management system (RDBMS) has a typical plan that shows number of tables and the connecting between these tables through MongoDB is document oriented. There is no understanding of relationship with schema less connections. There are few terms which are connected in both relational and non-relational databases. What's called table in RDBMS is known as a collection in MongoDB. MongoDB gives a default "_id" which is a twelve byte hexadecimal number which ensures the uniqueness of each archive. It resembles the primary key in RDBMS [22]. MongoDB Atlas gives an essentially technique to have and oversee information in the cloud. Dashboards are the foundation of MongoDB.

Inferential Statistical Measure: Pearson correlation coefficient (**PCC**), or the **bivariate correlation**, is a measure of linear correlation between two sets of data. It is the covariance of two factors, separated by the product of their standard deviations; thus, it is essentially a normalized measurement of the covariance, such that the outcome consistently has a reach somewhere in the range of −1 and 1 [4]. Visualization gives an easier understanding of the relation between the variables than the reading tabular data; heatmaps are normally used to imagine correlation matrices. This matrix tells a lot about the relationships between the variables involved. We could discover a correlation of 1.0 along **the diagonal of the matrix**. This is on the grounds that each variable is highly and positively correlated with itself [23].

5 Results and Discussion

Based on the above statistical approach, we visualized that in Indian green coffee export modeling average price per metric ton is steadily but slowly raising. Total export value reached the peak for some time till 2018 and need to analyze further for the post 2018. From Fig. 2, we found that quantity increases with the reduction in price from 2012 onwards. Fig. 3 shows the year-wise chart for the three countries based on rate. From Fig. 4, we have computed the inferential statistical measures through which we could find that there is a linear relationship between the explanatory variables.

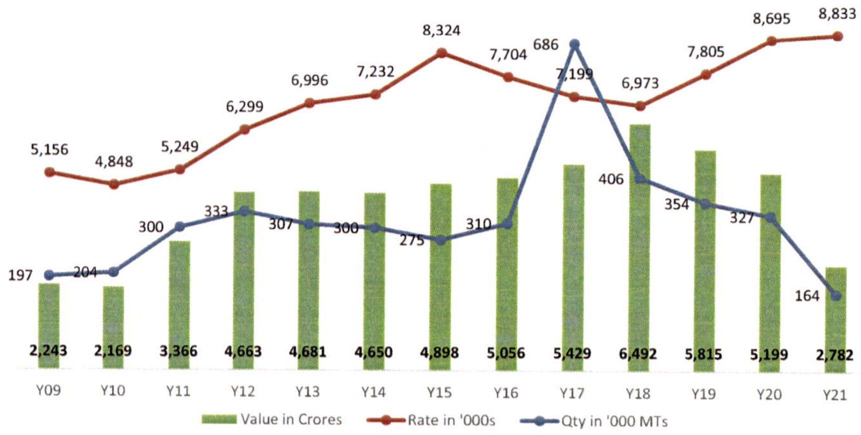

Fig. 2 Statistical charts for qty, rate, and value of exports in different years

Fig. 3 Year-wise chart for the three countries based on rate in Rs/M.Ton

Fig. 4 Inferential statistical measure: Pearson correlation coefficient (PCC) correlation matrix shows a relationship between the explanatory variables

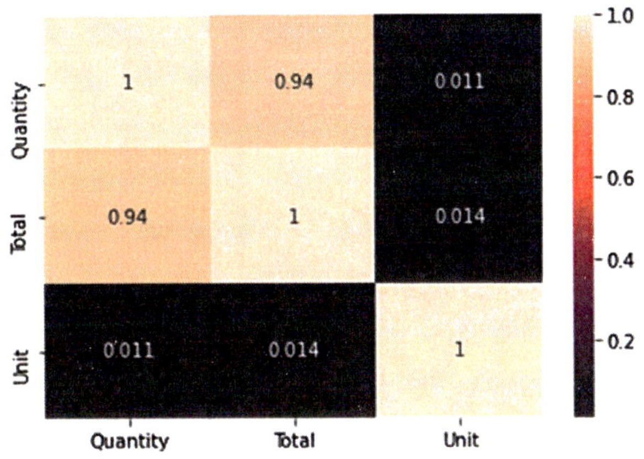

6 Conclusion

The present research provided some interesting knowledge into demand forecasting practices contributing toward the accomplishment of competitive advantage by organizations Furthermore, this study gives an structured methodology toward identifying and selecting the significant indicators for the forecasting model. In this paper, we introduced an outline on descriptive analytics tools and techniques that make it feasible for analytics to build classifiers more effectively. To date, just pilot studies have been executed during the initiative and results for some estimated situation still need improvement. Future works should focus on resolving more issues on demand modeling in big data, thus contributing to the feature engineering and big data analytics. We proposed to explore on deeper insights on factors influencing coffee export demand, demand forecasting/predictions. We have done bivariate analysis using Python with the overall demand of Indian green coffee with USA demand. Hence, the result is obtained compared with the traditional statistical models which give an accurate linear regression model.

References

1. S.A. Khan, Importance of measuring supply chain management performance. Indus. Eng. Manage. (2013)
2. Z. Sun, K. Strang, J. Yearwood, Analytics service oriented architecture for enterprise information systems. CONFENIS 2014, Hanoi, 4–6 December 2014, in *Proceedings of iiWAS2014*, pp. 506–518. ACM Press (2014)
3. R. Ganeshan, Managing supply chain inventories: a multiple retailer, one warehouse, multiple supplier model. Int. J. Prod. Econ. **59**, 341–354 (1999)
4. www.indiacoffee.org—Indian Coffee Portal. Available from http://www.indiacoffee.org
5. T.H. Davenport, J.G. Harris, *Competing on Analytics: The New Science of Winning* (Harvard Business Press, Boston, 2017)
6. M.J. Mortenson, N.F. Doherty, S. Robinson, Operational research from Taylorism to terabytes: a research agenda for the analytics age. Eur. J. Oper. Res. **241**(3), 583–595 (2015)
7. P. Mikalef, I. Pappas, J. Krogstie, M. Giannakos, Big data analytics capabilities: a systematic literature review and research agenda. Inf. Syst. e-Bus. Manage. **16**(3), 547–578 (2018)
8. B. Akerkar (ed.), *Advanced Data Analytics for Business Big Data Computing* (CRC Press, Boca Raton, 2013), pp. 373–397
9. J. Krumeich, D. Werth, P. Loos, Prescriptive control of business processes. Bus. Inform. Syst. Eng. **58**(4), 261–280 (2016)
10. L. Šikšnys, T.B. Pedersen, L. Liu, M. Özsu (eds.), *Prescriptive Analytics Encyclopedia of Database Systems* (Springer, New York, 2016)
11. R. Soltanpoor, T. Sellis, Prescriptive analytics for big data, in *27th Australasian Database Conference: ADC 2016. Databases Theory and Applications, LNCS*, Vol. 9877, ed. by A. Cheema, W. Zhang, L. Chang (Springer International Publishing, Sydney, NSW, 2016), pp. 245–325
12. C.P. Chen, C.-Y. Zhang, Data-intensive applications, challenges, techniques and technologies: a survey on big data. Inf. Sci. **275**, 314–347 (2014)
13. T. Mekuria, D. Neuhoff, U. Köpke, The status of coffee production and the potential for organic conversion in Ethiopia, in *Conference Paper* (2004)
14. C.L. Yang, H. Sutrisno, Short-term sales forecast of perishable goods for franchise business, in *2018 10th International Conference on Knowledge and Smart Technology: Cybernetics in the Next Decades, KST 2018*, pp. 101–105 (2018). https://doi.org/10.1109/KST.2018.8426091
15. L.X. Lu, J.M. Swaminathan, Supply chain management. Int. Encycl. Soc. Behav. Sci. (2015). https://doi.org/10.1016/B978-0-08-097086-8.73032-7
16. E. Hofmann, E. Rutschmann, Big data analytics and demand forecasting in supply chains: a conceptual analysis. Int. J. Logist. Manage. **29**(2), 739–66 (2018). https://doi.org/10.1108/IJLM-04-2017-0088
17. M. Arul Kumar, S. Gopalsamy, Coffee export from India. IJSDR **4**(5) (2019)
18. M.M. Najafabadi, F. Villanustre, T.M. Khoshgoftaar, N. Seliya et al., Deep learning applications and challenges in big data. J. Big Data **2**(1) (2015)
19. A. Gandomi, M. Haider, Beyond the hype: big data concepts, methods, and analytics. Int. J. Inf. Manag. **35**(2), 137–144 (2015)
20. S.C. Matz, O. Netzer, Using big data as a window into consumers' psychology. Curr. Opin. Behav. Sci. **18**, 7–12 (2017)

21. N. Ayyanathan, A. Kannammal, Combined forecasting and cognitive decision support system for Indian green coffee supply chain predictive analytics, in *2015 International Conference on Cognitive Computing and Information Processing (CCIP)* (2015)
22. www.geeksforgeeks.org/mongodb-an-introduction. Available from https://www.geeksforgeeks.org
23. www.statisticshowto.com—Pearson Correlation. Available from https://www.statisticshowto.com

Two-Tier Securing Mechanism Against Web Application Attacks

Vikas Matam, H. S. Shankaranarayana Hebbar, Prince Jha, Amruthanshu Bhat, S. Nagasundari, and Prasad B. Honnavalli

1 Introduction

Web applications continue to have a strong presence in a wide range of sectors, including online business applications to telecommunication, transport, e-commerce, and banking. As more and more services and applications move to the web, targeting the web application for exploits is increasing. Exploiting vulnerabilities of web applications can very likely compromise the data and privacy of its users. According to the most recent figures, upwards of 27 attacks per minute are attempted against web applications [1]. Out of all the web application attacks, well-documented code injection attacks like cross-site scripting (XSS) and Structured Query Language Injection (SQLi) have been repeatedly securing a spot in the OWASP Top Ten Web Application Security Risks for well over a decade now. According to the WordPress Security Learning Center, 65% of vulnerabilities are due to coding practices that introduce SQLi and XSS injection flaws [2]. The "State of the Internet" report from Akamai states that out of all the recorded web vulnerabilities from November 2017 till March 2019, 65% of them were SQLi attacks. Injection attacks top the list of critical vulnerabilities in 2021 as published by the OWASP foundation.

Many solutions have been proposed and implemented to avoid web application attacks. Because XSS and SQLi attacks are mostly triggered due to improper input sanitation, defensive coding approaches have been widely adopted to counteract. But this depends on the diligence and awareness of programmers to maintain code quality. A popular option to fend off SQLi and XSS attacks is to use a web application firewall (WAF). WAF detects possible attacks by analyzing HTTP requests/responses in real time. Legacy WAFs have a predefined rule set which is used to compare

V. Matam · H. S. Shankaranarayana Hebbar · P. Jha · A. Bhat · S. Nagasundari (✉) · P. B. Honnavalli
Department of CSE, PESU C-IS FCR, PES University, Bengaluru, India
e-mail: nagasundarisuresh@gmail.com

J. D. Peter et al. (eds.), *Disruptive Technologies for Big Data and Cloud Applications*, Lecture Notes in Electrical Engineering 905,
https://doi.org/10.1007/978-981-19-2177-3_73

against incoming requests to detect potential attacks. This works well for known web attacks but it fails for unknown or unseen attacks. Furthermore, the rules' creation and updation are heavily reliant on security experts and their experiences. Due to the obvious drawbacks of the legacy signature-based WAF, there is a growing interest in developing WAF which derives rules from actual web data. In the light of current trends, the proposed work uses AI concepts, deep learning to be exact, to detect web application attacks. Deep learning models can maximize the use of unstructured data and eliminate the need for feature engineering.

The proposed solution makes use of a recent deep learning model called Bidirectional Encoder Representations from Transformers (BERT) which will be attached with an artificial neural network (ANN) to further train the BERT model on a dataset comprising SQLi and XSS payloads. This acts as the first tier in attack detection. The second tier is a context-aware SQLi detection module that looks for a change in the structure of the intended query formed using user input to detect SQL injections.

The major contributions of this paper can be summarized as follows:

- A recently published unsupervised deep learning model, BERT, has been used, which has never been implemented to detect web application attacks.
- To train the deep learning model, a real-world dataset consisting of 51,604 unique payloads to detect XSS and SQL attacks has been constructed.
- A recursive decoder has been implemented which extracts plain text data from possible multiple encodings.
- A novel context-aware approach has been implemented to detect SQLi attacks.
- The proposed solution is benchmarked against existing solutions, on the publicly available dataset, HTTP DATASET CSCI 2010 [3].

The following is how the rest of the content is organized:

Section 1.1 summarizes all the relevant related works, and Sect. 3 details the working and mechanisms of the proposed solution that includes an overview, preprocessing of data, and explanation on classifiers. Section 4 describes experiments and results of the BERT model, and finally, Section 5 discusses the conclusions and further works intended to accomplish.

1.1 Related Work

Mokbal et al. [4] in their paper propose a two-step approach to detect XSS attacks. First, a dynamic feature extractor model is used to get a proper feature vector after removing irrelevant parts from the source web page. The generated feature vector is used as input to a multilayer perceptron (MLP) to detect malicious web pages. But the drawback is that this scheme cannot be used for real-time detection of attacks yet. Another approach to detecting XSS attacks makes use of ensemble learning [5]. This uses a group of Bayes network/decision trees, and each Bayes network is built including threat intelligence and domain knowledge. Further testing of the proposed scheme has to be done as authors explicitly state that their implementation needs to be tested on much more practical scenarios.

In another work, unlike regular rule-based WAF, Galbreath presented Libinjection [6] which uses lexical analysis to detect SQLi. The system was trained on tens of thousands of real SQLi attacks with minimal prior understanding of what SQLi is through semantic syntax. The authors [7] propose a real-time detection of web threats based on attention models. Here, supposedly malicious segments are extracted first from HTTP requests and then these segments are fed to the classifier. The authors state that by only focusing on the smaller section of large HTTP request data, the detection accuracy increased by 23% and reduced computation cost by 82.6% but as stated by authors they make use of supervised learning models and these tend to be inefficient while dealing with anomalous or obfuscated HTTP requests. Hidhaya and Geetha [8] in their paper proposed a reverse proxy-based server-side solution to mitigate the SQL injection attack and the XSS injection attacks. This solution does not require changing existing code and supposedly achieved 100% accuracy against the standard testbed. Makiou et al. [9] in their paper proposed a composite solution to detect complex SQL injection from raw HTTP requests. Their solution includes an ML classifier and a regex-pattern matching detection system. The aim was to reduce the number of detection cycles with the ML model predicting obvious legitimate requests from suspicious requests and checking the suspicious payloads with pattern matching detection engine rules. In another approach to detect SQLi attacks, Nagasundari et al. [10] make use of residual network (ResNet) model, a variant of convolutional neural network but as authors pointed out, the architecture requires considerable time to train and is not intended to be used for real-time attack detection.

2 Proposed Approach

This section explains the architecture and overview of the attack detection flow. This section also provides a detailed explanation of data preprocessing, training the deep learning classifier, and explaining the context-aware SQLi detection mechanism scheme.

2.1 Overview

Figure 1 represents the system architecture diagram of the proposed methodology, and here a flask application acts as a reverse proxy and accepts HTTP requests intended for the webserver. From the requested HTTP packet, path and input are extracted. The inputs are preprocessed, decoded, and converted into plain text. The fine-tuned BERT model and context-aware SQL model will detect for any attacks in

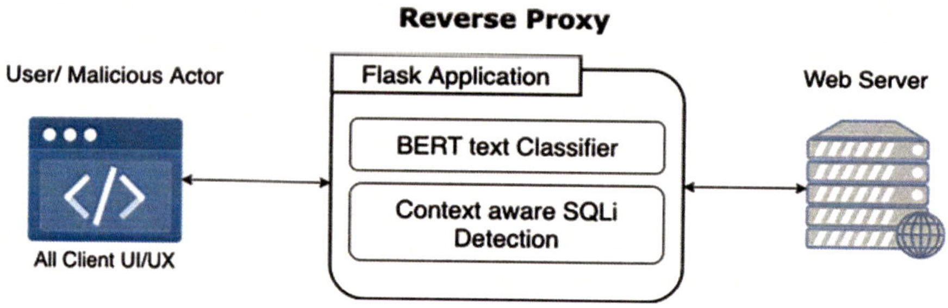

Fig. 1 System architecture

succession. If the request is found as benign, the packet is forwarded to the webserver else it is dropped.

2.2 *Preprocessing*

From the HTTP request packet, the URL path and user inputs are extracted. Multiple encoding attacks and mixed encoding attacks are some of the popular ways to bypass signature-based application firewalls, and thus in the reverse proxy, a decoding mechanism is implemented that recursively decodes the user input from different encoding schemes like URL encoding, Unicode, hex encoding into a plain text format.

2.3 *The Classification*

Figure 2 shows the flow of the HTTP packet in the reverse proxy. After the preprocessing, the user data is fed into the BERT-based model for potential attack detection. Here the saved weights of the fine-tuned BERT model are loaded and used for runtime classification of the HTTP payload. Based on the result of the BERT model classifier, if the payload is found benign, then it is forwarded to SQLi context-aware classifier; otherwise, the request is dropped immediately. SQLi context-aware classifier first checks the path of the request, if that path is pre-configured in the reverse proxy by the webserver, then the above-mentioned context-aware algorithm will detect for potential attacks. If the path is not configured, then the packet is considered as benign and forwarded to the web server.

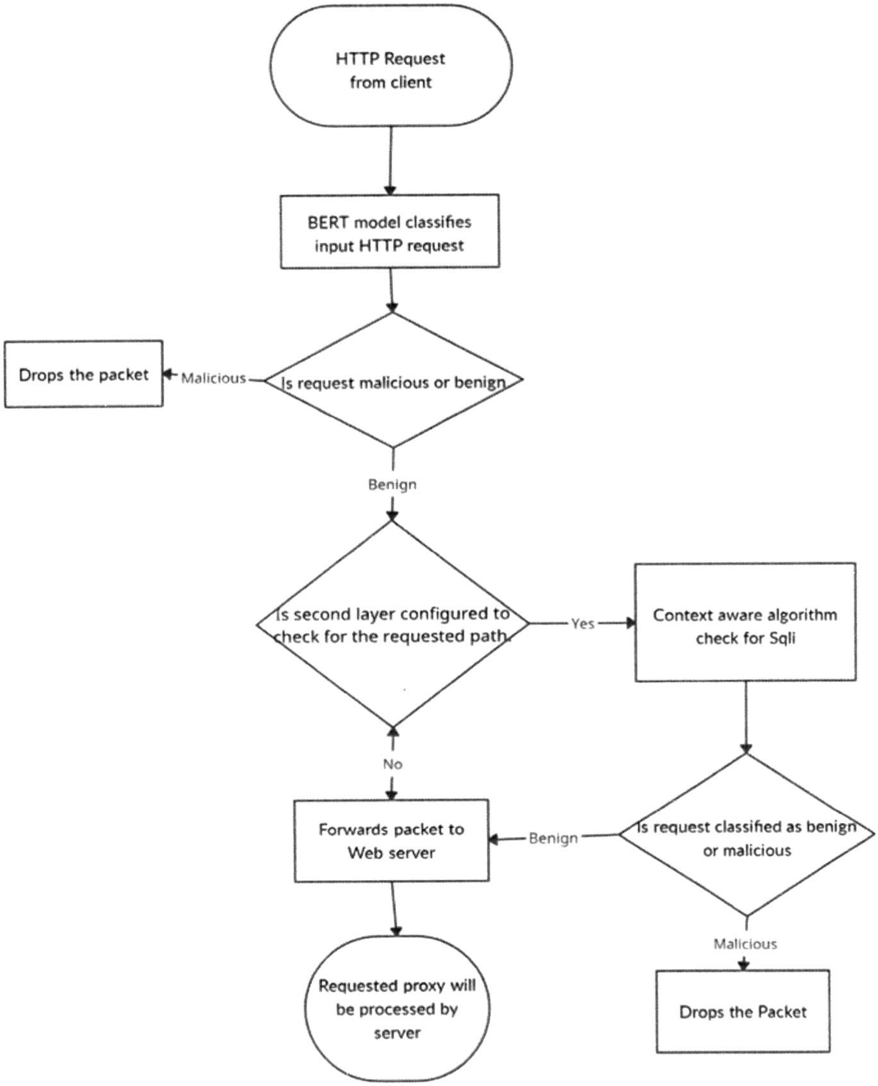

Fig. 2 Flowchart

2.4 Training of BERT Model

BERT is a language representational model [11]. BERT is considered truly bidirectional as it takes into consideration both left-side and right-side information of a token simultaneously during the training thus giving better word embedding. The BERT model, when published, produced the highest scores for natural language modeling tasks like GLUE, SQuAD, and SWAG. BERT is primarily a stack of transformer encoders. Initially, the BERT model had two variants: one is the BERT BASE which contains 12 encoder layers with 110 M parameters and another one is BERT LARGE which contains 24 encoder layers with 340 M parameters. Both of these models have been trained on very large amounts of English text corps, including BookCorps with 0.8 billion words and English Wikipedia with 2.5 billion words.

Since BERT has trained on such a massive dataset, it can be fine-tuned for specific tasks of supervised learning by adding a few extra output layers. There are few different ways of fine-tuning a BERT model depending upon the amount of resource consumed to train the model for downstream tasks [12] and one such method is to freeze the entire architecture, attach few neural network layers to the model then train it, the same has been carried out in our work. For the current work, two different variations for fine-tuning the BERT model have been tested out.

The first variant tried for the classification task is shown in Fig. 3, as depicted in the figure the output vector of length 768 from the BERT model which corresponds to [CLS] input token is fed to the input layer of ANN which is connected to a hidden layer of 512 nodes. The final output is an argmax of the softmaxed values from the output layer of the ANN. This model has a ReLU activation function with a dropout value of 0.1 to avoid overfitting of the ANN.

The other variant for fine-tuning the BERT model is depicted in Fig. 3. As shown in the Fig. 4, the model has an additional hidden layer of 198 nodes which is absent in BERT-1ANN, and in addition to this, BERT-2ANN uses a sigmoid activation function with a higher dropout value of 0.3.

The dataset used for training these deep learning models is an aggregation of different real-world datasets from the internet [13]. The dataset has 51,604 samples with 61.35% of them being benign. This dataset has sufficient variety in the samples and covers all possible known simple and sophisticated SQL and XSS attack vectors along with benign payloads.

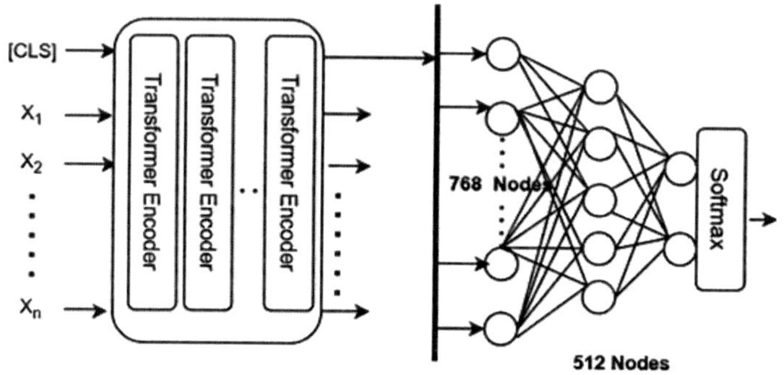

Fig. 3 BERT-1ANN

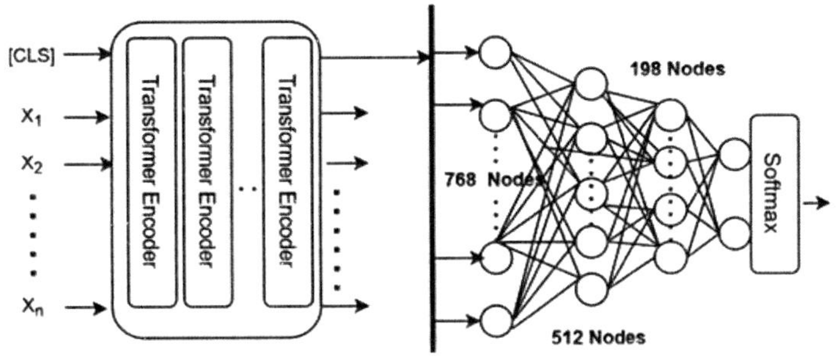

Fig. 4 BERT-2ANN

2.5 Context-Aware SQL Model

Context-aware models can sense and react based on the information available to them through the environment used to tailor the application's relevant data. In the application mentioned in the work, context-awareness is gained by using a configuration file, created/added by the webserver, which contains just the relevant SQL query structure and not the actual table/column names and its corresponding URI as shown in Fig. 5 The context-aware algorithm checks for any change in the intended logic of SQL statements using user input. Below is a detailed explanation of the implementation of this context-aware SQL injection detection algorithm.

```
{
"/products?category=userinput" : "SELECT * FROM table
                                  WHERE $attribute1  = userinput
                                  and $attribute2 = userinput ",
"/artist?id=userinput" : "SELECT * FROM table WHERE
                                  $attribute1 = userinput"
}
```

Fig. 5 Configuration example

Algorithm 1: Query Generator Algorithm

Algorithm 1 Query Generator

```
function QUERYGENERATOR(actualQuery, userInput)
    actualTokens = parser(actualQuery)
    userQuery = ""
    for token t in actualTokens do
        if t.type ==' userinput' then
            userQuery+ = userInput[t.value]
        else
            userQuery+ = t.value
        end if
    end for
    userTokens = parser(userQuery)
  ▷ Tokenize the query generated for userInput
    return ContextSQLClassifier(token_actualQuery,
  token_userQuery)
end function
```

Key-value pairs from the HTTP request and the intended SQL query from the configuration file are inputs to this algorithm. The actual tokens are collected by parsing the intended SQL query which is configured by the reverse proxy. The user input dictionary and the tokens of the actual query are available. Initially, the user query is empty. Iterate through the intended query tokens and add the tokens to the user query except when the intended token is a user input then the user input dictionary is indexed to obtain the corresponding attribute value and added to the user query. After forming SQL query from user inputs, the user query and intended query are both tokenized, and the classification is determined by using the ContextSqlClassifier algorithm (CSG algorithm), which is explained in detail below.

Algorithm 2: CSG Algorithm

Algorithm 2 CSG

```
function CONTEXTSQLCLASSIFIER(token_actual, token_user)
    token_actual = Removewhitespace(token_actual)
    token_user = RemoveWhitespace(token_user)
    if len(token_actual)! = len(token_user) then
        return"Malicious"
    else
        tokenLength = len(token_actual)
        for i in range (0, tokenLength) do
            if (token_actual[i].type ≠' userinput')and
                (token_acutal[i].type[i].type or
                    token_actual[i].value_user[i].value) then
                return"Malicious"
            else
                continue
            end if
        end for
    end if
    return"Malicious"
end function
```

The tokens of the intended query and the tokens of the user query of a particular path are the inputs. Initially, whitespaces are removed from both the token lists. For the user request to be benign, no modifications to the existing tokens or extra tokens addition should occur. So it is a necessary condition that token lengths of both the lists must be the same. To guarantee the query structure, iterate tokenwise comparing kth token in user tokens list to kth token in the intended token list, if the token type mismatches, then the algorithm detects it as a malicious user request.

3 Experiment and Results

This section explains the hardware used, time consumed to train and test variants of the BERT model fine-tuned for attack detection. This section also details benchmarking done on our solution against various published papers.

3.1 BERT Model Training and Comparative Analysis

Fine-tuning the BERT model with the dataset of benign and malicious XSS and SQLI payloads is done on Google Colab with 12 GB of RAM, 105 GB of virtual hard disk, and GPU hardware acceleration enabled. Two variations of BERT model fine-tuning have been carried out on the dataset, one with just one hidden layer in ANN and another one with two hidden layers. Fivefold cross-validation has been carried out for both models. Each fold needed almost an hour to train with the above-mentioned Google Colab configuration., Following are the results for both the models and represented in table 1.

From the above results for BERT model fine-tuning, inference can be made that there isn't much difference in accuracy, precision, recall, and F1-score values between these two models. Adding additional hidden layers, making the model complex offers very little advantage. This indifference was only achieved because the pre-trained BERT model does a really good job in producing accurate word embedding. And thus using pre-trained weights for BERT makes the fine-tuning for specific tasks much faster and less resource-consuming.

Figure 6 depicts the comparison of accuracies between the implemented BERT model with various other deep learning implementations for XSS and SQL injection detection, which include CNNSQL [14], DEEPXSS [15], MLPXSS [4]. None of

Table 1 Experimental result

Model	Accuracy	Precision	Recall	F1 Score
BERT-1ANN	0.9870	0.9830	0.9834	0.9832
BERT-2ANN	0.9898	0.9914	0.9822	0.9868

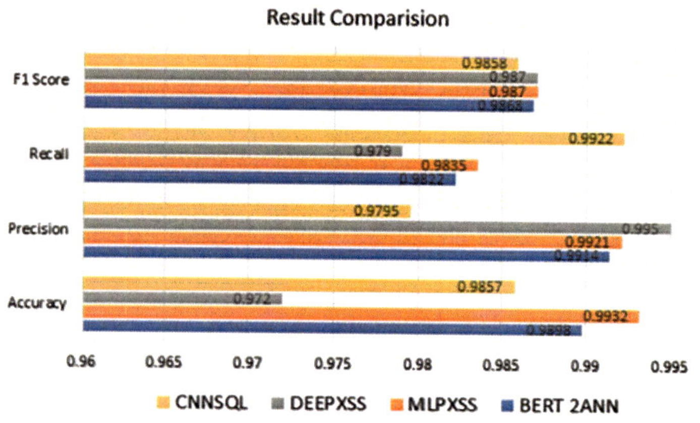

Fig. 6 Comparative analysis with first tier, BERT model

these solutions detect attacks in real time through HTTP packet examination unlike the BERT model yet accuracy scores are quite comparable.

3.2 *End-To-End Testing of Proposed Two Tire Solution*

HTTP CSCI 2010 [3] is a popular publicly available dataset to test web application firewalls. This dataset contains 36,000 benign and more than 25,000 malicious HTTP payloads. This dataset of HTTP requests includes attacks such as SQLi, XSS, buffer overflow, and parameter tampering. Thus, manually SQLi payloads were separated and a total of 900 SQL injection payloads were identified. A random 1100 benign HTTP requests were gathered from the same dataset to create a test dataset of 2000 payloads. The results were compared with Libinjection [6], an open-source real-time detection system, which was tested on the same dataset and depicted in the below given fig. 7.

The CSCI 2010 dataset is a popular dataset for benchmarking but it is really old and has been missing newer attack vectors so to truly assess the performance of the proposed solution in the event of zero-day attacks a new attack dataset with malicious payloads [16, 17] used in the pen-testing tools such as burp suite intruder and SQLMap is prepared. A total of 1000 SQL attack payloads and 1000 XSS payloads were gathered. Out of 1000 malicious SQL payloads, 98% of them were detected by the fine-tuned BERT model, the rest of the malicious payloads were detected by the context-aware model, thus producing zero false negatives. And out of 1000 malicious XSS payloads, 990 of them were correctly detected.

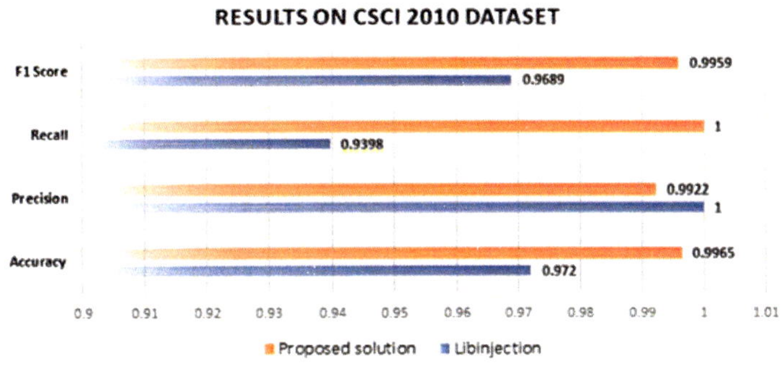

Fig. 7 Comparative analysis with proposed two-tier solution

4 Conclusions and Future Work

The proposed solution is a two-tier securing mechanism that uses both a deep learning model called BERT and a context-aware algorithm as classifiers to detect XSS and SQL injection attacks by examining HTTP requests in real time. Evaluation results show that fine-tuned BERT model achieves high performance with the precision of 99.14%, recall rate of 98.22%, and an F1-score of 98.68% after fivefold cross-validation. To our knowledge, the BERT model was never used before for web attack detection. The second classifier, the context-aware SQL detection mechanism produced zero false positives and false negatives while evaluating the solution on real-world attacks. Further work could extend the proposed work to detect other web application attacks such as code injection and file inclusion.

References

1. T. Beery, N. Niv, *Web Application Attack Report* (2013)
2. O.C. Abikoye, A. Abubakar, A. Haruna Dokoro, O. Noah Akande, A. Anthonia Kayode, A novel technique to prevent SQL injection and cross-site scripting attacks using Knuth-Morris-Pratt string match algorithm. EURASIP J. Inform. Secur. **2020**(1), 1–14 (2020)
3. https://www.tic.itefi.csic.es/dataset/
4. F.M.M. Mokbal, W. Dan, A. Imran, L. Jiuchuan, F. Akhtar, W. Xiaoxi, MLPXSS: an integrated XSS-based attack detection scheme in web applications using multilayer perceptron technique. IEEE Access **7**, 100567–110058 (2019)
5. Y. Zhou, P. Wang, An ensemble learning approach for XSS attack detection with domain knowledge and threat intelligence. Comput. Secur. **82**, 261–269 (2019)
6. N. Galbreath, Libinjection. Blackhat (2012). https://github.com/client9/libinjection
7. T. Liu, Y. Qi, L. Shi, J. Yan, Locate-then-detect: real-time web attack detection via attention-based deep neural networks. IJCAI 4725–4731 (2019)
8. S.F. Hidhaya, A. Geetha, Intrusion protection against SQL injection and cross site scripting attacks using a reverse proxy, in *International Conference on Security in Computer Networks and Distributed Systems* (Springer, Berlin, Heidelberg, 2012), pp. 252–263

9. A. Makiou, Y. Begriche, A. Serhrouchni, Improving web application firewalls to detect advanced SQL injection attacks, in *2014 10th International Conference on Information Assurance and Security*. IEEE (2014), pp. 35–40
10. S. Nagasundari, P.B. Honnavali, SQL injection attack detection using ResNet, in *2019 10th International Conference on Computing, Communication and Networking Technologies (ICCCNT)*. IEEE (2019), pp. 1–7
11. J. Devlin, M.W. Chang, K. Lee, K. Toutanova, *BERT: Pre-training of Deep Bidirectional Transformers for Language Understanding* (2018). arXiv preprint arXiv:1810.04805
12. https://www.analyticsvidhya.com/blog/2020/07/transfer-learning-for-nlp-fine-tuning-BERT-for-text-classification/
13. https://github.com/sn-hebbar99/Two-Tier-Securing-Mechanism-against-Web-Application-Attacks/tree/main/Dataset
14. D. Chen, Q. Yan, C. Wu, J. Zhao, School of Cyberspace Security, Chengdu University of Information Technology, Chengdu, China
15. Y. Fang, Y. Li, L. Liu, C. Huang, DeepXSS: cross site scripting detection based on deep learning, in *Proceedings of the 2018 International Conference on Computing and Artificial Intelligence* (2018), pp. 47–51
16. https://github.com/crawl3r/PortswiggerXSS/blob/master/payloads.txt
17. https://github.com/grananqvist/Machine-Learning-Web-Application-Firewall-and-Dataset/

IoT Enabled Detection and Notification System for Potholes and Road Cracks

N. Aishwarya, N. G. Praveena, C. Karthikeyan, and S. Priyanka

1 Introduction

Generally, cracks, potholes, ditches, and surface flaws are all examples of irregularities on the road surface. The potholes have two distinct aesthetic characteristics. Firstly, potholes areas are illuminated with low intensity, and they are darker than surrounding pavement. Secondly, the textures in potholes are coarser than in other places [1, 2]. All people need roads with good condition as the main mode of transportation is through roads. Even, if the small damages on the road are not immediately handled, then it becomes a life-threatening problem as the damages increases day by day due to the movement of vehicles. Further, the cost of repairing of such roads increases rapidly. Hence, pothole and road cracks detection play a crucial role to keep the roads in a better condition. Road damage detection cannot be done manually. So, the computer vision technology comes into picture which detects the road damages efficiently [3]. In [4], camera sensor is used to extract the characteristics of potholes. The accuracy is better but the method did not detect all the potholes. In [5], canny edge detector and Hough transform are applied to detect the potholes. The affected areas are then sent to officials via emails. Though the algorithm is implemented in real time, the detection results are not better in terms of visual quality.

A simple design using Raspberry Pi microcontroller is fixed in the vehicle to detect the road damages automatically using image processing techniques [6]. The hardware implementation suits the real-time scenario but the number of potholes in the road was not accurately predicted. Also, the authors fail to concentrate on the algorithmic perspective which may hinder the performance on different types of roads

N. Aishwarya
Department of ECE, Amrita Vishwa Vidyapeetham, Amrita School of Engineering, Chennai, India

N. G. Praveena (✉) · C. Karthikeyan · S. Priyanka
Department of ECE, R.M.K College of Engineering and Technology, Chennai, India
e-mail: drpraveenang@gmail.com

J. D. Peter et al. (eds.), *Disruptive Technologies for Big Data and Cloud Applications*,
Lecture Notes in Electrical Engineering 905,
https://doi.org/10.1007/978-981-19-2177-3_74

such as wet, rough, and dry roads. So, in [7] watershed-based thresholding algorithm is used to identify the potholes on flat and degraded road areas. The computational speed of the algorithm has to be improved, and the algorithm has to be tested for real-time scenario. As a number of pavements have been increased, pothole detection has become a great challenge in modern society. Hence, to improve the detection accuracy, deep learning models are developed to identify the potholes and cracks. Also, they are used to classify whether the road is a plain road or damaged one [8–10]. These models basically require a huge data set for training which increases the computational time, and practically, it is not feasible to collect such data sets in real-time scenario. Also, the success rate measured in terms of accuracy totally depends upon the number of layers, activation function, number of epochs, and other training parameters in the model. Based on these perspectives, a robust pothole detection algorithm is proposed to detect the road cracks and potholes. Then, the detected area is fed to an IoT device which is equipped with Wi-Fi and GPS module. These modules share the location of the affected areas, and complaints are registered simultaneously to highways department to improve the road safety.

The remainder of this work is arranged in the following manner. In Sect. 2, the proposed pothole detection algorithm is described. Section 3 explains the hardware implementation of the proposed method, and Sect. 4 concludes the paper.

2 Proposed Algorithm

Figure 1 shows the basic flow of the proposed process. As an initial step, image preprocessing is done using median filter to remove the noise and artifacts of the image. Then, histogram equalization is done to improve the overall contrast of the image. The resultant RGB image is then transformed into grayscale image. Next, the preprocessed grayscale image is subjected to segmentation by using Otsu's thresholding technique. This is an adaptive thresholding technique which separates the foreground pixels and background pixels using optimal threshold [11]. Though there are many techniques to perform the above segmentation process, this method is computationally fast and yields better results. The result of the segmentation process is shown in Fig. 2. From Fig. 2, it can be seen that the potholes and road cracks are properly segmented but the results have few defects which will degrade the performance of the algorithm. Hence, mathematical morphology technique [12] is applied to refine the segmentation process. Let us assume the output of Otsu's thresholding is denoted as OTH (x, y). Then, the refined output MM (x, y) is obtained as given below.

$$\mathrm{OTH}_1(x, y) = \mathrm{bwareaopen}(\mathrm{OTH}, T_1) \tag{1}$$

$$\mathrm{MM}_1(x, y) = \mathrm{bwareaopen}(1 - \mathrm{OTH}_1(x, y), T_1) \tag{2}$$

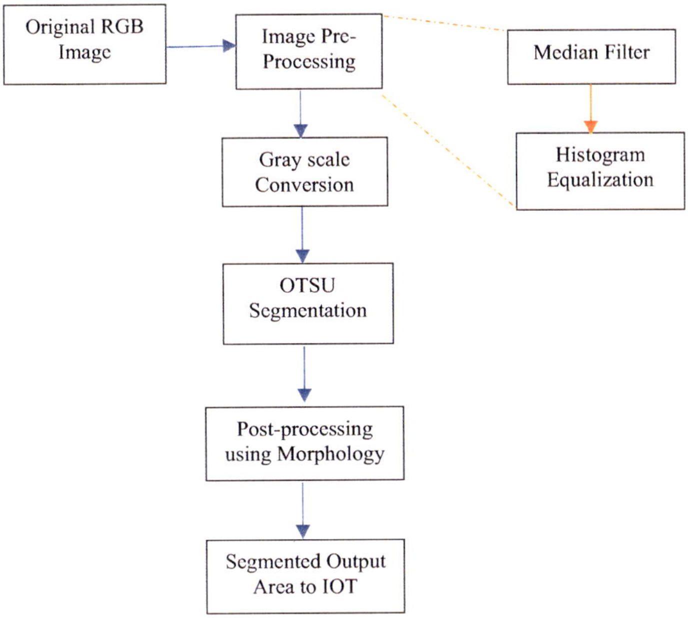

Fig. 1 Overview of the pothole detection algorithm

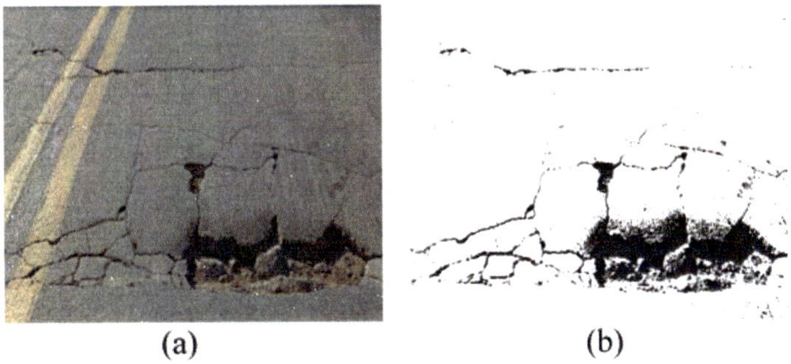

Fig. 2 Results of Otsu's thresholding. **a** Original image **b** Detected output

$$MM(x, y) = \sim MM_1(x, y) \tag{3}$$

Finally, the segmented output area is fed to IoT device which is equipped with Wi-Fi module and GPS module. Figure 3 shows the final results of detected potholes and cracks. From Fig. 3, it is visually evident that the road damages are properly detected by the proposed technique and thus useful to avoid the road accidents and other mishaps.

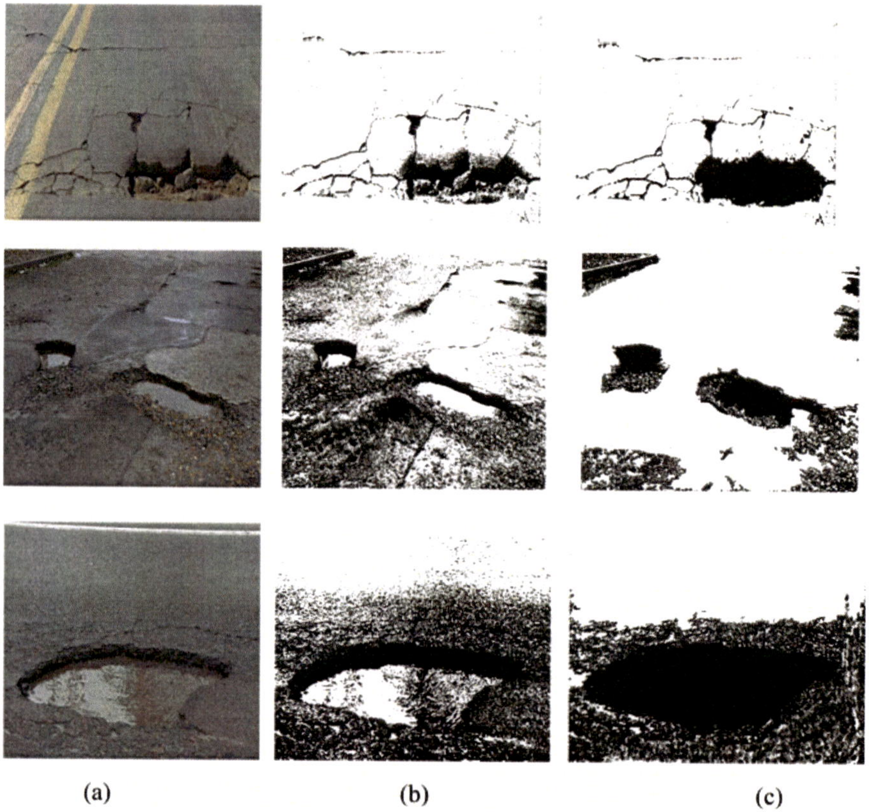

Fig. 3 Qualitative results of the proposed algorithm. **a** Original image **b** Otsu's output **c** Final segmented output after morphology

3 Hardware Implementation

The main aim of the system is to detect the potholes and to report the potholes to the highway department for the welfare of the public safety. The flow diagram is shown in Fig. 4. It consists of UART TTL logic, GPS module, and web server.

First, the prototype is switched on, the camera starts capturing the images of the road in video format, for each second per frame. It checks for potholes and if the pothole is detected, then the location is sensed using GPS module and the information is sent to the server. The complaints based on the potholes are registered using IoT device and then sent to the highway department which is stored in the database. Each and every time the complaints made by public; their location will be updated automatically to the highway department. Figure 5 shows the message format transported to the server. To further speed up the process, the location of the road damages is also intimated to the smart phone. Figure 6 shows the message received in smart phone after the detection of pothole.

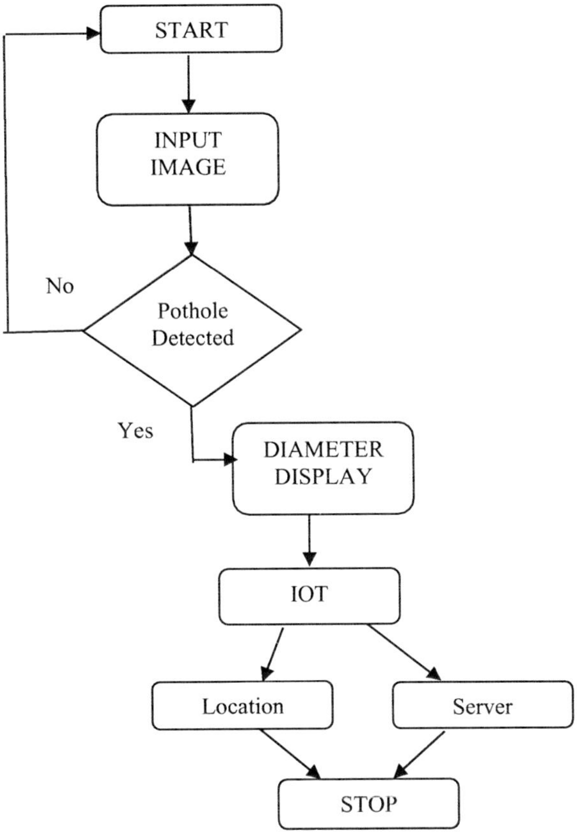

Fig. 4 Flow diagram of the proposed work

To verify the accuracy of the proposed system, two test cases are considered. One is the ideal setup and the other is the non-ideal setup. In the ideal setup, no vehicles, obstacles, and no pedestrians on the road are considered. In non-ideal setup, it is normal road condition where every obstacle is taken into account. The test was done to detect TP (true positive—correctly sensed as a pothole), TN (true negative—correctly sensed as a nonpothole), FP (false positive—incorrectly sensed as a pothole), and FN (false negative—incorrectly sensed as a nonpothole) manually. The recorded output, referred to as the confusion matrix, is shown in Table 1. The findings of the experimental trials, which included thirty trials at four different speeds, demonstrated that the system was successful at detecting potholes. Simultaneously, the system was effective in displaying the identified potholes on the web server, as well as their associated details such as time and location.

Based on Table 1, the accuracy is computed as,

$$\text{Accuracy} = \frac{\text{TP} + \text{TN}}{\text{TP} + \text{TN} + \text{FP} + \text{FN}} \times 100 \tag{4}$$

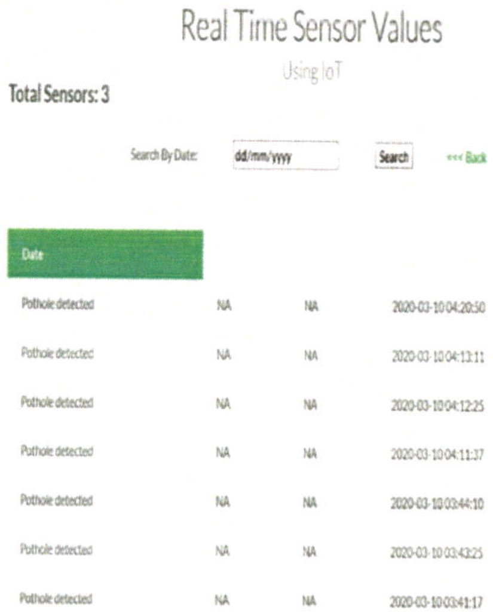

Fig. 5 Message format sent to server

Fig. 6 Message intimation in smart phone

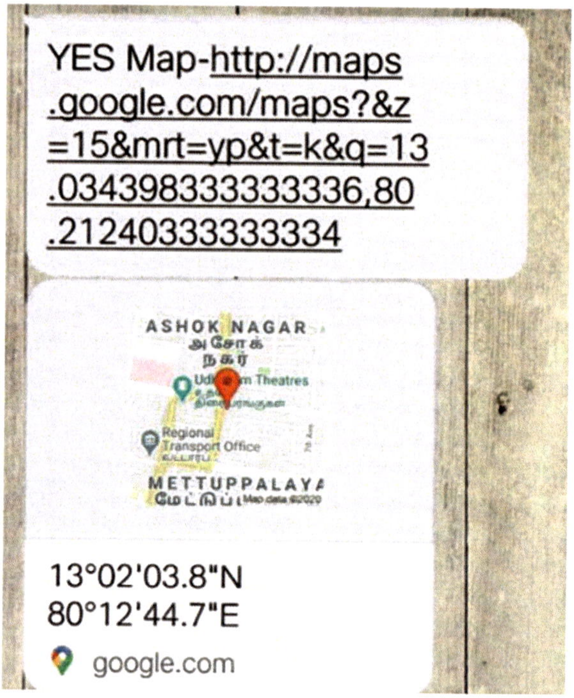

Table 1 Confusion matrix

TP	75
FP	3
FN	5
TN	45

Thus, the overall accuracy of the proposed system is 93.75%.

4 Conclusion

In this work, an effective pothole detection algorithm is proposed to identify the road cracks and potholes. Then a simple system using an IoT device is built which reports the damage and the location of the damaged area to the web server as well as to the smart phone. The implementation of automatic detection system did not use any laptop, desktop, and smart phones. According to the experiment results, the proposed method achieves an overall accuracy of 93.75%. As a result, the smart detection and reporting of pothole system were able to operate successfully. The proposed technique, however, has several limitations. Potholes might be misidentified depending on the type of shadow and the shape of the pothole. Thus, in order to detect potholes more reliably, images from not just a single sensor but also more sensors must be used, and more features for these sensors must be included in the suggested algorithm.

References

1. D.-W. Jang, R.-H. Park, Pothole detection using spatio-temporal saliency. IET Intell. Trans. Syst. Aug 2016
2. A. Rasyid, M.R.U. Albaab, M.F. Falah, Y.Y.F. Panduman, A.A. Yusuf, Pothole visual detection using machine learning method integrated with internet of thing video streaming platform, in *IES International Electronics Symposium* (2019)
3. P. Khumsap, N. Phisanbut, P. Watanapongse, P. Piamsanga, Novel feature extractions for reflection, alligator cracks and potholes road surface classification. IEEE (2018)
4. S.W. Lee, S.J. Kim, K.E. An, S.-K. Ryu, D. Seo, Image processing-based pothole detecting system for driving environment, in *IEEE International Conference on Consumer Electronics (ICCE)* (2018)
5. A. Bhat, P. Narkar, D. Shetty, D. Vyas, Detection of Potholes using image processing techniques, in *International Conference on Innovative and Advanced Technologies in Engineering* Mar 2018
6. M.M. Garcillanosa, J.M.L. Pacheco, R.E. Reyes, J.J.P. San Juan, Smart detection and reporting of potholes via image-processing using Raspberry-Pi microcontroller. IEEE (2018)
7. T.D. Chung, M.K.A.A. Khan, Watershed-based real-time image processing for multi-potholes detection on asphalt road, in *International Conference on System Engineering and Technology(ICSET)* Oct 2019

8. K.E. An, S.W. Lee, S.-K. Ryu, D. Seo, Detecting a pothole using deep convolutional neural network models for an adaptive shock observing in a vehicle driving, in *IEEE International Conference on Consumer Electronics (ICCE)* (2018)
9. N.D. Hoang, An artificial intelligence method for asphalt pavement pothole detection using least squares support vector machine and neural network with steerable filter-based feature extraction, in *Artificial Intelligence Applications in Civil Engineering* (2018)
10. Y.B. Aparna, R.R.V. Gupta, N. Aggarwal, A. Akula, Convolutional neural networks based potholes detection using thermal imaging, in *Journal of King Saud University—Computer and Information Sciences* (2019)
11. T. Kim, S.K. Ryu, Review and analysis of pothole detection methods. J. Emerging Trends Comput. Inf. Sci. **5**(8), 603–608 (2014)
12. Y. Yang, Y. Zhang, J. Wu, L. Li, S. Huang, Multi-focus image fusion based on a non-fixed-base dictionary and multi-measure optimization. IEEE Access **7**, 46376–46388 (2019)

Constructing Pixel Picture Languages Using Cell-Like SN^+P System

Y. Preethi Ceon, Hepzibah A. Christinal, S. Jebasingh, and D. Abraham Chandy

1 Introduction

Natural computing refers to modelling ideas from nature in order to perform computation and further developing computation model [7]. Membrane computing is a branch of natural computing that is inspired by the structure and functioning of biological cells. The computational model obtained is called P system, which is a distributed and parallel computational model.

The class of neural-like computing model is called spiking neural *P* system [4], where neurons communicate to each other by means of spikes and electric impulses of identical shape. It is a type of network model consisting of a number of neurons placed in nodes of a directed graph, where neuron sends signal along synapse. The neuron processes the information in the form of spikes by use of spiking and forgetting rules. The functioning of spiking rules is as follows: A neuron absorbs the current spikes and creates new ones. The generated spikes are conveyed to other neurons connected by synapses to the cell that embodies the rule. Similarly, the predefined number of spikes present in the neuron is removed using the forgetting rule. This rule is triggered when the amount of spikes in a neuron matches the regular expression linked with the rule. The output of the system is the set of spikes sent to the environment, a spike

Y. Preethi Ceon (✉) · H. A. Christinal · S. Jebasingh
Department of Mathematics, Karunya Institute of Technology and Sciences, Karunya Nagar, Coimbatore, Tamil Nadu 641114, India
e-mail: preeti.ceon@gmail.com

S. Jebasingh
e-mail: jebasinghs@karunya.edu

D. Abraham Chandy
Department of Electronics and Communication, Karunya Institute of Technology and Sciences, Karunya Nagar, Coimbatore, Tamil Nadu 641114, India

J. D. Peter et al. (eds.), *Disruptive Technologies for Big Data and Cloud Applications*, Lecture Notes in Electrical Engineering 905,
https://doi.org/10.1007/978-981-19-2177-3_75

train. Thus, the computation in SN P system is associated with the different types of spikes sent to the environment.

In the literature, several researchers studied P system as a generating device producing string languages, array languages and picture languages. Subramanian [10] studied P system and picture languages bridging membrane computing and picture generating grammars. In [6], the authors studied the chain code P system with using cycle grammar. In [9], the authors studied generation of cycle picture languages using sequential spiking neural P system. In [11], the generation of chain code picture language and space filling curves using context free parallel chain code P system were studied by Subramanian et al. Also in [3], the authors have studied the generation of cycle picture languages and fractal curves using cycle rewriting chain code P system.

Cell-like SN P with several types of spikes (cSN$^+$P System) [8] is a variant of cSN P system where the alphabet of spikes can contain more than one variable. Cell-like SN P system works in the same way of spiking neural p system. In this paper, we study cSN$^+$ P system to generate the pixel picture language. In Sect. 2, we briefly summarize the preliminaries of pixel picture language. In Sect. 3, we study the construction of one-dimensional and O-shaped pixel picture languages by cSN$^+$ P system.

2 Preliminaries

In this section, the preliminaries of pixel picture language are discussed.

2.1 *Pixel Pictures*

Pixel pictures are extension of drawn pictures with modified chain code interpretation of the alphabets $\{l, u, r, d\}$. Pixel pictures are defined as a collection of pixels in a square grid, starting from a point s and ending at a point q [1]. Pixel pictures are represented in a two-dimensional plane with connected set of unit cells over the alphabet $\{l, u, r, d\}$ where the interpretation of alphabets is as follows: l indicates one unit line movement to the left and a unit cell to the right of the line segment, r represents the one unit line movement to the right and a unit cell to the right of the line segment, u represents the one unit line movement up and a unit cell to the right of the line segment, and d represents the one unit line movement down and a unit cell to the right of the line segment. Pixel pictures can be represented using strings from the set of alphabets $\{l, u, r, d\}$. The collection of pixel pictures forms the pixel picture language.

2.2 Big Data

In the current digital information [2] world, huge amount of data is collected from various sources and useful information is obtained from that data. Huge data is processed and stored as either structured or unstructured facts as a result of processing images through various computation techniques. Recognizing a particular form of data from a huge volume of data is of interest to industry for business intelligence purpose. *P* system-based computational models can be used to recognize alphabets and other forms of digital pictures.

3 Constructing Pixel Picture Language Using Cell-Like Spiking Neural *P* System with Several Types of Spikes

In this section, we define a new computational model cell-like neural P system with several types of spikes by combining the two computational models cell-like spiking neural *P* system and spiking neural *P* system with several variables.

3.1 Cell-like Spiking Neural P System with Several Types of Spikes

A cell-like spiking neural *P* system [2] with several spikes of degree $m \geq 1$ is a construct in the form of

$$\Pi = (O, \sigma_1, \sigma_2 \ldots, \sigma_m, R_1, \ldots, R_m, i_0),$$

where

1. The limited set of alphabets is denoted by O.
2. $\sigma_1, \sigma_2 \ldots, \sigma_m$ are the neurons, it is in the form $\sigma_i = (W_i, R_i), 1 \leq i \leq m$, where
   (a) $.W_i \epsilon O^*$ is the initial multiset of spikes contained in σ_i;
   (b) The finite set of rules of the forms R_i is given by
   i. $\frac{E}{u} \rightarrow V$, where V is a sequence of pairs of the form (y_i, tar) and $\text{tar}\epsilon\{\text{here, out, in}_j\}$, E is a regular language over O.
   ii. $V \rightarrow \lambda$, with $V \epsilon O^*$ such that there is no rule $\frac{E}{u} \rightarrow v$ of type (a) and $V \epsilon E$.
3. The output region of Π is given by $i_0 \in \{0, 1, \ldots, m\}$.

A rule $\frac{E}{u} \rightarrow V$ is applied as follows: When a multiset w of spikes such that $w \in L(E)$ and $u \in \text{sub}(E)$ is contained in the neuron σ_i, then the firing rules get activated, which consume the spikes u, produces the multiset of spikes $y_i \in E$ and transmits it to the neuron indicated by the target. The forgetting rule $V \rightarrow \lambda$ is applied when exactly the spike V is present in the neuron, and it is removed from the neuron. Since the global clock is assumed, the working of the system is harmonized. The application of firing or forgetting rules in a neuron forms a transition of the system. Any sequence of evolutions starts from the initial configurations. When no more rule is applicable, it reaches a configuration where the computation stops.

The different types of spikes emitted from the output neuron in a halting configuration form the output of the cSN⁺P system. The sequence of spikes emitted from the output neuron forms the string language $L(\pi)$ generated by the cSN⁺P system.

The output string language $L(\pi)$ obtained from the SN P system is mapped with the alphabets $\sum = \{l, u, r, d\}$ which is defined by a homomorphism $\varphi : L(\pi) \rightarrow \sum$. The desired pixel picture language representation is obtained from the modified chain code interpretation [5] of the string language $\varphi(L(\pi))$.

3.2 *Pixel Picture Language Using Cell-Like Spiking Neural P System with Several Types of Spikes*

Theorem 3.1 *The one-dimensional pixel picture languages can be generated by cell-like spiking neural p system having one membrane and firing rule.*

Proof In the following, we consider the construction of one-dimensional horizontal and vertical strip pixel picture languages using cell-like spiking neural p system with several variables.

The language $L(\pi) = \{ r^n | n \geq 1\}$ represents 1D horizontal pixel picture language. The construction of language can be done by cell-like SN⁺P system with a neuron and firing rules in the sequential mode. The system is described as follows:

$$\pi = (O, \mu, \sigma_0, R_0, \sigma_0)$$

where

- $O = \{a\}$ is the alphabet of spike
- $\mu = [\,]_0$
- $\sigma_0 = \{a^2\}$
- Firing rules are given as

$$R_0 = \begin{cases} a^2/a \rightarrow (a, \text{out})(a, \text{here}) \\ a^2/a \rightarrow (a, \text{out}) \end{cases}$$

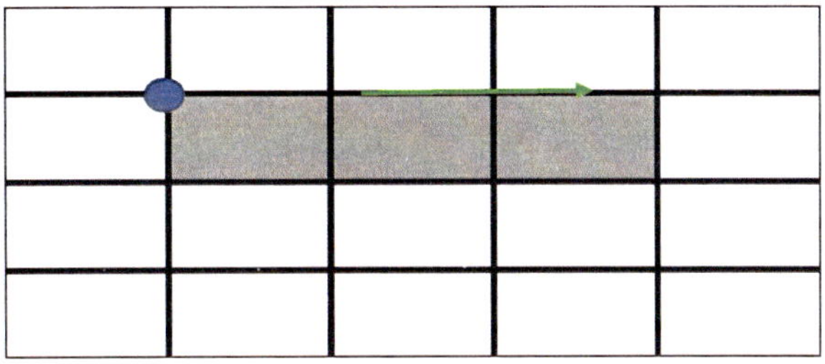

Fig. 1 One-dimensional horizontal pixel picture language

- $i_0 = [\,]_0$

When $n = 3$ in $\boldsymbol{L(\pi)}$, we have $w = rrr$ and the corresponding pixel picture is shown in Fig. 1.

Overview of the computation

The system has one membrane σ_0 having firing rules defined in the set R_0. The membrane σ_0 has the multiset a^2 spikes. The region σ_0 gets activated and the rules in R_0 are applied non-deterministically to the initial set of spikes.

Assume that the system non-deterministically choose the rule $a^2/a \rightarrow (a, \text{out})$ and applies it to the initial set of spikes a^2. The system will send a spike to the environment and stops the computation. Suppose that the system non-deterministically chooses the firing rule $a^2/a \rightarrow (a, \text{out})(a, \text{here})$. This rule consumes a spike a, produces spike a inside σ_0 and sends a spike to the environment. The process of sending the spike a to the environment takes place n number of times and when the system choose the rule $a^2/a \rightarrow (a, \text{out})$, the computation of the system stops. Therefore the sequence of the spikes emitted from the output neuron will be the language $L(\pi) = \{\, a^n | n \geq 1\}$.

Finally, let us define a homomorphism φ between the set of spikes and the chain code alphabets $\Sigma = \{r, l, u, d,\}$as $\varphi(a) = r$. The pixel picture interpretation of the words generated by the system gives the 1-D horizontal strip pixel picture language. Also when we define a homomorphism φ between the set of spikes and the chain code alphabets $\Sigma = \{r, l, u, d,\}$as $\varphi(a) = d$, we obtain the 1-D vertical strip pixel picture language $L(\pi) = \{\, d^n | n \geq 1\}$.

When $n = 2$ in $\boldsymbol{L(\pi)}$, we have $w = dd$ and the corresponding pixel picture is shown in Fig. 2.

Theorem 3.2 *The language of O-shaped pixel pictures can be generated by cell-like spiking neural p system with several variables having three membranes and firing rules.*

Proof The O-shaped pixel picture language is given by the expression $L(\pi) = \{\, r^n d^m l^n u^m | m, n \geq 1\}$. The pixel pictures have holes in the interior of the picture and connected in the boundary. In the following we define the cell-like SN+P system

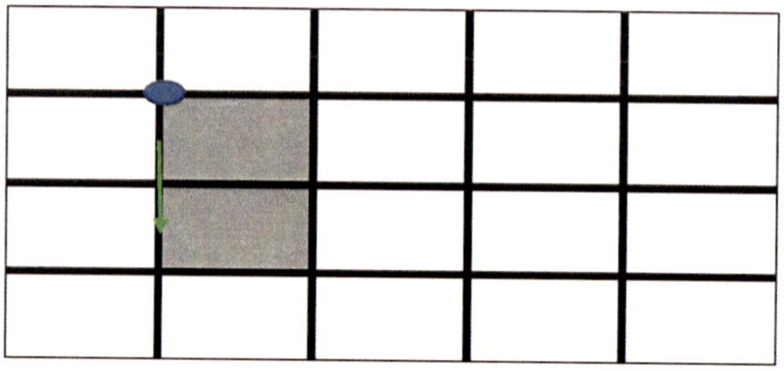

Fig. 2 The 1-D Vertical Pixel picture language

with several variables to construct the language in the sequential mode. The system is described as follows:

$$\pi = (O, \mu, \sigma_0, \sigma_1, \sigma_2, R_1, R_2, R_3, i_0)$$

where

- $O = \{a, b, c, e, f, g\}$ are the alphabets of spikes
- $\mu = [[]1[]2]_0$
- $\sigma_0 = \{a^2b^2fg\}, \sigma_1 = \{c^2\}, \sigma_2 = \{e^2\},$
- Rules are given as
- $R_0 = \begin{cases} a^2b^2fg/a \to (a, \text{out})(a, \text{here})(a, \text{in}_1) \\ a^2b^2fg/a^2 \to (a, \text{out})(a, \text{in}_1) \\ b^2fg/b \to (b, \text{out})(b, \text{here})(b, \text{in}_2) \\ b^2fg/fb^2 \to (b, \text{out})(b, \text{in}_2)(f, \text{in}_1) \\ a^+cg/a \to (c, \text{out}) \\ acg/acg \to (c, \text{out})(g, \text{in}_2) \\ e^+b/e \to (e, \text{out}) \\ eb/eb \to (e, \text{out}) \end{cases}$
- $R_1 = \begin{cases} a^+c^2f/a \to (a, \text{out}) \\ ac^2f/ac \to (a, \text{out})(c, \text{out}) \end{cases}$
- $R_2 = \begin{cases} b^+e^2g/b \to (e, \text{out}) \\ be^2g/be \to (b, \text{out})(e, \text{out}) \end{cases}$
- $i_0 = []_0$

When $n = 4$in $\boldsymbol{L(\pi)}$, we have $w = rrrrddddlllluuuu$ and the corresponding pixel picture is shown in Fig. 3.

Overview of the Computation

The computation of the system is sequential at the level of membrane and has firing present in the region. The system has 3 membranes σ_0, σ_1 and σ_2 each having firing rules defined in the sets R_0, R_1 and R_2. Initially the membrane region σ_0 has the

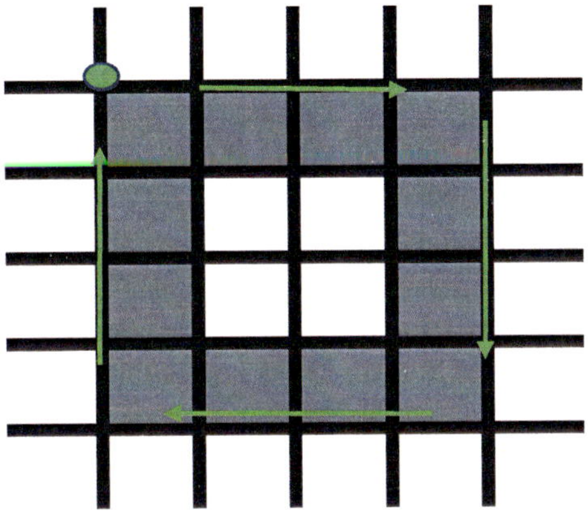

Fig. 3 The o-shaped pixel picture language

multiset a^2b^2fg spikes, region σ_1 has the multiset c^2 spikes and region σ_2 has the spike e^2. The region σ_0 gets activated and the firing rules from the set R_0 are selected non-deterministically to the multiset of spikes.

Assume that the system non-deterministically choose the firing rule $\boldsymbol{a^2b^2fg/a} \rightarrow (\boldsymbol{a}, \mathbf{out})(\boldsymbol{a}, \mathbf{here})(\boldsymbol{a}, \mathbf{in}_1)$ and applies to the multiset of the spikes $\boldsymbol{a^2b^2fg}$. In this case, it consumes a spikes, sends spike a outside and produce spike a and sends it to the neuron $\boldsymbol{n}_1$. The repeated application of the rule will produce a sequence of spikes a^{n} in the environment. This process will stop when the system applies the rule $\boldsymbol{a^2b^2fg/a^2} \rightarrow (\boldsymbol{a}, \mathbf{out})(\boldsymbol{a}, \mathbf{in}_1)$**.** The simultaneous action of sending a spike a outside and a spike to the neuron $\boldsymbol{n}_1$ will acts as register counter, which will be used to produce same number of spikes c. The subsequent application of the rules $\boldsymbol{b^2fg/b} \rightarrow (\boldsymbol{b}, \mathbf{out})(\boldsymbol{b}, \mathbf{here})(\boldsymbol{b}, \mathbf{in}_2)$ **and** $\boldsymbol{b^2fg/b} \rightarrow (\boldsymbol{b}, \mathbf{out})(\boldsymbol{b}, \mathbf{here})(\boldsymbol{b}, \mathbf{in}_2)$ **in** $\boldsymbol{n}_0$ will produce a sequence of spikes b^{m} in the environment.

Now the neuron $\boldsymbol{\sigma}_1$ gets activated and the application of the rule $\boldsymbol{a^+c^2f/a} \rightarrow (\boldsymbol{a}, \mathbf{out})$ and $\boldsymbol{ac^2f/ac} \rightarrow (\boldsymbol{a}, \boldsymbol{out})(\boldsymbol{c}, \boldsymbol{out})$ will produce the multiset of spikes $\boldsymbol{a^m}$ in the neuron $\boldsymbol{\sigma}_0$. The subsequent application of the rules $\boldsymbol{a^+cg/a} \rightarrow (\boldsymbol{c}, \boldsymbol{out})$ **and** $\boldsymbol{acg/acg} \rightarrow (\boldsymbol{c}, mathbfout)(\boldsymbol{g}, mathbfin_2)$ will produce a sequence of spikes c^{n} in the environment. Now the neuron $\boldsymbol{\sigma}_2$ gets activated and the application of the rule $\boldsymbol{b^+e^2g/b} \rightarrow (\boldsymbol{e}, \mathbf{out})$ **and** $\boldsymbol{be^2g/be} \rightarrow (\boldsymbol{b}, \mathbf{out})(\boldsymbol{e}, \mathbf{out})$ will produce the multiset of spikes $\boldsymbol{e^m}$ in the neuron $\boldsymbol{\sigma}_0$. The subsequent application of the rules $\boldsymbol{e^+b/e} \rightarrow (\boldsymbol{e}, \mathbf{out})$ **and** $\boldsymbol{eb/eb} \rightarrow (\boldsymbol{e}, \mathbf{out})$ will produce a sequence of spikes e^m in the environment.

At that time the sequence of spikes emitted from the output neuron $\boldsymbol{\sigma}_0$ will be the language $\boldsymbol{L}(\pi) = \{\, \boldsymbol{a^n b^m c^n e^m} | \boldsymbol{n}, \boldsymbol{m} \geq 1\}$. The output string emitted from $\boldsymbol{\sigma}_0$ is mapped to the chain code alphabet to the set {l, r, u, d}. The homomorphism Ø: $\boldsymbol{L}(\pi) \rightarrow \boldsymbol{\Sigma}$ is as follows,

$$\emptyset(\boldsymbol{a}) = \boldsymbol{r}, \emptyset(\boldsymbol{b}) = \boldsymbol{d}, \emptyset(\boldsymbol{c}) = \boldsymbol{l}, \emptyset(\boldsymbol{e}) = \boldsymbol{u}$$

Hence, the O-shaped pixel picture language obtained by cell-like SN + P system is given by.

$\emptyset(\boldsymbol{L}(\pi)) = \{\ \boldsymbol{r}^n \boldsymbol{d}^m \boldsymbol{l}^n \boldsymbol{u}^m | \boldsymbol{m}, \boldsymbol{n} \geq 1\}.$

4 Conclusion

In this article, we have studied the construction of the one-dimensional pixel picture languages and O-shaped pixel picture language using the spiking neural p system with several types of spikes. It remains to further investigate the recognition of pixel picture languages using SN P system and its application in big data. The study can be applied in recognizing particular form of a data from a reservoir of the big data system.

References

1. G. Costagliola, F. Ferrucci, C. Gravino, Adding symbolic information to picture models: definitions and properties. Theoret. Comput. Sci. **337**(1–3), 51–104 (2005)
2. Ezhilraman, S. Vahini, and Sujatha Srinivasan. "State of the art in image processing & big data analytics: issues and challenges." *International Journal of Engineering & Technology* 7.2.33 (2018): 195–199.
3. Freeman, Herbert. "Computer processing of line-drawing images." ACM 215 Computing Surveys (CSUR) 6.1 (1974): 57–97.
4. Ionescu, Mihai, Gheorghe P˘aun, and Takashi Yokomori. "Spiking neural P systems." Fundamenta informaticae 71.2, 3 (2006): 279–308.
5. Ionescu, Mihai, et al. "Spiking neural P systems with several types of spikes. Proceedings of the Ninth Brainstorming Week on Membrane Computing, 220183–192. Sevilla, ETS de IngenieríaInformática, 31 de enero-4 de febrero,2011 (2011).
6. Jebasingh. S, G.Johnsy and P.S.Divya "Chain Code P system using Cycle Grammar", International Journal of Recent Techonology and engineering, Vol-8 Issue 2, July 2019. 225
7. Leandro Nunes de Castro, Fundamentals of natural computing: an overview, Physics of Life Reviews, Volume 4, Issue 1, 2007, Pages 1–36, ISSN 1571–0645.
8. H.A. Maurer, G. Rozenberg, E. Welzl, Using string languages to describe picture languages. Inf. Control **54**(3), 155–185 (1982)
9. Preethi.C.Y, Hepzibah A. Christinal , S. Jebasingh, and A. Chandy. "Generation of Cycle Picture Languages Using Sequential Spiking Neural P System" Advances in Mathematics: Scientific Journal 2020, Vol. 9, No. 2: 583–590
10. Subramanian, K. G. "P systems and picture languages." *International Conference on Machines, Computations, and Universality*. Springer, Berlin, Heidelberg, 2007.
11. Subramanian, K. G., I. Venkat, and L. Pan. "P systems generating chain code picture languages." Proceedings of Asian Conference on Membrane Computing. 2012.

A Decision Support System for Restricted Movement with Semaphores During Lockdown in COVID-19 Pandemic

Bushra Haqqi and R. Joshua Samuel Raj

1 Introduction

The COVID-19 pandemic in India is part of the global coronavirus disease pandemic of 2019 (COVID-19), which is caused by coronavirus 2 that causes severe acute respiratory syndrome (SARS-CoV-2). On January 30, 2020, India announced the first case of COVID-19, which originated in China. The Indian government imposed a state-wide lockdown on March 24, 2020, restricting the movement of the country's 1.3 billion people. Lockdown has been identified as the sole way to stop the coronavirus from spreading [1]. The impacts of lockdown on illness are overlapping, and modeling implies that differentiating between them is critical for understanding the situation in India. The lockdown limits the spread of the infection by requiring imposing physical separation.

Second, it reduces the transmission of illness across geographical boundaries. Of course, neither impact is absolute. Poor lockdown planning resulted in fear and congestion, which likely sped up transmission for a short time. The abrupt announcement of a state-wide lockdown on March 24, 2020, sparked considerable concern about migrants' immediate, short-, and long-term futures in their destination countries. Multiple concerns linked to social, educational, economic, political, agricultural, psychological, and other levels have been identified amid the lockdown in Indian society, which has had a disastrous influence on the lives of the people.

Though lockdown is likely to be necessary [2], short-term lockdowns, quarantine, and social distance can lead to long-term impacts such as sleeplessness, anxiety, despair, and post-traumatic stress disorder symptoms and a proper scheduling of the E-pass is required to exploit the benefits which can be done with semaphores as it has been proven for real-time tasks [3]. The COVID-19 epidemic has resulted in a

B. Haqqi · R. Joshua Samuel Raj (✉)
Department of ISE, CMR Institute of Technology, VTU, Bengaluru, India
e-mail: joshuasamuelraj@gmail.com

J. D. Peter et al. (eds.), *Disruptive Technologies for Big Data and Cloud Applications*, Lecture Notes in Electrical Engineering 905,
https://doi.org/10.1007/978-981-19-2177-3_76

variety of mental health reactions, depending on the strengths and limitations of the person. The lockdown has demonstrated that "man is a social being," since there have been several proven cases of mental illness and despair as a result of the lockdown, which lasted nearly a year. Because India is known for its variety, COVID-19 has a wide and varied influence. Both rural and urban economies have suffered as a result of the economic downturn.

The lockdown had a significant influence on 63.4% of the population's job and income, which was favorably associated with persons aged 35–50 years, followed by 50–65 years, affecting their financial situation. During the lockdown, those at the top of the pyramid are likely to have had their earnings safeguarded and savings rates pressed up, putting more "fuel in the tank" for future consumption. Meanwhile, lower-income households are likely to have experienced long-term job and income losses, limiting their spending. The cleavages may already be seen.

The lockdown has been justified on the grounds that it prioritizes "life" above "livelihoods" and "health" over "economic well-being." Despite the tight quarantine, India's response to the epidemic has been underwhelming because of the misuse of E-pass given for travel as it was not customized and followed in a systematic manner.

With 26.7 million recorded cases of COVID-19 infection and 307,231 fatalities, India now has the world's second-highest number of confirmed cases (after the USA) and the third-highest number of COVID-19 fatalities (after the USA and Brazil). It was the first country to report over 400,000 new cases in a 24 h period on April 30, 2021, as shown in Fig. 1. Many governments have enforced lockdown measures to contain the second wave of COVID-19 and made it essential for people to obtain an e pass for travel movement due to an increasing number of cases, and the addition of intelligence can be done by implementing semaphores for E-passes.

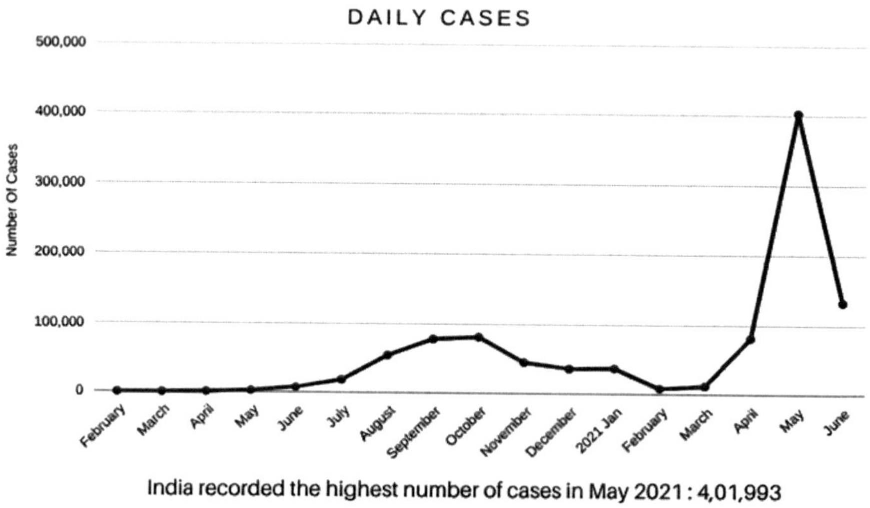

Fig. 1 History of COVID-19 cases count in India

The paper can be outlined as follows: the measures to combat COVID's growth are discussed in Sect. 2, and semaphores in Sect. 3. Section 4 elaborates on the methods and analysis done in our work. Conclusions and recommendations for future work are found in Sect. 5.

2 Measures to Combat COVID-19's Growth

Many means were used to control the spread of the fatal virus, such as social distance, quarantine, isolation, lockdowns, curfews, travel restrictions, schools/colleges closures, and so on, to control the spread of the COVID-19 virus, with varying results.

Social distancing is a method which was employed to curtail the number of people gathering at a particular place for any kind of functions or activities. It asked people to not risk their life, their family's lives, and the lives of others who are a part of those gatherings. Quarantine was a measure for individuals having been tested positive or those having symptoms to stay indoors away from non-affected people and reduce the number of people getting affected by this virus.

Screening was a method where the body temperature of people would be checked by means of thermometer at all possible places like airports, malls, and stations. Containment is used in the early phases of an outbreak, and it involves tracing and isolating persons who have been affected, as well as additional actions to prevent the illness from spreading.

When containing the illness is no longer viable, efforts shift to mitigation, which refers to actions attempted to reduce the illness's spread and minimize its impact on the healthcare system and society. Personal preventive measures such as hand hygiene, wearing face masks, and self-quarantine, community measures aimed at physical distancing such as closing schools and canceling mass gathering events, and community engagement to encourage acceptance and participation in such interventions are all non-pharmaceutical measures that help to curb the outbreak, as well as steps to protect the environment, such as surface cleaning. People will only be able to travel in an emergency, and owning an E-pass has become obligatory for inter-state travel. During the lockdown, this E-pass will allow you to roam about, but it should only be used in an emergency or urgent scenario.

3 Semaphores and Its Context

An event handler will "give" a semaphore each time an event happens (incrementing the semaphore count value), and a handler task will "take" a semaphore each time it processes an event in the counting events use scenario (decrementing the semaphore count value). The difference between the number of events that have happened and the number of events that have been handled is the count value. When the semaphore is generated, it is preferable for the count value to be zero [4]. Semaphore is also useful for implementing resource pools such as database connection pools.

4 Methods and Analysis

Many fraudulent examples of obtaining the E-pass have been documented, raising doubts about its validity. For example, it has been stated that brokers for obtaining E-passes may be found throughout the state of Tamil Nadu for a cost ranging from Rs. 500 to Rs. 2000 [5]. Getting E-passes on payments has been extensively reported in the media. Genuine candidates were unable to obtain an E-pass even after providing supporting documentation, prompting widespread criticism of the online application procedure.

We propose a model to use counting semaphores to issue E-pass for essential travel movements in lockdown as part of intelligence [6]. We will set an upper limit for each district considering the number of cases in that area and other lockdown guidelines issued by that particular state. This will not only help in preventing COVID-19 by restricting the travel movement but also enforce social distancing as a limited number of people will be allowed to move at any given time and will prove to be more authentic.

A semaphore is a synchronization tool to execute the critical section one by one. Semaphore can be thought of as a shared resource that is limited in quantity.

Example: Semaphore $S = 0$: no S resource available
Semaphore $S = 1$: Exactly one S resource is available
Semaphore $S = 2$: Two S resources are currently available

As long as the value of s is zero, the code that is waiting for this process is stalled. To get past the wait (S) command, the value of the semaphore must be greater than zero, indicating that the resource is available.

Grabbing the resource is achieved by decrementing the value of the semaphore by one. A signal (s) of semaphore indicates that one additional S resource has become available. The signal S command increments the value of s by one.

Counting semaphores are used when we need more than one process in the critical section at a given time. The value of counting semaphores indicates the maximum number of processes that can enter into the critical section at the same time.

This paper proposes to use counting semaphores to issue and keep track of maximum number of E-passes at any given time to curtail the spread of the virus by restricting the number of people allowed for movement. The E-passes will be issued to the people on a first come first serve basis and upon reaching the maximum limit, the program will display a limit reached message. A person can still apply for an E-pass but he'll be directed to the waiting queue and once a person submits back the E-pass by logging out of the system, the first person in the waiting queue will be allotted that E-pass.

Algorithm for the proposed solution:

(1) Start
(2) Setting upper limit for E-pass
(3) Person logged into the system
(4) Checking if limit has been reached or not
(5) If the limit is not reached, pass allotted to the person and count incremented. Go to step 7
(6) If the limit is reached, the person is in the waiting queue. Go to step 4
(7) Person logged out of the system and count decremented.
(8) End.

In our proposal of using counting semaphores to issue E-passes, we will be analyzing the population density accompanied with the number of active cases.

We have considered the daily average cases (7 days) and number of cases per 100,000 residents in the last 7 days (30/05/2021–05/06/2021) [7]. The E-passes will be generated according to the range of active cases in a particular district or state.

Table 1 shows the number of daily average cases taken over 7 days and number of cases per 100,000 of residents in that state. The upper limit of E-pass allowed is calculated by taking into account the range of cases per 100,000 and taking percent of population accordingly.

The extent of E-passes allowed is shown in Table 1.

We have considered the COVID-19 count per lakh cases and have suggested the following percentage of E-passes that can be sanctioned, and this is considered as the baseline to initialize the semaphore count as depicted in Table 2.

A visual statistics is generated based on the cases populated against the E-passes that can be generated by obtaining the semaphore in Fig. 2.

Taking into consideration the districts of Karnataka and analyzing the number of active cases [8], the maximum number of E-pass allowed at any particular time is taken as a percent according to the number of active cases, of the population of that district and is shown in Table 3.

A visual statistics is generated based on the active cases populated against which the maximum E-passes that are generated by obtaining the semaphore in Fig. 3.

Table 1 E-passes allowed statistics

State	Population	Daily avg. (7 days) cases	Per 100,000 cases	E-pass allowed
Maharashtra	12.43 cr	16,092	14	14,916,000
Tamil Nadu	7.85 cr	25,529	37	9,420,000
Karnataka	6.83 cr	17,528	29	8,196,000
Andhra Pradesh	5.44 cr	11,572	23	6,528,000
Kerala	3.6 cr	18,601	56	5,440,000
Lakshadweep	74,151	147	227	2966

Table 2 Upper ceil of E-passes in different states of India

COVID-19 cases count per lakh	Upper limit of E-pass in percentage of population (%)
0–49	12
50–99	10
100–149	8
150–199	6
200–249	4

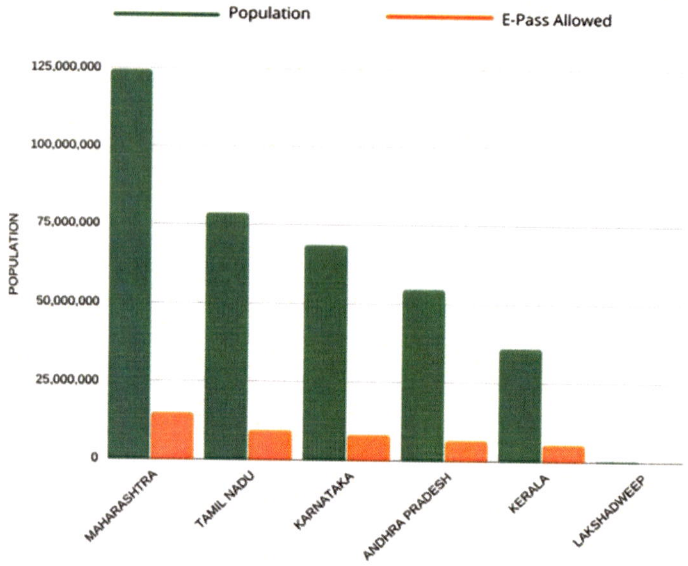

Fig. 2 Population versus E-passes allowed

By taking into account the number of active cases, the upper limit of E-pass is generated and the same is set as the value of the counting semaphore. The counting semaphore which is associated with the E-passes is made available for the people who would like to get an E-pass to travel, and based on the availability, it is approved. Whenever an E-pass is given, the semaphore count is decreased with the help of its wait operation, and whenever a person has completed his purpose of E-pass and surrenders it, the semaphore count is increased, thereby keeping a check on the people who are outside with the E-pass. The semaphore concept keeps a check on the upper limit and also does not give the pass if its value is reduced to zero. Using counting semaphores to generate the e-passes is better management as compared to the current approach. An upper limit will be set by analyzing the active cases daily, and the passes will be allotted to the people needing them, thus controlling the spread of the virus as can be seen with the dataset in Table 4 and the graphical output in Fig. 4.

Table 3 Upper ceil of E-passes in different districts of Karnataka

District	Active cases	Population	E-pass allowed
Bangalore Urban	0.0	10,960,871	1,315,304
Tumkur	139	3,051,894	366,227
Kalaburagi	200	2,923,559	292,355
Raichur	119	2,197,303	262,956
Davanagere	408	2,216,310	177,304
Dharwad	404	2,104,129	168,330
Hassan	153	2,023,699	242,843
Shimoga	283	1,996,736	199,673
Chitradurga	70	1,890,452	226,854
Kolar	88	1,750,268	210,032
Uttara Kannada	158	1,637,223	196,466
Active cases count based E-pass allowed percentage			
0–199—12%	200–399—10%	400–599—8%	600–799—6%

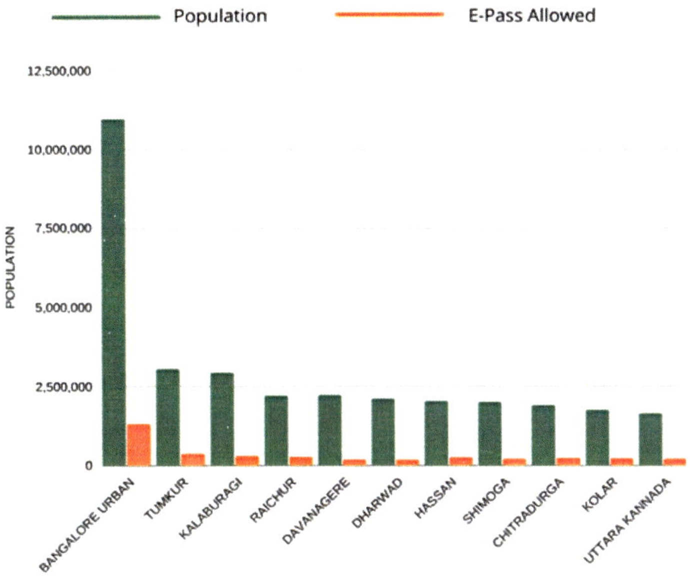

Fig. 3 E-passes allowed district-wise in Karnataka

Our proposed approach has served its purpose as can be seen from the comparative Table 5, wherein the random approach is not able to place an upper limit in terms of issuance of E-passes, whereas our proposed approach with semaphores serves the purpose of E-pass as shown in Fig. 4, wherein it exposes the maximum number of people who move out of their homes on emergency and it is the actual count of

Table 4 Random versus semaphore approach in E-pass

District	Active cases	Population	E-pass allowed (random approach)	E-pass allowed (with semaphores)
Bangalore urban	0	10,960,871	6,028,479	1,315,304
Tumkur	139	3,051,894	1,678,541	366,227
Kalaburagi	200	2,923,559	1,461,779	292,355
Raichur	119	2,197,303	1,208,516	262,956
Davanagere	408	2,216,310	997,339	177,304
Dharwad	404	2,104,129	946,858	168,330
Hassan	153	2,023,699	1,113,034	242,843
Shimoga	283	1,996,736	998,368	199,673
Chitradurga	70	1,890,452	1,039,748	226,854
Kolar	88	1,750,268	962,647	210,032
Uttara Kannada	158	1,637,223	900,472	196,466

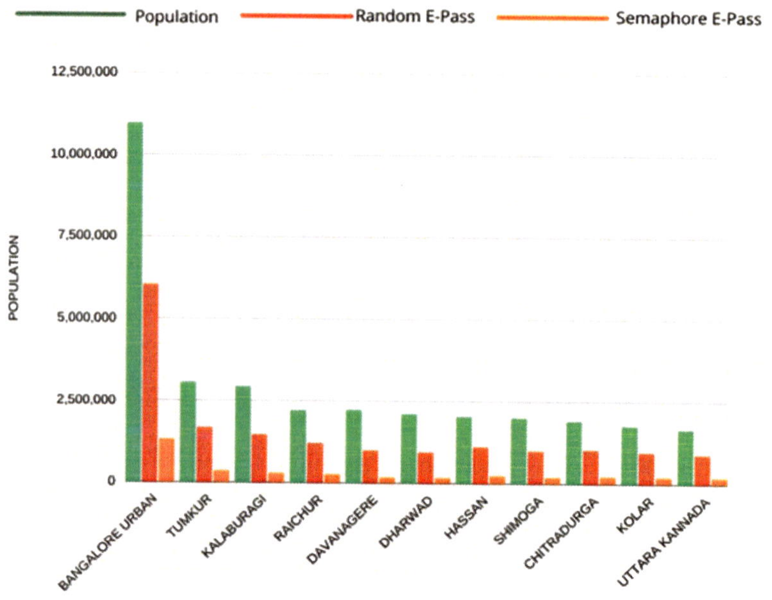

Fig. 4 Random and semaphore E-pass versus population

people outside as a person can return back and surrender the pass that can be taken by another person in need, exploiting the semaphore concept.

It has been reported that the spread of the coronavirus to this extent has mainly been because of ineffective measures taken to curtail it and by not following the proper guidelines issued by the WHO [9]. Another possible reason might be that

Table 5 Upper ceil of E-passes

COVID-19 cases count per lakh	Upper limit of E-pass in percentage of population in random approach (%)	Upper limit of E-pass in percentage of population with semaphores (proposed approach) (%)
0–199	55	12
200–399	50	10
400–599	45	8
600–799	40	6

about 80% of COVID-19 patients in India are asymptomatic (an internal analysis by the Integrated Disease Surveillance Programme (IDSP) has concluded after extensive discussions with states), and they travel around the streets coming in contact with several other persons transmitting the virus. So restricting the movement of people and maintaining social distancing is the need of the hour. Using counting semaphores we can restrict the movement of people at the particular point of time, in times of lockdown which will not only prove to be more effective but also helps in keeping track of the number of people with valid E-passes and as the issuance is transparent, it can also play a major role on the mental health.

5 Conclusion and Future Work

We concluded from the analysis of our proposed model that incorporating counting semaphores to generate the E-pass for travel around the districts will prove to be an effective way in controlling the spread of the virus. As seen from our decision support system by taking into consideration, the various factors such as the population of the people being considered for lockdown, alongside the history of COVID-19 cases count ensures that the purpose of lockdown is met more meaningfully and ensures a smooth and clean policy of getting E-pass when available. This decision support system with the help of semaphores ensures that there is an upper limit on the number of E-passes that are issued without any hidden agenda. In the future, we plan to extend this work to consider the deaths due to COVID-19 as a factor to issue E-pass and extend the issuance of E-pass as privilege for medical reasons.

References

1. D.W. Allen, COVID lockdown cost/benefits: a critical assessment of the literature (2021)
2. https://www.ncbi.nlm.nih.gov/pmc/articles/PMC7301781
3. X. Jiang, N. Guan, W. Liu, M. Yang, Scheduling and analysis of parallel real-time tasks with semaphores, in *2019 56th ACM/IEEE Design Automation Conference (DAC)*, pp. 1–6 (2019)
4. A. Silberschatz, P.B. Galvin, G. Gagne, *Operating System Concepts*, 9th edn. (Wiley, 2002)

5. https://www.health.harvard.edu/diseases-and-conditions/preventing-the-spread-of-the-coronavirus
6. J.S. Raj, V. Vasudevan, Smart bacterial foraging optimization algorithm for scheduling in grid. Eur. J. Sci. Res. **94**(2), 253–260 (2013)
7. https://www.nytimes.com/interactive/2021/world/india-covid-cases.html
8. https://covid19.karnataka.gov.in/covid-reports/report
9. https://www.outlookindia.com/website/story/india-news-demands-for-scrapping-tamil-nadus-e-pass-system-get-louder/358323

Comparative Analysis of Deep Learning Models for Cotton Leaf Disease Detection

X. Anitha Mary, Kumudha Raimond, A. Peniel Winifred Raj, I. Johnson, Vladimir Popov, and S. J. Vijay

1 Introduction

One of essential crop fiber is cotton, which are cultivated almost in 70 countries. In India, cotton is also called "White Gold" as India has a larger plantation area in cotton among other countries in the world. Cotton yarn and cotton fabrics account for roughly a quarter of India's overall textile and apparel exports. Cotton production in India was estimated to be 6.42 million metric tons in the year 2019–2020 [1]. This year, India is expected to grow 32.8 million bales (170 kg each), a decrease of 3.7 million bales from last year's crop of 36.5 million bales. This decline is due to various leaf diseases. A new pathogenic virus called leaf curl virus (CLCuV) is affecting a large proportion in northern India damaging nearly 15 lakh hectares [2]. So, timely and accurate diagnosis of leaf disease is important for sustainable agriculture. As most of the plant diseases show visible symptoms, experienced plant

X. A. Mary (✉) · A. P. W. Raj
Department of Robotics Engineering, Karunya Institute of Technology and Sciences, Coimbatore, Tamil Nadu, India
e-mail: anithamary@karunya.edu

K. Raimond
Department of Computer Science Engineering, Karunya Institute of Technology and Sciences, Coimbatore, Tamil Nadu, India
e-mail: k_raimond@karunya.edu

I. Johnson
Department of Plant Pathology, TNAU, Coimbatore, India

V. Popov
Additive Manufacturing, Technion–Israel Institute of Technology, Haifa, Israel

S. J. Vijay
Department of Mechanical Engineering, Karunya Institute of Technology and Sciences, Coimbatore, Tamil Nadu, India

J. D. Peter et al. (eds.), *Disruptive Technologies for Big Data and Cloud Applications*, Lecture Notes in Electrical Engineering 905,
https://doi.org/10.1007/978-981-19-2177-3_77

pathologist is needed to diagnose such diseases [3]. However, even a good pathologist fails in characterizing the diseases due to excess variety of plants, plant diseases which can vary due to climate and fast-spreading nature [4]. Hence, it is required to have an intelligent system that automatically diagnoses the diseases accurately. In this paper, the performance of three CNN architectures are evaluated to classify the disease leaf from the healthy cotton leaf.

2 Related Works

Today, Artificial Intelligence (AI)-based automatic leaf disease identification is often used [5]. Revathi and Hemalatha [6] proposed a method to detect six varieties of disease in cotton leaf. Particle Swarm Optimization (PSO) is used for feature extraction, and gain deep forward neural network is used to diagnose the disease. Zhang et al. [7] uses seven varieties of cotton diseases. Chan-Vese (C-V) algorithm and Local Binary Fitting (LBF) algorithm are used for segmentation. The model has achieved better accuracy than Geodesic Active Contour (GAC) algorithm. A new model named GMDH (Group Method of Data Handling) logistic algorithm is used by Chen et al. [8] to detect plant disease. The clustering technique is adopted with ANN to categorize different plant disease [9]. Support Vector Machine (SVM) methods are frequently used by many researchers [10, 11] for the detection of tomato and vine leaf disease and in sugar beet [12]. Naïve Bayes classifier is also being used for early diagnosis of wheat disease [13]. In [14], forward neural network is used for determining the injured leaf spot in cotton giving an accuracy of 95%. Zhang et al. [7] uses backpropagation neural network for identification of cotton leaf disease. Sladojevic et al. [15] used CNN for classifying different plant diseases with 30,880 images as testing and 2589 images as training data. CNN models like AlexNet and GoogLeNet [16], VGG 16 [17], Inception V4, ResNet50, ResNet101, Resnet152, and DenseNets 121 [18] are also used. Nine layer CNN is used for the plant village dataset and reported an accuracy level of 96.46% [19]. Apart from cotton leaf, many researchers use deep learning algorithms for disease identification for various crops like tomato, sugar beet, wheat, rice, etc. [20–26]. Table 1 shows the summary of CNN models used by different researchers in classifying different crops.

3 Materials and Methods

3.1 Dataset

In this study, cotton leaf datasets obtained using Kaggle, GitHub, plant village, and Internet sources are used containing 906 images, out of which 519 are healthy leaves and 387 are diseased leaves. The different diseases used are Alternaria leaf spot,

Table 1 Application of CNN Models in crop classification task

Author and year	Name of the crop	Architecture/Algorithm used	Accuracy in %
Revathi and Hemalatha [6]	Cotton leaf	Cross information Gain deep forward neural network	95
Zhang et al. [7]	Cotton leaf	BP neural network	92.72
Kawasaki et al. [20]	Cucumber leaf	CNN architecture-Customized	94.9
Mohanty et al. [21]	Various crop—Apple, blueberry, cherry, corn, grape, orange, potato, tomato	CNN-AlexNet and GoogleNet	99.34
Nachtigall et al. [22]	Apple	CNN-AlexNet	97.30
Brahimi et al. [23]	Tomato	AlexNet and GoogleNet	99.18
Ranganathan [24]	Tomato	AlexNet and VGG 16	97.49
Wang et al. [25]	Apple leaf	VGG 16 and VGG 19 Inception V3 Residual Network (ResNet) 50	90.4
Lu et al. [26]	Rice	AlexNet	95.48

Ascochyta Blight, bacteria blight, and Cercospora Leaf Spot and target leaf spot. Figure 1a shows images for healthy leaf, and Fig. 1b shows the image of Alternaria leaf spot which is characterized as circular and brown in the form of concentric rings and is found to be a common disease in cotton. Figure 1c shows the Ascochyta Blight which is characterized by light brown or white sport with different sizes. Bacterial Blight symptoms begin as small water-soaked spots and grow to distinctively angular shapes as bacterial movement is restricted by leaf veins. Lesions appear on the leaf's upper side and turn black as they develop, and defoliation can occur as shown in Fig. 1d. Figure 1e shows the Cercospora Leaf Spot. During the early stages, reddish lesions will appear. The lesions expand and turn white to light brown in the middle as the disease progresses [27]. Brown lesions, often up to 2 cm (1 inch) in diameter and exhibiting a pattern of concentric circles, are typical signs of Target Spot (Fig.1f). Apart from theses diseased leaf, healthy leaf is also taken for classification. Table 2 shows the details of dataset used.

3.2 Deep Neural Network—CNN Model

Deep learning has proven to be a very effective technique in recent decades due to its ability to handle vast volumes of data. Hidden layers have surpassed conventional techniques in popularity, especially in pattern recognition. CNNs is one of the most widely used deep neural networks [28]. The proposed method uses the ConvNet

Fig. 1 Images of cotton leaves **a** Healthy, **b** Alternaria leaf spot, **c** Ascochyta Blight, **d** bacteria blight, **e** Cercospora Leaf spot, **f** Target spot

Table 2 Details of dataset

Classes	Number of images
Healthy leaf	519
Alternaria leaf spot	102
Ascochyta blight	73
bacteria blight	80
Cercospora leaf spot	48
Target spot	84
Total	906

model to classify the healthy and diseased leaf as shown in Fig. 2. The ConvNet model is to reduce the images into a format that is easier to process while preserving features that are essential for obtaining a good prediction. The Pooling layer, like the Convolutional Layer, is responsible for shrinking the Convolved Feature's spatial scale. The computing power needed to process the data is reduced. It is also useful for removing rotational and positional invariant dominant characteristics, which helps to keep the molecular training phase moving forward. In this, max pooling layer is used, which returns the maximum value. Max Pooling works as a Noise Suppressant as well. It removes all noisy activations and performs de-noising as well as dimensionality reduction.

Many CNN state-of-the-art architectures have emerged and are being used for many applications on large benchmark datasets. Instead of building a model from

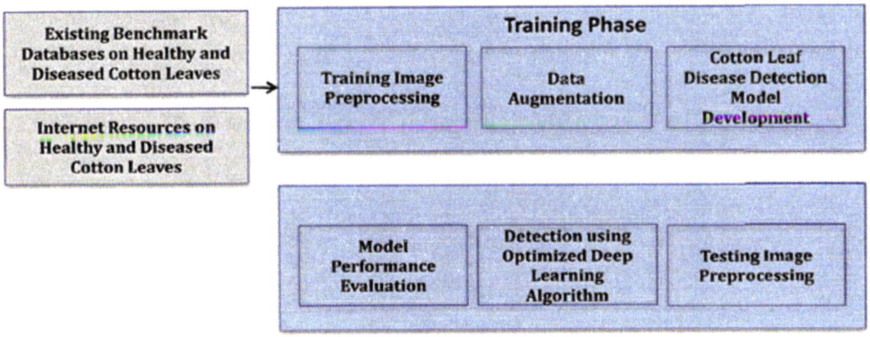

Fig. 2 Block diagram of the proposed method using deep CNN model

scratch for similar applications, the concept of transfer learning is used to accurately build new models using the pre-trained CNN models such as LeNet, ResNet, VGG 16, and VGG 19. In this work, AlexNet, InceptionV3, ResNet50, VGG 16, NASNet, and Xception models are used for the comparative analysis.

3.3 AlexNet

AlexNet is a CNN architecture as designed to recognize images and won ImageNet dataset challenge of size 14 million images of thousand categories. The model was built to accommodate colored images of size (224 × 224 × 3). The model consists of five convolutional layers, two normalization layers, and three fully connected layers. First two convolution layers follow a similar design, where the convolution operation is followed by max pooling and normalization operations, which is followed by three subsequent convolution layers and one max pooling layer. Each convolutional layer includes kernels and a nonlinear activation function Rectified Linear Unit (ReLU). The output layer consists of two subsequent fully connected layers and a soft max layer. The design has accelerated the training speed and prevented the model overfitting.

The image is first resized to 227 * 227 pixels. Following that, the image is sent to the first convolution layer. The first layer is made up of 96 kernels that are 11 * 11 in size with a stride of 4. These kernels detect the image's numerous edges. The image is then transmitted to the second convolution layer, where 256 kernels of 5 * 5 size are applied. With every kernel function, the max pooling layer reduces the picture dimensions by 3 * 3 pooling. The input to the third and fourth convolution layers are processed using 384 kernels of size 3 * 3. The fifth layer performed a convolution operation using 256 kernels of size 3 × 3. Finally, 4096 neurons were sent as input to three fully connected layers. The final layer produced an output of size 1000 × 1. After all convolution and the last two fully linked layers, a nonlinear and non-saturating activation function termed ReLu is applied. It increases the rate. To combat

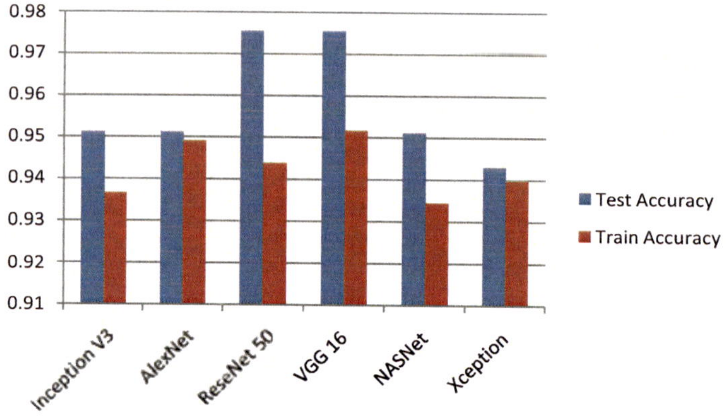

Fig. 3 Train and test accuracy

overfitting, the dropout is added to the network. The dropout will increase network performance during the test phase. Figure 3 shows the architecture of AlexNet.

3.4 VGG 16

It is a 16-layer deep CNN with thirteen convolution, five max pooling, and three dense layers. It was designed to improve the performance of AlexNet by using 3 × 3 kernels with a stride 1 and max pool layer of 2 × 2 filter of stride 2 for all layers instead of applying large filters of size 11 × 11 and 5 × 5 in the first and second convolutional layer‘s, respectively. The kernels of size 64, 128, 256, and 512 were used for convolution operations in subsequent layers. The output layer consists of two dense layers with 4090 nodes followed by the final dense layer with 1000 nodes will be used to classify 1000 image categories. The details are shown in Fig. 4.

3.5 Inception V3

The deep convolutional architecture, Inception V3, is commonly utilized for classification tasks. In the GoogleNet design, Szegedy introduced the model notion, and Inception V3 was offered by modifying the inception module [29]. Multiple symmetric and asymmetric construction components make up the Inception V3 network. With fewer parameters, Inception V3 is a 42-layer deep learning network. Factorizing convolutions are used to reduce the number of parameters. The model will be less overfit and thus more accurate with a reduced number of parameters. The Inception V3 pre-trained model is utilized to perform transfer learning on the problem of cotton leaf disease in this paper.

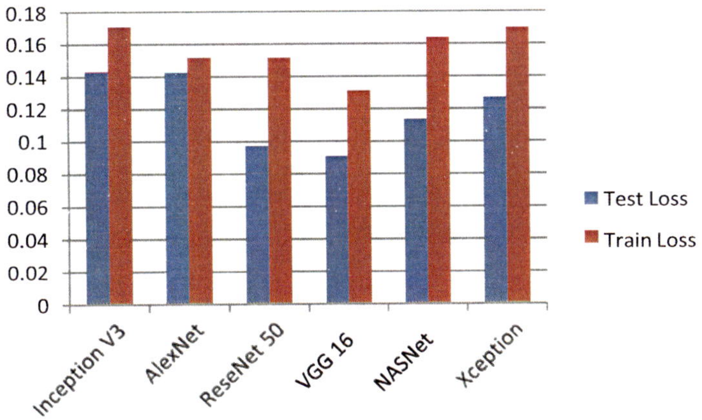

Fig. 4 Train and test loss

3.6 *Residual Network*

Single-layer architecture is useful to approximate any function, but it may overfit the data. Though deeper architectures are much useful, it has resulted in vanishing gradients problem [30] while training such a deep network. The gradient that is propagated back will slowly diminish and hinder the learning process.

But to solve the problem of vanishing gradients [30], ResNet is presented. Studies have discovered that VGGNets that use small kernels can only work to a certain depth due to the usage of small kernels. Skip connections are implemented in ResNet. Residual networks are presented as skip connections in this example. By adding the block's input to its output, the network is supposed to learn "the residue" of the networks. ResNet50 is implemented in the paper. Conv Block and Identity Block are stacked with the structure to create ResNet. In the beginning, the image data is passed through layers of a CNN. Five Conv Blocks and thirteen Identity Blocks make up ResNet 50. Before submitting it through soft max classification, average pooling is applied at the end.

3.7 *Xception*

Xception (extreme inception) extends the principles of operation of inception. It is a ConvNet with 71 layers. During the convolution operation on 3D images, the filter simultaneously considers both the depth (channels) and spatial dimensions (image region). In Xception, depth wise and spatial convolutions are performed independently for each channel which is followed by a pointwise convolution using a 1 × 1 convolution across channels.

3.8 NASNet

Google introduced NASNet, which framed the task of finding the ideal CNN architecture as a Reinforcement Learning challenge, thanks to its vast computer capacity and engineering talent. The basic idea was to find the optimal combination of filter sizes, output channels, strides, number of layers, and other characteristics in the specified search space.

4 Results and Discussion

The raw images with 906 images are divided into 80% training and 20% testing dataset. Further, to avoid overfitting, the training data is split into 80% training data and 20% validation data. The entire experiment was carried out using Keras and tensor flow.

The applicability of deep CNN is evaluated with six architecture, namely AlexNet, ResNet, Inception V3, NASNet, Xception, and VGG 16. Accuracy is one of the most commonly used metric to evaluate the performance of architecture. The accuracy was calculated for every 10 iteration till 50 epochs.

$$\text{Accuracy} = \frac{\text{TP} + \text{TN}}{\text{TP} + \text{TN} + \text{FP} + \text{F}} \quad (1)$$

where TP = True Positive; TN = True Negative; FP = False Positive; FN = False Negative

The comparison of training (Tr) and testing (Te) accuracy for six architecture is given in Table 3 for 10 to 50 epochs. Figure 3 shows the accuracy of various architecture up to 50th epochs, and Fig. 4 shows the training and testing loss of the above architectures. In terms of exactness and loss value, the ResNet50 model outperforms the other five models in the exploratory results.

Table 3 Performance measure of six pre-trained CNN models

Architecture	After 10 epochs		After 20 epochs		After 30 epochs		After 40 epochs		After 50 epochs	
	Tr	Te	Tr	Te	Tr	Te	Tr	Te	Tr	Te
Inception V3	85.25	84.55	88.93	93.5	87.95	95.12	90.69	94.31	93.67	95.12
AlexNet	82.62	93.5	90.18	95.93	90.17	96.75	91.58	94.31	94.92	95.12
ResNet50	83.22	88.62	86.94	95.12	90.59	95.12	89.7	92.68	94.39	97.56
VGG 16	81.26	90.24	89.07	95.93	90.28	89.43	92.6	95.93	95.17	97.56
NASNet	78.32	81.3	85.34	89.43	91.66	95.93	92.45	95.93	93.45	95.12
Xception	83.66	91.87	85.34	89.43	89.26	97.56	93.75	96.75	93.99	94.31

All the pre-trained models showed similar performance with small percentage variations in accuracy metric. The Xception model has resulted in lower accuracy with 94.31%, followed by Inception V3, AlexNet, and NASNet with 95.12%, VGG 16 with 96.56%, and the best classification model is obtained using ResNet50 and VGG 16 with 97.56% accuracy in classifying the cotton leaf as normal or diseased.

5 Conclusion

Convolutional neural networks' performance in object detection and picture classification has improved dramatically in recent years [31]. Based on deep CNN models, six architectures, namely AlexNet, ResNet50, VGG 16, NASNet, Inception V3, and Xception, were implemented and verified the exactness and loss value for cotton leaf disease images. It is observed that ResNet 50 and VGG 16 over-performs other models. Increased datasets can enhance classification accuracy since deep learning models can operate efficiently on them.

In future work, feature extraction strategies for data pre-treatment that are most suitable for the deep learning model for higher performance can be chosen. Using a fast computer device such as a GPU can increase performance even more.

References

1. Cotton-the white Gold of India, https://www.sutlejtextiles.com/blogs/blog-post.php?id=D21
2. S.G. Bhattacharyya, P. Kumar, D. Monga, K.K. Biswas, Recent status and distribution pattern of cotton leaf curl disease in Northwest India: An alarming situation in future cotton cultivation. Indian J. Agric. Sci. **87**(5), 624–633
3. S. Sankaran, A. Mishra, R. Ehsani, C. Davis, A review of advanced techniques for detecting plant diseases. Comput. Electron. Agric. **72**, 1–13 (2010)
4. S. Sladojevic, M. Arsenovic, A. Anderla, D. Culibrk, D. Stefanovic, Deep neural networks based recognition of plant diseases by leaf image classification. Comput. Intell. Neurosci. 11 (2016), https://doi.org/10.1155/2016/3289801
5. N. Jiang, Y. Jiang, H. Zhi, Y. Dong, H. Li, S. Ma, Y. Wang, Q. Dong, H. Shen, Y. Wang, Artificial intelligence in healthcare: past, present and future. Stroke Vasc. Neurol. **2**, 230–243 (2017), https://doi.org/10.1136/svn-2017-000101
6. P. Revathi, M. Hemalatha, Identification of cotton diseases based on cross information gain deep forward neural network classifier with PSO feature selection. Int. J. Eng. Technol. **5**, 4637–4642 (2014)
7. J.H. Zhang, L.J. Qi, R.H. Ji, Cotton diseases identification based on rough sets and BP neural network. Trans. Chin. Soc. Agric. Eng. **28**, 161–167 (2012)
8. Vaamonde, A.D. Navajas, A. Ortiz-Barredo, Automatic plant disease diagnosis using mobile capture devices, applied on a wheat use case. Comput. Electron. Agric. **138**, 200–209 (2017)
9. H. Al-Hiary, S. Bani-Ahmad, M. Reyalat, M. Braik, Z. Alrahamneh, Fast and accurate detection and classification of plant diseases. Int. J. Comput. Appl. **17**, 31–38 (2011)
10. U. Mokhtar, M.A.S. Ali, A.E. Hassanien, H. Hefny, *Identifying Two of Tomatoes Leaf Viruses Using Support Vector Machine* (Springer, New Delhi, 2015), pp. 771–782, https://doi.org/10.1007/978-81-322-2250-7_77

11. X.E. Pantazi, D. Moshou, A.A. Tamouridou, S. Kasderidis, *Leaf Disease Recognition in Vine Plants Based on Local Binary Patterns and One Class Support Vector Machines* (Springer, Cham, 2016), pp. 319–327, https://doi.org/10.1007/978-3-319-44944-9_27
12. T. Rumpf, A.-K. Mahlein, U. Steiner, E.-C. Oerke, H.-W. Dehne, L. Plümer, Early detection and classification of plant diseases with support vector machines based on hyperspectral reflectance. Comput. Electron. Agric. **74**, 91–99 (2010)
13. A. Johannes, A. Picon, A. Alvarez-Gila, J. Echazarra, S. Rodriguez-Vaamonde, A.D. Navajas, A. Ortiz-Barredo, Automatic plant disease diagnosis using mobile capture devices, applied on a wheat use case. Comput. Electron. Agric. **138**, 200–209 (2017)
14. J. Chen, H. Yin, D. Zhang, A self-adaptive classification method for plant disease detection using GMDH-logistic model. Sustain. Comput. Inform. Syst. **28**, 100415 (2020). https://doi.org/10.1016/J.SUSCOM.2020.1004
15. C. Zhou, H.B. Gao, L. Gao, W.G. Zhang, Particle swarm optimization (PSO) algorithm. Appl. Res. Comput. **12**, 7–11 (2003)
16. S. Sladojevic, M. Arsenovic, A. Anderla, D. Culibrk, D. Stefanovic, Deep neural networks based recognition of plant diseases by leaf image classification. Comput. Intell. Neurosci. 11 (2016). https://doi.org/10.1155/2016/3289801
17. G. Geetharamani, A. Pandian, Identification of plant leaf diseases using a nine-layer deep convolutional neural network. Comput. Electr. Eng. **76**, 323–338 (2019)
18. S.P. Mohanty, D.P. Hughes, M. Salathé, Using deep learning for image-based plant disease detection. Front. Plant Sci. **7**, 1–10 (2016). https://doi.org/10.3389/fpls.2016.01419
19. E.C. Too, L. Yujian, S. Njuki, L. Yingchun, A comparative study of fine-tuning deep learning models for plant disease identification. Comput. Electron. Agric. **161**, 272–279 (2019)
20. Y. Kawasaki, H. Uga, S. Kagiwada, H. Iyatomi, Basic study of automated diagnosis of viral plant diseases using convolutional neural networks, In *International Symposium on Visual Computing* (Springer, Cham, 2015), pp. 638–645
21. S.P. Mohanty, D.P. Hughes, M. Salathé, Using deep learning for image-based plant disease detection. Front. Plant Sci. **7**, 1419 (2016)
22. L.G. Nachtigall, R.M. Araujo, G.R. Nachtigall, Classification of apple tree disorders using convolutional neural networks, In *Proceedings of the 2016 IEEE 28th International Conference on Tools with Artificial Intelligence (ICTAI)* (San Jose, CA, 2016), pp. 472–476, 6–8 Nov 2016
23. M. Brahimi, M. Arsenovic, S. Laraba, S. Sladojevic, K. Boukhalfa, A. Moussaoui, Deep learning for plant diseases: Detection and saliency map visualisation, in *Human and Machine Learning* (Springer International Publishing, Cham, 2018), pp. 93–117
24. A.K. Rangarajan, R. Purushothaman, A. Ramesh, Tomato crop disease classification using pre-trained deep learning algorithm. Procedia Comput. Sci. **133**, 1040–1047 (2018)
25. J. Wang, L. Chen, J. Zhang, Y. Yuan, M. Li, W. Zeng, CNN transfer learning for automatic image-based classification of crop disease, in *Chinese Conference on Image and Graphics Technologies* (Springer, Singapore, 2018), pp. 319–329
26. Y. Lu, S. Yi, N. Zeng, Y. Liu, Y. Zhang, Identification of rice diseases using deep convolutional neural networks. Neurocomputing **267**, 378–384 (2017)
27. K.P. Ferentinos, Deep learning models for plant disease detection and diagnosis. Comput. Electron. Agric. **145**, 311–318 (2018)
28. Diagnosis and Management of Foliar Diseases in the United States › Cotton Production › Cotton Production Research › Plant Pathol. https://www.cottoninc.com/cotton-production/ag-research/plant-pathology/diagnosis-management-foliar-diseases-2/
29. E.C. Too, L. Yujian, S. Njuki, L. Yingchun, A comparative study of fine-tuning deep learning models for plant disease identification. Comput. Electron. Agric. **161**, 272–279 (2018)
30. M. Manav, Convolutional Neural Networks (CNN). https://www.analyticsvidhya.com/blog/2021/05/convolutional-neural-networks-cnn/#:~:text=In%20deep%20learning%2C%20a%20convolutional,applied%20to%20analyze%20visual%20imagery.&text=Now%20in%20mathematics%20convolution%20is,is%20mod ified%20by%20the%20other
31. A. Veit, M.J. Wilber, S. Belongie, Residual networks behave like ensembles of relatively shallow networks, In *Advances in Neural Information Processing Systems* (2016), pp. 550–558

Facility Recommendation Based on Trajectory Clustering

S. Sharmila and B. A. Sabarish

1 Introduction

Spatial data gained importance due to the rapid growth of geospatial systems and their impact on the field of smart infrastructure development. Spatial data analysis and mapping is a major challenge due to varying representation and sampling rates used for the data collection. Trajectory data is a trace of any moving objects traveling across. For the analysis of Geographic Information System (GIS) based on trajectory, the actual travel path of vehicles has to be analyzed. Trajectory information collected will be a three-dimensional information holding a pair of longitude and latitude values along with timestamp. Facility recommendation system is a multi-objective problem with numerous constraints including the travel distance, transportation cost, coverage, accessibility, and number of resources available. Trajectory clustering is an unsupervised learning methodology to group the similar trajectory information and use the patterns generated (knowledge) as an input prerequisite for the setting location and placement of facilities. Trajectory information can hold a little more information including shape, direction, and velocity. Many single facility location models are more complex when compared to having a model for multiple facility location models. Accessibility is a major measure for the facility location problem which is generally based on distance. Facility location system considers conflicting constraints for the design using multiple criteria decision-making (MCDM) technique. Identifying and fixing a candidate facility have been classified as a Generic Access Facility (GAF) problem which is solved using a clustering algorithm.

S. Sharmila · B. A. Sabarish (✉)
Department of Computer Science and Engineering, Amrita School of Engineering, Amrita Vishwa Vidyapeetham, Coimbatore, India
e-mail: ba_sabarish@cb.amrita.edu

S. Sharmila
e-mail: cb.en.p2cse19022@cb.students.amrita.edu

J. D. Peter et al. (eds.), *Disruptive Technologies for Big Data and Cloud Applications*, Lecture Notes in Electrical Engineering 905,
https://doi.org/10.1007/978-981-19-2177-3_78

The rest of this paper is organized with problem statement in Sect. 2, related works in Sect. 3, proposed framework in Sect. 4, dataset description, results, and conclusion from Sects. 5, 6, and 7.

2 Problem Statement

Facility location recommendation is the process of selecting the optimal set of facilities (F) for a bounded region (R) represented as set of trajectories $T = \left[(\mathrm{lat}_1, \mathrm{long}_1), (\mathrm{lat}_2, \mathrm{long}_2), \ldots (\mathrm{lat}_n, \mathrm{long}_n)\right]$ based on demand considering multiple constraints. Solution can be made by dividing regions into trajectory clusters based on similarity measures. Optimum number of clusters are identified using the elbow method, and the centroid locations or border locations are identified and recommended for various facilities including PHC, recycling centers (SWM), schools, banks, etc.

$$f(x) = \min(\mathrm{dist}\,(p_i, f))$$

where $p_i = (\mathrm{lat}_i, \mathrm{long}_i), \quad \forall i = 1, 2, ..n$ in T and $f = (\mathrm{lat}, \mathrm{long})$ of facility
For solid waste center facility,

$$f(x) = \max\,(\mathrm{dist}\,(p_i, f))$$

where $p_i = (\mathrm{lat}_i, \mathrm{long}_i), \quad \forall i = 1, 2, ..n$ in$T$ and $f = (\mathrm{lat}, \mathrm{long})$ of facility

3 Related Works

Spatial data are gathered from various means like transportation, logistics tracking, banking, health care. The huge volume of data gathered should be processed to extract valuable information. The extracted valuable information can increase the effectiveness and profitability of the system. Clustering is a general term used for grouping similar sets of objects or data [1–4]. Few real-time usage of clustering is outlier detection in a patient's physical activity, plagiarism detection in documents and used to recognize fraudulent activities, etc. Document plagiarism detection, patient's physical activity outlier detection, identifying fraudulent activities, etc. [5, 6].

The solid waste management facilities framework is proposed, where optimization algorithms can be applied using two methodologies either by exact or approximation algorithms. The exact solution faces a problem in terms of size of problem and approximate solution tries to identify near-optimal solution from the available option and find a local optimum from the available list. Facility location problem (FLP) consists of three major constraints including facilities to be located, number of locations available and demands. FLP can be either a single or multi-facility problem. Real-world application mainly focuses on the multi-facility problem where the objective is to identify location for various facilities in a bounded region [7]. A framework for the travel service recommendation using trajectory data mining is proposed. Trajectory data can help in generating patterns which can help in improving ratings of the services and locating facilities. Spatiotemporal features are taken into consideration, and critical points are identified based on the time spent on the location [8]. Zheng, Vincent proposed a framework for answering our daily routine questions based on historical travel of the user to choose based on the interest of the user and locations visited previously. Location features and activity correlation is carried out to gather more knowledge for better understanding for facility recommendation [9]. Facilities can be either undesired or desired facilities based on the impact of them over the environment. Distance either to be maximized or minimized based on the type of facilities [10].

Selection of facilities for the business objective of maximizing facilities is proposed as Multi-characteristic Information based Top-*k* Location Recommendation (MITLR) based on accessibility and profit maximization. Identifying behaviors based on historical and social relationships between the users travel and trajectory data in [11]. Tree-based mostly ranked clusters of visited locations are utilized to include the timely dynamic social trends and customized preferences of the users [12]. Traditional trajectory data mining algorithm does not consider the constraints of budgets, infrastructure development, or utilization. A solution using the primary factor coverage based on similarity of trajectory pattern generated by the user is proposed in [13]. Storage location problem which helps in storing and sharing the common pooled applications like rationing to pool and share across various locations. Warehouse application is solved using decomposition of cost matrix and reducing total cost [14]. In [15], a framework using domination set theory is proposed where each node represents the junction or critical points in the city and distance matrix is generated using Euclidean distance. Clustering is done using the distance matrix and validated using Davies–Bouldin index and Dunn index. Similarly, a clustering algorithm is proposed which uses domination set and *K*-means (hard clustering) clustering algorithm to choose location for facilities in [16]. Another variant of FLP is proposed using a nonlinear programming approach based on setup cost (capacity-based FLP) and other types of facilities can be located [17]. Generic access facility problem of network design also has been proposed based on density of people in the location and bandwidth availability constraints. Approach is based on cost-effective and selection of critical points based on density [18].

4 Proposed Framework

This framework applies the trajectory clustering algorithm-based facility recommendation system which identifies the crucial points to locate facilities and also helps in identifying the optimum number of facilities needed to cover the entire region.

Input: Trajectory data points
Output: Clusters and its corresponding centroids
Step 1: Preprocessing the trajectory data to remove redundant points.
Step 2: Calculate the distance between the trajectory points using the Haversine distance
$d(x, C) = 2R \sin^{-1} \sqrt{\sin^2\left(\frac{x_1 - C_1}{2}\right) + \cos(x_1)\cos(C_1)\sin^2\left(\frac{x_2 - C_2}{2}\right)}$
where d is the distance calculated, x the trajectory data point (x_1—latitude and x_2—longitude), centroid is denoted as C (C_1—centroid latitude, C_2—centroid longitude), and R is approximate radius of earth ($R \approx 6371$ km).
Step 3: Identify the number of clusters either based on the number of facilities available (count) or calculate optimum number of clusters and recommend number of facilities needed to cover the complete region.
Step 4: Identify the number of clusters using elbow method.
Step 5: Calculate the centroid of each cluster and recommended for facilities which need major access like PHC, hospital, schools, banks, ATMs.
Step 6: Identify the points which are farthest from central area but also have good accessibility as they can be recommended for the facilities like solid waste management.

5 Dataset

Coimbatore routes trajectory dataset [19] is used for the analysis. The points are selected to uncover complete travel routes of Coimbatore using Google Maps consisting of 4086 GPS points. Data will be preprocessed to remove the redundant points, which reduces the noise in the data. After the removal of redundant points, the dataset will contain about 2351 points representing trajectories of Coimbatore.

6 Results

The data spread of Coimbatore routes trajectory dataset is in Fig. 1.

This data set contains redundant data points. These duplicate data points are removed in the preprocessing stage such that 4086 data points are reduced to 2351 data points. There are two use cases using which the number of centroids for facility recommendation can be calculated. The first use case is using the elbow method when the number of centers that should be found for facility recommendation is unknown. The second use case is if the user knows the exact number of facilities points to be found. For the first case, the popular elbow method is used to foresee the number of clusters. Figure 2 shows the result of applying the elbow method on the Coimbatore routes trajectory dataset.

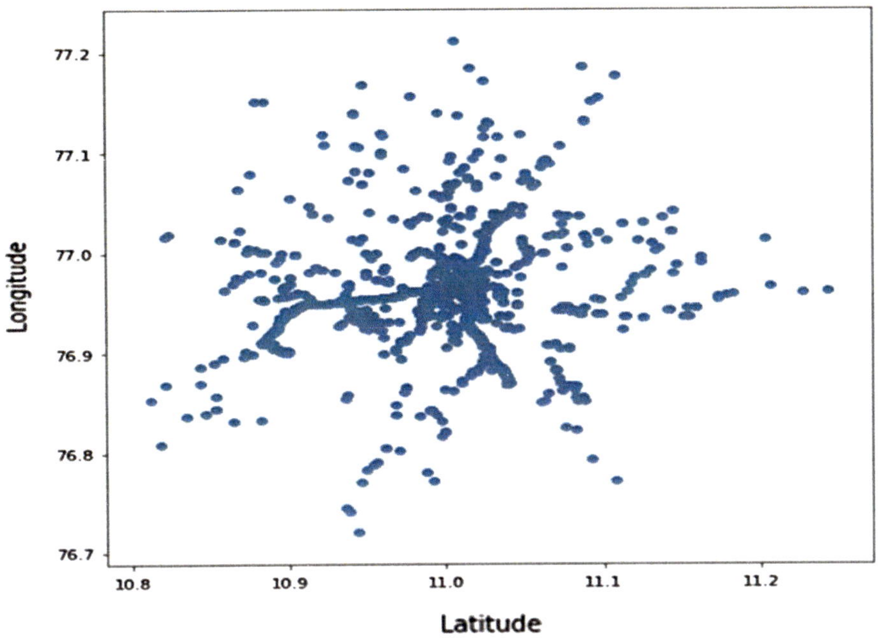

Fig. 1 Data spread of Coimbatore routes trajectory dataset

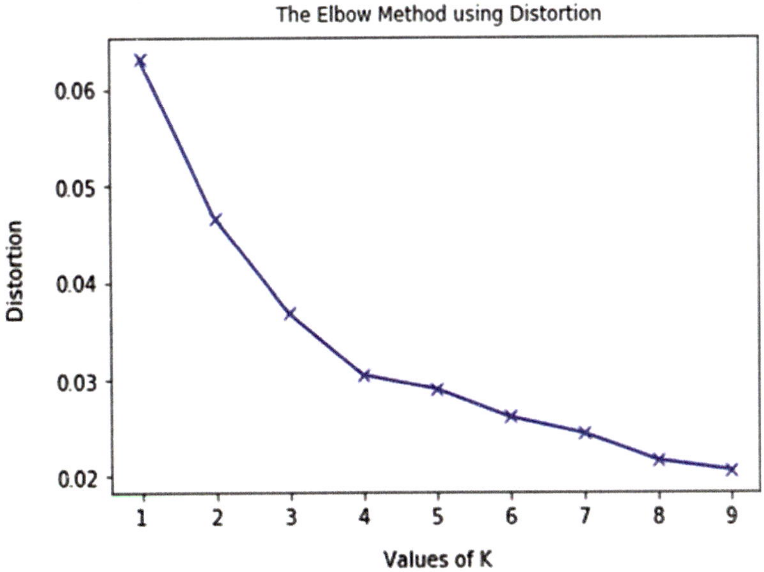

Fig. 2 Elbow method on Coimbatore routes trajectory dataset

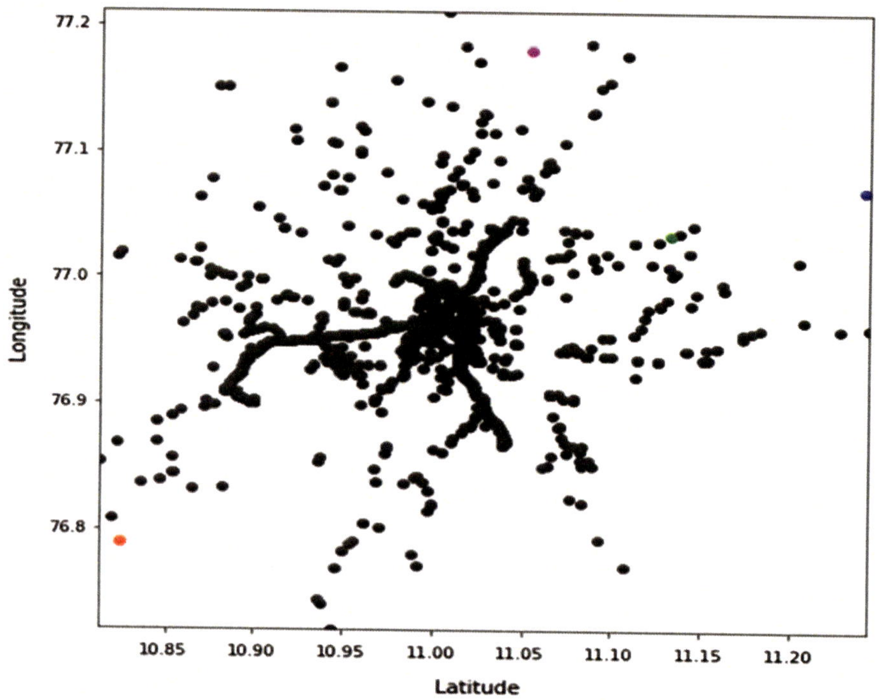

Fig. 3 Initial random centroids

In Fig. 2, the elbow can be clearly seen when the number of centers is 4 (i.e.) $K = 4$. Thus, in further implementation the number of centers is considered as 4. The initial centroids generated at random are highlighted in red, blue, green, and magenta color dots and represented using a scatter plot in Fig. 3.

The initial assignment of data points to specific clusters is made by calculating the Haversine distance between each data point and the random centroid. The data point is allotted to the cluster from which it is located the nearest. The result of the initial assignment stage is illustrated in Fig. 4 as follows.

After initial assignment, the centroid of each cluster is updated using the intra-cluster mean. The mean point is taken as the updated centroid for each cluster. Update stage updates the data point allocation to each cluster based on the updated centroid value. The result of the update stage is pictured in Fig. 5.

The steps assignment and updation are recursively carried out until there is no change in cluster allocation or centroids. When there is no update in cluster membership or centroid value, then the final clusters are formed. The final cluster formation with highlighted centroids is in Fig. 6.

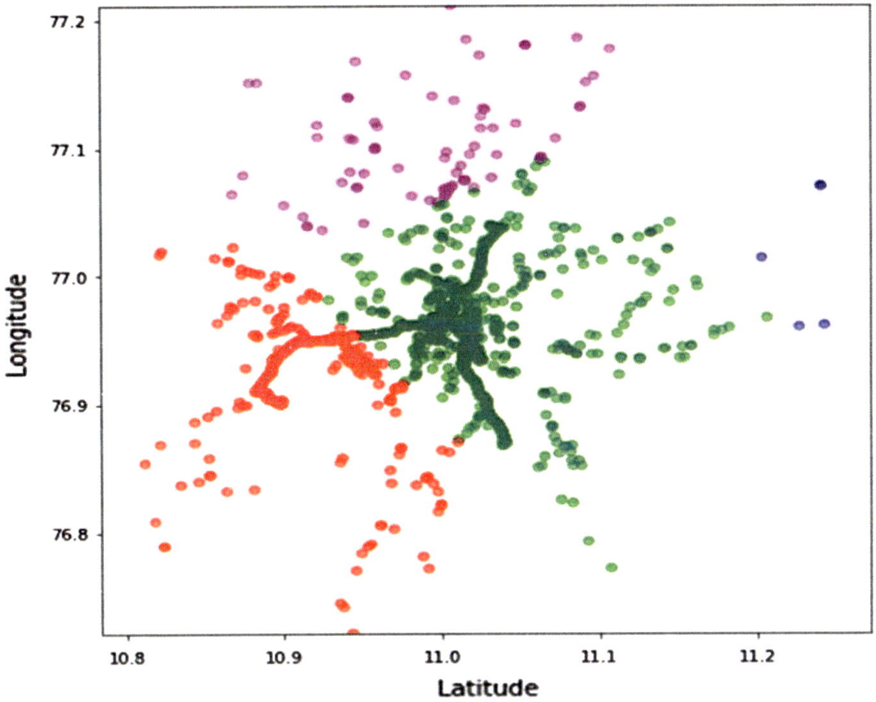

Fig. 4 Cluster formation after initial assignment

In Fig. 6, the highlighted stars are the final centroids which can be used as location to the facilities like banks, hospitals, ATMs, PHCs, and schools. The sample latitude longitude points which can be suggested for facility planning are [10.911139352159461, 76.93345363787373], [11.012163354735128, 76.9318751364365], [11.12502523076923, 76.97113338461537], [11.01597609589041, 77.02508413242018].

For solid waste management spots in the city, the central point in the city cannot be suggested. Instead, the farthest accessible point can be used for solid waste management processes. The outliers are detected in Coimbatore routes trajectory dataset, and those are suggested for solid waste management processes. Figure 7 shows the outlier points highlighted in different colors.

The outlier points detected for solid waste management are [10.81233, 76.85365999999998], [10.82257, 76.86846], [10.81908, 76.80830999999998], [11.20352, 77.01351], [11.24356, 76.96110999999998], [11.20672, 76.96693], [11.22731, 76.95978000000002], [10.82248, 77.01646], [10.82385, 77.01919000000002], [10.93614, 76.74434000000002], [10.93825, 76.74122], [10.94385, 76.72108], [10.94673, 77.1674], [11.10781, 77.17706], [11.087, 77.18576999999998], [11.02472, 77.17215], [11.00643, 77.21113000000003], [11.01644, 77.18454]. Figure 8a represents the centroids identified, and Fig. 8b shows border points using GPS visualizer for better visualization. The Coimbatore district data points for facility recommendation is chosen to verify and validate our proposed approach against the real-time facilities. The centroids and outliers identified are verified and validated against real points.

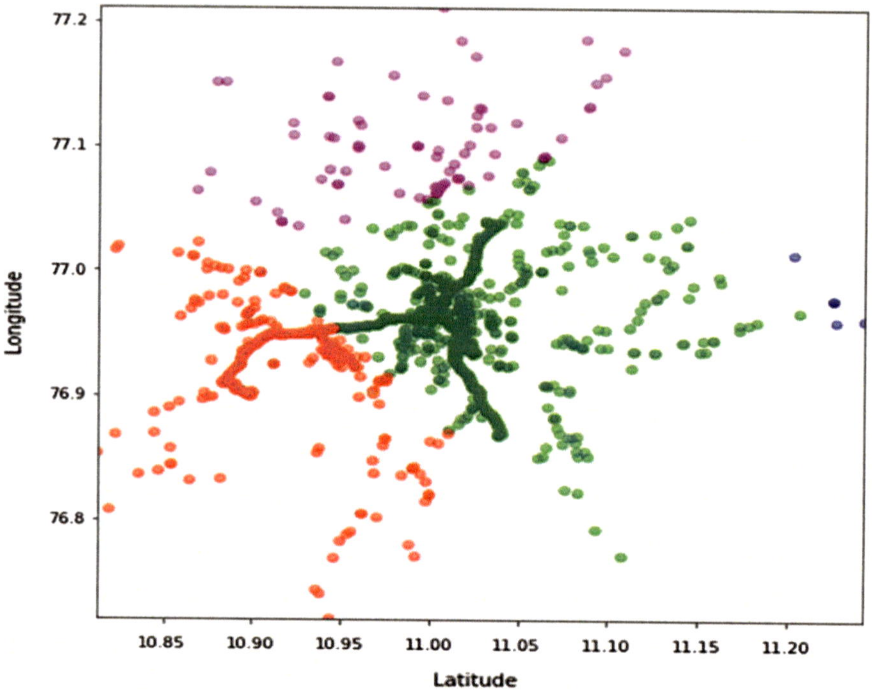

Fig. 5 Cluster formation after update stage

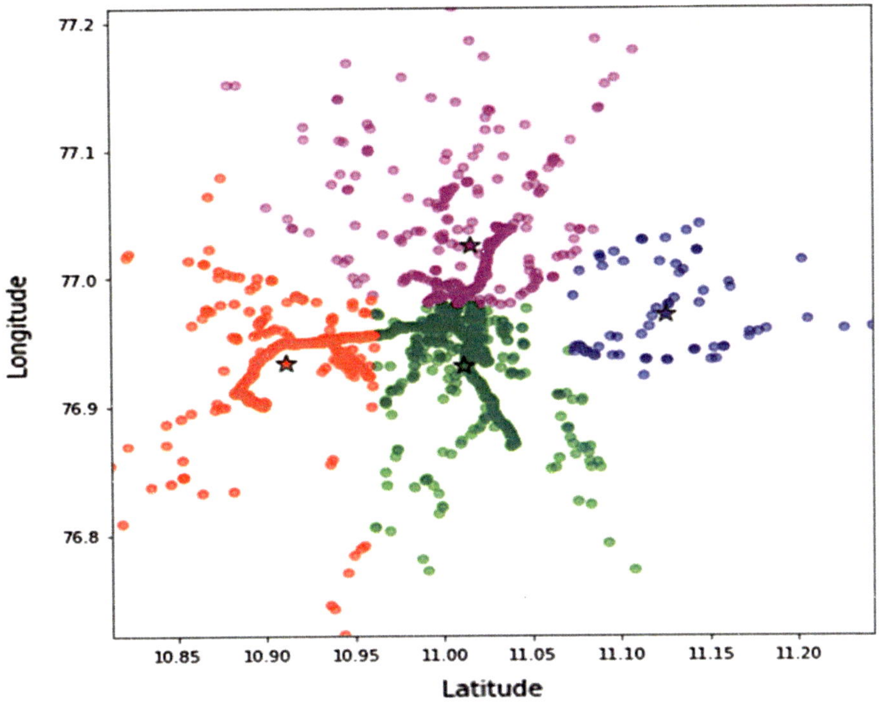

Fig. 6 Final clusters with centroids

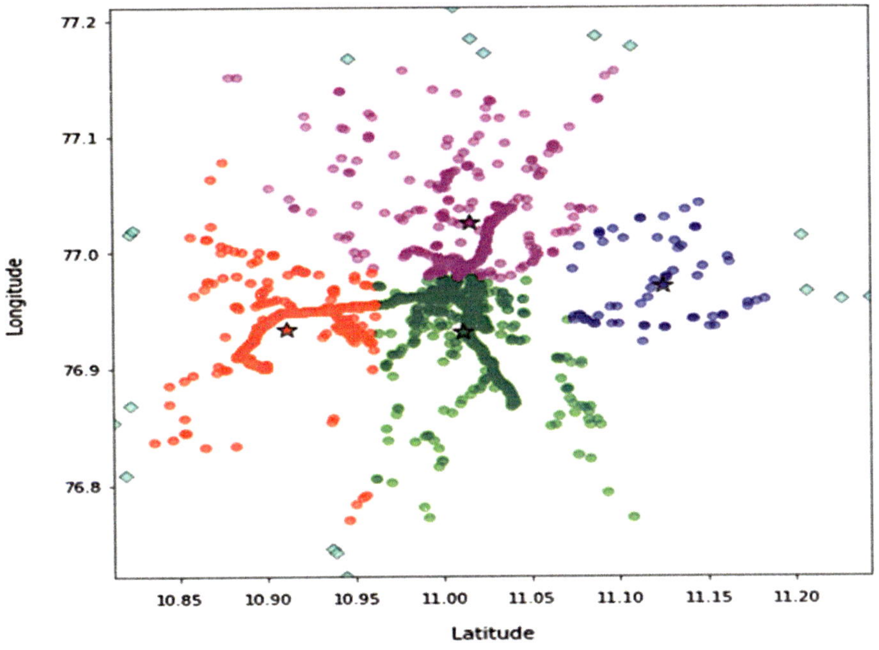

Fig. 7 Final clusters with efficient centroids and outliers

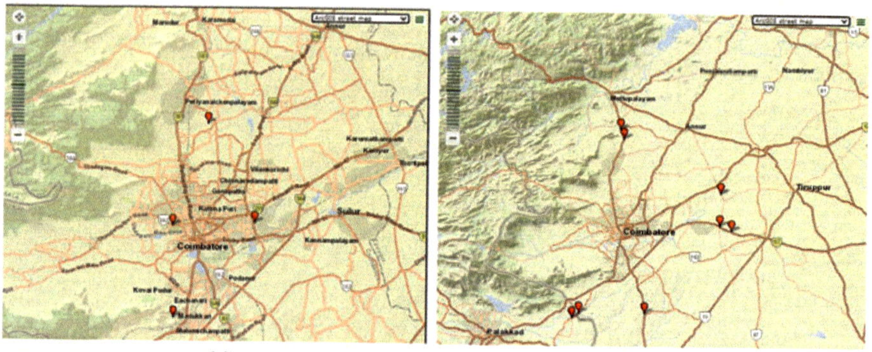

(a) (b)

Fig. 8 **a** Final centroids identified and **b** Border points for solid waste management shown using GPS visualizer

7 Conclusion

The proposed framework uses the trajectory clustering-based framework using the actual road map trajectories which has an advantage of considering only the points which are used for transportation. Optimum clusters are formed using the identifying the centroid points for general facility features and border points to recommend for the solid waste management facilities. This framework can play a major role in the infrastructure development of smart cities providing better accessibility and improving the quality of living. The results are verified in a known city called Coimbatore and which can be applied to any geographical region to place the facilities with better accessibility and reduced transportation cost.

References

1. M.K. Gupta, P. Chandra, A comprehensive survey of data mining. Int. J. Inf. Technol. **12**, 1243–1257 (2020)
2. A.C. Benabdellah, A. Benghabrit, I. Bouhaddou, A survey of clustering algorithms for an industrial context. Procedia. Comput. Sci. **148** (2019)
3. L. Rokach, A survey of clustering algorithms, in *Data Mining and Knowledge Discovery Handbook*, ed. by O. Maimon, L. Rokach (Springer, Boston, MA, 2009)
4. R. Xu, D. Wunsch, Survey of clustering algorithms. IEEE Trans. Neural. Netw. **16**(3), 645–678 (2005)
5. K. Vani, D. Gupta, Using *K*-means cluster based techniques in external plagiarism detection, in *2014 International Conference on Contemporary Computing and Informatics (IC3I)* (Mysore, India, 2014)
6. D. Gupta, K. Vani, C.K. Singh, Using natural language processing techniques and fuzzy-semantic similarity for automatic external plagiarism detection, in *2014 International Conference on Advances in Computing, Communications and Informatics (ICACCI)* (Delhi, India, 2014)

7. O.J. Adeleke, D.O. Olukanni, Facility location problems: models, techniques, and applications in waste management. Recycling **5**(2), 10 (2020)
8. Z. Duan et al., Personalized service recommendations for travel using trajectory pattern discovery. Int. J. Distrib. Sens. Netw. **14**(3), 1550147718767845 (2018)
9. V.W. Zheng, et al. Collaborative location and activity recommendations with GPS history data, in *Proceedings of the 19th International Conference on World Wide Web* (2010)
10. S.H. Tang, O. Boyer, A. Pedram, R.B. Yusuff, N. Zulkifli, A review on multiple criteria undesirable facility location problems. J. Basic. Appl. Sci. Res. **3**(8), 708–713 (2013)
11. P. Wang, W. Chen, L. Zhao, Towards effective top-*k* location recommendation for business facility placement, in *International Conference on Knowledge Science, Engineering and Management* (Springer, Cham, 2020)
12. M.A. Saleem, Y-K. Lee, S. Lee, Dynamicity in social trends towards trajectory based location recommendation, in *International Conference on Smart Homes and Health Telematics* (Springer, Berlin, Heidelberg, 2013)
13. J. Bao et al., Planning bike lanes based on sharing-bikes' trajectories, in *Proceedings of the 23rd ACM SIGKDD International Conference on Knowledge Discovery and Data Mining* (2017)
14. Z. Bagherpoor et al., Warehouse location problem, in *Facility Location* (Physica, Heidelberg, 2009), pp. 295–314
15. B.A. Sabarish et al., Recommendations of location for facilities using domination set theory, in *2017 International Conference on Communication and Signal Processing (ICCSP)* (IEEE, 2017)
16. B.A. Sabarish, S. Vidhya, Facility recommendation system using domination set theory in graph. Int. J. Innovative Technol. Explor. Eng. **8**, 313–317 (2019)
17. L.Y. Wu, X.S. Zhang, J.L. Zhang, Capacitated facility location problem with general setup cost. Comput. Oper. Res. **33**(5), 1226–1241 (2006)
18. P. McGregor, D. Shen, Network design: An algorithm for the access facility location problem. IEEE Trans. Commun. **25**(1), 61–73 (1977)
19. Coimbatore Routes Trajectory dataset. Available: https://www.kaggle.com/sharmilaselvaraj/coimbatore-routes-trajectory-dataset

COVID-19 Symptom Analysis and Prediction Using Machine Learning Techniques

S. Mahima, T. Mathu, and Kumudha Raimond

1 Introduction

Coronavirus disease 2019 or in short COVID-19 that began in Wuhan, China, has caused a worldwide respiratory disorder outbreak. Fever, cough, and headache are the most prevalent symptoms, whereas gastrointestinal symptoms are minimal. Until date, the coronavirus is responsible for almost 6 million infections around the world, with the number continuing to rise. More research is being done in the early diagnosis of COVID-19 leading to numerous AI-based intelligent models which are being discussed by specialists and researchers all over the world to make informed decisions and assist massive control efforts.

AI has emerged as a promising technology, with applications such as forecasting weather data, healthcare sector-related predictions, and trend analysis [1]. ML is a branch of AI that has recently been exhibiting quite promising results [2, 3] in the field of medicine and currently in works related to COVID analysis. For instance, in [4], by asking eight fundamental questions, an ML algorithm is employed to detect the positive COVID-19 case when RT-PCR test is performed. The input dataset for this work is acquired using [5] through the Israeli Ministry of Health. For this work, to make appropriate predictions, a gradient-boosting machine methodology with DT base-learners was also used. For this study, sex, age 60 or older, contact with an infected person, and the early symptoms such as cough, fever, sore throat, shortness

S. Mahima (✉) · T. Mathu · K. Raimond
Department of Computer Science and Engineering, Karunya Institute of Technology and Sciences, Coimbatore, Tamil Nadu, India
e-mail: smahima20@karunya.edu.in

T. Mathu
e-mail: mathu@karunya.edu

K. Raimond
e-mail: kraimond@karunya.edu

J. D. Peter et al. (eds.), *Disruptive Technologies for Big Data and Cloud Applications*, Lecture Notes in Electrical Engineering 905,
https://doi.org/10.1007/978-981-19-2177-3_79

of breath, and headache were all taken into account. The training-validation set was created using data from 51,831 tested participants obtained between March 22nd and March 31st, 2020. The test set included data from the following week, April 1st to 7th. The auROC (Area under the Receiver Operating Characteristics) was used to score the model on the test set.

A Boosted RF algorithm, which is a well-calibrated RF model that is enhanced by including the AdaBoost algorithm, has been proposed in [6]. In this paper, the "Novel Corona Virus 2019 Dataset" dataset is used, which was collected from Kaggle utilizing [7], John Hopkins University and the World Health Organization. The objective of the study was to accurately forecast a patient's result based on a variety of criteria, including but is not limited to journey record, statistics, and other characteristics. Multiple ML classification models, namely DT, SVM, Gaussian NB, and Boosted RF, were trained using the pre-processed dataset in this study. On the dataset used, the model has 0.86 $F1$-score and 94% of accuracy. To perform COVID-19 prediction, authors of [8] have applied ML techniques such as KNN + NCA, Multilinear Regression, and XGBoost Classifier. They have used two different datasets, a global dataset and an Indian dataset. R-Squared values were also used to compare the model and not just the accuracy value. According to their findings, the model XGBoost has the best accuracy of 42.5%. In [9], SVM, ANN, RF, DT, Logistic Regression, and KNN techniques have been addressed in order to estimate the death rate among people infected by COVID-19. The authors have used the confusion matrix and estimated the sensitivity and specificity of the classifier. The results show that ANN provides an accuracy of 93.75%.

Term frequency/inverse document frequency (TF/IDF), Bag of words (BOW), and report length are some of the classifications given to clinical reports. The authors of [10] have used this for COVID-19 prediction, where the classifications are fed into ML algorithms like Logistic Regression and Multinomial NB. They claim that models like Logistic Regression and NB have a 94% accuracy rate. The authors of [11] have employed Convolution Neural Network (CNN) in image categorization, which is useful in processing the X-ray images. Furthermore, Chest X-ray images are used to make predictions for detailed reporting. COVID-19 infection was indicated by anomalies in the chest. They developed COVID-Net, an architecture based on X-ray image prediction that has produced 98.9% accuracy.

In [12], COVID-19–confirmed instances are forecasted using Polynomial Regression, DT Regressor, and RF Regressor. The input data for this work is obtained using [13]. The dataset comprises confirmed, deaths, and recovered cases of COVID-19 as of May 20, 2020, with 7 fields and 27,166 records. The authors have discovered that polynomial regression has yielded the best accuracy of 90% for the COVID-confirmed cases. The authors of [14] have done a comparison of ML algorithms to predict COVID-19 death risk. RF Classifier, SVM, NB Classifier, DT Classifier, Linear Regression, Logistic Regression, k-Nearest Neighbor, and XGBoost are some of the models that are used to forecast. The prediction is then repeated using a DT and the $J48$ algorithm, as well as different regression types like Ridge, Lasso, and Elastic Net Regression. Additionally, XGBoost and Boruta are applied for feature selection. When all regression techniques on the Boruta filter are used, the accuracy rate was observed to have increased to 97.4%.

In [15], the results of the ML-based RF and deep learning-based CNN methods were compared. The dataset was gathered using Kaggle for X-ray dataset of COVID-19–affected people which comprise a total of 5600 images. The CNN model surpasses the RF approach, with an accuracy of 92.4% when 40% of training data and 60% of testing data are taken into account. The authors of [16] have performed an analysis on the impact of age and sex to find the recovery rate of COVID-19–affected patients from Israel. For this work, 5769 records of cured Israeli patients have been acquired using [5] that have been considered as one of the publically available datasets for COVID-19 analysis. In this public dataset, the patients whose age is between 20 and 59 years are allocated as they belong to 0–19 years or above 60 years. Additionally, for elderly and younger aged people, it is designated as belonging between 0 and 19 years or above 60 years of age. According to the observations by the authors, youth are less prone to have major COVID-19–related symptoms that require medical attention in an emergency ward, and they recover from SARS-CoV-2 infection faster on an average rate. Additionally, regarding SARS-CoV-2 infection, ladies are supposedly less susceptible to develop severe acute respiratory distress (ARDS) or mortality.

The goal of this study is to execute two works in accordance to the COVID crisis, with the symptoms experienced by a person affected by COVID as the primary focus. Initially, a correlation analysis utilizing Pearson and Spearman correlation coefficients is performed to identify the most significant symptoms detected in a person with COVID. Following that, six machine learning approaches were used to forecast the outcome using the most important attributes identified. The accuracy, precision score, $F1$-score, and ROC-AUC score have been applied to analyze the efficiency of these classification methods that includes RF, DT, Gradient Boosting, NB, SVM, and Bernoulli NB.

2 Materials and Methods

2.1 Data Collection and Description

The input dataset used for this work is obtained from [5]. The dataset was made available by the Israeli Ministry of Health after conducting the RT-PCR test using a nasopharyngeal swab in order to identify the presence of SARS-CoV-2 virus for several individuals. The dataset description illustrated in Table 1 with details about each attribute, their description, and the corresponding values is considered for our work. For this study, the data is collected from February–May 2021 with 103,0986 records. It was observed that the data for the symptoms was categorical data.

Table 1 Input data description

Attribute	Description	Values and their indications
Test_date	Arrival date of the test at the laboratory	Dates present in the format DD/MM/YYYY
Gender	Gender of the person taking the test	Male Female Null indicates unknown values
Corona_result	Results of the first COVID-19 test	Negative—not a carrier of COVID-19 Positive—carrier of COVID-19 Other—not tested/ in testing/ inconclusive
Age_60_and_above	Age indicator of the test taker	1–60 years and above 0—Below 60 years
Cough	Did cough symptoms appear before testing?	1—yes 0—no Null—unknown
Fever	Did fever appear before testing?	1—yes 0—no Null—unknown
Sore_throat	Did sore throat symptoms appear before testing?	1—yes 0—no Null—unknown
Shortness_of_breath	Did cough shortness of breath symptoms appear before testing?	1—yes 0—No NULL—Unknown
Head_ache	Did headache symptoms appear before testing?	1—yes 0—No NULL—Unknown
Test_indication	What is the indication for testing?	Abroad—arrived from abroad Contact_with_confirmed—contact with a confirmed case Other—other indication or not specified

2.2 Data Preprocessing

When the records for the "corona result" attribute were obtained, they were all in Hebrew. Google Translate was used to translate these into English. Further, the entries "Positive," "Negative," "Other," "Abroad," and "Contact with confirmed" were found in records for the "corona result" and "test indication" fields. Two additional rows, "corona result num" and "test indication num," were created to make them easier to interpret. In order to perform a classification task using the input data, the records with "Negative" and "Positive" values from the "corona result" attribute were considered for the present study. Table 2 shows the amended data description for the attributes included.

Table 2 Description of attributes included as a Preprocessing step

Attributes created	Original description	Modified description
corona_result_num	Negative—Not a carrier of COVID-19	0—Not a carrier of COVID-19
	Positive—Carrier of COVID-19	1—Carrier of COVID-19
test_indication_num	Abroad—arrived from abroad	0—Arrived from abroad
	Contact_with_confirmed—contact with a confirmed case	1—Contact with a confirmed case
	Other—other indication or not specified	2—Other indication or not specified

2.3 Feature Selection

With the objective of performing an analysis of COVID purely based on the symptoms observed, the attributes "Test_date," "Age_60_and_above," and "Gender" have not been considered in this work. Correlation analysis was performed using Pearson and Spearman correlation coefficients for all the attributes indicating symptoms with the "corona result" property in order to determine the most influential parameters. Cough, fever, headache, shortness of breath, and sore throat had positive correlation value for both Pearson and Spearman correlation. The attribute test indication num was observed to have negative correlation.

In this regard, the feature selection was performed for the further work. The findings of the correlation study are presented in Table 3. Based on the outcomes acquired using the correlation analysis, it is acceptable to infer that the most common symptoms in people who have been diagnosed with COVID are cough, fever, sore throat, shortness of breath, and headache. As a matter of fact, additional amount of care is required in order to stop the spread of these symptoms as soon as possible.

Table 3 Results for correlation analysis

Attributes created	Pearson correlation values	Spearman correlation values
Cough	0.411	0.411
Fever	0.322	0.322
Sore_throat	0.287	0.286
Shortness_of_breath	0.161	0.161
Head_ache	0.416	0.416
test_indication_num	−0.561	−0.582

2.4 *Machine Learning Algorithms*

2.4.1 Random Forest

RF is a technique called Bootstrap and Aggregation, which is traditionally referred to as bagging. With the application of multiple DTs, the RF technique is utilized for producing regression-based and classification-based models. Instead of depending on individual DTs, the primary idea behind this methodology is to merge many DTs in the process of producing final output. As a fundamental learning model, RF uses several DTs [12].

2.4.2 Decision Trees

DT is a non-parametric supervised learning approach for extracting knowledge from given data. It is termed non-parametric because it makes no assumptions about the dependent variable's distribution, the explanatory variables' distributions, or the functional form of the relationships between them. DTs are classified based on the type of target they must forecast. A classification tree differs from a regression tree in that it predicts a categorical response rather than a quantitative and, in most cases, continuous response. Any split, at any internal node, has the goal of generating two descendant child nodes with increasingly pure data. A simple visual representation of the notion could help to clarify it. A crucial phase in the tree-building process is deciding on the size. Because of the absence of independence between the training and test sets, a test-sample estimate is proposed as a solution [17].

2.4.3 Gradient Boosting

Gradient Boosting belongs to a class of sophisticated ML methodologies which have had a lot of success in a variety of applications. They are extremely configurable to the application's specific requirements, such as being taught with respect to various loss based functions. The learning strategy in gradient Boosting assigns newer models within a series to offer a better and accurate estimate of the dependent variable. The essential idea behind this strategy is to create new base-learners that are optimally associated with the ensemble's negative gradient of the loss related function [18].

2.4.4 Naïve Bayes (NB)

NB is an unsupervised ML technique widely used for classification tasks and is according to Bayes theorem with strong and naïve independence assumptions. It makes learning easier by presuming that features are unaffected by the class [19]. In the presence of feature dependencies, the success of NB can be described as: when observed in terms of zero–one loss (error in classification), the degree of the fit to a probability distribution is sometimes not linked to maximization. Alternatively, if the measured and forecasted distributions coincide on the most likely class, an efficient classification is determined [20].

2.4.5 Support Vector Machine (SVM)

The authors of [21] introduced SVM in 1995, and it has exhibited successful and efficient qualities for data analysis and data mining applications. SVM has emerged into powerful problem-solving tools in machine learning, with origins in Statistical Learning Theory (SLT) and optimization related methods. Most ML problems are reduced to optimization issues by SVMs, and optimization is at the core of SVMs [22].

2.4.6 Bernoulli Naïve Bayes

Bernoulli NB Classifier is according to the binary outcome generating concept of if a term or word appears in a document [23]. When trying to predict the polarity of a new newspaper story, we magnify the probability of appearance of all words in the article, as well as the probabilities of words which do not emerge in the article, for both polarities, and the one that is greater helps give the polarity of this news piece [24].

2.5 Methodology

2.5.1 Other Works with the Same Dataset

The Israeli Ministry of Health has made a dataset publicly available which is used in our study. The dataset is made up of a variety of attributes, which were described previously in Table 1. In this regard, many works have been performed using this as the input dataset. The works that were mentioned previously in the papers [4] and [16] have utilized this dataset, respectively. In [4], the authors have applied all the input parameters available from the dataset. Multiple evaluation criteria, such as auROC (Area under the receiver operating characteristics), sensitivity, specificity, positive predictive value, and negative predictive value, false-positive rate, false-negative rate, false discovery rate, and accuracy, were used to calculate the results. The authors of [16] used the dataset to conduct research that focuses on the influence of gender and age on the recovery rate of COVID patients. By analyzing these previous works and the authors' strategies, we want to highlight that a meaningful comparison of our work and the methodologies used in these two previous publications is impossible due to the distinctions in the approaches proposed.

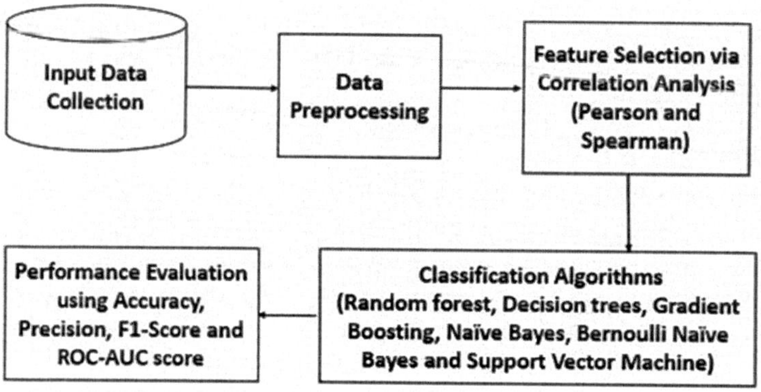

Fig. 1 Flow diagram for the methodology

2.5.2 Our Study

In this work, data linked to symptoms that are suggested in a person affected by COVID was analyzed using ML-based algorithms to accomplish a classification task. After the feature selection process via correlation analysis, the acquired symptoms data is loaded in the classification algorithms, namely RF, DTs, Gradient Boosting, NB, SVM, and Bernoulli NB. The input dataset is split such that 75% is training data while 25% is the testing data. The "corona_result" attribute is the target variable while the attributes signifying the symptoms experienced by a person affected by COVID, i.e., "cough, fever, head_ache, and sore_throat" are the predictors. The parameter tuning for the six algorithms is applied to enhance the results based on trial-and-error process. The flow diagram for the methodology carried out is depicted in Fig. 1.

3 Results and Discussions

Accuracy, precision score, $F1$-score, and ROC-AUC score were used to analyze the study's findings. The algorithms were fine-tuned through a trial-and-error procedure until they achieved the highest level of accuracy. The number of estimators, random state, and minimum number of samples split count, minimum samples leaf, minimum leaf weight fraction, maximum number of features, and maximum count of leaf nodes were all parameters for RF. The value 30 provided the best accuracy when the number of estimator parameters was changed from 10 to 30. Similarly, the random state was assigned a number between one and ten, with ten being the best. The DT algorithm employed the "entropy" criterion and the "best" splitter. For the SVM algorithm, the widely used "RBF" kernel function was used. When using the

Table 4 Results for the classification algorithms

Machine learning technique	Accuracy	Precision score	$F1$-score	ROC-AUC score
Random forest	96.35	0.774	0.58	0.728
Decision trees	96.35	0.774	0.579	0.727
Support vector machine	96.35	0.773	0.579	0.727
Naïve Bayes	96.1	0.77	0.57	0.72
Bernoulli Naïve Bayes	96	0.76	0.571	0.72
Gradient boosting	95.78	0.77	0.43	0.64

symptoms with the highest correlations, i.e., fever, cough, and headache combined, and then incorporating sore throat and shortness of breath, we found no significant differences in the results. Hence, we have considered that all the features obtained with a positive correlation could be taken into consideration as impactful symptoms.

With a score of 96.35%, the RF, DT, and SVM algorithms provided the best results, following with NB with 96.1%, Bernoulli NB with 96%, and Gradient Boosting with 95.78%. Precision scores for RF and DT were 0.774, SVM was 0.773, NB and Gradient Boosting was 0.77, and Bernoulli NB was 0.76. Regarding the $F1$-score, the RF has yielded 0.58, followed by the DT and SVM with a value 0.579. NB has acquired an $F1$-score of 0.57 with Bernoulli NB obtaining 0.571 and Gradient Boosting 0.43. In terms of the ROC-AUC scores, the RF has acquired 0.728, DT and SVM 0.727, NB and Bernoulli NB 0.72 and Gradient Boosting 0.64. As a result, it could be observed that all the six algorithms utilized in this work exhibited promising results with respect to the various performance metrics. The results of the six algorithms utilized in our study are summarized in Table 4.

4 Conclusion

In this paper, we have performed an analysis based on the symptoms observed in a person affected by COVID using the open-source dataset provided by the Israeli Ministry of Health. The inputs for this work consisted of records of patients who were tested for COVID in an RT-PCR test. Initially, to identify the most impacting symptoms and carry out a feature selection process, we have performed the correlation analysis using Pearson and Spearman correlation coefficients. According to the correlation analysis performed, it was seen that the most influential symptoms were cough, fever, sore throat, headache, and shortness of breath. Hence, it can be assumed that these are the most commonly observed symptoms in a person declared COVID Positive. As a result, utmost care should be exercised in order to eliminate these symptoms as quickly as feasible. Further, in order to further analyze the COVID cases by considering the features selected, we have applied six machine learning algorithms, namely RF, DTs, Gradient Boosting, NB, SVM, and Bernoulli NB to perform a classification task. To analyze the performance of these algorithms, we have applied the metrics accuracy, precision score, $F1$-score, and ROC-AUC score, respectively. When the deployed machine learning methods were evaluated, it was discovered that all six algorithms were effective in differentiating between positive and negative cases of COVID with excellent performance metrics.

References

1. S. Kathiresan, A.R.W. Sait, D. Gupta, S.K. Lakshmanaprabu, A. Khanna, H.M. Pandey, Automated detection and classification of fundus diabetic retinopathy images using synergic deep learning model. Pattern Recogn. Lett. **133**, 210–216 (2020). https://doi.org/10.1016/j.patrec.2020.02.026
2. R. Ahuja, A. Solanki, Movie recommender system using K-means clustering and K-nearest neighbor, in *Accepted for Publication in Confluence-2019: 9th International Conference on Cloud Computing, Data Science & Engineering*, vol. 1231(21), (Amity University, Noida, 2019), pp. 25–38
3. A. Tayal, U. Kose, A. Solanki, A. Nayyar, J.A.M. Saucedo, Efficiency analysis for stochastic dynamic facility layout problem using meta-heuristic, data envelopment analysis and machine learning. Comput. Intell. **36**(1), 172–202 (2019)
4. Y. Zoabi, S. Deri-Rozov, N. Shomron, Machine learning-based prediction of COVID-19 diagnosis based on symptoms. npj Digit. Med. **4** (2021). https://doi.org/10.1038/s41746-020-00372-6
5. https://data.gov.il/dataset/covid-19/resource/d337959a-020a-4ed3-84f7-fca182292308
6. C. Iwendi, A.K. Bashir, A. Peshkar, R. Sujatha, J.M. Chatterjee, S. Pasupuleti, O. Jo, COVID-19 patient health prediction using boosted random forest Algorithm. Frontiers. Publ. Health. **8** (2020). https://doi.org/10.3389/fpubh.2020.00357
7. Novel Corona Virus 2019 Dataset, (2020). Available online at: https://www.kaggle.com/sudalairajkumar/novel-corona-virus-2019-dataset/ (Accessed Apr 23, 2020)
8. B. Kolla, Analysis prediction and evaluation of COVID-19 datasets using machine learning algorithms. Int. J. Emerg. Trends. Eng. Res. 8 (2020). https://doi.org/10.30534/ijeter/2020/117852020
9. M. Pourhomayoun, M. Shakibi, Predicting mortality risk in patients with COVID-19 using artificial intelligence to help medical decision-making (2020). https://doi.org/10.1101/2020.03.30.20047308
10. A.M.U.D. Khanday, S.T. Rabani, Q.R. Khan, N. Rouf, M. Mohi Ud Din, Machine learning based approaches for detecting COVID-19 using clinical text data. Int. J. Inf. Technol. (2020). https://doi.org/10.1007/s41870-020-00495-9
11. L. Wang, A. Wong. COVID-Net: a tailored deep convolutional neural network design for detection of covid-19 cases from chest X-Ray images. arXiv:2003.09871
12. G. Monica, M. Bharathi Devi, Using machine learning approach to predict COVID-19 progress. Int. J. Mod. Trends. Sci. Technol, 6(8S), 58–62 (2020)
13. https://github.com/CSSEGISandData/COVID-19
14. P. Anandhanathan, P. Gopalan, Comparison of machine learning algorithm for COVID-19 death risk prediction (2021). https://doi.org/10.21203/rs.3.rs-196077/v1
15. N. Trivedi, S. Simaiya, S.K. Sharma, COVID-19 pandemic: role of machine learning & deep learning methods in diagnosis. Int. J. Curr. Res. Rev. 150–155 (2021). https://doi.org/10.31782/IJCRR.2021.SP192
16. I. Voinsky, G. Baristaite, D. Gurwitz, Effects of age and sex on recovery from COVID-19: analysis of 5769 Israeli patients. J. Infect. 81(2), e102–e103 (2020). https://doi.org/10.1016/j.jinf.2020.05.026
17. Casarin, Roberto & Facchinetti, Alessandro & Sorice, Domenico & Tonellato, Stefano. (2021). Decision trees and random forests*. https://doi.org/10.4324/9781003037903-2.

18. A. Natekin, A. Knoll, Gradient boosting machines, a tutorial. Front. Neurorobotics, 7(21). https://doi.org/10.3389/fnbot.2013.00021
19. P. Kaviani, S. Dhotre, Short survey on Naive Bayes algorithm. Int. J. Adv. Res. Comput. Sci. Manage. 4 (2017)
20. P. Domingos, M. Pazzani, On the optimality of the simple Bayesian classifier under zero-one loss. Mach. Learn. 29, 103–130 (1997)
21. C. Cortes, V. Vapnik, Support vector networks, in Proceedings of Machine Learning, vol. 20, (1995), pp. 273–297
22. Y, Tian, Y. Shi, X. Liu, Recent advances on support vector machines research. Technol. Econ. Dev. Econ. 18 (2012). https://doi.org/10.3846/20294913.2012.661205
23. Text Classification. https://monkeylearn.com/text-classification/
24. Bernoulli Naïve Bayes Classifier. https://mattshomepage.com/articles/2016/Jun/07/bernoulli_nb/

Segmentation of Streets and Buildings Using U-Net from Satellite Image

Jayasurya Subramanian, Senthil Kumar Thangavel, and Pasquale Caianiello

1 Introduction

There are more than three hundred thousand accidents that taken place in India every year, in that more than a hundred thousand people are dying. In those, 30% of accidents take place due to faulty roads and illegally constructed buildings. If all the illegally constructed buildings or detected early, we can avoid the danger of accidents and prevent human life from falling. To do that we are using the medium of the satellite to get the image of the urban and rural areas to find any illegally constructed building that is present in that area. The advantages of using satellite images (remote sensing) are its spatial resolution, spectral characteristics, temporal characteristics, and processing options. To do so convolutional neural network (CNN) and U-Net came to the rescue.

In today's world, CNN plays a key role in object recognition and computer vision [1]. There are several sub-branches of neural networks, including CNN and recurrent neural networks, which are similar to short attention span [2]. By learning from the training data, this neural network assists in categorizing the picture into multiple types. Apart from that, they learn from their layers which include their own max pooling, convolutional, and fully connected layers [3].

J. Subramanian · S. K. Thangavel (✉)
Department of Computer Science and Engineering, Amrita School of Engineering,Amrita Vishwa Vidyapeetham,Coimbatore, India
e-mail: t_senthilkumar@cb.amrita.edu

J. Subramanian
e-mail: cb.en.p2cse18007@cb.students.amrita.edu

P. Caianiello
Department of Mathematics and Computer Science, University of L'Aquila, L'Aquila, Italy
e-mail: pasquale.caianiello@univaq.it

J. D. Peter et al. (eds.), *Disruptive Technologies for Big Data and Cloud Applications*, Lecture Notes in Electrical Engineering 905,
https://doi.org/10.1007/978-981-19-2177-3_80

Mnih and Hinton [3] was the first to use deep learning method for road extraction, proposing restricted Boltzmann machines (RBMs) to identify street area using ariel photographs, then using a support vector machine (SVM) and K-nearest neighbor (KNN) for road extraction, with a precision of 0.8–0.9. Initially, U-Net was used in medical image segmentation but in recent years for gaining better accuracy U-Net was introduced to satellite image segmentation also.

It consumes more time and manpower to identify the buildings and roads that are illegally constructed or not through the field visit process. So we are moving to the satellite images. Also, it is not that easy, and sometimes we will get images with high noise and with low contrast which makes our process tougher. Not only that the artifact like vegetation and vehicles also should be taken care of to improve the object detection process. Motivated by deep learning [4] and U-Net [5], we use a deep residual U-Net that combines deep learning and the U-Net architecture. These classification methods are now evolving, and the findings are becoming more accurate. These techniques are also important in the remote sensing industry.

2 Literature Review

In the last few years, the computer vision field had advanced rapidly because of deep learning [6]. In that semantic segmentation is considered as the gem of all [7]. Even though there are a lot of ways to solve this problem, U-Net got a spatial place. U-Net is a fully convolutional network that uses convolutional operation and max pooling operation to solve the problem which gives an advantage over any other methods.

2.1 U-Net Architecture

U-Net architecture was initially developed by Olaf Ronnberger et al. [5]. He created it for the segmentation of biomedical images. Encoder or contraction path and decoder or symmetric expansion path make up the architecture. The encoder path captures the image background, while the decoder path allows for accurate localization using transposed convolutions. It just has convolutional layers, so we should call it a end-to-end fully convolutional network (FCN). It will tolerate any size image due to the lack of a thick layer. Figure 1 depicts the original architecture proposed by Olaf Ronnberger et al.

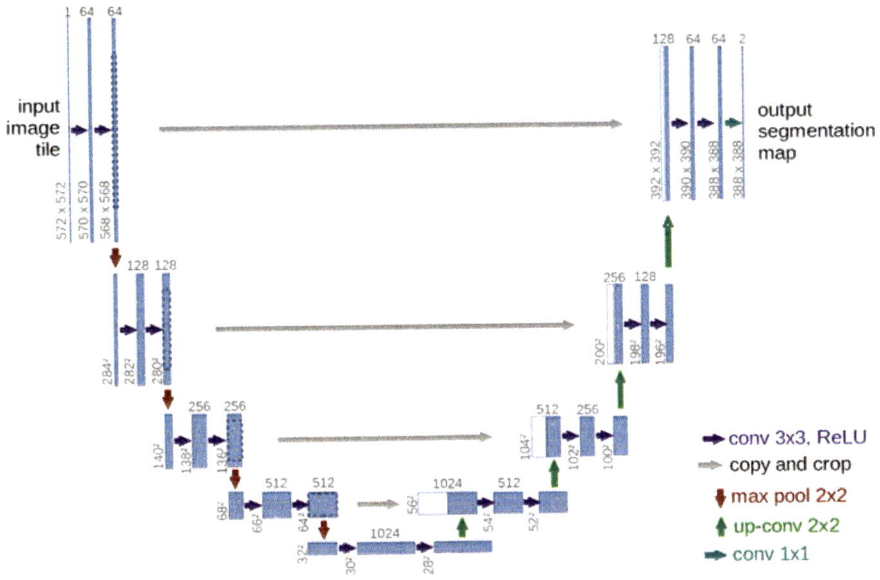

Fig. 1 Original U-Net architecture

3 Implementation

3.1 Dataset Description

We use the dataset of Massachusetts Benchmark dataset for roads and buildings extraction. In the image directory, there are 2300 satellite images representing rural and urban areas. In which we can find images of roads and buildings in the rural and urban areas. The training dataset contains three labeled images of 1300×1300 pixels. After the training, we run a prediction on 100 test images of 1300×1300 pixels. The predicted images are then cropped in patches of $256 \times 256 \times 3$ pixels and transformed into a CSV submission file containing the predicted output (22 for a building, 11 for a road, 00 for background) for each patch.

3.2 Preparing Dataset

After the preprocessing step, we need to digitize our image, for digitizing process we are using QGIS. QGIS is a cross-platform desktop geographic information system (GIS) program that allows you to browse, update, and analyze geospatial data for free.

3.3 Rasterization and Vectorization Process

The rasterization process is done in two ways: one is by using QGIS which is time-consuming, and the second method is by using python. In this, we will see how to rasterize the data using python. As a first process, we are going to create labels with shapefiles for the images. While doing that we need to take care of the projection of imagery and shapefile which should be the same. And another main constraint is the projection unit should be in meters if you want to buffer line features. It will take the input from the given file as the image and add the raster add vector data to those images and save those images in the destination file [8]. While doing so we are adding buffer length (–buffer) for line features. We are using line features for roads and polygon features for buildings. The polygon features do not need buffer length so we are ignoring that for those features. Then we are adding buffer length for the attributes (–buffer_atr) from the vector file. This attribute can be buffer width and multiples with the buffer of line. And this is also the same as—buffer which is not needed for polygon features [1]. After that the attributes form the vector file (labels_atr), pixels inside the polygons will be assigned by their attribute value. This is not required for the line and single class (polygon). This process includes

1. Creating the labels with shapefiles.
2. The projection of imagery and shapefiles should be the same.
3. Projection units should be in meters if you want to buffer line feature.

3.4 Patch Generation

After the rasterization and vectorization of the data, we need to generate patches for the input images [3]. Patches are the process of generating multiple images from a single image, and here we are making the patches for 256 × 256 pixel values. To generate the patches for train, test, and validate, the command must be run three times. The main constraint is that the name of the image and the label name should be the same.

3.5 CSV Generation

As a next step, we are going to save the directories of patches in .csv file. We are saving the patches in .csv because instead of reading patches from the folders directly it will be easy to read from the .csv file [9]. It includes the image folder which consists of image patches with the directory, image format, label_folder where label patches are get stored, label format, patch size, and output folder where the output .csv file gets saved.

3.6 Training the Models

After generating the patches, now we are ready to train our model. It consists of training model name; in our case, it is U_net model, the csv file name with the directory, consists of directories of image and label patches of the training set, the csv file name with the directory, consists of directories of images and labels patches of the validation set, input shape of models to feed (patch size and patch size channels), batch size depends on GPU/CPU memory; in our case, we are using a batch size of 1, the number of classes ion label data, and the number of epochs [10]. In our case, we are using the epoch size of 300.

3.7 Predicting Using Deep Learning Models

U-Net is a form of deep learning model, which is a completely convolutional network model. We use U-Net to train 18 layers in this experiment. Prior to training the files, preprocessing steps such as managing missed entries, inserting median values, segmenting, and resizing the images are completed. We get the accuracy as well as the failure after training. The predicting model is made up of the model's input form (patch size, patch size channels), a trained model with a directory, a folder of input images/tiles with a directory, and an output folder where the projected image/tiles are saved. Initially, we will have a $256 \times 256 \times 3$ input. Transpose and max pooling operations will be performed by each layer [11] [12]. As a result, the model's precision is 0.98, the total number of params is 31,379,075,000, the total number of trainable params is 31,379,075,000, and the total number of non-trainable params is 0. After convolution, the input layer provides the output of the convolutional layer, with the image's height and width decreased. The output size of the convolutional layer is limited to two in comparison with the input layer. After the convolutional layer, the data is fed into the max pooling layer. This layer reduces the image's dimension by downsampling the input. Finally, we have the encoding layer, which reconstructs the signal. Figure 2 depicts the input and output images. With the help of a segmented image, one can easily find the buildings which are illegally constructed along the roadside manually.

3.8 Accuracy

Following the prediction process, we calculate the model's accuracy. We calculate the model's accuracy using a variety of accuracy parameters, including IoU score, $F1$-Measure, precision, recall, mean IoU score, and mean $F1$-measure with the confusion matrix. It is done by taking the input shape of the model, weight which is trained model with the directory, .csvfile name with the directory, consists of the directories

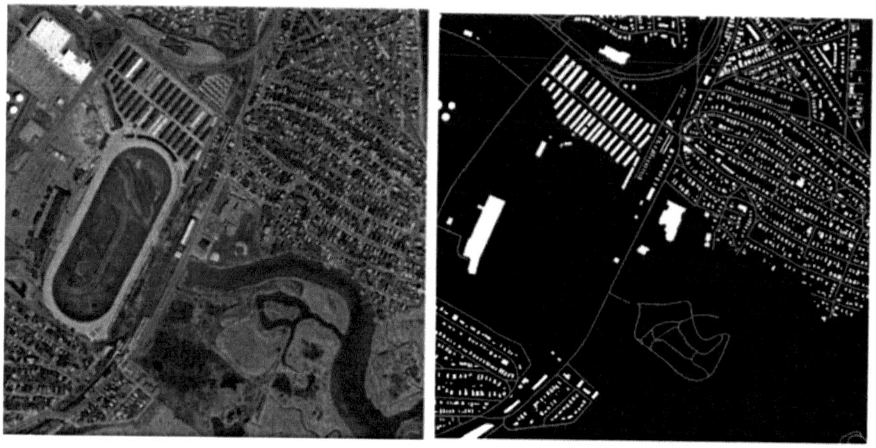

Fig. 2 Input and output of the model

of image and label patches of the test set. The overall accuracy we are getting is 0.98, the mean IoU score is 0.971, and the mean $F1$-Measure is 0.985.

3.9 Comparison of the Images

As the next step, we need to compare the timed images for checking if there are any illegally constructed buildings and roads that are present in that particular area. For the comparison of the images, we are taking the image from the repository which is checked by the government officials manually and say there are no illegally constructed buildings or roads in that area, with the timed image of the same area. And then we are converting both the images into NumPy arrays. After that we are converting all the 0's into 3's in both the arrays. We are doing that because the multiplicative inverse of zero will be zero itself; in that case, our total model will fail that is why we are converting the values 0's to 3's. There will be no change in the result by the 0's conversion. Now 1 will represent the roads, 2 will represent the buildings, and 3 will represent the backgrounds (other than roads and buildings). After that we are multiplying the first image array value (q) which is our checked image with "10" and adding the result with the checking image array (w). The following equation explains the above-mentioned operation.

$$\mathrm{Rt} = (q * 10) + w \tag{1}$$

where Rt = resultant array, q = checked image array value, w = comparing image array value. The result will be in the form of 11, 12, 13, 21, 22, 23, 31, 32, and 33. Table 1 gives a brief explanation of the representation of the outputs.

Table 1 Output array value description

Value	Description
11	No changes in the road
12	New buildings are constructed in the place of roads
13	Roads had been demolished or removed
21	New roads are constructed in the place of building
22	No changes in the buildings
23	Buildings had been demolished or removed
31	A new road is constructed in the empty area
32	New buildings are constructed in the empty area
33	No change in the empty area

(A) (B)

Fig. 3 **a** Image in the government repository, **b** image we obtain now

As the next step, we are extracting only the newly constructed buildings and roads if there are any. We are doing that by changing the value of the output array. We are converting the pixel value which denotes there is a change to 1 and the rest of the pixel value to 0. For example, consider there are new buildings have constructed in the empty area. In that case the pixel value for the place where a newly constructed building will be 32. Now we are converting that 32–1 and the rest of the pixel value to 0. While doing so we will get only the newly constructed building image [13]. Now we will open that image in the QGIS to know the coordinates. With the help of the coordinates, we can easily find the location of the buildings where it is constructed and the government officials can easily identify whether that building is constructed legally or illegally. Figure 3 explains the extraction of the particular area from the image.

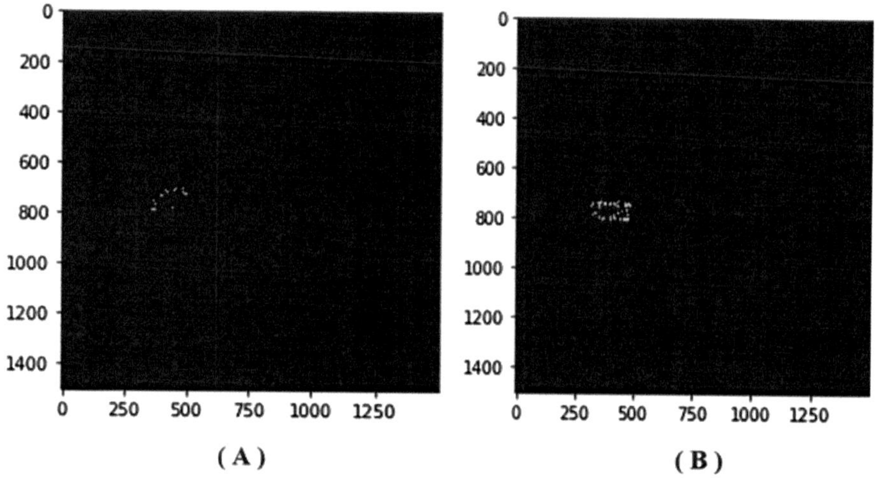

Fig. 4 **a** Construction of extra roads in that particular area **b** construction of an extra building in that particular area

Here figure A represents the image in the government repository which is already checked by the government officials and states each road and building in the image are constructed legally. Now figure B represents the newly obtained image of the same area. Now we need to check the presence of any newly constructed buildings and roads that are present in that area. Figure 4 shows the comparison results.

The above figure represents the output of the comparison of the two images. In the comparison process, we got to know that there are newly constructed roads and buildings in the comparison image. (A) represents that there are newly constructed roads in that particular area and (b) represents that there are newly constructed buildings in that particular area. After that, we obtain those images and we will upload those images in QGIS to get the coordinates of the newly constructed buildings and roads. Once we get the coordinates, it will be easy to find whether those roads and buildings are constructed legally way or not.

4 Conclusion and Future Work

We use satellite imagery to spot improperly built roads and buildings in this chapter and decided to automate the whole process in the future. Since here for the matching of the segmented image, we are manually uploading the image to QGIS for getting the coordinates of the buildings and roads. So instead of doing that we intend to do that automatically as our future work. In addition to that, we planned to merge two or more residual models for the prediction process to improve the processing time of the system in this paper. From high-resolution remote sensing results, we projected the U-Net model for street and building extraction.

To summarize, the expected solution incorporates residual learning and U-Net advantages, and the operation's good paradigm results in an average precision of 0.98. By analyzing the city database, we have developed a model that will assist in locating illegally built buildings in both urban and rural areas. Authorities can manually check unmatched buildings and roads which we get as the output against the city database to detect illegal structures and roads. The artificial intelligence for detecting patterns is enhanced by combining machine learning and deep learning techniques. As a result, we can conclude that CNN and other pre-training classifiers are useful to both the biomedical and remote sensing industries.

References

1. K.S. Gautam, L. Parameswaran, S.K. Thangavel, Computer vision based asset surveillance for smart buildings. J. Comput. Theor. Nanosci. **17**(1), 456–463 (2020)
2. I. Letteri et al., Performance of botnet detection by neural networks in software-defined networks, in *ITASEC* (2018)
3. J. Hormese, C. Saravanan, Automated road extraction from high resolution satellite images. Procedia Technol. **24**, 1460–1467 (2016)
4. K. He et al., Deep residual learning for image recognition, in *Proceedings of the IEEE Conference on Computer Vision and Pattern Recognition* (2016), pp. 770–778
5. O. Ronneberger, P. Fischer, T. Brox, U-Net: Convolutional networks for biomedical image segmentation, in *Medical Image Computing and Computer-Assisted Intervention—MICCAI 2015*, ed. by N. Navab et al. (Springer International Publishing, Cham, 2015), pp. 234–241. ISBN: 978-3-319-24574-4
6. P.N. Anil, S. Natarajan, A novel approach using active contour model for semi-automatic road extraction from high resolution satellite imagery, in *2010 2nd International Conference on Machine Learning and Computing* (IEEE, 2010), pp. 263–266
7. K.V. Sree, G. Jeyakumar, An evolutionary computing approach to solve object identification problem in image processing applications. J. Comput. Theor. Nanosci. **17**(1), 439–444 (2020)
8. K.S. Gautam, S.K. Thangavel, Video analytics-based intelligent surveillance system for smart buildings. Soft Comput. **23**(8), 2813–2837 (2019)
9. A. Krizhevsky, I. Sutskever, G.E. Hinton, ImageNet classification with deep convolutional neural networks. Communications of the ACM **60**(6), 84–90 (2017)
10. G. Kumar, P.K. Bhatia, A detailed review of feature extraction in image processing systems, in *2014 4th International Conference on Advanced Computing & Communication Technologies* (IEEE, 2014), pp. 5–12
11. R. Manjusha, L. Parameswaran, Design of an image skeletonization based algorithm for overcrowd detection in smart building, in *Interna- tional Conference on ISMAC in Computational Vision and Bio-Engineering* (Springer, 2018), pp. 615–629
12. L. Swati Nair, R. Manjusha, L. Parameswaran, Kernel based approaches for context based image annotation, in *Computational Vision and Bio Inspired Computing* (Springer, 2018), pp. 249–258
13. V. Mnih, G.E. Hinton, Learning to detect roads in high-resolution aerial images, in *European Conference on Computer Vision* (Springer, 2010), pp. 210–223

Jayasurya Subramanian is currently pursuing his master's in Computer Science and Engineering at Amrita School of Engineering, Amrita Vishwa Vidyapeetham, Coimbatore. His research interest includes video analytics, image processing, and deep learning. He completed his B.E (Computer Science and Engineering) from Kongu Engineering College, Perundurai. He then completed his M.S. (Informatics and Mathematics) at the University of L'Aquila, Italy. His research work includes satellite image processing and segmentation using convolutional neural networks.

Dr. Senthil Kumar Thangavel currently serves as an Associate Professor at the Department of Computer Science and Engineering at Amrita School of Engineering, Amrita Vishwa Vidyapeetham, Coimbatore. His research interest includes video analytics, big data analytics, intrusion detection systems, and deep learning. He completed his B. Tech. (Computer Science and Engineering) from Sethu Institute of Technology, Madurai. He then completed his M. Tech. (Distributed Computing Systems) from Pondicherry Engineering College, Pondicherry. He completed his Ph.D. in Information and Communication Engineering from Anna University, Chennai. He is part of the project titled "Malware detection using FPGA, Sandboxing and Machine Learning" under the IBM Shared University Research (SUR) scheme.

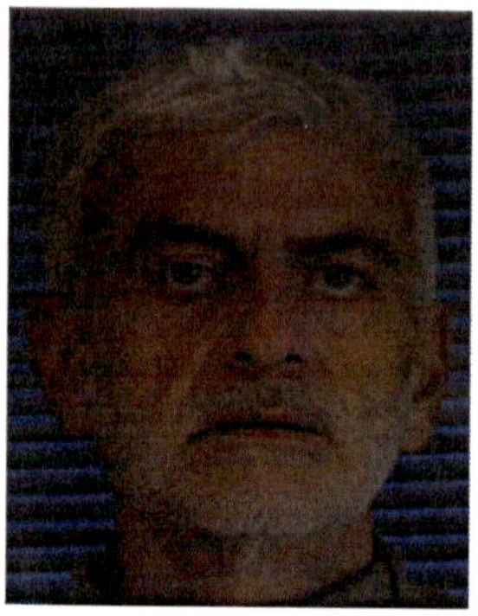

Dr. Pasquale Caianiello is a Senior Researcher and Assistant Professor at the Department of Information Engineering, Computer Science and Mathematics at Universitá degli Studi dell'Aquila. He graduated in Mathematics at the University of Rome "la Sapienza" and got his Ph.D. in Computer Science at the Courant Institute, New York University. His lifetime research project is to ground complex and intelligent behavior on Information Theory and is devoted to the application of machine learning to different scientific disciplines. His work can be found with the ORC ID, Pasquale Caianiello: 0000-0003-1006-326X.

Satellite Pose Estimation Using Modified Residual Networks

M. Uma Rani, Senthil Kumar Thangavel, and Ravi Kumar Lagisetty

1 Introduction

Estimating the relative attitude and position of uncooperative in-orbit objects is incredibly vital for varied autonomous operations in space. There are many space missions happening in the recent years; for example, ISRO is to conduct a space docking experiment known as SPADEX MISSION (Docking refers to connecting two flying objects in space to connect two structures to make a bigger one or to transfer men or material from one to the other), and there are missions like Active Debris Removal (ADR Mission), the 'RemoveDebris' mission deliberated to exhibit several technologies for removing debris present in space, which is led by Surrey Space Centre from the University of Surrey, Restore-L mission executed by National Aeronautics and Space Administration, etc., needs an accurate pose to do on-orbit servicing operations which are part of these missions. The determination of Space Vehicle's Pose for space-proximity operations of uncooperative target spacecraft using a monocular camera is recently drawing a lot of attention because it consumes low power and the requirements are simple. There are present strategies for determining spacecraft pose that utilizes image processing [1–3] algorithms which help with the identification of visible features of the target. Once the visible features get identified, which are probably the features which are hand engineered, they are matched against the known model of spacecraft to estimate pose. But there are quite

M. Uma Rani · S. K. Thangavel (✉)
Department of Computer Science and Engineering, Amrita School of Engineering, Amrita Vishwa Vidyapeetham, Coimbatore, India
e-mail: t_senthilkumar@cb.amrita.edu

M. Uma Rani
e-mail: cb.en.p2cse19010@cb.students.amrita.edu

R. Lagisetty
Indian Space Research Organization, Bangalore, India
e-mail: rkkumarl@ursc.gov.in

J. D. Peter et al. (eds.), *Disruptive Technologies for Big Data and Cloud Applications*, Lecture Notes in Electrical Engineering 905,
https://doi.org/10.1007/978-981-19-2177-3_81

some limitations to this method, as the preciseness gets degraded due to the presence of varied lighting conditions. The deep learning architectures, especially convolutional neural networks (CNN) based, are explored to address these problems. The interesting property of CNN is that it ignores the trivial features and learns more of significant features. This architecture supports transfer learning and works best by learning through the synthetic images of the space imagery dataset. The CNN model in its training phase automatically trains itself with the several synthetic images [3–5] generated of the satellite chosen, and in prediction phase, it takes in an image which is of type greyscale of the target spacecraft taken from close vicinity approximately at the distance of 10[m] between the two satellites. And then the trained convolutional neural network can now be used to estimate the pose of the spacecraft.

2 Background

In this section, some background information and the terms related to satellite pose estimation is provided. Pose determination of satellite is nothing but the calculation of the relative location and attitude of a target spacecraft in orbit relative to the chaser satellite. There are two main categories of pose determination: cooperative and non-cooperative. A target satellite is considered to be cooperative if it carries useful information, which can be helpful for pose estimation such as on-board sensors and communication links with chaser satellite. Uncooperative pose determination is further divided into unknown and known target objects. If prior data is available like positions of known features like solar panels [6], antennas, etc., then it comes under known target applications, and examples of such applications are debris removal or removal of defunct satellites. If the target is uncooperative and also we do not know the target model or we do not have prior data, then it comes under unknown, uncooperative targets and examples of such applications are asteroid exploration and debris removal. The position and orientation of a spacecraft S relative to world frame W_F is specified as $SF = (P, \phi)$, where P denotes the translation vector which indicates the spacecraft position from the origination in world frame W_F and ϕ denotes the satellite's orientation with respect to frame W_F. The 3D vector of Cartesian coordinates (x, y, z) denotes the space vehicle's position. And the orientation of satellite is represented with the help of a unit quaternion representation $\phi = (\phi w, \phi x, \phi y, \phi z)$, here ϕw is a real scalar and $\phi x, \phi y, \phi z$ is a 3D vector in complex space. Therefore, satellite pose is represented with the help of a 7D vector $[x, y, z, \phi w, \phi x, \phi y, \phi z]$.

3 Related Works

Pose estimation of an uncooperative satellite is a critical task, and a lot of research is happening in this area. Deep learning algorithms, especially CNNs, are being explored by most of the researchers for pose determination of spacecraft. The various methods for finding the pose of satellites have been demonstrated by many researchers. In the paper by Zeng et al. [7], a data augmentation-based deep learning approach has been proposed to space target recognition. Here they have built a nine-layer deep convolutional neural network model which has been trained on a simulated space target dataset to identify the space target. The dataset is prepared using the Satellite Tool Kit (STK), and further the images are degenerated by series of motion and out-of-focus blur to imitate the environment of space. There is another paper by Sharma et al. [5] which has demonstrated a new feature recognition approach based on weak colour progression filtering that is used to determine the target image's true edges. This method can easily identify the edges even if it contains the Earth image in the background. In order to reduce the search space and bring down the features correspondence problem, the features detected in the process are intermixed using the geometric constraints, which is solved using the EPnP method. The paper by Sharma et al. [8] introduces a novel method based on CNN which can give a predicted pose in real-time on board. This method trains the model with the set of images belonging to the particular pose labels which are obtained using the concept discretization of the pose space. Wu et al. [9] in his paper have demonstrated an artificial intelligence method for target identification in space, and this method has been called 'two-stage convolutional neural network' (T-SCNN). This method is composed of two stages: (1) targets locating and (2) target recognition. In the first stage, a concept called minimal bounding rectangle employing threshold method is used to detect all suspicious images from the complete image dataset, and after that all the regions that circumscribe target are cut to produce target images for training. In the second stage, all the target images are sent to the well-trained recognition network for identification. Harvard et al. [10] introduced a novel method for pose estimation using concept called key points that are specialized for the given target. A set of robust key points are created by examining the effectiveness of existing localization algorithms by simulating and testing different perspectives (Table 1).

4 Dataset

There are some dedicated datasets which are publicly available for training, validating and testing of pose estimation methods for space borne applications. Some of them are as follows.

1. The SPEED dataset(Spacecraft POSE Estimation Dataset): This publicly available dataset contains 15,000 augmented reality and 900 actual camera images of the Tango satellite. Figure 1 shows sample images from (A)SPEED dataset [14] and (B)URSO dataset [11]
2. URSO dataset: dataset prepared with the help of simulator built on Unreal Engine 4, named URSO
3. ImageNet dataset: Image Net ILSVRC 2010 dataset

Table 1 Summarization of recent research works

Paper title	Models used	Key takeaways
[5]	Robust monocular vision-based pose initialization architecture	[Dataset used: PRISMA] the architecture used WGW technique with the sobel operator, followed by the Hough transform, to detect small and large features of the target satellite
[10]	CNN	[Dataset used: ImageNet ILSVRC 2010 dataset] the architecture produces landmark positions for arbitrary key point detectors that can assess relative detectability and can better manage high relative dynamics between the target and the camera
[11]	a deep learning framework for estimating satellite's pose based on orientation and soft classification	[Dataset used: a simulator built on Unreal Engine4 named URSO] this framework has exhibited the effects of a variety of network hyperparameters and training options, as well as providing solutions to problems such as what is the best method for estimating orientation. It is also observed that soft categorization yields better results than direct regression
[8]	CNN	[Dataset used: ImageNet] better performance was observed on images containing high amounts of Gaussian white noise. Accuracy of this model was proved to be better compared to classical algorithms of feature detection
[12]	CNN-based HRNet architecture	[Dataset used: SPEED dataset] HRNet model gave precise predictions of two-dimensional landmarks. The SA-LMPE algorithm has improved pose refinement and removed more erroneous predictions
[7]	CNN	[Dataset used: Simulated space target dataset rendered by the Systems Tool Kit (STK)] This model automatically learns and efficiently represents relevant features using deep learning models
[9]	T-SCNN	[Dataset used: Images were collected from public database] this deep learning model has performed well in detecting and recognizing space targets from simple images of deep space
[13]	DCNN	[Dataset used: Simulated space debris dataset of Apollo spacecraft by TERRIER] this model is effective at learning features of high level and is resistant to transformations, like reflections, rotation and brightness changes

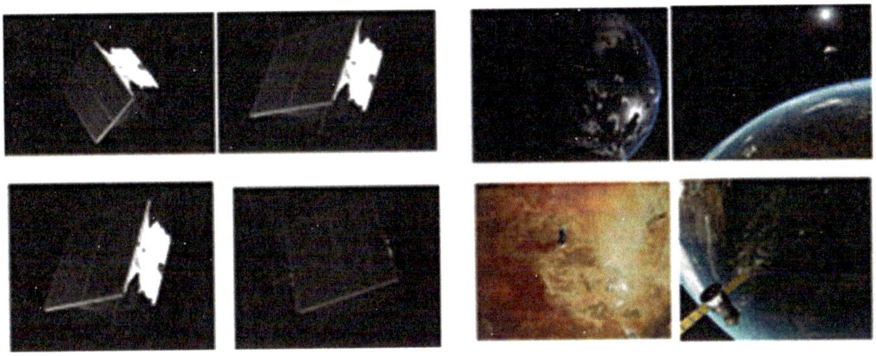

Fig. 1 Sample images from **a** SPEED dataset **b** URSO dataset

4. BUAA-SID Dataset: This dataset contains some 4600 grey images. These are the images of some 20 satellites and their corresponding images in binary format.

5 Methodology

The main goal is to train a convolutional neural network [15] using a synthetic satellite image dataset, and then the network should be able to estimate the seven-dimensional pose of the satellite on unseen images which are present in the test dataset. In this project, we started with convolutional neural network architectures, and then based on the results, we have experimented with residual networks by modifying it which is well explained in sect. 7 and applied regularization techniques further like image augmentation. While experimenting with different frameworks, the goal was to reduce overfitting [16] and to reduce computational time for training without compromising on the accuracy of results.

5.1 Convolutional Neural Networks

To overcome the challenges posed by the use of manually defined features, convolutional neural network (CNN) [17] is explored. The blocks which make up the CNN are as follows: convolution layer, pooling layer and fully connected layers. These blocks automatically learn the spatial features with the help of a back propagation algorithm (Fig. 2).

Convolutional Layer This is a very important layer, and its main function is to do feature extraction. Feature extraction has two operations: linear and nonlinear, which is nothing but a combination of convolution operation and activation function, respectively.

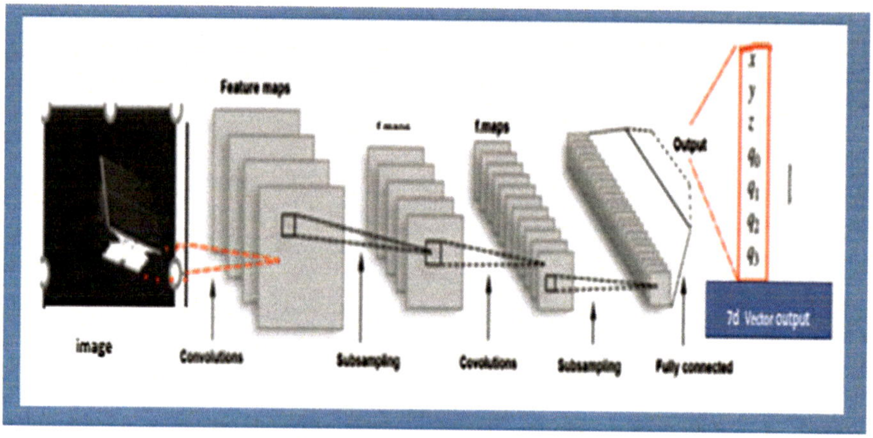

Fig. 2 Convolutional neural network architecture

Pooling Layer This layer's main function is to reduce the size of the volume by selecting the maximum value from the region covered by the kernel. Therefore, the output would contain the most important features only.
Fully Connected Layer(FC Layer) These layers are present at the end of CNN. In this layer, every node of the first layer is connected to every node of the second layer. The output of the last pooling layers of a CNN neural network serves as the input for the FC layer.

6 ResNet Architecture

It is observed that a considerable amount of depth in neural networks is important for critical tasks, but with increasing depth, there is a risk of vanishing gradient problem, i.e. when we transit through multiple levels, gradient details might be lost. Therefore, the ResNet architectures have been explored for this task (Fig. 3).

7 Modified ResNet Architecture to Reduce Overfitting

We started with training ResNet-18 model to determine position and orientation of satellite. This model has been pretrained on ImageNet weights using the concept of transfer learning so that it can work efficiently with limited dataset. By observing the results, we found that this model suffers from overfitting, so to bring down the overfitting, an extra block has been attached, where a dropout layer has been added with parameter 0.5 in between the two convolution layers as shown in the basic block

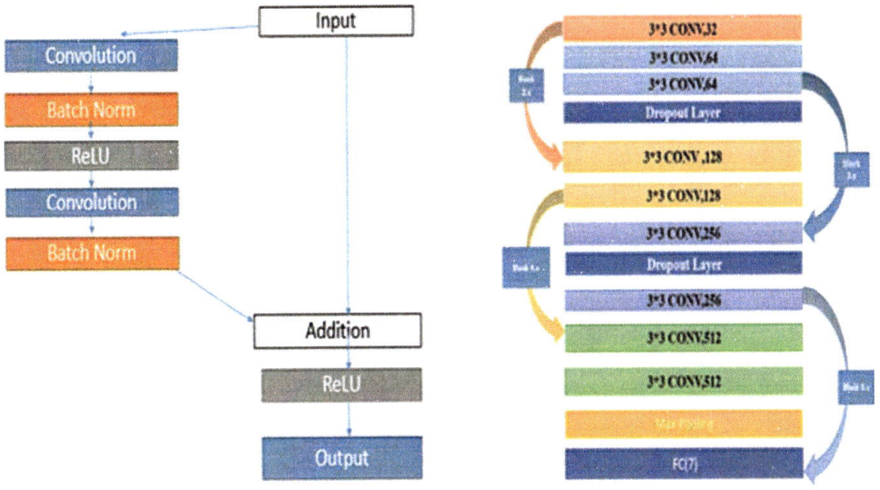

(A) (B)

Fig. 3 **a** Basic block of ResNet **b** sample ResNet-18 network for satellite pose estimation with dropout layer

diagram of ResNet architecture in Fig. 4 and also the softmax layer of ResNet block was replaced with fully connected layer to predict seven-dimensional vector pose of satellite.

7.1 Dropout Layer

Dropout [18]is one of the regularization techniques which reduces overfitting in neural networks. It helps reducing correlated learning amongst the neurons. This layer simply drops some of the units of neural network.

7.2 Data Augmentation

To overcome overfitting, we also used the data augmentation method to enlarge the dataset. This technique prepares new dataset from the training set of original dataset.

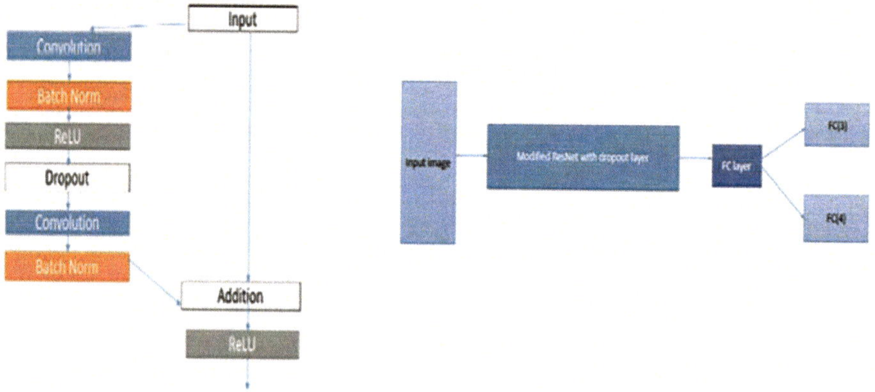

(A) (B)

Fig. 4 **a** ResNet architecture with dropout layer **b** block diagram with two fully connected layers for calculating position and orientation independently

8 Experiments

A lot of models have been implemented and experimented which has been mentioned in the methodology section to study the effects of these techniques on performance. In all the experiments, we used stochastic gradient descent with a learning rate of 0.001 using an Adam optimizer. The network structure has been modified for the regression task, and also the hyperparameters have been changed and analysed. The batch size considered is 32 as it produced efficient and precise results.

8.1 Evaluation Metrics

The total error of three-dimensional pose estimation is nothing but the calculated sum of orientation and position error. The total error on the synthetic image test set gives synthetic image error and the total error on the real images test set corresponds to the real image error, and this is presented in Table 3. For the designed experiments, the reports are given in Table 3.

For an image i, the position error is calculated using the formula shown in Eq. (1). It is the L2 normalization of the position error, which is nothing but the dissimilarity between the ground truth and the estimated position vectors is taken and normalized with the spacecraft's ground truth distance.

$$\mathrm{Err}_{\mathrm{pos}}^{(i)} = \frac{|P_{\mathrm{GT}}^{(i)} - P_{\mathrm{Est}}^{(i)}|_2}{P_{\mathrm{GT}}^{(i)}|_2} \tag{1}$$

The orientation error is calculated using the formula represented in Eq. (2). Orientation error is the angle of rotation that aligns the calculable with ground truth orientations.

$$\text{Err}^{(i)}_{\text{orient}} = 2.\arccos(| < \phi^{(i)}_{\text{EST}}, \phi^{(i)}_{\text{GT}} > |) \tag{2}$$

The pose error is calculated using the formula represented in Eq. (3). The sum of position error and orientation error is pose error.

$$\text{Err}^{(i)}_{\text{pos}} = \text{Err}^{(i)}_{\text{POS}} + \text{Err}^{(i)}_{\text{orient}} \tag{3}$$

The total error is calculated using formula shown in Eq. (4). Here 'N' denotes number of images present in the original dataset's test set. This error is found by taking the average of pose errors over all the images(N images) present in the test set.

$$\text{TotalError} = \frac{1}{N}\sum_{i=1}^{N} \text{Err}^{(i)}_{\text{POSE}} \tag{4}$$

8.2 *Loss Functions*

The loss function considered for the experiments is mean squared error(MSE). It is calculated using the formula shown in Eq. (5). For all the experiments, MSE has been selected as it gave efficient results especially for the regression tasks. MSE is nothing but the average squared dissimilarity in between the actual value and the estimated value given by the model. It penalizes even a small error as it squares the differences. This loss function is better compared to other loss functions as it is differentiable, therefore optimizes better (Table 2).

$$\text{MeanSquareError} = \frac{1}{N}\sum (x - \hat{x})^2 \tag{5}$$

Table 2 Hyperparameters used in our model

Momentum	Learning rate	Batch size	Activation function	Dropout	Optimizer
0.001	0.001	32	ReLu	0.5	Adam

Table 3 Error values for test set of real images and synthetic image dataset

Model name	Description	Real image error	Synthetic image error
A	ResNet with random initialized weights(baseline model)	4.6	3.3
B	ResNet-18	2.8	2.65
C	ResNet-50 + dropout	1.9	1.6
D	ResNet-50 + dropout + data augmentation	**1.5**	**1.7**

8.3 Hyperparameters Used in Our Model

8.4 Layers in ResNet

The performance of ResNet architecture showed improved results with an increased number of layers. The ResNet with 50 layers shows better results in comparison with ResNet-18. ResNet-50 is a deep network with 50 layers so it extracts more relevant features. The residual connections along with the added dropout layer have provided an added advantage in improving the performance which can be seen in Table 3 provided.

8.5 Dataset Used

SPEED dataset has been used in this project. There are 300 real images of the Tango satellite in this dataset and contain 12,000 synthetic images for training. These 8-bit monochrome images are in jpeg format, with a resolution of 1920 × 1200 pixels. Images are 8-bit monochrome in jpeg format, which are having resolution of 1920 × 1200 pixels. These are images of the Tango satellite of the PRISMA Mission.

9 Results

9.1 Comparison with Randomly Initialized Network

The ResNet model which was pertained on ImageNet (dataset) weights has given efficient results compared to the ResNet model which was randomly initialized. Table 3 lists the error values for various experiments carried out. The ResNet model

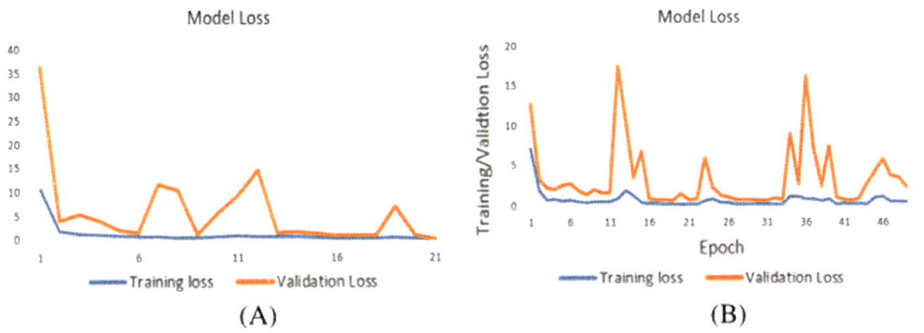

Fig. 5 **a** Graph representation for validation and training loss against epoch numbers for the base model **b** graph representing training and validation loss against epoch number for ResNet-50 model with dropout + image data augmentation and mean squared error loss function

along with data augmentation and with dropout techniques has shown improved results in terms of overfitting when compared with the base model. All the tests are performed on SPEED dataset (Fig. 5).

10 Conclusion

We did a deep study on satellite pose estimation using neural networks. We successfully have implemented modified residual neural networks which is a CNN architecture and also performed various experiments on this architecture. A lot of models have been implemented and experimented which have been mentioned in the methodology section to study the effects of these techniques on performance. The main aim of this project was to reduce overfitting and to train models with limited computational cost. These experiments have helped us to learn interesting results which can be further explored and used in enhancing precision in the establishment of future space navigation systems with less computational cost. First the ResNet architecture has been implemented with random initialized weights, but the performance levels increased dramatically when used the transfer learning with the help of ImageNet weights. The learning rate of the neural network has increased dramatically as ImageNet understands a number of low-level properties. Second, the modified ResNet with a dropout [19] layer has given good performance overall. The various regularization techniques used, like image data augmentation and dropout, have reduced the overfitting problem to quite an extent. Third, the dropout layer used in the ResNet has reduced the overall training time to a little extent but did not get significant results, but during the process, we got to learn some variants of residual networks like DRHNet [20], LeanResNet [21] and multi-residual networks which can be used and implemented to reduce the computational complexity and to speed up training the network. In future, we would like to further explore on more variants of residual networks and experiment on them with various techniques of regularization so that it could be useful in reducing, overfitting of neural networks, computational cost for training as well as rendering time of result as all these factors are very important for on-board pose estimation of satellite.

References

1. K. Gautam, L. Parameswaran, S.K. Thangavel, Computer vision based asset surveillance for smart buildings. J. Comput. Theor. Nanosci. **17**(1), 456–463 (2020)
2. R. Manjusha, L. Parameswaran, Design of an image skeletonization based algorithm for overcrowd detection in smart building, in *International Conference on ISMAC in Computational Vision and Bio-Engineering* (Springer, 2018), pp. 615–629
3. L.S. Nair, R. Manjusha, L. Parameswaran, Kernel based approaches for context based image annotation, in *Computational Vision and Bio Inspired Computing* (Springer, 2018), pp. 249–258
4. K. Gautam, S.K. Thangavel, Video analytics-based intelligent surveillance system for smart buildings. Soft Comput. **23**(8), 2813–2837 (2019)
5. S. Sharma, J. Ventura, S. D'Amico, Robust model-based monocular pose initialization for noncooperative spacecraft rendezvous. J. Spacecraft Rockets **55**(6), 1414–1429 (2018)
6. R. Lagisetty, P.K. Sinha, Multi-task dual arm space robot using co-operative control for planetary explorations, in *39th COSPAR Scientific Assembly*, vol. 39 (2012), p. 1021
7. H. Zeng, Y. Xia, Space target recognition based on deep learning, in *2017 20th International Conference on Information Fusion (Fusion)* (IEEE, 2017), pp. 1–5
8. S. Sharma, C. Beierle, S. D'Amico, Pose estimation for non-cooperative spacecraft rendezvous using convolutional neural networks, in *2018 IEEE Aerospace Conference* (IEEE, 2018), pp. 1–12
9. T. Wu, X. Yang, B. Song, N. Wang, X. Gao, L. Kuang, X. Nan, Y. Chen, D. Yang, T-scnn: a two-stage convolutional neural network for space target recognition, in *IGARSS 2019-2019 IEEE International Geoscience and Remote Sensing Symposium* (IEEE, 2019), pp. 1334–1337
10. A. Harvard, V. Capuano, E.Y. Shao, S.J. Chung, Pose estimation of uncooperative spacecraft from monocular images using neural network based keypoints (2020)
11. P.F. Proença, Y. Gao, Deep learning for spacecraft pose estimation from photorealistic rendering, in *2020 IEEE International Conference on Robotics and Automation (ICRA)* (IEEE, 2020), pp. 6007–6013
12. B. Chen, J. Cao, A. Parra, T.J. Chin, Satellite pose estimation with deep landmark regression and nonlinear pose refinement, in *Proceedings of the IEEE/CVF International Conference on Computer Vision Workshops* (2019)
13. J. Tao, Y. Cao, M. Ding, Z. Zhang, Visible and infrared image fusion for space debris recognition with convolutional sparse representation, in *2018 IEEE CSAA Guidance, Navigation and Control Conference (CGNCC)* (IEEE, 2018), pp. 1–5
14. M. Kisantal, S. Sharma, T.H. Park, D. Izzo, M. Märtens, S. D'Amico, Satellite pose estimation challenge: dataset, competition design, and results. IEEE Trans. Aerosp. Electr. Syst. **56**(5), 4083–4098 (2020)
15. Y. Xiang, T. Schmidt, V. Narayanan, D. Fox, Posecnn: a convolutional neural network for 6d object pose estimation in cluttered scenes. arXiv preprint arXiv:1711.00199 (2017)
16. N. Srivastava, G. Hinton, A. Krizhevsky, I. Sutskever, R. Salakhutdinov, Dropout: a simple way to prevent neural networks from overfitting. J. Mach. Learn. Res. **15**(1), 1929–1958 (2014)
17. S. Nandhini, R. Suganya, K. Nandhana, S. Varsha, S. Deivalakshmi, S.K. Thangavel, Automatic detection of leaf disease using cnn algorithm, in *Machine Learning for Predictive Analysis* (Springer, 2021), pp. 237–244
18. Y. Gal, J. Hron, A. Kendall, Concrete dropout. arXiv preprint arXiv:1705.07832 (2017)

19. P. Mianjy, R. Arora, R. Vidal, On the implicit bias of dropout, in *International Conference on Machine Learning* (PMLR, 2018), pp. 3540–3548
20. Y. Xu, Z. Fu, G. Xu, S. Zhang, X. Xie, Drhnet: a deep residual network based on heterogeneous kernel for steganalysis. Secur. Commun. Netw. **2020** (2020)
21. J. Ephrath, L. Ruthotto, E. Haber, E. Treister, Leanresnet: a low-cost yet effective convolutional residual networks. arXiv preprint arXiv:1904.06952 (2019)

M. Uma Rani is currently pursuing Master of Technology in Computer Science and Engineering at Amrita School of Engineering, Amrita Vishwa Vidyapeetham, Coimbatore. Her research interests are artificial intelligence and computer vision. She did this research work under the mentorship of Dr. T. Senthil Kumar and Dr.Ravi Kumar Lagisetty.

Dr. Senthil Kumar Thangavel currently serves as Associate Professor at the Department of Computer Science and Engineering at Amrita School of Engineering, Amrita Vishwa Vidyapeetham, Coimbatore. His research interest includes video analytics, big data analytics, intrusion detection systems and deep learning. He completed his B. Tech. (Computer Science and Engineering) from Sethu Institute of Technology, Madurai. He then completed his M. Tech. (Distributed Computing Systems) from Pondicherry Engineering College, Pondicherry. He completed his Ph.D. in Information and Communication Engineering from Anna University, Chennai. He is part of the project titled 'Malware detection using FPGA, Sandboxing and Machine Learning' under the IBM Shared University Research (SUR) scheme.

Dr. Ravi Kumar Lagisetty is Scientist 'SG' in ISRO and a doctorate from IISC in Aerospace Engineering. His research interest includes advanced optimal control, space docking, formation flying, space robotics, satellite control, image processing and AI for space and robotic application. He is currently working as Division head, Hardware in Loop Simulation, Mission Simulation Group at ISRO. He is having around 25 years of experience in spacecraft control and robotic control.

Opinion on Prediction Algorithms for Identifying Autism Spectrum Disorder

D. Darling Jemima, A. Grace Selvarani, and J. Daphy Louis Lovenia

1 Introduction

Brain is the most important part which is responsible for all kinds of activities likes talking, smelling, seeing, thinking, actions, etc., in human body. Brain comes under the central nervous system (CNS) [1]. The brain includes cerebrum, cerebellum, and the brain stem. The brain stem consists of three parts. They are mid-brain, pons, and medulla oblongata. Cerebrum is responsible for functions which include touch, sight and listening, speech, feelings, and movement control. Function of cerebellum is to order the movements of muscles and maintain balance and posture. Brain stem plays a vital role in linking the cerebrum and cerebellum with the spinal cord. Many functions such as inhalation and exhalation, heart rate, temperature, rise and shine cycles, digestion, vomiting, and coughing are performed by brain stem. Thalamus is founded on the top of the brain which is responsible to act as a switch board or the gate way to work with the cortex. People who are face difficulties in speech, having interactions with social world; any repetitive actions are diagnosed with autism spectrum disorder.

Usually, this can be found in kids during their first three years. This will impact their regular activities and will have behavioral issues such as reduced eye contact and

A. Grace Selvarani
Department of Computer Science and Engineering, Sri Ramakrishna Engineering College, Coimbatore, Tamil Nadu, India
e-mail: graceselvarani@srec.ac.in

J. Daphy Louis Lovenia
Department of Mathematics, Karunya Institute of Technology and Sciences, Coimbatore, Tamil Nadu, India
e-mail: daphy@karunya.edu

D. Darling Jemima (✉)
Department of Computer Science and Engineering, Sri Krishna College of Technology, Coimbatore, India
e-mail: darlingjemimai@gmail.com

J. D. Peter et al. (eds.), *Disruptive Technologies for Big Data and Cloud Applications*, Lecture Notes in Electrical Engineering 905,
https://doi.org/10.1007/978-981-19-2177-3_82

interacting with people [2]. Temporal lobe is associated with processing information and having a long-term memory. Parietal lobe is responsible for managing the sense of taste, hearing, sight, touch, and smell. Frontal lobe is the largest lobe in human body and responsible for managing high-level executive function and expressive language. At present, the cause of autism is unknown, a person can receive the support and services that they need in early stages. This survey is highly concentrated on diagnosis procedure for autism with the help of different methodologies. The different computing algorithms implemented on the autism dataset were discussed. The accuracy of correctly classifying the autistic patient from the healthy people was underlined in this survey. We have given the clear outline of which algorithm is best based on our predictions.

2 Literature Survey

The data assert for the initial intervention of autism spectrum disorder (ASD) was health records which contains symptoms and treatments which are collected from the patients regarding the activities of the kids, the gesture, their emotion through the gaming software, ontology features, repetitive patterns, phenotypic and genotypic information through virtual reality(VR), and haptic devices. As per the record of the World Health Organization (WHO), 1/160 children are affected to this neurological disorder, autism spectrum disorder (ASD). From the records, it is evident that autism is globally increasing in children. The diagnosis of ASD is improved due to advancement in technology and diagnostic criteria. Khalid Al-jabery et al. implemented various clustering algorithms on the samples of 208 probands from the Simon Simplex Collection Missouri Site patients.

Steps such as data pre-processing, correlation analysis, uni-dimensional clustering, *k*-dimensional clustering, and cluster evaluation have been followed. The feature representation and normalization of the data have been done in the pre-processing phase. With the help of pairwise Pearson's correlation, the correlation between the numerical features was examined. To cluster each input data, expectation and maximization algorithm were used. Once these features are clustered, they were ranked using feature ranking. Sub-space iteration has been done using K-dimensional clustering [3]. Finally, the cluster evaluation includes voting, univariate analysis, and multivariate analysis. After multivariate analysis result which includes 3 clusters were formed, with CER 8.65% for cluster 1, 15.4% for cluster 2, 34.6% for cluster. Liu et al. [4] used Naïve Bayes algorithm to classify autistic and non-autistic records. The electronic health records (HERs) are taken as the input where communication and eye contact features are extracted from the records. Each feature is given a weight, and each record is treated as the feature vector. The independency between features with the probability distribution was estimated to determine the class label. The positive class and the negative class are segregated by marginal hyperplane in support vector machine (SVM).

Only 50 out of 10,000 inputs from electronic health record were used for the study purpose. F-score is used as the evaluation metrics with SVM classifier as 0.71. This study by Liu et al. [4] uses initial early intervention (EI) assessment records to detect children with autism spectrum disorder (ASD) much earlier to benefit from the funds aided by the state EI systems. The dataset contains 9777 initial El assessment records in which 1711 are ASD negative, 57 are ASD positive, 2998 are speech and language impaired. (SLI), and 5011 as unspecified development delayed (DD). Data pre-processing has been done by applying lemmatization for unigrams and removing stopwords for bigrams. Unified ontology has also been used to remove all irrelevant features and construct a compressed feature set. The effectiveness of the three feature set representations (Boolean, term frequency (TF), and term frequency times inverse document frequency (TFIDF)) [5] has been compared. Three classification methods, such as Naive Bayes (NB), Bayesian logistic regression (BLR), and support vector machine (SVM), have been used. Out of the three classification methods, SVM model trained using unigrams proposed by manually developed ontologies seems to give a better accuracy of early autism spectrum disorder identification. F-score for the all three models was analyzed, and finally, SVM [6] outperforms other two algorithms with 0.071.

3 Different Prediction Algorithms to Identify ASD

The existing work in the area of autism detection is done manually by collecting the various features. The proposed work is to automate the autism detection using the machine learning approaches. The CSV files which have the autistic and non-autistic features are given as an input for the autism detection. There are two classes, namely 0 and 1 (0-autistic, 1-non-autistic). As per the analysis taken, the accuracy was recorded based on the training the various models. The various models which have been implemented are

1. Voted perceptron
2. K-nearest neighbor (KNN)
3. Logistic regression
4. Hoeffding tree
5. Random forest
6. Support vector machine (SVM)
    - 6.1. Linear kernel
    - 6.2. Sigmoid kernel
    - 6.3. Poly kernel
    - 6.4. Radial basis function (RBF) kernel
7. Bayesian network classifier
8. Decision stump
9. Adaptive boosting.

3.1 Voted Perceptron

Rosenblatt's perceptron model with Helmbold and Warmuth's leave-one-out technique, both combine together to form voted perceptron. This is used for high-dimension spaces using kernels. It has the supremacy over the data which is separated linearly with great margin. The voted perceptron algorithm performs leniently while deploying in the autism spectrum disorder dataset with accuracy of 83.94%. So voted perceptron algorithm is not much efficient when compared to other models for the autism detection.

3.2 K-Nearest Neighbor (KNN)

KNN is a straightforward machine learning algorithm which follows supervised learning. It does not learn from the trained data of ASD. Instead, it stores the data and performs action while doing the classification. The KNN algorithm performs up to the mark while deploying in the dataset with accuracy of 94.88% in the detection of ASD. So KNN is efficient when compared to voted perceptron for the autism detection.

3.3 Logistic Regression

Logistic regression was proposed by Joseph Berkson. This algorithm is used to predict the outcome based on the group of independent variables. It calculates the probability of binary event's occurrence and the deal with the issues of classification. In a linear separable dataset, it works in finest manner. Logistic regression outwitted the above algorithm with accuracy 96.01% for the dataset. So logistic regression algorithm is much efficient when compared to other models for the autism detection.

3.4 Hoeffding Tree

The Hoeffding tree is a decision tree learner for huge datasets that considers that the data allocation is not varying regulation time. Hoeffding tree builds decision tree on the basis of theoretical assurances of the Hoeffding bound. Hoeffding tree algorithm is effective with the dataset when compared to the above algorithm showing accuracy 97.85%. So Hoeffding tree algorithm is compatible when compared to other models for the autism detection.

3.5 *Random Forest*

Random forest is algorithm which has more decision trees which uses feature randomness and bagging. While constructing, each tree combine together to create an unrelated forest of trees which give more accuracies than any individual tree. It is suitable for very large datasets, and interpretability is not a major concern, because it consists of multiple decision trees. Random forest is a flexible algorithm that produces even without hyperparameter tuning. It is a popular algorithm for its greatest accuracy. Since it is simple and diverse, it is been used in classification and regression task. Random forest algorithm outperformed with the dataset when compared to the above algorithm showing accuracy 98.50%. So random forest algorithm is best fitted when compared to other models for the autism detection.

3.6 *Support Vector Machine (SVM)*

A support vector machine (SVM) is a supervised machine earning model that are used for classification, regression, and outliers detection. The main objective of SVM is to find the data points with maximum margin. "Kernel" is a result of mathematical functions used in support vector machine. So, kernel generally transforms the training set of data into a linear equation in a higher number of dimension spaces. Using four kernels, autism detection has been implemented, and their accuracies have been recorded. The four kernels are

1. Linear kernel
2. Sigmoid kernel
3. Poly kernel
4. Radial basis function (RBF) kernel.

3.6.1 Linear Kernel

In linear dataset, preferred and common SVM kernels are linear kernel. It is fastest and easiest to implement while the dataset has large number of features. Linear kernel has a effective high-dimensional space, even while the dimensions are greater than the samples. Since it utilize a subset of training points in the decision function, it is been declared as memory efficient. Linear kernel performs on top notch as it gives 98.70% accuracy when implemented on the dataset. Therefore, linear kernel is much more efficient and works with high computation when compared to other kernels in SVM.

3.6.2 Sigmoid Kernel

The hyperbolic tangent kernel is also known as the sigmoid kernel and as the multilayer perceptron (MLP) kernel. The sigmoid kernel comes from the neural networks field. Bipolar sigmoid function is used as an stimulate function for artificial neurons. Sigmoid kernel function is similar to a two-layered, voted perceptron neural network. The sigmoid kernel in SVM gives an accuracy of 73.11% when implemented on the dataset. Sigmoid kernel will not be very much effective when compared to other SVM kernels as it gives accuracy of 73.11% when it is implemented in the dataset.

3.6.3 Poly Kernel

Polynomial kernel is common kernelled model, and it always allows learning of non-linear models. In this method, it is not determining the given features also the similarity and combinations are determined. Poly kernel works fine when compared to RBF kernel as it gives accuracy of 83 0.96% while implemented in the dataset.

3.6.4 Radial Basis Function (RBF) Kernel

RBF is one of the most powerful kernel among the kernels of SVM. It is mainly used in case of non-linear data. It helps to make proper separation when there is no prior knowledge about the data. RBF is not advisable as it gives very less accuracy of 74.05% when implemented in the dataset.

3.7 Bayesian Network Classifier

Bayesian network is a graphical portrayal of different probabilistic relationships among the random variables in a set. It is a classifier where it will not depend on any features. Bayesian network algorithm is not preferable as it gives only 80% of accuracy when it is applied in the dataset.

3.8 Decision Stump

Decision stump tree known as 1-rules, because of it prediction method, based on single input feature. It is one level tree due to one internal node connecting to terminals. Decision stump tree is preferable as it gives 90.01% of accuracy when it is implemented in the model.

3.9 *Adaptive Boosting*

AdaBoost stands for adaptive boosting which is used as an ensemble learning in machine learning. It works based on the principle where learners are grown chronologically. AdaBoost works in an acceptable manner as it gives accuracy of 92% while the algorithm was implemented in the dataset.

4 Discussion

Different machine learning algorithms have been implemented for the application of predicting autism spectrum disorder in the human being. According to the results obtained by applying different predicting algorithms on autism spectrum disorder dataset, support vector machine with linear kernel outperforms the other predicting algorithms. Support vector machine algorithm with different kernels such as linear, sigmoid, Poly, and radial basis function network have been implemented and analyzed the results. Since linear kernel with SVM worked faster and efficient in understanding the features such as ethnicity, gender, Jaundice affected history, and answers for all the queries with Q-chat reply in the toddler autism dataset. It utilize a subset of training points in the decision function, and it is been considered to be one of the efficient process in terms of memory utilization as well. Whereas KNN is supervised learning algorithm which is comparatively effective than voted perceptron. Random forest and adaptive boosting pick up the features randomly, and accuracy is better than Bayes classifier of finding out the probabilistic features in the autism spectrum disorder dataset.

5 Conclusion

Various machine learning algorithms were implemented in the autism spectrum disorder dataset, and their accuracy and loss were recorded for the study. Based on the opinion, SVM linear kernel algorithm has performed well when compared to the other predicting algorithms. Thus, the autism spectrum disorder can be identified, and preventive measures can be taken in early stage with the help of machine learning algorithms.

References

1. M.-Y. Hsiao, D.-Y. Chen, J.-H. Chen, Constructing human brain function association models from fMRI literature, in *IEEE Xplore, 29th Annual International Conference of the IEEE Engineering in Medicine and Biology Society* (2007)

2. G. Noriega, Restricted, repetitive, and stereotypical patterns of behavior in Autism—an fMRI perspective, in *IEEE Transactions on Neural Systems and Rehabilitation Engineering* (2018)
3. K. Al-jabery, T. Obafemi-Ajayi, G.R. Olbricht, T.N. Takahashi, S. Kanne, D. Wunsch, Ensemble statistical and subspace clustering model for analysis of Autism spectrum disorder phenotypes, in *IEEE Xplorer* (2016)
4. M. Liu, Y. An, X. Hu, D. Langer, C. Newschaffer, L. Shea, An evaluation of identification of suspected autism spectrum disorder (ASD) cases in early intervention (EI) records, in *IEEE Xplorer* (2013)
5. M.A. Miskam, S. Shamsuddin, H. Yussof, I.M. Ariffin, A.R. Omar, A questionnaire-based survey: Therapist's response on emotions gestures using humanoid robot for autism, in *IEEE Xplorer International Symposium on Micro-NanoMechatronics and Human Science (MHS)* (2015)
6. X.-A. Bi, Y. Liu, Q. Sun, X. Luo, H. Tan, J. Chen, N. Zeng, The genetic-evolutionary random support vector machine cluster analysis in autism spectrum disorder. IEEE Access **7**, 30527–30535, Mar 2019

Classification of Music Genres Based on Mel-Frequency Cepstrum Coefficients Using Deep Learning Models

Manoj Preetham, Jemimah Beulah Panga, J. Andrew, Kumudha Raimond, and Hien Dang

1 Introduction

The word 'genre' means to interpret. Genres usually won't have any music theoretic foundations, e.g.,—Indian genres are geographically defined [1]. At present, genre classification is performed manually by humans by applying their personal ideas, knowledge, and understanding of music. The drawback in this manual approach is finding the candidates having good field expertise in classifying the music. The proliferation of music streaming services such as Spotify, Amazon Music, Apple Music, Pandora, and SoundCloud in recent years has necessitated the need for these apps to arrange and categorize these audio songs by genre [2]. As a result, it is critical to create a robust and accurate music genre classification system to aid in the automated organization of this content. However, because of the imprecise definition of music genres, categorizing music genres remains a tough job [3]. Although categories like

M. Preetham · J. B. Panga · K. Raimond
Department of Computer Science and Engineering, Karunya Institute of Technology and Sciences, Coimbatore, India
e-mail: manojpreetham@karunya.edu.in

J. B. Panga
e-mail: jemimahbeulah@karunya.edu.in

K. Raimond
e-mail: kraimond@karunya.edu

J. Andrew (✉)
Department of Computer Science and Engineering, Manipal Institute of Technology, Manipal Academy of Higher Education, Manipal, India
e-mail: andrewj.research@gmail.com

H. Dang
Department of Computer Science, University of Massachusetts Boston, Boston, MA, USA

Faculty of Computer Science and Engineering, Thuyloi University, Hanoi, Vietnam

J. D. Peter et al. (eds.), *Disruptive Technologies for Big Data and Cloud Applications*, Lecture Notes in Electrical Engineering 905,
https://doi.org/10.1007/978-981-19-2177-3_83

hip-hop, country, and disco are commonly used, there is no specific definition to differentiate them. These issues, as well as the demand for a music genre categorization system, have opened opportunities to the scholars all around the world and have recently attracted a large number of them [4]. The construction of this system requires the extraction of acoustic features, which are good estimators for the type of genre we are interested in, followed by the classification of labels or the regression stage. Generally, feature extraction relies on front-end signal processing to evaluate relevant features from the time-domain or frequency-domain audio representation [5].

Machine learning (ML) and DL techniques are currently used in music genre classification [6]. Tzanetakis and Cook [7] in 2002 developed a music genre classification approach by integrating EM clustering and KNN, as well as three types of features reflecting timbral texture, rhythmic content, and pitch content, and they achieved 61% accuracy. As a matter of fact, for this type of genre categorization job, human accuracy is approximately 70% [8]. In recent years, there has been an increase in audio classification methods using ANN [9]. However, to implement such a classification model for this task, the main problem lies in the decision of the frame rate in feature calculation and extraction. The decision of the frame rate is important because the classifier varies greatly over time in an audio file, which means that the audio may lean toward one genre at the beginning and another genre at the end. So, the average of it is what matters the most. So, the extraction of features which detect the subtle differences between genres is very important to increase the accuracy of the classifier. Cepstrum or Mel-frequency cepstrum coefficients (MFCCs) [10] are the most common inputs used in these models. Cepstrum is a logarithmic operation that occurs after a signal's Fourier transform, followed by an inverse Fourier transform to produce a spectrogram.

DL models such as convolutional neural network (CNN) [11], multilayer perceptron (MLP) [12], and long short-term memory (LSTM) [13] are employed in this study to identify songs depending on genre. In this study, the GTZAN [14] dataset is utilized, which comprises audio data from several genres.

The remaining part of this study is arranged as follows: Sect. 2 is 'literature survey' in which we have discussed a few previous studies in this area. In Sect. 3, 'Methods', we will explore the project's architectural diagram and the dataset, particularly on how the audio dataset works and how to handle it by learning a few key concepts which helps in data preprocessing and visualization of the audio dataset. Theoretical considerations will be discussed later in this chapter. In Sect. 4, we will look at how to put the music genre classifier together by learning about the modules, features, how to design a neural network models, and to train them. The output, training accuracies, and accuracy plot of these models will be described in Sect. 5. Finally, in Sect. 6, the paper's conclusion and the scope of future study linked to it are presented. Finally, the work is concluded by a list of references.

2 Literature Survey

Most musical functions are based on the spectrogram, which have been proved to be efficient in music genre classification tasks based on deep learning [15–17] Many deep learning models used in computer vision, notably CNN-based models, are employed similarly for music genre classification tasks due to the similarities between audio spectrograms and RGB pictures. Zhang et al. [15] suggested an outstanding convolutional neural network design based on residual blocks, drawing inspiration from image categorization [18] Costa et al. [17] proposed a classification framework involving CNN and support vector machine to make better use of acoustic information in music (SVM). Furthermore, recurrent neural networks (RNN) are popular due to their ability to capture time information. In [19], Dai et al. have used LSTM and recurrent neural networks in music genre classification tasks and got good results. In addition, Choi et al. in [20] utilized both CNN and RNN to collect spatial data from the spectrogram using CNN.

Recent research indicates that spectrograms (e.g., STFT spectrograms, MFCC spectrograms) converted from audio signals may be successfully employed on music genre classification tasks [21–23] because of their capacity to depict temporal variations in energy distribution across frequency. More and more CV approaches, such as local binary pattern (LBP) [21], Gabor filter [22], and Weber local descriptor (WLD) [24], are being used to music genre classification systems. Deep learning techniques, particularly CNNs, should not be overlooked. Sigtia and Dixon [16] have previously validated that ReLU [25], dropout [26], and Hessian-free optimization may increase feature learning effects in the music genre classification area utilizing deep neural networks. Chan et al. [27] trained a CNN as a feature extractor first and subsequently as a classifier based on features to take advantage of CNN's feature learning capacity. They produced amazing results in the GTZAN [14] dataset when combined with the majority voting methodology. In contrast to the preceding design, Zhang et al. [15] incorporated the residual blocks described by He et al. in [18] into their model to achieve better results. Despite the widespread availability of spectrograms, Costa et al. [17] evaluated the impact of acoustic information and provided a framework that included acoustic features and an audio frequency spectrogram.

As a result of their recent success, many studies have applied deep neural networks to speech and other types of audio data. Because of the high sample rate of the audio output, it is difficult to represent the audio in the time domain for neural network input. Van Den Oord et al., on the other hand, have found a solution to this difficulty [28]. Audio generation operations are performed with this library. A spectrogram of a signal, which incorporates time and frequency information, is a typical alternative representation. The spectrogram can be used to train a convolutional neural network (CNN) as an image [29]. To classify musical genres, a CNN was created using the original MFCC matrix from Li et al. as input [27]. To achieve the same goal, Schindler and Lidy [30] used a constant Q transform spectrogram (CQT) as CNN input.

For classification, most of the previous work relies on content-based characteristics and typical machine learning techniques. This study will help in further exploration of using audio signal waves transformed into spectrogram images as feature sets for a deep learning technique. In order to categorize the GTZAN data, a convolutional neural network was used. The proposed hypothesis is that for a similar given audio dataset, the general music type grouping exactness of a deep learning convolutional neural network, utilizing spectrogram images as data, will be greater than the order precision of conventional machine learning approaches, utilizing content-based highlights as a contribution. Unlike previous works, we have classified the audio files into ten various genres. The goal of this research is to perform a comparison of deep learning-based models that only require spectrograms as input. We also examined at the relative significance of various features.

3 Methods

3.1 Music Genre Classification Architecture

Firstly, the dataset is downloaded and imported into the Spyder IDE. Then, from the audio data, MFCCs are extracted. We divided the dataset in a 7:3 ratio, with 70% of the data going to the training phase and 30% going to the testing phase. The testing phase is used to evaluate the findings, whereas the training phase is further separated into training data and validation data. The validation phase receives 10% of the data from the training phase, which contains 70% of the data. The training data are then fed into three different architectures: CNN, MLP, and LSTM. Here, we firstly pass the data into the convolution layer followed by the max pooling layer, and then, we normalize the data to prevent overfitting. This is repeated several times before flattening the data and passing it to the dense layers. The dropout function is then used to prevent overfitting. Finally, the output is passed to the SoftMax function as it standardizes the output. Validation data are used to determine whether or not the training data are making valid predictions. The data are then tested to predict the model's correctness. Finally, we compare and assess all of the models' accuracies in order to predict the model with the best accuracy. The block diagram of the proposed work is shown in Fig. 1.

3.2 Dataset

The MARYSAS Web site's GTZAN [14] dataset is utilized. This provides ten distinct music genres, each with hundred audio samples in.au form. Among the genres are country, reggae, jazz, rock, classical, metal, pop, blues, and disco. In order to reflect a diversity of recording settings, these tracks have been obtained in 2000–2001 via a

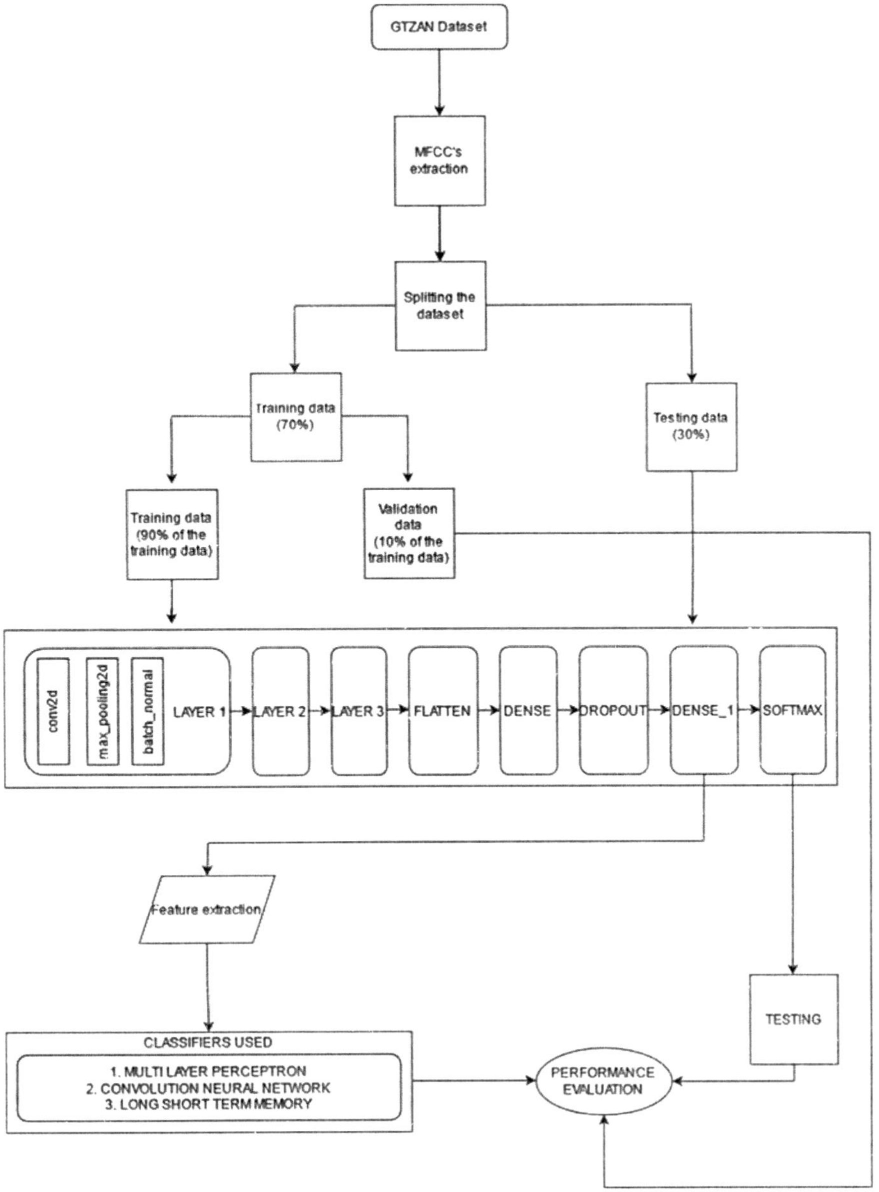

Fig. 1 Block diagram of the study

Table 1 Overview of dataset

S. No	Genre Name	Count
1	Pop	100
2	Hip-hop	100
3	Rock	100
4	Classical	100
5	Blues	100
6	Country	100
7	Metal	100
8	Reggae	100
9	Disco	100
10	Jazz	100
	Total	1000

range of sources, such as CDs, radios, and microphones tapes. The dataset contains 1000 audio recordings, each lasting 30 s. These audio tracks are 16-bit mono-audio files at 22050 Hz. The following Table 1 gives the detailed of the dataset.

3.3 Data Preprocessing

Analog–Digital conversation: Analog sound waves like pianos are continuous waves. It must be digitized by analog-to-digital conversion or ADC. This involves two steps: sampling and quantification. The wave is sampled into multiple time intervals in sampling, whereas the amplitude is quantized with a finite number of bits in quantization. The interval is given by the sample rate, which is the number of samples we have in one second. The number of bits used to store the amplitude of a digital sound wave will determine its quality.

Fourier Transformation: The Fourier transformation is the process of converting a complex periodic acoustic wave into a sum of sine waves vibrating at various frequencies. If we consider the red sound wave is 's', it can be computed as the sum of the decomposed sine waves, i.e.,

$$s = A_1 \sin(2\pi f_1 t + \varphi_1) + A_2 \sin(2\pi f_2 t + \varphi_2) \tag{1}$$

Through Fourier transform, we move from time domain to frequency domain that means the time information is being lost. This can be observed from the spectrum plot after transforming.

Short-Time Fourier Transform (STFT): In STFT, we compute several FFT at different intervals; it also preserves the time information. It will have a predetermined frame size. Following STFT, we will be provided a spectrogram, which will provide knowledge about the magnitude as a function of frequency and time.

Mel-frequency Cepstral Coefficients (MFCCs): MFCCs capture the timbral or textural aspects of sound. For extracting MFCCs, we perform FFT, where we move from time domain to frequency domain. The great advantage of MFCCs is they approximate the human auditory system. The applications of MFCCs are speech recognition, music genre classification, and music instrument classification.

3.4 Data Visualization

Data visualization is an important component of data science. It aids in the comprehension of data as well as the explanation of data to others. The visualization of each value from the dataset is as shown below (Fig. 2).

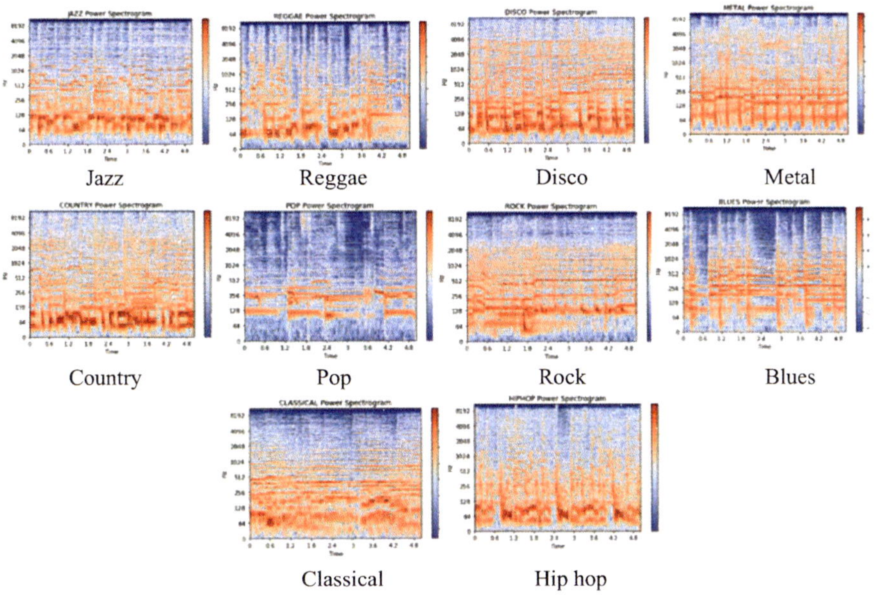

Fig. 2 Visualization of different genres of the dataset

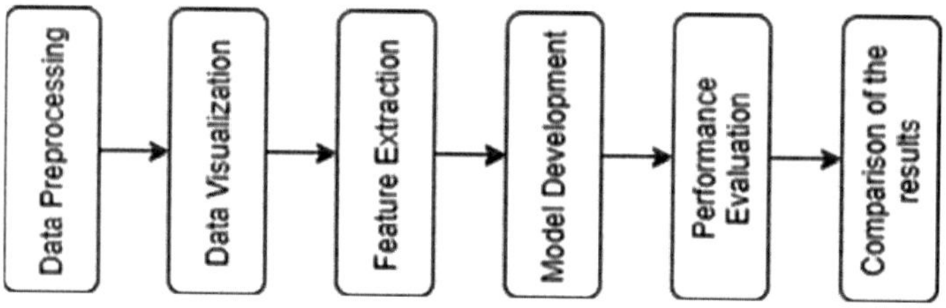

Fig. 3 Flow diagram of the process

4 Implementation

4.1 Flow Diagram

The flow diagram of the proposed work is shown in Fig. 3.

4.2 Feature Extraction

The dataset used contains 1000 30-s songs in in.au form from 10 different genres. Python is used to read this file. Audio tracks that are damaged or less than 28 s long will be removed, and this audio tracks are then converted into Mel spectrogram [31]. First, the audio signal is processed with a sample rate of 22,050. The size of the input sample window is 2048 (n ft). 512 jumps (jump length) are carried out simultaneously to scan the next window. Second, when calculating the fast Fourier transform (FFT) of each window, the time domain is converted to the frequency domain. Later, the Mel scale is created by dividing the entire frequency spectrum into 128 (n Mel) evenly spaced frequencies. Lastly, a spectrogram is made by dividing the amplitude of each window signal into components according to the frequency of the Mel scale.

4.3 Splitting the Dataset

The dataset is separated into three sections, namely training data, testing data, and validation data. Firstly, we split the dataset in the ratio of 7:3, where 70% data is given to the training phase and 30% data is given to the testing phase. The testing phase is used to assess the findings, the training phase, on the other hand, is further subdivided into training data and validation data. The validation phase is given 10% data from the 70% data of the training phase.

4.4 Neural Network Classification Models

MLP: This is a specific type of neural network that is very common and is one of the first neural networks engineered back in the day. We have started using MLP because perceptron's couldn't learn linearly inseparable functions. In neural networks, it is necessary to build the network topology by providing information about the layers, namely input, hidden, and output layers. Here, we use Keras to specify the same. Firstly, we have flattened the audio data and sent it to the dense layer which is using ReLU as an activation function. We used ReLU to minimize the gradient descent problem. In this model, we used three hidden layers which are followed by an output layer. We use the SoftMax function for the output layer as it normalizes the output. The model was assigned with a batch size of 32 and was iterated 50 times.

CNN: CNNs are an advanced type of neural network which are mainly used for processing images than MLPs [32–34]. They have the advantage in the fewer parameters as they use dense layers. To apply convolution or pooling layers to the audio file, we can consider the audio file itself as an image and use spectrogram, MFCCs as images. To create a CNN network topology model, we must first pass the audio data to the convolution layer, then to the max pooling. ReLU is used to reduce the gradient descent issue. To avoid overfitting, we apply batch normalization afterward. This set of layers is called the feature learning phase, and we have performed it three times in our model. The output must next be flattened and sent to the dense layer, where the ReLU activation function is used once more. We also utilize a dropout function that drops 30% of the results to prevent overfitting. This is then passed on to the final output layer for classification, where the SoftMax function is employed for normalizing the data. A batch size with 32 was allocated to the model, and it was iterated 30 times. The network details of the MLP model is shown in Fig. 4.

LSTM: LSTM solves the vanishing or exploding gradient problem—which basically extends their memory. It was designed to combat vanishing gradients through a gating mechanism. The solution is a memory cell, which is updated at each step in the sequence. To build a network topology for the LSTM model, we must first feed the data into a few LSTM layers that are sequence to sequence. The current sequence is sent onto the next layer through the sequence-to-sequence layer. To avoid overfitting, we immediately pass the output through dense layers with a ReLU activation function and a dropout function. The output layer comprises ten nodes, one for each of the ten genres in the dataset. The SoftMax activation function is being used here to effectively normalize and classify the output. A batch size of 32 was allocated to the model, and it was iterated 30 times. The neural network summary of CNN and LSTM are shown in Figs. 5 and 6.

```
Data succesfully loaded!
Model: "sequential"
_________________________________________________________________
Layer (type)                 Output Shape              Param #
=================================================================
flatten (Flatten)            (None, 1690)              0
_________________________________________________________________
dense (Dense)                (None, 512)               865792
_________________________________________________________________
dropout (Dropout)            (None, 512)               0
_________________________________________________________________
dense_1 (Dense)              (None, 256)               131328
_________________________________________________________________
dropout_1 (Dropout)          (None, 256)               0
_________________________________________________________________
dense_2 (Dense)              (None, 64)                16448
_________________________________________________________________
dropout_2 (Dropout)          (None, 64)                0
_________________________________________________________________
dense_3 (Dense)              (None, 10)                650
=================================================================
Total params: 1,014,218
Trainable params: 1,014,218
Non-trainable params: 0
_________________________________________________________________
```

Fig. 4 Model summary of MLP

5 Results and Discussion

5.1 MLP-Based Music Genre Classification

During the testing phase, the MLP model has outperformed with an accuracy of 96%. However, we discovered that the testing accuracy was just 50%. As a result, there is about a 40% difference in testing and training accuracy. This leads to a problem called overfitting (Fig. 7).

Now, we have to solve overfitting to achieve our goal. So, we have come up with two solutions, namely L2 regularization and dropout. We chose L2 regularization because it is excellent for complicated datasets, and dropout allows us to assess accuracy by randomly omitting certain units. Finally, with a loss of just 2.87, we were able to attain an accuracy of 0.63 (Fig. 8).

5.2 CNN-Based Music Genre Classification

To avoid the overfitting problem in the CNN model, we used batch normalization and dropout. With a loss of just 0.8425, we were able to attain an accuracy of 0.70.

```
Model: "sequential_1"
_________________________________________________________________
Layer (type)                 Output Shape              Param #
=================================================================
conv2d (Conv2D)              (None, 128, 11, 32)       320
_________________________________________________________________
max_pooling2d (MaxPooling2D) (None, 64, 6, 32)         0
_________________________________________________________________
batch_normalization (BatchNo (None, 64, 6, 32)         128
_________________________________________________________________
conv2d_1 (Conv2D)            (None, 62, 4, 32)         9248
_________________________________________________________________
max_pooling2d_1 (MaxPooling2 (None, 31, 2, 32)         0
_________________________________________________________________
batch_normalization_1 (Batch (None, 31, 2, 32)         128
_________________________________________________________________
conv2d_2 (Conv2D)            (None, 30, 1, 32)         4128
_________________________________________________________________
max_pooling2d_2 (MaxPooling2 (None, 15, 1, 32)         0
_________________________________________________________________
batch_normalization_2 (Batch (None, 15, 1, 32)         128
_________________________________________________________________
flatten_1 (Flatten)          (None, 480)               0
_________________________________________________________________
dense_4 (Dense)              (None, 64)                30784
_________________________________________________________________
dropout_3 (Dropout)          (None, 64)                0
_________________________________________________________________
dense_5 (Dense)              (None, 10)                650
=================================================================
Total params: 45,514
Trainable params: 45,322
Non-trainable params: 192
_________________________________________________________________
```

Fig. 5 Model summary of CNN

This indicates that the training and testing data are at the same pace. As a result, our model is prepared to provide accurate predictions.

5.3 LSTM-Based Music Genre Classification

We used dropout to mitigate overfitting concerns in an LSTM model using Adam as an optimizer. So, after the training, testing, and validation phases, we have an accuracy of 0.61 and a loss of just 1.123. This means our model doesn't have any overfitting problems (Figs. 9 and 10).

```
Model: "sequential_2"
_________________________________________________________________
Layer (type)                 Output Shape              Param #
=================================================================
lstm (LSTM)                  (None, 130, 64)           19968
_________________________________________________________________
lstm_1 (LSTM)                (None, 64)                33024
_________________________________________________________________
dense_6 (Dense)              (None, 64)                4160
_________________________________________________________________
dropout_4 (Dropout)          (None, 64)                0
_________________________________________________________________
dense_7 (Dense)              (None, 10)                650
=================================================================
Total params: 57,802
Trainable params: 57,802
Non-trainable params: 0
_________________________________________________________________
```

Fig. 6 Model summary of LSTM

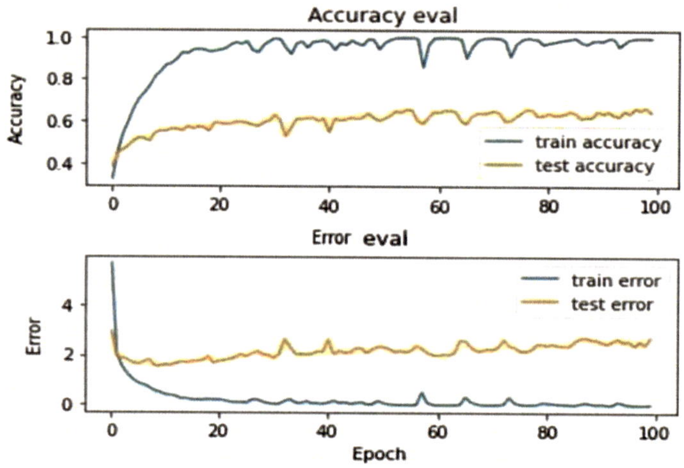

Fig. 7 Accuracy plots of MLP

5.4 *Comparative Analysis*

We used three deep learning neural network models, namely convolutional neural networks, multilayer perceptron, and long short-term memory networks for classifying the music genres. We found that CNN produces the best accuracy of 70.42% which is greater than the human accuracy and over any other neural network models, meanwhile MLP gives 63.28% accuracy and stands second in the Table 2. The comparative analysis of various neural network models are shown in Fig. 11.

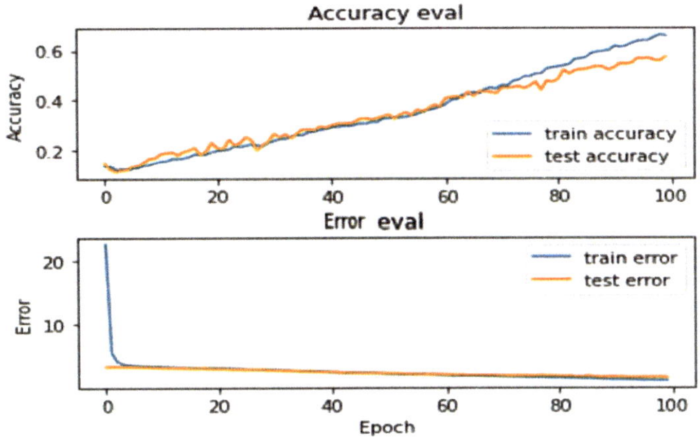

Fig. 8 Accuracy plots after solving overfitting problem of MLP

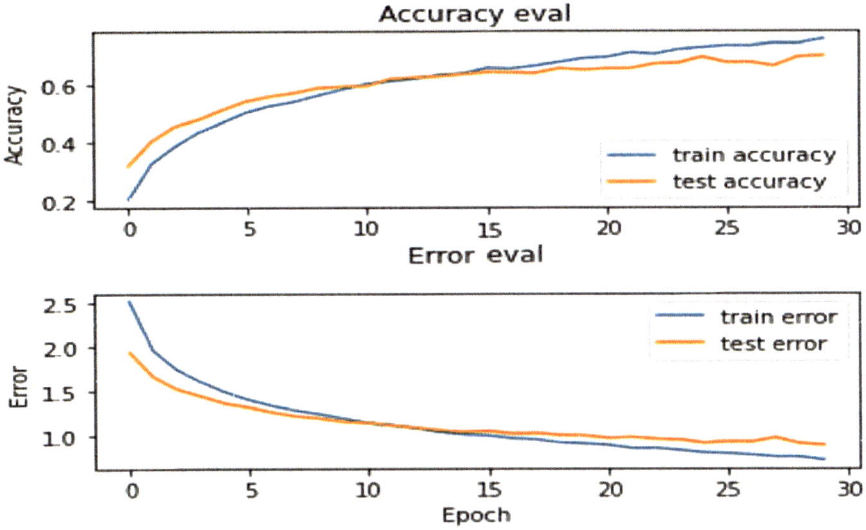

Fig. 9 Accuracy plots of CNN

6 Conclusion and Future Scope

According to this research, the CNN model outperforms the MLP and LSTM models regardless of the dataset used; we found that CNN produces the best accuracy of 70.42% which is greater than the human accuracy and over any other neural network models, meanwhile MLP gives 63.28% accuracy and stands second in the table indicating that CNN is the best neural network model for identifying music genres. Combinatorial models have been studied in depth. In this report, CNN's model

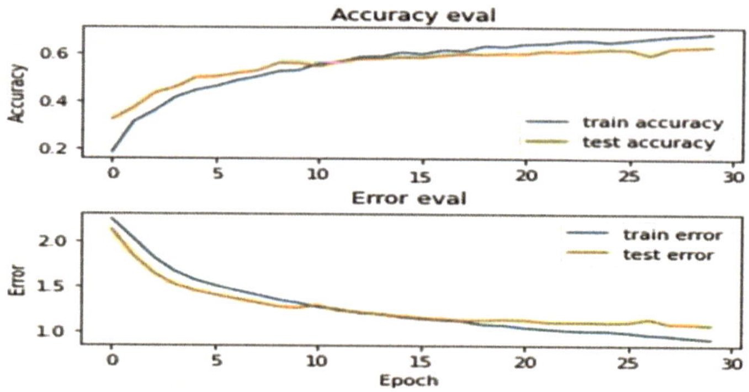

Fig. 10 Accuracy plots of LSTM

Table 2 Comparative analysis overview

Approach	Accuracy (%)	Loss (%)
MLP	63.28	1.082
CNN	**70.42**	**0.842**
LSTM	61.50	1.123

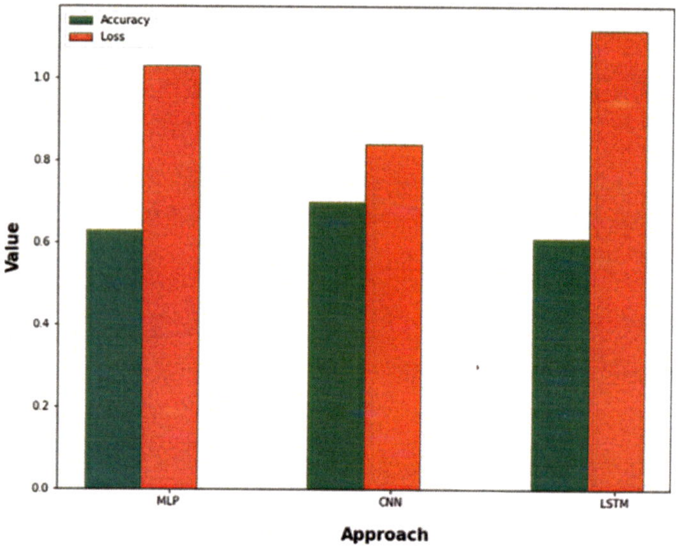

Fig. 11 Comparative analysis of all the approaches

performed better, and concluded that it may be a useful model for building more complicated models. Given that their classification is also more constant, it is clear that finding patterns in music is easier. It is certainly feasible to detect patterns in a song that are linked to a particular genre, but implementing only the CNN, MLP, or LSTM models may not be sufficient. This is true for any dataset. In general, estimating the high accuracy of automated systems is difficult since even people fail to classify music. It may, however, be beneficial for systems that rely on manual confirmation since it decreases burden.

This study only examines the impact of CNN model over MLP and RNN models when it comes to predicting song genre, but it is also fascinating to observe how they perform when it comes to identifying song emotion. Knowing whether a song is energetic or joyful, for example, will give another approach to help categorize songs when genres start to sound alike. Further study comparing these models with various attributes will aid in determining which model works best based on the data available. Nowadays, publishing and browsing have become so simple, resulting in an ever-increasing diversity of music, as well as a steady evolution of our understanding of genre. As more forms of music are generated, categorizing music may grow increasingly challenging.

References

1. T. Li, M. Ogihara, Q. Li, A comparative study on content-based music genre classification 282 (2003), https://doi.org/10.1145/860435.860487
2. T. Johnson, Analyzing genre in post-millennial popular music. City Univ. New York, p. 206, Sep 2018, Accessed: 12 Jun 2021. [Online]. Available: https://academicworks.cuny.edu/gc_etds/2884
3. Y.E. Kim et al., Music emotion recognition: A state of the art review. Proc. Ismir **86**, 937–952 (2010)
4. Z. Fu, G. Lu, K.M. Ting, D. Zhang, A survey of audio-based music classification and annotation. IEEE Trans. Multimed. **13**(2), 303–319 (2011). https://doi.org/10.1109/TMM.2010.2098858
5. C. McKay, I. Fujinaga, P. Depalle, jAudio: A feature extraction library, in *Proceedings of the International Conference on Music Information Retrieval* (2005), pp. 600–603
6. A. Karatana, O. Yildiz, Music genre classification with machine learning techniques, 1–4, Apr 2017, https://doi.org/10.1109/siu.2017.7960694
7. G. Tzanetakis, P. Cook, Musical genre classification of audio signals. IEEE Trans. Speech Audio Process. **10**(5), 293–302 (2002). https://doi.org/10.1109/TSA.2002.800560
8. M. Dong, Convolutional neural network achieves human-level accuracy in music genre classification, Feb 2018
9. A.R. Rajanna, K. Aryafar, A. Shokoufandeh, R. Ptucha, Deep neural networks: A case study for music genre classification, in *Proceedings—2015 IEEE 14th International Conference on Machine Learning and Applications, ICMLA 2015*, pp. 655–660, Mar 2016, https://doi.org/10.1109/ICMLA.2015.160
10. B. Logan et al., Mel frequency cepstral coefficients for music modeling, in *Ismir*, vol. **270**, (2000), pp. 1–11
11. S. Lawrence, C.L. Giles, A.C. Tsoi, A.D. Back, Face recognition: A convolutional neural-network approach. IEEE Trans. Neural Netw. **8**(1), 98–113 (1997). https://doi.org/10.1109/72.554195

12. J. Tang, C. Deng, G. Bin Huang, Extreme learning machine for multilayer perceptron, IEEE Trans. Neural Networks Learn. Syst. **27**(4), 809–821, Apr 2016, https://doi.org/10.1109/TNNLS.2015.2424995
13. S. Hochreiter, J. Schmidhuber, Long short-term memory. Neural Comput. **9**(8), 1735–1780 (1997). https://doi.org/10.1162/neco.1997.9.8.1735
14. P.C.G. Tzanetakis, GTZAN dataset
15. W. Zhang, W. Lei, X. Xu, X. Xing, Improved music genre classification with convolutional neural networks, in *Proceedings of the Annual Conference of the International Speech Communication Association, INTERSPEECH*, 2016, vol. 08-12-Sept, pp. 3304–3308, https://doi.org/10.21437/Interspeech.2016-1236
16. K. Choi, G. Fazekas, M. Sandler, K. Cho, Transfer learning for music classification and regression tasks. Proc. 18th Int. Soc. Music Inf. Retr. Conf. ISMIR 2017, 141–149, Mar 2017
17. Y.M.G. Costa, L.S. Oliveira, C.N. Silla, An evaluation of convolutional neural networks for music classification using spectrograms. Appl. Soft Comput. J. **52**, 28–38 (2017). https://doi.org/10.1016/j.asoc.2016.12.024
18. K. He, X. Zhang, S. Ren, J. Sun, Deep residual learning for image recognition, in *Proceedings of the IEEE Computer Society Conference on Computer Vision and Pattern Recognition*, vol. **2016-Decem** (2016), pp. 770–778, https://doi.org/10.1109/CVPR.2016.90
19. J. Dai, S. Liang, W. Xue, C. Ni, W. Liu, Long short-term memory recurrent neural network based segment features for music genre classification, 2017. https://doi.org/10.1109/ISCSLP.2016.7918369
20. K. Choi, G. Fazekas, M. Sandler, K. Cho, Convolutional recurrent neural networks for music classification, in *ICASSP, IEEE International Conference on Acoustics, Speech and Signal Processing—Proceedings*, Jun 2017, pp. 2392–2396, https://doi.org/10.1109/ICASSP.2017.7952585
21. Y.M.G. Costa, L.S. Oliveira, A.L. Koerich, F. Gouyon, J.G. Martins, Music genre classification using LBP textural features. Signal Process. **92**(11), 2723–2737 (2012). https://doi.org/10.1016/j.sigpro.2012.04.023
22. Y. Costa, L. Oliveira, A. Koerich, F. Gouyon, Music genre recognition using gabor filters and LPQ texture descriptors, in *Lecture Notes in Computer Science (including subseries Lecture Notes in Artificial Intelligence and Lecture Notes in Bioinformatics), 2013*, vol. **8259**, LNCS, no. PART 2, pp. 67–74, https://doi.org/10.1007/978-3-642-41827-3_9
23. M.-J. Wu, J.-S.R. Jang, Combining acoustic and multilevel visual features for music genre classification. ACM Trans. Multimed. Comput. Commun. Appl. **12**(10) (2015), https://doi.org/10.1145/2801127
24. L. Nanni, Y.M.G. Costa, A. Lumini, M.Y. Kim, S.R. Baek, Combining visual and acoustic features for music genre classification. Expert Syst. Appl. **45**, 108–117 (2016). https://doi.org/10.1016/j.eswa.2015.09.018
25. V. Nair, G.E. Hinton, Rectified linear units improve restricted Boltzmann machines
26. G.E. Hinton, N. Srivastava, A. Krizhevsky, I. Sutskever, R.R. Salakhutdinov, Improving neural networks by preventing co-adaptation of feature detectors, Jul 2012
27. A.B. Chan, A. Hon, W. Chun, T.L. Li, A.H. Chun, *Automatic Musical Pattern Feature Extraction Using Convolutional Neural Network Image/video restoration and people tracking in outdoor environments View project SmartPalette View project Automatic Musical Pattern Feature Extraction Using Convolutional Neu.* (2010)
28. A. van den Oord et al., WaveNet: A Generative model for raw audio, Sep 2016
29. L. Wyse, Audio spectrogram representations for processing with convolutional neural networks, Jun 2017
30. A. Schindler, T. Lidy, *Parallel Convolutional Neural Networks for Music Genre and Mood Classification Europeana Sounds View Project SCAPE Project View Project Parallel Convolutional Neural Networks for Music Genre and Mood Classification* (2017)
31. F. Gouyon, Y.M.G. Costa, L.S. Oliveira, A.L. Koericb, Music genre recognition using spectrograms, in *IEEE Conference Publication, IEEE Xplore*

32. J. Andrew, S.S. Mathew, B. Mohit, A Comprehensive analysis of privacy-preserving techniques in deep learning based disease prediction systems 0–9 (2019), https://doi.org/10.1088/1742-6596/1362/1/012070
33. J.A. Onesimu, J. Karthikeyan, An efficient privacy-preserving deep learning scheme for medical image analysis. J. Inf. Technol. Manag. **12**(Special Issue: The Importance of Human Computer Interaction: Challenges, Methods and Applications), 50–67, Dec 2021, https://doi.org/10.22059/jitm.2020.79191
34. J. Andrew, R. Fiona, H. Caleb Andrew, Comparative study of various deep convolutional neural networks in the early prediction of cancer, in *2019 International Conference on Intelligent Computing and Control Systems, ICCS 2019*, May 2019, pp. 884–890, https://doi.org/10.1109/ICCS45141.2019.9065445